# *GEOHAZARDS*

# *GEOHAZARDS*

## *Natural and Human*

**Nicholas K. Coch**
Queens College, City University of New York

Prentice Hall
Englewood Cliffs, New Jersey 07632

**Library of Congress Cataloging-in-Publication Data**
Coch, Nicholas K.
Geohazards : natural and human / Nicholas K. Coch. — 1st ed.
p. cm.
Includes bibliographical references and index.
ISBN 0-02-322992-6
1. Natural disasters. 2. Disasters. I. Title.
GB5014.C63 1995
363.3'4—dc20 94-23581
CIP

Cover: Hurricane Elena, Sept. 1, 1985, photographed from *Discovery* at an altitude of 236 nautical miles; Richard O. Covey, pilot. Inset: Eruption at Kilauea Volcano, 1983; photo by Ken Sakamoto/Black Star. Back cover: Collapse of freeway, Loma Prieta earthquake, 1989; photo by James A. Sugar/Black Star. Aerial view of Davenport, Iowa, during the 1993 flooding of the Mississippi River; photo by Chris Stewart/Black Star.

Editor: Robert A. McConnin
Developmental Editor: Fred C. Schroyer
Production Editor: Mary Harlan
Photo Editor: Chris Migdol
Photo Researchers: Julie Tesser, Clare Maxwell
Text Designer: Robert Vega
Cover Designer: Robert Vega
Production Buyer: Pamela D. Bennett
Illustrations: BioGraphics

This book was set in Galliard by The Clarinda Company and was printed and bound by Von Hoffmann Press, Inc. The cover was printed by Von Hoffmann Press, Inc.

Printed in the United States of America

10 9 8 7 6 5 4 3 2 1

ISBN: 0-02-322992-6

Prentice-Hall International (UK) Limited, *London*
Prentice-Hall of Australia Pty. Limited, *Sydney*
Prentice-Hall of Canada, Inc., *Toronto*
Prentice-Hall Hispanoamericana, S. A., *Mexico*
Prentice-Hall of India Private Limited, *New Delhi*
Prentice-Hall of Japan, Inc., *Tokyo*
Simon & Schuster Asia Pte. Ltd., *Singapore*
Editora Prentice-Hall do Brasil, Ltda., *Rio de Janeiro*

# Preface

*Geohazards* was written in order to provide a more inclusive treatment of geologic hazards than is presently available in environmental geology texts. The text deals with both natural geologic processes that can be harmful to people (natural geologic hazards) and with human modifications of natural geologic systems that now make them harmful (human geologic hazards). In addition, the term *geologic* is used in its broadest sense to include all aspects of the Earth—the hydrosphere and atmosphere as well as the lithosphere. Whereas current environmental geology books emphasize the lithosphere, *Geohazards* includes a wider varietZy of topics at a level suitable for both non-majors and beginning geology students. This book includes the major topics found in all environmental geology books (volcanoes, earthquakes, flooding, landslides, wastes). Hopefully it will treat these topics with a fresher approach. However, in order to include more information on *hazards,* topics such as land use, esthetics, medical geology, resources, and environmental law are not considered in detail in this book. This was done in order to include detailed discussions of geologic hazards not discussed in presently available environmental geology texts. These topics include a more detailed treatment of air pollution problems (acid rain, the greenhouse effect, radon, the depletion of the ozone layer); new discussions of such severe weather hazards as extra-tropical storms, hurricanes, and tornadoes; and coastal problems including reef and mangrove destruction, estuarine problems, and wetlands elimination, as well as a detailed treatment of coastal erosion and the effects of sea-level rise. In addition, the discussion of mass movements (lateral and vertical) is detailed because each of these topics is treated in a separate chapter.

*Geohazards* is based on almost twenty years experience teaching environmental geology courses on all levels from the non-major to the graduate levels. The text assumes no prior geologic knowledge and is designed to be used in a beginning course for non-majors or majors that meets for three hours a week. Each chapter explains the *principles* as simply as possible, describes the *problems* (with examples), and suggests *solutions.* Wherever possible, citations, quotes, and diagrams are based on recently published papers.

Chapter 1 deals with why geologic hazards are causing more damage and loss of life each year, the way in which society can deal with hazards and problems, and some basic concepts about environmental problems. Chapters 2 (Inside Earth) and 3 (Outside Earth) provide a basic review of geologic information pertinent to discussions of geologic hazards in later chapters. Readers with a geology background can skip these chapters and move directly to the later chapters. I have placed this basic material at the beginning of the book, rather than distribute it among many chapters, so that readers will have a coherent discussion available before they get into the detailed information in later chapters. In addition, there is the question of where to introduce certain information (porosity, plate tectonics, bedding, shear forces, etc.). Beginning with coherent units of basic geologic information is not an experimental approach; it has worked well in my courses for many years. The choice of topics included in Chapters 2 and 3 was based on the requirements of subsequent chapters, the suggestions of reviewers, and the need for a coherent discussion in each of those chapters. The selection of topics and the geologic terminology in Chapters 2 and 3 were kept at a beginning to intermediate level so that this book could be useful also for students and workers in other fields, such as environmental biology and chemistry, forestry, engineering, architecture, disaster preparedness, and city planning. Chapter 2 describes the source of energy for internal processes, the materials (rocks and minerals) formed, the way that earth materials react to stress (structural geology), and the concepts of plate tectonics and earth cycles. Chapter 3 provides an overview of major surface processes. Although some repetition is unavoidable between Chapters 2 and 3 and some of the later chapters, I have tried to keep this to a minimum. The structure of each of the remaining chapters enables an instructor to add more advanced concepts and details easily and to utilize the chapters in a different order if desired. Chapters 4 to 16 each begin with an additional discussion of the geologic principles relevant to the topic. These principles are applied to as many different actual examples as possible in the remainder of the chapter. Wherever possible, I have cross-referenced related material in different chapters.

The boxed discussions in most chapters deal with a fascinating aspect of the subject material. New terms are boldfaced in the text when they are first defined. They are defined again in a glossary at the end of the book. Many new and detailed conceptual figures and abundant photographs, many taken by the author specifically for this book, are used to illustrate concepts in the text. A summary in text form, review questions of different difficulties, and selected readings at the end of each chapter will help you to review the material.

## ACKNOWLEDGMENTS

My thanks to the many chapter reviewers for their constructive criticisms, additions and suggestions:

☐ Fred P. Wolff, Hofstra University
☐ Robert I. Tilling, U.S. Geological Survey
☐ Frank W. Fletcher, Susquehanna University
☐ Raymond Pestrong, San Francisco State University
☐ Winton Cornell, University of Tulsa
☐ Hobard M. King, West Virginia University
☐ Hannes K. Brueckner, Queens College, CUNY
☐ Jeffrey S. Haner, Louisiana State University
☐ Joseph F. Schreiber, Jr., University of Arizona
☐ William J. Wayne, University of Nebraska, Lincoln
☐ Ann G. Harris, Youngstown State University
☐ Peter H. Mattson, Queens College, CUNY
☐ John E. Sanders, Columbia University
☐ Marjorie J. Clarke, New York, NY
☐ Samuel C. Snedaker, University of Miami
☐ Stephen P. Leatherman, University of Maryland
☐ Richard A. Davis, University of South Florida
☐ Orrin H. Pilkey, Jr., Duke University
☐ John C. Kraft, University of Delaware
☐ Otto H. Muller, Alfred University
☐ H. Robert Burger III, Smith College
☐ Carol L. Ekstrom, Rhodes College
☐ Guerry H. McClellan, University of Florida
☐ Sherry D. Oaks, Colorado State University
☐ Robert Dolan, University of Virginia
☐ Katherine V. Cashman, University of Oregon
☐ John Hope, The Weather Channel, Atlanta
☐ Alina M. Szmant, University of Miami
☐ Frederic R. Siegel, George Washington University
☐ Rona J. Donahoe, University of Alabama
☐ Anthony F. Randazzo, University of Florida
☐ Ellen E. Wohl, Colorado State University
☐ George W. Fisher, Johns Hopkins University
☐ Parker E. Calkin, University of Buffalo, SUNY
☐ Barry F. Beck, University of Central Florida
☐ David C. Locke, Queens College, CUNY
☐ Henry Bokuniewicz, SUNY Stony Brook
☐ Barbara W. Murck, University of Toronto
☐ Judson L. Ahern, University of Oklahoma

My special thanks to the people who provided critiques of specialized aspects of the book, provided additional material, identified new information sources, and/or showed the author geologic hazards in the field. These include Carol Coch, U.S. Army Corps of Engineers; Fred Wolff, Hofstra University; Robert Tilling, U.S. Geological Survey; Hannes Brueckner and Patrick Brock, Queens College; Ray Pestrong, San Francisco State University; Frank Fletcher, Susquehanna University; Ervin Otvos, Gulf Coast Research Laboratory; Dean Malouta, Shell Production Research, Houston; Orrin Pilkey, Duke University; Stephen Leatherman, University of Maryland; Jerre Johnson, College of William and Mary; Michael Katuna, College of Charleston; Paul Gayes and Doug Nelson, Coastal Carolina College; John Hope, The Weather Channel; Robert Sheets and Brian Jarvinen, National Hurricane Center; William Gray, Colorado State University; Robert Dolan, University of Virginia; Barry Beck, P. E. LaMoreaux Associates; Henry Bokeniewicz, SUNY at Stony Brook; Cliff Zimmerman, New York City; Matilda Fitton, Clearwater, Florida; Mel Nishihara, Hawaii Civil Defense Agency; and Marcia Brockbank, San Francisco Estuary Project. Finally, I would like to thank collectively all those geologists who have taught me a great deal about environmental geology in their lectures, field trips, and publications over the years.

Many people have helped in the production of this book. Robert McConnin, Senior Editor, recognized the need for this book and oversaw its development. I am deeply indebted to Fred Schroyer, Developmental Editor, for his expertise and organizational skills that contributed significantly to the molding of the final version of this text. Mary Harlan, Production Editor, did a superb job in bringing together all the diverse components of this book and meticulously supervising its final production. Chris Migdol and Julie Tesser obtained many additional photos for the text. Dave Carlson and his crew at Biographics rendered my crude sketches into fine art.

I thank my colleague, Allan Ludman, for his encouragement and for allowing me to use in *Geohazards* some of the diagrams he conceived for our text *Physical Geology*. Leila Woolley was of great help in obtaining permissions and in the production of the index and glossary. Henry Mesa, Randy Levy, and Sara Woolley provided significant support help. Katherine Andrade deserves special recognition for her hard work on many aspects of the production of the text, as well as keeping my research program going while I was writing this book. Finally, I reserve special thanks for my wife, Carol Coch, for her encouragement and understanding, help, and technical expertise during the writing of *Geohazards*.

*To Carol*

# *Contents in Brief*

# Contents

## 2

## Earth's Interior 17



## 3

## Earth's Surface 49

## 6

## Soil Erosion and Sediment Pollution 155



## 7

## Streams 177

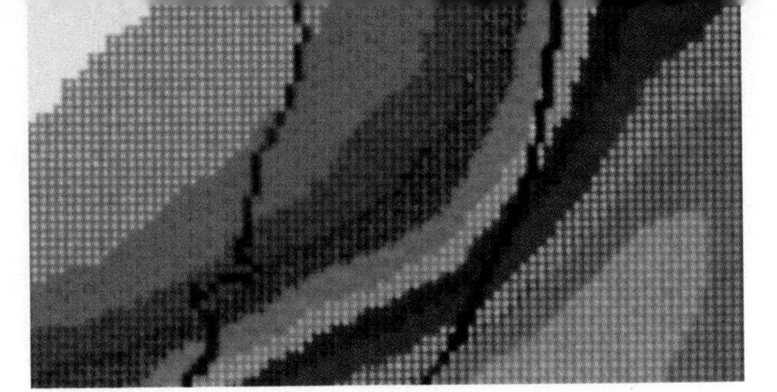

# 12

## Waste Disposal 311



# 13

## Estuarine and Wetland Problems 345

# 14

## Problems of Mangrove Wetlands and Coral Reefs 367



# 15

## Coastal Problems 395

## 16

## Severe Weather Hazards 419

# 1

# *The Geologic Hazards of Living on Earth*

How hazardous is it where you live? We are not talking about hazards from crime, careless driving, or unsafe working conditions. This book is about **geologic hazards,** or **geohazards,** which are Earth processes that are harmful to humans and their property. Do you live in an earthquake-prone area, near a volcano, along a coast prone to major storms and hurricanes, in an area subject to groundwater pollution, in a low-lying place that often floods, or in an area where the land is being undermined? These are the kinds of geohazards we will explore in this book. Geohazards include two broad classes, natural and human.

## GEOHAZARDS—NATURAL

Each year seems to bring more media reports of volcanic eruptions, earthquakes, hurricanes, floods, coastal erosion, landslides, and surface collapse. However, these are *natural* geologic processes that have been operating on Earth for several billion years. Why do we hear more about them now? Part of the reason is that we live in a more environmentally conscious society with better communications than ever before.

Another reason is that Earth's expanding population creates more situations that are vulnerable to harmful geologic processes. For example, the island of Hawaii actually is built of several volcanoes that have harmlessly spewed lava for millennia. But as Hawaii's population has grown and spread, new housing developments are threatened by inundation with lava from renewed volcanic activity (Figure 1–1).

This natural process, volcanism, is a geologic hazard because it *adversely affects humans and their property.* Sometimes human activity can even trigger a geologic hazard where it might not naturally occur. For example, expand-

**The Washington, D.C., metropolitan area. This region has undergone explosive growth since 1965. (NASA).**

**FIGURE 1–1**
Lava flowing into a housing development on Hawaii. (Photo by J. D. Griggs, U.S. Geological Survey.)

ing a housing development by cutting into a hillside steepens the slope and may initiate a landslide (Chapter 9). The great weight of water impounded behind a newly built dam in a seismically active area may trigger an earthquake (Chapter 5). In a geologic hazard, some Earth process causes a safety hazard or economic risk for people.

Geological events like volcanic eruptions, earthquakes, and floods happen, regardless of whether people are exposed to them. But it is only when these events *affect people* that they become geologic hazards. Geologic hazards can cause injuries, death, and property destruction.

Some hazards have beneficial aspects as well as destructive ones. For example, a flooding river may destroy structures and injure people, but it also deposits nutrient-rich sediments over the area, greatly increasing soil fertility. Similarly, destructive volcanic lava and ash form very fertile soils when they weather and break down. Lava deposits form new land as they pour into the ocean on volcanic islands such as Iceland and Hawaii, creating new real estate (Chapter 4).

Geologic hazards vary in their power. **Magnitude** is the power of a destructive event such as an earthquake or hurricane; magnitude indicates how much energy is released. **Frequency** is how often events of a given magnitude occur or are expected to occur. All other factors being equal, larger magnitude events are less frequent but cause far more damage than smaller magnitude events. For instance, the Charleston, South Carolina, area was struck by one of the largest earthquakes in U.S. history in 1886. While earthquakes are uncommon in the coastal plain of the southeastern United States, this high-magnitude, low-frequency event was highly destructive. Of course, even a very small-magnitude event in a densely populated area can cause tremendous destruction and many deaths.

Hazards also differ in the size of the area affected. Some hazards such as landslides (Chapter 9) and surface collapse (Chapter 10) affect local areas. Others, like volcanoes (Chapter 4), earthquakes (Chapter 5), and hurricanes (Chapter 16) may affect entire regions.

Geologic hazards may give advance notice, or they may occur without warning. Hurricanes can be tracked for days and landfall sites predicted accurately several hours before the storm passes over a coast. Earthquakes, on the other hand, may occur without detectable warning.

Many hazards are preceded by **precursors,** minor events that indicate an impending major event. For example, cracks in the ground are common precursors of surface collapse (Chapter 10). Steam venting and ground bulging are typical precursors of volcanic eruptions (Chapter 4). Surface cracking and creeping are precursors of landslides (Chapter 9). Geologists have identified precursors for many hazards. Unfortunately, however, these events seldom predict when the major event (the "big one") will occur.

A natural geohazard, then, is a natural geologic process that adversely affects humans and their property.

## GEOHAZARDS—HUMAN

Sometimes people's activities interact with natural geologic systems and result in a hazard. Pollution is one example of a hazard in which human activities increase the concentration of undesirable substances in air and water, creating a hazard. Such human-generated effects are called **anthropogenic** in this book (*anthropo* for "human," *genic* for "generated by"). In a human geohazard, *the harm results from anthropogenic activities that accelerate or alter a normally benign process to cause a problem.*

Underground water flow is a natural geologic process that among other things, supplies drinking water to many communities (Chapter 8). However, when anthropogenic activities allow liquid waste to enter the ground, the underground water becomes polluted, the contamination spreads to water wells in the area, and a geohazard is created.

Another example is the movement of sediment along a shoreline. This natural process replenishes beach sand that is eroded by storms. However, in many places, human manipulation has altered coastal processes, causing a variety of coastal problems (Chapter 15).

Human geohazards differ from natural geohazards in several ways. In a human geohazard, humans both *affect* the process and then are *adversely affected by* its results. In general, human geohazards develop slowly, and in most cases they can be stopped or greatly reduced. Although we can do little to control most natural geohazards, there is greater promise for control and elimination of human ones.

A human geohazard, then, is a normally benign Earth process that is made dangerous by human activity.

## POPULATION AND POLLUTION

The dramatic increase in human population (Figure 1–2) is degrading the environment, promoting development of geologic problems, and increasing injuries, property damage, and loss of life.

Human population increased slowly but steadily until the Middle Ages (roughly A.D. 1100–1500). A sharp drop in population occurred when the Black Death (bubonic and pneumonic plagues) struck Europe in the mid-1300s. By 1700, however, world population had rebounded to about 700 million and the growth rate was increasing. Since that time, an exponential increase in population growth had created a world population of about 5.6 *billion* by 1994. It is estimated that world population will reach 6 billion before the year 2000.

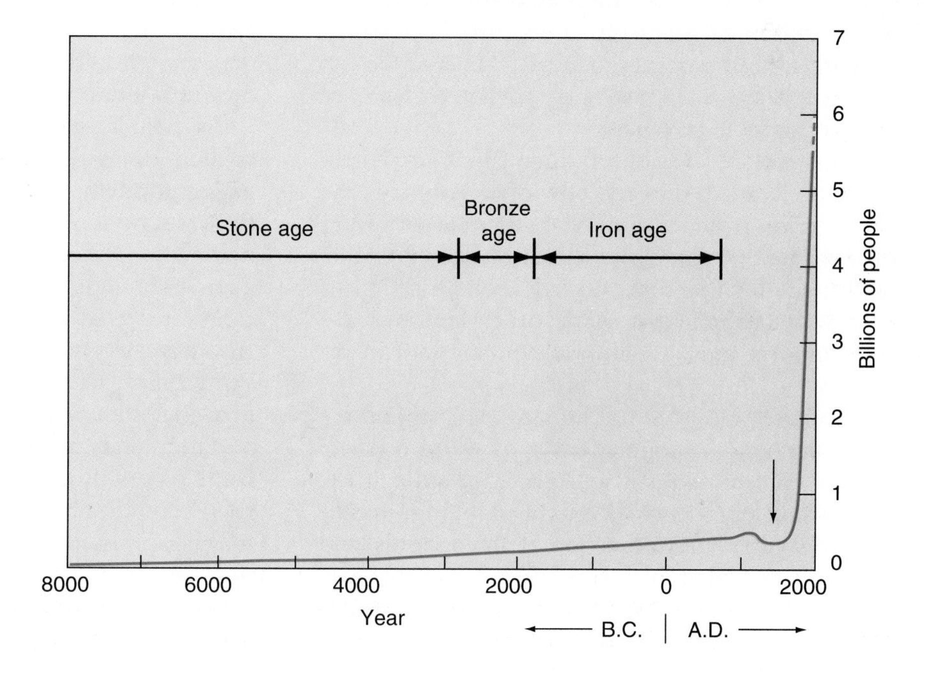


**FIGURE 1–2**
Graph showing the growth of human populations. The arrow marks a sharp population drop caused by the Black Death, which first struck Europe in 1348. In 1992, the United Nations Earth Summit Conference in Rio de Janeiro confirmed that Earth's environmental problems are intimately tied to our mushrooming population. (Data from the Population Reference Bureau, 1994.)

This rapid rate of population growth is causing serious shortages of resources, including food, and is degrading the environment. The unfavorable alteration of our surroundings, wholly or largely as a by-product of human action, is called **pollution.** Pollution of surface and underground water and air is described in detail in subsequent chapters.

## POPULATION GROWTH STRAINS EARTH'S RESOURCES

The rapid increase in human population (Figure 1–2) is straining Earth's ability to provide food, clothing, shelter, and energy. We already use more than 500 kilograms (1,100 pounds) of steel, 25 kilograms (55 pounds) of aluminum, and 200 kilograms (440 pounds) of salt per person on average each year. As today's 5.6 billion multiply in future years, where will the additional resources come from? The problem is even more serious because technologically advanced societies consume greater quantities of resources than do less developed cultures. However, only part of this consumption is essential, whereas the rest is for convenience or luxury. Thus, as we achieve higher standards of living, our resource consumption increases per person.

A **resource** is any natural material that is both useful to people and available in sufficient amounts. Some resources, such as wood and water, are **renewable resources,** because they can be replenished within a lifetime. Other resources are called **nonrenewable resources** because they can be replenished only on a geologic scale of *millions* of years, too slow to help us.

Resources are distributed unequally over Earth. The more varied the geology of a country, the greater is its resource potential. In general, larger countries such as the United States, Canada, China, and Russia, include more varied geology and thus are richer in resources. Many other nations have more uniform geology and are notably poor in resources.

The *basic* resources for human civilizations are land and water. As populations grow, so grows the need for land to use as living space, for agriculture, and for *resource extraction* (mining, drilling, pumping, and so on). Only a fraction of any nation's land is suitable for growing crops, because fertile soil *and* water both must be present. As homes replace fields in flat areas, farming is displaced to hilly or mountainous regions. In these areas, steeper slopes accelerate soil loss, causing not only loss of farmable land but also pollution of streams with sand and silt (Chapter 6).

As human societies developed, they first used **nonmetallic resources,** such as wood and stone, to make tools (the "Stone Age"). Later, people developed the technology to use **metallic resources** for the same purposes in the Bronze Age and the Iron Age (see Figure 1–2). As local metal deposits were depleted, new sources were found in more distant locations. These resources are extracted from Earth by underground mining, surface mining, and drilling. This extraction, as well as processing, purification, and use, has created serious problems of both air and water pollution (Chapters 10, 11, and 12).

## THE URBAN CONCENTRATION FACTOR

Pollution of our environment has been occurring since the dawn of civilization. For several thousand years, most people lived in sparsely settled rural areas; urban centers were small and widely dispersed. Earth's entire human population was small. Therefore, the quantities of solid, liquid, and gaseous waste produced per person also were small. These substances could be disposed of effectively through burial or be diluted and dissipated by wind and flowing water. However, as human population skyrocketed and urban centers grew (Figure 1–3), diverse pollution problems developed.

As people concentrated in cities, pollution increased dramatically and waste disposal became a major problem. The urban concentration of harmful waste no longer could be dispersed effectively or diluted by winds and rivers. For example, the organic-rich sludge produced from sewage treatment can be recycled into fertilizer in small communities. In a large city such as New York, however, sewage sludge is too rich in metals to be safely used for growing edible crops.

The metals can be removed, and this is done in some places. But the huge volume of sludge makes this economically impractical for New York City. You may be surprised to learn that the high metal content does not come primarily from industrial sources. Instead, it is from *domestic* uses (household cleaners and deodorants) and weathering of metallic surfaces such as galvanized fences (zinc-coated), structural iron, aluminum window frames, and metal roofs.

Let us examine this further, for it illustrates pollution problems in general. In each domestic use or weathering event (for example a rain storm or winter-time freezing and thawing), only minuscule amounts of metal are dissolved. They enter sewage treatment plants via sewers. The problem is that *millions* of domestic uses and weathering events occur daily in large metropolitan areas. As a result, the total metal content entering sewage treatment plants may be considerable.

A small amount of iron may be dissolved from a structure, or a minute amount of zinc may be dissolved from a galvanized fence each day. However, multiply each occurrence by the great surface areas of steel or the many miles of galvanized fence in a large city, and the metal input to sewage plants each day can be significant. Such a multiplication of small amounts of pollutants by large populations is referred to in this book as the **urban concentration factor.**

## EXPANSION OF POPULATIONS INTO HAZARDOUS AREAS

Housing and industry require land for buildings, roadways, recreation, and services. This demand for land has caused us to extend development into areas that are potentially hazardous. Some of the most scenic, agriculturally productive, and pleasant places to live and work always have been associated with geologic hazards. Farmers cultivate the floodplains of river valleys because they are fertile and offer plenty of water. However, they are called *floodplains* for a reason: they flood periodically, as we learned during the great Missouri-Mississippi floods of 1993. Similarly, those who fish for a living dwell and work on barrier islands and in coastal wetlands, near the fish—and where hurricanes (or at least strong storms) are a danger.

Additionally, human population growth has encouraged development of urban population centers in formerly inhospitable, but inexpensive, areas. Las Vegas, in the Nevada desert, grew up because it was possible to make the area more livable with cheap electricity (for air conditioning) and water derived from underground sources. Ironically, Las Vegas, in dry desert country, experiences flash flooding from storms because the steep slopes around the city are sparsely vegetated.

**FIGURE 1–3**
The New York–New Jersey metropolitan area. The Hudson River runs vertically through the center of the photo. The darker areas on land represent parks and open spaces. The lighter areas are densely settled. (NASA photo.)

Development has increased even into areas *known* to be hazardous. Consider the explosive population growth in earthquake-prone San Francisco and Los Angeles. Look also at the growing cities of Miami Beach, Florida; Galveston, Texas; Atlantic City, New Jersey; and Long Beach, New York—all built on barrier islands, which are especially vulnerable to hurricane damage.

In addition, outward expansion from relatively safe major population centers has pushed suburbs into adjacent areas that are potentially hazardous. Charleston, South Carolina, is somewhat sheltered a few miles inland from the Atlantic. Despite this location, Charleston suffered severe damage from Hurricane Hugo in 1989. Some of its coastal suburbs, however, that are built on wetlands and barrier islands, sustained much worse damage from wind and flooding (Figure 1–4).

**FIGURE 1–4**
Damage caused by the storm surge and high winds of Hurricane Hugo (1989) on Sullivans Island, a coastal suburb of Charleston, South Carolina. (Photo by author.)

## HAZARD MANAGEMENT

A geologic hazard, current or potential, can be managed in several ways. The approach used in any specific situation is the result of (1) environmental awareness and the residents' perception of the problem, (2) commercial considerations, and (3) fiscal restraints placed on local government. In this section, we will consider five approaches: accepting the loss, restrictive zoning, structural solutions, hazard warning and evacuation, and abandonment and public use of hazardous areas.

### Accepting Hazard Losses

In some areas, neither people nor their government take strong action against a potential hazard. Sometimes this lack of action is because of lack of funds. In other cases, it is because the hazard is incorrectly perceived as being slight. A good example is zoning and construction practice in two coastal states, South Carolina and New York. In South Carolina, newly constructed beach homes must be elevated above the expected hurricane storm-surge level (Figure 1–5). In fact, this practice greatly reduced struc-

**FIGURE 1–5**
Homes in coastal zone of South Carolina elevated to minimize storm flooding and wave damage. (Photo by author.)

**FIGURE 1–6**
Homes on south shore of Long Island, New York, built directly at ground level. This construction provides no protection in a storm. (Photo by author.)

tural damage during Hurricane Hugo in 1989. By contrast, building codes in many New York coastal communities do not require elevation of newly built beach homes (Figure 1–6). Many prudent homeowners, however, have done so on their own.

Why the difference between South Carolina and New York? It is in part because the frequency of powerful hurricanes in New York is far less than in South Carolina. Consequently, South Carolinians are more aware of the hurricane hazard. In the northeast, major coastal storms (called *nor'easters*), some with local winds approaching hurricane force, struck in 1992 and 1993. These storms caused significant damage, but not nearly as much as a hurricane would.

**FIGURE 1–7**
The Rockaway Peninsula in New York City. This highly urbanized coastal segment has the greatest year-round residential population density in America. (Photo by author.)

The last major hurricane to hit coastal New York was more than half a century ago, in 1938, so the danger is forgotten and seems remote (Chapter 16). However, the northeast is long overdue for a major hurricane. The damage that such a storm will do to this highly developed and ill-prepared coastal region (Figure 1–7) will far exceed that caused by Hurricane Hugo in South Carolina in 1989 or Hurricane Andrew in Florida and Louisiana in 1993.

## Restrictive Zoning

Zoning laws can reduce loss of life and structural damage in two ways. Zoning can limit population density in a potentially hazardous area. Zoning also can specify the kind of structures permitted in a hazardous area. Examples include (1) denial of a permit to locate a nuclear reactor in an area with a known geologic fault under the land surface, (2) prohibition of development on the floodplain of a river, or (3) prohibition of development in front of the dune line in a coastal area.

In some earthquake-prone parts of California, the ground is cut by active fault lines. Despite this, housing developments, fuel storage tanks, utilities, hospitals, and schools still sprawl across the land (Figure 1–8). This construction occurred because the danger was not realized accurately at the time or because the risk was glossed over; now corrective measures are too late. Some newer areas being developed have used zoning to avoid areas of seismic risk.

## Engineering Solutions

Hazards are mitigated in some communities through engineering solutions. These include building codes that specify materials and construction techniques. Such codes minimize damage to structures during earthquakes (San Francisco) or hurricane winds (Miami).

Sometimes whole communities are protected by engineering structures. For instance, New Orleans is built along the Mississippi River, and much of the city is nearly 2 meters (over 6 feet) below sea level. Whole areas of New Orleans are protected by pumping stations, floodgates, and locks, and by floodwalls built above river level (Chapter 7).

**FIGURE 1–8**
Trace of the San Andreas Fault from Silicon Valley (in the distance) into the Pacific Ocean south of San Francisco. Most of this area was not developed before 1956. Note the proximity of part of the development (Daly City) to the delineated landslide area. (Photo from U.S. Geological Survey.)

A community that needed such an engineering solution, but constructed it too late, was Galveston, Texas. This city is built on a barrier island that is highly vulnerable to storms in the Gulf of Mexico. In 1900, a hurricane destroyed the city and claimed 6,000 lives. Had a seawall been in place, it might have prevented this great loss of life, the largest in the history of U.S. disasters. Today, a massive seawall (Figure 1–9) protects most of Galveston from coastal flooding by future storms. Unfortunately, it would be financially impossible today to build such massive seawalls to protect other vulnerable cities built on barrier islands.

## Hazard Warning and Evacuation

Some communities rely on combinations of warning systems and evacuation plans. For example,

**FIGURE 1–9**
West end of the seawall on Galveston Island, Texas. Note the beach erosion along the developed coastal segment beyond the end of the seawall. (Photo by author.)

Hawaii uses sirens to warn coastal inhabitants of approaching catastrophic sea waves *(tsunami)* that are generated by distant earthquakes. The warning gives the population a chance to evacuate to higher elevations.

The Florida Keys, a string of islands off southern Florida, have been repeatedly devastated by hurricanes. Indeed, the most powerful hurricane in American history devastated the central Florida Keys in 1935. Further, the islands are difficult to evacuate, because there is only one practical evacuation route, U.S. Highway 1, along which the evacuation route and locations of hurricane shelters are marked (Figure 1–10). This system successfully guided tens of thousands of people off the Florida Keys and to mainland shelters before Hurricane Andrew hit the northernmost islands in August 1992 (Chapter 16).

**FIGURE 1–10**
Evacuation route–shelter location sign in the Florida Keys. Signs with this hurricane symbol guide motorists to safety away from threatened areas. (Photo by author.)

## Abandonment and Public Use of Hazardous Areas

Each hazard-mitigation approach described so far provides some protection for people and structures. In the long run, however, the best solution may be to abandon the use of certain hazardous areas as population centers, and to adapt them for communal use. An example is to permit only parks and athletic fields to be built on river floodplains (Figure 1–11). Another example is prohibiting or severely restricting home construction in earthquake-prone areas. Unfortunately, these actions are usually taken *after* disasters have struck.

Public recreation areas and wilderness areas, with minimal permanent structures, are a prudent use for hazardous areas (Figure 1–12). This land usage obviously reduces the risk to people, avoids the great expense of evacuating communities before a river flood or hurricane hits, and minimizes reconstruction cost afterward. In addition, this approach affords great numbers of people access to beautiful recreational areas.

**FIGURE 1–11**
Floodplain area zoned for public use. This flooding event caused minimal damage to the park facilities. (FEMA Floodplain Management.)

## SOME ENVIRONMENTAL PRINCIPLES

To manage geohazards and to minimize their effect, we must understand some basic environmental principles. We also need to understand how we have altered our environment to suit our needs.

### Earth Is a Closed System

A common misconception is that flushing pollutants or blowing them away somehow gets rid of them. This is the "out of sight, out of mind" or "the solution to pollution is dilution" strategy. However, pollutants only *appear* to go away. In actuality, they may relocate from one place to another or from the land to the atmosphere or ocean. They may even become absorbed and hidden, or chemically altered. But they still are in our environment.

As mentioned, this strategy was acceptable when Earth's population was much smaller than today.

**FIGURE 1–12**
Robert Moses State Park on the south shore of Long Island. (Photo by author.)

Geologists understand the fallacy of such thinking today, because Earth neither gains nor loses significant material to space—it is merely transferred and later recycled. Scientists refer to this internal cyclic transformation as a **closed system.*** In short, if you dump something toxic, sooner or later it affects you or others.

For years, wastes were dumped in rivers, which carried them to the sea. At the same time, ocean dumping of wastes (e.g., sewage sludge) and toxic materials (e.g., acids) was permitted. These practices resulted in the transport and dispersal of harmful substances elsewhere by ocean currents. In recent years, toxic waste dumping has become a great concern. The practice is being eliminated to an increasing degree in many countries.

The concentration of heavy metals and toxic chemicals in fish and animals in oceans, rivers, and lakes is increasing in many areas. These contamination problems are most severe in water bodies with restricted circulation, such as the Mediterranean Sea or the Great Lakes. The harmful elements, coming from a variety of sources, are incorporated into plants and animal tissues and are successively concentrated in higher and higher organisms in the food chain. We humans, at the top of the chain, must keep this in mind as we enjoy our presumably "healthy" seafood dinners.

## Throwing the Trash Over the Fence: Displaced Problems

In some cases, the solution to a geologic environmental problem in one area results in development of the same problem, or a different one, in another area. This is a **displaced problem.** For example, taller smokestacks in midwestern U.S. industrial areas have greatly reduced local industrial pollution. However, industrial gases, which are injected into high-altitude winds, now are carried much farther eastward and have increased acid rain in the northeastern United States and Canada (Chapter 11).

Along shorelines, migrating sand replenishes eroding beaches. Where erosion has been severe, structures called groins are built perpendicular to the coast in order to trap migrating sand and build up the beach. Groins have been locally successful in their intended purpose, but they often create a displaced problem by causing erosion problems on the *other* side of the last groin. This is because they prevent the flow of sand from replenishing those eroding beaches, resulting in marked shoreline erosion and structural damage (Figure 1–13).

**FIGURE 1–13**
Shoreline erosion induced by coastal engineering structures in Erie County, Pennsylvania. Sand moves along the shoreline from upper right to lower left and is trapped by the groins (structures extending perpendicular to the shore). The area to the left of the last groin is sand starved and the shore is being actively eroded, placing the house at risk. (Photo from Pennsylvania Department of Environmental Resources.)

Few geologic environmental problems are ever truly solved, but some have been reduced in severity. To reduce severity and to prevent displaced problems, it is essential to understand the *dynamic and interrelated* nature of Earth processes. Unless we do, the solution to a problem in one area may lead to the development of another problem in an adjacent area.

## Recurrence Interval: It Hasn't Happened in Years, So . . .

Even the worst disasters, such as a major river flood or an earthquake, begin to fade from people's memories after five to ten years. After one generation (about twenty-five or thirty years), recollections of the event become mostly anecdotal, fondly offered by older people with sentences that begin,

*Earth receives some material from space (meteors, asteroids, comets, dust) and loses some material to space (lightweight gases in the atmosphere), so it is not a pure closed system. As far as pollutants are concerned, however, Earth is a closed system.

**FIGURE 1–14**
Dauphin Island, Alabama, near Mobile, seven years after Hurricane Frederick (1979) destroyed most of the structures and rebuilding was underway. By 1994 the area had been almost completely rebuilt. (Photo by author.)

"Why, I remember the great flood that . . ." The attitude becomes "it hasn't happened in years, so why worry?"

Geologists encounter this kind of wishful thinking in people who promote property development in hazardous areas. However, geologists know that this is an unrealistic way to assess the probability of occurrence of a major hazard.

Geologists are trained to consider the *long-term history* of an area. They understand that if a destructive geologic event has occurred here before, it *undoubtedly* will occur again. The problem is that we cannot predict *exactly when* a hazard will recur.

For example, coastal Mississippi was hit by two exceedingly powerful hurricanes in a ten-year period (Hurricane Camille in 1969 and Hurricane Frederick in 1979). Usually, the expected time between such powerful storms—the **recurrence interval**—is much more than ten years. Recurrence intervals are statistical probabilities based on *averages over a long period of observation.* Does this mean that a very long time must elapse before the next major hurricane hits the Mississippi coast? Probably not. Should development on the barrier islands (Figure 1–14) continue at all in the meantime? If so, under what restrictions?

Less-intense hazards can lull a population into complacency. For example, consider the history of earthquake activity along the San Andreas Fault System in California. Every year, numerous minor earthquakes, and some significant ones, shake communities near faults in this system. People have learned to accept this phenomenon because most have experienced only minor quakes during their lifetimes.

The last "great earthquake" in southern California (Los Angeles) was in 1854, and the last in northern California (San Francisco) was in 1906 (Chapter 4). It has been estimated from the geologic record (Chapter 4) that the recurrence interval for a major quake on the San Andreas system is 130 to 200 years. Thus, these areas are primed for a major seismic event. Because recurrence intervals are averages, this event may not recur for many more years, or even decades. On the other hand, it could be happening as you read this sentence! In such highly urbanized and populated areas, these statistical probabilities should be given greater credence before additional urban or suburban developments (Figure 1–9) are implemented.

## Disturbing Dynamic Equilibrium

You may have heard the phrase "the balance of nature." In geology, this means that natural systems maintain a balanced state over long periods. We call this **dynamic equilibrium.** Note that it is not a *static* equilibrium, but a *dynamic* equilibrium, in which a change in one part of the system tends to become balanced by change in another part to preserve overall system equilibrium.

Even catastrophic events are "absorbed" and adjusted for by dynamic equilibrium, over the long term. For example, a severe hurricane may seem to "destroy" a coral reef. However, reefs have evolved to be tough, wave-resistant structures *as long as they have time to regenerate and if the surviving corals have favorable living conditions.* If these conditions are met, the damaged reef will regenerate by growing a new layer of live coral over the storm-torn rubble.

Other factors can disrupt dynamic equilibrium, however. Natural regrowth may be slowed or prevented if pollution inhibits new coral growth (Chapter 14). This is one of many ways in which people interfere with geologic systems that have been in dynamic equilibrium for millions of years.

## LOOKING AHEAD

Before we contemplate modifying geologic systems, it is important to understand how they work and how they interrelate. Chapters 2 and 3 provide a general overview of geologic systems. Then, Chapters 4 through 16 present the geohazards of living on Earth, both the natural hazards and the human ones that result when we interfere with geologic processes.

## SUMMARY

Geohazards (geologic hazards) are natural Earth processes that adversely affect humans and property. These dangers include two broad classes, natural and human. Our interference with geologic processes has increased the occurrence of geohazards and the devastation from them. In some cases, human activity can trigger a geohazard.

Magnitude is the *power* of a destructive event, indicating how much energy was released. Frequency is *how often* events of a given magnitude occur or are expected to occur. Large-magnitude events cause greater devastation but fortunately they occur with far lower frequency. Some hazards, such as landslides, have a local effect; others, including earthquakes and hurricanes, have a regional effect; and some, such as volcanic dust clouds (Chapter 4), affect weather worldwide and thus have a global effect. Many hazardous events are preceded by warning precursors.

Human geohazards result from activities that accelerate or alter normally benign natural processes into problems. Human geohazards differ from natural ones in that humans both *affect* a geologic process and then are *adversely affected* by its results. In general, human geohazards develop slowly and there is greater promise for control.

Rapid population increase creates even greater demand for such resources as land, water, and metallic and nonmetallic materials. The unfavorable alteration of our environment, wholly or partially resulting from human actions, is called pollution. As population becomes increasingly concentrated in cities, the effects of pollution are accentuated. Population growth has forced expansion of housing and industrial areas into potentially hazardous areas.

Potentially hazardous situations can be dealt with in several ways. Each situation depends on (1) environmental awareness and hazard perception, (2) commercial consideration, and (3) fiscal restraints. In some cases, the hazard is incorrectly perceived to be small and little action is taken.

Various means are used to address hazards. Some communities use zoning laws to limit population and development in potentially hazardous areas. In other places, hazard warning systems and evacuation plans are used to protect lives. Structural reinforcement through building codes also is used to minimize damage and provide more protection to occupants in case evacuation is not possible. Finally, potentially hazardous areas may be abandoned or used for low-density public-use development (parks, beaches, athletic fields).

Managing geohazards requires us to keep four basic principles in mind:

1. Earth is a closed system (if you dispose of a toxic substance in the ground, water, or air, sooner or later it affects you or others).
2. An attempt to solve a geologic problem in one area commonly creates a displaced problem in another area.
3. Most people have a poor perception of hazard frequency and magnitude in their area.
4. Most geologic systems are in dynamic equilibrium. Attempts to modify the natural system may result in a hazardous situation.

Continuing population growth is forcing the development of hazardous areas that once were remote. It is straining Earth's mineral, food, water, and energy resources. It is increasing anthropogenic hazards as humans adversely affect geologic systems. Pollution problems long have ceased to be local; now they are more regional or international in scope, because harmful substances are carried great distances by wind, river flow, and ocean currents. To ensure quality of life in the future, it is essential to take remedial action now.

## KEY TERMS

anthropogenic
closed system
displaced problem
dynamic equilibrium
frequency
geologic hazard (geohazard)
magnitude
metallic resources
nonmetallic resources
nonrenewable resources
pollution
precursors
recurrence interval
renewable resources
resource
urban concentration factor

## REVIEW QUESTIONS

1. a. Distinguish between a *natural geohazard* and a *human geohazard*.
   b. Provide three examples of each.
2. Explain this statement: Geohazards are natural processes, but sometimes they are triggered by humans.
3. a. List geohazards you have heard about, read about, or observed in your area.
   b. In each case, describe attempts, if any, made to reduce the risk to people and structures.
4. Distinguish between the *magnitude* and *frequency* of a geohazard.
5. Describe an example of each of the following:
   a. a displaced problem.
   b. a geohazard triggered by human activity.
   c. an urban concentration problem.
   d. a geohazard that is being solved.
   e. a problem caused by manipulating a geologic system.
6. Choose a place where people could live without being subjected to a geologic hazard. Where would it be? Explain why no hazards exist there. Discuss with the class.
7. From the standpoint of geologic hazards or geologic environmental problems, where would you consider to be the most dangerous place to live in the United States? In Europe? Asia? South America? Africa? Antarctica? On Earth? Discuss with the class.

## FURTHER READINGS

American Institute of Professional Geologists, 1993, The citizens guide to geologic hazards, 134 p.

Coch, N. K., and Ludman, A., 1991, Physical geology: New York, Macmillan, 678 p., Chapter 21 (Resources).

Keller, E. A., 1992, Environmental geology, sixth edition: New York, Macmillan, 521 p., Chapter 1 (Philosophy and Fundamental Principles).

Skinner, B. J., and Porter, S. C., 1987, Physical geology: New York, Wiley, 750 p., Chapter 1 (The Human Planet).

U.S. Geological Survey, 1981, Facing geologic and hydrologic hazards, Professional Paper 1240-B, 108 p.

U.S. Geological Survey, 1978, Nature to be commanded, Professional Paper 950, 95 p.

# 2

# *Earth's Interior*

A major earthquake reduces an urban center to ruins. A volcanic eruption buries nearby villages in ash and mud. A coastal area is destroyed by a hurricane before the storm cuts a path of destruction 100 miles wide as it moves far inland. Rapid melting of ice and snow results in massive river flooding that destroys structures and injures people. What do all of these large-scale hazards have in common? Each of them releases great amounts of energy.

**Energy** is the ability to do work, which can be constructive, like heating a home or generating electricity, or destructive, like an earthquake or lightning strike. Energy exists in two states, stored (potential) or active (kinetic). It can be stored, like the **potential energy** in gasoline. When gasoline is burned in a car's engine, this energy is converted to **kinetic energy** that moves the car.

Energy takes six basic forms: mechanical energy (for example, a landslide), heat energy (anything hot), chemical energy (from chemical reactions), electrical energy (electricity), nuclear energy (nuclear power), and electromagnetic energy (ultraviolet, light, radio). These forms of energy are constantly transformed from one into another. For example, the sun's nuclear energy is transformed into electromagnetic energy waves that travel to Earth. When they reach Earth, they are transformed into heat energy, which stirs the atmosphere (mechanical energy).

Energy performs work on *matter*. Whether we are looking at individual atoms and molecules of matter, or mineral grains, or large rocks, or a mountain, or the entire Earth, we are watching energy at work on matter.

Earth processes are powered by two major energy systems. The *internal energy system* powers earthquakes and volcanoes and is derived from Earth's internal heat. The *external energy system* powers surface processes such as winds, rivers, and glaciers and is driven by gravity and by the sun's energy. To a lesser degree, Earth's rotational energy affects the external energy system. This chapter describes Earth's *internal* energy system and how it works on matter to produce minerals, rocks, and major movements in Earth. In Chapter 3, we will look at Earth's *external* energy system.

**Folded sedimentary rock layers. (Photo by Michael Collier/Condit.)**

*Special Note:* This chapter and the following one summarize those aspects of physical geology most relevant to geohazards. Some of it will review familiar material that you learned earlier in your education. There also will be many new terms and concepts. If you spend a little extra time on this material, it will help you understand and enjoy the rest of the book. Following these two chapters, we will return to the study of geohazards in Chapter 4 to look at the problems of living with volcanoes.

## MATTER

All Earth materials exist as solids, liquids, or gases, depending on temperature and pressure. A substance in any state can be transformed (at the same pressure) into another state, simply by adding or removing enough heat energy. The most familiar example is water, which we routinely convert to ice by removing heat energy or convert to water vapor by adding heat energy. The fact that water changes state so readily is especially significant because water affects the development of many geologic hazards.

All substances are made of *atoms,* which in turn consist of subatomic particles (Figure 2–1). Two types of particles form the central *nucleus* of an atom: *protons* have a mass of one and a positive electrical charge; *neutrons* have a mass of one and no electrical charge. Orbiting the nucleus at various atomic distances are *electrons,* particles having a negligible mass and a negative electrical charge.

The number of protons in the nucleus is the *atomic number.* The total number of neutrons and protons in the nucleus is the *atomic mass.* The more particles in an atom, the heavier it is. For example, the lightest atom, *hydrogen* (H), has only a single proton, but extremely heavy *uranium* (U) packs 238 protons and neutrons into its nucleus.

The electrical charge on an atom is very important. When the number of protons in the nucleus equals the number of orbiting electrons, the overall electrical charge of an atom is zero, or neutral.

A chemical *element* is a substance composed exclusively of atoms having the same atomic number. (A

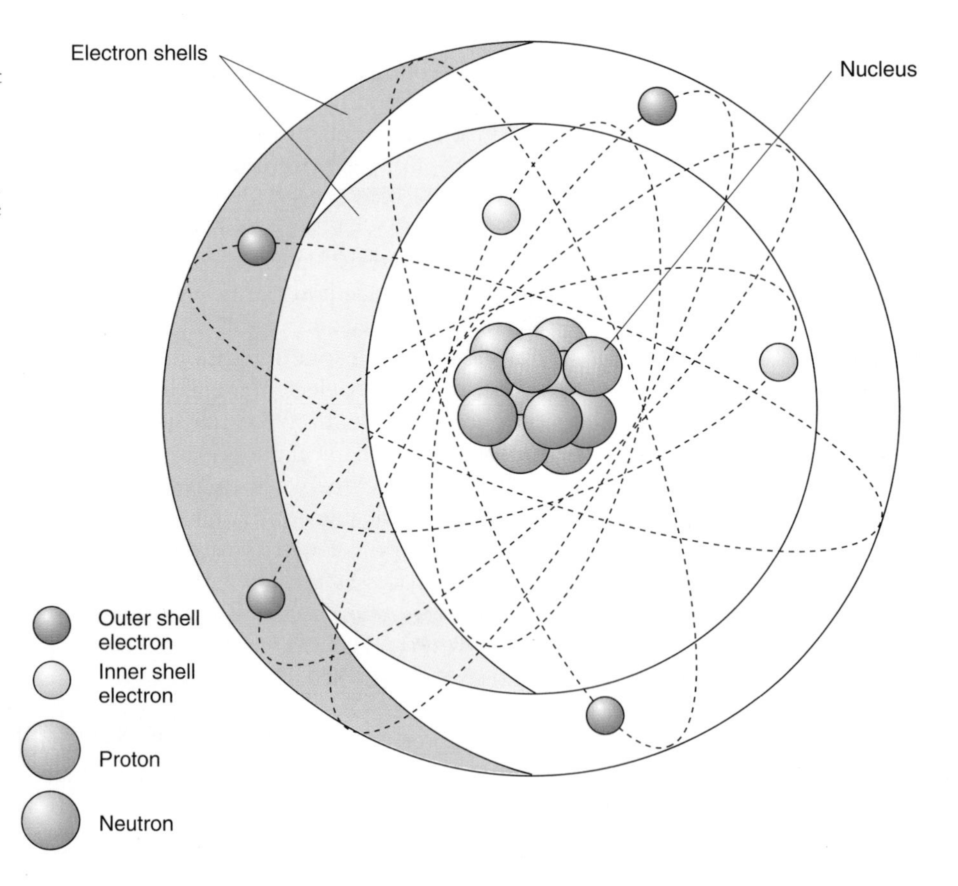


**FIGURE 2–1**
The structure of a carbon atom. Not all of the six protons and six neutrons can be shown because of perspective. The electrons orbit at different distances from the nucleus, two at the same distance close to the nucleus and four farther away. (Reprinted with the permission of Macmillan College Publishing Company from *Physical Geology* by Nicholas K. Coch and Allan Ludman. Copyright © 1991 by Macmillan College Publishing Company, Inc. Courtesy of Allan Ludman.)

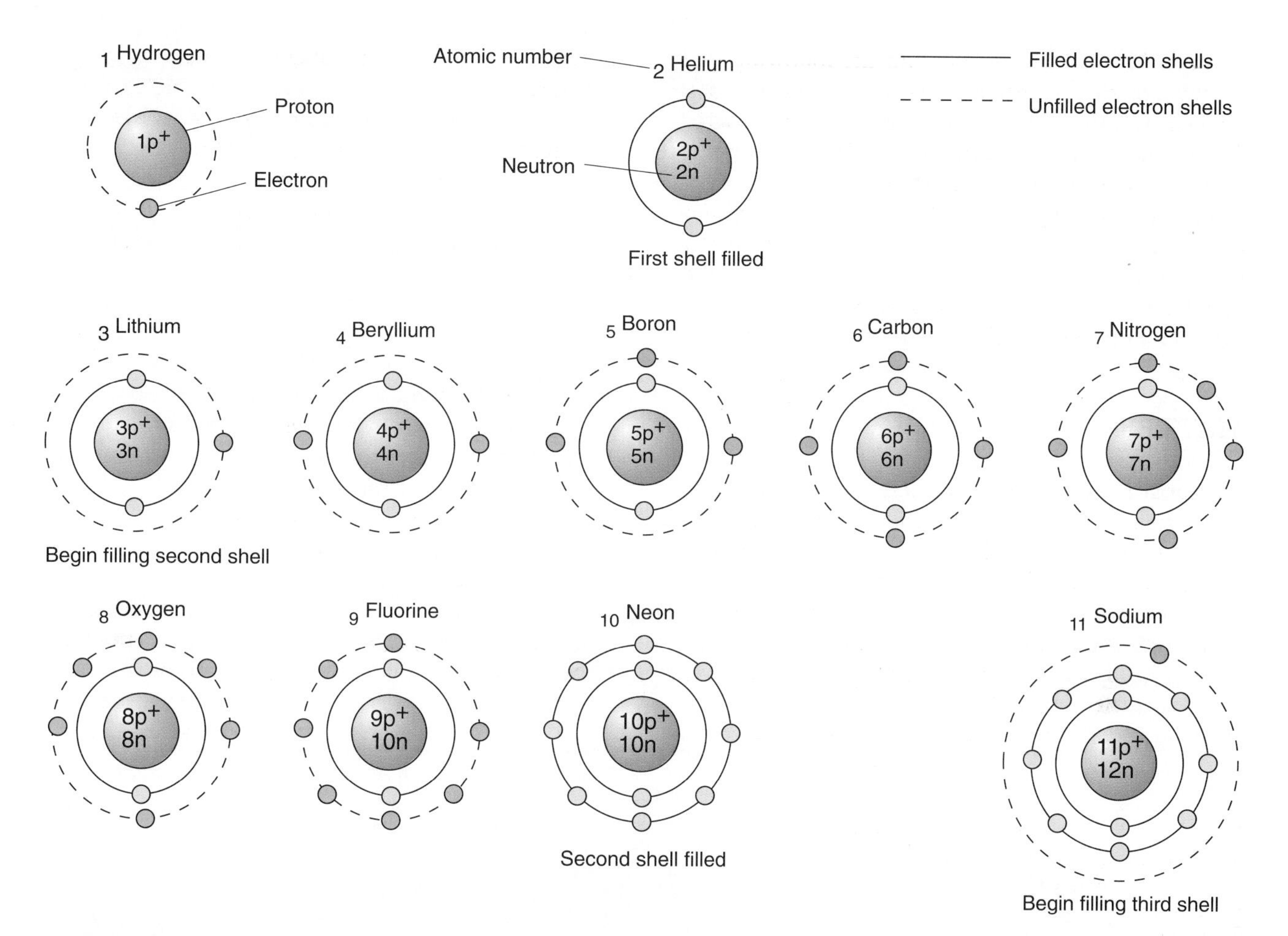


**FIGURE 2–2**
Of the 106 known chemical elements, 11 are depicted here, showing their atomic structures. (Reprinted with permission of Macmillan College Publishing Company from *Physical Geology* by Nicholas K. Coch and Allan Ludman. Copyright © 1991 by Macmillan College Publishing Company. Courtesy of Allan Ludman.)

chemical *compound* combines two or more elements to form a different substance.) One hundred and six different elements exist because the number of protons, neutrons, and electrons varies (Figure 2–2).

For example, the element *carbon* (C) has an atomic number of 6, and its structure contains six protons and six neutrons in the nucleus, giving it a mass of 12 (Figure 2–1). All atoms of carbon have six protons, but some atoms have more neutrons, giving them different atomic masses. Atoms of an element having different masses are called **isotopes** of that element. Carbon has three major isotopes: carbon-12 (six neutrons), carbon-13 (seven neutrons), and carbon-14 (eight neutrons).

Of the 106 known elements, 92 occur naturally and the remainder have been synthesized in the laboratory. *Oxygen* (O) and *silica* (Si) are by far the most abundant elements in Earth's crust (47% and 28% respectively), so more rocks contain these elements. The next-most-abundant elements in the crust include *aluminum* (Al), *iron* (Fe), *calcium* (Ca), *sodium* (Na), *potassium* (K), and *magnesium* (Mg). The remaining 98 elements together make up only 1% to 7% of Earth by weight, depending on how deep within Earth you look.

## NUCLEAR REACTIONS

Most elements are stable throughout time. They may combine with other elements to form compounds, but the elements themselves do not change. However, some elements have isotopes (variant forms) that are radioactive. *Radioactive isotopes* have unstable nuclei that break down by releasing energy or particles. Such changes in the nucleus are called *nuclear reactions.*

In a nuclear reaction, some material or energy is lost from the nucleus of an atom, converting the nucleus into the nucleus of a different element. We call this process *radioactive decay.* For example, an unstable uranium nucleus eventually decays to become a *lead* (Pb) nucleus. Some of the energy that held the original nucleus together is released as heat, which is why nuclear reactors and nuclear weapons give off so much heat.

Earth contains many radioactive isotopes, which decay steadily, regardless of the conditions around them, releasing heat. (There is no risk of a nuclear explosion inside Earth, because these elements are not concentrated enough.) These radioactive isotopes have undergone radioactive decay since their formation, and the amount of heat generated over geologic time has been tremendous. It is this heat that is responsible for the bulk of internal geologic processes: earthquakes, mountain building, and plate tectonics.

## HEAT ENERGY AND ITS MOVEMENT INSIDE AND OUTSIDE EARTH

Earth's internal and external energy systems are powered by heat energy. The heat energy moves from hotter areas to cooler ones by three processes: conduction, convection, and radiation (Figure 2–3). All three are equally important in moving heat energy, but they operate differently.

### Conduction

**Conduction** transfers heat energy directly from atom to atom in solids. If you hold a metal spoon in a flame, you will soon drop it, because heat is *con-*

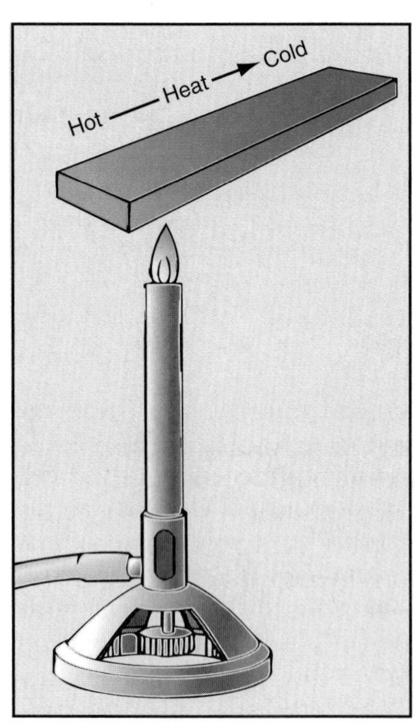


(a) Conduction

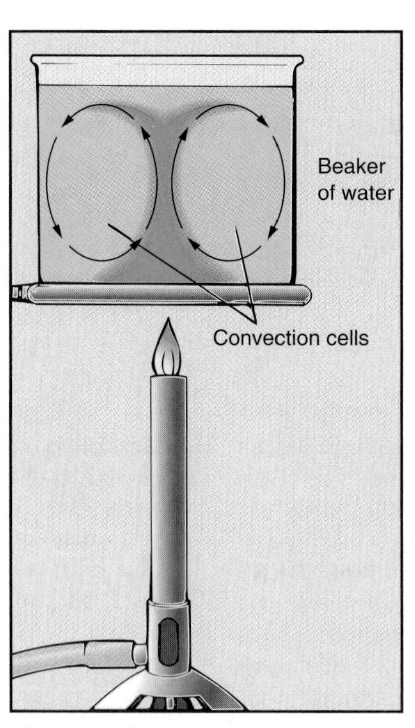


(b) Convection

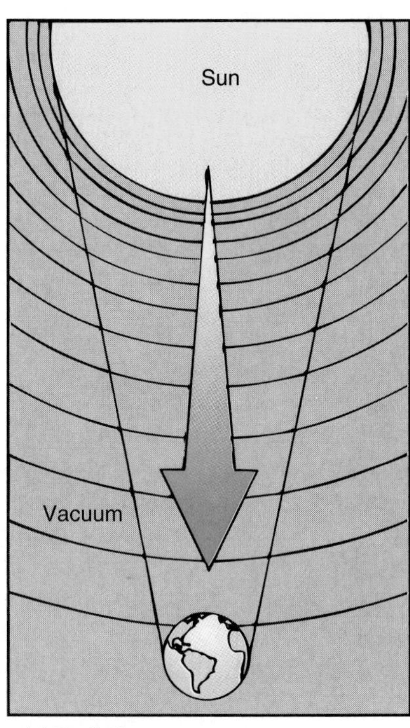


(c) Radiation

*ducted* rapidly from atom to atom along the spoon to your hand.

Each substance has a different heat *conductivity.* Metals are highly conductive, so they heat and cool rapidly. Rocks are much less conductive, so they are much slower to heat and cool. This low conductivity of rock is important, for it helps heat to build up inside Earth. The heat generated from radioactive decay within Earth occurs faster than it can be conducted away. Over time, this has led to a heat buildup so great that it has exceeded the melting point of the rock, changing its state from solid to liquid and making a molten rock called **magma.**

### Convection

**Convection** transfers heat energy through liquids and gases by moving *masses* of the heated substances (Figure 2–3b). You probably have watched this happen when heating soup; it is the slow turbulence you see before the soup actually boils. As a gas or liquid is heated, the atoms absorb this energy and grow more active, moving farther apart. As a result, the density of the substance decreases. The less-dense (warmer) material rises upward.

For example, when a fire is lit in a stove in the center of a cold room, the less-dense heated air rises and the denser, cooler air moves in to take its place, creating convection, which eventually warms the whole room. Conduction is involved here, too—the heat of the fire is *conducted* through the metal of the stove where it heats the air touching the stove. The warmed air then rises to the ceiling as a warm, moving *convection current.* The air eventually cools, grows denser, and returns to the floor to be heated again.

Eventually, this circulating movement, called a **convection cell,** distributes air throughout the room, warming it. Convection is an extremely important method of moving fluids in and on Earth. Inside Earth, molten magma moves very slowly (over thousands and millions of years) by convection. On the surface, the ocean waters are slowly stirred by convection currents. In the atmosphere, convection currents influence our daily weather (Chapter 3).

### Radiation

*Radiation* transfers heat energy by electromagnetic waves, even through the vacuum of space. You feel heat radiation by standing near an open fire or stove or in sunshine. Radiation is an essential part of Earth's external energy system, described in Chapter 3. Nuclear reactions in the sun produce radiant electromagnetic energy that travels to Earth.

## CHEMICAL REACTIONS AND COMPOUNDS

In the nuclear reactions described previously, changes occur only in the nucleus of atoms, which gain or lose nuclear particles. In contrast, in a **chemical reaction,** changes occur *only in the electrons* orbiting the nucleus. By means of these electrons, atoms become *bonded* to each other without altering their nuclei.

Recall that electrons orbit their nucleus at different atomic distances. Electrons that orbit at the same distance are said to belong to the same *electron shell.* (This is not a real, physical shell, but an energy level that is called a "shell" for easy visualization.) In the carbon atom (Figure 2–1), electrons orbit the nucleus at two levels ("shells"). Rigid physical laws limit the number of electrons that can fill each electron shell. The innermost shell can hold two, the second shell can hold eight, the third can hold eight, and so on. You can see this structure in the elements shown in Figure 2–2.

Atoms of elements that have *filled* outer shells are called *inert* because they are electrically balanced and cannot enter into chemical reactions. Examples are the elements *helium* (He, atomic number 2), and *neon* (Ne, atomic number 10), shown in Figure 2–2.

However, atoms that have *unfilled* outer electron shells are electrically unbalanced, or unstable. When such an atom comes near another atom that has an incomplete outer electron shell, it tries to bond one of the other atom's outer electrons. The other atom does the same thing. This bonding is a *chemical reaction.* The result is that the two atoms bond to each other by means of their outermost electrons. This can happen in so many different combinations that thousands of different chemical reactions exist.

When elements combine in a chemical reaction, they form a *compound.* Consider the chemical reaction between the elements sodium (Na) and chlorine (Cl), illustrated in Figure 2–4. The sodium atom has a single electron in its outermost shell, giving it a negative charge (surplus electron, like the

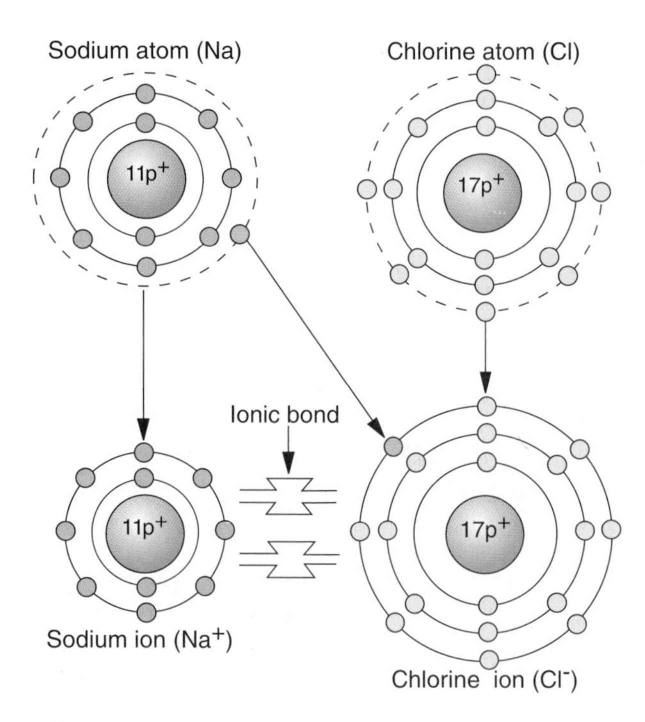


**FIGURE 2–4**
Ionic bonding of sodium ($Na^+$) and chlorine ($Cl^-$) to form NaCl, sodium chloride or common salt. (Reprinted with permission of Macmillan College Publishing Company from *Physical Geology* by Nicholas K. Coch and Allan Ludman. Copyright © 1991 by Macmillan College Publishing Company. Courtesy of Allan Ludman.)

negative end of a battery). It can become stable if it can lose the electron. In the case of the chlorine atom, a single electron is needed to complete its outer shell, which has seven, giving it a positive charge (missing an electron).

When the chemical reaction occurs, and the two atoms suddenly are joined to share the sodium's extra electron, the configuration for both becomes stable. The charge of each atom changes, however. The sodium atom lost an electron, giving it a net *positive* charge (because its nucleus has 11 protons, but only 10 electrons). The chlorine atom gained an electron and now has a net *negative* charge (17 protons in its nucleus, but 18 orbiting electrons). This type of bonding, in which atoms gain or lose electrons to fill their outermost shells, is called *ionic bonding.*

When an atom gains or loses an electron in a chemical reaction, the charged particles that result are call **ions.** Ions are distinguished from atoms with a + sign or – sign ($Na^+$ and $Cl^-$).

## MINERALS

The solid earth is made up of minerals. To be called a **mineral,** a substance must meet several criteria:

- It must be a solid.
- It must occur naturally (even though it also may be made in a laboratory).
- It must be inorganic.
- It must have an orderly arrangement of its atoms (Figure 2–5a).
- It must have a specific chemical composition (like quartz, $SiO_2$) or vary within a well-established range of composition.

A substance must meet *all* these criteria to be a mineral. For example, quartz meets all these criteria, and therefore is a mineral. But coal is organic in origin, being composed of once-living fossil plant material, so it is not a mineral, even though it is a naturally occurring solid.

Under favorable growth conditions, minerals form regular geometric shapes called **crystals** (Figure 2–5b). Crystalline structure is a consequence of the orderly internal arrangement of atoms (Figure 2–5a). The number of faces on a crystal and the angles between them are characteristic of each mineral and can be used in identifying them.

Only a few minerals occur in the pure state—not chemically combined with other elements. A notable example is gold. Most minerals, however, are compounds of two or more elements. Because silica and oxygen are the most abundant elements in Earth, most minerals are silica-oxygen compounds called *silicates.* These minerals are combinations of silica and oxygen, a common example being quartz ($SiO_2$). Often other elements are incorporated, too. More than 3,000 distinct minerals have been identified by geologists.

In the chemical reactions that form a compound, the component elements lose their individual characteristics, and the new compound has its own. A fine example is the chemical reaction of sodium and chlorine to form the mineral *halite* (NaCl), or common table salt (Figure 2–5b). By itself, sodium is a violently reactive metal that must be stored away from air (it is kept in kerosene). Chlorine is a poisonous green gas. However, the chemical reaction that bonds these two dangerous elements produces a benign mineral, salt, which we eat every day.

Minerals form under a wide variety of conditions below and on Earth's surface:

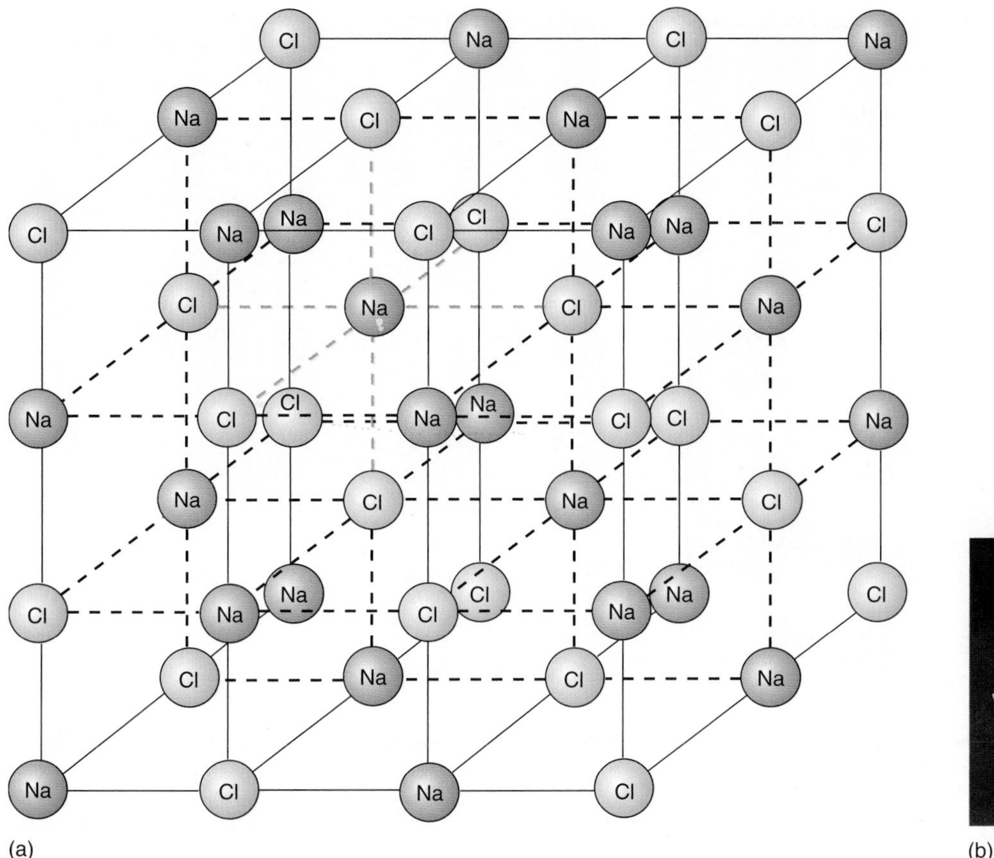

**FIGURE 2–5**
(a) The internal atomic structure of halite or sodium chloride. The structure is composed of an array of alternating Na+ and Cl– ions. (b) Crystals of halite. (Part (a) reprinted with permission of Macmillan College Publishing Company from *Physical Geology* by Nicholas K. Coch and Allan Ludman. Copyright © 1991 by Macmillan College Publishing Company. Courtesy of Allan Ludman.)

1. Minerals crystallize from *magmas and hot water solutions* as they slowly cool beneath Earth's surface. As the liquid cools, some minerals form from chemical reactions; with further cooling other minerals may form (Figure 2–6).
2. Minerals form *under great pressure* below the surface. Changes in pressure and/or high temperatures transform some minerals into different ones, which are more stable at the higher pressure or temperature (Figure 2–7).
3. Minerals form *wherever more surface water evaporates than can be replenished.* Examples are in desert basins (Death Valley, California) or around the edges of salt flats and ponds along hot, dry coastlines such as the Red Sea (Figure 2–8).
4. Minerals form *through the activities of aquatic organisms and plants.* Clams and other animals extract simple elements and compounds from seawater or lake water and combine them to form the mineral *calcite* (calcium carbonate, $CaCO_3$), which is used in building their shells. Some algae in shallow tropical seawater remove calcium carbonate from the water to form calcite. The calcite crystals along with organic material make up the mass of the algal structure. Upon the algae's death, the organic material decomposes and the calcite crystals sink to the seafloor to form calcium carbonate mud or sand. (This calcium carbonate is not an exception to the definition of a mineral; it is considered a mineral because it has all the properties of one, even though its crystallization was stimulated by organisms.)

FIGURE 2–7
These garnet crystals in a rock called schist formed as a result of tremendous pressure on shale.

FIGURE 2–6
These crystals of quartz formed from hot solutions underground.

FIGURE 2–8
Evaporite deposition in Death Valley, California. (Photo by author.)

## ROCKS

Rocks are composed of minerals. Whereas a mineral has an orderly arrangement of its atoms and a definite chemical composition (or a specific range), a rock commonly is a *mixture* of minerals. **Granite** is a good example, typically containing three principal minerals: quartz, mica, and feldspar. In a few cases, a rock may contain a single mineral, as in pure halite "rock salt" (NaCl), pure quartzite ($SiO_2$), or pure *limestone* ($CaCO_3$).

There are three types of rock, each formed in a different way:

1. *Igneous* rocks form from the cooling of molten rock underground and on Earth's surface (examples: granite, basalt).

2. *Metamorphic* rocks form from high pressures and temperatures within Earth (examples: slate, marble).
3. *Sedimentary* rocks form from compaction of sediment on or just below Earth's surface (examples: sandstone, shale).

We will now look more closely at each type.

## Igneous Rocks

**Igneous rocks** form by the cooling of molten rock. (*Igneous* means "fire-formed"; this is easy to remember if you think of "igniting" a fire.) Liquid magma inside Earth is less dense than the solid rock around it, so it slowly rises. Two things can happen: the magma may cool underground and solidify to form **plutonic (intrusive) rock.** Or, the magma may break through Earth's surface as **lava** from a volcano. The cooling of lava forms **volcanic (extrusive) rocks.**

Plutonic and volcanic rocks derived from the same magma have similar *chemical* compositions, as you would expect. However, they are different in their *physical* appearance. They also may contain the same minerals, but their crystal sizes are different. The main reason for such differences is the amount of time that the molten rock had to cool. If it cools very fast, as in a volcanic eruption, crystals will be tiny, microscopic, or even nonexistent. If it cools very slowly, crystals have time to grow large, commonly in the 1 to 2 centimeter (.4 to .8 inch) range, and rarely to a meter (3.3 feet) or more in length.

The crystal size in an igneous rock also depends on how viscous (thick) the magma or liquid is. The **viscosity** (stiffness) of a fluid indicates its resistance to flow. Fluids like water are not very viscous, but others, like poured concrete, are highly viscous.

As molten rock cools, small "seed" crystals form and grow larger as they bond to ions in the fluid, which are compelled by their electrical charges to migrate toward the crystals. If the magma cools slowly, there is plenty of time for seed crystals to grow and for ions to migrate to join them. But if the magma cools rapidly, seed crystals have less time to grow. In addition, a fast-cooling magma or lava becomes viscous more quickly and the ions cannot migrate as easily to the seed crystals. Consequently, plutonic rocks have larger crystals than volcanic rocks.

What happens if molten rock cools *very* quickly (is quenched)? This commonly happens with explosive volcanic eruptions. In this case, no crystals can develop and filaments or fragments of **volcanic glass** form. Like window glass, volcanic glass is actually a super-cooled liquid and, given enough time, it will gradually flow. (Old window panes may have a wavy surface and be thicker at the bottom as a result of very slow flow under gravity.)

Igneous rocks can be divided into three major families on the basis of their composition: granite, andesite, and basalt. These groups are distinguished by their composition, color, density, and where they occur (Table 2–1). These facts about igneous rocks will become important in Chapter 4, when we look at the effect that these properties have on volcanic hazards.

Each of the three rock families has an *volcanic* form and a *plutonic* form, listed at the top of the table. For example, if a granitic magma cools within Earth, it is called a *granite.* If the same magma is extruded as lava onto the surface, the lava cools quickly there to form a *rhyolite.*

## Sedimentary Rocks

Sedimentary rocks are produced by processes acting on Earth's surface. As such, they logically could be presented in the next chapter, "Earth's Surface." However, we treat them here to keep the discussion of the three rock types in one place and to emphasize the relation among the three types.

**Sedimentary rocks** are produced from the deposition and compaction of *sediment* at or near Earth's surface. **Weathering** by rain, snow, wind, freezing, and thawing relentlessly breaks down existing rocks to produce particles of **sediment**—sand grains, tiny silt and clay particles, crystals formed by evaporation of surface water, and sediments produced by plants and animals (Figure 2–9).

Sediment particles are *eroded* (removed) from their original site and transported by rivers, glaciers, wind, and downslope movement across Earth's surface, often for long distances. Often, they are deposited and eroded repeatedly. Eventually, the sediment reaches a *basin of deposition* from which it cannot be eroded any more. Examples of basins of deposition include low places like lakes, oceans, and desert basins that have no outlet.

Over time, sediments accumulate layer by layer in the depositional basin. Such layering or **bedding**

**TABLE 2–1**
Characteristics of the three major igneous rock families.

| | Granite Family | Andesite Family | Basalt Family |
|---|---|---|---|
| *Location* | | | |
| If volcanic (extrusive, or above ground): | Rhyolite | Andesite | Basalt |
| If plutonic (intrusive, or below ground): | Granite | Diorite | Gabbro |
| *Property* | | | |
| Silica ($SiO_2$) content | High | Intermediate | Low |
| Viscosity of magma/lava | High | Intermediate | Low |
| Major elements (excluding silica) | High in Al, K, Na Low in Ca, Mg, Fe | Mixture of elements common in basalt and granite families | High in Ca, Mg, Fe Low in Al, K, Na |
| Color | Light | Intermediate | Dark |
| Density | Lowest | Intermediate | Highest |
| Explosivity (volcanic rocks only) | Highest | Intermediate | Lowest |
| Occurrence | Continents only | Island arcs; continent edges | Primarily ocean basins; also on continents |

is the single most significant characteristic of sedimentary rocks (Figure 2–10). As sediments become buried, the pressure on the grains increases. This pressure forces the grains together and expels the water and air that was trapped between the particles in a process called **compaction** (Figure 2–11a).

At the same time, fluids containing dissolved ions circulate through the sediments. They deposit minerals in the open spaces between these grains in a process called **cementation** (Figure 2–11b). The combination of compaction and cementation eventually produces solid sedimentary rocks from the original unconsolidated sediment layers (Figure 2–10b).

Some sediment particles are produced from the breakdown of older rocks. For example, the breakdown of granite frees crystals of quartz, mica, and feldspar. When these particles are buried, compacted, and cemented, they produce a *clastic* sedimentary rock. Clastic rocks are divided into major groups and named based on particle size, as listed in Table 2–2.

The size names are preceded by modifiers that describe the composition of the particles—see examples in last column. Thus, a clastic rock composed of quartz particles between 1.0 and 2.0 millimeters (.04 and .08 inch) is called a *quartzose sandstone*. An iron-bearing shale is called a *ferruginous shale* (after *ferrum*, Latin for iron).

Not all sedimentary rocks are formed of clastic particles. Other sedimentary rock types are biogenic and chemical.

*Biogenic rocks* are sedimentary rocks produced by the action of plants and animals (Figure 2–9). A good example is *coal*, a rock produced from the burial and compaction (Figure 2–11a) of swamp vegetation. Over millions of years, the buried vegetation is transformed into a carbon-rich rock. A more wide-

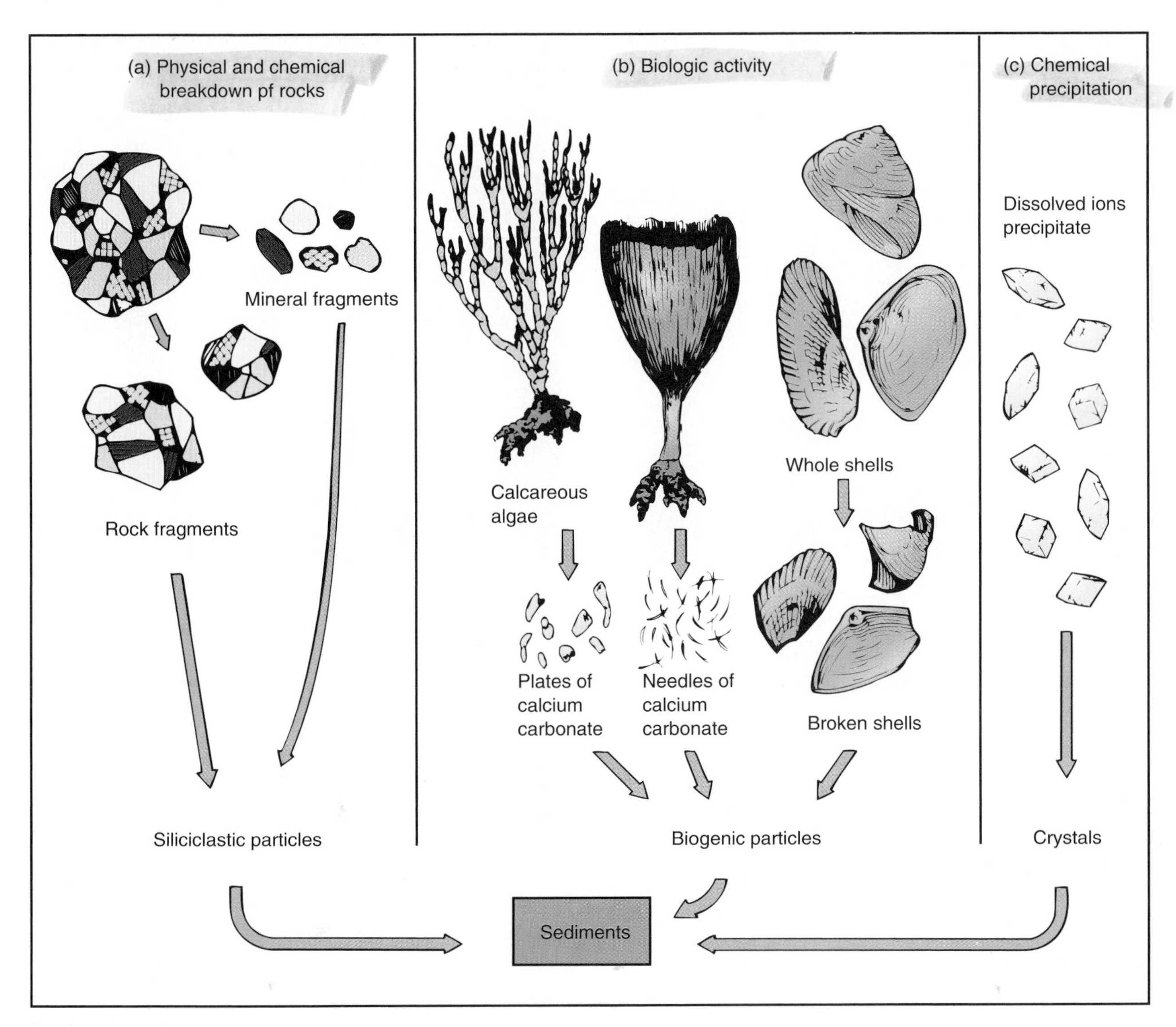


**FIGURE 2–9**
The origin of sediments. Sediments are composed of various percentages of (a) clastic particles from the physical and chemical breakdown of older materials, (b) debris from biological activity, and (c) crystals formed from concentrated solutions. These are the components that form sedimentary rocks when sediments are compacted. (Reprinted with permission of Macmillan College Publishing Company from *Physical Geology* by Nicholas K. Coch and Allan Ludman. Copyright © 1991 by Macmillan College Publishing Company.)

spread biogenic rock is **limestone,** produced by the deposition of *aragonite* (calcium carbonate) crystals and their subsequent compaction. The calcium carbonate is produced by the activities of plants such as calcareous algae (Figure 2–12a) and shellfish such as clams (Figure 2–9). Compaction and cementation of calcareous mud and shell material produces the rock *limestone* (Figure 2–12b).

*Chemical rocks* are produced by crystals deposited from concentrated solutions. Chemical sediments are deposited wherever water containing dissolved ions becomes concentrated by evaporation or where the supply of fresh water to dilute the ion concentration is inhibited. For example, beds of salt (halite) are produced by evaporation of shallow coastal bays.

(a)

(b)

**FIGURE 2–10**
(a) Layering in modern beach sand, Long Island, New York. (b) Layering in ancient sand, now sedimentary rock. (Photo (a) by author; (b) courtesy of Fred Wolff.)

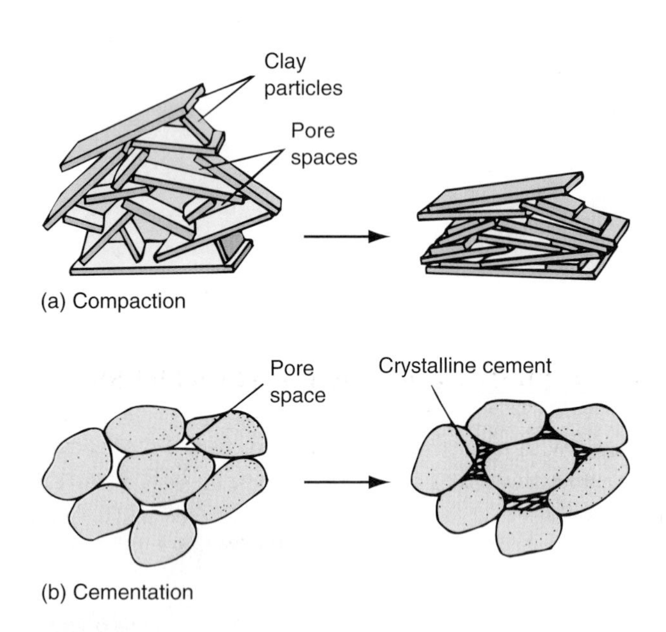


**FIGURE 2–11**
(a) Compaction of sediments and (b) cementation. These processes typically require thousands or millions of years. (Reprinted with permission of Macmillan College Publishing Company from *Physical Geology* by Nicholas K. Coch and Allan Ludman. Copyright © 1991 by Macmillan College Publishing Company.)

## Metamorphic Rocks

**Metamorphic rocks** have undergone change in their characteristics. The name is easily understood: *meta* means "change," *morphos* means "form." The changes happen through *recrystallization* of other rocks when they are subjected to extreme changes in temperature, pressure, the presence of hot solutions, or any combination thereof. These processes generally occur deep within the Earth.

The process of **metamorphism** may significantly alter the properties of the original rock. This may entail change in crystal size, shape, and composition, as well as an alignment of the minerals (Figure 2–13).

The largest volume of metamorphic rocks is produced where temperature, pressure, and hot solutions all play significant roles over a wide area (hundreds of square kilometers) to produce *regional metamorphism.* Regional metamorphism requires large-scale forces such as mountain-building (discussed later in this chapter.)

A common example of metamorphic change is the transformation of shale into *schist.* Shale is made of clay-sized particles (Table 2–2) of silicate miner-

**TABLE 2–2**
Names of clastic sediment particles and the clastic rocks they form.

| Particle Size | Particle Name | Clastic Rock Name | Sample Name with Composition |
|---|---|---|---|
| Larger than 2 mm (.078 in) | Gravel | Conglomerate | Quartzose conglomerate (gravel comprises the quartz fragments) |
| 2–0.063 mm (.078–.0025 in) | Sand | Sandstone | Quartzose sandstone (sand comprises the quartz grains) |
| 0.063–0.004 mm (.0025–.00016 in) | Silt | Siltstone | Calcareous siltstone (contains calcium carbonate) |
| smaller than 0.004 mm (.00016 in) | Clay | Shale | Ferruginous shale (contains iron) |

als (Figure 2–13a). Under the great stress of metamorphism, these particles must change to accommodate their new environment. They do so by converting to layers of platy (flat) mica minerals. The mica minerals tend to form perpendicular to the direction of compression, creating a distinctive texture that gives the metamorphosed rock its new name, **schist** (Figure 2–13b).

In some rocks, metamorphism produces separate layers of platy and granular minerals. Such a layered rock is called **gneiss** (pronounced "nice"). A quartz-rich sedimentary rock, such as a quartzose sandstone, is metamorphosed into **quartzite.** A limestone is recrystallized into **marble.**

## REACTIONS OF ROCKS TO STRESS

Rocks within Earth are subjected to forces that tend to make them deform. These forces include gravity and pressure. If rocks are not free to move, they experience internal forces called **stress.** Stress is measured as the force applied to a unit area, such as kilograms per square meter ($kg/m^2$) or pounds per square inch ($lb/in^2$).

Three types of stress are common in Earth processes (Figure 2–14). **Compressional stress** (Figure 2–14a) occurs when a rock is squeezed by forces acting *toward* one another, like a piece of wood clamped in a vise. **Tensional stress** develops

(a)

(b)

**FIGURE 2–12**
(a) Calcareous mud in the Dry Tortugas, Florida. The dark green plants in the field of grass are the calcareous algae that produce the mud which is deposited in tropical and subtropical environments. (b) Fine-grained (lithographic) limestone produced by compaction of calcareous mud. (Photo (a) by author.)

(a)

(b)

**FIGURE 2–13**
Under certain conditions shale (a), a sedimentary rock, was slowly transformed into schist (b), a metamorphic rock, by the process of metamorphism. The agents causing the metamorphism were great heat and pressure, which forced the minerals in the shale to recrystallize. Active mountain building in the past produced the heat and pressure sufficient for metamorphosis.

from *opposite* forces that tend to pull a rock apart (Figure 2–14b). **Shear stress** acts along planes *parallel to* the deforming forces, tending to twist the rock in opposite directions (Figure 2–14c). The reaction to all three types of stress is called *strain.*

## Strain

If you apply any type of stress to an inflated balloon—compression, tension, or shear—the balloon shows the strain by changing shape and volume. Although it is harder to see, rocks similarly express

FIGURE 2–14
Types of stress: (a) compressive, as in cracking a walnut; (b) tensional, as in a tug-of-war; (c) shear, as in sliding a deck of cards between your fingers.

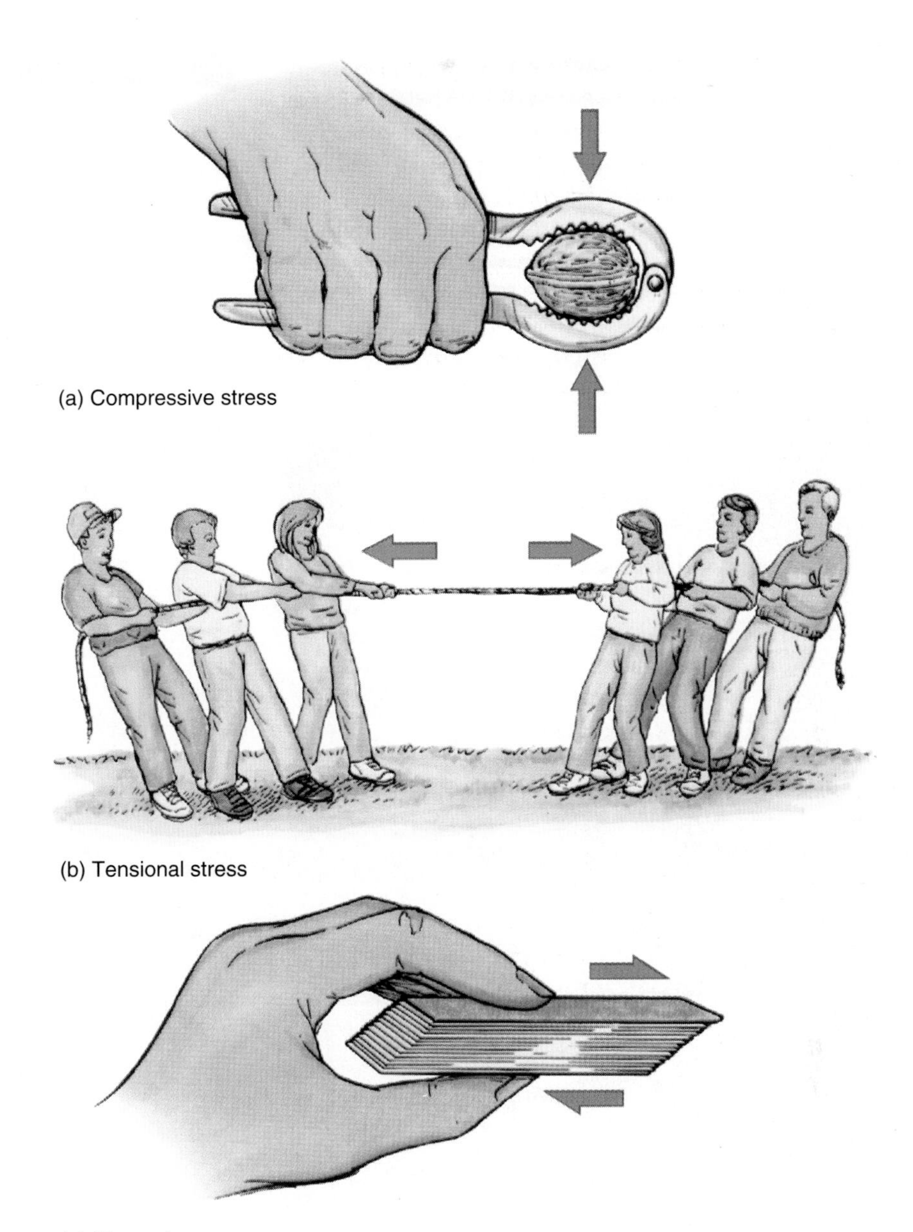


**strain** as a change in shape or volume. The behavior of a stressed rock depends on whether it is **brittle** or **ductile** (capable of flowing).

A brittle rock eventually breaks when the stress applied exceeds its strength. Such breaking under stress is called **failure** by engineers. On slopes such as hillsides, landslides occur when stresses on the rocks and soil exceed their strength (see Chapter 9). Tensional stress from the force of gravity produces cracks in rocks and soil, and the material roars down the slope (again, from the force of gravity), creating a geologic hazard.

Brittle materials can become ductile as conditions change (temperature and/or pressure). For example, consider pulling on a bar of taffy, causing tensional stress. Chilled taffy is brittle and will break instead of stretching. But let the taffy warm to room temperature and stress it again. The warm taffy deforms ductilely and can be twisted or pulled into different shapes.

Similarly, rocks at shallow depths (hundreds of meters) are brittle and will fail when stressed. These same rocks at greater depth are under far greater pressure (the weight of overlying rocks) and are deeper in Earth, where it is hotter than at the surface. In this environment, rocks tend to deform ductilely, without breaking. Thus, the depth of burial and the temperature determine to a large degree the type of deformation (brittle or ductile) that will result when a rock is stressed.

Conversely, if a rock that deforms ductilely at depth is brought to the surface and stressed, it will be brittle and break.

## Relationship Between Stress and Strain

To understand how stress affects rocks, consider the example of a stretched spring. As stretching (stress) begins, the strain is proportional to stress, a condition called **elastic strain.** When the spring is released, it elastically returns to its original shape, so there is no permanent deformation (Figure 2–15a). You may not think of rocks as being elastic, but they are. For example, the passage of a tractor-trailer truck over a thick concrete highway stresses the concrete. It elastically strains but returns to its original shape after the truck passes.

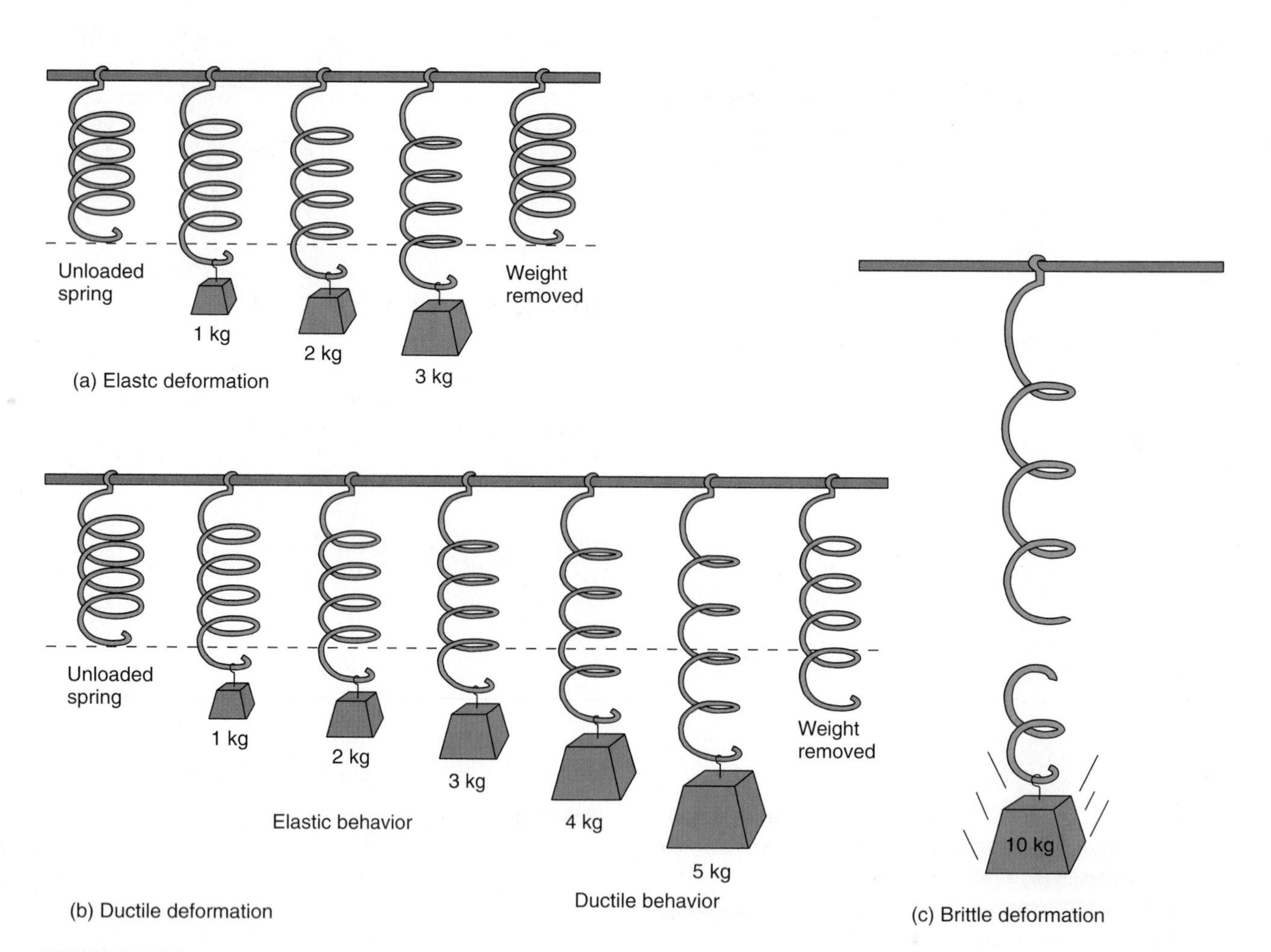


**FIGURE 2–15**
Effects of increasing strain on a spring. (a) Elastic deformation—spring returns to original length when weight is released. (b) Ductile deformation—stress exceeds strength and spring is permanently deformed. (c) Brittle deformation—spring breaks from excessive stress. (Reprinted with permission of Macmillan College Publishing Company from *Physical Geology* by Nicholas K. Coch and Allan Ludman. Copyright © 1991 by Macmillan College Publishing Company. Courtesy of Allan Ludman.)

Stretching the spring too far will exceed its elastic limit, permanently deforming it so it can no longer return to its original shape (Figure 2–15b). Stretching the spring still farther will break it—in other words, it fails (Figure 2–15c).

Rocks behave the same way, except for one step. As stress increases in rock, it first deforms elastically, but when the stress exceeds the strength of the rock, it fails without any **plastic deformation** (which is a permanent change in rock shape or volume). The strain that results from stress may be either recoverable or permanent. Permanent plastic deformation is shown in Figure 2–16.

### Strength

The ability of a rock to resist failure is its **strength.** A rock's strength depends on its mineral composition, cementation, pores (small openings) in the rock, structural features (such as bedding), and the arrangement of the mineral grains. A rock's strength varies even further, depending on its temperature and the type of stress applied (tensional, compressional, or shear). For example, most rocks are stronger under compression than under tension (in other words, rocks are harder to crush than to pull apart). Thus, when we talk about a rock's strength, we must specify the conditions under which its strength is measured.

Rock strength is affected by the *confining pressure* to which it is subjected at great depth within Earth. Therefore, rocks are stronger within Earth than at its surface. One reason surface rocks become broken and jointed so readily is their reduced strength at the surface and their subsequent failure under tension.

Sometimes a rock fails at an unexpectedly low stress level. This can occur if the rock has experienced repeated episodes of low-level stress. The *accumulated strain* may weaken the rock and cause it to break, like strong metal that has been repeatedly flexed until it breaks easily. Such failure due to accumulated low-level strain is called *fatigue.*

To understand many geohazards, it is essential to know about the strength of rocks and sediments, the stresses imposed on them by geologic processes, and how they react. This knowledge is especially important in understanding earthquakes (Chapter 5), landslides (Chapter 9), and surface collapse (Chapter 10).

## ROCK STRUCTURES

The phrase "solid as a rock" usually is inaccurate because most rocks are not all that solid. They contain fractures, pore spaces, and other openings that are important in enhancing rock weathering

(a)

(b)

**FIGURE 2–16**
(a) Sandstone (arkose) and (b) gneiss, its plastically deformed equivalent.

and erosion and in triggering geologic hazards such as landslides (Chapter 9) or surface collapse (Chapter 10).

## Pore Spaces in Rocks

You may not have realized that small open spaces exist between mineral grains in rocks, but they do. These open spaces are important in allowing oil and gas to accumulate and in permitting water to migrate underground through the rocks. Most clastic sedimentary rocks are composed of particles separated by open spaces called **pores** (Figure 2–17a). The amount of pore space in a rock is its **porosity.**

In the compaction and cementation of clastic sediment into a clastic rock, the compaction (Figure 2–11a) greatly reduces the original porosity. It is further reduced by cementation when water circulating through the sediment's pore spaces leaves deposits of crystals in them (Figure 2–11b). The cemented rock has greatly reduced porosity (Figure 2–17b).

Most igneous and metamorphic rocks and many chemical sedimentary rocks have a different kind of porosity than clastic rocks, because they are composed of tightly interlocking crystals instead of clastic particles.

## Solution Cavities in Rocks

Some chemical sedimentary rocks, such as many types of limestones, have a low initial porosity because they are composed of interlocking crystals of calcite (calcium carbonate). However, with time,

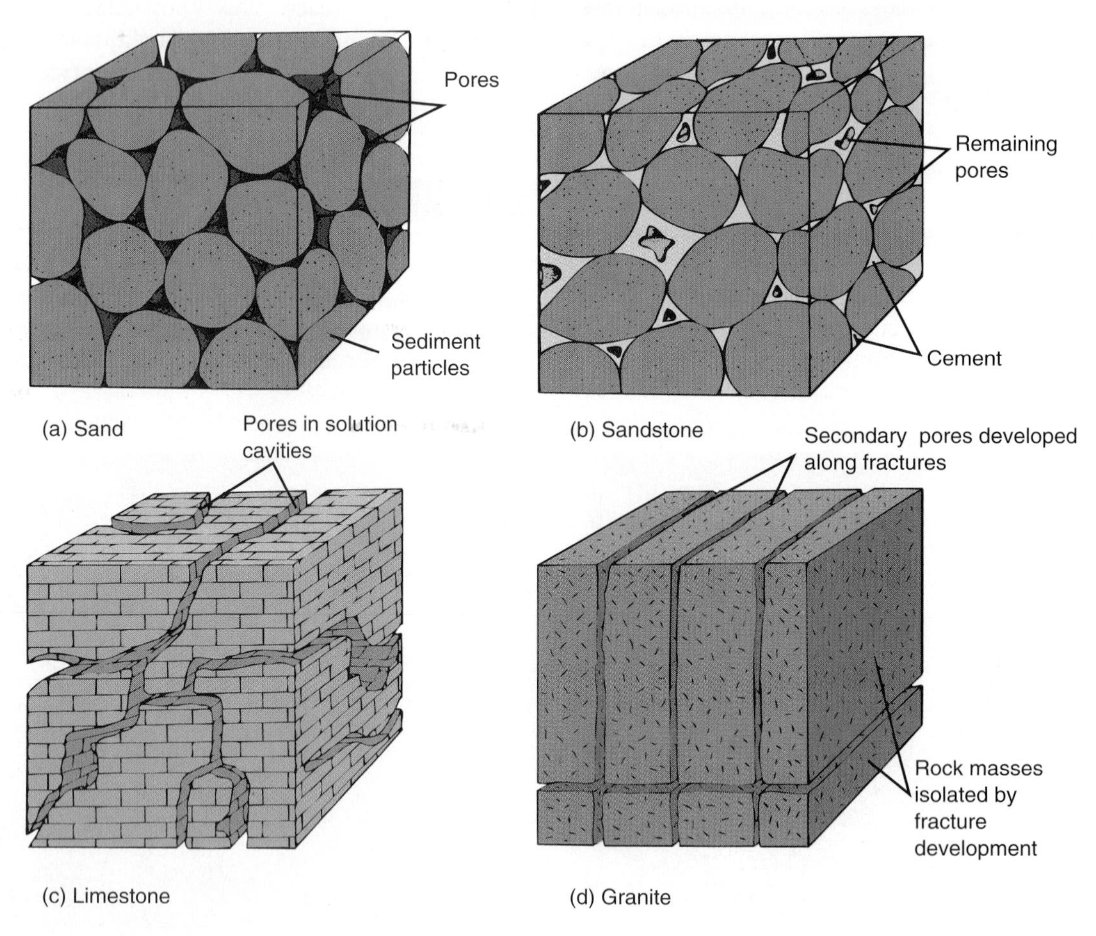


**FIGURE 2–17**
Types of porosity developed in (a) sandy sediment, (b) sandstone, (c) limestone, and (d) granite. (Reprinted with permission of Macmillan College Publishing Company from *Physical Geology* by Nicholas K. Coch and Allan Ludman. Copyright © 1991 by Macmillan College Publishing Company.)

acidic underground water circulating through limestone can dissolve cavities in it, making it far more porous (Figure 2–17c). Taken to a large scale, this is how limestone caves and caverns are formed, like Mammoth Cave in Kentucky and Carlsbad Caverns in New Mexico.

Underground limestone solution in some parts of Florida has created large voids (measured in meters) beneath the surface. These voids decrease surface support, leading to its collapse, a well-known problem in Florida that is described in Chapter 10.

## Joints in Rocks

Many rocks have **joints,** which are fractures along which no movement has occurred. In some igneous rocks, which formed from cooling magma, joints are caused by tensional stresses during cooling.

However, most joints are caused when stress on buried rocks causes more strain than a rock can accommodate by ductile flow. When the confining pressure is reduced, the rock responds by breaking, as we noted before. It does not break at random but systematically, into *sets* of joints. As the overlying rocks erode, these joint sets slowly become exposed. Joint sets intersect, dividing the rock into blocks (Figure 2–17d).

Joint development in rocks is very important because it facilitates subsequent weathering and transport of fluids (Chapter 3). Joint development also breaks the rock into blocks, which are more prone to downslope movement (Figure 2–18).

## Faults in Rocks

Recall that deeply buried rocks may respond ductilely to stress. But when rocks are stressed near Earth's surface, the much lower pressures there allow brittle deformation (breaking) rather than ductile deformation. Consequently, the rock fractures along a planar surface called a **fault,** along which there is movement. An example is California's well-known San Andreas Fault.

Fault movement can be vertical, horizontal, or a combination. Different types of stress produce different types of faults, as shown in Figure 2–19:

- ☐ *Tensional* stress pulls apart rocks, producing a **gravity fault** or *normal fault.* As tension pulls apart the rock, one block *drops* under the force of gravity. It moves relative to the other block, down the plane of the fault. This forms a high-angle fault surface.
- ☐ *Compressional* stress produces a **thrust fault** or *reverse fault.* Here the rock on one side of the fault is pushed upward relative to the other side. This forms a low-angle fault surface.
- ☐ *Shear* stress along a vertical section within a rock can produce a rupture along which there is no vertical movement but rather a horizontal displacement. When the fault happens, rocks on

**FIGURE 2–18**
Jointing in light-colored cap rock on the Colorado Plateau. The joints allow water to enter the rock, which promotes weathering. As blocks of rock are loosened by weathering, they fall to the slope below. (Photo by author.)

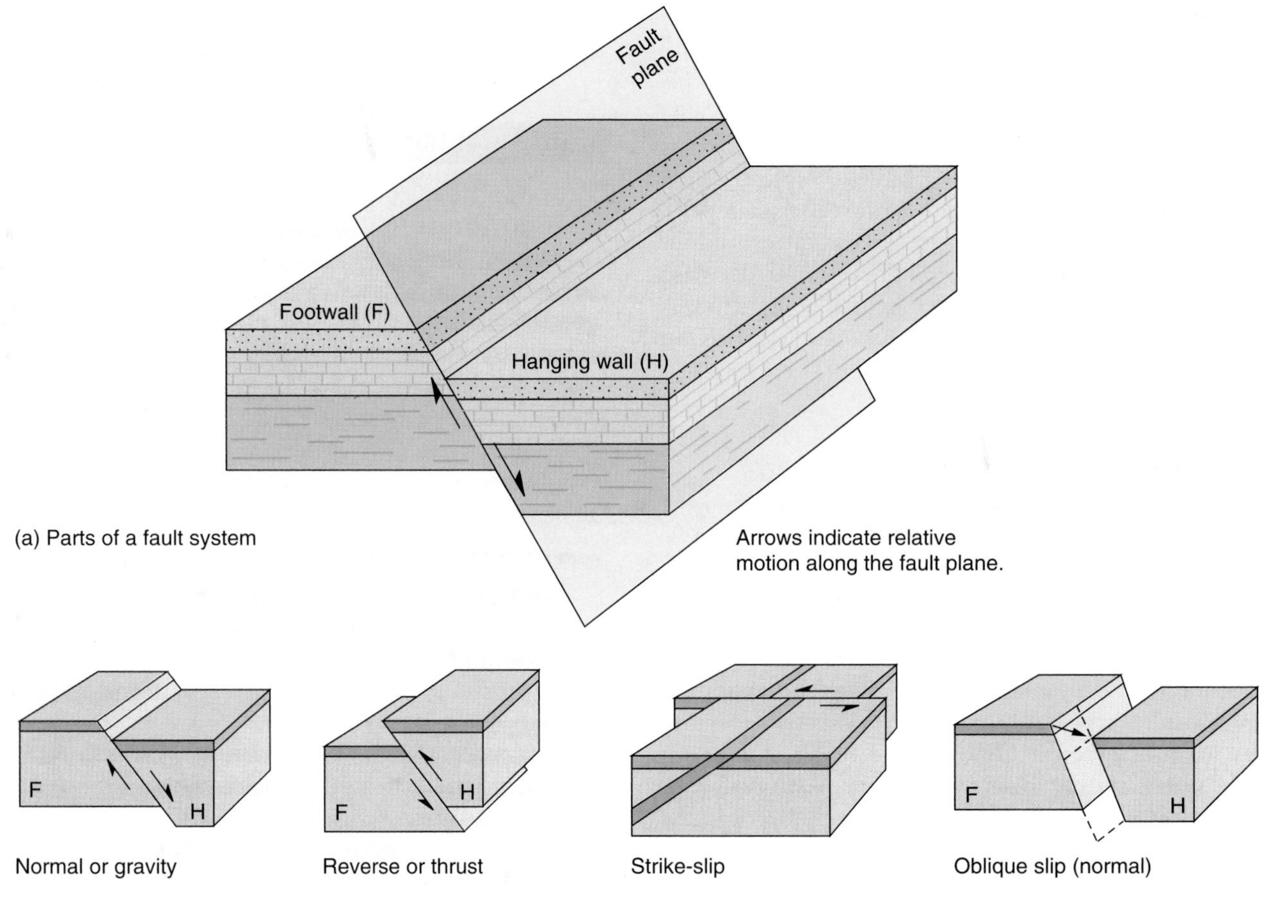


**FIGURE 2–19**
(a) Parts of a fault. Years ago, miners who worked underground along faults called the lower block on which they stood the *footwall* (F), and the upper block on which they hung their lanterns the *hanging wall* (H). This terminology remains in use today. (b) Types of faults. (1) Normal or gravity fault. (2) Reverse or thrust fault. (3) Strike-slip fault. (4) Oblique-slip fault.

either side of the fault plane slide past each other in a *strike-slip fault.*

- ☐ Some faults have displacement both vertically and horizontally and are called *oblique-slip faults.*

Faulting is a major cause of earthquakes and is described further in Chapter 5.

## INTERNAL STRUCTURE OF EARTH—CORE, MANTLE, CRUST

Now that many of the components and processes of Earth have been presented, let us put them together and take a larger-scale view of Earth's interior.

Earth is about 6,370 kilometers (3,959 miles) deep from surface to the center, but we have been able to drill wells only a few kilometers deep. So, how can we tell what is inside Earth? Several kinds of evidence have been used:

- ☐ study of once-deeply buried rocks that have been uncovered by erosion;
- ☐ study of interior material that emerges as lava and gases from volcanoes;
- ☐ analysis of earthquake waves that pass through Earth;
- ☐ study of the curious configuration of Earth's continents and ocean basins;
- ☐ study of Earth's movements through space; and
- ☐ study of meteorites.

Collectively, this research has provided evidence of Earth's structure and composition. Much of the evidence is indirect and inferred, but it generally fits together—in other words, one piece of information generally confirms another. One of the greatest achievements of science has been to assemble these puzzle pieces into a coherent picture of Earth's interior.

Earth's interior is composed of three major regions, on the basis of chemical composition and physical state (liquid or solid). From the center outward, these regions are the core, mantle, and crust (Figure 2–20).

The **core** is the innermost region of Earth (Figure 2–20). It is ball-shaped, about 3,570 kilometers (2,220 miles) thick from the center to its approximate outer edge. (This "edge" is not precise, but is a thick zone that grades into the mantle.) Indirect evidence indicates that it is predominantly metallic nickel and iron, not rock. Geologists subdivide the core into two zones of very different properties: the *inner core* is solid and the *outer core* is believed to be liquid.

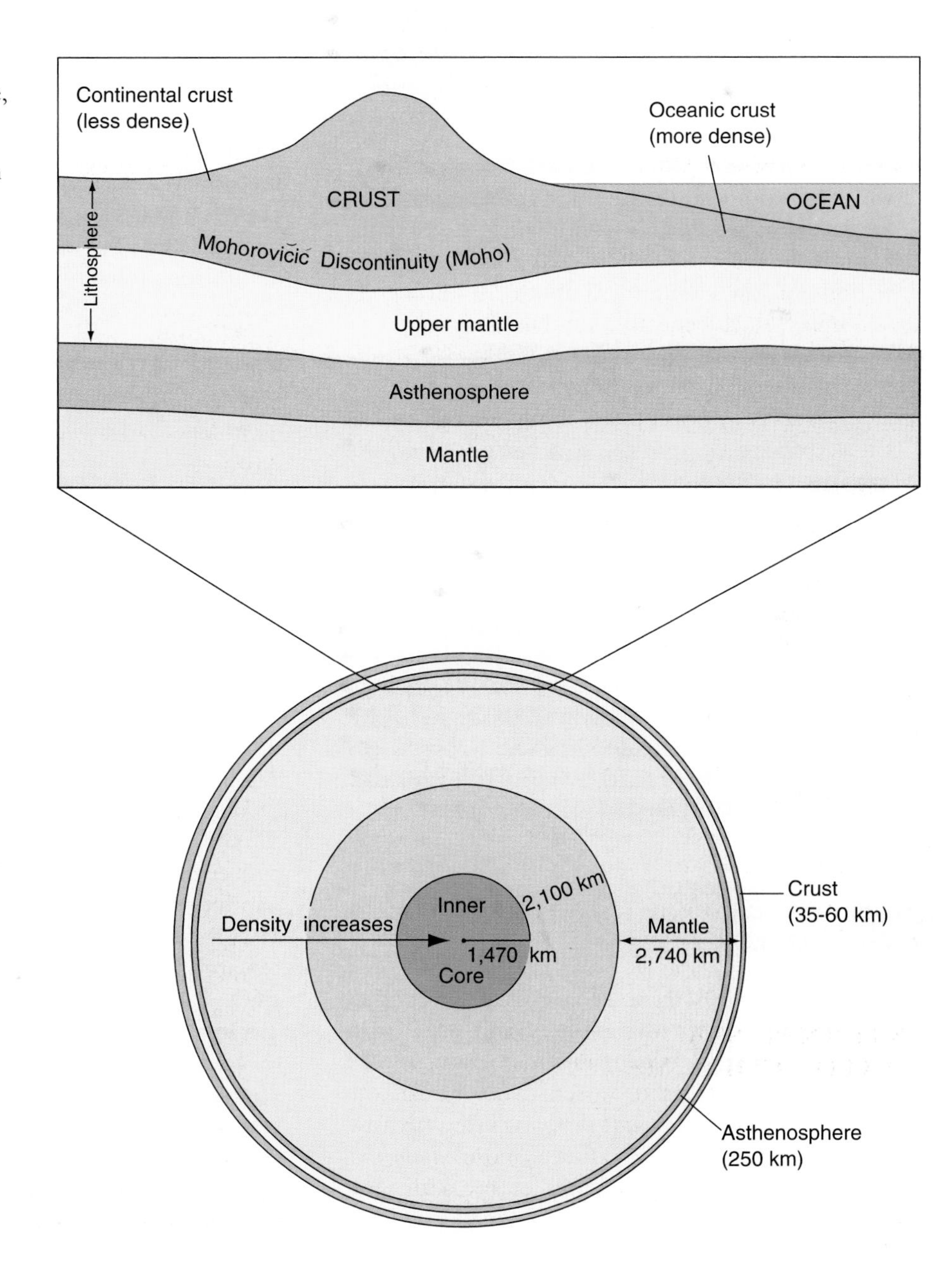


**FIGURE 2–20**
Earth's three major regions—core, mantle, and crust. The cross section describes the outer 300 kilometers (185 miles) of Earth in more detail.

The **mantle** is the middle layer of Earth (Figure 2–20). It is about 2,740 kilometers (1,700 miles) thick and very dense. In the uppermost part of the mantle is a thin but very important layer, the **asthenosphere.** This layer is more ductile or plastic than the mantle rock above and below it (Figure 2–20). The asthenosphere begins about 50 kilometers (30 miles) below the top of the mantle and is about 250 kilometers (155 miles) thick. Earthquake waves that pass through this zone provide evidence that the asthenosphere is less rigid (more ductile or plastic) than the rest of the mantle.

The **crust** is the outermost and thinnest layer of Earth. Two different types of crust exist, *continental* and *oceanic* (Figure 2–20). This distinction is extremely important, as you shall see shortly.

- **Continental crust** makes up the continents. It varies in composition but has the gross overall composition of granite. Continental crust ranges from about 35 kilometers (20 miles) thick in regions of low elevation to about 60 kilometers (35 miles) thick beneath mountain chains.
- **Oceanic crust** makes up the vast seafloor. It differs markedly from continental crust. Oceanic crust averages only about 5 kilometers (3 miles) thick. Instead of granite, it is composed of the dark, dense rock **basalt.**

The contact zone between the crust and the mantle is called the *Mohorovičić discontinuity* (mo-hor-oh-VEE-chik), commonly shortened to **moho.**

The rigid outer part of Earth, composed of solid crustal rock and the solid portion of the mantle above the asthenosphere, is called the **lithosphere** (Figure 2–20). The combination of the rigid lithosphere over the non-rigid asthenosphere has extremely important geologic and environmental significance. It is the basis for plate tectonics.

## PLATE TECTONICS—HOW CONTINENTS "DRIFT"

Up until the 1960s most geologists believed that the continents remained stationary and grew with time, surrounded by oceans underlain by very old rocks. Although voices of dissent were heard, the scientific community was not ready to listen because too little evidence existed for a more *dynamic,* mobile Earth. However, a series of breakthrough discoveries in the 1960s led to a dramatic revision of thinking and general acceptance of the remarkable idea that continents "drift."

### Plate Tectonics Theory

**Plate tectonics** is a theory that explains the movement and deformation of parts of the outer Earth. Its basis is the movement of rigid lithospheric slabs called *plates* over a less-rigid layer (asthenosphere) in the upper mantle (Figure 2–20). The word *tectonics,* which means to build, refers to the construction of Earth's surface features such as mountains and ocean basins.

Alfred Wegener, a German meteorologist, proposed in 1912 that the continents had once been joined in a supercontinent called *Pangaea.* It subsequently broke apart, resulting in continents migrating slowly across Earth's surface. His hypothesis became known as **continental drift.** He presented several lines of evidence to support his hypothesis, all of which remain valid today.

For example, you can observe on a globe the striking "fit" of the Atlantic coastlines of South America and Africa. They resemble two matching puzzle pieces. This suggested to Wegener that these continents once were joined and subsequently drifted apart.

Wegener also noted the marked similarity of fossil plants and animals found on different continents that today are separated by vast oceans. How could similar land animals have developed identically on separate continents? An obvious explanation was that these continents once were together as a single giant land mass. Long after the animals and plants lived, breakup of the continent would create the distribution of fossils that we find today (Figure 2–21).

Wegener's continental drift hypothesis had few supporters for many years because no one could conceive of a *mechanism* by which massive, rigid continents could move long distances. By the 1970s, however, advances in seafloor and on-land research led more geologists to accept the hypothesis.

One of the important discoveries was the existence of the asthenosphere (Figure 2–20). This zone of ductile rock provides a plastic surface on which the lithosphere can slide, albeit at the ponderous rate of a few centimeters per year. The masses of lithosphere that move over the asthenosphere are called **lithospheric plates.** The theory

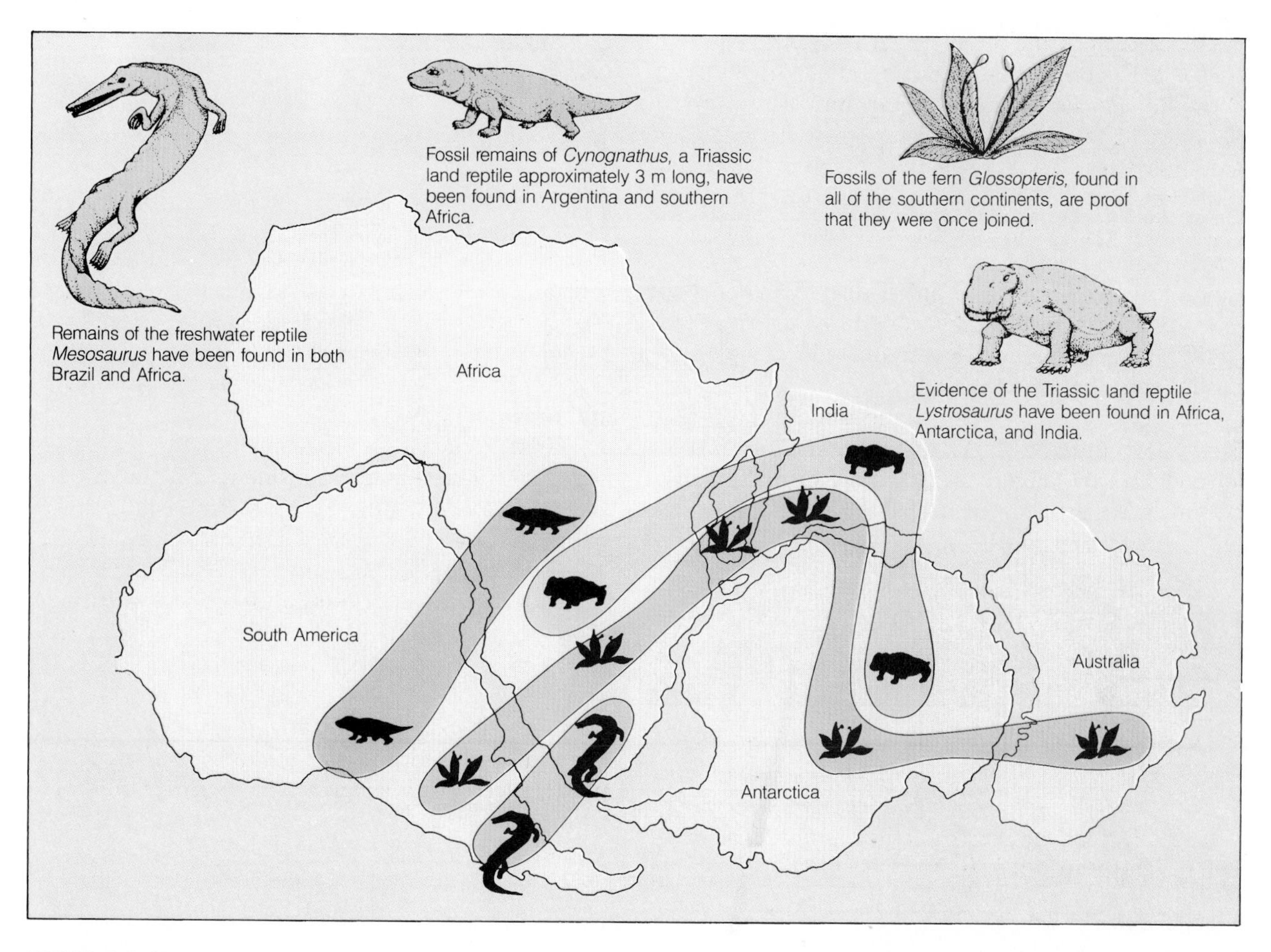


**FIGURE 2–21**
Fossil evidence of continental drift. Fossils of certain species of animals and plants are distributed on continents now separated by oceans. Fossils of *Mesosaurus,* a freshwater reptile, are found in both Brazil and South Africa. *Glossopteris,* a fossil fern, is found in all of the southern continents. *Lystrosaurus,* a land reptile, is found in Africa, India and Antarctica. *Cynognathus,* a reptile, is found in Argentina and southern Africa. (Reprinted with permission of Macmillan College Publishing Company from *Earth's Dynamic Systems* by W. Kenneth Hamblin. Copyright © 1992 by Macmillan College Publishing Company.)

that explains the movement of lithospheric plates is called *plate tectonics theory.*

So much evidence now supports this explanation that it has been elevated from the level of hypothesis (proposed explanation) to the level of a theory (greater acceptance). Unlike their counterparts of only forty years ago, most geologists today accept the plate tectonics theory.

## Ocean Basins and Seafloor Spreading

In the mantle, radioactive decay of certain elements (mostly uranium, thorium, and potassium) produces great heat. This *radiogenic heat* may drive slow-moving convection currents within the mantle. These currents, combined with gravitational forces, plus the expansion of the overlying rocks due to the heating, create tension that spreads the lithosphere apart. This spreading is called *rifting* (Figure 2–22a), and it forms *rift basins* (Figure 2–22b).

Continued subsidence of the rift basins settles them below sea level. Eventually they fill with saltwater from existing oceans to form narrow seas that steadily widen (Figure 2–22c). Along the lines of rifting, slowly rising basalt magma from the mantle

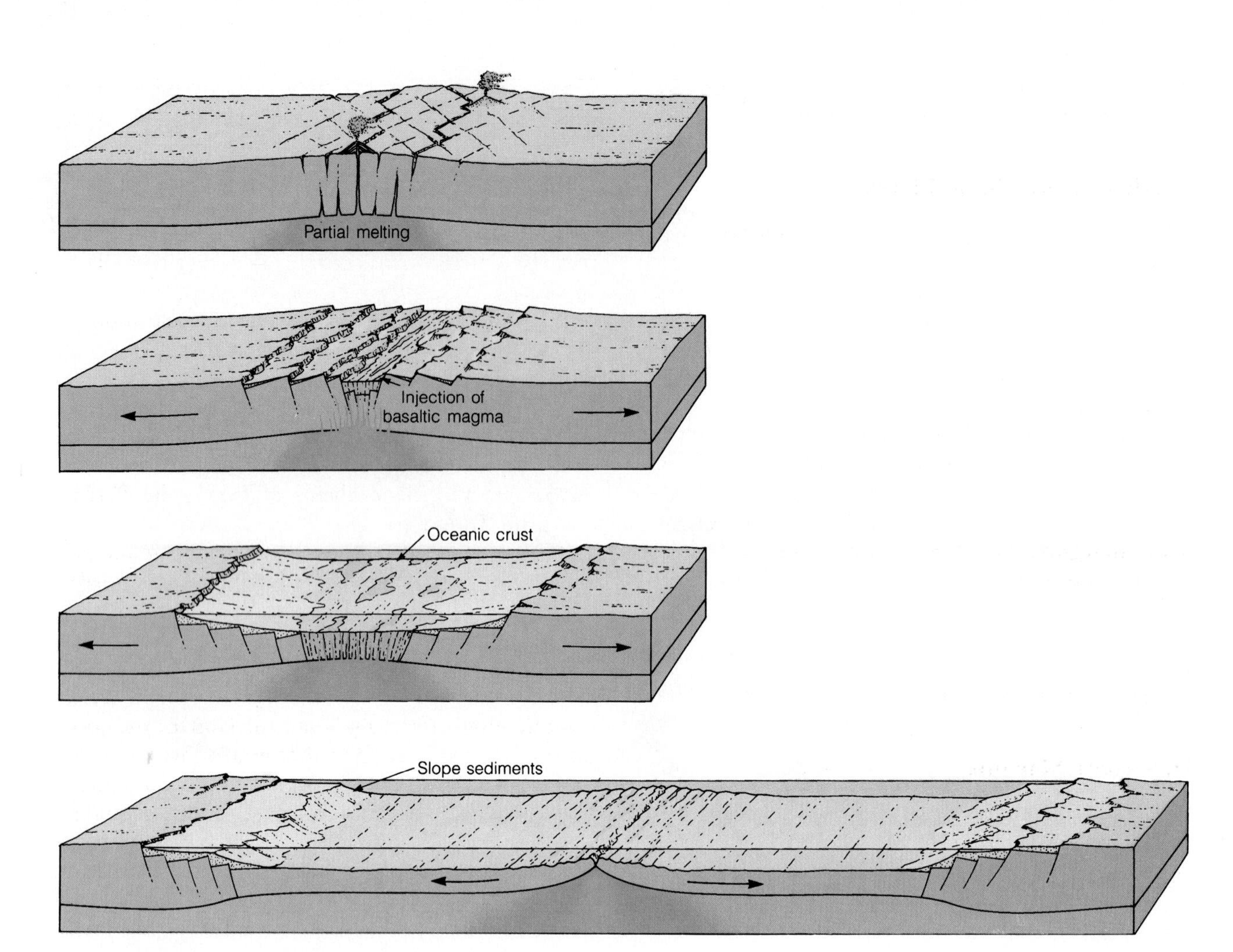


**FIGURE 2–22**
Formation of an ocean basin by continental rifting. (a) Rifting of the continental crust is accompanied by extrusion of basaltic magma. (b) As the rift zone subsides, an arm of the ocean encroaches. (c) The new ocean basin becomes wider and deeper. (d) The plates continue to move apart in the process of seafloor spreading. (Reprinted with permission of Macmillan College Publishing Company from *Earth's Dynamic Systems* by W. Kenneth Hamblin. Copyright © 1992 by Macmillan College Publishing Company.

reaches the surface and is extruded onto the seafloor. Here the seawater quickly cools it to form new seafloor.

As the convection currents slowly spread apart the seafloor, and fresh extrusions add more lava along the rift, the lava piles up to form a long *mid-ocean ridge.* (Worldwide, 64,000 kilometers [40,000 miles] of these ridges form Earth's longest continuous mountain chain, almost entirely underwater.) Each fresh extrusion splits the previous one down the middle. The halves of the previous flow spread apart, in opposite directions on either side of the rift line. This is called **seafloor spreading** (Figure 2–22d).

The slow convection currents in the mantle propel *entire plates* in opposite directions away from the spreading center, somewhat like conveyor belts carrying material in opposite directions. As new, less-dense lava is extruded into the rift, the older and denser crust sinks into the asthenosphere. Because every ocean is flanked by continents, the continental plates on either side of the widening ocean are slowly forced apart, too. If the asthenosphere were not plastic, the crustal plates would simply rumple. Instead, the plastic asthenosphere makes possible their sliding movement.

## Earth's Lithospheric Plates

Like an egg with a cracked shell, Earth's lithosphere is broken into eight giant slabs or *plates* and several smaller ones (Figure 2–23). Most plates include both continental and oceanic crust (note the North American, African, and Australian plates). Others, such as the Pacific plate, are composed largely of oceanic crust.

Each plate moves at a different rate (generally a few centimeters a year) and in a different direction, as you can see from the arrows. Each plate is continually in contact with those adjacent to it. At the **plate margins,** or edges where the plates meet, potential hazards exist for humans. The reason is that, at their margins, plates interact in three different movements: they diverge, converge, or slide past one another. We will now examine each of these movements.

## Divergent Margins

Where tensional stress moves plates away from each other, **divergent margins** are created. The mid-ocean ridges in Figure 2–22d are divergent margins. Along the Mid-Atlantic ridge today, the North American and Eurasian plates are moving apart from each other at an average rate of about 2.5 centimeters (1 inch) each year.

Divergent margins generally are regions of non-explosive eruptions of basalt lava and moderate earthquake occurrence. Because these earthquakes and eruptions are of relatively low intensity and because they mostly occur underwater, they present little risk to people, and usually happen unnoticed.

## Convergent Margins

Where compressional stress moves plates toward each other, **convergent margins** are created. Convergent margins are more complex than divergent margins because different things can happen, depending on whether the convergence of plates is between two oceanic plates, between an oceanic and a continental plate, or between two continental plates.

When two oceanic plates converge, one usually moves beneath the other, a process called **subduction.** The upper plate simply rides over the other one. Gravity pulls the lower (subducted) plate down into the mantle, carrying oceanic crust toward the asthenosphere (Figure 2–24). The subducting plate is partially melted by the tremendous heat, generating magma that slowly rises to the surface. If it reaches the surface, it is extruded from volcanoes.

If one oceanic plate subducts beneath another oceanic plate, the rising magma forms a chain of underwater volcanoes, some of which may grow to form island volcanoes in an **island arc** chain (Figure 2–24, left side). Well-known examples of island arcs in the Pacific Ocean include the Aleutian Islands of Alaska and the island chains of Japan, the Philippines, and Indonesia.

If an oceanic plate pushes against a continental plate, the denser (basaltic) oceanic plate is subducted under the less dense (granitic) continental plate (Figure 2–24, right side). The magma, formed from the partial melting of the subducted plate, rises to the surface where it erupts to form an active volcanic mountain chain along the edge of the continent. Examples are the Cascade Mountains of the northwestern United States (including Mount Saint Helens) and the Andes Mountains of South America (Figure 2–24). The hazards of volcanic eruptions along convergent margins are considered in Chapter 4.

If two continental plates converge, neither plate is subducted. Instead, the force of collision forms a mountain range at the point of juncture. For example, the Himalayas formed when the sub-continent of India collided with the Eurasian plate.

## Transform Margins

At some plate margins, the plates neither diverge nor converge but *slide* horizontally past each other under shearing stress. This is called a **transform-fault margin** (Figure 2–23). Transform margins are different in several ways. Typically, neither volcanism nor transfer of material between the asthenosphere and lithosphere occurs at a transform boundary. Earthquakes, however, are common along them.

The most famous example of a transform boundary is the San Andreas fault system in California (Figure 2–23). This fault system marks the boundary between the northward-moving Pacific plate on the west and the westward-moving North American

FIGURE 2–23
Earth's lithospheric plates showing names, directions of movement, and types of margin. (Reprinted with permission of Macmillan College Publishing Company from *Physical Geology* by Nicholas K. Coch and Allan Ludman. Copyright © 1991 by Macmillan College Publishing Company. Courtesy of Allan Ludman.)

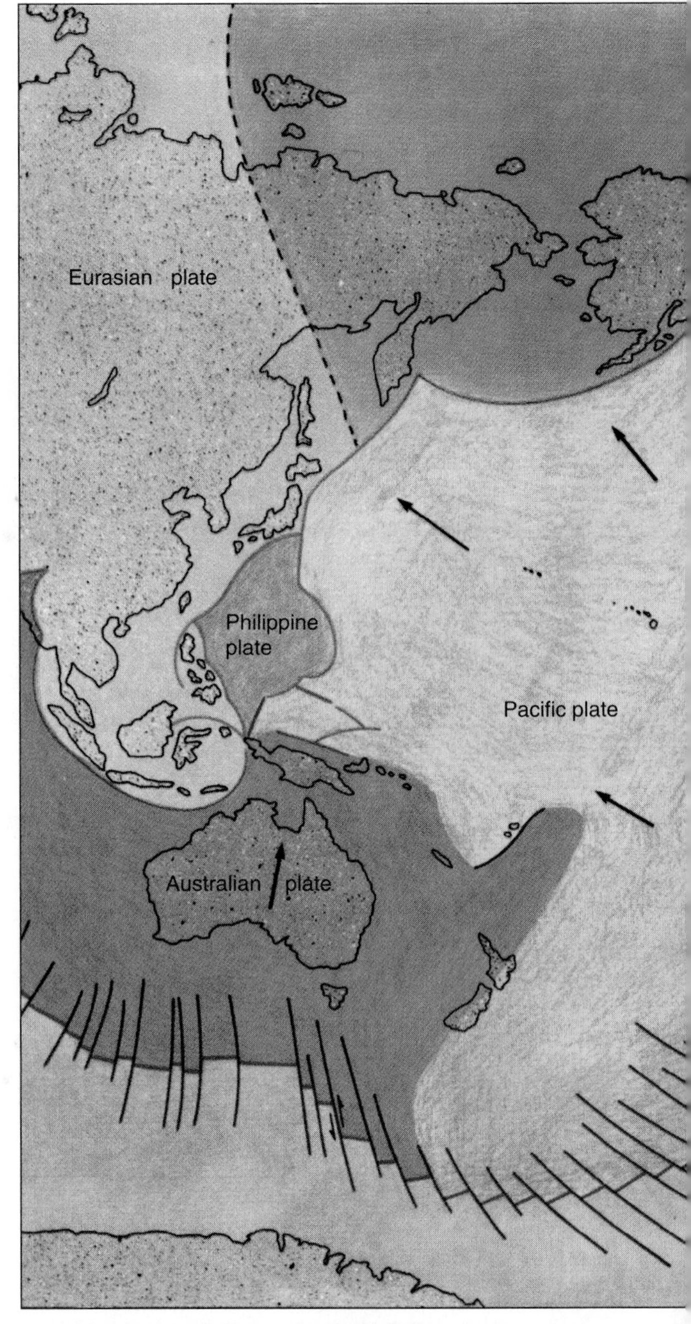


plate on the east. Great shear forces act along this boundary. The environmental consequences of fault movement here are serious, as you will see in Chapter 5.

## PLATE TECTONICS AND OTHER EARTH CYCLES

You can see that material subducted at converging margins ultimately is recycled into new magma. Thus Earth has been recycling itself for several *billion* years. A proof of the recycling of seafloor by plate tectonics is that no ocean crust exists that exceeds 200 million years in age—despite Earth's great age of about 4.6 billion years! This surprising fact is simply explained: newly formed seafloor at the rifts moves away from them at an average few centimeters a year. In the largest ocean, the Pacific, 200 million years of such movement transports the seafloor from the rifts to the converging margin with another plate, where the recycling occurs.

Earth's recycling of its own materials proceeds in several ways. In addition to recycling by plate tectonics, early in Earth's history rocks were transformed from one type to another in the geologic **rock cycle** (Figure 2–25).

Other Earth cycles are important. The *water cycle* involves the journey of water from the ocean, into the atmosphere, and back to the ocean via rainfall and streams. The *atmospheric cycle* involves the cycling of atmospheric gases through water, rocks, animals, and plants. The *carbon cycle* is crucial to life on Earth, moving carbon among the atmosphere, rocks, and living things.

Geologists believe that the meteorites colliding with Earth today formed at or near the time of Earth's formation. Radioactive isotopes have been used to date these meteorites at about 4.6 billion years. However, the oldest Earth rock found so far is 3.96 billion years old. Will we ever find a "genesis" rock that is as old as Earth? It is highly doubtful, because it is probably long gone, recycled into new rock by Earth's internal and external processes.

The generalized picture of the rock cycle in Figure 2–25 shows the variety of rock transformations that are possible. Under one part of Earth, deeply buried sedimentary rocks are being transformed by

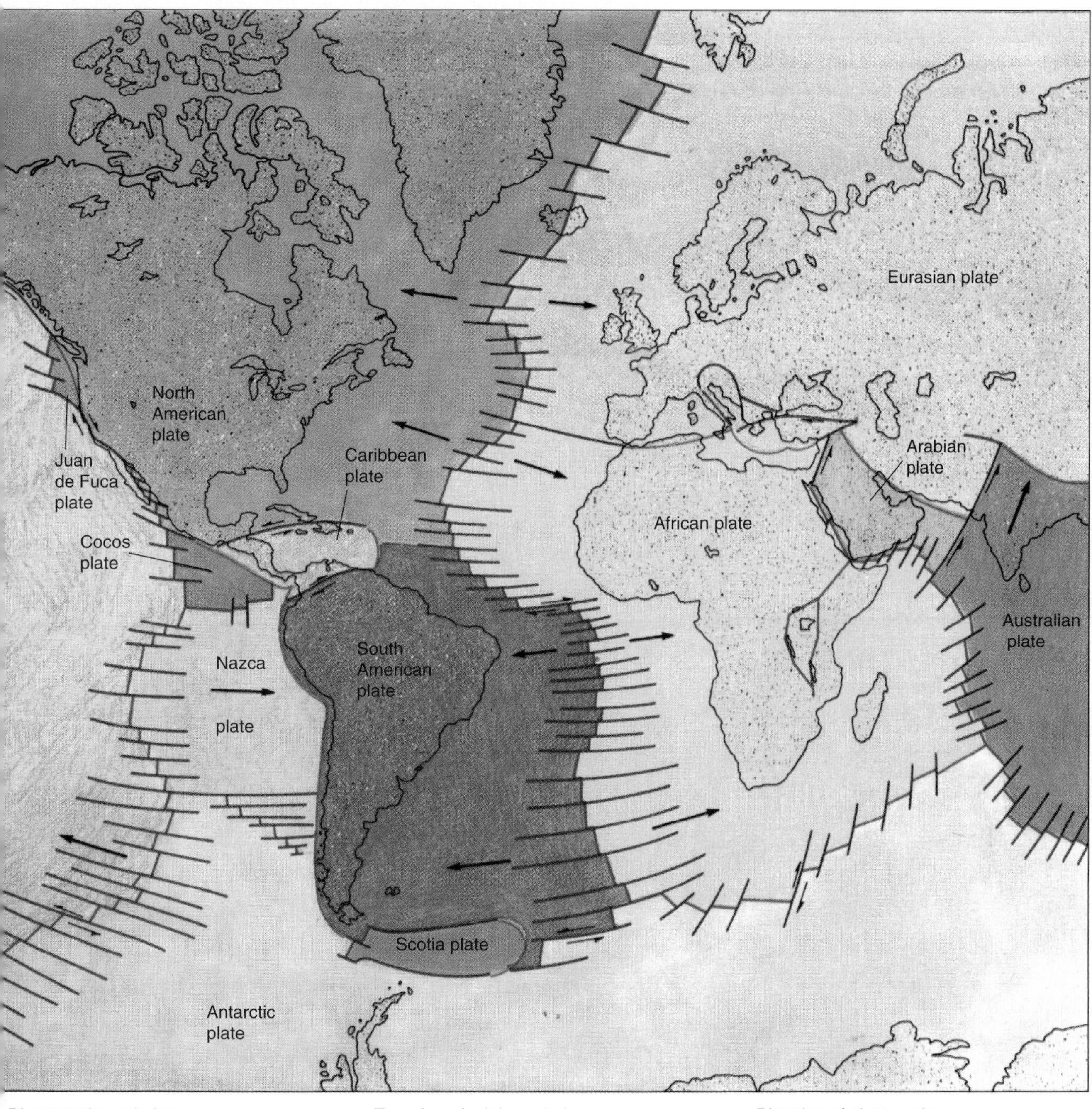


Divergent boundaries ——— Transform fault boundaries ——— Direction of plate motion ——→

heat and pressure into metamorphic rocks. At depth elsewhere, metamorphic rocks are melting to form magma that will solidify into new igneous rocks. At the same time, surface rocks are being weathered to form sediments that become buried, compacted, and cemented into new sedimentary rocks . . . and so on, probably for billions more years.

## LOOKING AHEAD

This completes our short tour of Earth's interior matter, energy, and processes. In the next chapter we will consider the role of air, water, ice, and living things in surface processes, and the energy source that drives it all—the sun.

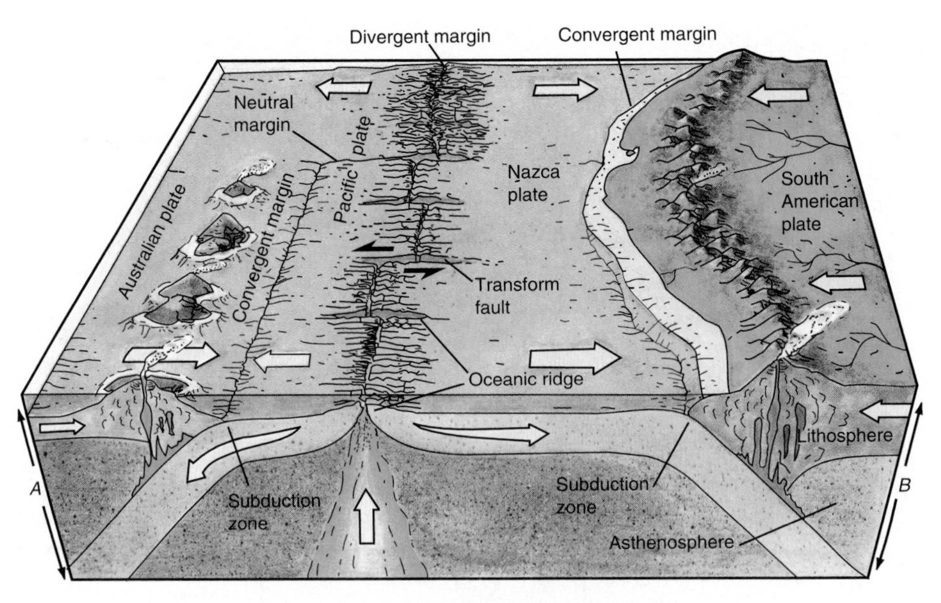


**FIGURE 2–24**
Geologic block diagram showing examples of plate margins from the South Pacific Ocean. Note the oceanic/oceanic plate convergence at left and the oceanic/continental plate convergence at right. (Reprinted with permission of Macmillan College Publishing Company from *Physical Geology* by Nicholas K. Coch and Allan Ludman. Copyright © 1991 by Macmillan College Publishing Company. Courtesy of Allan Ludman.)

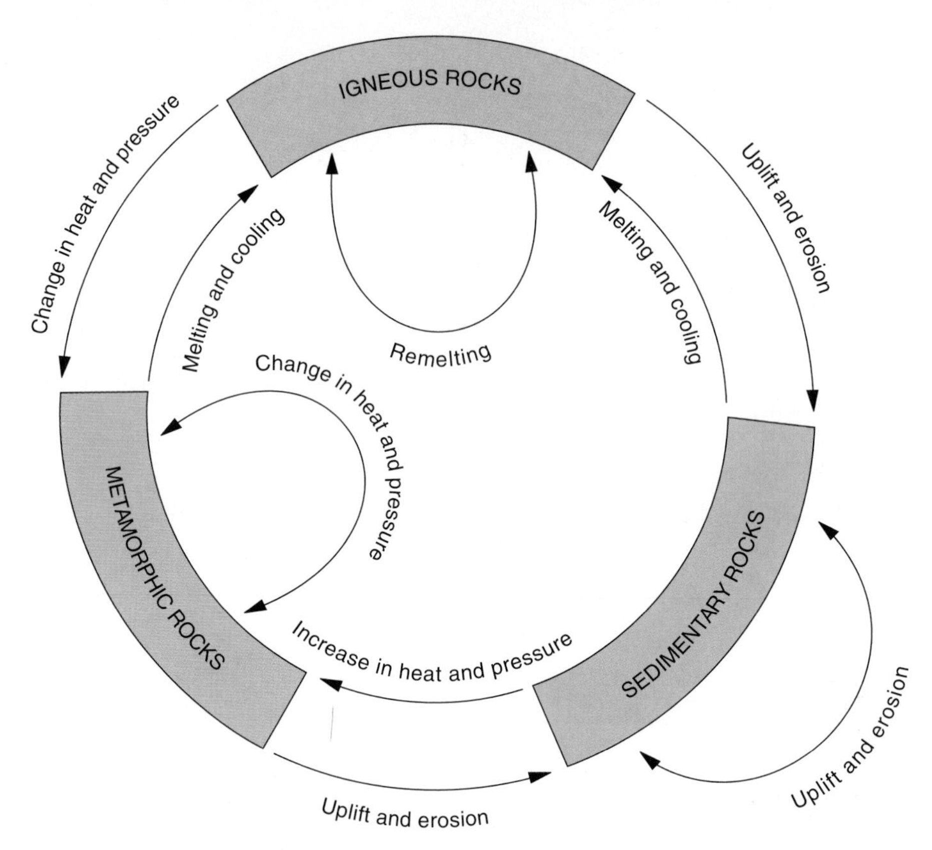


**FIGURE 2–25**
The rock cycle. Note how each type—igneous, sedimentary, metamorphic—sooner or later is converted into another type by Earth's continuing geologic processes. (Reprinted with permission of Macmillan College Publishing Company from *Physical Geology* by Nicholas K. Coch and Allan Ludman. Copyright © 1991 by Macmillan College Publishing Company. Courtesy of Allan Ludman.)

## SUMMARY

Energy is the ability to do work. Potential energy is stored, whereas kinetic energy is in action. Earth's geologic processes are powered by two energy systems: (1) internal systems (derived from gravity and heat generated within Earth) that drive mantle convection currents and tectonic plate movement and (2) external systems (powered by gravity and solar energy) that drive surface processes like rivers, glaciers, and wind.

All matter (solid, liquid, or gas) is made of atoms, which in turn are made of three major subatomic particles. An atom's nucleus contains electrically neutral neutrons and positively charged protons. Orbiting around the nucleus are negatively charged electrons (electrons equal protons in number). An element is composed entirely of atoms of a single kind. Isotopes of an element have the same atomic number but a different atomic mass (number of neutrons). Of 106 known elements, oxygen and silicon are the most abundant in Earth's crust. Other abundant elements are aluminum, iron, sodium, calcium, potassium, and magnesium.

Some elements have radioactive isotopes with unstable nuclei that decay and release heat. Because rock is a poor heat conductor, heat from radioactive decay in Earth's interior builds up and melts rocks into magma. Less-dense magma rises slowly and either cools to form plutonic (intrusive) rocks below the surface or is erupted as lava on the surface. Heat moves by three means: conduction, convection, and radiation.

Reactions between elements that involve their orbiting electrons are chemical reactions. Minerals are naturally occurring, inorganic solids with specific chemical compositions and orderly atomic structures. Most minerals are formed of elements that combine into compounds. Minerals form from hot solutions, pressure, deposition from concentrated surface waters, and through activities of plants and animals.

Rocks are made of one or more minerals. Igneous rocks form when magma cools underground (plutonic rocks) or on Earth's surface (volcanic rocks). Intrusive igneous rocks cool slowly and have time to grow larger crystals; extrusive equivalents cool faster and have finer crystals. The major igneous rock families are granite, andesite, and basalt.

Sedimentary rocks are surface and near-surface products of sediment accumulation, burial, compaction, and cementation. Sediments include clastic mineral particles from weathered rocks, biologically produced debris, and crystals deposited from evaporating surface water bodies. Sediments are transported and accumulate in a low topographic area or deposition basin (lake, ocean, desert basin). Sedimentary rocks are clastic (mineral and rock fragments), biogenic (formed through the plant and animal activities), and chemical (produced by crystals deposited from concentrated solutions).

Metamorphic rocks are produced by the recrystallization of other rocks by intense temperature, pressure, hot solutions, or any combination thereof. Each type of rock is transformed into others in the rock cycle.

Earth processes produce compressional, tensional, and shear stress. The reaction to stress is strain, a change in shape or volume. Ductile materials flow under stress; brittle materials break. Elastic strain is recoverable when stress is relieved; plastic deformation is a permanent change in shape or volume of rocks. Joints are systematic breaks in a rock that occur without movement. Faults are breaks in the rocks with movement. Tensional stress produces normal faults; compressive stress produces thrust faults; shear stress produces strike-slip faults.

Earth is divided into three layers: core, mantle, and crust. The lithosphere includes the crust and solid mantle portion above the asthenosphere. Slabs of lithosphere called plates slowly move over the plastic asthenosphere. How they move is explained by plate tectonics theory. Plates move away from each other at divergent margins, move toward each other at convergent margins, and slide horizontally past each other along transform margins.

## KEY TERMS

### Physical Processes

compressional stress
conduction
convection
elastic strain
energy
kinetic energy
potential energy
shear stress
strain
stress
tensional stress
viscosity

### Chemical Processes

chemical reaction
ions
isotopes

### Igneous Rocks

basalt
crystal
granite
igneous rocks
lava
magma
mineral
plutonic (intrusive) rock
volcanic glass
volcanic (extrusive) rock

### Sedimentary Rocks

bedding
cementation
compaction
limestone
pores
porosity
sediment
sedimentary rocks
weathering

### Metamorphic Rocks

gneiss
marble
metamorphic rocks
metamorphism
quartzite
schist

### Rock Behavior Under Stress

brittle
ductile
failure
faults
gravity fault
joints
plastic deformation
strength
thrust fault

### Earth Structure

asthenosphere
continental crust
core
crust
lithosphere
mantle
moho (Mohorovičić discontinuity)
oceanic crust

**Plate Tectonics**

continental drift
convection cell
convergent margins
divergent margins
island arc
lithospheric plates
plate margins
plate tectonics
rock cycle
seafloor spreading
subduction
transform-fault margin

## REVIEW QUESTIONS

1. Distinguish between conduction and convection as ways to transfer heat energy within Earth.
2. a. Define plutonic (intrusive) and volcanic (extrusive) igneous rocks.
   b. Name a plutonic-volcanic equivalent in igneous rocks.
   c. How do plutonic and volcanic equivalents vary in appearance? Why?
3. How do nuclear and chemical reactions differ?
4. a. What are radioactive isotopes?
   b. What are their roles in driving Earth's internal processes?
5. Describe four different ways in which minerals form, and give examples.
6. Distinguish among the formation of igneous, metamorphic, and sedimentary rocks.
7. Distinguish between joints and faults.
8. What types of faults are generated by (a) tensional, (b) compressional, and (c) shear stress?
9. Rocks subject to stresses nearer the surface are strained very differently than the same rock under the same stress at depth.
   a. Why are there differences in the strain under these two conditions?
   b. What might the two rocks look like after the stress is relieved?
10. Contrast plate movements at (a) divergent, (b) convergent, and (c) transform-plate margins.
11. Describe the inner structure of Earth and the characteristics of its various layers.
12. What is the rock cycle? What is the energy source driving it?

## FURTHER READINGS

Coch, N. K., and Ludman, A., 1991, Physical geology: New York, Macmillan, 678 p., Chapters 2 (Matter and Energy), 3 (Minerals), 4 (Igneous Processes and Rocks), 5 (Volcanoes), 6 (Sedimentation and Sedimentary Rocks), and 7 (Metamorphism and Metamorphic Rocks).

Costa, J. E. and Baker, V. R., 1981, Surface geology—Building with Earth: New York, Wiley, 498 p., Chapter 5 (Rock: Its Strength, Durability, and Uses).

Hamblin, W. K., 1992, Earth's dynamic systems, sixth edition: New York, Macmillan, 647 p. Chapter 18 (Plate Tectonics).

# 3

# *Earth's Surface*

Two main external processes shape Earth's surface: weathering (physical and chemical) and the action of streams, glaciers, and wind. These processes require energy. The energy comes from two sources: Earth's own gravity and a giant nuclear reactor about 150,000,000 kilometers (93 million miles) away: the Sun.

Gravitational energy, or **gravity,** is very important in driving many surface processes such as stream erosion, glaciation, and landslides. The force of gravity *(G)* between two objects depends on their masses and the distance between them. The greater the mass, or the closer they are, the greater is the gravitational force between them.

For example, consider the Sun, Moon, and Earth. Is the gravitational force greater between Earth and the Sun, or between Earth and the Moon? The Sun's mass is 333,000 times that of Earth, whereas the Moon's is only 0.01 that of Earth, so you might think the force between Earth and Sun would be far greater. However, in this case, distance is more significant. The Sun is nearly 400 times farther from Earth than the Moon, greatly reducing the gravitational attraction between them. In this case, the gravitational force between Earth and the Moon is greater.

The gravitational force of the Moon on the oceans along with Earth's rotation, causes the ocean surfaces to rise and fall as **tides.** The total mass of Earth is much greater than that of any individual object, such as a stream, glacier, or rock cliff, on its surface. Consequently, Earth's gravity "pulls" on the water in a stream, causing it to flow downslope.

## EARTH'S EXTERNAL ENERGY SYSTEM

In the solar nuclear reaction, four hydrogen atoms are forced together under extremely high temperature and pressure, to form two atoms of helium (see Figure 2–2 in Chapter 2). Such a nuclear reaction, in which nuclei of two atoms are fused to make a new element, is called **nuclear fusion.**

**View of the Nile Delta (triangular dark area), looking west along the south coast of the Mediterranean Sea, as seen from the Gemini 4 spacecraft. (NASA.)**

During nuclear fusion, some mass from the hydrogen atoms is converted into radiant energy (Figure 2–3c). This energy radiates from the Sun in many wavelengths (mostly light). Part of it travels through the vacuum of space to Earth. Upon reaching Earth, this radiant energy is converted to heat. It warms the top few meters of the solid Earth **(lithosphere),** the upper part of water bodies **(hydrosphere),** and Earth's gaseous envelope **(atmosphere).**

Radiant energy from the Sun that heats the upper portion of water bodies (Figure 3–1) makes water molecules so active that some of them rise into the atmosphere as vapor, against the pull of Earth's gravity. This process is **evaporation.** The heated air and water vapor are less dense than surrounding cooler air, so they rise. This carries moisture upward into the atmosphere, where temperatures become progressively cooler with altitude.

As the rising air cools, it has less capacity to hold moisture. The water vapor begins to change back into microdroplets of water, a process we call **condensation.** Depending on air temperature, the condensed water vapor forms **precipitation**—liquid particles of rain or solid ones of snow.

Rain that falls to Earth's surface replenishes **surface water** (oceans, lakes, and streams) and also infiltrates into the ground to supply water to wells. Snow that falls in constantly frigid regions accumulates and eventually becomes compressed into glacial ice. Gravity pulls on water, snow, and ice, converting their potential energy into the kinetic energy of flowing streams and creeping glaciers. This is the energy that enables streams and glaciers to erode the surface.

A similar mechanism drives Earth's wind systems (Figure 3–1). Radiant energy heats Earth's surface and the air in contact with it. The heated air, which is less dense, rises by convection (Figure 2–3b). Cooler, denser air moves in horizontally to replace the rising air. Coupled with Earth's rotation, this results in the winds that sweep the land, drive the surface currents in the oceans, and create waves on the ocean surface.

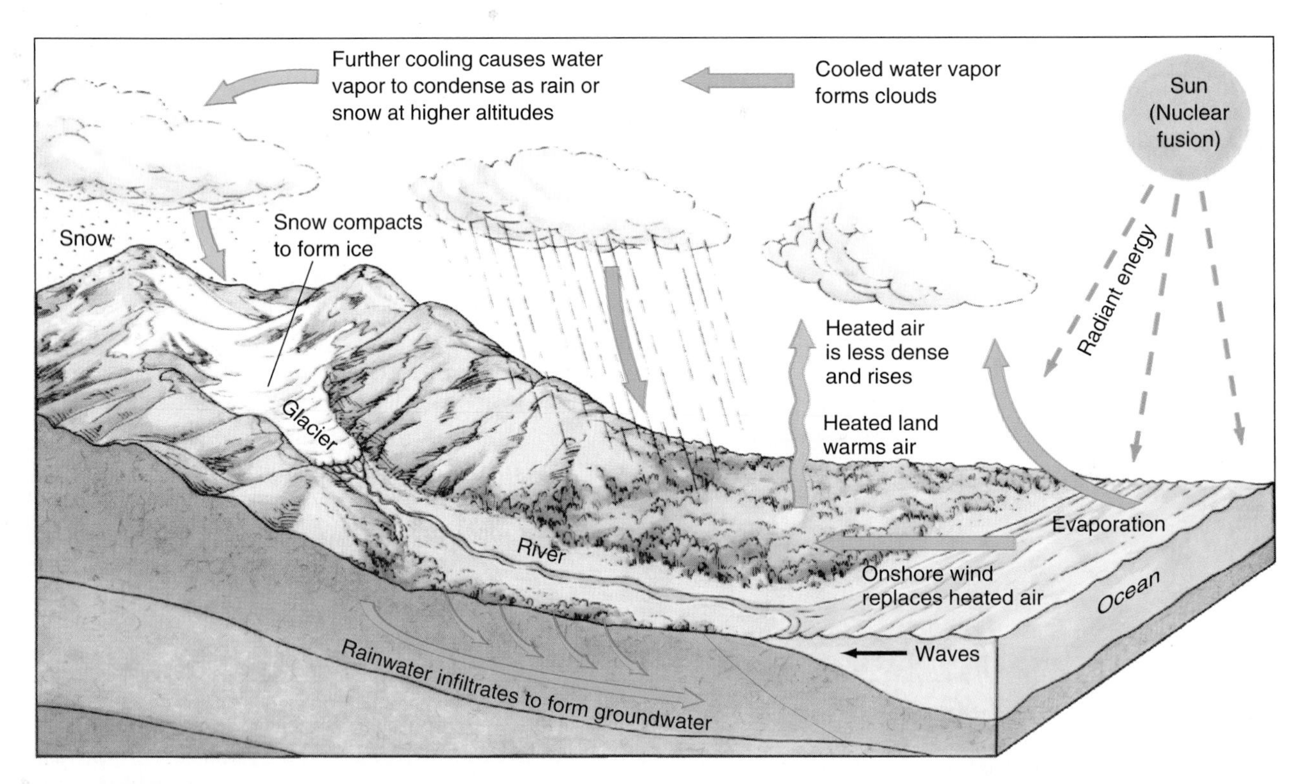


**FIGURE 3–1**
The mechanisms by which solar energy drives Earth's surface processes.

## THE HYDROLOGIC CYCLE

Water exists in three states on Earth: liquid (rain, streams, lakes, oceans, and groundwater); vapor (steam, clouds, fog, and invisible humidity); and solid (ice and snow). Water constantly changes from one state to another, depending on temperature.

For example, *right now,* all around Earth, surface water is evaporating to form water vapor, water vapor is condensing to form liquid water, water is freezing to form ice crystals and snow, and glaciers are melting to form liquid water. This continual circulation of water in its three states through the atmosphere, hydrosphere, and lithosphere is called the **hydrologic cycle** (Figure 3–2).

The distribution of water in the hydrosphere is as follows:

| | |
|---|---|
| 97.30% | in the ocean (saltwater) |
| 2.14% | frozen into glaciers |
| 0.54% | groundwater |
| 0.02% | in streams, rivers, freshwater and saltwater lakes, and soil moisture |
| 100.00% | |

Most of Earth's water resides in the oceans as saltwater. The freshwater that people rely on for drinking, washing, and growing crops is the surprisingly tiny percentage in the ground, streams, lakes, and soil.

The total water on Earth in all forms (solid, liquid, gas) probably has remained nearly constant

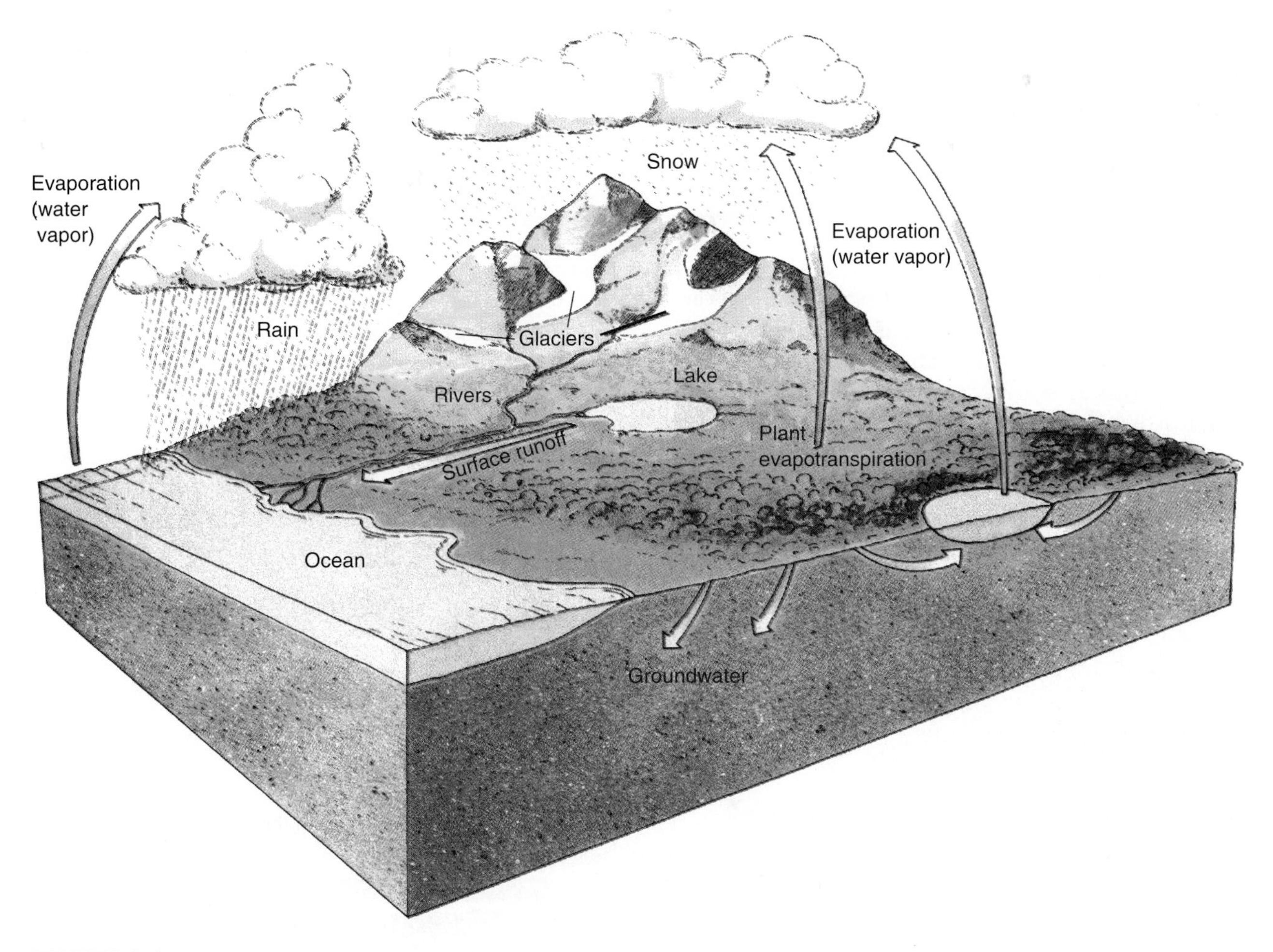


**FIGURE 3–2**
The hydrologic cycle. (Reprinted with permission of Macmillan College Publishing Company from *Physical Geology* by Nicholas K. Coch and Allan Ludman. Copyright © 1991 by Macmillan College Publishing Company.)

since water first formed early in Earth's history. However, the *relative* percentage of water in each form has varied over time. For example, during the most recent Ice Age, which peaked around 18,000 years ago, more water was stored in glaciers compared to the amount today.

The growth of extensive glaciers on the continents required water, which came from the water vapor evaporated from the oceans (Figure 3–1). This withdrawal of seawater to form glaciers caused a substantial lowering of sea level, approximately 125 meters (410 feet) worldwide, at the peak of the last glacial age.

### Humans and the Hydrologic Cycle

As you have seen, only a small amount of Earth's water is available on land for industrial, agricultural, and domestic use. Worldwide population increase and industrialization are creating greater demand for this limited resource. Many scientists feel that a crisis is rapidly approaching unless water conservation is practiced more widely.

The problem periodically becomes acute in dry regions, such as the Sahel of Africa and even parts of the U.S. Southwest and California. After several years of drought in California, that state began water rationing in 1991. Although California's situation is not as severe as that in Africa, it demonstrates that even wealthy nations are vulnerable to drought.

One possible solution to our potential water crisis is desalination of ocean water. The most effective method is distillation (evaporating seawater and condensing the vapor to collect freshwater). However, distillation consumes considerable energy and creates environmentally damaging saline waste water. Technological advances in desalination could become one of the major environmental breakthroughs of the next century.

### Evaporation, Infiltration, and Runoff

Rain that falls to the surface can do three things: it can evaporate back into the atmosphere, soak into the ground, or run off the slope into streams. Some water absorbs enough heat energy to evaporate back into the atmosphere. Some water is absorbed by plants, and they return a portion of it to the atmosphere through their breathing-like process of **transpiration.** The remaining water either soaks into the ground or flows along the surface. Movement of water into the ground is called **infiltration.** Water that flows downslope across the surface is called **runoff.**

The portion of water that infiltrates compared to the portion that runs off varies with land slope, vegetation density, type of vegetation, soil type, rock type, rainfall rate, and the degree of soil saturation. In general, the steeper the slope, the greater the runoff. Vegetation promotes infiltration because it slows surface water flow and because plant root systems open the soil to promote infiltration. The more openings in the surface—cracks, pores, joints—the more infiltration occurs.

The volume of water that seeps into the ground per unit of time and area is the *infiltration rate.* If rain falls faster than the infiltration rate, the excess water flows over the surface in a thin layer called **sheetflow.** Thus, during especially heavy rainfalls, a greater danger of flooding exists on slopes.

### Groundwater

The portion of rainfall that infiltrates the ground surface to migrate underground under the influence of gravity is called **groundwater** (Figure 3–2). Groundwater moves much more slowly than surface water, because of frictional effects as it flows through pore spaces and rock structures.

Some groundwater eventually seeps into stream channels at the surface, thus augmenting their flow. The portion of a stream's flow contributed by the underground water supply is the **base flow;** this component can maintain streamflow during a drought.

## WEATHERING

Minerals form deep within Earth, where they are stable at temperatures and pressures far greater than those at Earth's surface. These minerals eventually become exposed at the surface because overlying rocks become weathered and eroded. At the surface, these minerals are exposed to dramatically different conditions, which change them into different minerals that are more stable under surface conditions.

The physical and chemical processes that cause this transformation are called **weathering.** They result in the formation of soil, which is Earth material

**FIGURE 3–3**
Weathering processes and products. (Adapted from *Physical Geology* by Nicholas K. Coch and Allan Ludman. Copyright © 1991 by Macmillan College Publishing Company.)

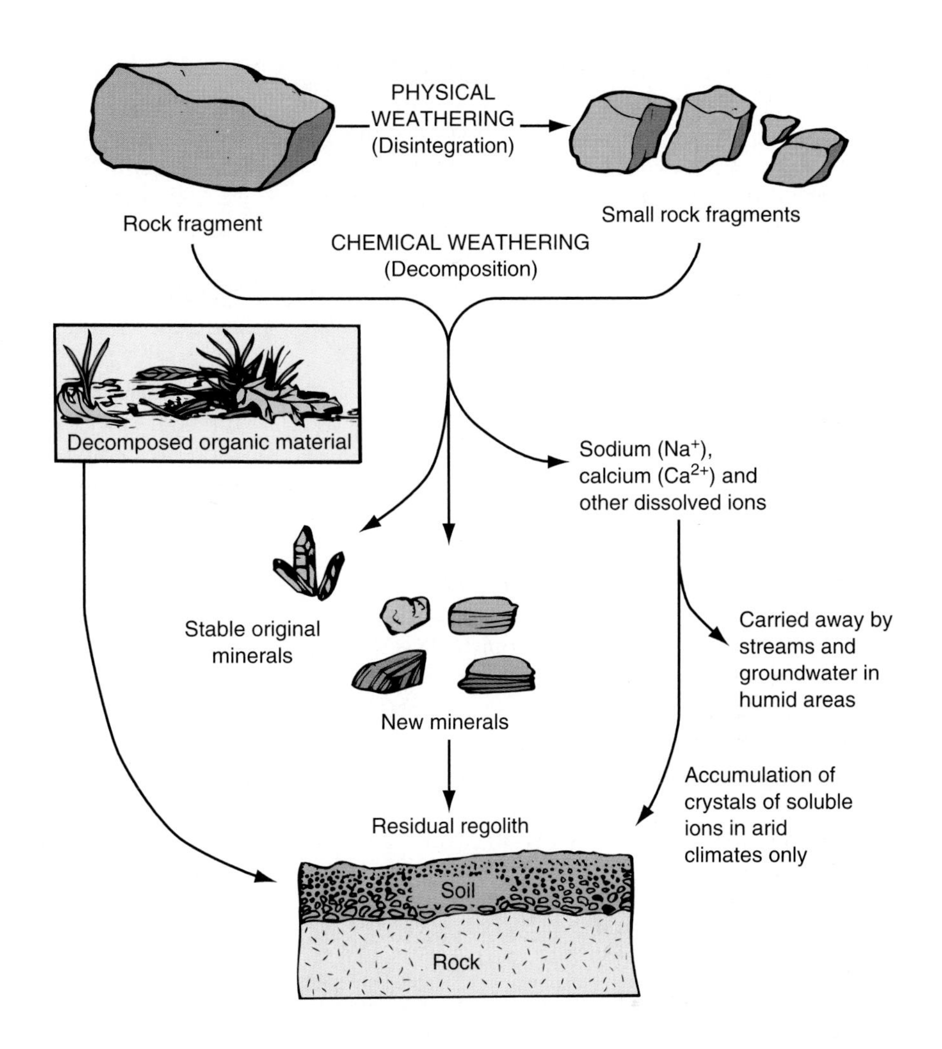


at equilibrium with surficial temperatures and pressures. Weathering, therefore, is an equilibrium adjustment from internal to surficial conditions. Figure 3–3 shows major weathering processes and their products.

## Physical Weathering

**Physical weathering** is the disintegration of rock into smaller pieces, with no change in the chemical makeup of the minerals (Figure 3–3). The process is analogous to smashing a rock into tiny pieces—slowly. The agents of this type of weathering include gravity, the growth of plant roots within rock openings, the growth of crystals (ice, salt) within rock openings, wind, and burrowing animals.

A major physical weathering process is **frost wedging,** in which water enters rock openings (joints, pore spaces) and freezes. Freezing water expands about 9%, exerting powerful tensional forces sufficient to wedge the rocks apart (see Chapter 2).

Frost wedging forms angular pieces and particles of the minerals that make up the parent rock (Figure 3–4). This process is most effective in temperate and subarctic climates where the temperature *repeatedly* fluctuates between freezing and thawing.

Physical weathering greatly increases the surface area of rock. To see how this happens, examine the rock cube in Figure 3–5. The rock measures 2 meters (6.5 feet) on a side. Thus, its total surface area is 24 square meters (2 meters × 2 meters × 6 sides), or 253.5 square feet. Frost wedging along joints breaks the rock into eight cubes, each measuring 1 meter (3.3 feet) on a side. This doubles the area *exposed to weathering,* to 48 square meters (1 meter × 1 meter × 6 sides × 8 cubes), or 523 square feet. In this manner, physical weathering exposes more rock surface to chemical weathering.

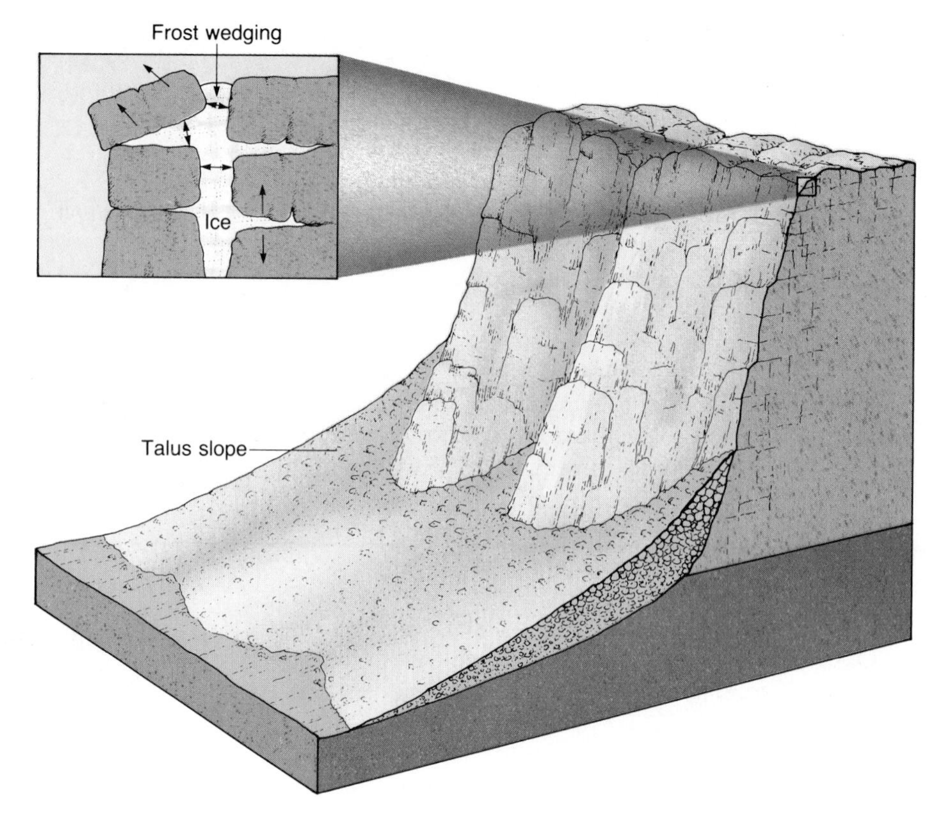


**FIGURE 3–4**
Frost wedging. When water in rock openings freezes, it expands and fractures the rock. The accumulation of rock debris at the base of a cliff is called talus. (Reprinted with permission of Macmillan College Publishing Company from *Earth Science* (seventh edition) by Edward J. Tarbuck and Frederick K. Lutgens. Copyright © 1994 by Macmillan College Publishing Company.)

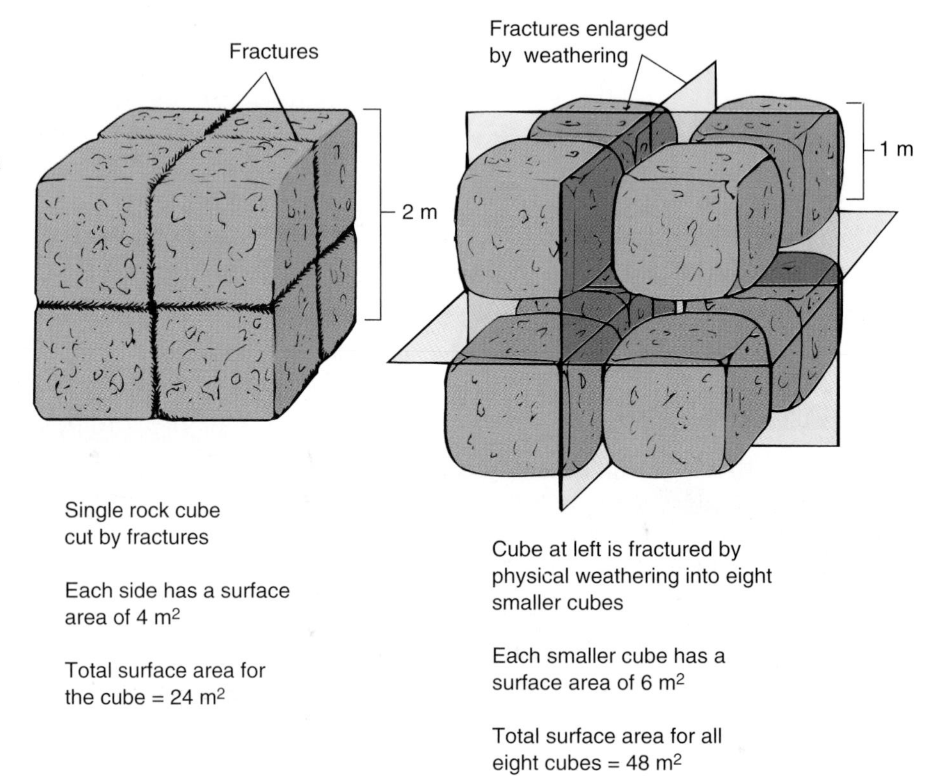


**FIGURE 3–5**
Physical weathering increases the surface area of a rock. (Reprinted with permission of Macmillan College Publishing Company from *Physical Geology* by Nicholas K. Coch and Allan Ludman. Copyright © 1991 by Macmillan College Publishing Company.)

## Chemical Weathering

**Chemical weathering** is the compositional change that occurs in the minerals of a rock that is under *chemical* attack (Figure 3–1). An example is the formation of rust on an iron pipe exposed to air and moisture. Major agents of chemical weathering include oxygen, water, carbon dioxide, and organic acids from decaying plants.

Reaction between these chemical weathering agents and the minerals in a rock alter the rock's minerals. This usually changes the rock's appearance, because different minerals have different properties of color, shape, hardness, and so on (Figure 3–6). For example, unweathered basalt is a dark, grayish-black extrusive rock (Chapter 2). However, the chemically weathered basalt shown in Figure 3–6 has undergone chemical weathering, changing its surface to a rust-colored, fine-grained weathering product.

One of the most important chemical weathering processes is **oxidation,** in which oxygen and water react with minerals in a rock. A piece of iron left outside quickly oxidizes and becomes covered with iron oxides collectively called "rust." Similarly, in Figure 3–6, iron oxides develop from the original iron-rich silicates in the basalt, creating a rust-colored weathering residue on the rock surface.

**FIGURE 3–6**
Weathered basalt on the island of Kauai, Hawaii. Chemical weathering has reduced the original dark gray rock to a rust-colored clay. (Photo by author.)

Another important chemical weathering process is **solution.** Here, carbon dioxide combines with water to form carbonic acid. Over time, this weak acid very effectively dissolves limestone. The unique atomic structure of water ($H_2O$) makes it very effective in weathering. In a water molecule, positively charged hydrogen atoms are on one side and negatively charged oxygen atoms are on the opposite side. This creates a molecule that is polar—a negative end (oxygen) and a positive end (hydrogen)—Figure 3–7. (The overall molecule is neutral, with a net charge of zero.)

These polar water molecules can react with oppositely charged ions in a mineral structure, dissolving the mineral. This process of **hydrolysis** is a very important reaction in the weathering of silicate minerals, which make up most igneous and metamorphic rocks (Chapter 2).

## Products of Weathering and Their Significance

Some products of weathering are carried away in solution as **dissolved ions** (Chapter 2), transported in both surface and underground water. They eventually are deposited as chemical sediments on land or in the ocean (Figure 3–3). Weathering also leaves a residue of resistant minerals that weather much more slowly.

Weathering leaves loose rock material on the surface, called **regolith** (Greek, "blanket of stones"). Regolith that remains in place above the underlying material from which it derived is **residual regolith** (Figure 3–3). Regolith that has been moved and deposited by streams, wind, or glaciers is **transported regolith.**

Regolith that has weathered enough to become enriched in the elements needed to support plant life is called **soil** (Figure 3–3). Residual regolith, soils, and problems of soil erosion are discussed in Chapter 6.

When solid rock has disintegrated and decomposed through weathering, the resulting regolith moves downslope more easily under the influence of gravity. Thus, weathering is very important in the development of a common geologic hazard: landslides (Chapter 6).

# EROSION

**Erosion** is the removal of regolith and rock by the action of streams, glaciers, wind, and coastal waves.

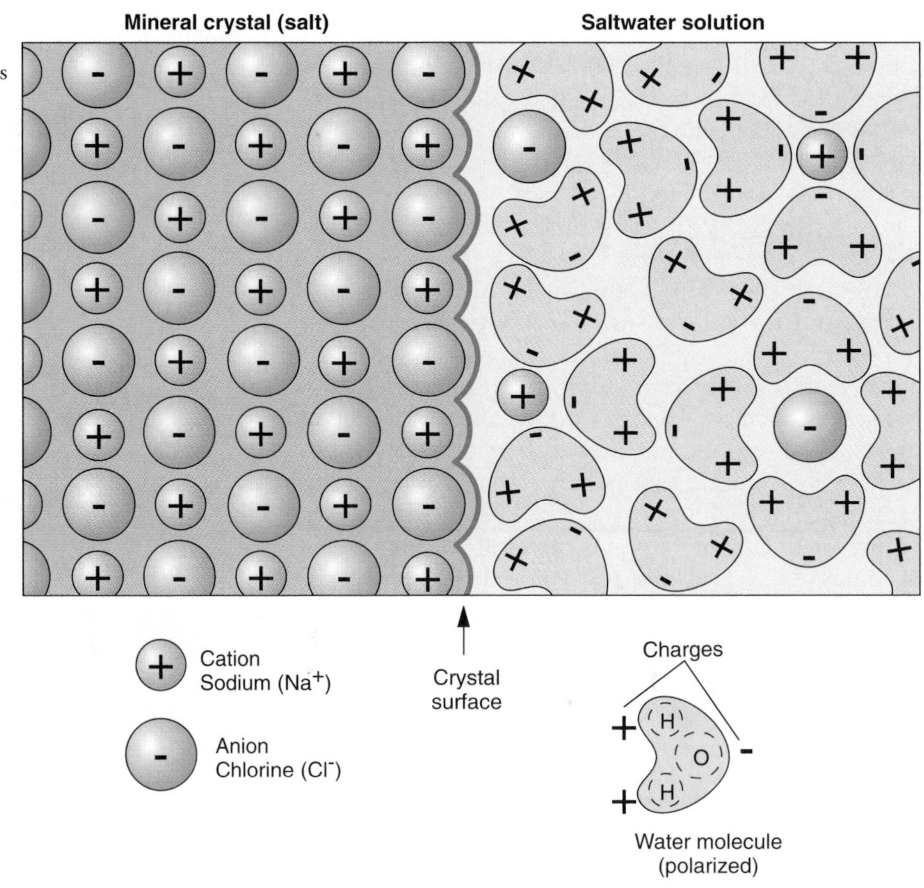


**FIGURE 3–7**
Role of polar water molecules in weathering. Because water molecules are polar, with a negative end (oxygen) and a positive end (hydrogen), they can react with oppositely charged ions in a mineral structure, dissolving the mineral. (Reprinted with permission of Macmillan College Publishing Company from *Physical Geology* by Nicholas K. Coch and Allan Ludman. Copyright © 1991 by Macmillan College Publishing Company.)

The relationship between erosion and transportation is shown in Figure 3–8. The cliff in Figure 3–8a has been eroded by storm waves, which cut a notch into its base (Figure 3–8b). This removed the support for the slope, resulting in a landslide—the mass erosion of a great volume of sediment from the cliff.

The landslide material then is reworked by waves and transported from the area by coastal currents. Waves move sand along the beach, while smaller particles are deposited offshore. Only the coarsest particles are left to make up the beach at the base of the new cliff face (Figure 3–8c).

## SEDIMENTS

Sediment particles that compose regolith can have very different characteristics from place to place. For example, when you walk along an ocean beach, you feel the coarse sand that is very different from the fine sand in the dunes inland from the beach or the dark-colored mud in the bay still farther inland, behind the dunes. Similarly, the fine-grained, uniform-sized sediment in the downstream part of a river is very different from the pebbles and boulders found in its upstream reaches.

Why are sediments so different from place to place? We will answer this question by first considering sediment properties and then showing how these properties give clues to the transporting agent—water, ice, or wind—that deposited them.

### Composition

The composition of clastic sediment is determined by the minerals in the source rocks. For example, abundant crystals of quartz, mica, and feldspar in a regolith suggest that the source rock was probably a granite. The original mineral composition may be changed significantly by chemical weathering. The feldspar in the original rock, for example, may have been chemically weathered into the clay minerals in the regolith.

**FIGURE 3–8**
Erosion in coastal cliffs. (a) Coastal cliff before a storm. (b) Breaking storm waves erode the cliff base, undermining it and causing its collapse. (c) Collapsed material is reworked by the waves and the cliff recedes landward in the process. (Reprinted with permission of Macmillan College Publishing Company from *Physical Geology* by Nicholas K. Coch and Allan Ludman. Copyright © 1991 by Macmillan College Publishing Company.)

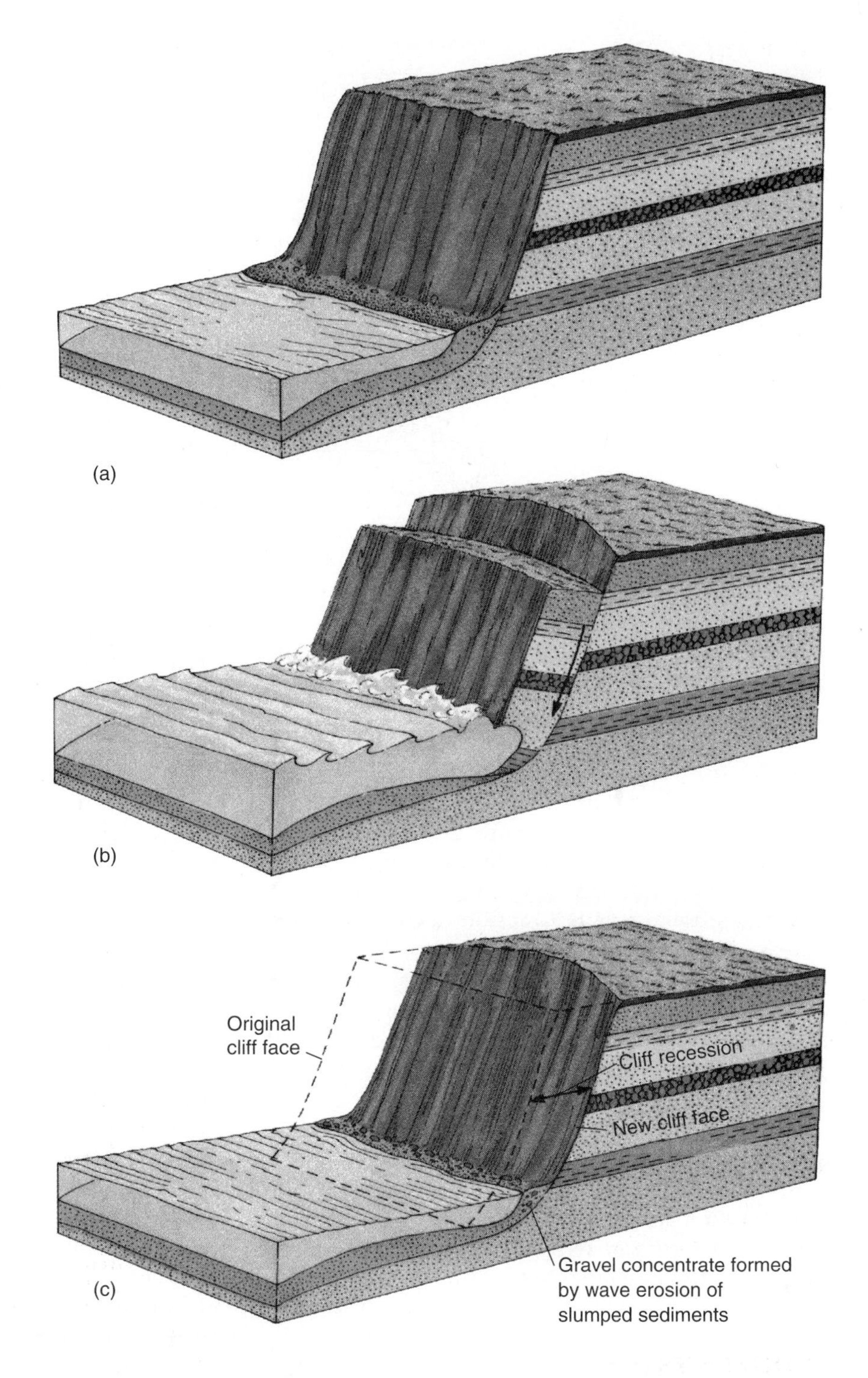


Chemical sediments are crystals that precipitate from a solution that became concentrated. Evaporation of surface waters in arid climates results in deposition of salt crystals in the regolith. Biogenic sediments, produced by the action of plants and animals, are made up largely of calcium carbonate or silica.

## Size

Sediment particles are labeled by size, and the most commonly used size terms are defined in the Wentworth classification (Table 3–1). Note that, in this context, **sand, silt,** and **clay** refer only to *particle size* and have *nothing* to do with composition.

**TABLE 3–1**
The Wentworth Size Scale for particles.

| Clast or Particle Size Range (mm) | Clast or Particle Name | Name of Sediment |
|---|---|---|
| Larger than 256 | Boulder | Boulder gravel |
| 256–64 | Cobble | Cobble gravel |
| 64–4 | Pebble | Pebble gravel |
| 4–2 | Granule | Granule gravel |
| 2–0.063 | Sand | Sand |
| 0.063–0.004 | Silt | Silt |
| Smaller than 0.004 | Clay | Clay |

Particles are described more specifically by a modifier. For example, sand-sized particles could be labeled *quartzose sand* (indicating a quartz composition). Silt-sized particles could be labeled *river silt* (indicating the transporting agent). Well-worn pebble-sized particles could be labeled *rounded pebbles* (indicating a physical characteristic).

Several factors determine particle size. The initial size of particles and crystals in the source rock affects sediment size. For example, chemical weathering of a coarse-grained granite initially produces larger crystals than weathering of a fine-grained granite. Given sufficient time and favorable conditions for further weathering, particles are continually reduced in size.

The dominant type of weathering also is important in determining particle size. For example, dominant *physical* weathering can form large rock fragments, even if the rock has small mineral crystals.

## Kinetic Energy and Viscosity

Recall from Chapter 2 that kinetic energy is the energy of motion and viscosity is the resistance to flow. The kinetic energy and viscosity of a transporting medium (water, air, ice) are two other important interactive factors in moving sediment.

A fast-moving stream has greater kinetic energy than a slow-moving stream and thus can transport larger pieces and particles. Similarly, fast-moving air (wind) can lift and move sand grains, whereas slow-moving air cannot. Both the fast-moving air and water have high kinetic agency and low viscosity (resistance to flow).

Glacial ice appears to have little kinetic energy, so you might expect this slow-moving mass to transport only fine sediments. However, glacial ice has a much greater viscosity (resistance to flow) than air or water. Just as viscous tar can pick up pebbles when it flows, the great viscosity of flowing ice enables it to carry all sediment particles it encounters, even car-sized boulders. In fact, ice is the most viscous transporting agent.

Another example of high viscosity is a saturated mass of regolith that flows down a slope. This flowing mixture of water and sediment has a high viscosity and therefore is able to transport a wide range of large particles (Figure 3–9). The geologic hazards posed by these movements are discussed in Chapters 4 and 9.

## Sorting

**Sorting** is the degree of uniformity of particle sizes. In a deposit of sediment, if particles are the same size, the deposit is considered *well-sorted.* If they are a mixture of very different sizes (as in Figure 3–9), the deposit is *poorly sorted.* Sorting also is a function of the kinetic energy, viscosity, and density of the transporting medium.

For example, air is the least viscous of all transporting agents. In addition, its density is only 1/1000 that of water, so it transports only the finer sediments (clay, silt, and some sand-sized particles). Consequently, wind-blown deposits generally are well-sorted. Glacial deposits, in contrast, are the most poorly sorted because ice is the most viscous of transporting agents and can thus transport a wide range of particle sizes.

FIGURE 3–9
Poorly sorted mudflow deposits in Death Valley, California. (Photo by author.)

In a similar fashion, sorting by flowing water separates (sorts) sediments of different sizes. Waves breaking on the landslide mass in Figure 3–8b have high kinetic energy that separates sand-size sediments and moves these particles along the shore to form a sandy beach. Currents that return the water slowly offshore carry finer silt and clay particles and deposit them in deeper water. The largest rock particles remain near the cliff until a severe storm or hurricane moves them.

## Roundness

**Roundness** describes the angularity of particles in a deposit, ranging from sharp to smooth. Crystals in rocks and in freshly weathered regolith are quite angular. As regolith is eroded and transported, the edges of the crystals become more rounded as the particles collide with each other and their environment. Rounding decreases as the viscosity of the transporting agent increases because viscous fluids allow few grain-to-grain impacts. On the other hand, rounding increases in low-viscosity media (flowing air, for example) because more grains are thrown into contact with each other.

Particles frozen in ice have little opportunity to have their edges rounded, so sediments deposited directly by ice have the poorest rounding. In contrast, wind-blown particles hit each other and the ground with great force and their edges become well-rounded. Sediments deposited by mud (mudflows) and water (streams) have rounding intermediate between glacial and wind deposits.

## Color

Sediment color reflects the color of component minerals, color of weathering products, and the percentage of organic material. For example, many beach and dune sands are rich in quartz, and so they commonly are white or tan. Some sediments are colored by weathering products. A common example is the red and brown produced by iron oxidation (Figure 3–6).

Organic material imparts a dark color (gray or black) to sediment. The darkness has two causes: the dark organic material itself and indirectly from the organic material's inhibiting effect on oxidation.

Decomposing organic material requires oxygen, and by consuming the available oxygen it inhibits oxidation of any iron minerals present. Thus, when organic material is present, we generally do not see the reddish weathering products characteristic of oxygen-rich deposits (Figure 3–6). To illustrate, mud in a lake or swamp may be black, but sediments in streams that feed them may be much lighter-colored because they have a higher oxygen level.

## STREAMS

Surface water that flows unconfined over the land is *sheetflow,* but flowing surface water confined to a channel is a **stream.** This section describes geologic characteristics of streams. Hazards and problems associated with human use and interference with stream processes are detailed in Chapter 7.

### Stream Energy at Work

Water on a slope has potential energy. Gravity pulls the water downstream and the potential energy is

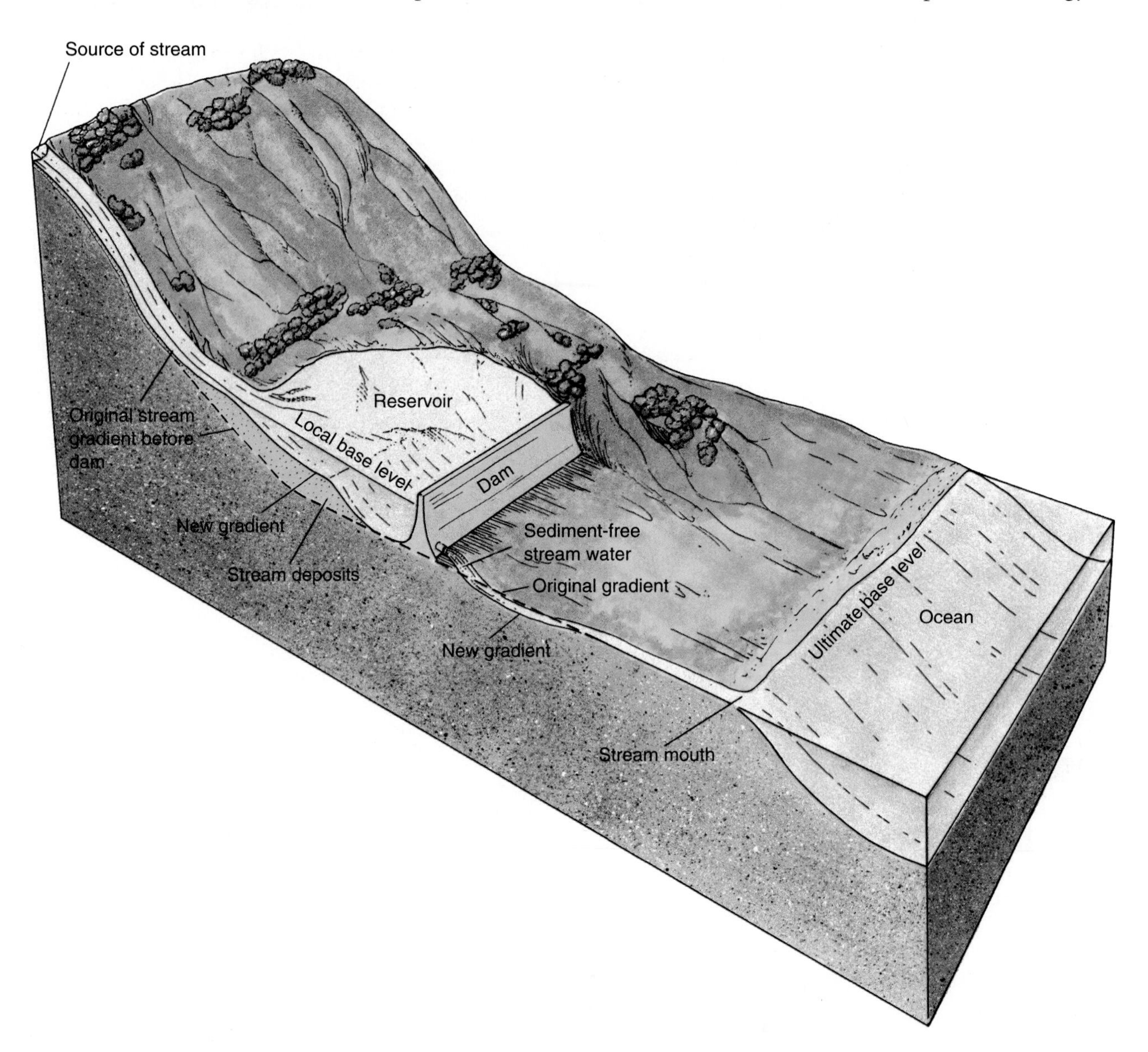


**FIGURE 3–10**
Effects of local base levels on a stream. Here, the reservoir acts as a local base level for the stream. Streamflow decreases in velocity as it enters the reservoir, resulting in deposition of its sediment. The deposition of sediment load immediately upstream from the reservoir decreases the stream gradient locally. Sediment-free water exiting the base of the dam has sea level as a base level; it erodes its bed, steepening its stream gradient locally. (Reprinted with permission of Macmillan College Publishing Company from *Physical Geology* by Nicholas K. Coch and Allan Ludman. Copyright © 1991 by Macmillan College Publishing Company.)

converted to the kinetic energy of flow. The amount of a stream's potential energy is proportional to its gradient. **Gradient** is the vertical drop of a stream channel over a horizontal distance. For example, a stream cascading from the mountains might have a gradient of 100 meters per kilometer, meaning that an elevation drops 100 meters (328 feet) over the distance of 1 kilometer (0.62 mile). A near-level prairie stream might have a gradient of only 0.5 meter per kilometer (1.6 feet per 0.62 mile). The steeper the gradient, the greater the potential energy in a stream.

Stream energy performs work—eroding regolith and rock and transporting sediment—until it reaches its lowest elevation, called **base level.** At the base level, the stream has little potential energy, and therefore little kinetic energy, so the sediment load in the water drops to the bottom.

Most streams have a *local base level,* the point at which they enter a larger stream, river, or lake. For most streams, the *ultimate base level* is the ocean, or sea level.

Streamflow can become temporarily slowed or blocked by natural obstacles (lakes, landslides, fallen trees) and artificial ones (human dams, beaver dams). This creates a *temporary base level* (Figure 3–10). A stream encountering such a temporary base level loses energy. Its ability to erode and transport sediment therefore is reduced until it can flow past the obstacle via a lake outlet, spillway, or overflow. This is why lakes and reservoirs, with their negligible gradients, fill with sediment over time (Figure 3–10). This problem is detailed in Chapter 6.

Although most streams attain their ultimate base level where they enter the ocean, some streams reach theirs on land. Numerous streams that flow into closed desert basins in the western United States are examples. Streams reaching their base level in California's Death Valley deposit clastic sediment as their flow infiltrates into the ground, and chemical sediment as the water evaporates into the dry desert air (Figure 2–8).

## Stream Sediment Transport and Deposition

Some of a stream's kinetic energy is expended in eroding and some in sediment transport. Streams erode by the force of flowing water. They move the resulting load of sediment in several ways (Figure 3–11).

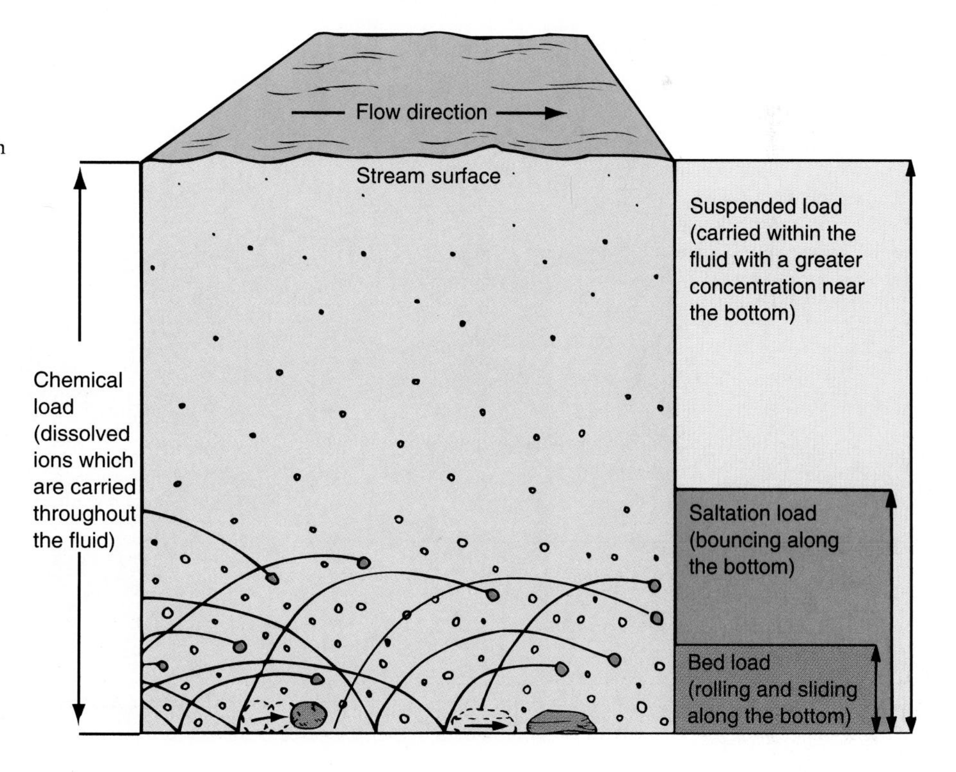


**FIGURE 3–11**
Types of stream load. (Reprinted with permission of Macmillan College Publishing Company from *Physical Geology* by Nicholas K. Coch and Allan Ludman. Copyright © 1991 by Macmillan College Publishing Company.)

The largest particles and pieces roll or slide along the bottom and make up the **bed load.** Sand-sized particles bounce along the stream bed and make up the **saltation load** (from the Latin word for "jump"). The finer particles (fine sand-size, silt-size, and clay-size) are the **suspended load,** which makes a stream appear cloudy. As the stream physically carries sediment particles, it also carries ions dissolved in the water. This is the **dissolved load.**

Bed load, saltation load, and suspended load are deposited as the velocity of the stream decreases—in that order, because the heaviest particles tend to drop out first. Thus, where a stream loses velocity, you can observe the pebble-sized particles dropping out, then the sand farther downstream, and finally silt and clay, all *sorted* by the stream.

The dissolved load is deposited only when chemical conditions change. This can occur where a stream enters the ocean or where evaporation concentrates the dissolved salts, making them precipitate as chemical sediments (Figure 2–8).

## Stream Channels

Under non-flood conditions, a stream flows within the channel it has cut into the land surface. Its channel is defined by the **width** (measured across the water surface at right angles to the flow) and the **depth** (averaged from several places across the stream). The width multiplied by the mean depth gives the stream's **cross sectional area** at that point. The volume of water carried past a point in a given time is called the stream's **discharge.**

Over time, each stream erodes a channel. The channel continually adjusts, seeking an **equilibrium** of all conditions along its length. When people disturb the channel or the area along it, this natural equilibrium is upset. When stream discharge increases markedly from rain or snow melt, the stream not only tries to change its channel but also floods more frequently. Hazards resulting from these problems are discussed further in Chapter 7.

## Floodplains and Levees

The **floodplain** is the flat area along a low-gradient stream channel. (High-gradient channels, such as steep mountain ravines, may not have a floodplain). When a stream **floods,** the volume of water exceeds the channel's capacity. The excess water overflows the channel and spreads across the floodplain.

Flood waters rapidly decrease in velocity, and the stream's energy becomes spread over a broad area.

**FIGURE 3–12**
Aerial view of the Mississippi River showing the ridges (levees) along the channel and the floodplain sloping away from the channel. Note the development on the floodplain. (Photo from Bureau of Reclamation.)

This loss of kinetic energy releases sediment from the water, causing deposition over the floodplain. Along the edges of the channel, the greatest amount of sediment is deposited. This creates slightly elevated areas that parallel the stream, called **levees** (Figure 3–12).

Natural levees grade gently outward from the channel across the floodplain. In many areas, people add height to natural levees to reduce the frequency of flooding, as along the Mississippi and Missouri Rivers. The more distant portions of the floodplain, called *backswamps,* receive the finer flood sediments (silt, clay).

## Deltas

When a stream enters a standing body of water (a lake or the ocean), its velocity decreases rapidly. The abrupt reduction of kinetic energy causes deposition of its sediment load on a shallow-water depositional plain called a **delta.** Figure 3–13 shows the Nile River delta, named for its resemblance to the Greek letter delta (Δ).

The delta grows thicker and wider over time (Figure 3–14). The weight of accumulating sediment slowly compresses the layers below, causing land subsidence beneath the delta surface. This can result in significant environmental problems if the delta surface is urbanized, as the Mississippi delta in southern Louisiana. Such problems are considered in Chapter 10.

**FIGURE 3–13**
Satellite view of the Nile River delta. (Photo from NASA.)

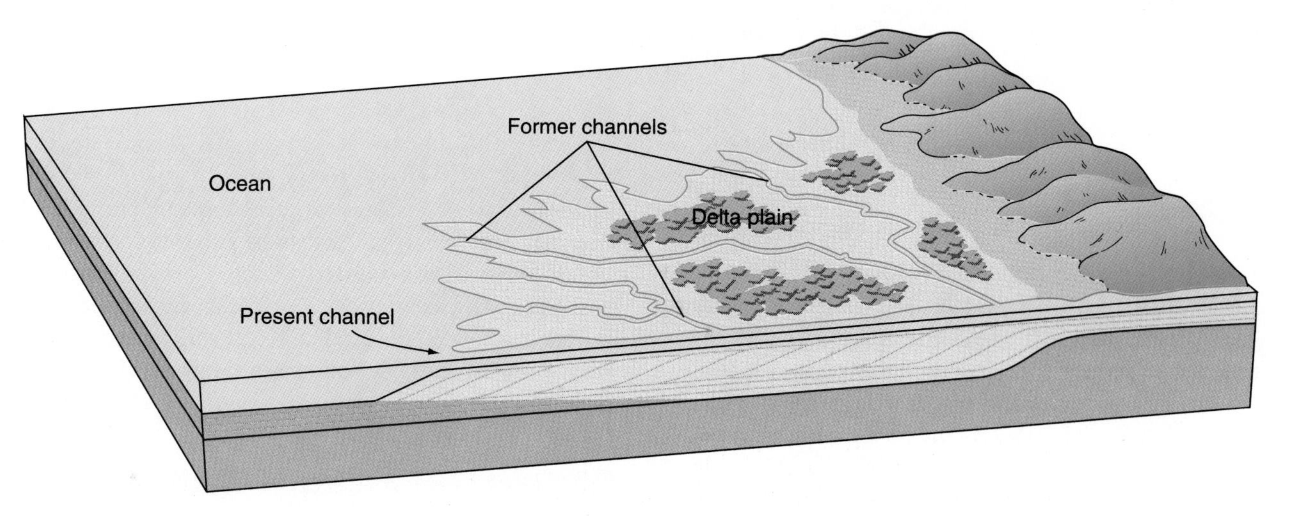


**FIGURE 3–14**
Cross section of a delta.

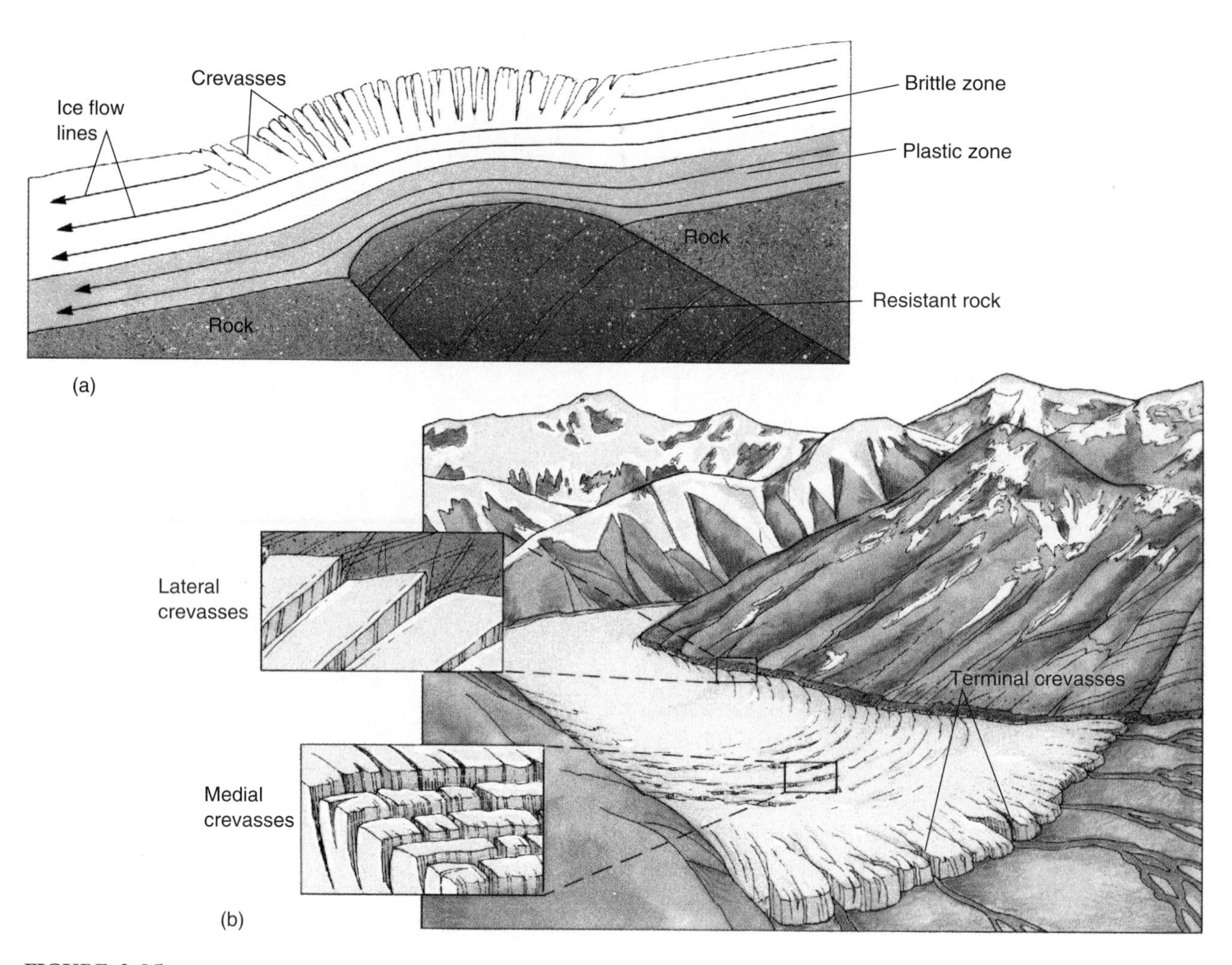


**FIGURE 3–15**
Valley glacier. (a) Cross-sectional view, showing brittle zone and plastic zone. (b) Map view, showing formation of crevasses as the glacier flows over an irregularity in the bedrock. (Reprinted with permission of Macmillan College Publishing Company from *Physical Geology* by Nicholas K. Coch and Allan Ludman. Copyright © 1991 by Macmillan College Publishing Company.)

## GLACIERS

A **glacier** is a long-lived mass of ice that forms from accumulated snow on land. It flows very slowly, like thick plastic material, under the force of gravity. As it flows, it scours the underlying surface, plucking and eroding particles and pieces from large boulders down through clay size. As the lower edge or *terminus* of a glacier enters a warmer area, the edge melts, releasing its load of unsorted sediment.

Glacial erosion and deposition have shaped the land surface in northern latitudes and in mountains at high altitudes. Glaciers have deposited much of the regolith that today exists in subarctic and temperate areas of the United States, Canada, Europe, and Asia. Glacial deposits provide significant economic resources of sand and gravel, and they are very porous, capable of storing large volumes of groundwater.

### Growth and Movement of Glaciers

Glaciers form wherever snow can accumulate year after year, continually adding to its thickness. This

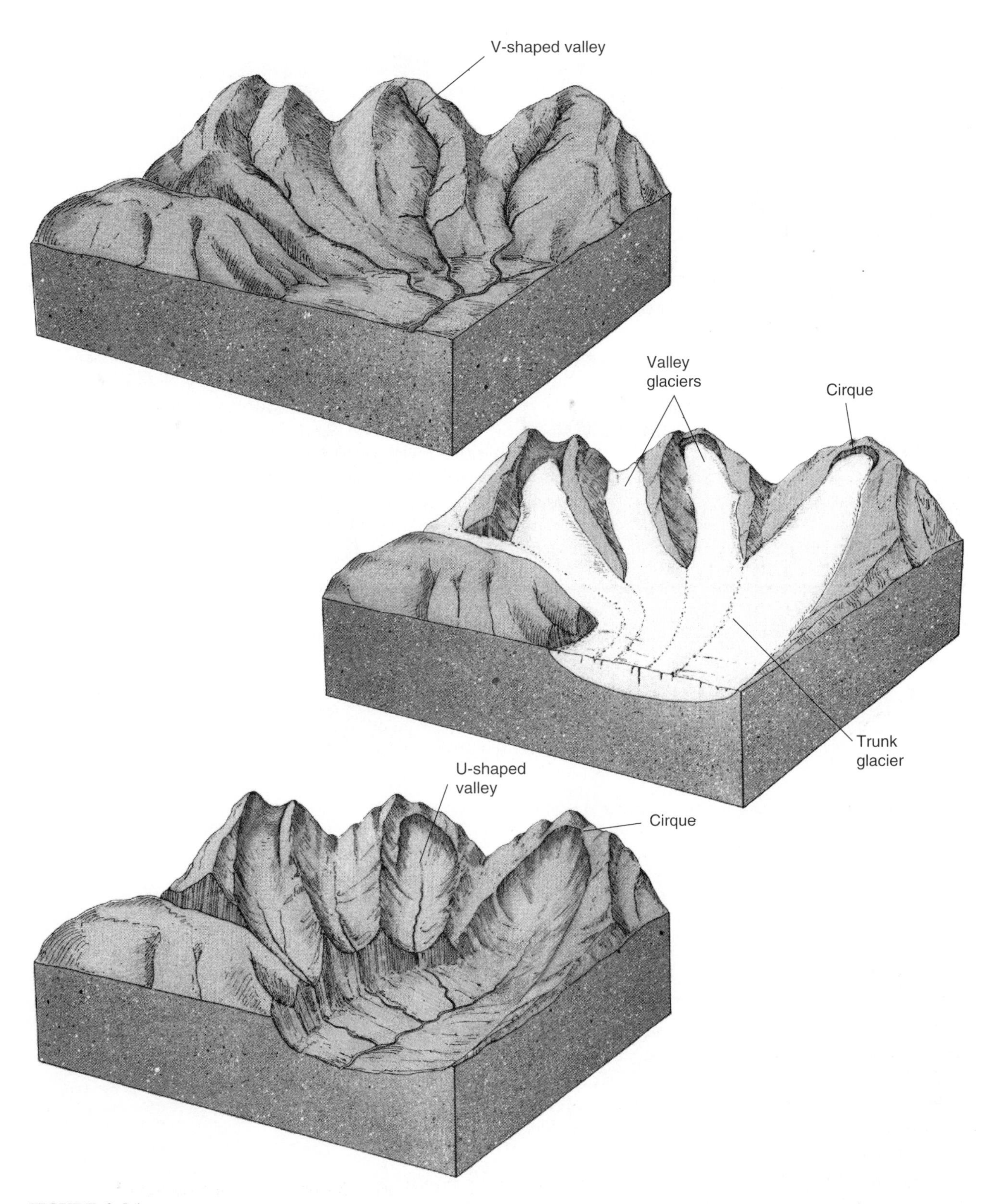


**FIGURE 3–16**
Evolution of alpine glacial features. (a) Preglacial mountain topography. (b) Mountain area during glacial episode. (c) After several glacial episodes, enlargement of cirque basins forms a variety of glacial features. (Reprinted with permission of Macmillan College Publishing Company from *Physical Geology* by Nicholas K. Coch and Allan Ludman. Copyright © 1991 by Macmillan College Publishing Company.)

occurs in Earth's frigid regions, at the poles and in high mountain elevations at any latitude. The great pressure of the accumulated snow compresses the lower portion of the permanent snow field into ice. Many glaciers form above the snowline on mountains. The **snowline** is the altitude above which snow is preserved from year to year.

Under ordinary conditions, ice is hard and brittle, like an ice cube. But under the pressure of its own great weight, glacial ice actually *flows* slowly, like extremely viscous plastic. The lower part of a glacier, which is under this great pressure, is called the **plastic zone** (Figure 3–15a; also see "Relationship Between Stress and Strain" in Chapter 2). Above this lower plastic zone, ice behaves in its more familiar fashion in the **brittle zone,** only about 60 meters (200 feet) thick, which extends to the surface (Figure 3–15a).

When the glacier passes over an irregularity on the bedrock below, such as the resistant rock shown in Figure 3–15a, the surface layer becomes bent and stretched. Tensional forces break it into deep **crevasses.** Crevasses pose a serious hazard to skiers and hikers because they can become spanned by thin "snow bridges" that look sound, but are not. Crevasses can form in the center of a glacier, at the sides, or at the front (Figure 3–15b).

## Glacier Types

Mountain glaciers begin on high slopes above the snowline in basins called **cirques** (Figure 3–16c). As they grow, these glaciers extend out of the basins as **valley glaciers** that flow down existing stream valleys (Figure 3–16c). Glaciers from adjoining valleys combine to form larger **trunk glaciers** (Figure 3–17). Trunk glaciers merge at the base of mountains to form **piedmont glaciers** that spread over the lowlands beyond the mountains.

Glaciers that form in polar regions are vast (hundreds of square kilometers) and thick (some over a kilometer). These glaciers cover all but the highest topographic features in a region. Smaller ones are

**FIGURE 3–17**
Aerial view of several mountain glaciers merging to form a trunk glacier in Alaska. (Photo courtesy of Austin Post.)

FIGURE 3–18
Maximum extent of Pleistocene glaciers in the Northern Hemisphere during the last glacial advance. Arrows denote local flow directions. Sea level shown is 100 m (330 ft) lower than today. (After R. F. Flint and B. Skinner, 1971, *Physical geology* (second edition). Used with permission of John Wiley & Sons, Inc.)

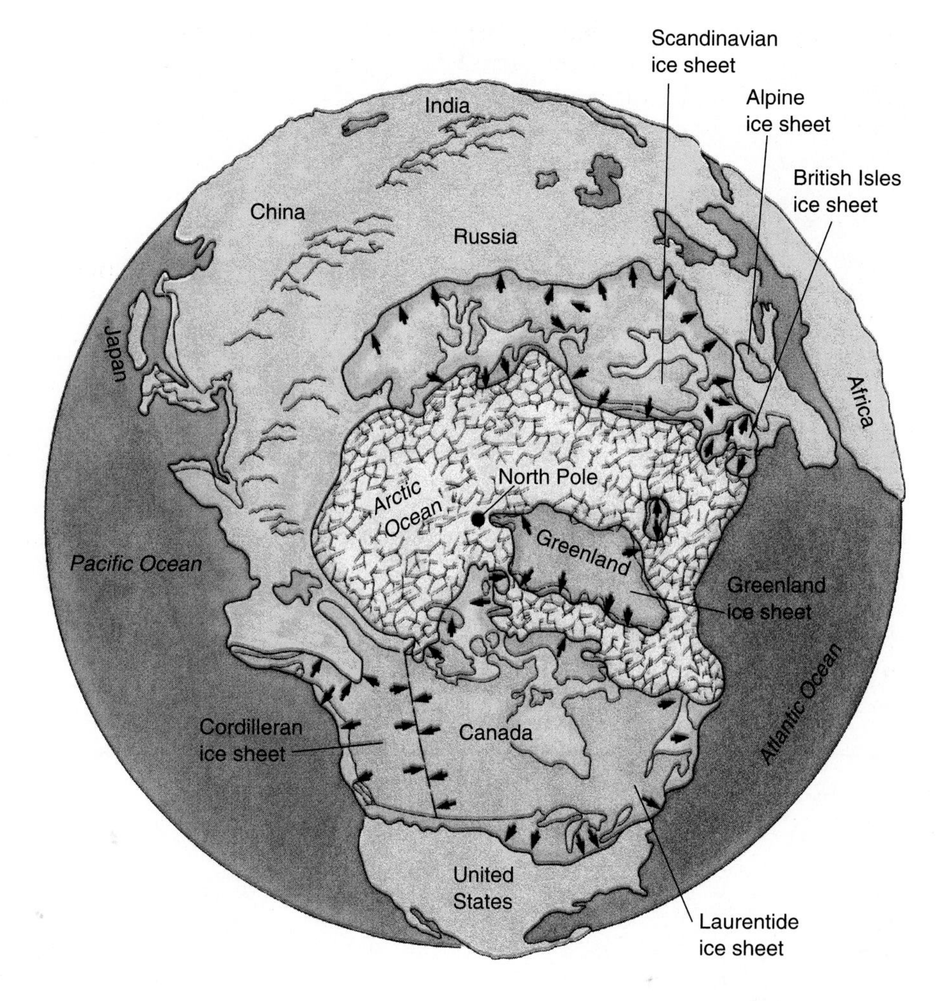


called **ice caps.** The largest are the **continental ice sheets,** thick masses that cover areas of subcontinental or continental size.

Today, continental ice sheets exist only on Greenland and Antarctica. However, about 18,000 years ago, during the most recent Ice Age, continental glaciers extended over large areas. They crossed regions that today have temperate climates (Figure 3–18). In these areas, most of today's landscape features and transported regolith result from continental glaciation.

## Glacial Erosion and Transport

Glaciers erode the land in several ways. Meltwater from glaciers infiltrates the rock beneath them and freezes, causing frost wedging, which loosens bedrock. This makes it easier for the glacier to incorporate this loose material. The moving glacier plucks the fractured rock and removes (erodes) it. Debris frozen into the ice abrades and polishes the bedrock, like sandpaper.

Glaciers transport material both on their surfaces (Figure 3–17) and frozen into the ice. Erosion is greatest at the bottom of the ice, but it also is important along the sides. This erosion typically reshapes V-shaped stream valleys into U-shaped glacial valleys (compare Figure 3–16a and c).

## Glacial Deposition

Deposition of a glacier's load happens in two ways: by direct action of the ice and by melting and transport of the load in meltwater. The differences between these two mechanisms explain distinctive characteristics of two types of glacial deposits.

**FIGURE 3–19**
Glacial till deposit. The till shows no stratification and is poorly sorted. (Photo by author.)

**FIGURE 3–20**
Glacial outwash deposit. The outwash shows well-developed bedding, and individual beds are well sorted. (Photo by author.)

Sediment eroded and deposited directly by moving ice is called **till.** The high viscosity of ice enables it to carry all of the material it encounters. Consequently, till is poorly sorted, usually does not show bedding (distinct layering), and displays angular, unsmoothed particles (Figure 3–19).

Sediment released from melting ice and carried away by meltwater streams is called glacial **outwash.** Large fluctuations in meltwater stream volumes and velocities cause deposition of different particle sizes at different times.

Consequently, outwash deposits show bedding (layering) of moderately sorted to well-sorted and particles that have been rounded (Figure 3–20). The most extensive outwash deposits are in an **outwash plain** that extends away from the terminus of a glacier.

### Moraines

A ridge of glacial deposits left by a melting glacier is called a **moraine** (French, "stone pile"). A *lateral moraine* is a deposit along a glacier's flanks (Figure 3–21). A *medial moraine* is the combined lateral moraines of two valley glaciers that have flowed together to form a trunk glacier (Figure 3–21). A *terminal moraine* is a deposit at the end of a glacier. A *ground moraine* is till that is deposited directly under the ice as it melts.

The wide range of deposits and landforms produced by continental glaciation is shown in Figure 3–21. The moraines and *drumlins* (streamlined ground moraine) are composed mainly of till. The *eskers* (subglacial stream deposits) and *kames* (crevasse fillings) are composed of outwash.

## WIND

Wind is an important agent of erosion and deposition in dry areas and along shorelines. It can move clay, silt, and sand-sized particles hundreds of kilometers, shaping the dunes of deserts and seashores. Wind erosion is becoming a serious concern be-

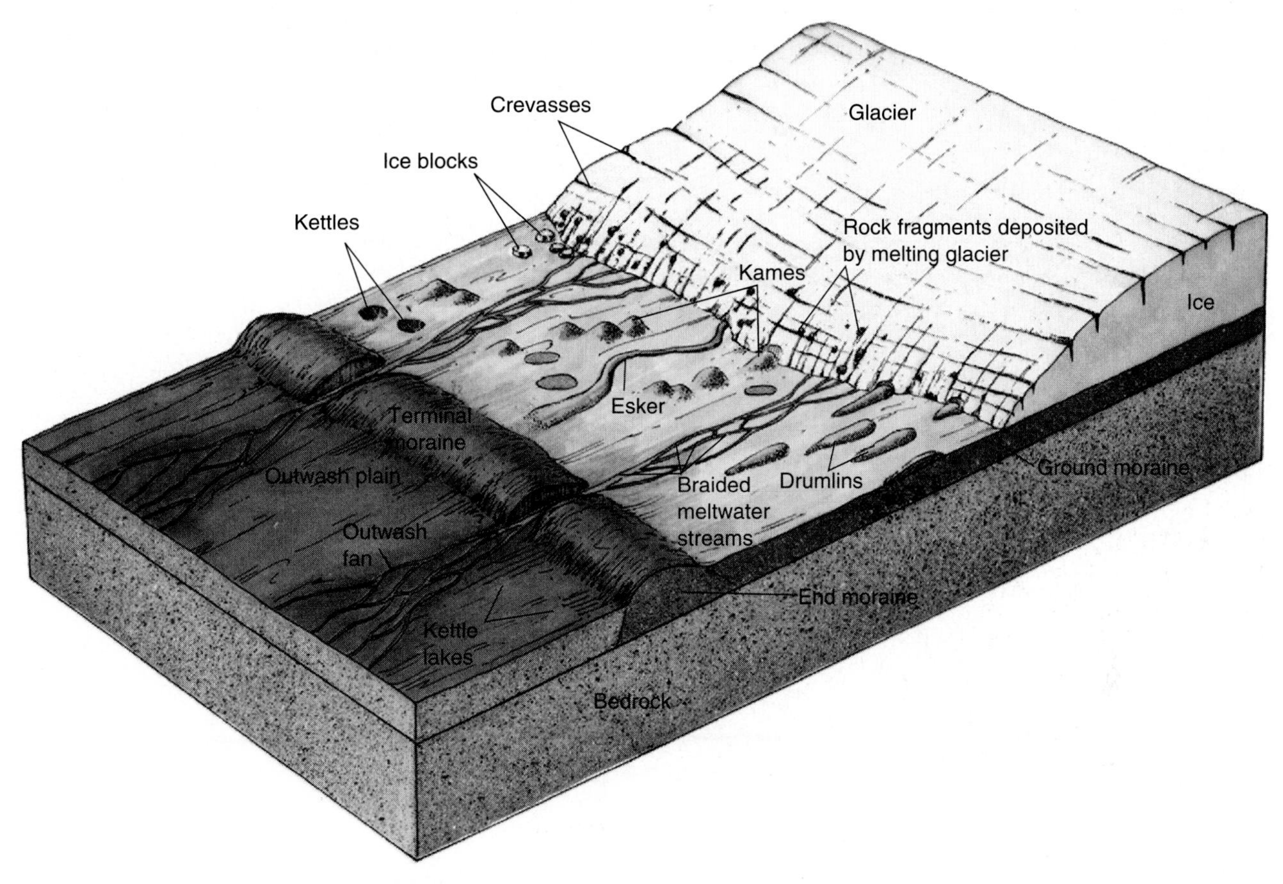


**FIGURE 3–21**
Landforms and sediment deposits resulting from continental glaciation. (Reprinted with permission of Macmillan College Publishing Company from *Physical Geology* by Nicholas K. Coch and Allan Ludman. Copyright © 1991 by Macmillan College Publishing Company.)

cause desert areas appear to be expanding. Long-term climatic change may result in the drying of presently humid areas, enabling wind erosion to increase. The hazardous consequences of wind erosion are considered in Chapter 6.

## Wind Erosion

Wind erodes sediment by the shear force (Chapter 2) that it exerts on loose particles on the land surface. Wind erosion of sediment is called **deflation.** In many cases, the wind-borne eroded particles strike the surface, eroding additional particles.

The wind picks up particles and moves them by saltation, just like moving water does (Figure 3–11). The saltating grains dislodge others as they bounce along, and soon there is a layer of saltating grains moving across the surface. Very little sand transport occurs more than 2 meters (6.5 feet) above the ground.

## Wind Deposition

Wind-carried particles are deposited when the wind's velocity decreases or where its flow is blocked, slowed, or deflected by a fence, rock, or other obstruction. Sand accumulates as sheets or as streamlined forms called **dunes.** Dune shapes form under different conditions of wind direction, sand supply, and vegetation. Figure 3–22 shows the four major types of dunes and their controlling conditions.

The finer particles (silt-sized and clay-sized) are transported in the air above the ground as suspended sediment. They may be carried far from the source area (Figure 3–23). In fact, silt eroded in African dust storms has been carried across the At-

| Dune type | Remarks | | |
|---|---|---|---|
| Barchan | 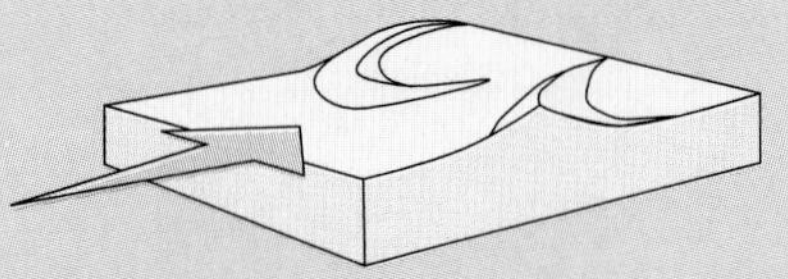 Maximum size: 30 m (65 ft) high 300 m (650 ft) point to point | 8-15 m (25-80 ft) annual movement | Most common. Generally in groups in areas of constant wind direction. |
| Transverse | 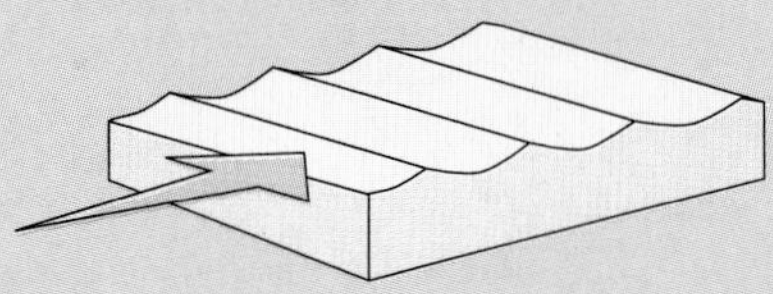 Similar to barchans but not curved. Form in areas with strong winds where more sand is available. | | |
| Parabolic | 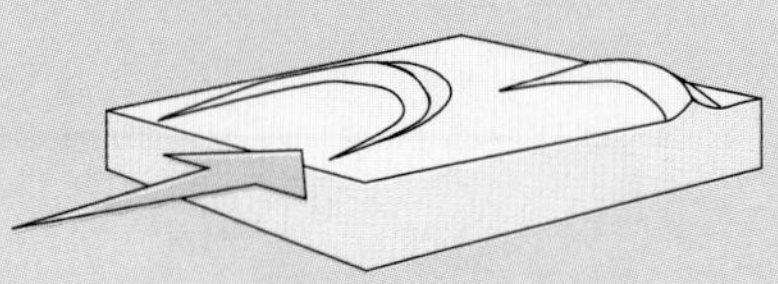 Maximum size: 30 m (65 ft) high. | Form in areas with moderate winds and some vegetation. | Extremely curved types called hairpin. Common at seacoast. |
| Longitudinal (Seif) | 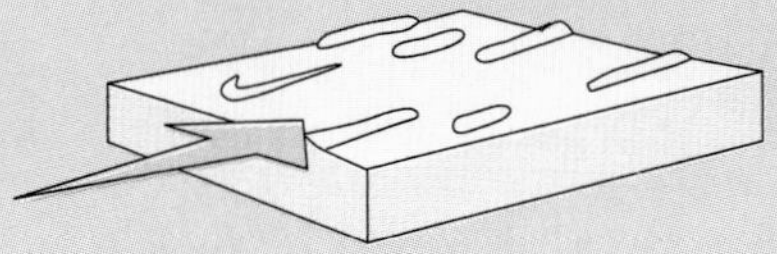 Maximum size: 90 m (300 ft) high, 100 km (60 mi) long. Avg. 3 m (10 ft) high, 60 m (200 ft) long. | Form in areas of high, somewhat variable winds, where little sand is available. | |

**FIGURE 3–22**
The four major dune types and their controlling conditions. (a) *Barchan dunes* develop best with constant wind direction, limited sand supply, and little vegetation. (b) *Transverse dunes* form at right angles to moderate winds where the sand supply is large. (c) *Parabolic dunes* resemble barchans but are "backwards." One type forms where wind erodes coastal dunes into *deflation basins.* (d) *Longitudinal dunes* form parallel to prevailing higher-velocity winds where the sand supply is large and wind direction is more variable. These dunes are common in the Sahara and other sandy deserts. (Reprinted with permission of Macmillan College Publishing Company from *Environmental Geology,* sixth edition, by Edward A. Keller. Copyright © 1992 by Macmillan College Publishing Company.)

lantic Ocean and identified in air samples taken in the Caribbean.

Deposits of windblown silt are called **loess** (*LOW-ess* or *luss,* Swiss for "loose"). The Missouri, Mississippi, and Ohio valleys contain widespread loess deposits, which form good farm land. Interestingly, the silt in these deposits is of glacial origin. The silt was deposited by strong winds that deflated the finer particles from glacial outwash plains to the north. The winds deposited the silt in sheets of loess, far from the original glacial margins. Today, *non-glacial* loess is being deposited by winds that blow silt from dry areas such as the Gobi Desert in China.

## SURFACE PROCESSES AND PEOPLE

People long have modified natural surface systems to improve agriculture and to facilitate travel. Population growth and industrialization have severely strained these natural surface systems. The resulting geologic hazards and environmental problems are increasing in severity and cost. The surface process-

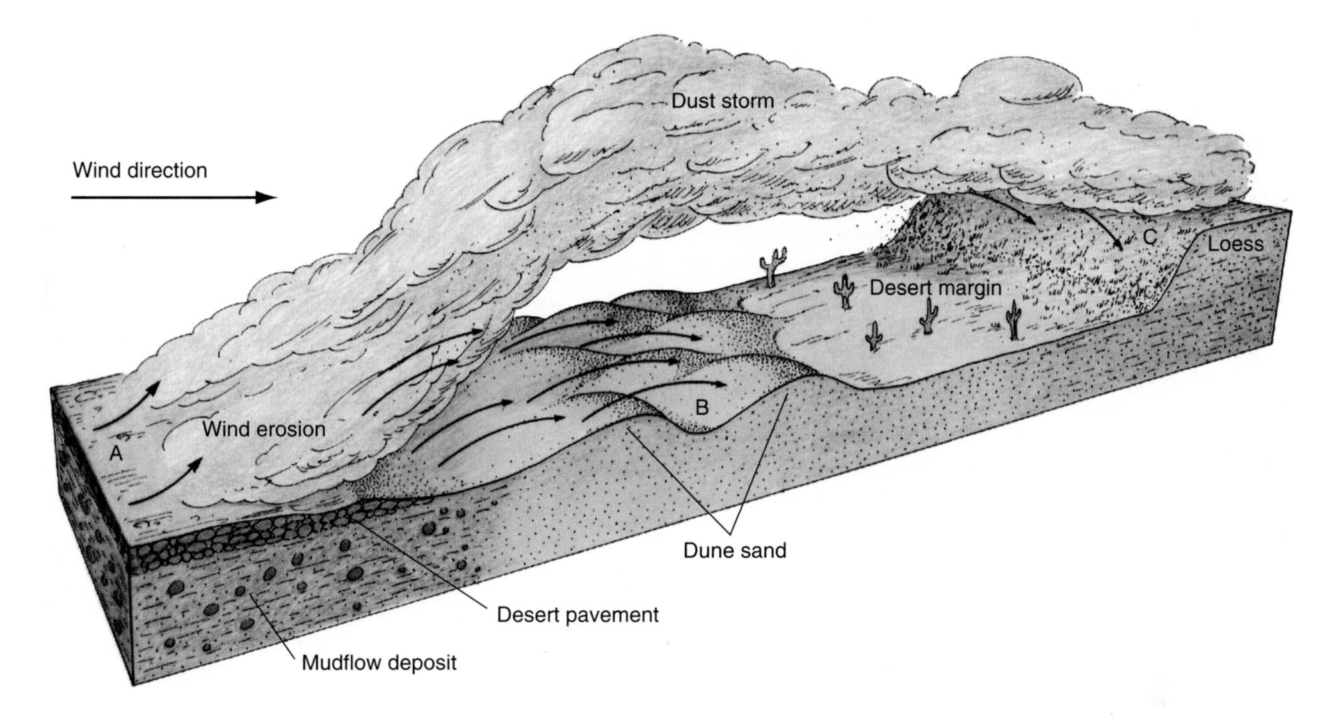


**FIGURE 3–23**
Deflation of a poorly sorted desert sediment deposit (at A) results in formation of three separate sediment accumulations: desert pavement, dune sand, and loess. Each has a different average size but each is better sorted than the original deposit. A, coarse rock material is concentrated as desert pavement as wind erodes the finer sand and silt from the mudflow. B, sand particles accumulate downwind to from dunes. C, the finest particles are suspended in the wind as dust and deposited at or beyond the desert margins. (Reprinted with permission of Macmillan College Publishing Company from *Physical Geology* by Nicholas K. Coch and Allan Ludman. Copyright © 1991 by Macmillan College Publishing Company.)

es reviewed in this chapter provide the background for understanding the hazards and problems presented in the rest of this book.

## LOOKING AHEAD

This completes our look at Earth's surface processes, and our two-chapter overview of physical geology. You now have the general background needed to appreciate the rest of this book. In the next chapter, we'll look at one of Earth's most interesting geohazards—volcanic eruptions. The chapter explains why and where volcanoes occur and discusses lava types, nuées ardentes (glowing clouds), lahars, the Ring of Fire, Hawaii's volcano observatory, the science of predicting eruptions, hazard control, and some famous volcanic disasters.

## SUMMARY

The Sun's nuclear fusion radiates energy that heats Earth's atmosphere, hydrosphere, and lithosphere, driving the complex, interrelated surface processes of air and water circulation, erosion, transportation of pieces and particles, and deposition.

Heat evaporates surface water into the atmosphere, where cooler temperatures condense the water vapor to form rain or snow. Rain falling on land either evaporates, runs across the surface as sheetflow and into stream channels back to the ocean, or infiltrates the surface to form groundwater. Snow falling on land melts and follows a similar course, unless it falls in constantly frigid areas where it accumulates to form glaciers. As glaciers melt, the meltwater follows the same pattern as rainfall. The hydrologic cycle describes water movement in its three states (solid, liquid, gas) among the atmosphere, hydrosphere, and lithosphere.

Conditions at the surface are very different from those in Earth's interior, where most rocks form. Rocks exposed on the surface become physically and chemically weathered. Physical weathering breaks a rock into smaller particles but does not change its chemical composition.

Chemical weathering changes a rock's composition, color, and other characteristics. Major chemical weathering processes include oxidation, hydrolysis, and solution. Chemical weathering forms different minerals that are more stable under surface conditions. It also dissolves elements that travel from the weathering area as dissolved ions in surface and underground waters.

Regolith is the loose, weathered rock material on the surface. If weathered sufficiently to support plants, regolith is called soil.

Sediment characteristics include size (clay, silt, sand, pebbles), sorting, roundness, and color. These attributes reveal the transporting agent that deposited the sediment particles.

Generally, the smallest, best-sorted, most-rounded particles are deposited by wind. The largest, most poorly sorted and least-rounded are carried by glaciers and mudflows. Stream deposits have characteristics between those of glaciers and wind.

Erosion is the removal of regolith by streams, glaciers, waves, or wind. Deposition is the dropping out of sediment from its transporting medium—water, air, or ice.

Streams continually erode their channels as they flow toward their base level, which can be permanent (the ocean and landlocked continental basins) or temporary (lakes and reservoirs). As streams approach base level, they lose velocity and deposit their bedload and saltation load to form a delta. Their dissolved load remains in solution until the water chemistry changes or evaporation occurs.

Many streams are flanked by natural levees, areas where the flooding stream deposits coarser sediment. Finer sediments spread over floodplains that extend away from the channels.

In frigid areas, accumulated snow becomes compressed into glacial ice, which flows to lower elevations where it melts. As glaciers flow slowly, they erode unsorted particles and pieces, from clay-size to boulder-size. Moving glaciers deposit till (in place) and meltwater deposits (outwash) over an area in front of the ice. Till forms ground moraines (under the ice), lateral moraines (along the flanks), and terminal moraines (at the end).

In the most recent glaciation, continental glaciers spread toward the equator from the polar regions. They reached into temperate areas of the United States, Europe, and Asia, modifying the topography and depositing till and outwash to create much of the surface we see today.

As solar energy heats air, making it expand and rise, cooler air moves in to replace it, creating wind. Wind erodes sediment by deflation. Sand-sized grains saltate and are deposited as wind velocity drops or when the grains hit an obstruction, sometimes accumulating in streamlined dunes. Finer silt and clay travel in suspension, sometimes to be deposited far away as loess.

## KEY TERMS

**Streams**

base level
base flow
bed load
condensation
cross sectional area
delta
depth
discharge
dissolved load
equilibrium
erosion
evaporation
floodplain
floods
gradient
groundwater
hydrologic cycle
hydrosphere
infiltration
levees
precipitation
runoff
saltation load
sheetflow
stream
surface water
suspended load
transpiration
width

**Glaciers**

brittle zone
cirques
continental ice sheets
crevasses
glacier
ice caps
moraine
outwash
outwash plain
piedmont glaciers
plastic zone
snowline
till
trunk glaciers
valley glaciers

**Weathering**

chemical weathering
dissolved ions
frost wedging
gravity
hydrolysis
lithosphere
nuclear fusion
oxidation
physical weathering
regolith
residual regolith
soil
solution
transported regolith
weathering

**Wind**

atmosphere
deflation
dunes
loess
tides

**Sediment**

clay
roundness
sand
silt
sorting

## REVIEW QUESTIONS

1. a. Describe the ultimate energy source that drives the surface processes of Earth.
   b. How does it differ from the energy source that drives Earth's internal processes (Chapter 2)?
2. Describe the formation of each of the following as a result of radiant energy falling on the land and ocean surface of Earth: (a) winds, (b) glaciers, and (c) streams.
3. How do sediment particles differ when deposited by glaciers, streams, and the wind?
4. What conditions favor infiltration of rainwater instead of runoff?
5. Distinguish between temporary and permanent stream base levels and give an example of each.
6. How do the products of chemical and physical weathering differ?
7. What effect does geographic latitude have on the relative efficiency of physical versus chemical weathering?
8. Distinguish between the movement of sediment particles as part of a stream's bed load, saltation load, and suspended load.
9. Contrast the characteristics of glacial sediments deposited directly by the ice (till) and by glacial meltwater (outwash).

## FURTHER READINGS

Coch, N. K., and Ludman A., 1991, Physical geology: New York, Macmillan, 678 p. Chapters 6, 9, 11, 13, and 14.

Costa, J. E., and Baker V. R., 1981, Surficial geology—Building with Earth: New York, Wiley, 498 p.

Ritter, D. F., 1986, Process geomorphology: Dubuque, W. C. Brown, 579 p.

# 4

# Volcanic Hazards

Volcanoes prove the existence and power of Earth's great internal heat. This heat melts rocks at depth (Chapter 2), and the resulting magma rises slowly toward the surface. Magma that does not reach the surface gradually cools and solidifies underground as plutonic (intrusive) igneous rock (Chapter 2). Magma that does reach the surface erupts as lava or ash from volcanic vents, creating new landforms—and geohazards. Despite the risk, however, people long have lived on or next to volcanoes because soils developed on volcanic deposits are among the most fertile. In addition, many volcanic areas are near the sea coast, where numerous people live.

## VOLCANIC ERUPTIONS: SOME EXPLOSIVE, SOME GENTLE

A **volcano** is a mountain formed by the accumulation of erupted lava and/or volcanic ash. The name comes from *Vulcan,* Roman god of the forge. We usually think of volcanoes as tall, conical mountain peaks, but that is only volcanism's most familiar style. Volcanism occurs in three basic forms: (1) relatively quiet lava flows that create broad plains instead of mountains; (2) lava flows that build massive, mound-like mountains; and (3) explosive eruptions from mountain peaks that produce ash, lava, and headlines.

The quiet flows have formed the Snake River Plain in Idaho, the Deccan Plateau in India, and Earth's largest landform: the entire seafloor, which is built of lava extruded from submarine mid-ocean ridges (Chapter 2). The massive mountains built of lava include those of Hawaii and Iceland. Their bountiful lava flows are impressive but usually are not explosive. The explosive category includes Mount Saint Helens (Washington State), Mount Pinatubo (Philippines), Mount Unzen (Japan), and many others that you hear about in the news. This chapter explains why these different types exist, and the geohazards surrounding each.

**Ash eruption from Mount Saint Helens. (Photo from U.S. Geological Survey.)**

Volcanoes occur on all continents, even Antarctica. However, they are not equally distributed over Earth. About 80% of active above-sea volcanoes occur within the Pacific Basin and near the edges of the continents that border it (Figure 4–1). With few exceptions, volcanoes are concentrated at tectonic plate boundaries (see Figure 2–23). About 80% occur at convergent boundaries where subduction is in progress, and about 15% occur where plates are diverging (spreading apart). The remainder are caused by heat sources *within* plates, such as the volcanic Hawaiian Islands (in the middle of the Pacific Plate) and the Yellowstone National Park area (in the North American Plate).

Some volcanoes are far more explosive than others. The energy released by different volcanoes can be compared by using the *Volcanic Explosivity Index (VEI)* rating, developed in 1982 by the U.S. Geological Survey and the University of Hawaii. The VEI is based on the volume of material erupted,

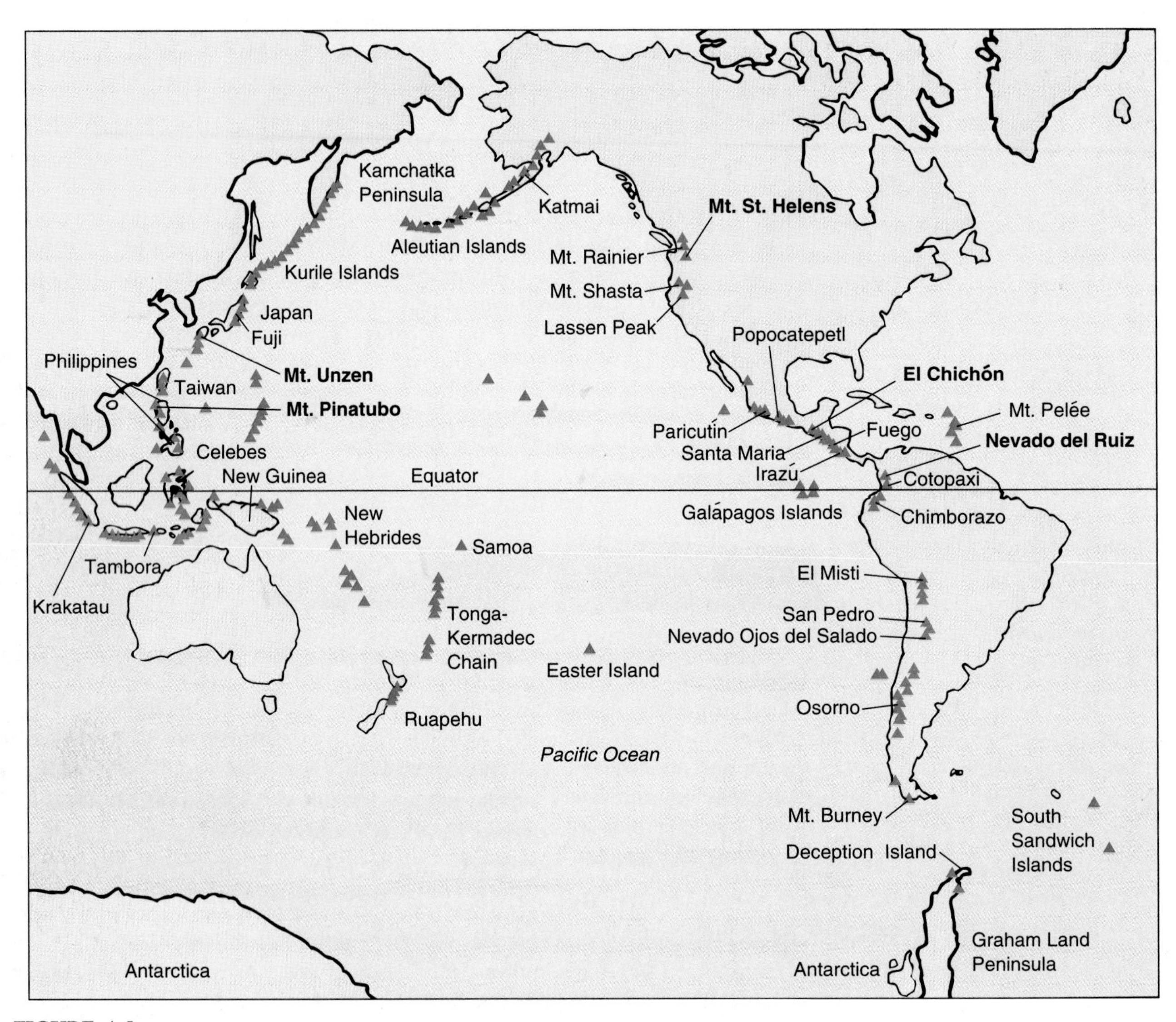


**FIGURE 4–1**
Location of active volcanoes in the Pacific Ring of Fire. The major eruptions occurring between 1980 and 1993 are highlighted. (Reprinted with permission of Macmillan College Publishing Company from *Physical Geology* by Nicholas K. Coch and Allan Ludman. Copyright © 1991 by Macmillan College Publishing Company. Courtesy of Allan Ludman.)

height of the ash plume, and other observations. It has a lower limit of 0 (non-explosive), but no upper limit. A VEI of 5 is considered very large (for example, the May 1980 eruption of Mount Saint Helens). The 1815 eruption of Tambora in Indonesia—the largest in recorded history, ejecting about 150 kilometers$^3$ (36 miles$^3$) of material—only rates a VEI of 7. Some enormous prehistoric eruptions would have rated a VEI of 8 or higher (for example, the eruption at Yellowstone in Wyoming, about 2 million years ago).

## ERUPTING MAGMA INTO LAVA, TEPHRA, AND GASES

Melting of solid rock occurs inside Earth where heat energy is sufficient to break the bonds that hold minerals together. Pure substances have distinct melting points, but rocks melt over a range of temperatures, depending on their mineralogic composition and the amount of water present. Melting produces *magma,* a hot mixture of molten rock, dissolved gases, and mineral crystals (Chapter 2). The temperature of magma is around 650–1,350°C (1,170–2,430°F).

Magma rises and breaks through either oceanic or continental crust at a point called a **vent.** Abruptly, the magma is exposed to the dramatically different environments of ocean or atmosphere. It undergoes changes to produce various solid, liquid, and gaseous products. The liquid product—molten lava—does not remain liquid for long, but quickly cools and solidifies. Erupted lava, either in fluid or solid form, can do great damage.

How the magma emerges—as relatively quiet flows, or explosively—depends on its composition. We will look first at "gentler" eruptions of lava.

### Lava

Magma that reaches the surface non-explosively spreads out as **lava flows.** There is wide variation in lava, and its characteristics determine how it flows and solidifies. In general, the less silica ($SiO_2$) in the lava and the hotter the melt, the lower its **viscosity** (stiffness) will be. This is similar to the difference between cold syrup (very viscous) and hot syrup (very fluid). Two general types of lava, pahoehoe and aa, are distinguished by their manner of flow and appearance once solidified.

**Pahoehoe** (pa-hoy-hoy) is a Hawaiian term for low-viscosity, "runny" lava that solidifies into a distinctive "ropy" texture (Figure 4–2). The surface of a pahoehoe flow cools rapidly to form a glassy surface while the lava below is still cooling and fluid enough to flow. Pahoehoe generally forms in hot, low-silica basaltic lavas (Table 4–1).

**Aa** (AH-ah) is a Hawaiian term for more viscous and slow-moving lava flows that have a jagged, blocky surface (Figure 4–3). The cooled, rubbly surface is dragged along and pulled under as the flow advances, in a manner similar to the caterpillar treads on a moving bulldozer. Like pahoehoe, aa also forms in basaltic lavas, but in mixtures that are less gaseous and therefore more viscous.

### Tephra

Explosive volcanoes eject *pyroclastic material* (Greek for "fire-broken"). This includes both airborne material that can travel some distance (ash, cinders) and material that travels near the ground surface as a mixture of hot gas and fragments.

Airborne lava fragments are called **tephra.** Some tephra forms in masses or drops of magma that quickly cool and solidify in the air before they can return to the ground. In other cases, pyroclastic fragments include pieces of older rock that were

**FIGURE 4–2**
Pahoehoe lava flowing through a structure on the island of Hawaii. (Photo by J. D. Griggs, U.S. Geological Survey.)

TABLE 4–1
Materials deposited from lavas of different compositions.

| Property | Basaltic composition (Hawaii, Iceland, seafloor, mid-ocean ridges) | Andesitic composition (Mount St Helens, Mount Pinatubo) | Rhyolitic composition (Long Valley Caldera, Ca.—prehistoric) |
|---|---|---|---|
| Temperature | hottest | intermediate | cooler |
| Silica content ($SiO_2$) | least | intermediate | greatest |
| Water content | least | intermediate | greatest |
| Viscosity (stiffness) | lowest (very fluid) | intermediate | highest (stiffest) |
| Products | mostly flows | flows and tephra | mostly tephra |
| Volcano type | shield volcano | stratovolcano | lava dome |
| Representative hazards | lava flows | lahars, tephra, nuées ardentes | violent eruption of gas and ash |

FIGURE 4–3
Flow of aa lava advancing across a field in Hawaii, Kona District. (Photo from U.S. Geological Survey, Menlo Park.)

broken from the volcano interior by the eruptive force. Tephra can include glassy particles.

Tephra varies from sand-sized particles **(volcanic ash)** up to boulder-sized pieces of rapidly cooled and solidified lava **(volcanic bombs)** (Figure 4–4). Larger pieces of tephra fall near the eruptive vent, whereas the finest particles (volcanic ash) may be dispersed hundreds of kilometers by wind. The consequences of this dispersal are considered later in this chapter.

## Lava Flows or Tephra?

Some volcanoes produce large lava flows but little tephra (Hawaii, Iceland). Others are the opposite, producing more ash than lava (volcanoes along island arcs like Japan, the Aleutians in Alaska, the West Indies, and ranges bordering ocean basins like the Cascades and the Andes). What accounts for this difference? The proportion of lava flow to tephra depends on the type of lava. Lava type in

**FIGURE 4–4**
Tephra eruption from Pu'u O'o volcano, Kilauea East Rift zone, Hawaii. (Photo by J. D. Griggs, U.S. Geological Survey.)

turn depends on silica content, water content, and magma temperature (Table 4–1).

The relative violence of volcanic eruptions also depends on the lava type. Some Hawaiian and Icelandic eruptions may appear explosive, with spectacular fire fountains exceeding a hundred meters (330 feet) in height. But these are relatively gentle compared to eruptions along island arcs, which can blast tephra thousands of meters into the atmosphere. Later we will study how different lavas produce different volcanoes.

### Volcanic Gases

Expanding gas drives eruptions, so gas is emitted during all eruptions, whether gentle or explosive. Most of the gas released is water vapor (steam), mixed with several other gases. The steam is created as the hot magma vaporizes surface water or groundwater in fractures and pores of near-surface rocks. Other components of volcanic vapors include carbon dioxide ($CO_2$), sulfur dioxide ($SO_2$), sulfur trioxide ($SO_3$), carbon monoxide (CO), and hydrogen sulfide ($H_2S$). The distinctive "rotten egg" odor of hydrogen sulfide is prominent during some eruptions. Sulfur dioxide ($SO_2$) and trioxide ($SO_3$) combine with water vapor to form acids (sulfurous acid and sulfuric acid, respectively). Hydrochloric acid (HCL) and hydrofluoric acid (HF) also may form.

Carbon dioxide is *heavier than air* and thus creates a special hazard if it is emitted in great concentrations during an eruption. This heavy gas hugs the ground, moves downslope, and tends to fill low areas (valleys and depressions). It may asphyxiate animal life by displacing normal air, thus depriving the animals of oxygen.

## VOLCANO FORMS

The mountainlike landforms we call volcanoes are built of lava, or ash, or both, and within that variation lies the key to their different forms. Major volcano forms are *shield volcanoes, stratovolcanoes, cinder cones,* and *lava domes.* They differ in size, slope steepness, the materials of which they are composed, explosivity, and hazardousness. These differences are related to the properties listed in Table 4–1. We will now examine each type.

### Shield Volcanoes

Highly fluid basaltic lava (Table 4–1) builds broad, massive, gently sloping volcanoes. These **shield volcanoes** are named for their resemblance to the curved shield of a warrior (Figure 4–5). Their gentle slope results from the greater distances that the more fluid basaltic lava can flow downslope before it solidifies.

Shield volcanoes are built of numerous overlapping lava flows extruded from a **central vent.** Smaller vents may develop on the volcano's flanks as **satellite volcanoes.** Tensional (spreading) forces on the solidified rocks form linear fissures (called *rift zones*) on the volcano's flank, and **fissure eruptions** may occur along these (Figure 4–6).

**FIGURE 4–5**
Mauna Loa, background, a shield volcano on the island of Hawaii. (Photo by Swanson, U.S. Geological Survey.)

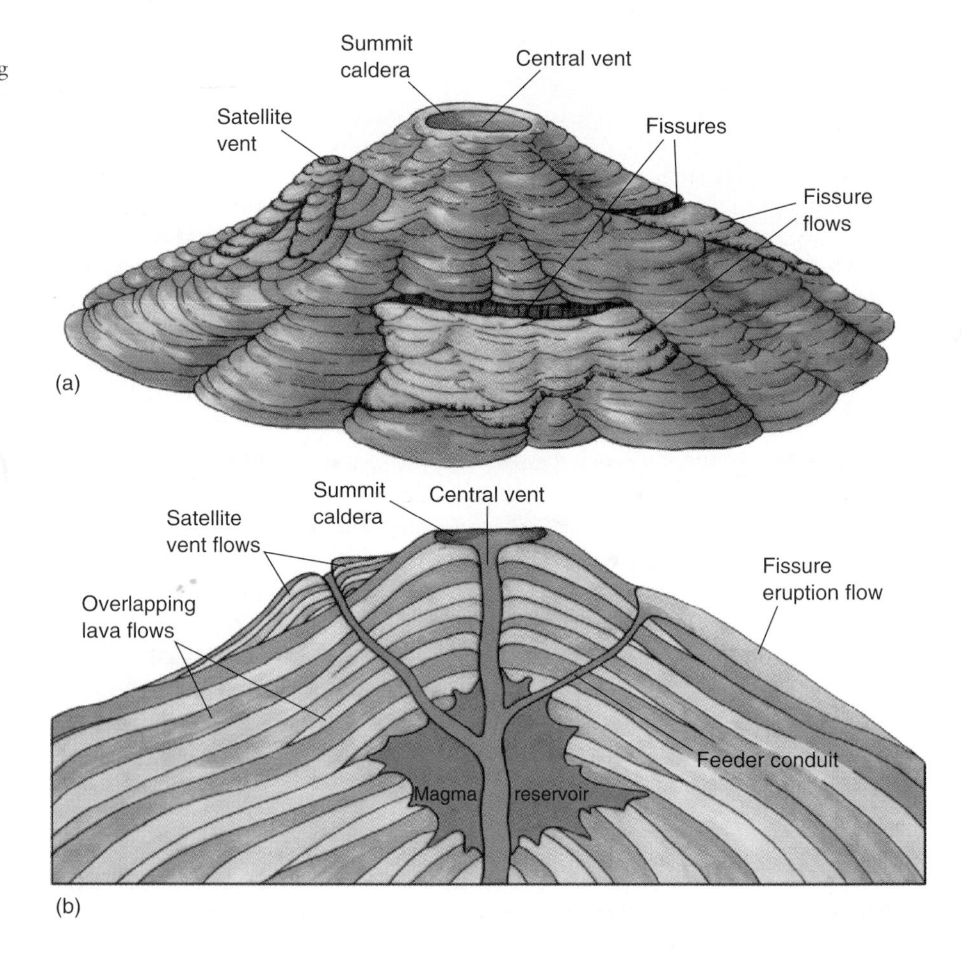


**FIGURE 4–6**
Diagram of a shield volcano showing variety of eruptive sites. (a) Volcano surface showing eruptive sites and flow sequence. (b) Generalized geologic section through (a), showing internal relations of flow sequences. Vertical scale is greatly exaggerated.

In surface area and lava volume, shield volcanoes are by far the largest volcanic landforms on Earth. The volcanoes in the Hawaiian Islands and many others built up from the seafloor are of this type. Interestingly, the planet Mars has shield volcanoes, too—and one of them (Olympus Mons) is larger than the entire Hawaiian Islands chain.

Shield volcanoes on Hawaii form when magma rises from a stationary heat source called a **hot spot** beneath the lithosphere (Figure 4–7). The magma migrates up through fractures in the overriding Pacific plate and extrudes onto the ocean floor, gradually building a mountain of successive lava flows. Over time, the mass accumulates until it emerges above sea level, thereby becoming an island volcano.

Mauna Loa volcano, on the island of Hawaii, illustrates the immense lava volume extruded to form a shield volcano (Figure 4–8). The portion of Mauna Loa that you can see rises over 4,000 meters (13,000 feet) above sea level. However, Mauna Loa extends another 6,000 meters (19,700 feet) from sea level down to the Pacific Ocean floor. This makes Mauna Loa about 350 meters (1,150 ft) higher than Mount Everest when measured from its seafloor base to its summit. Mount Everest is Earth's highest point, but Mauna Loa is Earth's tallest mountain!

The Hawaiian Islands are arranged in a chain, and the key to understanding why lies in the relation between the Hawaiian hot spot and the movement of the Pacific plate. The hot spot is a stationary heat source over which the Pacific plate slowly moves toward the northwest. Imagine a conveyor with hamburger patties slowly moving over a stationary gas flame, cooking the burgers one after another (Figure 4–7).

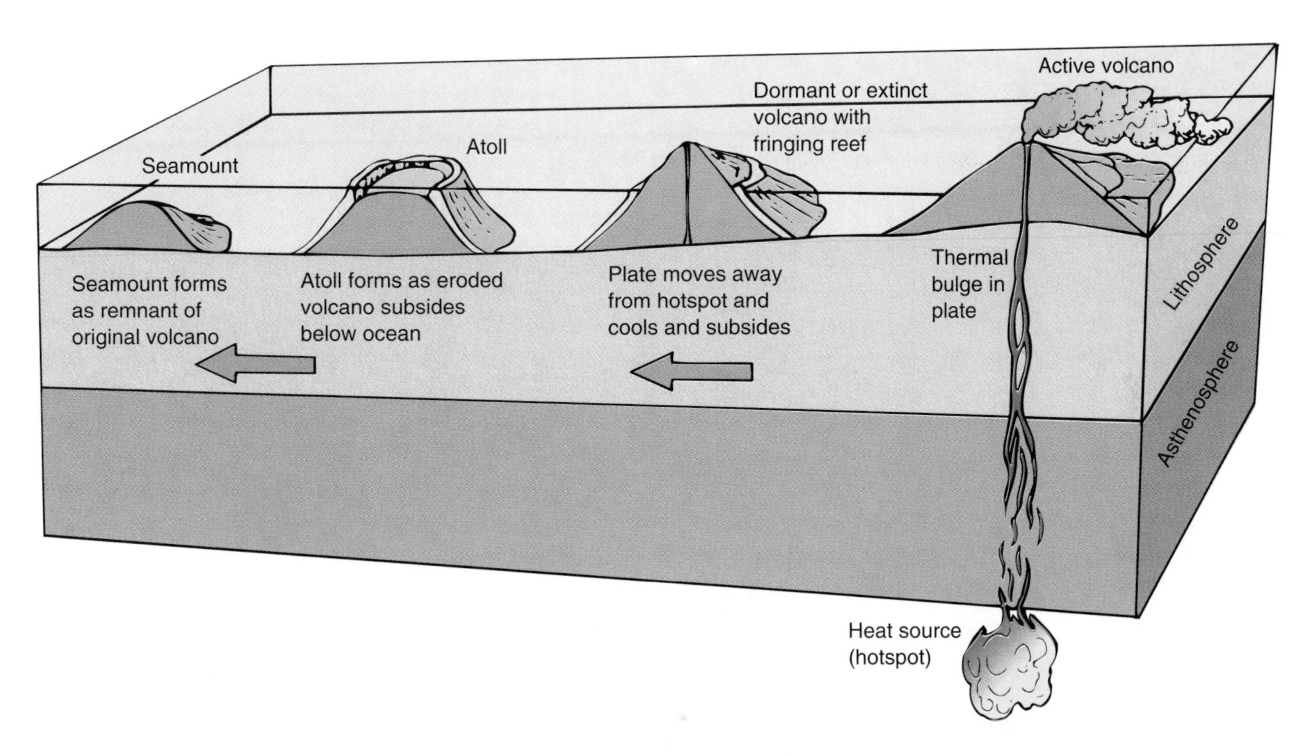


**FIGURE 4–7**
Origin of Hawaiian-type oceanic islands. Volcanic islands form over a hot spot and are moved away from it by continual plate movement. As the islands move away from the hot spot, they cool, and the rock density increases, resulting in subsidence of the island. This subsidence plus surficial erosion reduces the islands in height more and more with time. Fringing reefs grow upward around the subsiding islands to form walls. When the islands eventually sink below the surface, they evolve into seamounts. (Reprinted with permission of Macmillan College Publishing Company from *Physical Geology* by Nicholas K. Coch and Allan Ludman. Copyright © 1991 by Macmillan College Publishing Company. Courtesy of Allan Ludman.)

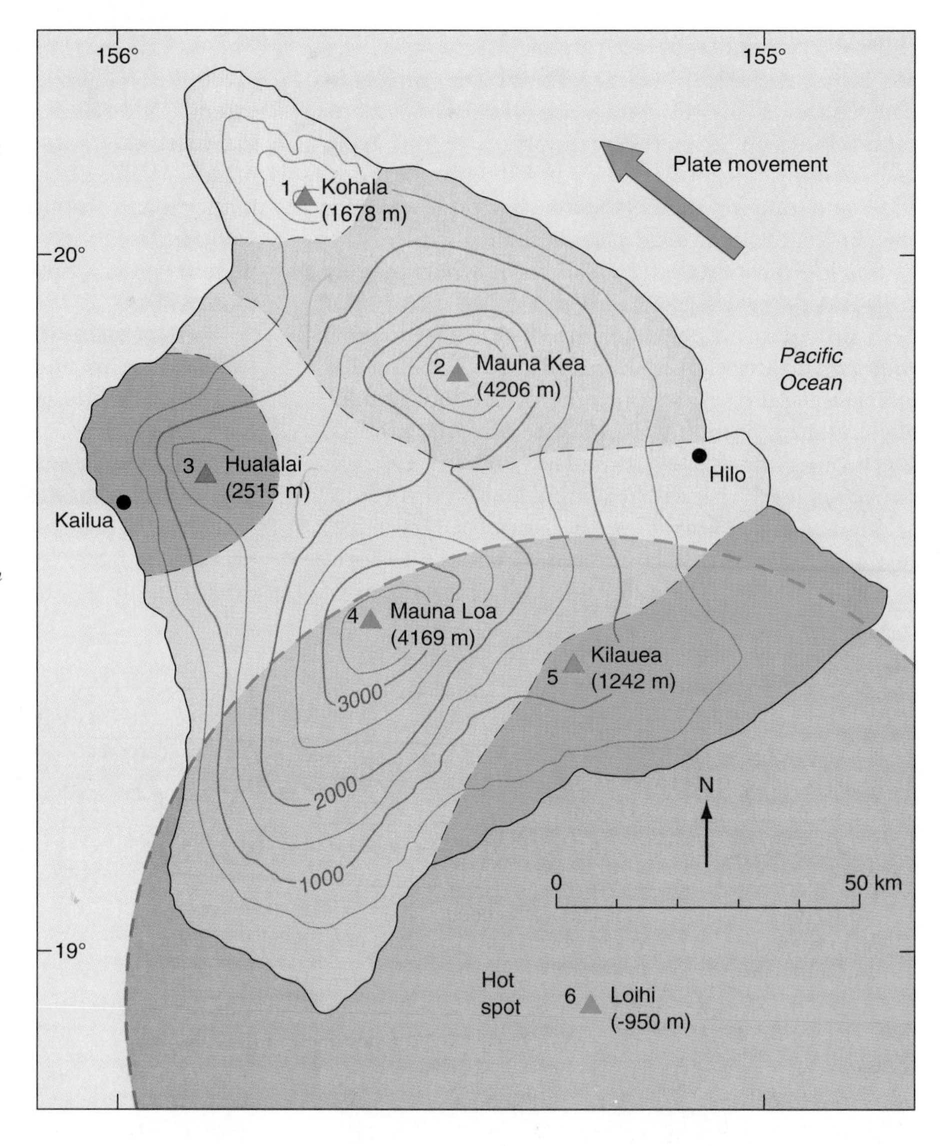


**FIGURE 4–8**
Volcanic centers on the island of Hawaii show the island's evolution. The Pacific plate is moving northwestward over a stationary hot spot. The hot spot currently is beneath the southern part of the island, as shown. Numbers indicate the relative ages of volcanic centers as the Pacific floor moved over the hot spot (1 = oldest, 6 = youngest). The currently active volcanoes are Mauna Loa, Kilauea, and Loihi. At present, Loihi is a submerged seamount—note its elevation, –950 meters (–3,117 feet) below the ocean surface. Elevations are in meters. (After Malahoff, "Geology of the Summit of Loihi Submarine Volcano," Chapter 6 in Decker, Wright, and Stauffer, eds., *Volcanism in Hawaii,* U.S. Geological Survey Professional Paper 1350, 1987.)

An island volcano forms over the hot spot until the gradual migration of the Pacific plate positions another part of the lithosphere above it. The volcano formed previously slowly migrates away from the magma source and becomes dormant. Weathering and erosion then slowly reduce the island that lava built. Coral reefs develop on the flanks of the subsiding volcano, forming a circular reef called an **atoll.** Eventually, the extinct volcano is worn down to below sea level, and is called a **seamount.**

You might think that steady movement of the plate over the hot spot would produce a continuous volcanic ridge, rather than a series of islands. It appears that Pacific plate movement may be intermittent rather than continuous, and magma generation from a hot spot also is believed to be intermittent as well. If the time intervals of plate movement and hot spot activity are long, their joint effect would produce the individual volcanic islands we see today. If the time intervals are short, however, a new **eruptive center** may grow near the former one. In fact, the island of Hawaii includes several volcanic centers, overlapping one another (Figure 4–8).

Most volcanologists consider a volcano to be **active** if it has erupted at least once in historical time, and a volcano to be **extinct** if it is not expected to erupt again. An active volcano may remain **dormant** before erupting again, for periods ranging from a few years to several centuries. In the case of Hawaii (Figure 4–8), Mauna Loa and Kilauea are considered active, while Hualalai and Mauna Kea

FIGURE 4–9
Mount Shasta, California, a stratovolcano. (Photo by Freeburg, U.S. Geological Survey.)

are currently dormant but potentially active; Kohala is regarded as extinct.

Where will the next Hawaiian island develop? Scientists have observed a submarine vent at the Loihi Seamount on the seafloor, about 30 kilometers (19 miles) off the southern coast of Hawaii (Figure 4–8).

## Stratovolcanoes

Most people visualize a volcano as a steep-sided conical mountain like Mount Fuji in Japan or Mount Shasta in California. These are called **stratovolcanoes** because they are composed of strata (layers) of pyroclastic debris and lava flows (Figure 4–9). Their lava is much more viscous (stiff) and gassy than that of shield volcanoes, and this combination of viscous lava and gas pressure typically makes stratovolcanoes more explosive than shield volcanoes.

Stratovolcanoes are most commonly composed of *andesite,* named for its common occurrence in the volcanically active Andes Mountains of South America. Andesite forms where oceanic lithosphere, with its overlying silica-rich oceanic sediments, is subducted under other oceanic lithosphere (Figure 2–23) to form volcanic island arcs (like Japan, Indonesia, and the Aleutians). This subduction into the mantle results in a magma that rises to the surface to erupt as andesite.

Andesite and stratovolcanoes also form where oceanic lithosphere is subducted under continental lithosphere to form volcanic mountain ranges—for example, the Andes Mountains and the Cascades of the northwestern United States (Figure 2–23).

One of the most devastating eruptions in recorded history involved a stratovolcano named Krakatau on the Indonesian island arc. After being dormant for two centuries, the volcano exploded violently in August 1883 in a series of eruptions that were about four times greater than the Mount Saint Helens eruption of 1980.

The culminating eruption was heard thousands of kilometers away in Japan and Australia and created an atmospheric pressure wave that was recorded on barometers around the world. The 800 meter-high (2,600 feet) volcano was largely blown away. Part of it collapsed into the ocean, creating seismic sea waves tens of meters high (called *tsunami,* Chapter 5). These waves inundated nearby villages, drowning about 33,000 people. Volcanic ash was ejected over 25 kilometers (15 miles) into the atmosphere and 3,000 people were killed by ash and gas flows on the ground. The ash thrown into the atmosphere circled the globe and caused slight global cooling for a year and colorful sunsets for several years afterward.

## Cinder Cones

Some volcanoes called **cinder cones** consist entirely of tephra. No lava flows are involved. They have a well-developed central vent and symmetrical flanks.

**FIGURE 4–10**
A cinder cone on Mount Lassen, California. (Photo from U.S. Geological Survey.)

The tephra erupted at the vent falls nearby and builds up the flanks to form a slope between the size of a shield volcano and a stratovolcano (Figure 4–10).

Cinder cones are the smallest volcanoes, but their eruptions can be quite explosive and they can grow rapidly. For example, the Paricutín (puh-ree-koo-TEEN) cinder cone in central Mexico erupted from a corn field in February 1943, forming a cone 300 meters (1,000 feet) high *within a month.* However, most cinder cones are smaller features.

## Lava Domes

Some volcanoes are built up of viscous lava that does not flow far from the vent before it cools and solidifies. This builds a landform called a **lava dome.** The Mono Craters of California are an excellent example (Figure 4–11). Lava domes are particularly dangerous because their viscous magma forms a "plug" in the vent. Gas pressure then can build tremendously behind this plug until it is released explosively in a violent eruption.

## Calderas

Rare but extremely violent eruptions can produce immense craters up to tens of kilometers across called **calderas** (Spanish for "cauldron"). Most calderas are produced when a particularly violent, large-volume eruption literally blows off the top of a volcano, leaving only the lower part. Calderas form after an eruption, as the volcano collapses into the emptied magma chamber below it. For example, the eruption of Krakatau in 1883 formed a huge caldera that extended 300 meters (1000 feet) below sea level.

Violent eruptions are typical of viscous, silica-rich magmas such as andesite or rhyolite that form "plugs" in the volcano. (Rhyolite is the extrusive equivalent of granite—see Table 4–1) The gases are confined in the magma below the plug. When the eruption starts, pressure is released suddenly, and the gases separate quickly and violently.

**FIGURE 4–11**
Mono Craters, Nevada, an example of lava domes. The snow-capped peaks in the distance are the Sierra Nevada Mountains. (Photo by C. Dan Miller, U.S. Geological Survey.)

***Crater Lake, Oregon.*** Mount Mazama was an ancient, huge stratovolcano that existed at present-day Crater Lake in southwestern Oregon. About 6,800 years ago, a cataclysmic explosion of viscous andesitic lava blew off the top of the mountain, which collapsed to leave a large caldera. Tephra from the caldera-forming eruption spread across the countryside, accumulating a thickness of 15 centimeters (6 inches) some 900 kilometers (560 miles) downwind to the east. The crater rim collapsed into the void created by the eruption. Later, a cinder cone grew and the caldera filled with rainwater. The cinder cone today is Wizard Island in Crater Lake (Figure 4–12). Mount Mazama's caldera is substantial—about 52 kilometers$^2$ (20 miles$^2$).

***Mammoth Lakes, California.*** This particularly worrisome caldera is an elliptical depression, formed about 730,000 years ago from a violent rhyolitic eruption (Figure 4–13). Volcanism within the old caldera has built a lava dome. Around 1980, the caldera floor slowly began to deform upward. By 1982, the uplift increased and was accompanied by medium-sized earthquakes (magnitudes 5 and 6). The points of origin of these earthquakes moved closer to the surface between 1980 and 1982, and changes occurred in the activity of natural hot springs. All of these signs of volcanic unrest indicate that a magma body exists about 5 to 8 kilometers (3 to 5 miles) beneath the caldera's center. A future eruption is a distinct possibility. Authorities are concerned because the popular resort town of Mammoth Lakes and the ski slopes of nearby Mammoth Mountain are located within the caldera.

In fact, a low-level volcano alert was issued by geologists of the United States Geological Survey, in consultation with local authorities, after the first indications of a possible eruption appeared. The economic impact on the area (loss of tourist revenue) was significant and the alert was lifted when the activity quieted down. Many local citizens believe the alert was unnecessary.

## VOLCANIC HAZARDS

Volcanoes can pose serious risks to people and property. The relative severity depends on the volcano and lava type, climate, topography, and the population density. Humans and property may be affected by lava flows, tephra falls, blasts of gassy incandescent pyroclasts (nuées ardentes), volcanic mudflows (lahars), and poisonous gases.

### Lava Flows

Lava flows rarely are life-threatening because they move slowly enough for people to escape. However, they can cause great environmental and property damage because they are so difficult to stop. The

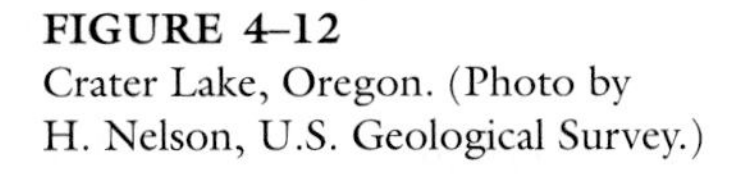

**FIGURE 4–12**
Crater Lake, Oregon. (Photo by H. Nelson, U.S. Geological Survey.)

**FIGURE 4–13**
The Long Valley Caldera and nearby features. The Long Valley Caldera measures 18 by 32 kilometers (11 by 20 miles). The eruption that formed it is estimated to have injected 584 kilometers$^3$ (140 miles$^3$) of ash into the atmosphere (greater than the volume of Lake Erie). The resulting deposit is today a distinctive buried ash layer in adjacent states called the Bishop Tuff. The resurgent dome is a recent feature that has built up within the caldera. (After S. L. Lamb, April 1991, Magma exploratory well, Long Valley Caldera: California Geology, Fig. 1, p. 85).

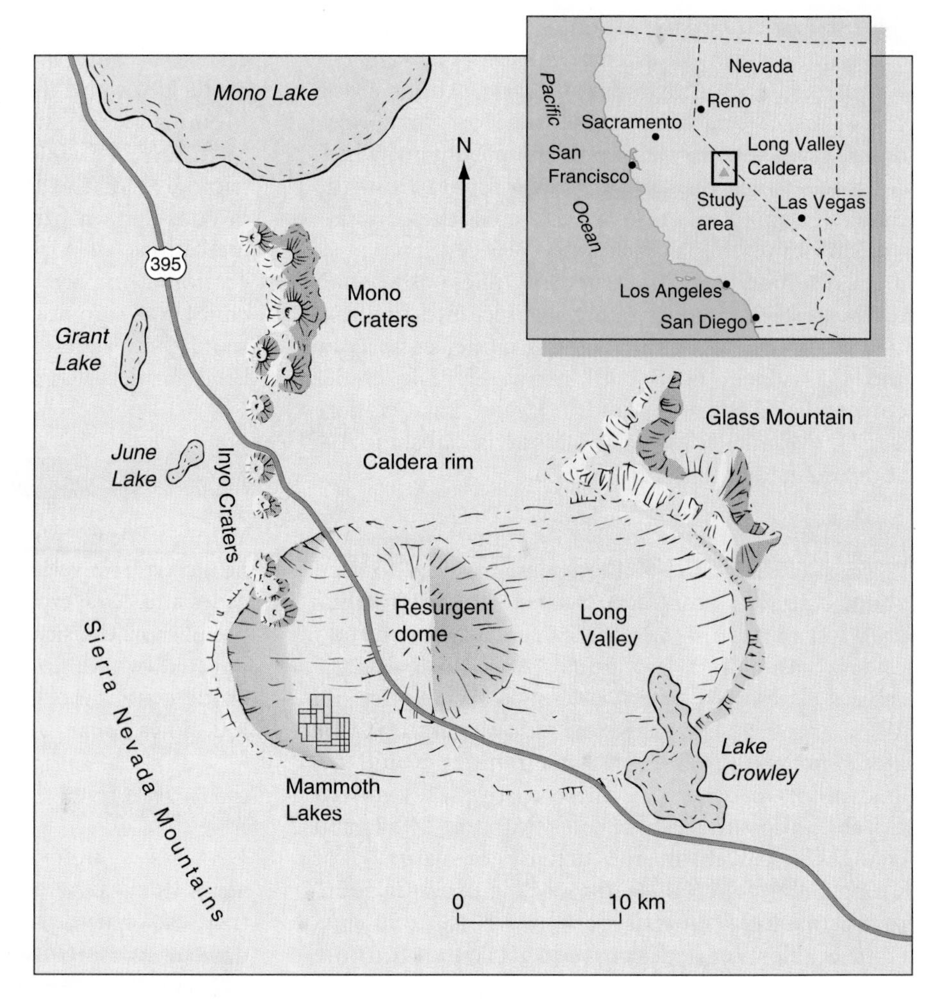

**FIGURE 4–14**
Lava flows over a coastal highway in Hawaii. (Photo by J. D. Griggs, U.S. Geological Survey.)

momentum of an advancing lava flow can demolish buildings and other structures. The lava can cover highways, and its heat can roast trees, houses, and crops.

Since 1983, lava flows have been particularly destructive in southern Hawaii (Figure 4–8). Lava erupting from satellite craters and along fissures has buried roads (Figure 4–14) and destroyed structures (Figure 4–15) in the Kalapana area on the southern coast of the island of Hawaii. When these flows reached the ocean, they formed clouds of acidic steam (Figure 4–16).

## Tephra

Most tephra is composed of fine-grained fragmented debris and ash. It contains bits of volcanic glass and is very abrasive. Explosive eruptions that produce great amounts of tephra are very hazardous.

**FIGURE 4–15**
Lava flows engulf structures in the Kalapana area, Hawaii. (Photo by J. D. Griggs, U.S. Geological Survey.)

**FIGURE 4–16**
Lava entering the sea in the Kalapana area. (Photo by J. D. Griggs, U.S. Geological Survey.)

They can create structural damage (roof collapse from the weight), mechanical failures of engines in vehicles and airplanes (clogging of air intakes and grit in mechanical parts), and breathing problems (see Box 4–1).

The damage caused by tephra is both immediate (burial under ash and collapsed structures) and long-lived. The extremely violent explosion in 1815 of the volcano Tambora, on the island of Sumbawa in Indonesia, provides an example. The eruption killed 12,000 people outright, but another 80,000 in the area died from starvation and disease as the year progressed. Crops were almost completely destroyed and livestock died because they ground down their teeth on grass coated with the glassy ash and starved.

During the May 1980 eruption of Mount Saint Helens, a largely andesitic volcano in the Cascade Range of the northwestern United States, volcanic ash formed thick deposits (Figure 4–17). Ash covered the surrounding area and the ash plume from the volcano was dispersed by high-altitude winds across the United States.

The Mount Saint Helens ashfall caused massive disruptions in the region and posed environmental problems in distant urban areas. Ash clouded the skies, reduced visibility, clogged automobile carburetors, fouled water systems, and made breathing difficult. A prior eruption of Mount Saint Helens almost 4,000 years ago generated three times the material (Figure 4–18), but fortunately the area was relatively unpopulated then.

Mount Saint Helens is one in a chain of largely andesitic volcanoes that make up the Cascade Range (Figure 4–19). These volcanoes all have erupted repeatedly through history and several are close to urban areas, posing a potential tephra hazard when they erupt. Mount Rainier, for example, is about 100 kilometers (60 miles) from Seattle, Washington.

The 1973 eruption on Heimaey Island off the coast of Iceland nearly buried the village of Vestmannaeyjar (Figure 4–20). Houses collapsed when the weight of the tephra and lava flows covered 3.2 $km^2$ (1.2 $miles^2$) of the surface and over 300 structures were destroyed.

## BOX 4–1 MAGMA TO MAGMA: TEPHRA HAZARD ALOFT!

One geohazard associated with tephra was unheard of as little as fifty years ago—dangers to planes flying through volcanic ash clouds. Although many volcanoes are in relatively unpopulated areas, some lie beneath commercial air routes. An eruption of a volcano along one of these routes can inject tephra into the air at altitudes used by airliners. This problem first gained wide attention when several aircraft suffered damage after flying through volcanic ash clouds from the 1980 eruptions of Mount Saint Helens (Figure 4–17).

In 1982, eruptions of Galunggung Volcano in Java, Indonesia created hazards to two Boeing 747 jets that flew through its ash clouds. In both cases, ash entered the jet engines and reduced power in all four of them. The pilots handled powerless descents of 7,600 meters (25,000 feet) and managed to restart the engines in the more oxygen-rich air at the lower altitude. Both planes landed safely at Jakarta airport in Indonesia. However, both planes experienced extensive damage to their engines and exterior surfaces.

The eruption of Redoubt Volcano in Alaska from December 1989 to April 1990 was modest compared to Mount Saint Helens and the Galunggung Volcano, but it caused a serious hazard to commercial and military jets flying the Trans-Alaska route to Asia. The tephra plume from Redoubt eruptions attained altitudes of 7 to 12 kilometers (4 to 7 miles) (Figure 1).

**FIGURE 1**
Ash plume from the eruption of Mount Redoubt in Alaska. (Photo from Alaska Volcano Observatory, U.S. Geological Survey.)

On December 15, 1989, a new Boeing 747 encountered an ash cloud at 7,600 meters (25,000 ft), some 240 kilometers (150 miles) northeast of Redoubt (Figure 1). Attempting to climb above the ash cloud, the pilot ascended about 900 meters (3,000 feet) when all four engines failed. The aircraft then descended 4,000 meters without power until the engines could be restarted. The plane landed safely, but the repair bill for its engines, aviation electronics, and exterior exceeded *$80 million!*

Tephra is an aircraft hazard in a number of ways. Volcanic ash particles are hard and angular, so they abrade compressor and turbine blades in the engines as well as windows, fuselage, and control surfaces. The extremely fine ash particles (measuring 1 to 1000 micrometers) penetrate even tiny openings in the fuselage and are deposited inside the plane (Figure 2).

The gas concentration associated with the tephra clouds may remain in the stratosphere for years after the tephra particles have settled. Sulfur dioxide in the clouds absorbs water vapor to form sulfuric acid. The acid droplets adhere to the exterior of the aircraft fuselage and penetrate it through microcracks. The acid corrodes plastics and rubber used in seals and makes small cracks in the acrylic windows on airliners, reducing visibility and requiring replacement after landing.

The most serious tephra problem results from *melting* of the ingested fine ash as it makes contact with the very hot combustor section of the engine (Figure 3). The ash is volcanic glass, so it remelts and flows onto the rear turbine blades. It builds up, blocking air flow and making the engine stall.

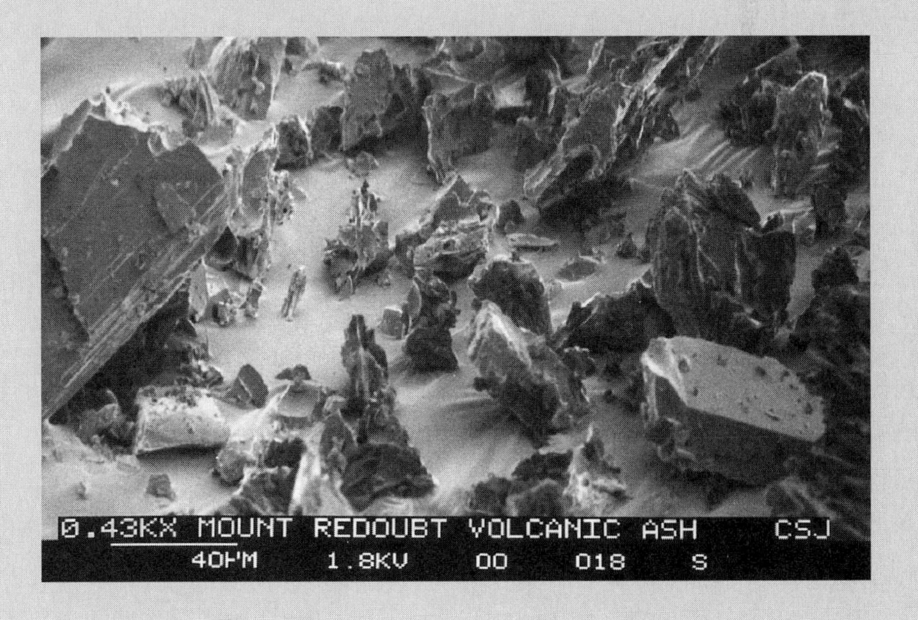


**FIGURE 2**
Photomicrograph of volcanic ash from Mount Redoubt, Alaska. Note the angularity of the particles. (Courtesy of C. St. John, Boeing)

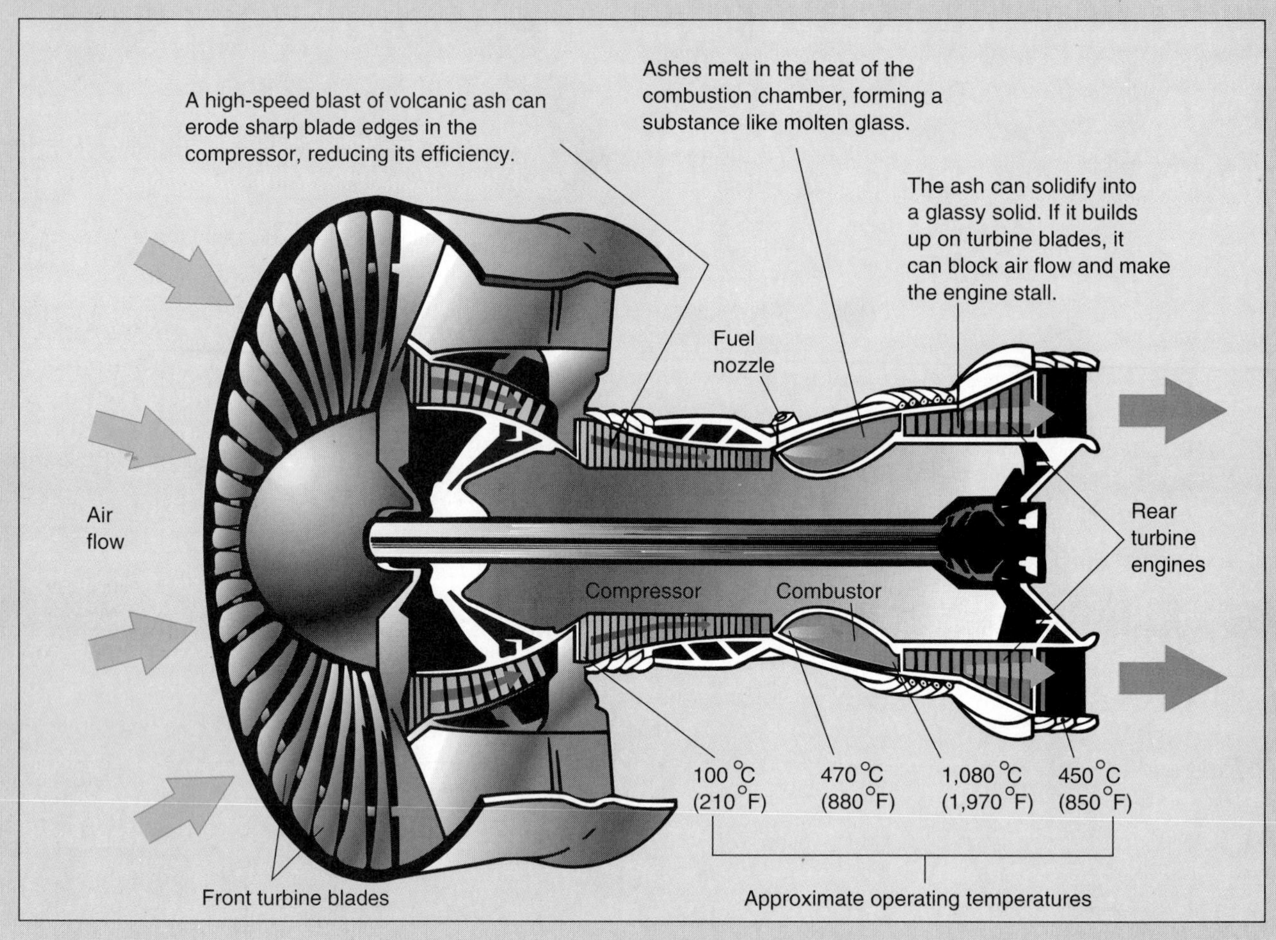


**FIGURE 3**
Cross section of jet engine showing mechanism by which ingested volcanic ash is melted, coating turbine blades and causing engines to lose power and stall. (After FAA. Detail from Henry Mesa.)

Thus, we have a remarkable situation: molten magma to molten magma. Magma from Earth's interior, part of which crystallizes into glass particles after extrusion through a volcano, remelts in the engines of planes passing above. The original magma forms a hazard when it erupts and a second hazard when it melts in aircraft engines!

This box is adapted from "Volcanic hazards and aviation safety: Lessons of the past decade," *FAA Aviation Safety Journal* 2:3. Reprints are available from Safety and Promotion Projects Division ASF-20, Federal Aviation Administration, 800 Independence Avenue SW, Washington, DC 20591.

**FIGURE 4–17**
Ash eruption from Mount Saint Helens. (Photo from U.S. Geological Survey, Menlo Park.)

## Pyroclastic Flows (Nuées Ardentes)

Near explosive volcanoes, a major danger is an incinerating mixture of gas and volcanic debris called a **pyroclastic flow.** Its temperature ranges between 700 and 1,000°C (1,300–1,800°F). Many geologists call it a **nuée ardente** (new-ay ar-DAWNT), French for "glowing cloud." This geohazard is most commonly associated with violent eruptions of andesitic volcanoes. Pyroclastic flows are ground-hugging but can move very fast; velocities to 150 kilometers/hour (90 miles/hour) have been recorded.

Pyroclastic eruptions may extend vertically from a volcano's summit, or they may be directed laterally from a weak area on the flank of the cone. A laterally directed pyroclastic flow is far more dangerous because the hot gas cloud is concentrated in one direction, engulfing and burning everything in its path as it roars down the volcano's slope. A lateral blast during the Mount Saint Helens eruption of 1980 leveled entire forests and arranged the trees in flow lines parallel to the movement of the blast (Figure 4–21).

Of the many pyroclastic flows in recorded history, the most famous occurred in A.D. 79 when Mount Vesuvius destroyed the city of Pompei, Italy.

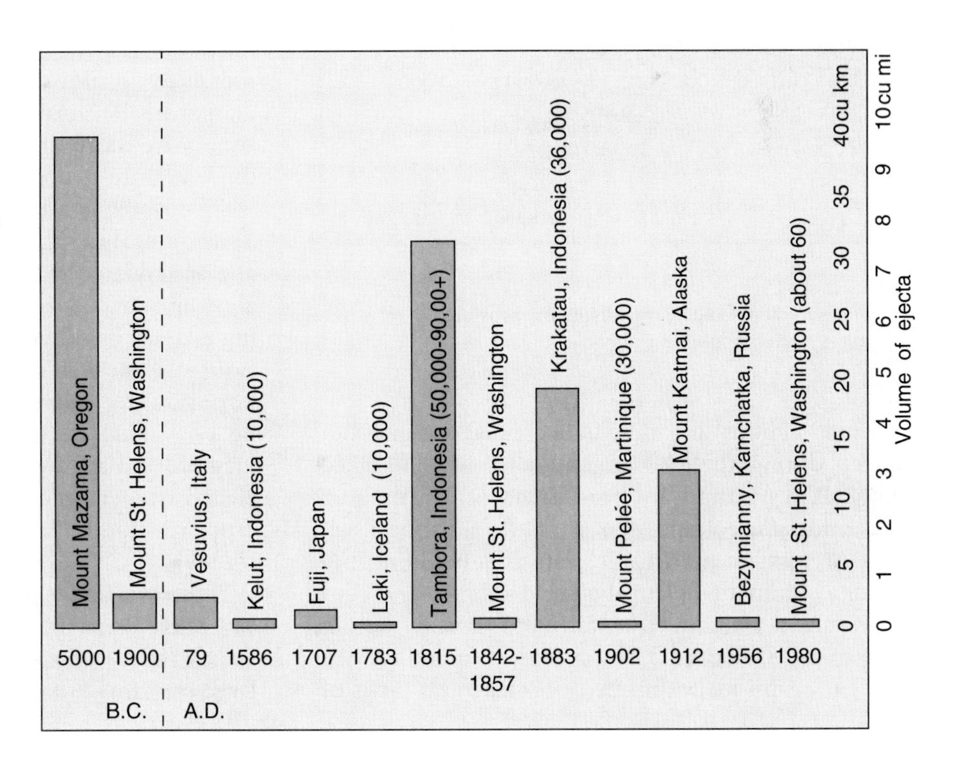


**FIGURE 4–18**
Comparison of the amount of pyroclastic material ejected during major volcano eruptions. The numbers of casualties are shown in parentheses. The 1980 explosion of Mount Saint Helens seemed massive because of media coverage, 57 deaths, and damage to agriculture, but it is dwarfed by numerous other eruptions throughout history. (Data from U.S. Geological Survey.)

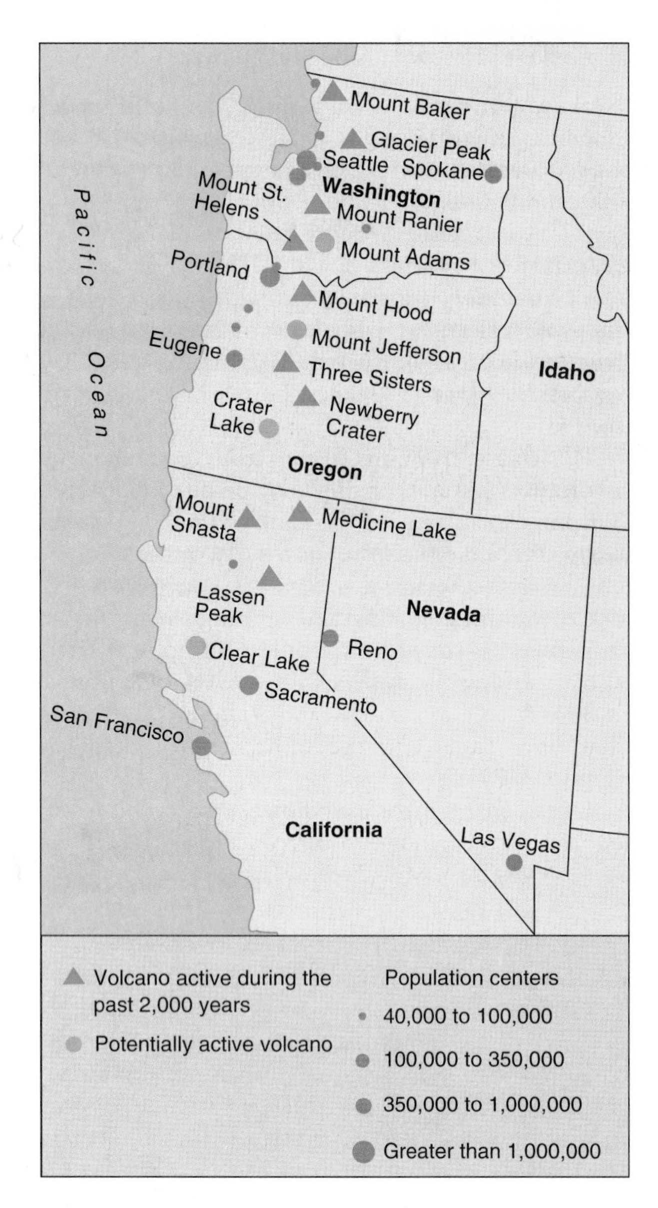


**FIGURE 4–19**
Volcanoes in the Cascade Range. Note the proximity of many of the volcanoes to population centers.

Another famous and destructive nuée ardente occurred during the 1902 eruption of Mount Pelée (puh-LAY) on the Caribbean island of Martinique. This andesitic volcano is part of the Caribbean Island Arc, where an oceanic plate is subducting beneath a continental plate (Figure 4–22).

Mount Pelée gave clear warning for weeks before it erupted violently. The precursors included ash and gas eruptions within its crater lake and flows of hot mud down river valleys. Local politicians assured the population that they were safe. However, a plug of very viscous magma blocked the crater, preventing release of the gas pressure building below the plug.

Shortly after dawn on May 8, 1902, the hot gas and debris escaped violently between the lava plug in the crater and a notch in the crater rim. Unfortunately, the lateral eruption aimed straight at the city of St. Pierre. The nuée ardente hugged the ground as it roared down the steep mountain slope, taking only two minutes to travel 5.5 kilometers (3.4 miles) to the city at an estimated 160 kilometers/hour (100 miles/hour). The momentum of the flow was sufficient to drive tree trunks into stone walls!

The blast flattened structures and trees and burned everything in its path, both across the land and over ships in the harbor. Some 30,000 awaking inhabitants of St. Pierre died instantly from the superheated gases. Among the two or three survivors was a prisoner, safe in his windowless dungeon.

## Lahars (Volcanic Mudflows)

**Lahar** is the Indonesian name for a mudflow—a mixture of water and volcanic debris that sweeps down a volcano's slope. A lahar's velocity depends on the slope and water content in the debris. The more watery the flow, the lower its viscosity, and the faster it flows. These low-viscosity lahars are dangerous because of their speed. More viscous lahars move more slowly, but their greater density can remove large structures, trees, and vehicles.

The water in a lahar comes from various sources. Some volcanoes have summits at high elevations and are covered with snow and ice, so the heat from an eruption may rapidly melt this cover, precipitating a lahar. In other cases, the violence of the explosion expels water from a summit crater lake. Groundwater in surface layers of volcanic rock may be converted to steam by the heat of the rising magma, and ash from the eruption may serve as nuclei to trigger precipitation of steam rising from an eruptive cloud. Rain, of course, is another water source.

Stratovolcanoes are effective lahar generators because of their steamy, explosive eruptions, steep slopes, and layers of easily eroded ash that cover large portions of their sides.

**FIGURE 4–20**
Homes buried by ash in the Heimaey Island eruption of 1973. (Photo by W. A. Benjamin, U.S. Geological Survey.)

**FIGURE 4–21**
Trees broken and aligned by a nuée ardente on the slope of Mount Saint Helens. (Photo from U.S. Geological Survey, Cascades Volcano Observatory.)

A good example of a lahar's destructive power is the eruption of the Nevado del Ruiz Volcano in Colombia, South America, in November 1985. The volcano was capped with a glacier and snow, which melted quickly in the eruption. Lahars swept down the slopes, killing about 25,000 people and causing great damage to the surrounding towns (Figure 4–23).

Lahars also caused great destruction during the 1980 eruption of Mount Saint Helens. They moved down slopes and along stream valleys, transporting trees, structural debris, and great volumes of sediment (Figure 4–24). These lahars choked local channels and dumped huge volumes of sediment into tributary streams of the Columbia River, a major transportation route. Extensive dredging by the U.S. Army Corps of Engineers was necessary to re-open and maintain navigation.

The U.S. Geological Survey has studied prehistoric lahars associated with Mount Rainier in Washington State. Past lahars reached within 16–25 kilometers (10–16 miles) of the location of Tacoma, Washington (Figure 4–25). What if a future eruption is more severe? What if the snow and ice cover on Mount Rainier is greater in the future? According to the USGS, the risk from Mount Rainier ashfalls is restricted mainly to the east side of Mount Rainier because prevailing westerly winds would disperse more ash on the east side of the volcano. This is fortunate for the cities of Seattle and Portland, located on the upwind (western) side of the Cascades Range.

Experience with Cascades volcanic processes and detailed studies of deposits from previous eruptions have enabled the U.S. Geological Survey to produce **hazard maps** that illustrate the areas affected by a particular geohazard. These maps have proven very useful to emergency-management authorities and the public. Hazard maps for the Mount Shasta volcano in the Cascade Range, for example, show

**FIGURE 4–22**
The island of Martinique is midway along the Caribbean Volcanic Island Arc. Mount Pelée, at the northern end of the island, has a number of knife-like ridges and valleys radiating out from its vent. Several lead toward the city of St. Pierre. This enabled the lateral blast from the volcano to destroy the city of St. Pierre to the south.

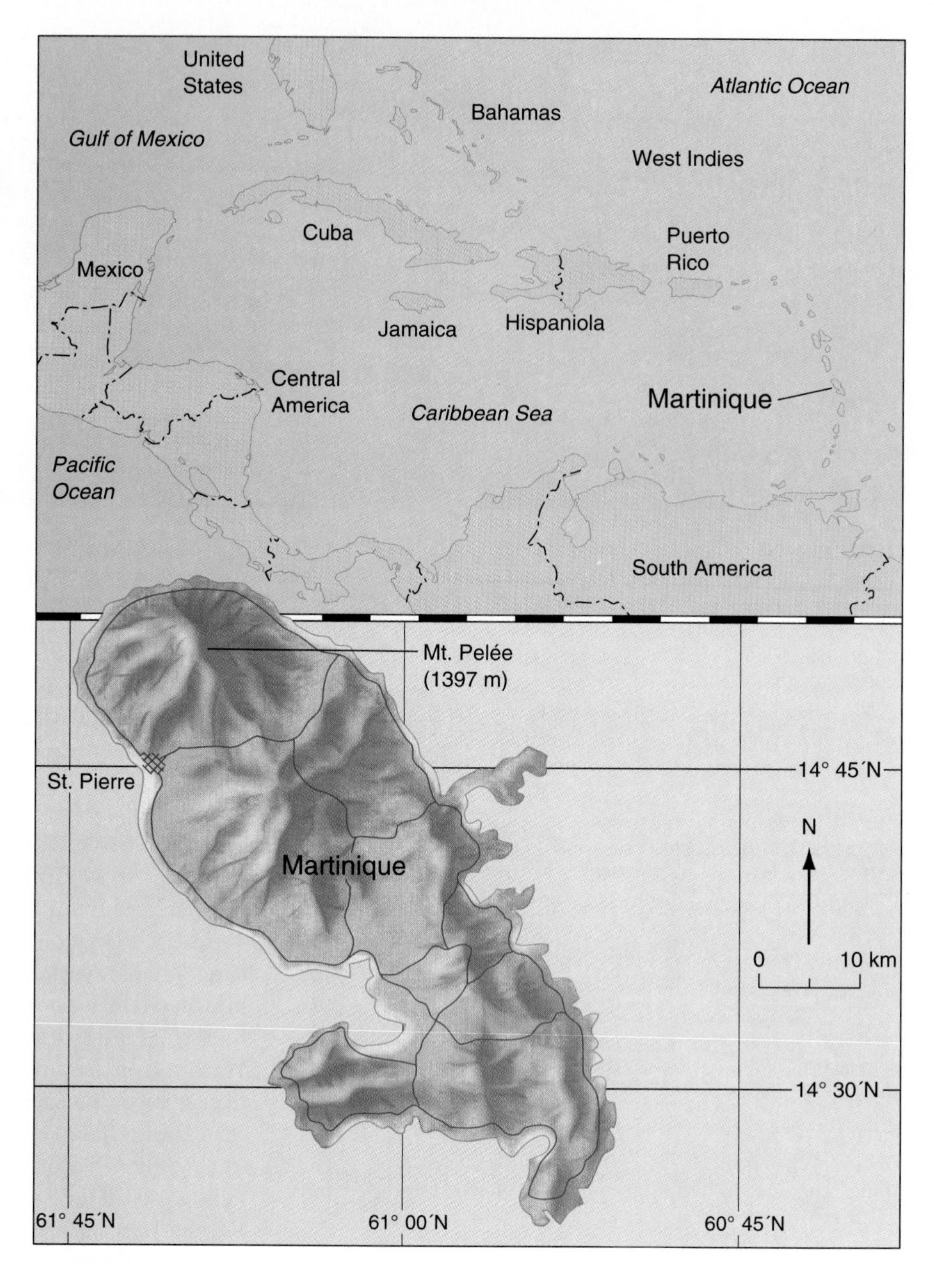


the danger from pyroclastic flows and lahars (Figure 4–26).

## Poisonous Gases

Although the major component of volcanic gases is water vapor, they also contain various amounts of carbon dioxide ($CO_2$), sulfur dioxide ($SO_2$), carbon monoxide (CO), hydrogen sulfide ($H_2S$), sulfuric acid ($H_2SO_4$), hydrochloric acid (HCl), and hydrofluoric acid (HF). These gases can constitute a geohazard under certain conditions.

For example, continuing eruptions along the east rift zone of Kilauea volcano in Hawaii emit toxic gases that drift over the agricultural and urbanized areas of the island of Hawaii. According to the U.S.

**FIGURE 4–23**
Lahar damage in the Nevado del Ruiz eruption in Armero, Colombia. The eruption of this volcano killed almost 25,000 people. The lahar swept down the area at center. The streets and block patterns are still visible. The structures are gone. (Photo by Tom C. Pierson, U.S. Geological Survey.)

**FIGURE 4–24**
Mudflows and debris from the May 18, 1980, eruption of Mount Saint Helens piled up against a bridge on Interstate 5 between Portland and Seattle. Fortunately the bridge was not destroyed. (Photo by Dan Dzurisin, U.S. Geological Survey.)

**FIGURE 4–25**
This map of prehistoric lahars on Mount Rainier, Washington, shows the extent of mudflows in the White River Valley and the Puyallup River Valley. (After Crandell and Mullineaux, 1967, U.S. Geological Survey Bulletin 1238.)

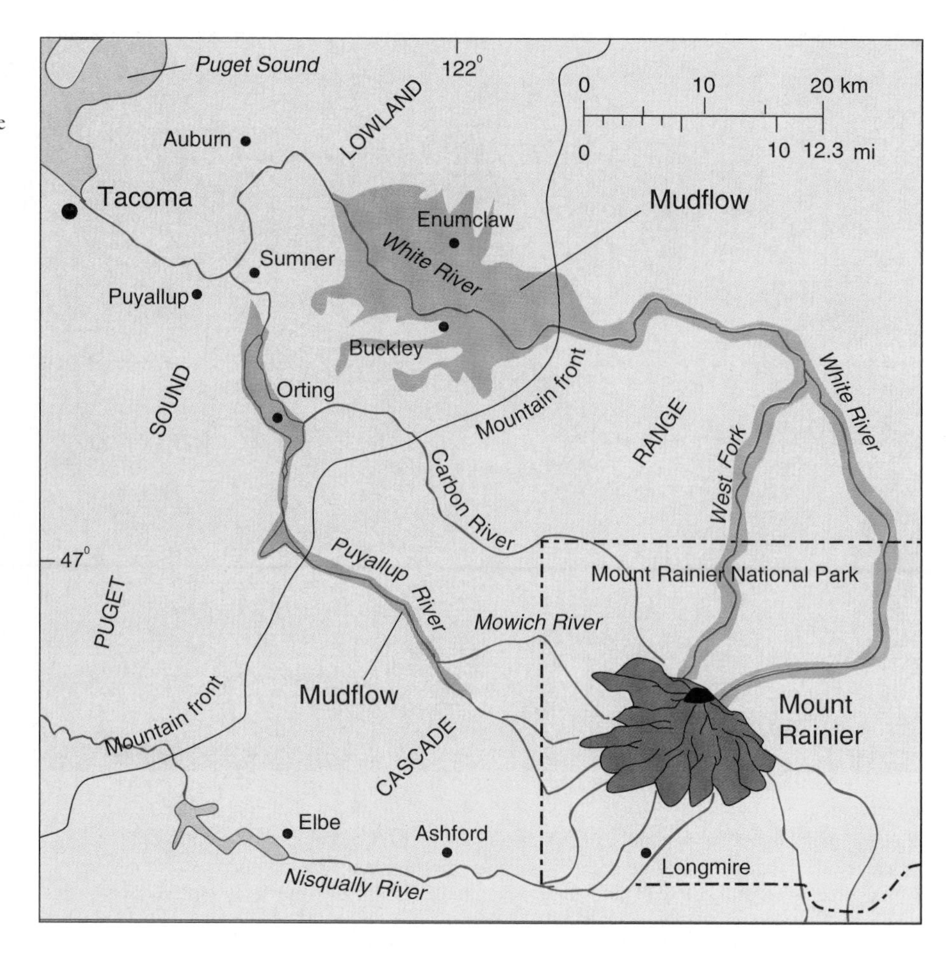


Geological Survey, these toxic gases are produced both at active vents and where lava flows enter the ocean and react with seawater.

At the eruption site, sulfur dioxide gas ($SO_2$) is converted to sulfuric acid ($H_2SO_4$) a short distance downwind to produce **vog** (volcanic fog). Lava entering the sea reacts with seawater to produce hydrochloric acid-bearing **laze** (lava haze). The plumes of laze may also contain particulates (volcanic glass particles). Vog and laze obscure scenic vistas, lower agricultural yields for certain crops, and affect those with respiratory or heart conditions.

Carbon dioxide also can be a geohazard, as is made clear in these accounts:

> During the [Mount] Katmai eruption [in Alaska], in 1912, acid rain fell [on the towns of] Seward and Cordova, 400 and 575 km. [250 and 360 miles] from the volcano, burning the skin of some persons and damaging both vegetation and metals. Near Cape Spencer, 1,100 km [700 miles] away, fumes tarnished brass. Drifting slowly southward, a whole month later the gases reached Vancouver [British Columbia], and acid rains damaged clothes hung out to dry. . . . Most insidious of the volcanic gases are $CO_2$ and CO because they are colorless and odorless. . . . During the 1947 eruption of Hekla [Volcano in Iceland], $CO_2$ accumulated as shallow pools in hollows, so that sheep entering the hollows were drowned [in the carbon dioxide gas], but men's heads were above the surface of the $CO_2$, and they walked through the hollows unharmed.[1]

In the central African nation of Cameroon, 1,700 people died one night in 1986. Clouds of carbon dioxide were released from Lake Nyos, a

[1]From B. A. Bolt, W. L Horn, G. A. Macdonald, and R. A. Scott, 1977, *Geologic hazards,* (second edition), New York: Springer-Verlag, p. 113.

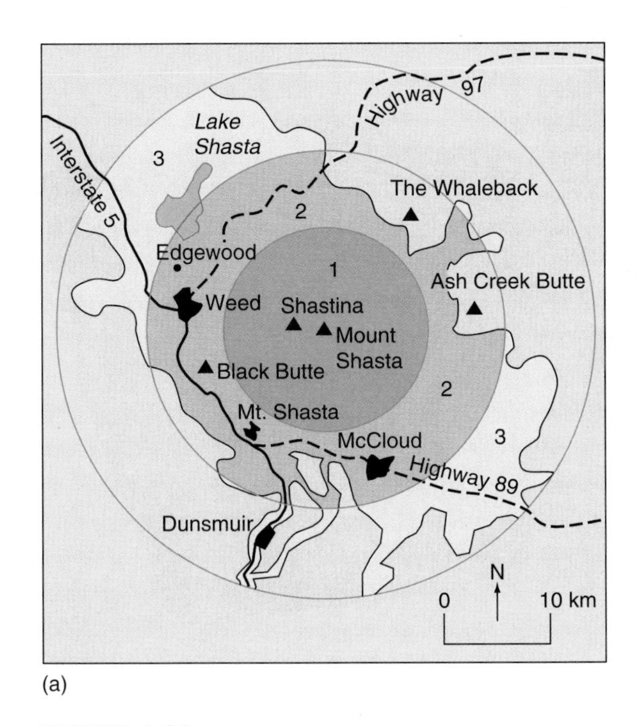


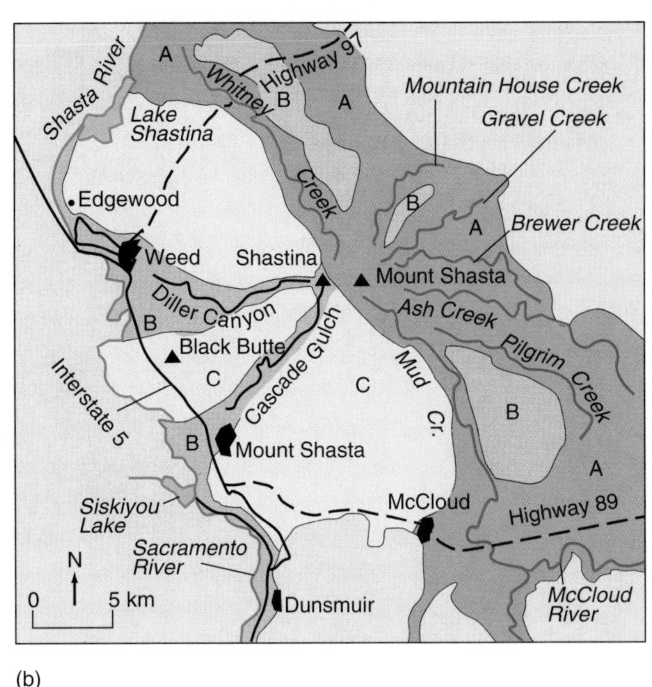


**FIGURE 4–26**
Hazards maps for Mount Shasta in the Cascade Range. (a) Pyroclastic flow and lateral blast hazard map. (b) Lahar (mudflow) hazard zones map. Zones designated by letters show relative likelihood of being affected by future mudflows. Zone A is most likely and zone C is least likely to be affected. (After Crandell and Nichols, 1989, U.S. Geological Survey.)

crater lake filling a dormant volcano, and spread downslope over nearby villages. The people died from oxygen starvation, inhaling mostly carbon dioxide that had displaced air. Geologists believe the $CO_2$ originated in magma beneath the lake bed (Figure 4–27(a)). As much as *a billion cubic meters* (35 billion cubic feet) of carbon dioxide were expelled from Lake Nyos in a few hours. Dead animals found up to 100 meters (330 ft) higher than the lake helped to determine the volume of the gas cloud.

Geologists have reconstructed the event like this: gas from the magma seeped upward into the lake. It dissolved in the lower lake water, just like $CO_2$ is dissolved in soda, and was maintained in solution by the pressure of the water above. Some disturbance of the water overturned it, quickly reducing the pressure and releasing the gas (Figure 4–28(b)). The dynamics are similar to quickly opening a bottle of warm soda.

What overturned the water? Possibilities include a submarine landslide, a normal seasonal overturning of the waters, an underwater eruption, or an earthquake. Whatever the cause, the gas quickly escaped from solution and broke the surface, creating huge waves. The waves destroyed nearby vegetation and the heavy gas cloud spread out through low areas. The gas killed—not by poisoning people, as might happen with other volcanic gases, but by depriving them of oxygen. In essence, the people and animals were quickly asphyxiated.

Although these events are dramatic, it is important to note that only a small percentage of volcano-related deaths have been caused by poisonous gases. Volcanic gases are not a major geohazard.

## PREDICTING VOLCANIC ERUPTIONS

A variety of precursors—minor events that warn of a major one to come—may indicate a potential volcanic eruption. Unfortunately, at present, no single

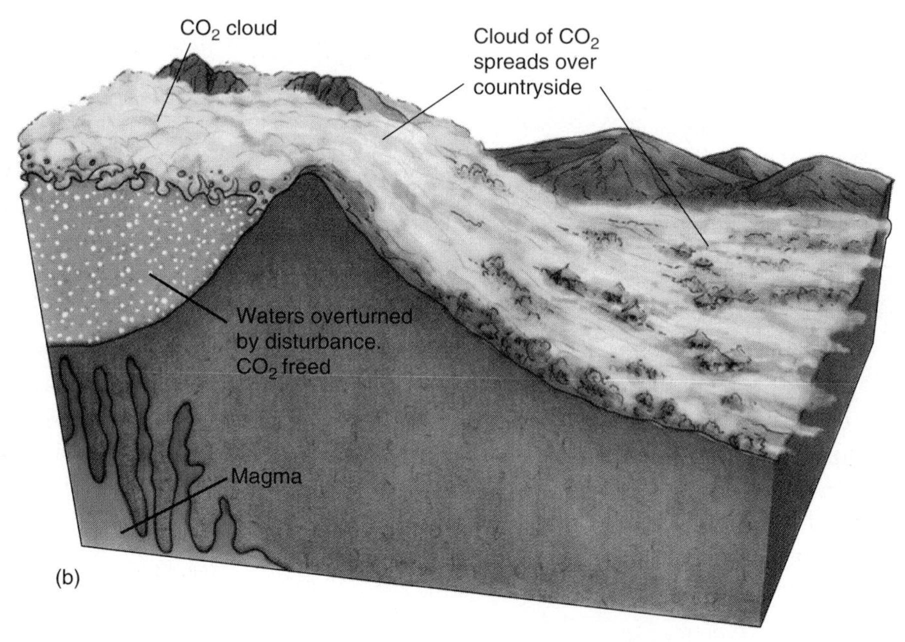


**FIGURE 4–27**
Diagram showing the origin of the Lake Nyos eruption. (a) Carbon dioxide of magmatic origin builds up in the bottom waters of the lake and is held there by the pressure of the overlying water. (b) Bottom waters are overturned by a disturbance. Carbon dioxide is freed rapidly, rises to the lake surface, and spills out across the surrounding countryside.

precursor can predict accurately *when* a volcano will erupt. Clearly, the more precursors that are evident simultaneously, the greater is the probability of an eruption. Several methods are being developed to provide sufficient warning for evacuation. These include detecting changes in the ground surface (bulging and tilting), detecting an increase in surface temperature, monitoring the occurrence of earthquakes caused by magma movement and/or hydrothermal pressure effects, and changes in volcanic gas composition.

## Ground Tilt and Displacement

Magma rising in a volcano causes the surface to bulge and change slope. In addition, fluid and gas pressures in the magma may cause the surface to bulge (Figure 4–28). These phenomena can be de-

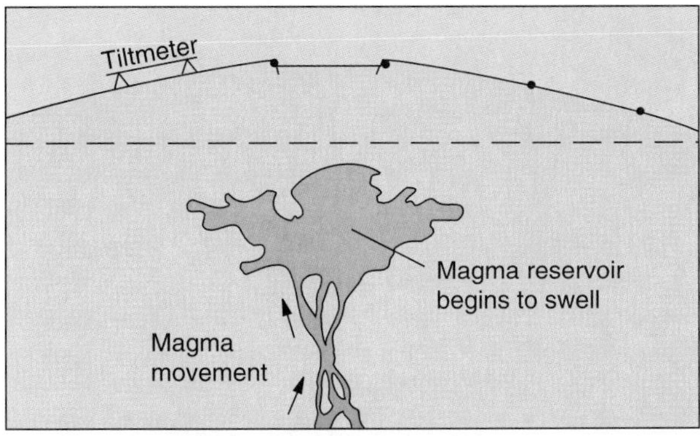


(a) Inflation begins

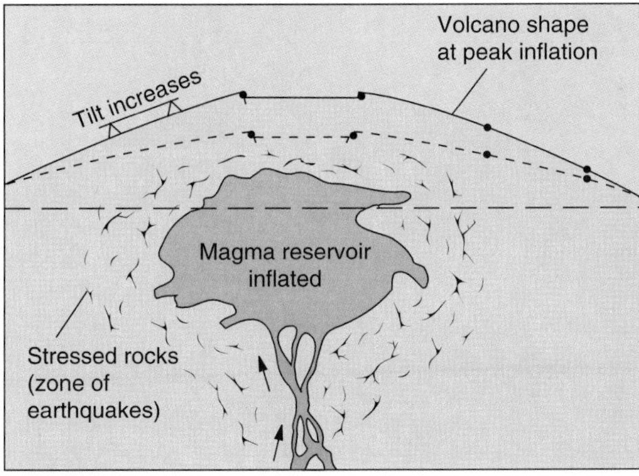


(b) Inflation at peak

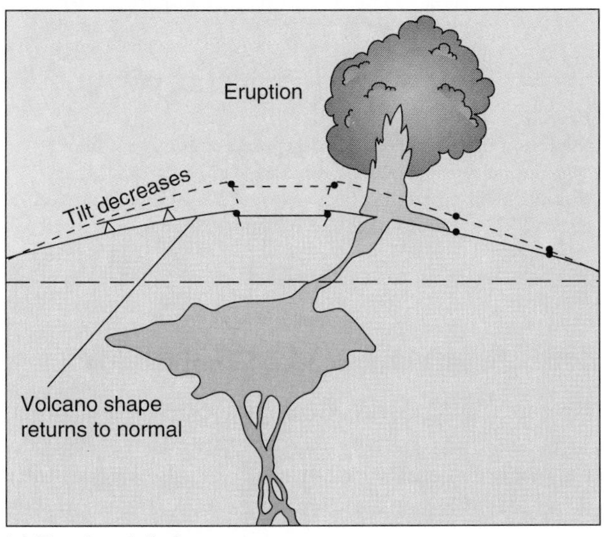


(c) Eruption deflation

**FIGURE 4–28**
Schematic model showing ground deformation of Kilauea, Hawaii, volcano summit, related to filling and emptying of magma reservoir. (a) Inflation begins. (b) Tiltmeters on the summit record amount of inflation. (c) Kilauea summit subsides as magma drains laterally to feed an eruption on a rift zone and the magma reservoir deflates. (After Tilling, R. I., 1983, as reproduced in U.S. Geological Survey, 1986, Earthquakes and volcanoes, v. 18, no. 1, p. 18.)

tected by measuring changes in elevation, measuring changes in the tilt of reference stations, and measuring changes in the distance between two points over time. Observations of Mount Saint Helens prior to its eruption in 1980 revealed pronounced bulging on its north side. The volcano eventually erupted on that side (Figure 4–20) and devastated the area nearby.

## Increase in Surface Temperatures

Magma nearing the surface conducts heat to the outer rocks of a volcano. This anomalous heat can be detected with thermal probes inserted into the surface and used with other evidence for eruption prediction. Aerial surveys using infrared (heat) sensing devices can map temperature differences on the volcano surface. However, temperature variations in crater lakes give better information because the water body absorbs heat from the volcano with time (for example, Kelut in Indonesia and Poás in Costa Rica).

## Monitoring Earthquakes

Eruptions often are preceded by small earthquakes (volcanic tremors). They sometimes number in the thousands but generally are too small to be felt (less than Richter magnitude 3, explained in Chapter 5). These volcanic earthquakes are closely related to magma movement or pressure changes in the reservoir, so they provide valuable clues to possible eruption times and locations. Pattern analysis of these earthquakes has helped reveal the depth of magma reservoirs and the migration of their point of origin.

Beneath Kilauea volcano in Hawaii, earthquake points of origin have been analyzed. This work revealed distinctive characteristics of earthquakes coming from three different areas (Figure 4–29):

1. Long-period (LP) earthquakes define the magma conduit (open circles on Figure 4–29).
2. A general absence of seismic activity outlines the magma storage area.
3. Shallow, high-frequency earthquakes above the magma storage region reflect the brittle fracturing of rocks as the magma intrudes them (solid circles in Figure 4–29). If long-term monitoring registers a rise in the magma storage area, an eruption may be more imminent.

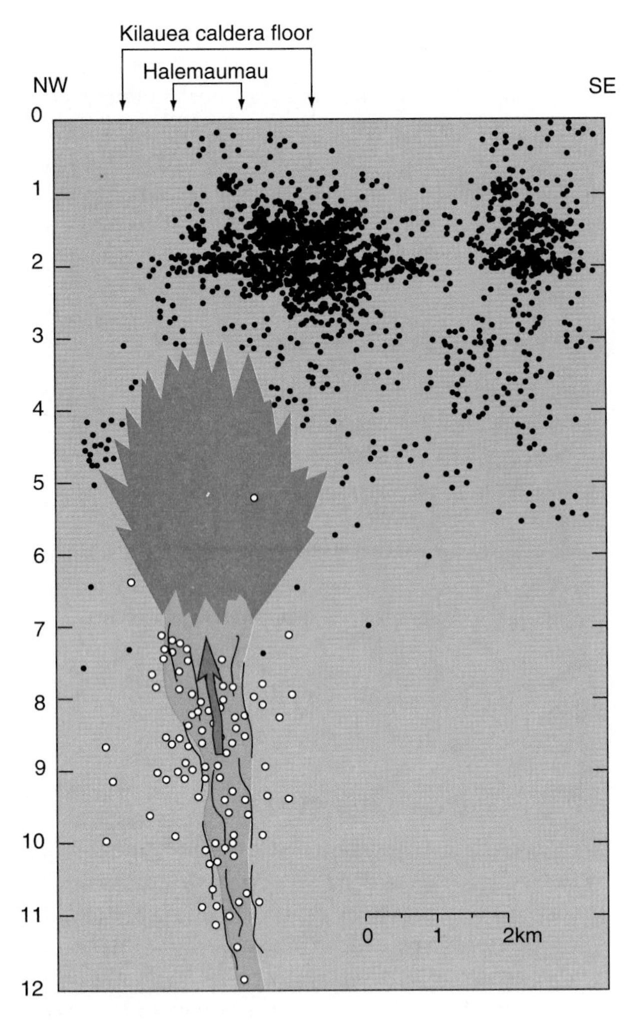


**FIGURE 4–29**
Cross section of Kilauea Volcano, Hawaii. Long-period earthquakes (open circles) outline the conduit along which the magma travels from its origin. The shaded area, where little earthquake activity occurs, outlines the magma storage area. Shallow, high-frequency earthquakes (solid circles) occur above the magma storage region, reflecting brittle fracture of the roof rocks. (After Koyanagi and others, 1974, as reproduced in U.S. Geological Survey, 1986, Earthquakes and volcanoes, v. 18, no. 1, p. 17.)

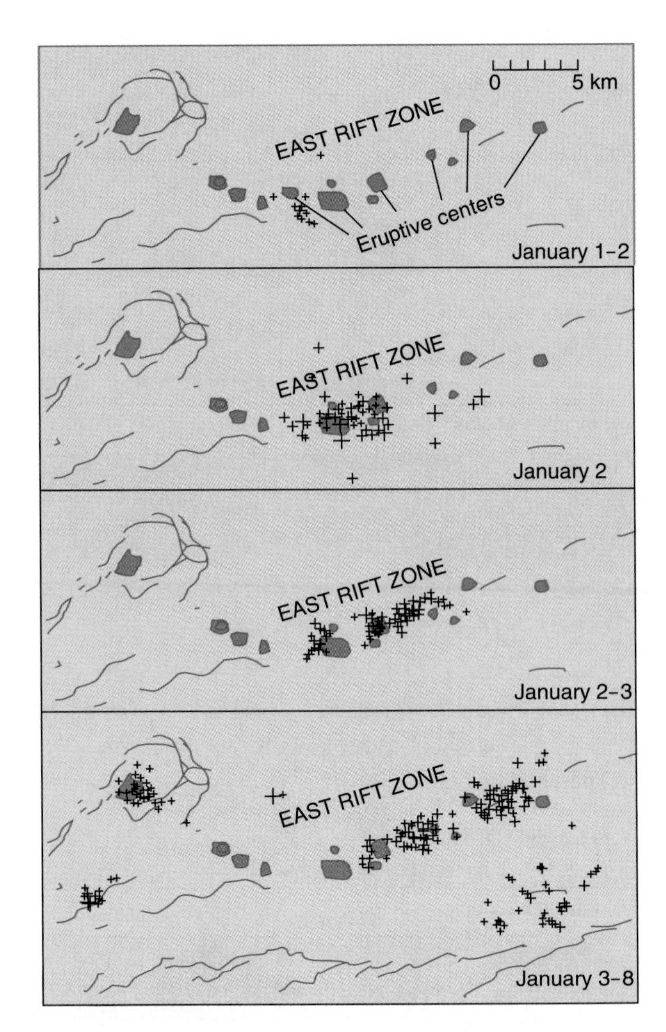


**FIGURE 4–30**
Migration of earthquakes (crosses) preceding and accompanying an eruption in Kilauea's east rift zone, January 1983. (After Koyanagi and others, 1986, as reproduced in U.S. Geological Survey, 1986, Earthquakes and volcanoes, v. 18, no. 1, p. 30.)

Migration of earthquake origins has been observed prior to eruptions. Such a migration is shown in Figure 4–30 for an eruption in Kilauea's east rift zone. This zone is defined by a string of eruptive centers. Note the increasing development of earthquake activity within the east rift zone prior to the eruptions.

## Changes in Volcanic Gas Composition

Research by the U.S. Geological Survey and others suggests that changes in gas emitted by a volcano may be a precursor to an eruption. Levels of sulfur dioxide ($SO_2$) and other gases may be related to change in magma supply rate, change in magma type, or change in the pathways through which the gas escapes.

Prior to the major eruption of Mount Pinatubo in the Philippines in June 1991, changes were noted in its $SO_2$ emission. Activity on the volcano

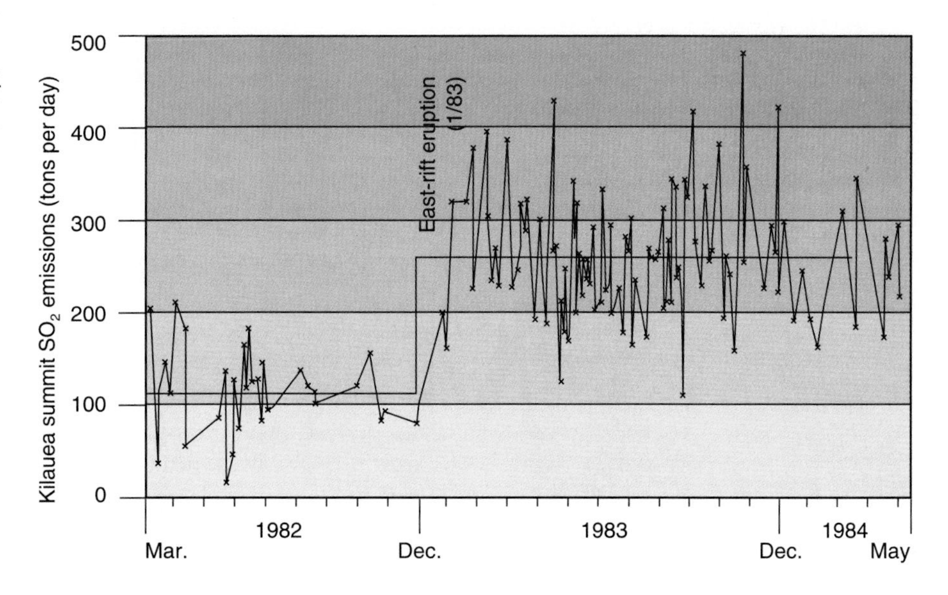


**FIGURE 4–31**
Rise in $SO_2$ concentration at Kilauea volcano, Hawaii, at the beginning of the 1983 eruption. (U.S. Geological Survey, 1986: Earthquakes and volcanoes, v. 18, no. 1, p. 64.)

began on April 2 with small steam explosions and earthquake activity. On May 13, its $SO_2$ emission was 500 tons per day; only 15 days later, $SO_2$ emission had increased tenfold, to 5,000 tons per day. A similar rise in $SO_2$ emission was noted at the summit of Kilauea at the beginning of its January 1983 eruption (Figure 4–31).

## Prediction Success: Mount Pinatubo, 1991

Predicting volcanic eruptions can be quite difficult because some volcanoes display plentiful precursor activity and fail to erupt, whereas others may explode violently with few precursors. In recent years, volcanologists have greatly increased their knowledge, have developed new techniques, and have improved their ability to predict when a volcano might erupt. These new techniques, combined with a knowledge of the volcano's eruptive history, were used successfully to reduce loss of life in the 1991 eruption of Mount Pinatubo.

Mount Pinatubo, an andesitic volcano on the Philippine Island of Luzon, last erupted 600 years ago. The volcano again stirred on April 2, 1991, with small steam explosions. Would it erupt? If so, *when?* Experts from the Philippine Institute of Volcanology (PHIVOLCS) immediately installed earthquake-monitoring systems on the mountain. On April 23, scientists from the U.S. Geological Survey joined the study. There was a sense of urgency because Mount Pinatubo was close to Philippine agricultural and population centers plus several major U.S. military bases (see Box 4–2). The following account is based on USGS descriptions.

The PHIVOLCS-USGS team installed earthquake-monitoring stations and tiltmeters (Figure 4–28) and used U.S. Air Force planes to sample the volcano's steam plumes. By May 28, all studies indicated that magma was rising within the volcano.

They established a scale of alert levels from 1 (low level unrest) to 5 (eruption underway). A geohazard map showing the expected hazards from a Pinatubo eruption was completed and distributed by May 23.

By June 5, the seismic activity increase prompted PHIVOLCS to raise the alert level to 3 (eruption possible within two weeks). Areas threatened by nuées ardentes (pyroclastic flows) were evacuated following this alert. By June 7, a small dome extruded on the north flank of Pinatubo. This prompted a level 4 alert (explosive eruption possible in 24 hours).

On June 9, PHIVOLCS raised the alert level to 5 (eruption underway). Clark Air Force Base was evacuated on June 10. On June 12, the first of several large volcanic explosions occurred. The major eruption on June 15 occurred coincidentally during a typhoon (Pacific hurricane—see Chapter 16). This greatly increased the mudflow (lahar) hazard. When the weather cleared on June 16, the top of the

BOX 4–2

## THE RESTLESS "RING OF FIRE" IN THE SUMMER OF 1991

Most of Earth's active volcanoes lie around the edge of the Pacific Ocean (Figure 4–1), an area appropriately called the "Ring of Fire." In 1991, three of them, well-separated geographically, erupted during a short period. These eruptions threatened the lives, health, homes, and livelihoods of thousands. One of them even affected diplomatic relations between the United States and the Philippines.

### MOUNT UNZEN, JAPAN

In 1792, this andesitic volcano erupted to cause the worst volcanic disaster in Japanese history, in which over 15,000 perished. Unzen then slumbered for two centuries. A small steam-blast explosion in November 1990, followed by intermittent increased earthquake activity, signalled the volcano's reawakening.

This andesitic volcano had shown intermittent precursors of eruption for several months. Hot springs commonly are associated with volcanoes, and it

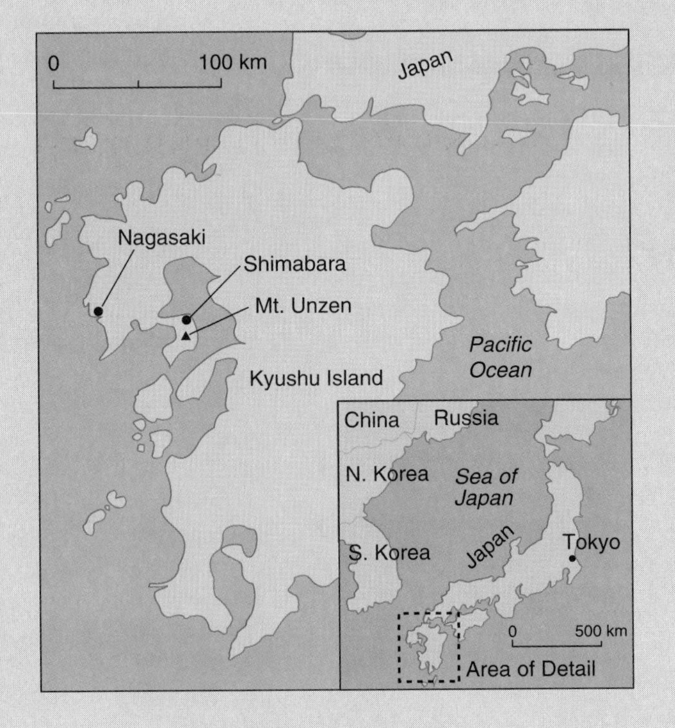


FIGURE 1
Mount Unzen, Japan, is about 64 kilometers (40 miles) east of Nagasaki.

**FIGURE 2**
Devastation by mudflows from Mt. Unzen, Japan. (Courtesy of Shigeo Aramaki.)

was near a popular hot-springs resort on the mountain that the 1991 eruption occurred. On June 3, Mount Unzen exploded (Figure 1), subjecting nearby residents to a devastating pyroclastic flow, flying tephra particles, hot ash that burned clothing and lungs, and intense heat that set homes afire from a distance (Figure 2). This event claimed 43 lives. Heavy seasonal rains are typical, so further damage was expected from lahars.

## MOUNT PINATUBO, PHILIPPINES

Only days later and 2,400 kilometers (1,500 miles) away in the Philippines, Mount Pinatubo erupted violently (Figure 3). The geohazard was great because half a million people live within 40 kilometers (25 miles) of this andesitic volcano. The volcano also was dangerously close (100 kilometers, or 62 miles) to a major U.S. western-Pacific military installation, Clark Air Base. At the time, lease renewal was pending for U.S. military bases in the Philippines. (The United States subsequently abandoned the bases.)

The climactic eruption on June 15 showered inhabitants with rocks, and heavy ashfalls that collapsed roofs (Figure 4). The ash became mud when a typhoon reached the islands shortly after the eruption began. Air traffic soon was halted at Manila International Airport because the ash-laden air damaged jet engines and wet, slippery ash made landings difficult.

Massive evacuation began a few days before the June 15 eruption because the Philippine Institute of Volcanology and Seismology and colleagues from the U.S. Geological Survey warned that pressure within the volcano and new fissures near its southern side indicated an imminent and violent eruption. Heavy rains increased the danger of lahars on the mountain flanks.

**FIGURE 3**
Mount Pinatubo is on the island of Luzon, 90 kilometers (55 miles) northwest of Manila, the Philippine capital.

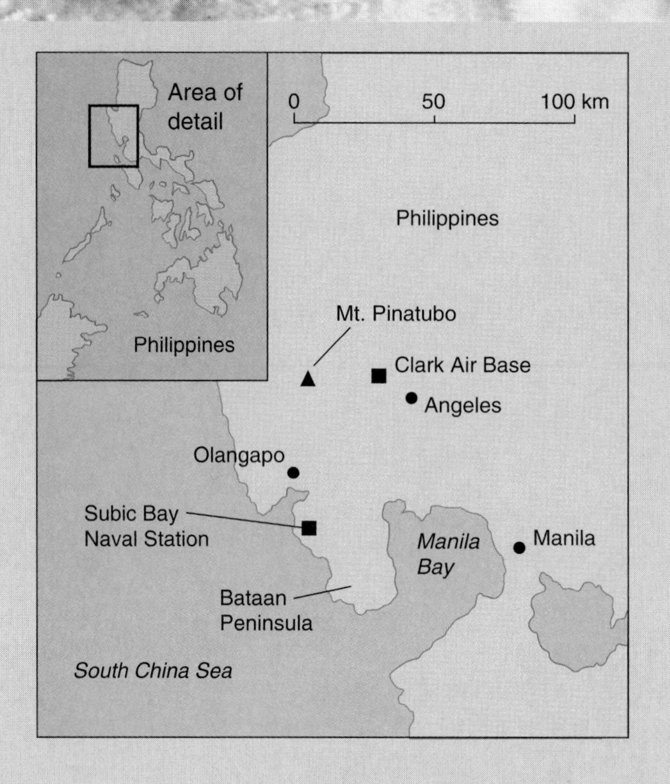


**FIGURE 4**
Spectators watching the eruption of Mount Pinatubo. (Photo by Raymond Pestrong, San Francisco State University.)

Approximately 600 people were killed and 200,000 acres, including Clark Air Base, were covered with thick volcanic ash. Rain caused massive avalanches and lahars from the volcano's flanks. The debris covered all of some towns and parts of others, changed river courses, and buried vast tracts of rice and sugar cane. Clark Air Base became useless as runways and roads became covered with wet ash and buildings collapsed under its weight.

Temporary shelters housed 60,000 refugees, and more homes and land were predicted to be lost with each storm and hurricane-like typhoon. Experts cautioned that avalanches would recur during each monsoon season for the next 7 to 10 years and that up to half the material deposited by the eruption would be washed onto surrounding areas.

## KILAUEA, HAWAII

The volcanoes of Hawaii (in the mid-Pacific) are basaltic rather than the andesitic type that make up the Pacific Ring of Fire. However, these volcanoes are frequently active, fed by lava from a hot spot below. An eruption had begun in January 1983 and continued to affect the southern coast of the island of Hawaii, near Kalapana. Basaltic lava poured from Kilauea's east rift zone (Figure 4–8) and lava flows cut off roads and caused significant property damage. By 1989, the eruption had destroyed 76 homes and the Visitor Center at Hawaii Volcanoes National Park. In 1990, lava flows moving toward Kalapana buried another 100 homes under 15 to 25 meters (50 to 80 feet) of lava.

The Kilauea eruptions were observed closely by the staff of the U.S. Geological Survey's Hawaiian Volcanoes Observatory, established in 1912 at the summit of Kilauea. The observatory staff uses instruments and field surveys to monitor volcanic and seismic activity on the island of Hawaii.

Observatory scientists cautioned that Kilauea's present activity would be dwarfed by potential eruptions from neighboring Mauna Loa Volcano. A Mauna Loa eruption would be very serious, for its eruption rates are much greater and its slopes much higher than Kilauea's, plus the volcano is closer to the resort areas of Kona and Hilo. Following a decade of dormancy, Mauna Loa erupted briefly in 1984. In only three weeks, its lava flows covered half the area that it took Kilauea eight years to blanket.

## WERE THESE ERUPTIONS RELATED?

No. The reason is simple: the plate tectonic setting for each is different (Figure 2–23). The Mount Unzen eruption was caused by subduction of the Philippine plate under part of the Eurasian plate, and the Mount Pinatubo eruption was caused by subduction of the Philippine plate under another part of the Eurasian plate. The Hawaiian eruptions are fed by basaltic magma from a hot spot presently located under the southernmost part of the island of Hawaii (Figure 4–8).

Not only were these three eruptions notable events in their own right, but they also demonstrate that similar Earth events are not necessarily directly related and can occur together in time purely by chance. What they have in common is that all three were governed by plate tectonic processes.

## LIVING WITH THE THREAT

Why do people persist in living so close to volcanic geohazards? The answer is a combination of the area's attractiveness and the human perception of hazard:

- ☐ Mount Unzen is a significant hot-spring resort that thrived around the volcanic source of heat—and it had not erupted in 200 years, so the risk appeared negligible.
- ☐ The fertile soil on Mount Pinatubo's slopes was a major agricultural resource—and it had been dormant for 500 years, so to most, the mountain appeared harmless.
- ☐ The Hawaiian Islands were populated sparsely until World War II. Their spectacular beauty spurred a post-war boom, and population and development extended even onto the flanks of active volcanoes on the island of Hawaii. However, volcanic geohazards are well known throughout Hawaiian history and the warnings are there.

These developments in such hazardous areas are good examples of a principle cited in Chapter 1: *people have short memories as far as geohazards are concerned.* The Japanese have a proverb for it: "Natural calamity strikes at about the time when one forgets its terror."

volcano was gone, pyroclastic deposits and lahars filled valleys, and an ash deposit covered the region to a thickness of 30 centimeters (12 inches) some 40 kilometers (25 miles) from the volcano.

The eruption death toll was about 600, most of whom were killed in collapsing buildings. That certainly constitutes a tragedy, but—given the fact that the eruption was the second or third largest in the twentieth century (VEI 6)—the death toll could have been far greater in this densely populated area (see Table 4–2). The monitoring, alert level scheme, and evacuation program based on PHIVOLCS-USGS studies was a great success.

## VOLCANIC HAZARD MITIGATION

For those living in a densely populated area near a volcano, whether it is a nonexplosive Hawaiian/Icelandic type or an explosive andesitic type like Mount Saint Helens or Mount Pinatubo, *evacuation* is the only effective safety action. People simply must relocate well before a major eruptive event. If evacuation is delayed until an area is partially inundated with lava or until valleys and streams are choked with pyroclastic or mudflow debris, exit routes may be blocked.

Violently explosive andesitic and rhyolitic volcanoes allow no room for control or mitigation because of the rapidity of their eruption and their explosive power. Evacuation at the first sign of new activity is the only course of action. However, in the case of relatively non-explosive basaltic volcanoes, such as those in Hawaii and Iceland, volcanic hazards have been mitigated to some extent. Evacuation also is essential, but some measures may reduce damage to property. These measures are described in this section.

### Lava Diversion by Explosion and Barriers

In times past, barriers have been built hastily to try to divert oncoming lava flows from populated areas. Their purpose is not to stop the lava, because lava can flow easily over the tops of barriers. The intent is to *divert* the lava sideways, possibly protecting a town, but at least allowing more time for escape.

Diversion barriers must resist two forces. One is the *mass pressure* of the lava against the barrier, which increases with flow thickness. The other force results from the *momentum* of the advancing flow. This force increases with the velocity of the flow, which is governed by the lava's viscosity and the steepness of the slope down which it is flowing. Both stresses must be considered in the design of a diversion barrier.

Another technique is to use explosives to breach the solidified wall of the lava and divert its flow.

Both barriers and explosives were used effectively to protect critical areas during a 1983 eruption of Mount Etna on the island of Sicily in Italy. Massive aa flows from a vent on the south side of the volcano threatened the areas of Monte Vetore and Sapienza (Figure 4–32). Sapienza, a major skiing and hiking area, lay directly in the path of the flows.

To divert some of the flow toward Monte Vetore and ease the pressure on Sapienza, explosives were tried. Lava diversion was minimal, but the cool debris blown from the explosion created blockages in the lava flow downstream, slowing its progress.

**TABLE 4–2**
Volcanic Explosivity Index (VEI) for selected volcanoes discussed in this chapter.

| VEI | Year | Location | Estimated Deaths |
|---|---|---|---|
| 2 | 1792 | Mount Unzen, Japan | 15,000 |
| 3 | 1985 | Nevado del Ruiz, Colombia | 25,000 |
| 4 | 1902 | Mount Pelée, Martinique (West Indies) | 29,000 |
| 5 | 1980 | Mount Saint Helens, Washington State | 57 |
| 6 | 1883 | Krakatau, Indonesia | 36,400 |
| 7 | 1815 | Tambora, Indonesia | 92,000 |

SOURCE: Derived from data in Barberi et al., 1990, Reducing volcanic disaster in the 1990s: Bulletin Volcanological Society of Japan, v. 35, p. 80–95.

**FIGURE 4–32**
Map of lava flows from Mount Etna's south flank in the 1983 eruption. (After Lockwood, J. P., and Romano, R. R., May 1985, Lava diversion proved in 1983 test at Etna: Geotimes, p. 9–12. Used with permission.)

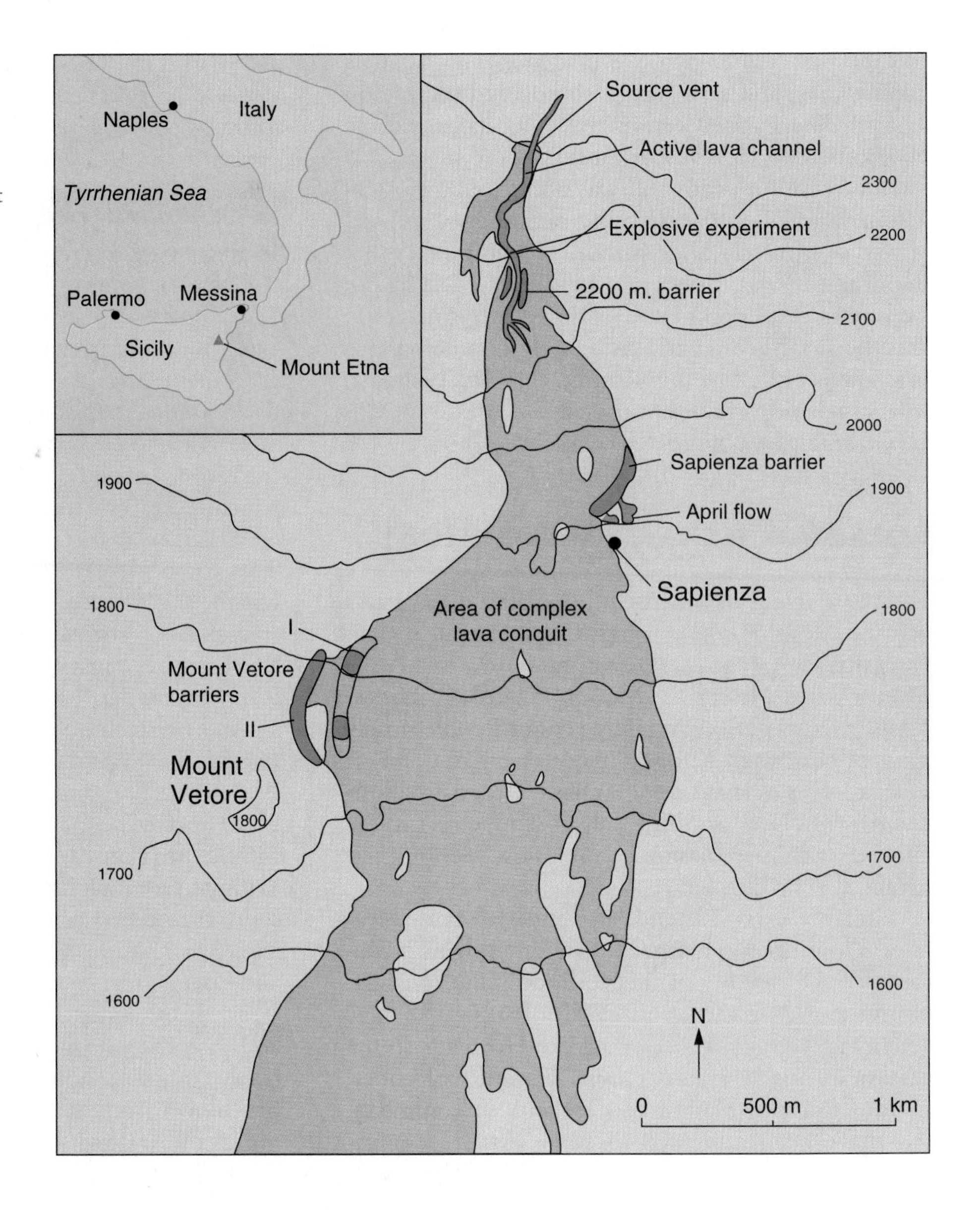


More than 100 workers then used bulldozers and trucks to build a massive diversion wall upstream of Monte Vetore. The goal was to divert flows that threatened a large hotel and an astronomical observatory. The barrier was largely overtopped and another one was built 100 meters (330 feet) to the west. The second barrier effectively diverted the flow.

A barrier 14 meters (45 feet) high successfully blocked the flow and diverted the lava around critical facilities in Sapienza (Figure 4–33). Although barrier construction cost was $3 million, losses would have been an estimated $5 to $25 million had the barriers not been built. The barrier diversions clearly were successful.

Lava barriers are most effective when constructed properly. However, as the Mount Etna case illustrates, barriers typically are improvised hastily because authorities rarely know where and when an eruption will occur, leaving too little time to construct a well-engineered diversion wall. In actual situations, the diversion walls are constructed in haste *after* an eruption starts and the likely path is known. These diversions usually are made of available rock that can be stacked quickly or of loose surface material that can be bulldozed quickly into a ridge.

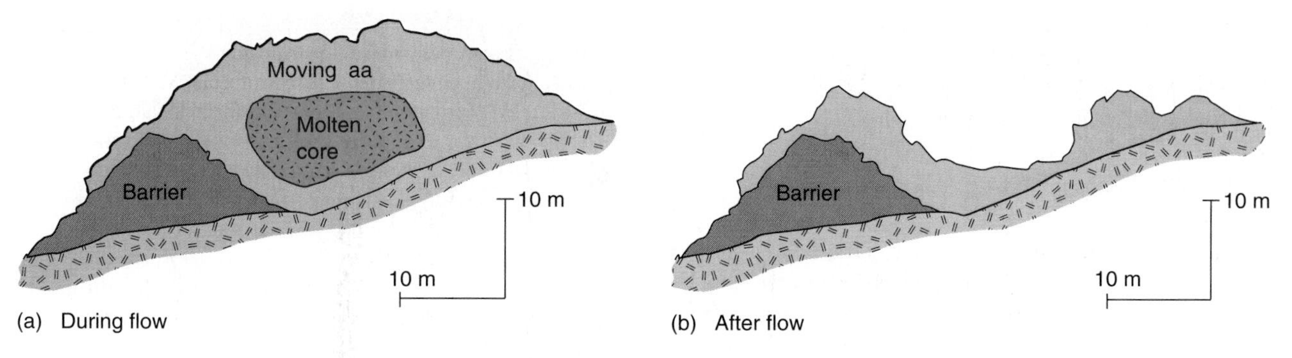


**FIGURE 4–33**
Sapienza, Italy, lava barrier and lava flow in cross section. (a) Only the leading edge of the lava flow overrode the earthen barrier that had been built to divert it. The lava mounded up while cooling. (b) After the flow cooled, lava subsided, leaving a depression. The barrier thus had served its purpose in protecting facilities at Sapienza. (After Lockwood, J. P., and Romano, R. R., May 1985, Lava diversion proved in 1983 test at Etna: Geotimes, p. 9–12. Used with permission.)

## Cooling Lava Fronts with Water

Stopping an approaching lava front by rapidly cooling it from molten to solid is another strategy. Rapid cooling creates a "stone wall" to slow the flow and may divert the flowing lava laterally. The success of this technique depends on being able to deliver abundant volumes of cooling water to the lava front.

The island nation of Iceland consists almost entirely of basalt lava from the Mid-Atlantic Ridge (see Figure 2–23). Volcanism continues in Iceland today, and periodic non-explosive eruptions threaten communities. Icelanders used the cold-water method with apparent success in stopping lava flows during the 1973 eruptions on the island of Heimaey, off the southwestern coast (Figure 4–34). Icy North Atlantic seawater, plentiful and close at hand, was pumped onto the basaltic flows from ships and pipelines running inland. Cooling the lava front appears to have worked well. However, some

**FIGURE 4–34**
Water-cooling of lava flows in Heimaey, Iceland. Cold sea water was pumped to the lava front in an attempt to stem the advance. (Photo from U.S. Geological Survey, Menlo Park.)

scientists feel that the lava-cooling strategy was not very effective and that the eruption and flows were ceasing as the water was applied.

## LOOKING AHEAD

It is no accident that the next chapter explores earthquakes, for they have intimate ties to volcanoes, as you saw in the discussion of volcano precursors. Both phenomena are expressions of Earth's tectonic movements.

Earthquakes are far more widespread than volcanoes, however, and can occur in areas that have experienced no volcanic activity for millions of years.

## SUMMARY

A volcano is a mountain formed by the accumulation of erupted lava and/or volcanic ash and other pyroclastic deposits. Volcanism occurs as quiet lava flows that create broad plains and plateaus, lava flows that build massive lava mountains, and explosive eruptions.

Volcanic eruptions produce lava flows, pyroclastic debris, and gases. Pahoehoe is a highly fluid basaltic lava, whereas aa lava is more viscous and occurs both in basaltic eruptions and those of more siliceous volcanoes. Airborne lava fragments, called tephra, can include pieces of older volcanic debris as well as solidified masses of lava from sand size to boulder size. The eruption of lava flows is favored in magmas of higher temperature, lower silica content, and lower water content. Water vapor is the major component of volcanic gas, with lesser volumes of carbon dioxide, sulfur dioxide, carbon monoxide, and hydrogen sulfide.

Shield volcanoes have gently sloping sides and are formed of fluid basaltic lavas (Hawaiian Islands). The Hawaiian volcanoes form as the Pacific plate moves over a stationary hot spot, which provides magma from the mantle.

Stratovolcanoes, built of lava layers and pyroclastic materials, have much steeper slopes. They are formed of lavas having greater silica and gas content. Stratovolcanoes occur along island arcs (Philippines and Aleutians) and near the edges of continents where oceanic plates are being subducted (Andes Mountains and Cascade Mountains). Cinder cones are small, steep-sided accumulations of tephra. Calderas are immense, cauldron-shaped craters resulting from a violent explosion and collapse of the original volcano.

Volcanic hazards include lava flows, tephra falls, pyroclastic flows (nuées ardentes), mudflows (lahars), and poisonous gases. Lava flows cause minimal loss of life but considerable damage to property and crop lands. Lahars occur when rain or rapidly melting snow or ice mobilizes loose volcanic debris on volcano slopes. Their high velocity and viscosity enables them to destroy anything in their path. They also cause major dislocations of transportation as they cut roads, wash out bridges, and clog rivers.

Pyroclastic flows (nuées ardentes) are by far the most deadly volcanic geohazard because of their speed and incinerating temperatures; they destroy all living things and structures in their path. The weight of accumulated ash on roofs can result in their collapse. In addition, fine dust causes health hazards, reduces visibility, clogs mechanical systems and water systems, and poses a danger to aircraft.

Volcanic gases can form a serious hazard in some situations. Carbon dioxide is especially dangerous because it is odorless, colorless, and heavier than air. Consequently, it can spread over low areas and asphyxiate humans and animals by depriving them of oxygen.

Several precursors are clues to potential volcanic eruptions. Ground surfaces bulge and tilt as magma rises, deforming the volcano. Thermal sensors can detect newly heated rock and suggest whether an eruption may be imminent. Earthquakes and other seismic processes can occur as the rock below the volcano surface is stressed and ruptured by the rising magma.

Attempts have been made to control volcanic flows by cooling lava fronts with water (Iceland) and building diversion barriers to channel advancing flows away from populated areas (Italy, Hawaii). These techniques may minimize damage to structures, but evacuation is the only way to minimize loss of life.

## KEY TERMS

aa
active volcano
atoll
calderas
central vent
cinder cones
dormant volcano
eruptive center
extinct volcano
fissure eruptions
hazard maps
hot spot
lahar (mudflow)
lava dome
lava flows
laze
nuée ardente (pyroclastic flow)
pahoehoe
pyroclastic material
satellite volcanoes
seamount
shield volcanoes
stratovolcanoes
tephra
vent
viscosity
vog
volcanic ash
volcanic bombs
volcano

## REVIEW QUESTIONS

1. What factors determine whether lava flows or pyroclastic debris predominate in an eruption?
2. What factors determine whether a shield volcano or a stratovolcano will develop?
3. Describe the plate tectonic conditions that determine whether Hawaiian-type islands or volcanic arc-type islands will develop in an ocean basin.
4. What geologic conditions favor the development of a nuée ardente?
5. What is a lahar? Why are they such a common geohazard associated with many volcanoes? What makes them so dangerous?
6. In what ways were the eruptions of Mount Pelée (1902) and Mount Saint Helens (1980) similar?
7. Describe four different precursors that occur before many volcanic eruptions. Explain each of them and show how they are monitored by geologists.
8. What special problem occurs in the release of carbon dioxide during volcanic activity? Give an example of such a geohazard.
9. Describe three different ways in which volcanic flows may be controlled.

## FURTHER READINGS


Bolt, B. A., Horn, W. L., MacDonald, G. A., and Scott, R. F., 1977, Geologic hazards (rev. second edition), Chapter 3 (Hazards from Volcanoes): New York, Springer-Verlag, 330 p.

Coch, N. K., and Ludman, A., 1991, Physical geology: New York, Macmillan, 678 p., Chapter 5 (Volcanoes).

Crandell, R., and Mullineaux, D. R., 1978, Potential hazards from future eruptions of Mount Saint Helens Volcano, Washington. U.S. Geological Survey Bulletin 1383 C, 26 p.

Crandell, R., and Nichols, D. R., 1989, Volcanic hazards at Mount Shasta, California, U.S. Geological Survey, 21 p.

Editors of Time-Life Books, 1982, *Volcano:* Alexandria, Va., Time-Life Books. (An excellent introduction in popular style, with good coverage of the 1980 Mount Saint Helens eruption, Krakatau, Mount Pelée, Kilauea.)

Findley, R., 1981, St. Helens, Mountain with a death wish: National Geographic, v. 159, no.1, p. 1–65.

McPhee, J., 1989, The control of nature, Chapter 2 (Cooling the lava): New York, Farrar Strauss Giroux, 272 p.

Miller, C. D., 1989, Potential hazards from future volcanic eruptions in California: U.S. Geological Survey Bulletin 1847, 17 p.

Parks, N., 1994, Exploring Loihi: The next Hawaiian island: Earth, v. 3, no. 5 (Sept.), p. 56–63.

Parks, N., 1994, The fragile volcano: Earth, v. 3, no. 6 (Nov.), p. 42–49.

Pendick, D., 1994, Under the volcano: Earth, v. 3, no. 3 (May), p. 34–39.

Tilling, R. I., 1989, Volcanic hazards and their mitigation: Progress and problems: Reviews of Geophysics, v. 27, p. 237–269.

Tilling, R. I., Heliker, C., and Wright, T. L., 1987, Eruptions of Hawaiian volcanoes—Past, present, and future: USGS General Interest Publication, 54 p.

Tilling, R. I., Topinka, Lyn, and Swanson, D. A., 1990, Eruptions of Mount Saint Helens: Past, present, and future (rev. edition): U.S. Geological Survey General Interest Publication, 56 p.

Wright, T. L., and Pierson, T. C., 1992, Living with volcanoes: U.S. Geological Survey Circular 1073, 57 p.

# 5

# *Earthquake Hazards*

Everything around you is moving and you become totally disoriented. The very ground beneath your feet, on which you depend for security and stability, is suddenly moving. This is an earthquake, an event that occurs with little or no warning. Earthquakes are perhaps the most frightening of all hazards, for once they start, *there is no time to escape*—you can only ride it out.

There are precursors of earthquakes, but they usually are too subtle to provide timely warning. Earthquakes cause considerable damage in rural and suburban areas, but the damage can be catastrophic in major population centers. Particularly vulnerable to the shaking ground are high-rise buildings, elevated highways, utility systems, poorly constructed dwellings common in third-world nations, and dense populations. Many of Earth's major metropolitan areas—Tokyo, San Francisco, Los Angeles, Mexico City—are in seismically active areas where "great" earthquakes have struck—and will again.

## ORIGIN OF EARTHQUAKES

The Greek word *seismos* means to quake, and **seismology** is the study of earthquakes. An **earthquake** is a shaking of the ground, usually caused by rocks rupturing under stress (Chapter 2). The rocks on opposite sides of the rupture move with respect to each other, typically distances ranging from millimeters to many meters. Energy released from the ruptured rock travels in waves, sometimes felt locally as an earthquake, but in any case measurable by sensitive instruments called seismometers.

The place within Earth where the rock breaks is the earthquake's **focus** (plural, *foci*), also called its *hypocenter*. Earthquake foci occur anywhere from the surface to about 700 kilometers (435 miles) depth. However, more than 75% of earthquakes are shallower than 60 kilometers (37 miles); seismic events occurring within this depth are called *shallow-focus earthquakes*. The point on Earth's surface directly above the focus is called the **epicenter.** Damage is usually, though not always, most severe at or near the epicenter.

**Cars crushed by the collapse of a Northridge apartment building into the adjacent parking area. (Courtesy of Peter W. Weigand, California State University, Northridge.)**

The strongest earthquakes commonly are caused by rocks rupturing in response to tectonic stresses at active plate margins, where one moving plate overrides another, grinds past another, or converges with another. Stress acting upon a body of rock builds up, being stored as strain energy (or energy of deformation). When the stress exceeds the strength of the rock, it breaks abruptly along a *fault* (see Chapter 2). The released energy radiates rapidly from the rupture point, shaking the surrounding ground as an earthquake. This explanation of earthquakes is the **elastic rebound theory,** proposed following the great San Francisco earthquake of 1906.

Smaller earthquakes occur in association with volcanic eruptions. They result from the brittle fracture of solid rock that surrounds rising magma in a volcano before it erupts (see Figures 4–29 and 4–30). Minor earthquakes also have been caused by human activities, such as the filling of a reservoir or the injection of fluids into deep fractured rocks (both are explained in a later section).

## TYPES OF EARTHQUAKES

Over 90% of earthquakes occur where tectonic plates move against one another in some manner (Figure 5–1). A different type of stress builds along each type of margin: divergent-plate margins (tensional stress), convergent-plate margins (compressional), and transform-plate margins (shear); see Figure 2–23. Because of the nature of each type of contact, seismic events occur at different focus depths and different power levels for each type.

It is interesting to compare Figure 5–1 with Figure 4–1 and note the similar global pattern of vol-

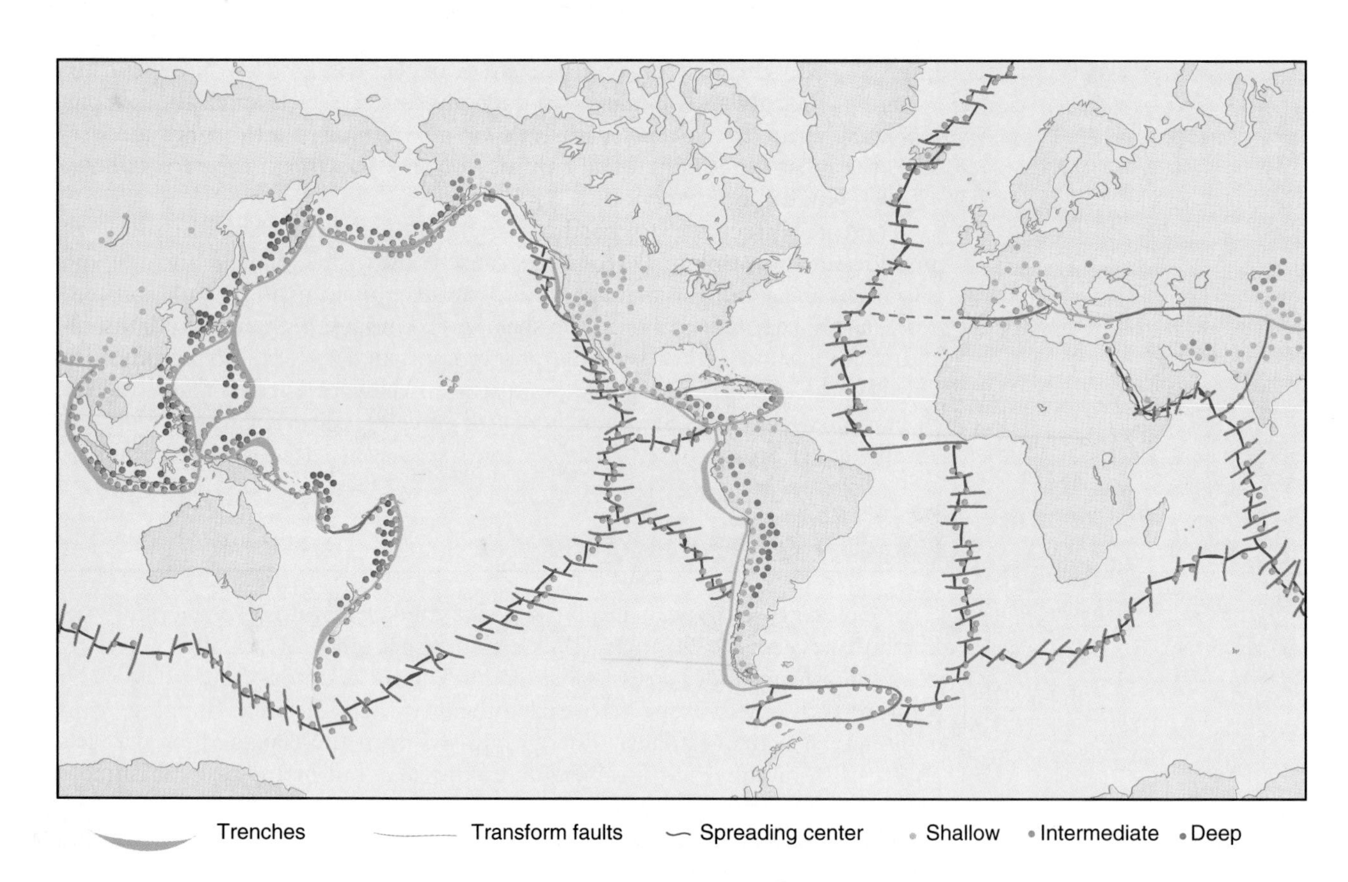


**FIGURE 5–1**
Computer-generated relief map of worldwide earthquake epicenters (red) and plate boundaries (yellow) for large earthquakes, 1980–1990. (Reprinted with permission of Macmillan College Publishing Company from *Earth's Dynamic Systems,* 6th ed., by W. Kenneth Hamblin. Copyright © 1992 by Macmillan College Publishing Company.)

canic activity and earthquake activity. Both activities outline the active margins of tectonic plates.

## Divergent-Margin Earthquakes

Wherever continental or oceanic crust is under sufficient tension to break, seismic waves are generated. Such earthquakes occur in **rift valleys** along mid-oceanic ridges such as the Mid-Atlantic ridge (see Figure 2–23), where oceanic plates are moving apart and basalt is extruding to form new oceanic crust. Rift valleys also occur *within* plates on continents, such as the huge African Rift Valley in East Africa and the Rio Grande Rift in the western United States. The origin of these continental rift zones is discussed later in this chapter.

Generally, rocks under compression are extremely strong. Subjected to shear force, they remain very strong, although less so. Under tensional force, however, rocks are comparatively weak. Thus, earthquakes at divergent (tensional) margins tend to be less powerful and shallower because rocks break more easily under tension. Rupture and faulting usually occur before great stress can build in the rocks.

## Convergent-Margin Earthquakes

Wherever an oceanic plate is subducting beneath another plate, powerful forces along their large area of contact generate continual stress that produces earthquakes. This occurs along a steeply dipping fault zone called a **Benioff zone** or subduction zone (Figure 5–2). The subducting (sinking) oceanic plate generates earthquakes that have progressively deeper foci in the direction of subduction. The *shallow focus* earthquakes are near the oceanic trench, whereas the *deep focus* earthquakes occur far beneath the island arc or continent (Figure 5–2).

Subduction-zone earthquakes are more powerful because compressive forces dominate. As explained, rocks typically are stronger under compression and thus can store greater strain energy before they rupture. Such earthquakes usually have deeper foci, reflecting the depth of the downward moving plate.

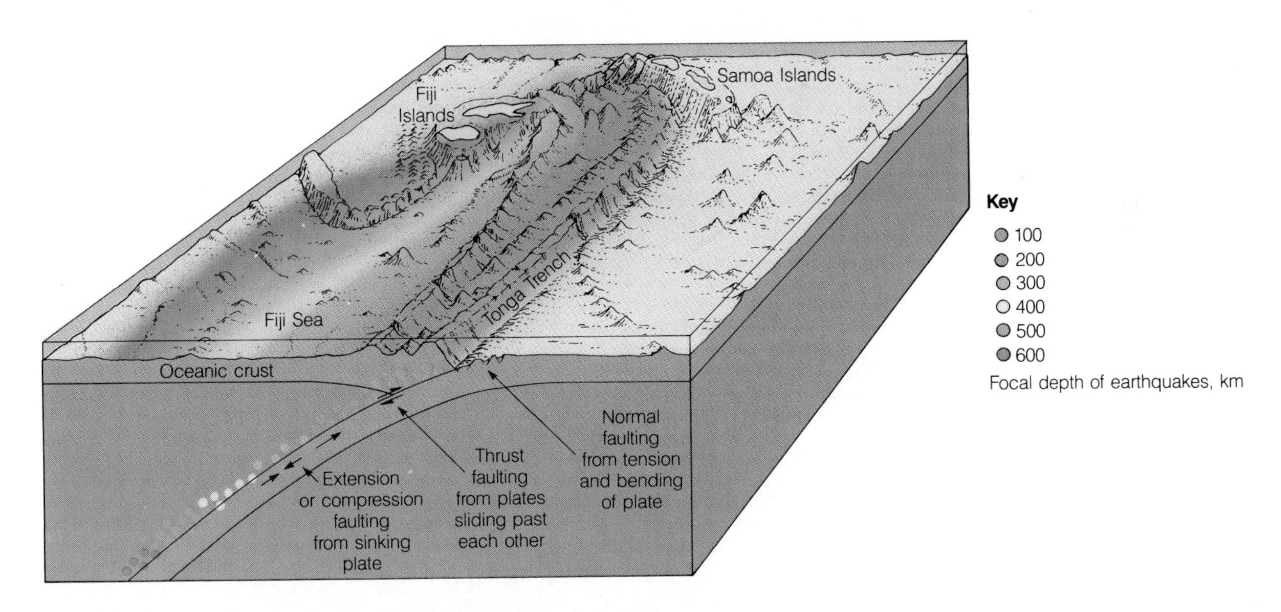


**FIGURE 5–2**
Cross section showing a Benioff zone in the Tonga region of the South Pacific. The top of the diagram shows the aerial distribution of foci, with focal depths represented by different-colored dots. The cross section on the front of the diagram shows how the seismic zone is inclined away from the trench. This seismic zone accurately marks the boundary of the sinking plate in the subduction zone. (Reprinted with permission of Macmillan College Publishing Company from *Earth's Dynamic Systems,* 6th ed., by W. Kenneth Hamblin. Copyright © 1992 by Macmillan College Publishing Company.)

Subduction causes most of the earthquakes along island arcs (Japanese, Philippine, and Aleutian).

Earthquakes in the Puget Sound area near Seattle, Washington, are believed to result from subduction of the Juan de Fuca plate beneath the North American plate (Figure 5–3). This subduction also causes the characteristically explosive volcanism in the Cascade Range (described in Chapter 4). Geologists believe that the Juan de Fuca plate and North American plate are locked together. Strain is accumulating along their boundary. U.S. Geological Survey geologists reported in 1991 that this region

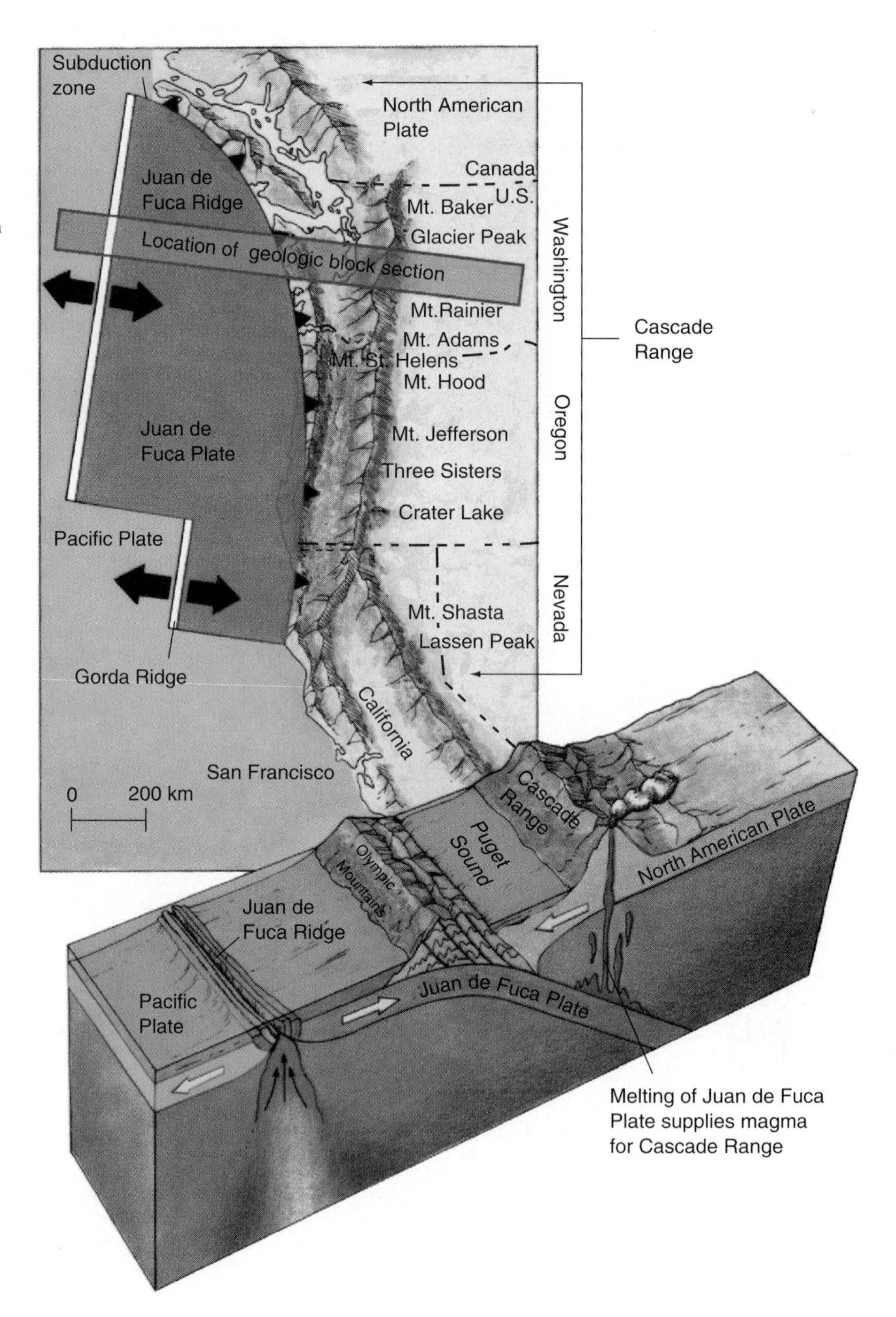


**FIGURE 5–3**
Origin of Puget Sound earthquakes and volcanism by subduction of the Juan de Fuca plate beneath the North American plate. (Reprinted with permission of Macmillan College Publishing Company from *Physical Geology* by Nicholas K. Coch and Allan Ludman. Copyright © 1991 by Macmillan College Publishing Company. Courtesy of Allan Ludman.)

is in much greater danger of a major earthquake than previously thought because of this strain accumulation. Laser-beam measurements of distances between mountaintop benchmarks over eight years reveal that the coastal mountains are being compressed or depressed and that the coastal area is rising. The increasing strain could trigger major earthquakes sometime in the future.

## Transform-Margin Earthquakes

Along transform faults, the plate edges are subject to great shear stress. The plates apparently move horizontally past each other in a bind-and-lurch pattern. Strain builds as the plates are pressured to move past one another, but friction binds them. When the strain level exceeds the friction and/or the strength of the rock, the plates abruptly lurch past one another, commonly by a few centimeters or meters, depending on conditions.

Elastic rebound theory explains how rock deforms elastically (Figure 5–4). When the shear stress exceeds the rock strength, rupture occurs and seismic waves are generated (Figure 5–4c). As the plates continue moving, elastic deformation resumes and stress again accumulates along the fault (Figure 5–4d).

This mechanism demonstrates one way in which stress is accommodated along the San Andreas fault zone in California, which separates the Pacific plate and North American plate (Figure 5–5). Some sections of the fault, as in the San Francisco and Los Angeles areas, are virtually stationary. Unfortunately, stress builds to very high levels along these *locked segments* until it eventually is released, causing a major earthquake. The longer the fault segment remains locked, the greater is the strain energy that can accumulate, and the greater is the potential for a major earthquake.

Along other portions of the fault, such as in the wine-grape region near Hollister, continuously sliding or *creeping segments* exist. This continual movement releases stress before it can build to dangerous levels. Along the creeping sections, frequent small earthquakes occur (the "window-rattlers" you often hear about in the news) rather than infrequent major ones. The consequences of locked faults and creeping faults are contrasted in the photos in Figure 5–5.

Earthquakes on transform faults tend to have shallow foci and thus can cause more damage than deep-focus events of comparable energy release. If they originate from locked sections, they can be very powerful, like the San Francisco (1906) and Loma Prieta (1989) earthquakes. Rocks generally have significant shear strength, so high levels of strain energy can be stored until the rock ruptures.

Some faults close to transform margins, such as those near a part of the San Andreas fault in California, are not strike-slip faults but thrust faults. Where transform faults make abrupt turns, such as north of Los Angeles, the plates on either side do not simply slide past each other in a shearing motion. The opposing forces near these fault bends set up zones of compression that eventually generate thrust faults. This is discussed in more detail in Box 5.1.

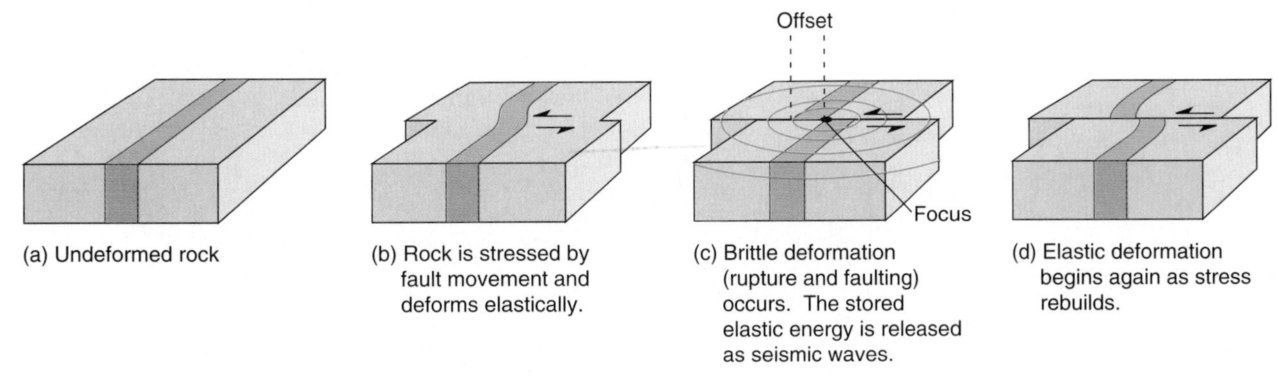


**FIGURE 5–4**
The elastic rebound theory is demonstrated by deformation and rupture along a transform fault. (Reprinted with permission of Macmillan College Publishing Company from *Physical Geology* by Nicholas K. Coch and Allan Ludman. Copyright © 1991 by Macmillan College Publishing Company. Courtesy of Allan Ludman.)

**BOX 5–1**

## THE 1994 NORTHRIDGE FAULT: A NEW TYPE OF SEISMIC HAZARD IN THE LOS ANGELES BASIN

The millions of southern Californians jolted awake at 4:31 A.M. on January 17, 1994, thought that the "Big One," a powerful earthquake on the San Andreas fault, had finally occurred. What actually happened was very different. The earthquake registered only 6.6 on the Richter scale, rather than the 7.5 plus expected for the "Big One." Additionally, it was *not* located on the San Andreas fault system but had an epicenter tens of kilometers to the west in Northridge (Figure 1). Finally, the earthquake did not result from movement on a strike-slip fault, characteristic of the San Andreas, but from displacement on a *thrust* fault (see Figure 2.19) with a focus at a depth of 15 kilometers (9 miles) below the surface.

Along most of the length of the San Andreas fault, the Pacific and North American plates slide past each other in a shearing motion (Figure 1), generating strike-slip faults (see Figure 2.19). However, the bend in the San Andreas fault line north of Los Angeles (Figure 1) changes the stress conditions in that area. The curve in the fault forces the two plates to push around each other, generating *compressional* stress where the two plates oppose each other (see Figure 2.14). These compressional forces generated thrust faults in the materials under the northern margin of the Los Angeles basin. Because they do not reach the surface, they are referred to as "blind" or "hidden" thrust faults. While they are not visible on the surface, the thrusting moves material upward, resulting in an elevation of the surface (Figure 1). Recent compilations of subsurface (drilling) records show that there are a number of thrust faults under the Los Angeles basin. These faults are not believed to be capable of generating earthquakes above 6.7 on the Richter scale. In comparison, a major 8.0 event is expected eventually on the San Andreas system.

Thrust faults of the Northridge type may be less powerful than San Andreas strike-slip faults, but they can cause *equal* destruction. This is because buildings constructed from the 1950s to the 1980s were designed to withstand lateral motion (San Andreas type), not the vertical motion associated with thrust faults. In southern California, the San Andreas fault generally runs through less developed areas, whereas the hidden thrust fault systems underlie heavily populated areas to the north and northwest of the Los Angeles basin (Figure 1). For example, the Wilshire fault, a thrust fault underlying Hollywood and Beverly Hills, has the potential to cause serious damage to some of the most expensive real estate in the world, as well as injuring or killing many inhabitants of that area.

Damage estimated at ten billion dollars was caused in the Northridge earthquake. Interstates 5 and 210 were cut by overpass failures, structural

**FIGURE 1**
Map of California showing San Andreas fault and Northridge fault.

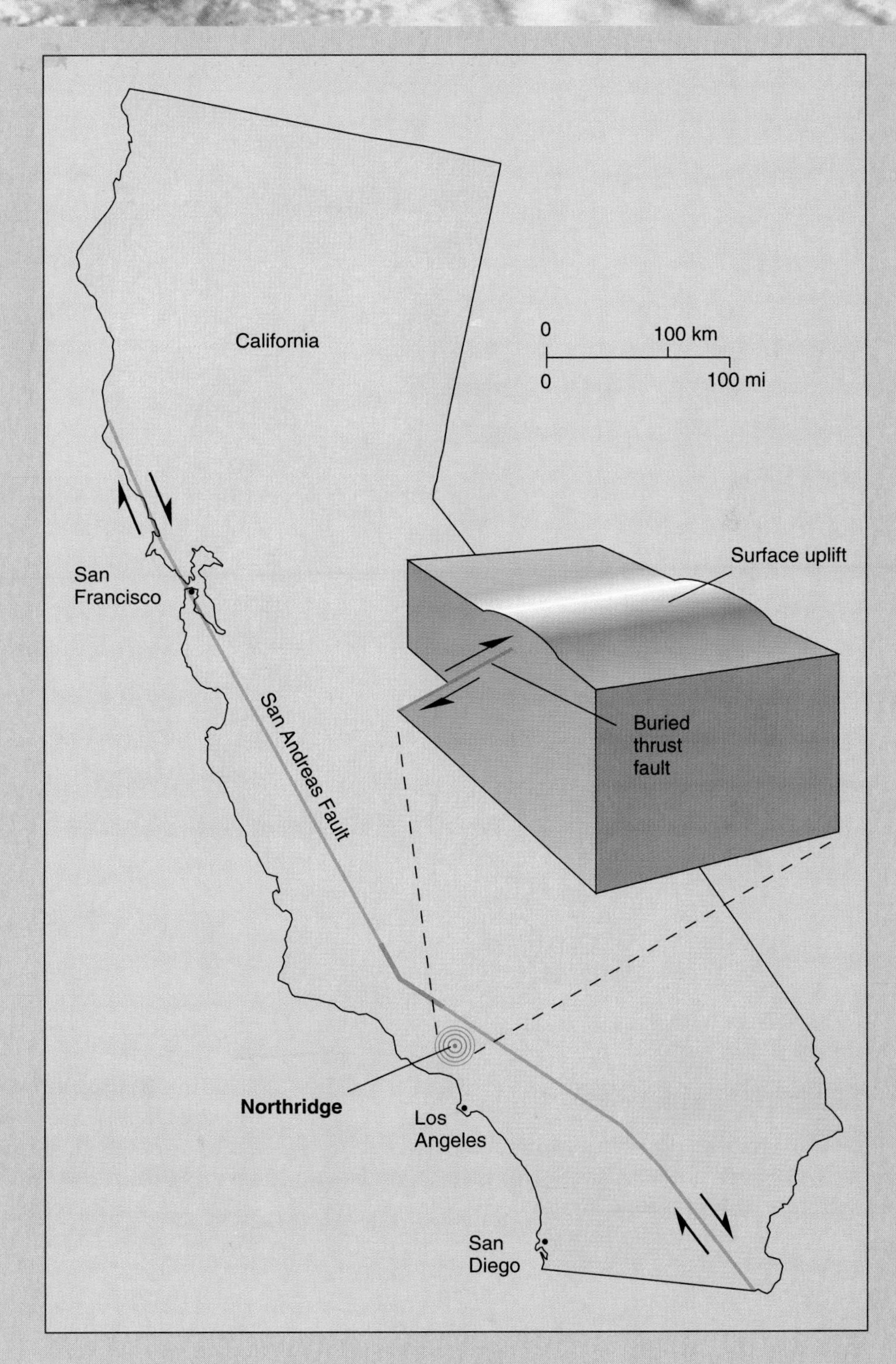

**FIGURE 2**
Structural failure in the Northridge earthquake of January 1994. (Photo by Les Stone/Sygma.)

## Intraplate Earthquakes

The geographic distribution of earthquakes shows that some are not associated with plate boundaries but occur *within* plates (Figure 5–6). These seismic events are called mid-plate or **intraplate earthquakes.**

***Basin and Range Earthquakes.*** Some intraplate earthquakes occur where continental crust is under tension. Tensional forces literally have pulled apart the North American plate in the western United States, breaking it into a series of blocks (Figure 5–7). Some blocks are lowered *basins* and others are pushed-up *ranges,* together forming a topography known as the *Basin and Range.* Continuing tension leads to recurring *vertical* movements along normal faults (Chapter 2). Contrast this displacement with the *lateral* sliding along transform faults like the San Andreas.

damage was extensive (Figure 2), broken gas lines caused explosions and fires, and homes on bluffs at Pacific Palisades and Malibu collapsed down the slopes. Communications were cut and transportation was severely disrupted around the greater Los Angeles area. The earthquake also caused a public health hazard: more than three dozen people came down with a flu-like sickness called "valley fever" (coccidioidomycosis). They inhaled dust containing the fungal spores that cause the illness. The dust had been stirred up by the earthquake, the frequent after shocks, and the cleanup process itself.

The Northridge earthquake provided a number of lessons. Knowledge from past earthquakes had led to a program to retrofit highway overpasses to make them seismically resistant. Those that had been retrofitted fared much better than those that had not. The most heavily damaged structure, the Santa Monica Freeway (I-210), had been scheduled for structural reinforcement to take place the month following the earthquake. The existence of hidden thrust faults with vertical displacements in the Los Angeles Basin requires rethinking of structural supports and some innovative engineering. Analysis of the failed structures showed that future designs must include safeguards against both horizontal *and* vertical shaking. Structures built over open parking areas proved to be particularly weak in the ground motions typical of a blind thrust fault. The upper floors collapsed into the open space, injuring and killing people on the lower floors and crushing parked cars (chapter opening photograph). In the Northridge Meadows complex alone, 16 people were killed in such a building collapse.

The discovery that the Los Angeles basin is in peril both from strong earthquakes on the San Andreas fault *and* from moderately strong earthquakes along numerous blind thrust faults has significantly increased the seismic hazard for area inhabitants. The development of building codes that address both horizontal and vertical shaking, delineation of hazardous areas, and effective disaster preparedness are even more important than ever.

One product of this basin/range faulting is the Wasatch fault zone, which extends from Utah into Idaho (Figure 5–8). Just west of the fault zone are Utah's largest urban centers—Provo, Salt Lake City, and Ogden—with a combined population of more than 1 million. Earthquakes of destructive magnitude have occurred along the fault periodically in the past 10,000 years. The fault zone has experienced repeated major earthquakes every several hundred to several thousand years.

Salt Lake City spreads westward across a basin from the Wasatch Range (Figure 5–9). The dominant Wasatch fault flanks the western edge of the Wasatch Range and is conspicuous as a scarp, about 22 meters (75 feet) high. This scarp is a vertical cliff created by repeated fault displacements that accompany significant earthquakes. No movement along the Salt Lake City segment of the Wasatch fault has been noted since 1857. However, when movement recurs along the fault, the large population could

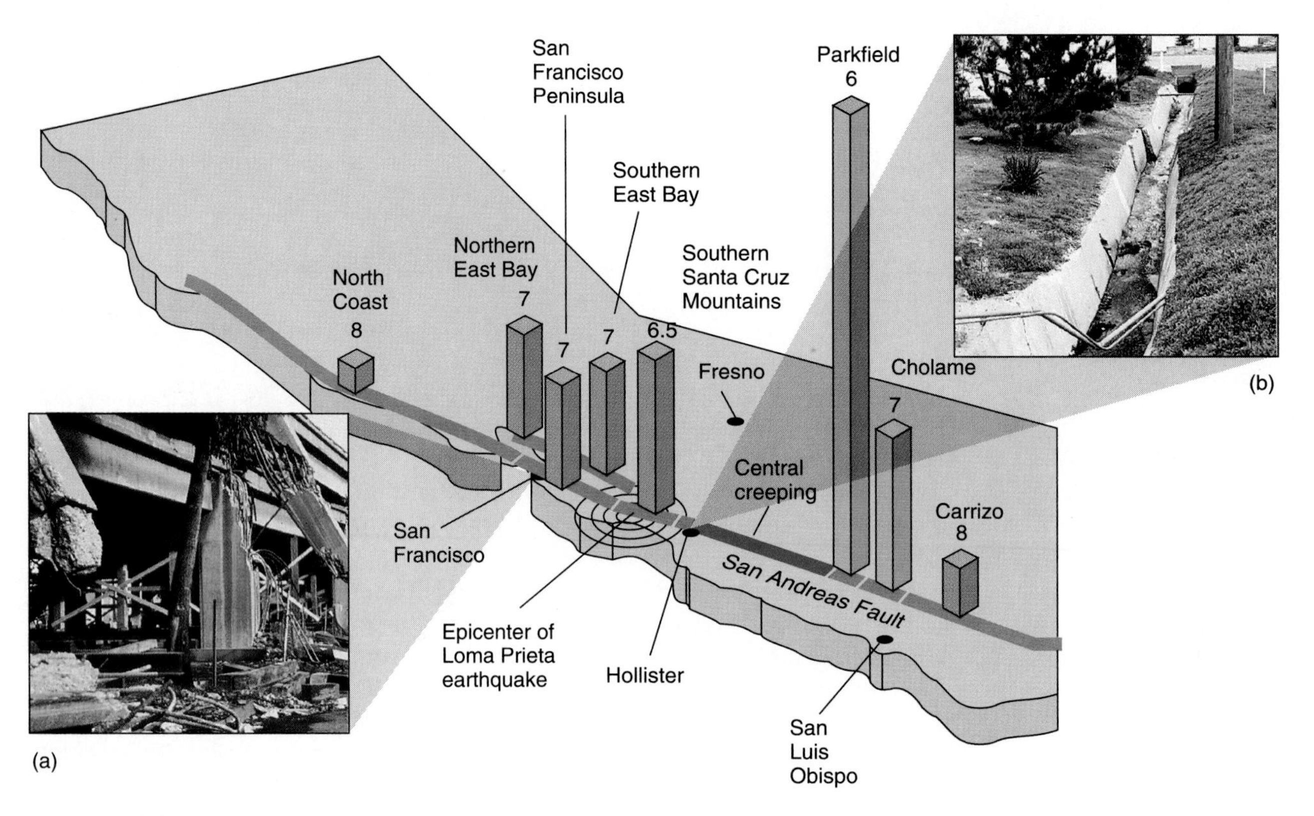


**FIGURE 5–5**
Two types of damage along California's San Andreas fault. (a) Collapsed upper deck of the Cypress Street Viaduct, San Francisco, in the Loma Prieta earthquake illustrates the consequences of a locked section releasing its energy. (b) Offset in drainage ditch in Hollister results from creep along the fault. (Diagram modified from USGS, 1989, "The Loma Prieta Earthquake": Earthquakes & Volcanoes, v. 21, no. 6, p. 221. Photo (a) by H. G. Wilshire, U.S. Geological Survey; (b) by Raymond Pestrong, San Francisco State University.)

experience loss of life and serious injury, and the modern structures could suffer major damage.

***Mid-Continent Earthquakes.*** Some intraplate earthquakes are more puzzling because they are not associated with mountains and fault scarps such as the Wasatch fault in the Basin and Range area. In fact, the greatest seismic event ever recorded in North America was the devastating 1811–1812 series of intraplate earthquakes around New Madrid, Missouri. Another intraplate earthquake, as strong as most plate-margin earthquakes in the western United States, devastated Charleston, South Carolina, in 1886. And then there are the earthquakes in New York and Boston. You probably do not think of these cities as being in earthquake country, but large intraplate earthquakes have occurred in there as well (1737 and 1884 in New York, and 1755 near Boston).

The cause of mid-continent earthquakes is not known. However, their occurrence implies that the North American plate is neither completely rigid nor uniform, and that it may be undergoing differential stress between its upper and lower portions as it migrates westward from the Mid-Atlantic ridge (Chapter 2). Seismic profiling recently has revealed some evidence for this: deep, old fault systems in the New Madrid area of Missouri (Figure 5–6).

Stresses within the plate may reactivate these old faults or cause new ruptures, triggering new intraplate earthquakes (Figure 5–10). A recurrence of such faulting could have hazardous consequences for nearby cities in Missouri (St. Louis and Springfield), Tennessee (Memphis, Chattanooga, and

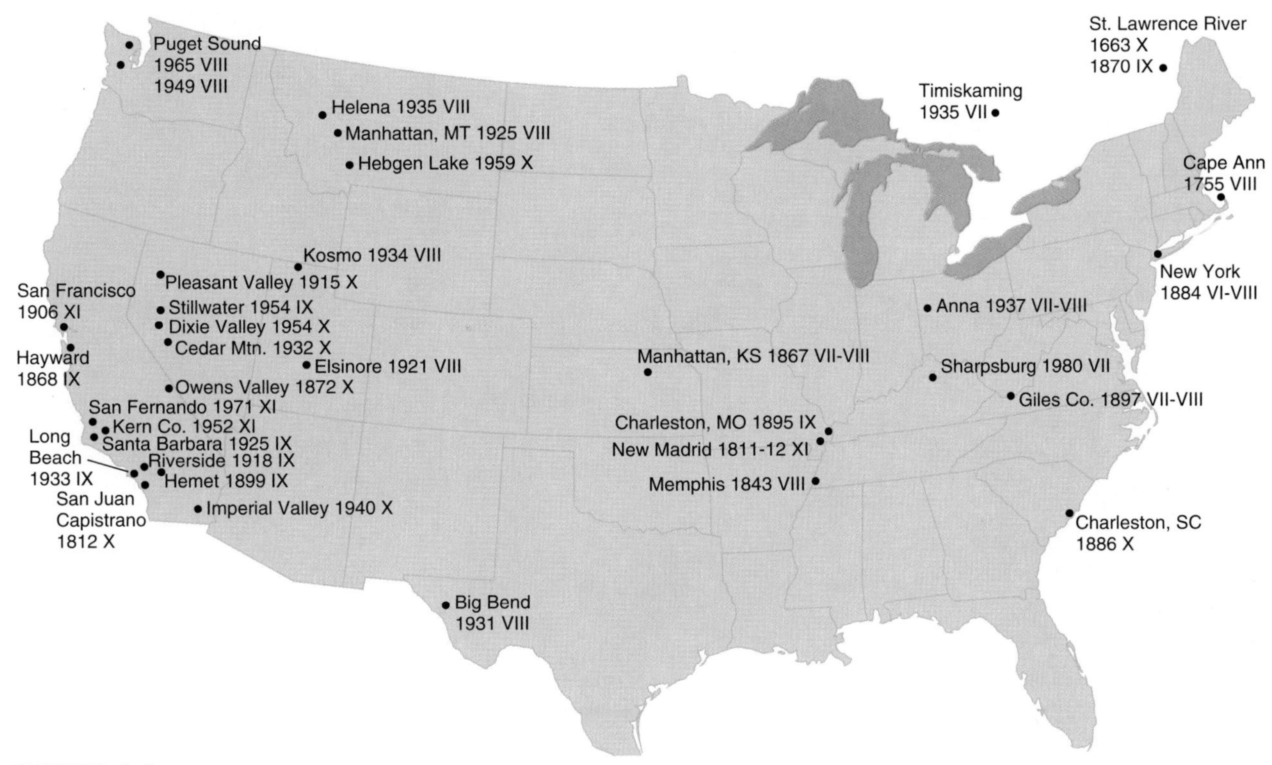


**FIGURE 5–6**
Distribution of U.S. earthquakes. The Roman numerals indicate earthquake intensity levels on the Mercalli scale. (Modified from U.S. Geological Survey, 1990, Earthquakes and Volcanoes, v. 22, no.1, p. 6–7.)

Nashville), Alabama (Birmingham), and Kentucky (Louisville), encompassing a total population of about 12 million. People in these cities are living in a major seismic zone, based on the power of the last major earthquake in 1811–1812.

The tremendous energy released in the New Madrid earthquakes created seismic waves powerful enough to ring church bells 1,800 kilometers (1,100 miles) distant *in New England!* Locally, seismic displacements *diverted the Mississippi River* and created a new lake, caused both subsidence and lifting of the land surface, triggered landslides, and formed sand boils (fast-rising groundwater that creates turbulent puddles in sand). Imagine the damage today if the same size earthquake were to occur!

## Earthquakes Triggered by Human Activity

Some earthquakes are triggered by human activities. These seismic events are less powerful than those generated by plate tectonic processes, but can cause minor damage.

Filling a large reservoir with water can induce small earthquakes. In 1935, Boulder Dam (formerly Hoover Dam) was built to impound the Colorado River, creating Lake Mead reservoir. In 1936, many moderate earthquakes (window-rattlers) started to be felt in nearby Las Vegas, Nevada. Thousands were recorded by 1973.

These earthquakes were triggered in part by the tremendous pressure exerted by the mass of water in Lake Mead. The pressure forced water into existing cracks and faults in underlying rock, allowing the faults to move and release energy, causing earthquakes. Adding this much weight to a small area of Earth's surface depresses the underlying rock, possibly creating new faults and certainly modifying existing ones. Thus, minor earthquakes—called "induced seismicity"—have occurred beneath Lake Mead for 60 years.

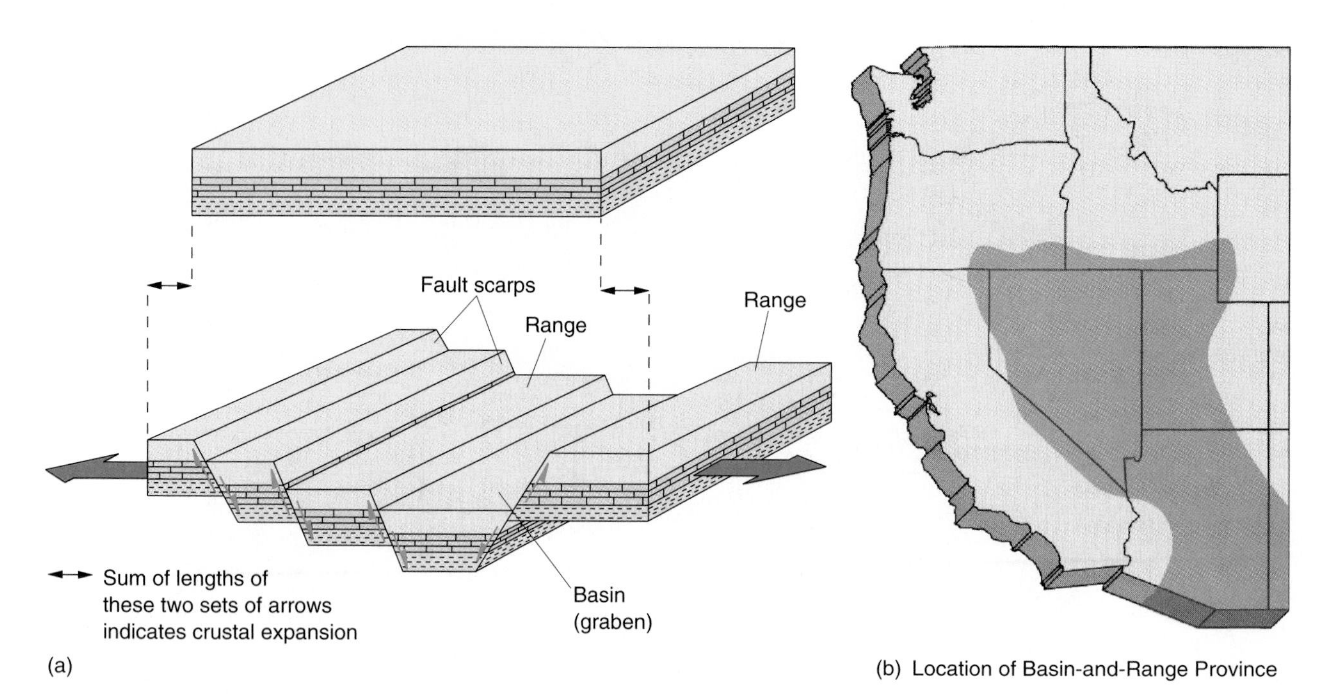


**FIGURE 5–7**
(a) Block diagrams of the horst (uplifted) and graben (downdropped) structure of the Basin and Range region of the southwestern United states. (b) Location of Basin and Range region. (Reprinted with permission of Macmillan College Publishing Company from *Physical Geology* by Nicholas K. Coch and Allan Ludman. Copyright © 1991 by Macmillan College Publishing Company. Courtesy of Allan Ludman.)

Small earthquakes near Denver, Colorado, in the mid-1960s led to recognition that other human activities can trigger earthquakes. At the Rocky Mountain Arsenal, wastewater was pumped under pressure deep underground for disposal. This injection caused movement along existing faults. When the pumping stopped, so did the earthquakes, demonstrating human influence on seismic activity. (This is described later in this chapter.)

## SEISMIC WAVES

We all know that earthquakes shake the ground, but how? Rocks rupturing from great stress generate seismic waves that radiate from the earthquake focus in all directions, similar to waves rippling in a pond when a rock is tossed in.

Not just one but four types of waves are generated when faulting triggers an earthquake (Figure 5–11). Two types of **surface waves** move along the surface of the ground (top of block in Figure 5–11). The other two types penetrate Earth and travel through it, and are called **body waves** (side of block in Figure 5–11). All four types are generated simultaneously, but they travel in different ways and at different speeds. The result is that they arrive at distant points *at different times, with different energies.* Therefore, at any particular point on Earth's surface and at any particular moment, the earthquake effects reflect the combination of wave types arriving there at that time.

### Types of Seismic Waves

The faster of the two body waves is the *primary* or **P wave.** In a P wave, the ground is shaken in a push-pull, or *compressional,* motion (Figure 5–11) in the same direction in which the wave is traveling. This type of movement is like bumping a table very hard, with predictable effect on the bowls of soup, tall beverage containers, and stacks of textbooks on

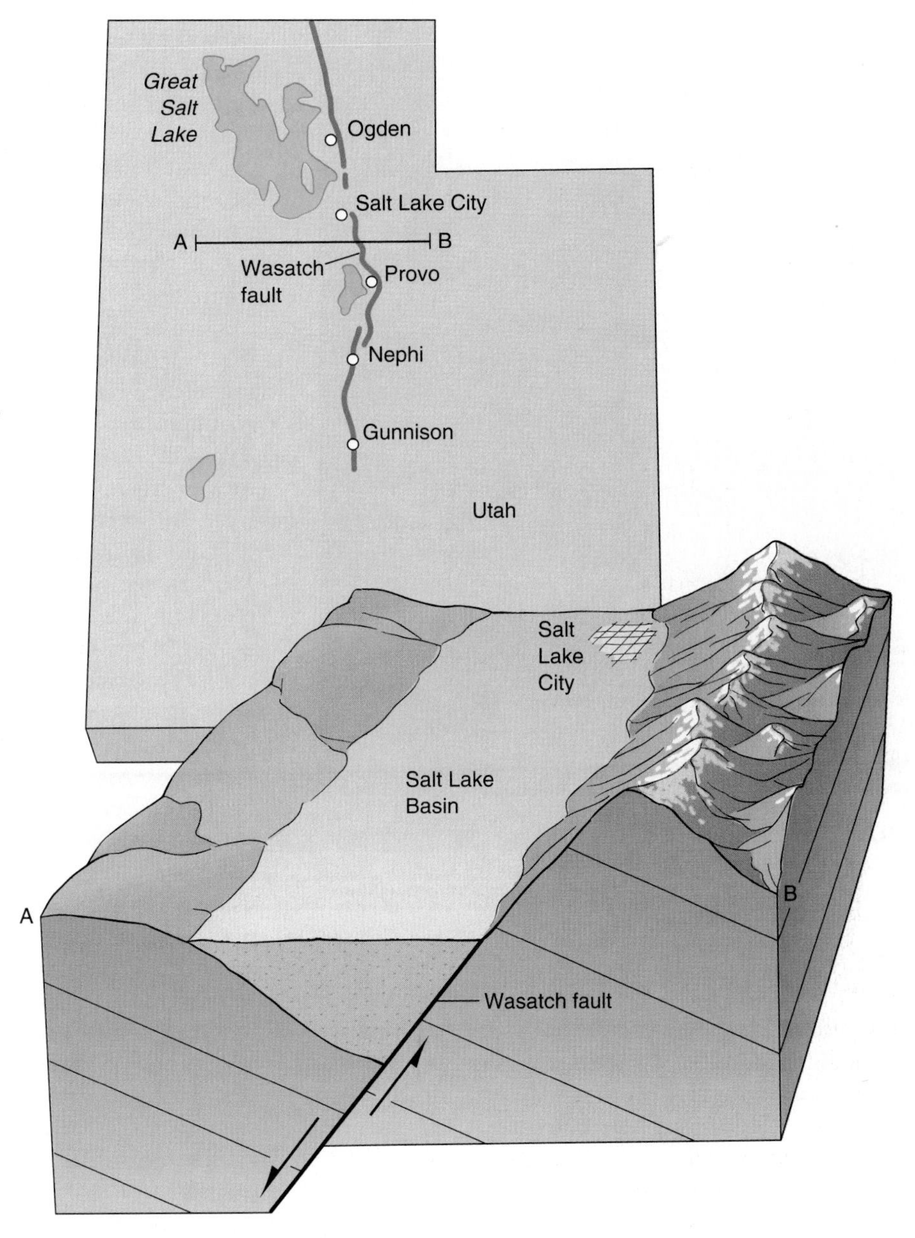


**FIGURE 5–8**
The largest urban centers in Utah are located along the 360-kilometer (220-mile) long Wasatch fault zone, Utah. Geologic evidence shows that earthquakes of Richter magnitude 7.0 or greater have occurred along the fault in the past 10,000 years. A generalized geologic section through the Wasatch Mountain-Salt Lake Basin area shows the structure of the Wasatch fault and relative movement along the fault. (After U.S. Geological Survey, 1988, Earthquakes and Volcanoes, v. 20, p. 210.)

the table—try it! Imagine this movement on a large scale in a city. P waves can travel through all media (solids, liquids, and gases).

The slower body wave is the *secondary* or **S wave.** In an S wave, the ground moves at right angles to the direction of travel of the wave through Earth. Imagine a rope tied to a tree. Flipping the rope up and down or side-to-side generates a series of waves that move along the rope. The rope moves *perpendicularly,* or *transversely,* to the direction of wave movement. Try lifting the end of a table rapidly up and down and note that this motion is less disturbing to objects resting on the table. Interestingly, S waves can travel only through solids.

The surface waves generated by an earthquake are called Love waves and Rayleigh waves (both named for English mathematicians). **Love waves** are a special kind of transverse wave. **Rayleigh**

**FIGURE 5–9**
View from Salt Lake City toward the Wasatch Range. The Wasatch fault runs along the base of the mountain front. (Photo courtesy of Utah Travel Council.)

**waves** cause an elliptical (rotating) motion in Earth materials (Figure 5–11). The combination of these waves produces a motion that is devastating to most structures. (Try lifting the end of a table up and down *and* simultaneously rotating it!).

## DETECTING, MEASURING, AND LOCATING EARTHQUAKES

Several thousand earthquake-monitoring stations operate worldwide at universities, observatories, and in government facilities. Earthquakes are detected when their seismic waves reach these stations. The detecting instrument is a **seismograph,** which times and records the incoming waves.

A seismograph has two parts, a *seismometer* to sense the ground vibrations and a recording device. The seismometer usually is anchored to bedrock so that it shakes in unison with the rock when disturbed by seismic waves (Figure 5–12). The recording device takes various forms, but to illustrate the principles simply, we will use an older design, a rotating drum covered with chart paper.

A stationary pen is suspended in isolation from ground motion, so it holds perfectly still as it touches the recording paper. When seismic waves arrive, they shake the bedrock and the seismometer frame that is attached to it, thus moving the recording paper against the stationary pen to trace the ground movement. The chart paper with the tracing is called a **seismogram.** In this manner, the instrument records the pattern of incoming waves. Typically, three separate seismometers record ground movement in three components: north-south, east-west, and vertical (Figure 5–12a, b).

The apparatus described is a traditional model. Modern seismographs are based on the same principles but use more sophisticated recording techniques. One model records electrical signals from the seismograph onto magnetic tape, providing a more flexible way to analyze the data by computer processing.

How sensitive is a seismograph? A seismometer easily can sense the rumble of a passing truck or storm waves pounding a shoreline tens of kilometers away. This interference is undesirable and usually is filtered out electronically. Underground nuclear tests thousands of kilometers away also show up clearly with a distinctive wiggle on the seismogram (in fact, this is a standard way of detecting them). Earthquakes large and small from all points on the globe show clearly. A seismogram also records countless, ever-present "microseisms" from minor fault movements and other causes too slight to be noticed by people as earthquakes.

### How Big Was That Earthquake?

Some way is needed to categorize earthquake power. This is done in two ways: *intensity* perceived by humans (Mercalli Scale) and *magnitude* measured by instruments (Richter Scale).

### Earthquake Intensity Perception

The **Mercalli Scale** describes the **intensity** of a seismic event, based on the damage caused and its *effects perceived by people* (Table 5–1). The Mercalli Scale values are qualitative, derived from field inter-

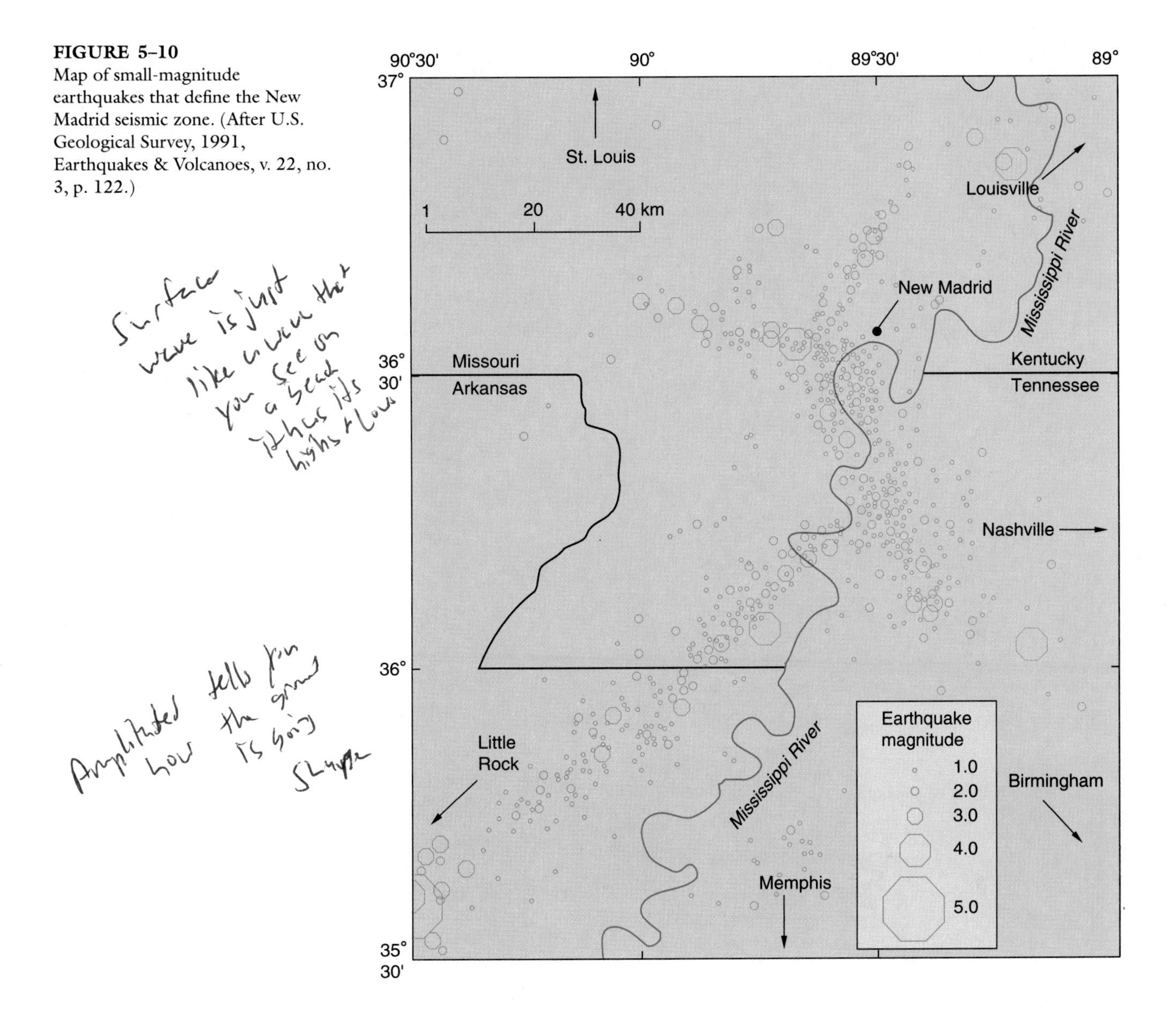


**FIGURE 5–10**
Map of small-magnitude earthquakes that define the New Madrid seismic zone. (After U.S. Geological Survey, 1991, Earthquakes & Volcanoes, v. 22, no. 3, p. 122.)

views and observations of damage, but are quite accurate and useful. Plotting Mercalli intensity values on a map gives an informative *pattern of the destructive effects* as one moves away from the epicenter.

The Mercalli intensity map of the New Madrid earthquake graphically shows the effects of this powerful seismic event outward from its epicenter in the Mississippi Valley (Figure 5–13):

- ☐ In the New Madrid area, most structures were destroyed (intensity of 9–10).
- ☐ In Cincinnati, 450 kilometers (280 miles) away, it was difficult to remain standing (intensity of 7).
- ☐ In Pittsburgh, the seismic waves awakened sleepers (intensity of 5).
- ☐ In southern New Hampshire, 1,800 kilometers (1,100 miles) away, the earthquake was felt by people on the upper floors of homes (intensity of 2–3).

## Earthquake Magnitude Measurement

A more quantitative way to gauge earthquake power involves measuring the amplitude (height) of the seismic waves on a seismogram. The **amplitude** is the vertical distance between the crest and the

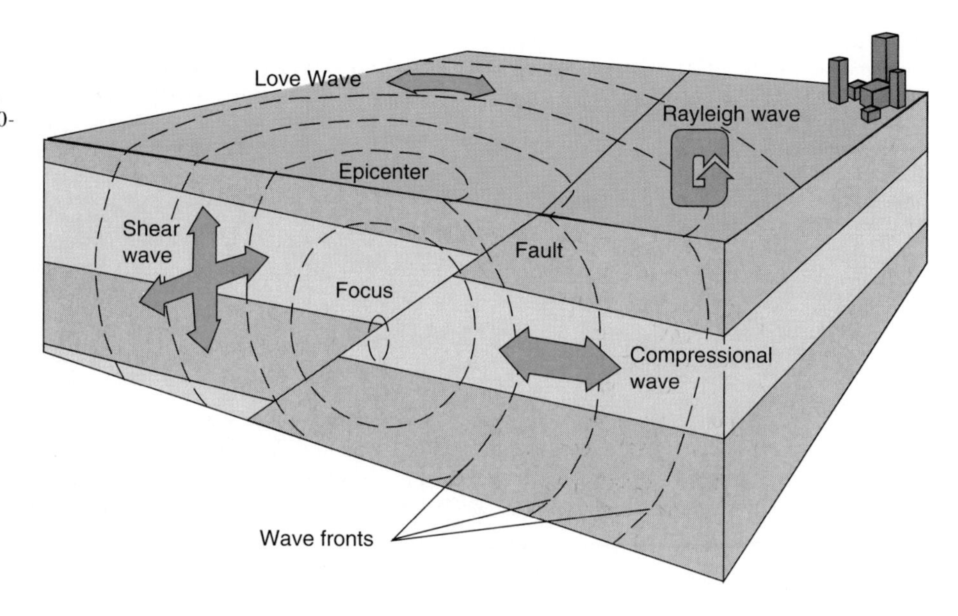


FIGURE 5–11
Ground motion associated with different seismic waves. (After U.S. Geological Survey Prof. Paper 1240-B, p. B9.)

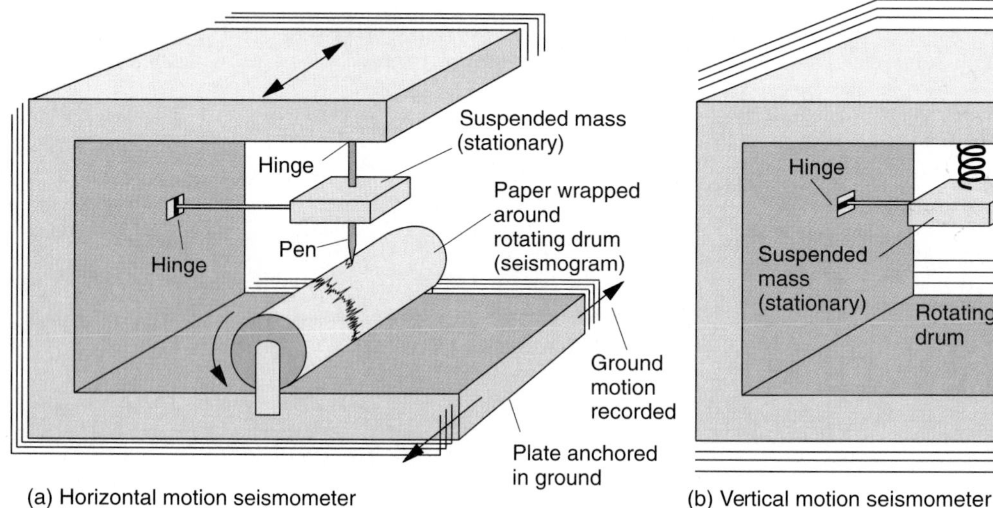


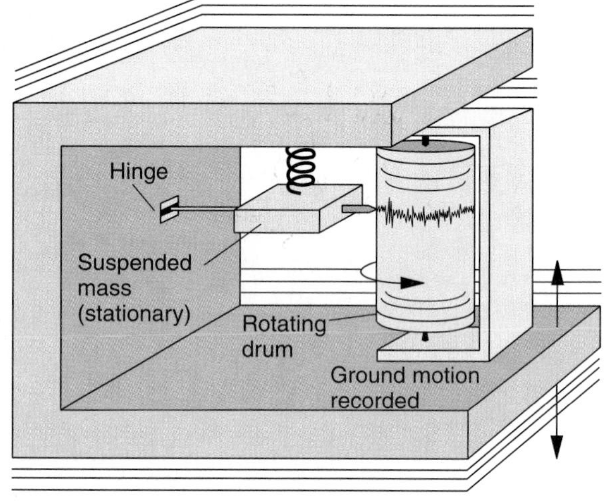


FIGURE 5–12
Seismograph. When seismic waves arrive, seismometers shake in unison with bedrock vibrations, including the seismograph drum with its chart paper. The stationary pen marks the moving paper, creating a seismogram record. (a) Horizontal motion seismometer. (b) Vertical motion seismometer. (Reprinted with permission of Macmillan College Publishing Company from *Physical Geology* by Nicholas K. Coch and Allan Ludman. Copyright © 1991 by Macmillan College Publishing Company. Courtesy of Allan Ludman.)

trough of a waveform, a measurement that generally indicates the **magnitude** of the seismic energy released (Figure 5–14). Consequently, the greater the amplitude of a wave on a seismogram, the greater the energy released by earthquake.

The magnitude is described by the numerical **Richter Scale,** named for American seismologist C. F. Richter, who developed it in 1935. Today's instruments are quite different from the one used by Richter, but all earthquake magnitude determi-

TABLE 5–1
Modified Mercalli Intensity Scale. Italian seismologist L. Mercalli developed the scale in 1902.

| Intensity Value | Description |
|---|---|
| I | Not felt. Marginal and long-period effects of large earthquakes. |
| II | Felt by persons at rest, on upper floors, or favorably placed. |
| III | Felt indoors. Hanging objects swing. Vibration like passing of light trucks. Duration estimated. May not be recognized as an earthquake. |
| IV | Hanging objects swing. Vibration like passing of heavy trucks, or sensation of a jolt like a heavy shell striking the walls. Standing motor cars rock. Windows, dishes, doors rattle. Glasses clink. Crockery clashes. In the upper range of IV, wooden walls and frames creak. |
| V | Felt outdoors; direction estimated. Sleepers wakened. Liquids disturbed, some spilled. Small unstable objects displaced or upset. Doors swing, close, open. Shutters, pictures move. Pendulum clocks stop, start, change rate. |
| VI | Felt by all. Many frightened and run outdoors. Persons walk unsteadily. Windows, dishes, glassware broken. Knickknacks, books, etc., off shelves. Pictures off walls. Furniture moved or overturned. Weak plaster and masonry D[1] cracked. Small bells ring (church, school). Trees, bushes shaken visibly, or heard to rustle. |
| VII | Difficult to stand. Noticed by drivers of motor cars. Hanging objects quiver. Furniture broken. Damage to masonry D, including cracks. Weak chimneys broken at roof line. Fall of plaster, loose bricks, stones, tiles, cornices, also unbraced parapets and architectural ornaments. Some cracks in masonry C[1]. Waves on ponds; water turbid with mud. Small slides and caving in along sand or gravel banks. Large bells ring. Concrete irrigation ditches damaged. |
| VIII | Steering of motor cars affected. Damage to masonry C; partial collapse. Some damage to masonry B[1]; none to masonry A[1]. Fall of stucco and some masonry walls. Twisting, fall of chimneys, factory stacks, monuments, towers, elevated tanks. Frame houses moved on foundations if not bolted down; loose panel walls thrown out. Decayed pilings broken off. Branches broken from trees. Changes in flow or temperature of springs and wells. Cracks in wet ground and on steep slopes. |
| IX | General panic. Masonry D destroyed; masonry C heavily damaged, sometimes with complete collapse; masonry B seriously damaged. General damage to foundations. Frame structures, if not bolted, shifted off foundations. Frames cracked. Serious damage to reservoirs. Underground pipes broken. Conspicuous cracks in ground. In alluviated areas, sand and mud ejected, earthquake fountains, sand craters. |
| X | Most masonry and frame structures destroyed with their foundations. Some well-built wooden structures and bridges destroyed. Serious damage to dams, dikes, embankments. Large landslides. Water thrown on banks of canals, rivers, lakes, etc. Sand and mud shifted horizontally on beaches and flat land. Rails bent slightly. |
| XI | Rails bent greatly. Underground pipelines completely out of service. |
| XII | Damage nearly total. Large rock masses displaced. Lines of sight and level distorted. Objects thrown into the air. |

[1]Key to 1956 revision prepared by Charles F. Richter, *Elementary Seismology,* W. H. Freeman, San Francisco, 1958, p. 137–38.
**Masonry A—**Good workmanship, mortar, and design; reinforced, especially laterally, and bound together by using steel, concrete, etc.; designed to resist lateral forces.
**Masonry B—**Good workmanship and mortar; reinforced, but not designed in detail to resist lateral forces.
**Masonry C—**Ordinary workmanship and mortar; no extreme weaknesses such as failing to tie in at corners, but neither reinforced nor designed against horizontal forces.
**Masonry D—**Weak materials, such as adobe; poor mortar; low standards of artisanship; weak horizontally.

SOURCE: After H. O. Wood and F. Neumann, "Modified Mercalli Intensity Scale of 1931," *Seismological Society of America Bulletin* 21(4): 277–88. Used with permission.

nations are still calibrated against the characteristics of his original instrument.

The Richter Scale is not linear. If one earthquake shows twice the amplitude of another on a seismogram, that does *not* mean the earthquake was twice as strong. It actually was *many times* stronger. Each whole unit on the Richter Scale represents a *ten-fold* increase in wave amplitude and about a *thirty-two-fold* increase in the energy released by an earthquake. This can be confusing, but it is the best way to scientifically compare earthquake magnitudes.

You can appreciate the difference in magnitudes by comparing an earthquake of magnitude 5.0, which causes minimal damage, to one of magnitude

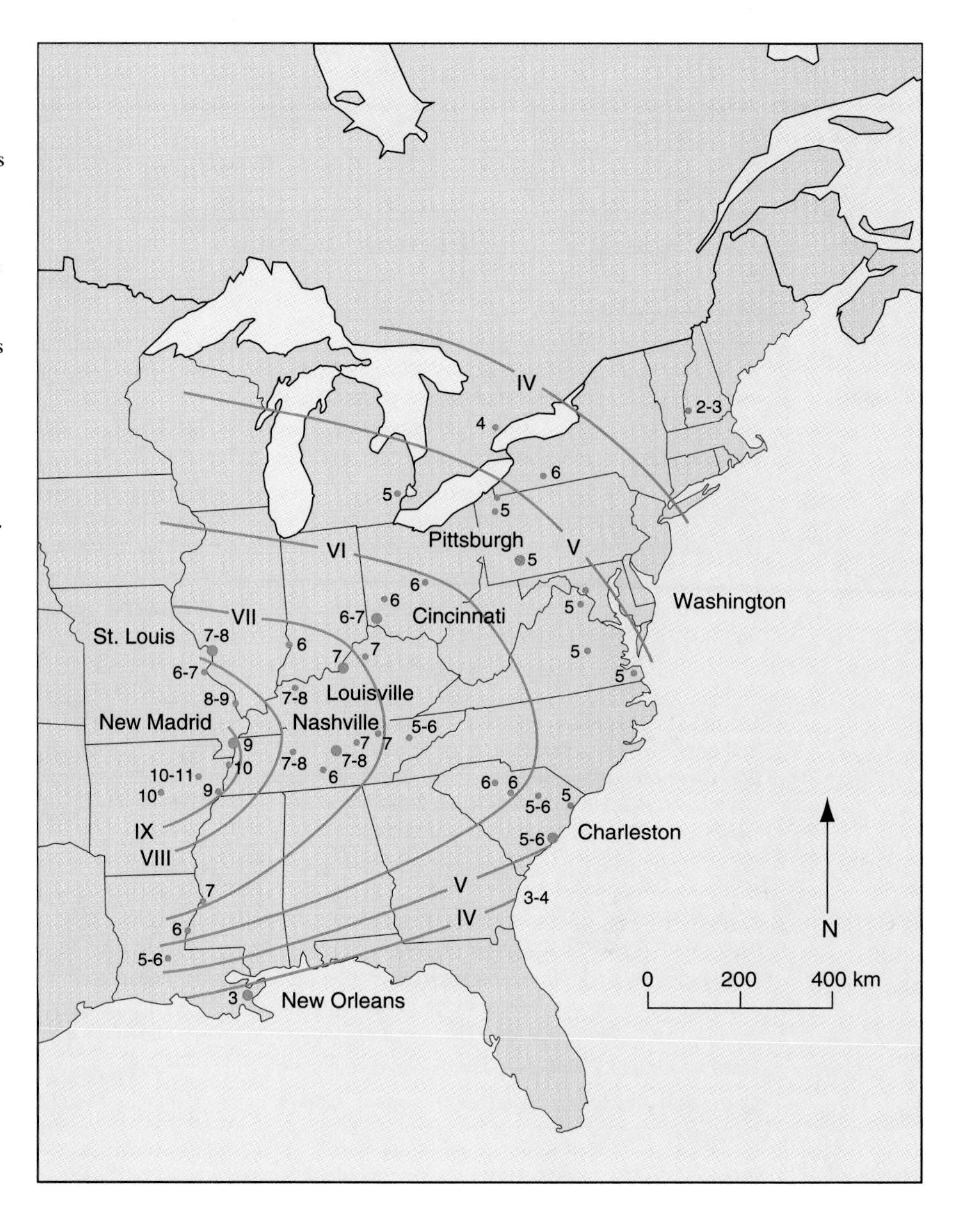


**FIGURE 5–13**
Mercalli intensity map of the New Madrid, Missouri, earthquake of December 1811. Notice that although the powerful seismic waves radiated in all directions, the intensity estimations are mostly east of the Mississippi. The reason is simple: In 1811, most people to the west were Native Americans, who relied on oral heritage rather than written records to recall events. This map was constructed from accounts in eastern newspapers, diaries, and government reports. (After Nuttli, 1973, "The Mississippi Valley Earthquakes of 1811 and 1812: Intensities, Ground Motion, and Magnitude," *Seismological Society of America Bulletin* 63(1): 227–248. Reprinted with permission of Macmillan College Publishing Company from *Physical Geology* by Nicholas K. Coch and Allan Ludman. Copyright © 1991 by Macmillan College Publishing Company. Courtesy of Allan Ludman.)

7.0, which causes great damage. The 7.0 earthquake has 100 times (10 × 10) the amplitude and 1,024 times (32 × 32) the energy released by the magnitude 5.0 earthquake.

Fortunately, the average yearly frequency of earthquakes worldwide decreases with their magnitude. More than 20,000 tremors of magnitude 1–3 occur each year, whereas shocks of magnitude 5–6 number about 4,000. There are about 300 magnitude 6–7 earthquakes and only a few earthquakes of magnitude 7.0 or greater annually.

## Locating Earthquakes

The key to locating an earthquake's epicenter is *the differences in arrival times of P and S waves.* The faster P wave arrives first at a recording station and is followed, after a **lag time,** by the S wave. This difference is proportional to the *distance* from the epicenter. Data from at least three seismic stations are required. Seismologists measure the difference in P wave and S wave arrival times at each of the three stations, and these data are used to determine

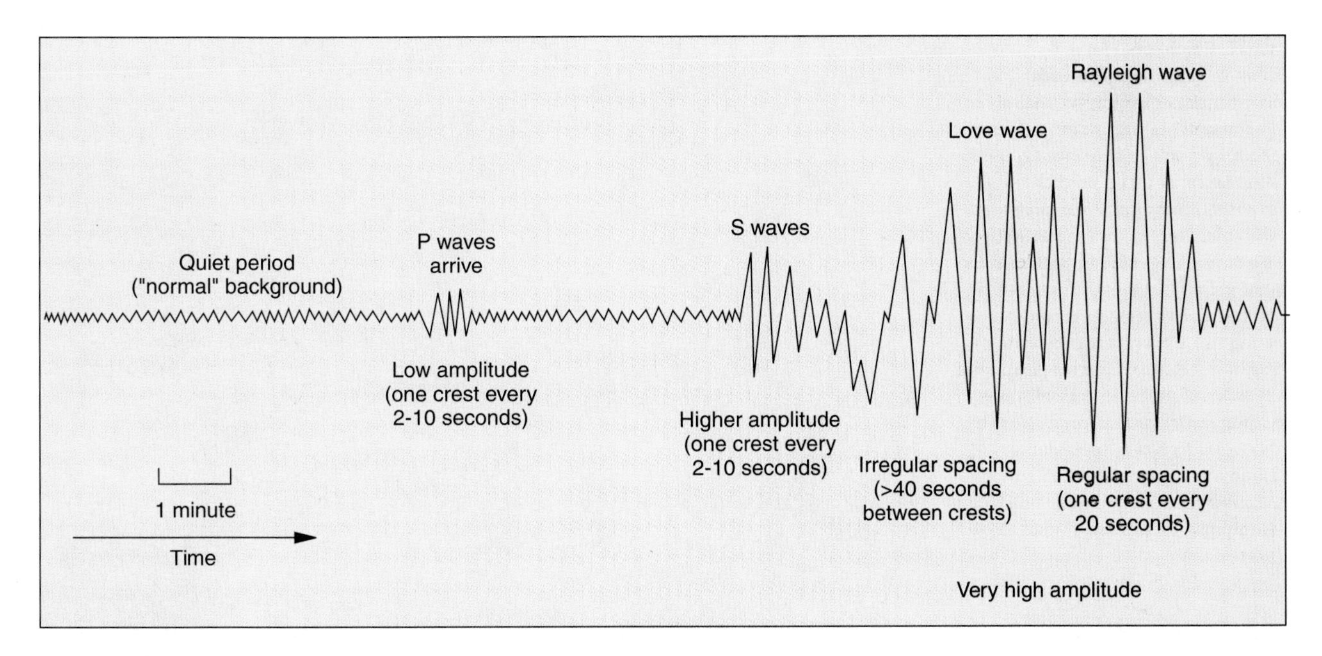


**FIGURE 5–14**
Simplified seismogram showing the distinctive characteristics of each of the four types of seismic waves. (Reprinted with permission of Macmillan College Publishing Company from *Physical Geology* by Nicholas K. Coch and Allan Ludman. Copyright © 1991 by Macmillan College Publishing Company. Courtesy of Allan Ludman.)

the epicenter (Figure 5–15). Data from at least four stations are required to determine the location of the focus.

The lag time increases with distance between the epicenter and the recording station. A familiar analogy helps to explain this: the light of a lightning flash travels 300,000,000 meters per second (984,000,000 feet/second), but the associated thunder travels at only about 330 meters per second (1,100 feet/second). Thus, to a distant observer, the sound of thunder lags way behind the lightning flash by many seconds. Similarly, the slower-moving seismic S wave lags behind the faster P wave. In both cases, the farther from the lightning (or earthquake epicenter), the greater is the lag time between the flash and the thunder (or P wave and S wave).

The time lag between seismic wave arrivals indicates the distance, *but not the direction,* from the recording station to the epicenter. The direction of the epicenter is easily determined by using seismograms from three or more stations. Figure 5–15 shows three seismic stations, each with a circle that has a radius equal to the distance to the epicenter. The epicenter is located within the area where the three circles intersect.

In practice, records from dozens of seismic stations are used to precisely locate an epicenter. In fact, data from more than three stations are required to determine the depth of the earthquake's focus. The graphical method illustrated here worked well for decades, but, as in all areas of geology, it has given way to sophisticated computer techniques for fast, precise calculation of earthquake epicenters and foci.

## Checking Earth's Pulse

In Boulder, Colorado, the National Earthquake Information Center (NEIC) monitors earthquake activity 24 hours a day. Seismic stations placed strategically around the United States transmit earthquake data to a satellite, which then retransmits the data to NEIC. The system is funded by the Nuclear Regulatory Commission and the U.S. Geological survey.

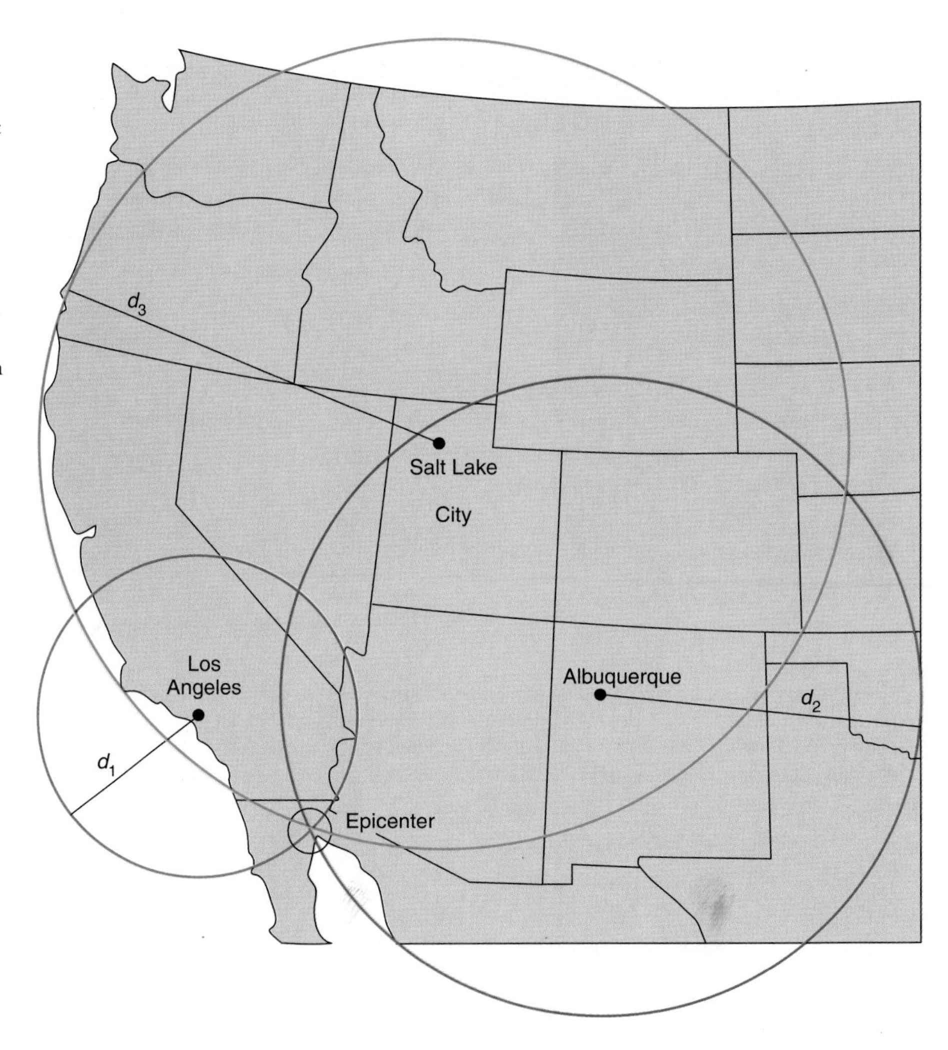


**FIGURE 5–15**
Map showing location of an earthquake epicenter by analysis of P–S arrival lag times at three seismic recording stations. Although this illustration shows a perfect intersection of the three circles at the epicenter, in reality the method is less precise, yielding a small area that includes the epicenter. The precise epicenter is pinned down by using more than three recording stations. (Reprinted with permission of Macmillan College Publishing Company from *Physical Geology* by Nicholas K. Coch and Allan Ludman. Copyright © 1991 by Macmillan College Publishing Company. Courtesy of Allan Ludman.)

## EARTHQUAKE DAMAGE SUSCEPTIBILITY

The structural damage and loss of life from an earthquake obviously vary with its magnitude. Beyond that, the essential factors in damage susceptibility are:

- surface geology (the nature of the rocks and sediments in the area);
- integrity of structures (whether they have been reinforced to withstand earthquake ground motion);
- population, building density, and time of day; and
- integrity of utilities.

### Surface Geology

Any building (dwelling, shopping mall, high-rise office complex) with its foundation on unfractured bedrock is more stable than a building built on sediment, or on fractured or altered rock. The reason is that solid rock compresses very little, whereas sediment tends to become more compact under pressure and to subside in doing so.

In general, when buildings must be built on sediment, well-drained sand and gravel are best. Less stable foundations are water-rich, fine clayey sediments; sandy fill that covers water-rich sediments; or landfill material such as trash and garbage. Thus, in a severe earthquake, buildings survive best if built on bedrock and least if built on water-rich sediment.

Structural stability also depends on how underlying materials affect the seismic waves passing through them. Studies of earthquakes in California, Japan, and Mexico show that the effects of seismic waves can be *increased* by thick sequences of sediment, causing buildings to shake more violently. An example of this wave amplification occurred in the Mexico City earthquake of September 19, 1985. The earthquake's epicenter actually was hundreds of kilometers to the west, but Mexico City was heavily damaged because the effects of seismic waves were increased as they passed through ancient lake sediments, 50 meters (164 feet) thick, which underlie the city (Figure 5–16).

The Loma Prieta, California, earthquake of October 17, 1989, also demonstrated how amplification increases damage. San Francisco's Marina District is built on a fill above a marshy area—in other words, water-rich sediment, the least-stable building surface. The Marina District is about 80 kilometers (50 miles) from the Loma Prieta epicenter, yet it sustained significant damage compared to more stable areas closer to the epicenter (Figure 5–17). The damage resulted when the underlying soggy fill became unstable and was sheared by the seismic waves. Ironically, this area had been filled in part with debris from the 1906 San Francisco earthquake! Much of this poorly emplaced material failed during the Loma Prieta Earthquake.

Another example of intensification of earthquake vibration in areas underlain by wet, fine sediment (bay mud) during the Loma Prieta earthquake was the collapse of a portion of Interstate 880 (Figure 5–18). Seismic waves passing through bedrock had a low amplitude, whereas those passing through the muds in which the expressway was anchored had a much higher amplitude (see seismograms in Figure 5–18). The intensified shaking in the area collapsed a one-mile segment of the interstate, causing two-thirds of the 62 deaths attributed to the earthquake. Passengers were crushed in their cars as the upper deck of the freeway collapsed onto the lower deck (Figure 5–19). Subsequent studies also revealed inadequate reinforcing of the freeway columns.

## Integrity of Structures

How a building is constructed is important to its stability in an earthquake. Brickwork fails easily because the cement between the bricks is weaker than the bricks (you can see this in the stairstep-crack patterns common in brick buildings that have settled). Properly built and braced wood frame structures are more flexible and thus more elastic. They may flex without breaking in a earthquake. Ornamental decorations on buildings, such as overhanging cornices and stone facing panels, are hazardous

**FIGURE 5–16**
Damage resulting from the Mexico City earthquake of September 19, 1985. (Photo by D. Aguilar, U.S. Geological Survey.)

**FIGURE 5–17**
Damage in the Marina district of San Francisco, resulting from the Loma Prieta earthquake. (Photo by C. E. Meyer, U.S. Geological Survey.)

because they are easily shaken loose and can fall on pedestrians below.

Seismic waves act like low-frequency sound waves, passing through the ground and causing it to vibrate at a certain rate (waves per second). Earthquake waves have much slower rates than sound waves. The number of waves passing a point per second is the **period** of the earthquake.

Every object has its own natural vibration period, including buildings. Buildings vibrate like a super-low-frequency tuning fork, each with its own period, depending on size and height. If the period of the earthquake ground waves happens to match the period of a building, the seismic waves make the building vibrate and, if intense enough, literally may shake it apart. Even if a well-designed building remains intact, its windows will shatter, creating a broken-glass hazard for people in the building and on the streets below.

Innovative designs are being developed by Japanese and American engineers to minimize the vibration effects on buildings. These *active control*

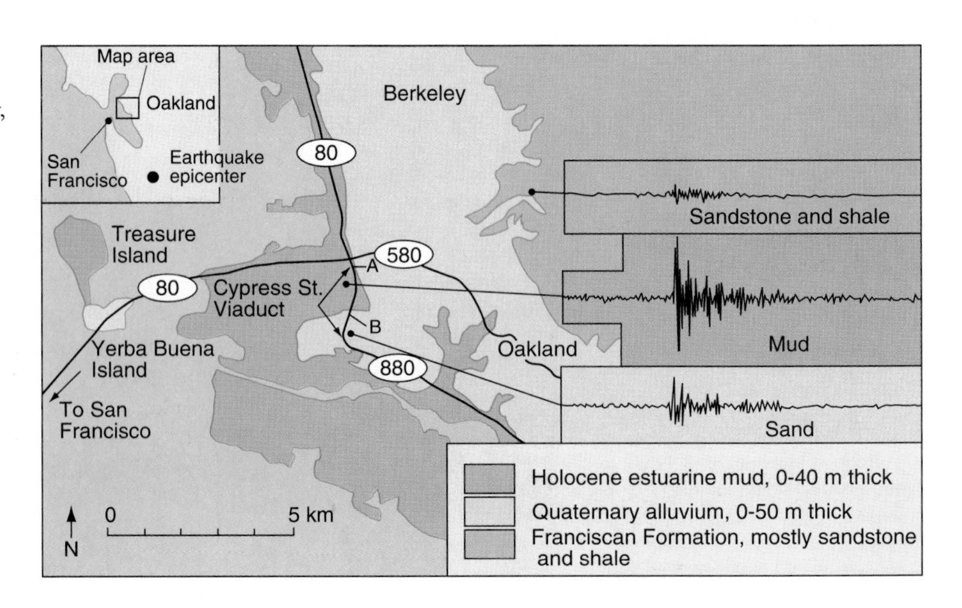


**FIGURE 5–18**
Map of the eastern San Francisco Bay region showing surface geology, location of Cypress Street Viaduct, and seismograms revealing the relative amplitude of waves passing through different material. (U.S. Geological Survey Circular 1079, "USGS Earthquake Reduction Program," Fig. 1, p. 7)]

**FIGURE 5–19**
The collapsed Cypress Street Viaduct of Interstate 880 in San Francisco during the Loma Prieta earthquake. (Photo by H. G. Wilshire, U.S. Geological Survey.)

*systems* move structural elements in the building to counteract motion detected by sensors. One system uses a movable weight atop the building. If sensors detect the building starting to vibrate, computers calculate the counterforce needed to reduce the vibration. A hydraulic system then moves the weight to the best position to dampen the vibration, much like balancing a vibrating wheel on a car by adding lead weights.

Another method uses movable internal bracing. If sensors detect building vibration, hydraulic devices shorten or lengthen the bracing sections to compensate for building motion. If these systems prove effective, they also may protect buildings from motion induced by high winds, such as during a hurricane.

## Population, Building Density, and Time of Day

The more people and the more buildings, the greater is the potential for death, injury, and structural damage in an earthquake. In 1981, the Federal Emergency Management Agency (FEMA) estimated losses in future major earthquakes along the San Andreas fault. Maximum death and injury would occur in all cases if the earthquake hit during daylight hours when most people are on the road, at work, or at school.

For example, FEMA describes the results of a *hypothetical* earthquake occurring on the Newport-Ingleside fault segment, southeast of Los Angeles, during the start of evening rush hour. Results: 23,000 dead, 91,000 requiring hospitalization, nearly one million less severely injured, and total losses to structures and contents approaching $69 billion.

Although such a seismic event has a *low probability of occurring in any given year, it eventually will happen* when sufficient stress builds along this locked section of fault. These consequences are so staggering that preparedness planning and implementation of disaster-mitigation measures *must* be taken far more seriously than they are now.

## Integrity of Utilities

Maintenance of electricity, gas, and water are essential for disaster recovery. However, the underground and surface conduits that carry these utilities are easily disrupted in an earthquake. The problem is not only the loss of the utilities, but the *secondary* damage it may cause. For example, electric sparks from downed power lines and open gas pipes pose explosion, fire, and electrocution hazards.

The 1906 San Francisco earthquake claimed more than 500 lives and did considerable structural damage. However, most of the damage resulted not from the direct shaking by seismic waves or the collapse of structures, but from *fires* that started from snapped gas lines and overturned wood and coal stoves. The problem was compounded when the water mains broke, making fire fighting nearly impossible. Whole sections of the city had to be blown up to create a fire break (an area that contains little that can burn, thus halting the spread of the fire).

Certain types of utilities did not even exist in 1906, such as nuclear and oil-fired power plants, high-voltage transmission lines, major water aqueducts, and high-pressure pipelines carrying natural gas, gasoline, and oil. Today we have these additional potential earthquake casualties, plus a far greater density of all utilities. These two factors make the potential catastrophe of a "big one" in San Francisco and Los Angeles truly frightening to contemplate.

## Aftershocks and the Reliability of Evacuation Routes

**Aftershocks** are smaller seismic events that persist for days or weeks following the main event. The fault movement that produces the large shock typically leaves areas of unrelieved stress in the rocks. Subsequent failures in these areas produce the aftershocks. Although they generally are lower in magnitude than the main event, they can produce significant damage because they repeatedly shake already weakened structures. Following the Loma Prieta earthquake of October 17, 1989, strong aftershocks continued for several days (Figure 5–20). In only three weeks, 4,760 aftershocks had been recorded in the area.

Because of aftershocks, many people leave an area that has experienced an earthquake. Reliable travel routes are essential to evacuate people, to treat the injured, to bring in fire-fighting equipment, and to bring relief supplies to the stricken area.

Reliable evacuation routes are especially important in a seismically high-risk city such as San Francisco. Its location on a peninsula bounded by San Francisco Bay and the Pacific Ocean limits access, and these routes become especially critical. Although the Loma Prieta earthquake epicenter was 80 kilometers (50 miles) south of San Francisco, it caused severe damage to the transportation system in the Bay region. Ground vibration collapsed a 15-meter (50-foot) section of the Bay Bridge over San Francisco Bay, along with a 2.5-kilometer (1.5-mile) section of the Cypress Street Viaduct of the Nimitz Freeway (Interstate 880) in Oakland, across the bay from San Francisco.

An important lesson from this earthquake is that distance from the epicenter may be less important than the ***nature of the foundation and method of construction.*** Better-built structures on stable ground closer to the epicenter fared much better than poorly built structures on unstable ground farther away!

**FIGURE 5–20**
Plot of aftershocks following the Loma Prieta earthquake.

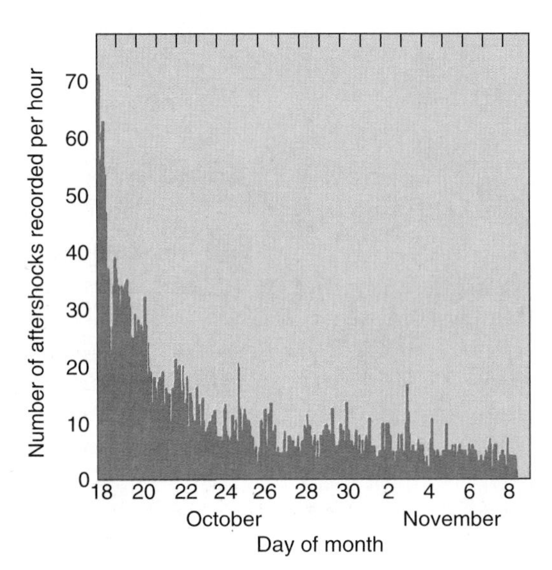


Loma Prieta aftershocks

| Magnitude | Number | Effect |
|---|---|---|
| 5 | 2 | Damaging |
| 4 | 20 | Strong |
| 3 | 65 | Perceptible |
| 2 | 384 | Not felt |
| 1 | 1,855 | Not felt |
| <1 | 2,434 | Not felt |
| Total | 4,760 | |

In our concern for the integrity of the bridges, tunnels, elevated freeways, and subways, we must not overlook the *connecting routes* to those structures. A well-designed bridge is useless if the roads leading to it are damaged by subsidence, cracking, or faulting, or if they are blocked by landslides. Consider the chaos during the aftershock period if the approaches to evacuation routes are made impassable by earthquake damage.

## EARTHQUAKE DAMAGE

Earthquake damage is of two kinds. *Primary* damage results directly from ground movement. *Secondary damage,* also called *collateral damage,* results when the ground movements cause other disruptions, such as broken gas lines, fires, or changed soil characteristics that affect the stability of building foundations.

### Buildings—Swaying and Pancaking

As seismic waves pass through the ground they set the buildings above into different kinds of movement (Figure 5–11). Exactly how a building will react to ground movements is a function of its construction, height, size, underlying strata, and distance from the epicenter.

The intensity of building vibration increases with height. Closely spaced buildings may sway so much that their tops collide. Internal damage may result when interior furnishings smash into partitions, walls, and windows. During an earthquake, in high-rise buildings, poured concrete floors can separate from their corner fastenings and fall, floor by floor, onto each other in a process called **pancaking** (Figure 5–21).

Pancaking caused serious damage to buildings in the Mexico City earthquake of September 15, 1985. Virtually all of Mexico City is built on ancient lake beds. As noted, these form an unstable building foundation because the thick sediments amplify ground motion. However, many older buildings survived where modern office buildings did not, suggesting that the problem lay in the construction methods, the quality of workmanship, or the degree of adherence to building codes in newer buildings.

### Fire and Explosion from Pipelines and Storage Tanks

Fire and explosion from ruptured gas lines have been discussed, but another fire hazard is rupture of high-pressure fuel transport lines. These can release great volumes of combustible material such as natural gas or aviation fuel very quickly. Chemical and

FIGURE 5–21
Pancaking. The upper floors of this building pancaked during the Mexico City earthquake of September 19, 1985. (Photo by M. Celebi, U.S. Geological Survey.)

fuel storage tanks often are located near waterways, built on fill laid over water-rich marine or bay sediments. Rupture of pipelines in such unstable soil will contaminate underground water and may contaminate nearby waterways.

### Shearing and Subsidence of Sand Fills

When surface seismic waves disturb (shear) a sand-filled area, the surface may subside. This results from a rearrangement of the sand particles. The arrangement of particles in a deposit is called the **packing.** A fill is created by dumping sand and bulldozing it over an area. When sediments are dumped, they have a very open **cubic packing** with a high volume of pore space (Figure 5–22a). However, when sand is deposited naturally by wind or flowing water, the particles assume a tighter **rhombohedral packing,** with a lower volume of pore space (Figure 5–22b).

Thus, when a sandy fill (cubic packing) is later shaken by seismic waves, the grains reorient themselves into a more compact rhombohedral arrangement. This results in decreased pore space, expulsion of part of the water in the pores, compaction of the layer, and subsidence of the overlying land and any structures on it.

You can test this phenomenon by partially filling a jar with any granular material—such as marbles or rice. On the jar, mark the upper surface of the fill. Now shake the jar up, down, and sideways to simulate the shearing caused by the four types of seismic waves (Figure 5–11). Note the movement of the particles and the subsidence of the original surface. You can see that any structure built on loosely packed sandy fill will sustain significant damage when the supporting surface drops.

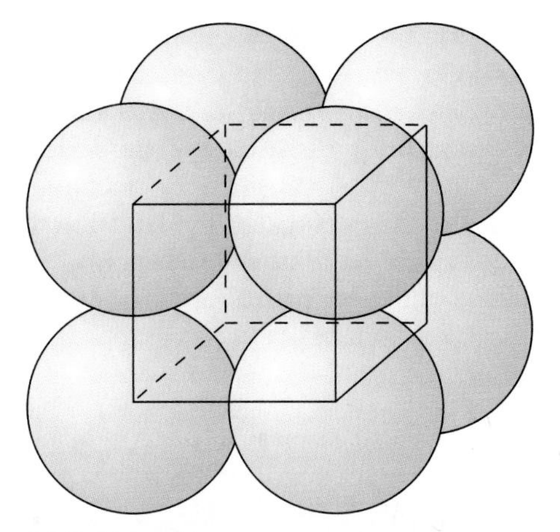

(a) Cubic packing

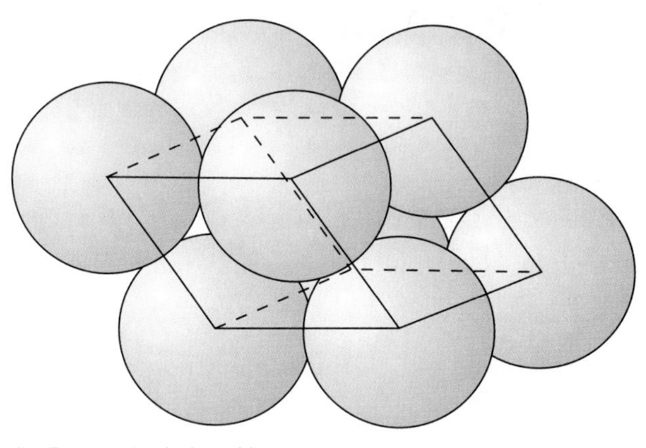

(b) Rhombohedral packing

**FIGURE 5–22**
Packing types in sediments. (a) Cubic packing. (b) Rhombohedral packing. (Reprinted with permission of Macmillan College Publishing Company from *Physical Geology* by Nicholas K. Coch and Allan Ludman. Copyright © 1991 by Macmillan College Publishing Company.)

### Quicksand, Sand Boils, and Sand Volcanoes

Most sandy fills are not dry, but are cubically packed sediments with pores full of groundwater. If these deposits subsequently are sheared, quicksand may develop. **Quicksand** is a sand/water mixture that is *fluid* because water flows *upward* through a deposit and exerts pressure on sand grains, keeping them from touching each other. This converts the sediment to a flowing sediment/water mass, through a process called **liquefaction.** (Note that liquefaction has no relation to melting, which is a *heat-caused* change of state.) Quicksand occurs where groundwater is forced to flow upward through sand because local geologic conditions prevent it from moving laterally.

Seismically induced quicksand forms when wet sandy fill (Figure 5–23a) is sheared and becomes more tightly packed. The reduced pore volume squeezes out part of the pore water, which migrates upward, exerting pressure on overlying sand grains and keeping them from making contact (Figure 5–23b). Some of this fluidized sand/water flows upward, disrupting the strata above and forming ir-

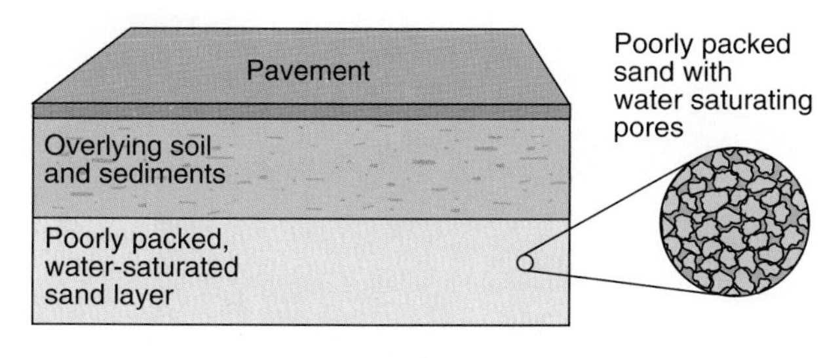


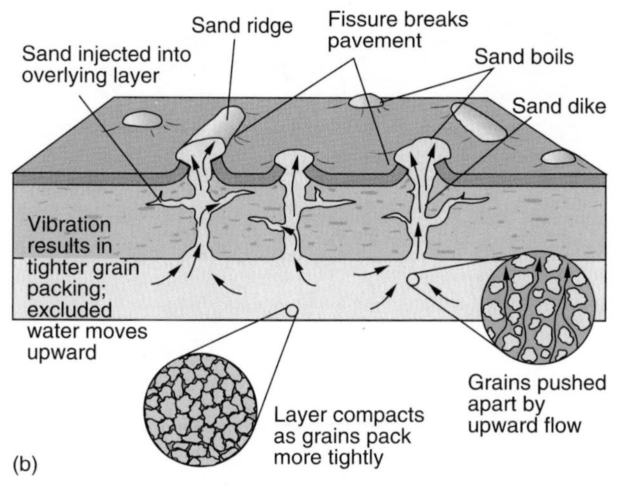


**FIGURE 5–23**
Cross section showing seismically induced quicksand formation and the origin of sand fountains on the surface. (a) Before the earthquake, sand grains make poorer contact and contain more water in spaces between them. (b) During the earthquake, grains are tightly packed with little water between grains.

**FIGURE 5–24**
Earthquake-induced liquefaction in surface deposits of the Pajaro River Plain, Watsonville, California. Numerous sand volcanoes formed along a fissure 6–7 meters (20–23 feet) long. (Photo by J. C. Tinsley, U.S. Geological Survey.)

regular sediment wedges called **sand dikes** within the overlying layer.

Some of the fluidized sand appears to "boil" through surface cracks, forming **sand boils** or sand volcanoes (circular or conical features) or **sand ridges** (linear features). These round and linear features appear on the surface after many earthquakes (Figure 5–24). Later in this chapter you will see how these features have provided vital clues in determining the recurrence rates of earthquakes.

## Quickclays

The 1964 Alaska earthquake caused massive ground failure in the Anchorage area (Figure 5–25). Part of Anchorage is built on a clay layer originally deposited in salty marine water. The ions in the salty pore water held the clay minerals together as aggregates in an open "house of cards" structure (Figure 5–26).

Hundreds of years ago, when erosion exposed the clays, freshwater flushed the saltwater from the pores and the clay particle structure became unstable. When seismic waves sheared this mass in 1964, it liquified almost instantly and began to flow as a **quickclay.** This removed the support for surface structures and caused widespread subsidence and landsliding in Anchorage.

## Landslides

Landslides are examined in Chapter 9, but we mention them here because the seismic vibration is a common triggering mechanism for landslides. Af-

**FIGURE 5–25**
Surface failure in Anchorage in southern Alaska in the 1964 earthquake. Prior to the earthquake, the sidewalk in front of the stores was at the level of the adjacent street; the earthquake lowered it nearly 4 meters (13 feet). (Photo from U.S. Geological Survey.)

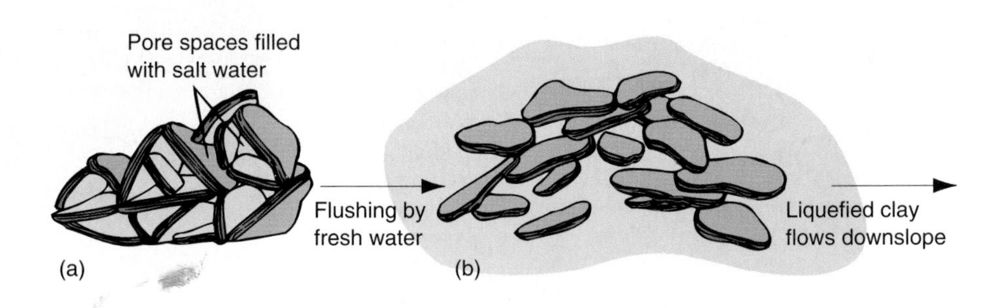


**FIGURE 5–26**
Formation of quick clays as fresh water flushes saltwater from pore spaces between clay particles. (a) Ions in salty pore waters hold clay together in an open structure. (b) Collapse of clay structure results in an excess of water, and liquefied clay starts to flow. (Reprinted with permission of Macmillan College Publishing Company from *Physical Geology* by Nicholas K. Coch and Allan Ludman. Copyright © 1991 by Macmillan College Publishing Company.)

tershocks can continue to trigger subsequent landslides. This is an added problem following a earthquake, because slides can close highways and railroads needed to evacuate survivors or bring aid to the stricken area.

## Regional-Scale Tectonic Deformation

Earthquakes can cause surface uplift or subsidence on a regional scale. The 1964 Alaska earthquake resulted in vertical crustal deformation over an area of 170,000 to 200,000 kilometers$^2$ (66,000 to 77,000 miles$^2$) in coastal Alaska (Figure 5–27). Interior portions subsided as much as 2 meters (6.5 feet), whereas coastal and offshore sections rose up to 4 meters (13 feet). Such movements also caused surface faulting.

Earthquake-related movement also displaced ocean water, generating a *seismic sea wave* (or *tsunami*, described in the next section). The wave swept the Alaska coast, its initial crest striking the Kenai Peninsula within 18 minutes and Kodiak Island within 34 minutes (Figure 5–27). The wave later reached wide areas of the Pacific.

Large-scale surface changes are especially serious in low-lying coastal areas because of the danger of rapid submergence and flooding by sea waves. Fortunately, in Alaska in 1964, the subsidence and tsunami struck a relatively unpopulated area. A similar event in a populated coastal region would be far

**FIGURE 5–27**
Map showing uplift and subsidence along the Alaska Coast during the 1964 Alaska earthquake. (After Plafker, G., 1965, Tectonic deformation associated with the 1964 Alaska earthquake: Science, v. 148, no. 3678, p. 1675–1687, Figure 2, p. 1677. Used with permission.)

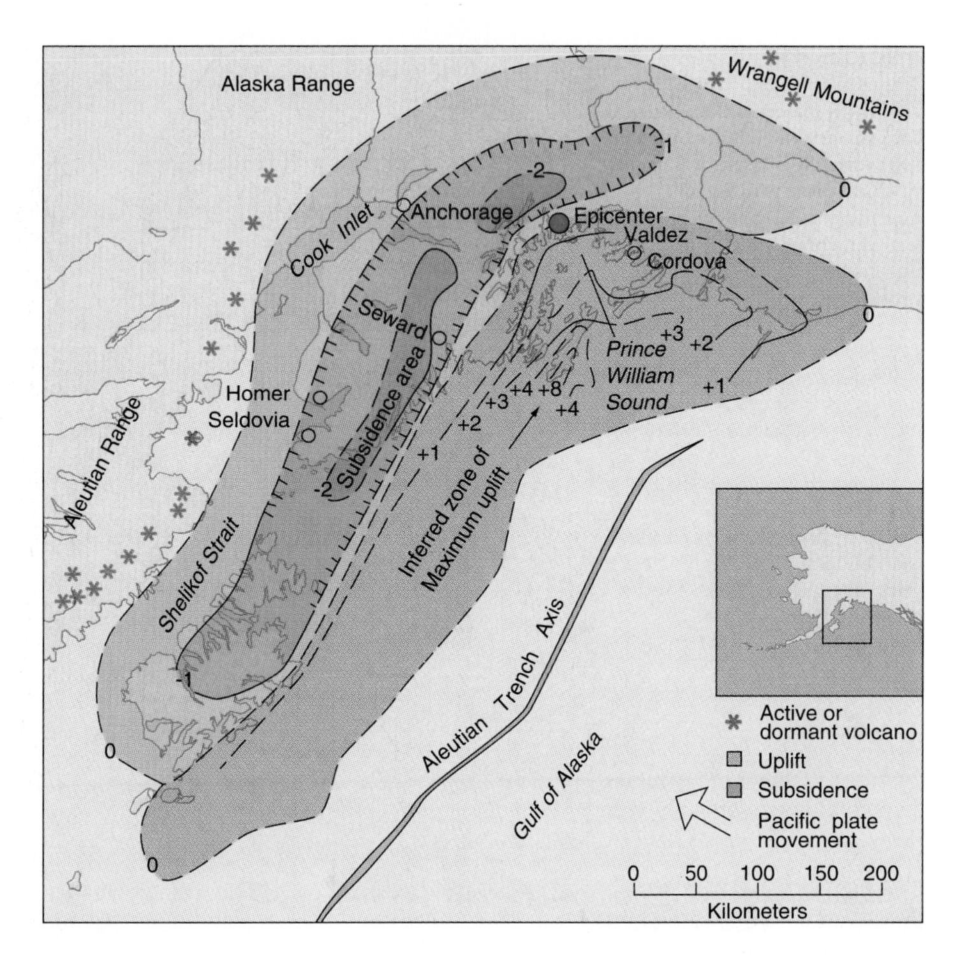


more serious. This possibility is considered in the following section.

## TSUNAMI

In the news you sometimes hear of a giant "tidal wave" that damages a coastal area. The name is very misleading, for the daily rise and fall of the tides has nothing to do with generating these waves. The waves are seismic sea waves, or tsunami.

### Generating Tsunami

A **tsunami** (soo-NAH-me) is a sea wave generated by a major disturbance of the seafloor and overlying water. As the water is displaced, it surges outward in all directions in a large wave. Tsunami are caused by faulting associated with earthquakes, volcanic eruptions, and submarine landslides. The most common cause of major tsunami is vertical displacement of the seafloor along faults (Figure 5–28).

Very large tsunami can be triggered by major volcanic eruptions. In 1883, the Indonesian volcanic island of Krakatau literally was blown out of the water by a violent eruption. Thousands of cubic meters of seawater rushed into the hot cavity on the seafloor and then surged back upward as hot water and steam. This great energy and violent movement generated tsunami that moved out in all directions, with devastating results. The Krakatau tsunami attained *30 meters (100 feet) in height* in places as they crashed ashore on Java and Sumatra, killing more than 30,000 people.

An infrequent cause of tsunami is submarine landslides that move masses of sediment from oceanic slopes onto the deep seafloor. This movement draws the ocean surface downward, and it rebounds to form a tsunami. Submarine landslides are orders of magnitude greater in volume than terres-

**FIGURE 5–28**
Origins of tsunami produced by faulting on the ocean floor. Vertical faulting starts the water movement that eventually becomes a tsunami. (a) Ocean crust with line of impending fault. (b) Sudden vertical displacement of seafloor causes a momentary drop in local sea level. (c) Water rushes into the depression, but overcorrects, locally raising the sea level. (d) Sea level locally oscillates before stabilizing. These oscillations are transmitted as long, low waves that travel thousands of kilometers. (Reprinted with permission of Macmillan College Publishing Company from *Physical Geology* by Nicholas K. Coch and Allan Ludman. Copyright © 1991 by Macmillan College Publishing Company. Courtesy of Allan Ludman.)

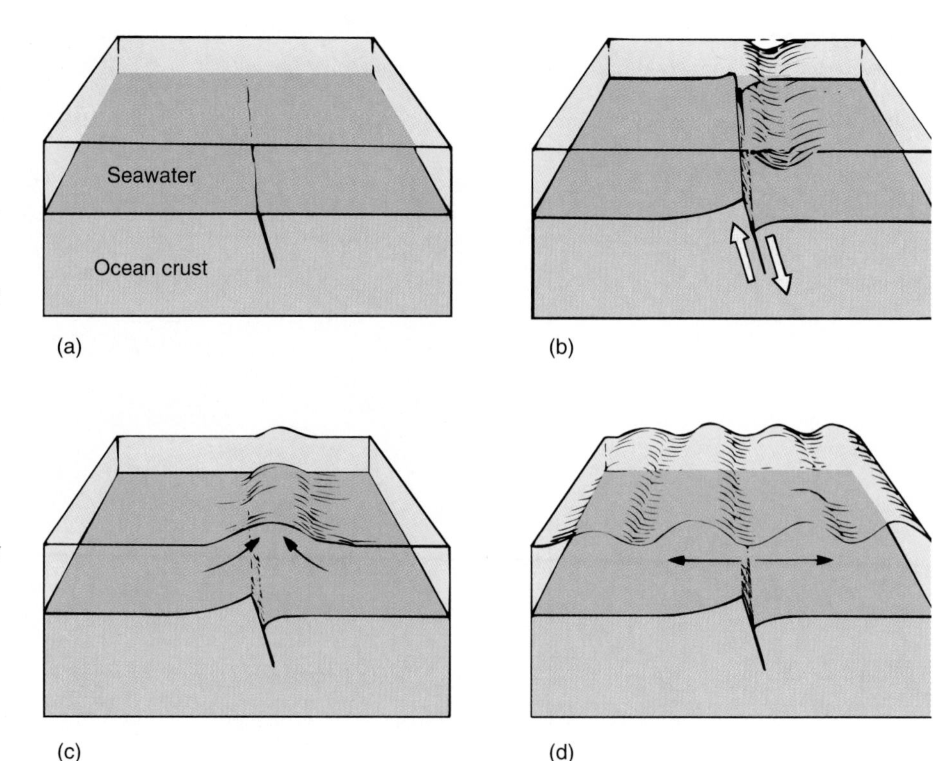


trial landslides. This process can generate rare but catastrophic "mega-tsunami."

## Characteristics and Movements of Tsunami

Tsunami waves have exceptionally long wavelengths (crest-to-crest distance) measuring tens of kilometers, even up to 100 kilometers (60 miles). (Imagine a wave as wide as Lake Michigan!) Tsunami waves also are very low in the open ocean, less than 0.5 meter (1.5 feet) high. Consequently, tsunami cannot destroy a ship at sea—in fact, these waves are so low and broad that they pass unnoticed beneath ships. However, if the ship is tied up at a dock, that is another story, as you shall see.

The speed of a tsunami is greater in deep water. Consequently, in a deep basin like the Pacific, tsunami can travel up to 700 kilometers/hour (435 miles/hour)—an aircraft speed. For perspective, consider that the tsunami generated by the 1960 southern Chile earthquake took only 22 hours to traverse the Pacific to Japan, where it caused great damage and over 100 deaths.

As a tsunami enters shallow coastal waters, its speed decreases but its height dramatically increases. (The transformation of wave speed and height as waves enter shallow water is discussed more fully in Chapter 15.) Commonly, water withdraws quickly from the shore before the tsunami hits. This water rises up offshore to form the first tsunami wave to strike the coast. Successive tsunami may attain several tens of meters height, depending on local conditions.

Japan often experiences these waves and has bestowed the name tsunami, meaning "harbor wave," because the waves rise in harbors to do great damage.

Successive tsunami may hit a coast at time intervals of 15 minutes to an hour, because of their extremely long wavelengths. In many cases, the first tsunami wave is not the biggest. At Crescent City, California, the first two waves of the 1964 Alaska earthquake tsunami swept along the coast 23 minutes apart, about four hours after the earthquake (Figure 5–29). These waves caused only minor flooding and gave a false sense of security to some residents, who returned to their places of business to clean up or save their merchandise.

However, some of these people, and a significant part of the town's waterfront, then later were

FIGURE 5–29
Tsunami travel-time map for the 1964 Alaska earthquake. Numerals indicate time in hours for the wave front to reach the points indicated. (After U.S. Geological Survey.)

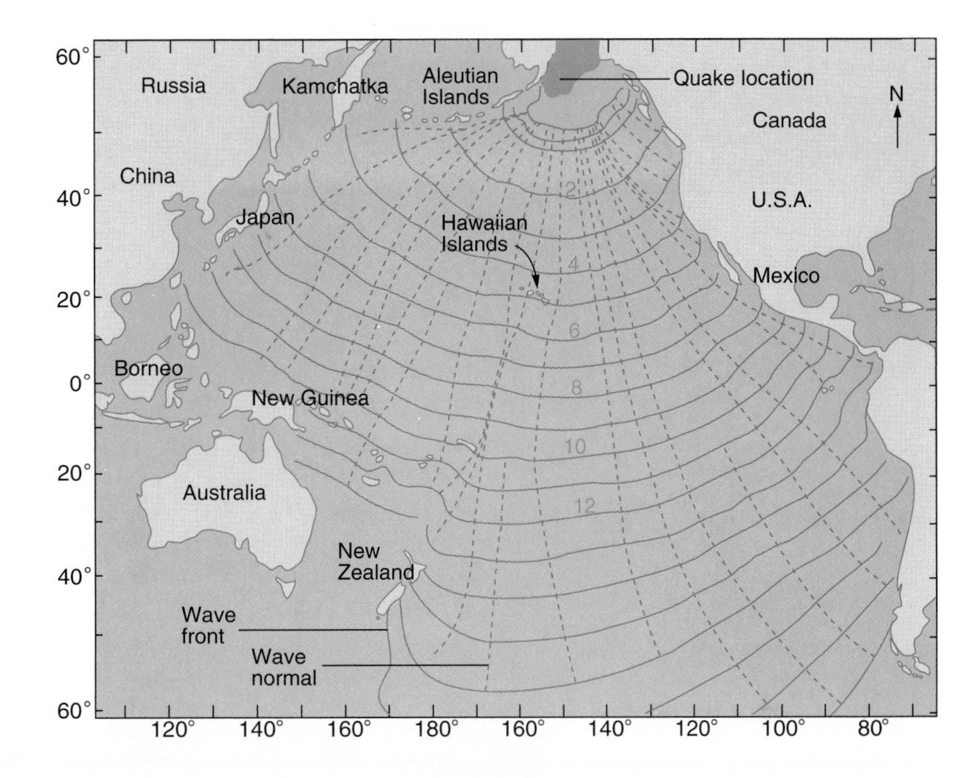


smashed by successive waves, which were much higher. These damaging waves reached 6.5 meters (21 feet) in height. Eleven people were killed and the damage was about $7.5 million.

## Damage from Tsunami

For tsunami caused by earthquakes, the maximum wave height of a tsunami is related to the earthquake's magnitude and the depth of its focus. The larger the earthquake's magnitude the higher the wave height. Local offshore topography also can affect wave height (see Chapter 15). Differences in tsunami damage along a coast commonly are related to variations in offshore topography.

Tsunami can vary greatly in the damage they cause. The orientation of the coast relative to the incoming wave front is one factor. If the tsunami is moving perpendicular to the coast (head-on collision), more damage is done than if it sweeps past, parallel to the coastline (glancing blow). Any intervening islands help shield part of the coast from the full force of the approaching waves.

A tsunami hitting at high tide is more dangerous than one that hits at low tide because the water level already is elevated, allowing the tsunami to reach even farther inland from shore. The nature of the coast is important: low-lying coastal plains can be swamped by the waves, whereas coastal cliffs are not. Populations in low-lying areas obviously are at risk.

Tsunami from *nearby* seafloor earthquakes, volcanic eruptions, or marine landslides can be especially devastating because of the lack of time to warn coastal inhabitants. On the night of September 1, 1992, a major earthquake (Richter magnitude of 7.0) occurred on the Pacific seafloor 56 kilometers (35 miles) off the coast of Nicaragua. The resulting tsunami, some more than 9 meters (30 feet) high, devastated a 240 kilometer (150 mile) stretch of the Nicaraguan coast. The death toll was high, not only from the wave size but because some residents lay asleep, unwarned when the tsunami hit. The waves obliterated coastal structures and spread as far as 0.8 kilometer (0.5 mile) inland. The Nicaraguan Red Cross put the death toll at 86—mostly children swept from their beds—with 101 missing and 3,700 homeless.

## Evidence for Ancient Tsunami

Until recently, ancient tsunami were known only if they had been recorded in the history, oral or written, of a region. However, recent studies show that

## BOX 5–2 LAUPAHOEHOE: A NEEDLESS TRAGEDY

Laupahoehoe is a small community on the northeast coast of Hawaii, about 40 kilometers (25 miles) north-northwest of the city of Hilo (Figure 5–30). A portion of the town, including its former school, is built on a low-lying coastal area at the base of a steep stream valley (Figure 1).

The community was devastated by the 1946 tsunami that resulted from a magnitude 7.5 earthquake in the Aleutian Islands. It destroyed the school,

**FIGURE 1**
The town of Laupahoehoe, Hawaii. (Photo by author.)

ancient tsunami can be recognized and dated by using the carbon-14 dating method on organic materials, by vegetative characteristics, and by type of sediment deposits.

Geologists recently described evidence for rapid submergence of the Washington State coast in the past. Fossil stumps of Western red cedar and Sitka spruce had characteristics indicating that they were killed by rapid salt water inundation. Preservation of stems and leaves of salt marsh grass also suggested rapid burial. In some places, the stems and leaves (dated as 300 years old) were surrounded by thinly layered sand beds, indicating rapid deposition in a series of pulses.

Taken together, the evidence suggests abrupt subsidence of the coast, resulting from tectonic deformation accompanying an earthquake about 300 years ago. This is similar to what happened to the Alaska Coast in the 1964 earthquake (Figure 5–27). The tsunami caused by the 300-year-old Washing-

killing twenty-five students and their teacher (Figure 2). The only school-age survivors in the community were those home sick that day. The area now is mapped as a category 4 zone on the Tsunami Hazard Map (Figure 5–30).

The tsunami-caused loss of life here and elsewhere led to the development of the Pacific Tsunami Warning System. The system alerts local civil defense personnel to sound warning sirens, which should warn inhabitants in time to evacuate to higher ground when tsunami threaten. The Laupahoehoe tragedy will hopefully never be repeated.

**FIGURE 2**
Laupahoehoe marker explaining loss of life in tsunami of 1946. (Photo by author.)

ton earthquake inundated the coast, depositing the layered sand in the process. This documentation of a major earthquake 300 years ago in this region is extremely important because this area is far more populated (Figure 5–3) than coastal Alaska. Detection and dating of former tsunami deposits give a general frequency of past tsunami inundation for a region. However, the deposits do not provide any information on the location of the epicenter of the past earthquakes.

## Tsunami Hazard Mitigation

The state of Hawaii is probably the most obvious tsunami target in the Pacific Basin. The Hawaiian Islands are in the middle of the Pacific, which is surrounded by the volcanically active and earthquake-prone "Ring of Fire." Hawaii is vulnerable to any tsunami generated in the seismically active Pacific (Figure 5–29). The Aleutian Islands earthquake of April 1, 1946 caused a 10-meter (33-foot) tsunami in the Hawaiian Islands. Over 150 people were killed, with widespread property damage.

The 1946 tsunami disaster led to the Pacific Tsunami Warning System, based at the Pacific Tsunami Warning Center (PTWC) in Honolulu, Hawaii. Data from seismological stations around the Pacific Rim are analyzed at PTWC and provide information on the position and magnitude of each earthquake. PTWC staff use that data to determine travel times to islands and mainland coastal areas that are threatened and notify civil defense personnel in these places.

Local officials implement their own warning system to evacuate low-lying coastal areas subject to imminent tsunami inundation. The first advance warnings using this system were issued for a tsunami expected from a 1952 earthquake in the Russian peninsula of Kamchatka. Because of the warning, damage in Hawaii was decreased and no lives were lost. In the Hawaiian Islands at present, sirens mounted on poles throughout the area are used, along with media warnings, to alert inhabitants of impending danger such as a tsunami or hurricane. In the front of the telephone directory for each island is the sequence of warnings and a map of locations and heights of coastal flooding.

Tsunami-prone areas are reducing potential property damage by restricting building in low coastal areas. A good example is Hilo, the largest city on the island of Hawaii, whose harbor area was devastated by tsunami in 1946 and 1960. Old commercial structures nearest the harbor have been removed and the area transformed into an attractive waterfront park that serves as a natural buffer against tsunami for the rest of the city. A tsunami hazard map (Figure 5–30) for residents indicates areas susceptible to tsunami damage.

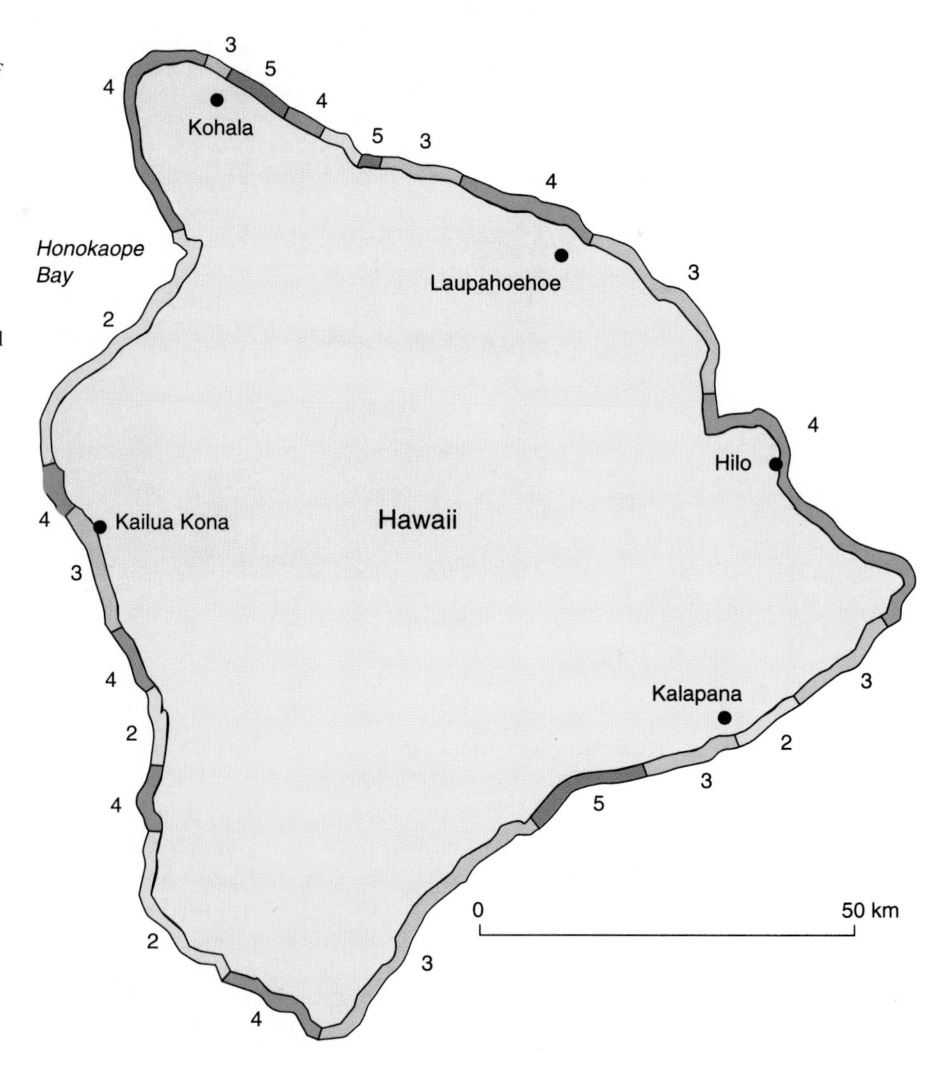


**FIGURE 5–30**
Tsunami hazard map of the island of Hawaii. The zone numbers indicate the areas in which the wave heights were at these levels: Zone 2—1.5 to 4.6 meters (5 to 15 feet); Zone 3—4.6 to 9 meters (15 to 30 feet); Zone 4—9 to 15 meters (30 to 50 feet); Zone 5—15 meters or greater (50 feet or greater). The island of Hawaii does not have a Zone 1. (U.S. Geological Survey Professional Paper 1240-B, Facing geologic and hydrologic hazards, p. B36).)

## EARTHQUAKE FORECASTING

Earthquake forecasting is a young but advancing endeavor. Several measurable **precursors,** which may indicate impending earthquakes, are being examined in seismically active areas: changes in the rocks, seismic gaps, and unusual animal behavior. The study of recurrence rates also holds promise in predicting probable occurrences.

### Changes in Rocks

The properties of a rock change as it is stressed (Chapter 2), and most are easily measured. The most fundamental change in rock is that, as stress increases, *minute cracks* develop. These cracks increase the rock volume, a phenomenon called **dilation.** The effects of dilatancy (die-LATE-un-see) may be expressed as *surface bulging,* an important earthquake precursor (Figure 5–31). Minor breaks in strained rocks, especially along tributary faults, may produce small seismic events called **foreshocks.** These release some stress and may be followed by a quiet period during which strain resumes building toward rupture levels on the main fault.

The level of strain that builds along potential rupture zones can be measured by instruments called *strain gauges.* Seismologists measure strain buildup as a useful precursor with which to forecast seismic activity.

Cracks in rocks lead to other precursors. *Decreases in groundwater level* may result from water draining into the cracks in the rocks, causing the local water table (Chapter 3) to drop. The addition of water draining into the cracks can *change the electrical conductivity* of the rocks, which also is easily measured.

Local minor changes in Earth's electromagnetic field immediately preceding some earthquakes also have been noted.

*Increases in radon gas levels* have been observed in groundwater before some seismic events. Radon is a radioactive gas produced by the natural decay of uranium-238 (Chapter 2), which is common in several rock types. Normally, the radon slowly accumulates within the rock and remains there. If the rock is strained so that cracks form, however, the radon may escape into groundwater. The gas is detected by analyzing water samples from wells or by sensors placed into the wells and connected to a recording station.

Another precursor involves measuring the speeds of P and S waves from earthquakes as they pass through the area being studied. A *decrease in the P wave/S wave speed ratio* has been detected before some earthquakes. This change may occur as cracks develop, changing the rock's ability to transmit the seismic vibrations.

It is important to understand that, to date, none of these precursors have enabled successful prediction of an earthquake.

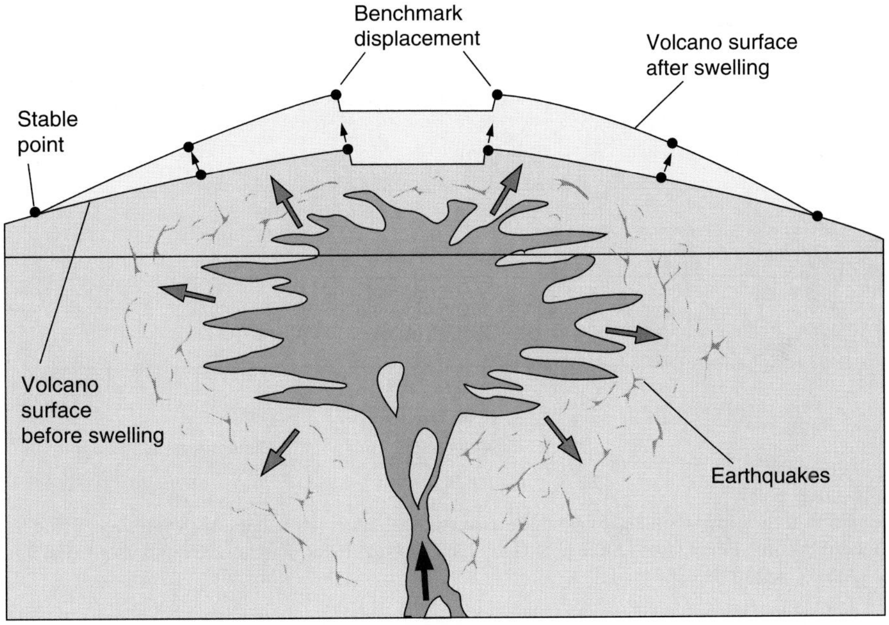

**FIGURE 5–31**
Geologic section showing earthquake generation prior to an eruption of Kilauea Volcano, Hawaii. Pressure from the rising magma cracks the rocks and generates earthquakes. (After U.S. Geological Survey, 1991, Living with volcanoes, Circular 1073, p. 23.)

## Seismic Gaps

As explained earlier, along a fault like the San Andreas, some segments creep, whereas others are locked. Fault segments along which large strain-releasing seismic events have occurred in the past, but not recently, are called **seismic gaps.** Recent research indicates that seismic gaps are places where the fault is locked and, therefore, accumulating strain. Because the strain is not relieved by minor seismic events, it can build to extreme levels. When the strain in the rocks finally causes them to rupture, a major earthquake occurs. According to the seismic gap theory, fault segments that long have been quiet are the ones to watch for the next major earthquake.

The seismic gap forecasting method is well illustrated by the Loma Prieta earthquake of 1989. In this seismic event, the Pacific plate moved northwestward and upward relative to the North American plate. The U.S. Geological Survey had identified the Loma Prieta area as one of the three seismic gaps along this section of the San Andreas fault (Figure 5–32a) and therefore one of three areas to watch. Although they correctly identified the place, the key problem remains: predicting *when* a major earthquake will occur.

The main Loma Prieta event and its thousands of aftershocks "filled" the seismic gap along that portion of the San Andreas fault (Figure 5–32b). But which of the remaining gaps will be the site of the next major earthquake?

## Anomalous Animal Behavior

It has long been noted, particularly in China, that erratic animal behavior precedes some volcanic eruptions and earthquakes. These activities have included horses bolting from their stables, snakes emerging in winter (and freezing to death), dogs barking, chickens refusing to lay eggs, and fish leaping from ponds. What do these animals sense?

What warning signals these animals receive is unknown. We do know that animals are adapted to their individual environments, and so, while they have the same basic sensors we do, some operate in different ranges or at different sensitivity levels. Well-known examples are that dogs and bats hear higher-frequency sounds than we do; dogs have a far keener sense of smell than humans; some snakes

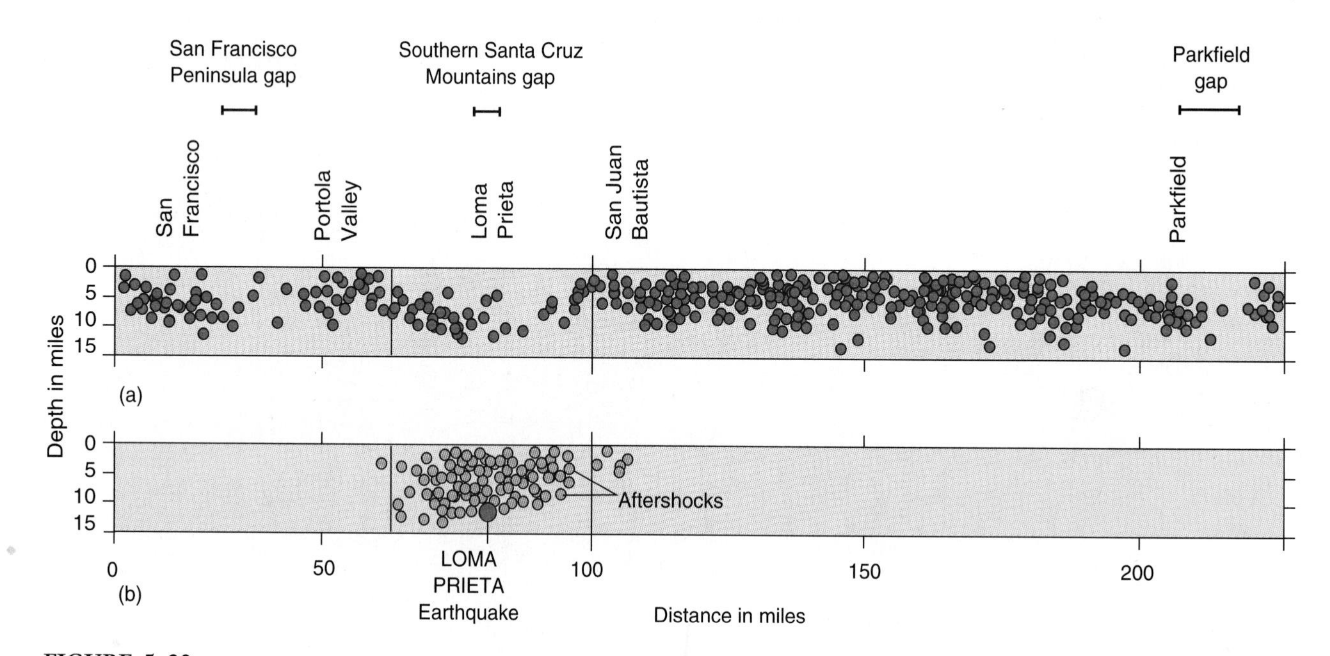


**FIGURE 5–32**
(a) Section showing seismic gaps along the San Andreas fault. (b) Section showing shocks that filled the Loma Prieta Gap. (After U.S. Geological Survey, 1989, Earthquakes and volcanoes, v. 21, no. 6, p. 218.)

sense infrared heat waves like we see light; big-eyed owls can see in deep darkness; and many animals feel sound vibrations and ground vibrations too slight for humans to notice.

These precursors are hard to quantify and apply to forecasting earthquakes, but millennia of experience in earthquake-prone China suggests that anomalous animal behavior should not be dismissed as a possible earthquake precursor, even though no scientific evidence has been obtained to date.

## Statistics

Statistics are an important *general* predictive tool. The probability of future earthquake occurrence can be calculated from long-term experience. As in any statistical effort, the longer the period of record, the better the prediction. The longest-term record comes from areas that have been settled for a long time—for example, Chinese records go back well over a thousand years.

In the eastern United States, earthquake records go back several hundred years (starting around 1638), but in the seismically active areas of the west, the historical record is much shorter (see Figure 5–13). However, the state of knowledge is advanced enough that analysis of past history, as well as present-day monitoring, provides the means for constructing maps showing probabilities of earthquake occurrence in a given area. Figure 5–33 shows such a probability map for the San Francisco Bay area.

As always, statistics show only probabilities. It is like forecasting a 50% chance of rain: should you carry an umbrella or not? If seismologists forecast a 50% chance of an earthquake where you live over the next six years, how should you prepare?

## Dating Faults and Disturbances in Sediment

A promising tool for determining or refining earthquake recurrence intervals is emerging: examination of underlying sediment layers to find datable evidence of a seismic disturbance. Earlier in this chapter, we looked at liquefaction of sand with water and its disturbing effect on sediment layers (Figure 5–23). The liquefied sand flows upward, cutting

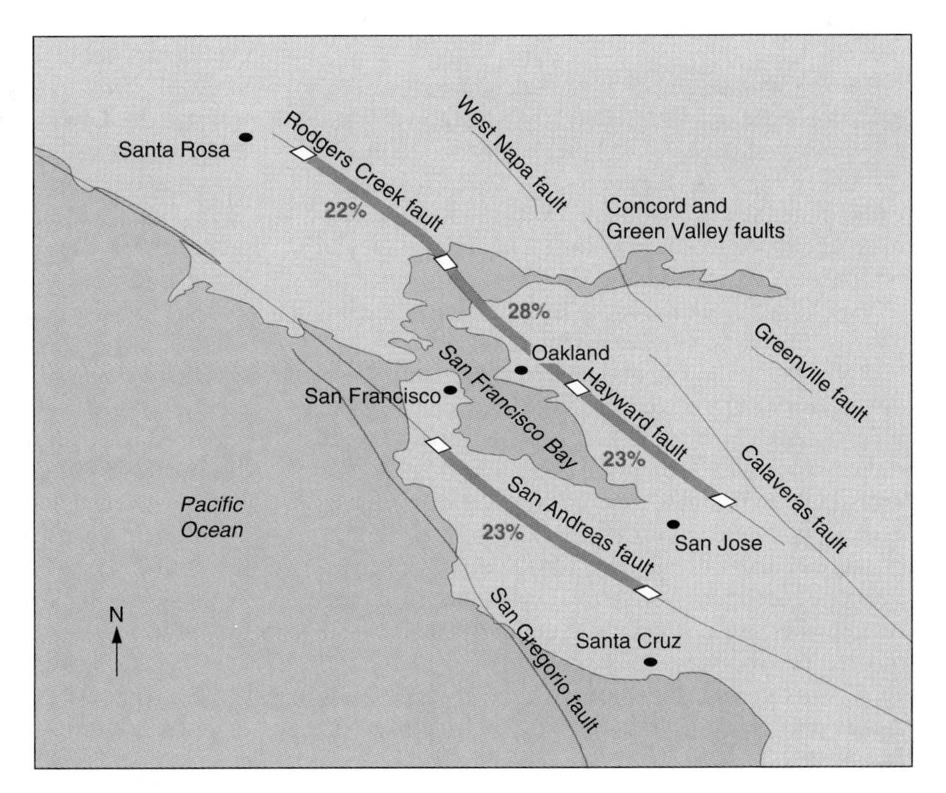


**FIGURE 5–33**
Map of 30-year probabilities for fault segments in the San Francisco Bay area. The total 30-year probability of one or more earthquakes for the entire region is 67%.

(a)

(b)

**FIGURE 5–34**
(a) Liquefaction features seen in a trench at New Madrid, Missouri, probably related to the 1811–1812 earthquakes. (b) Zone of liquefaction is shown in bright blue in diagram. Dating these layers can determine the age of past seismic events. (Photos by David Russ, U.S. Geological Survey.)

across and disturbing younger layers and emerging on the surface as sand boils (Figure 5–24). These sand boils are then covered by newer sediments, but their recognizable pattern remains in the sediment. Later, another seismic event may repeat the process.

Ideally, this produces a stacked sequence of liquefaction episodes. Hopefully, they include organic materials (wood, peat, etc.) that can be dated by the carbon-14 method (Figure 5–34). By dating the organic materials, it is possible to judge the approximate age of liquefaction events. The dates then can be analyzed to determine the average recurrence intervals between earthquakes in that area and thus forecast the likely next occurrence.

Another method is dating layers that have been offset (broken) by vertical fault movements and then overlain by organic-rich sediments that can be carbon-14 dated. Along one section of the San Andreas fault at Cajon Creek, California, for example, geologists found ruptures that suggest six earthquakes occurred in the past 1,000 years. This averages one earthquake every 166 years, but of course earthquakes are not evenly timed, so a reasonable recurrence interval would be roughly 100 to 200 years. This should cause concern, because the study area is only about 100 kilometers (60 miles) east of Los Angeles.

## FLUID INJECTION TO TRIGGER EARTHQUAKES

An interesting event in Denver suggests a possible earthquake-mitigation method. Denver began to experience minor earthquakes (magnitude 1.0-4.3) in April 1962. Their epicenters clustered around wells at the Rocky Mountain Arsenal, where wastewater was being pressure-injected into rocks greater than 4 kilometers (over 2 miles) below.

Injection had begun one month prior to the appearance of the earthquakes, and the timing of water injection showed a strong correlation with seismic activity (Figure 5–35). Geologists hypothesized that the pressurized water was releasing accumulated stress on old faults beneath the city. The water pressure acted to decrease the pressure on the fractured rock surfaces, enabling them to slip more easily.

Controlled fluid injection was tried in the Rangely Oil Field in northwestern Colorado, not to trigger earthquakes, but to stimulate more production from oil-bearing layers. Detailed monitoring showed that epicenters of local earthquakes were near the fluid-injection sites, and their magnitude related directly to the rate of pumping.

These studies suggest that fluids might be injected to trigger faulting—carefully. Periodic water in-

**FIGURE 5–35**
Plots of earthquake frequency versus volume of contaminated water injected at Rocky Mountain Arsenal. (Modified from Evans, D.M., 1966, Man-made earthquakes in Denver: Geotimes, Fig. 3, v. 10, no. 9, p. 11–18.)

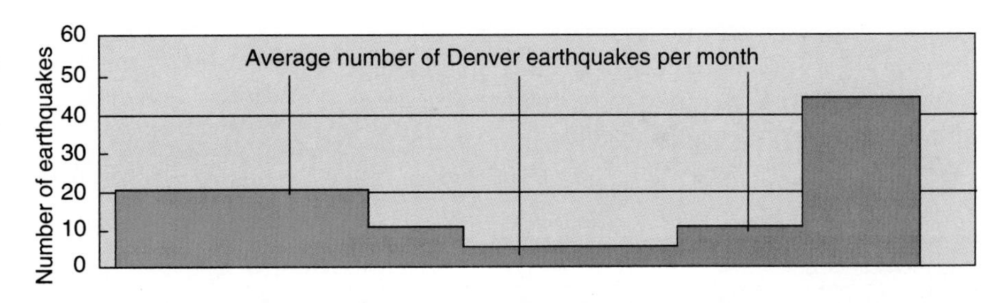


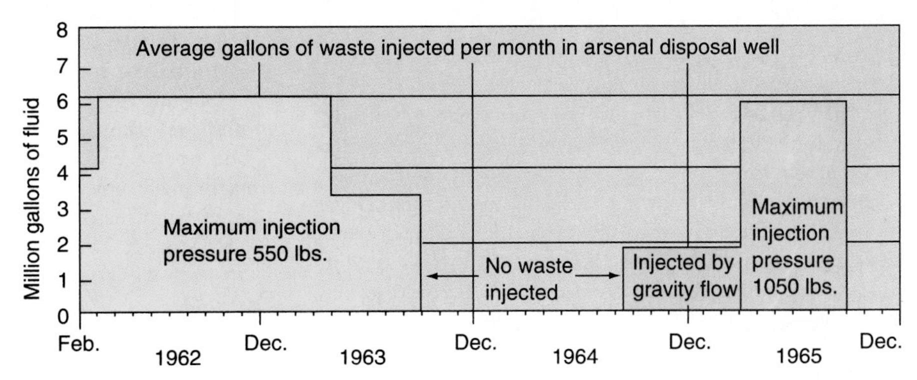


jection might trigger small earthquakes to relieve strain on a fault, perhaps preventing or reducing the intensity of a larger seismic event.

Applying this method to "locked" sections of the San Andreas fault could be very dangerous, however; stress levels on these segments already are so high that injection could trigger a large-magnitude earthquake. This could cause record-setting liability suits against the government and negative public perception ("if they had left the fault alone, we wouldn't have had an earthquake"). However, controlled fluid injection is a potential scientific tool for reducing strain buildup on other fault segments.

Controlled release of earthquake energy is not without its problems, however. Consider an area where a magnitude 8.0 earthquake is expected every 100 years. One would have to produce nearly 33,000 earthquakes (32 × 32 × 32) of magnitude 5.0 to release this same energy in small increments. This would require a magnitude 5.0 event *almost every day* for 100 years! Would this be more desirable than the rare magnitude 8.0 event, especially if the 8.0 event could be predicted and the area evacuated? (Of course, no such prediction is possible at this time.)

Magnitude 5.0 seismic events do little damage to well-engineered structures typical of developed countries, so a daily 5.0-magnitude temblor might be tolerable in the United States. The situation would be very different in a less-developed nation where widespread damage would occur because of poor construction, lack of building codes, or lack of code enforcement.

## LOOKING TOWARD THE "BIG ONE"

Research continues for ways to predict and prevent earthquakes along faults. At the same time, community education and disaster preparedness programs must increase. However, our only hope to reduce deaths, injuries, and structural damage is to develop prudent land-use policy, protective zoning, structural reinforcement, and adequate evacuation plans and routes. This must be done in locked areas of the San Andreas fault zone as well as the other areas mentioned in this chapter before the inevitable "big one" occurs.

## LOOKING AHEAD

You have now seen how Earth's tectonic forces generate two major hazards—volcanoes and earthquakes. The next chapter examines soils and sediments. They lack the drama of flowing lava and swaying buildings, but they cause many more common hazards for many more people.

## SUMMARY

Earthquakes are generated when highly stressed rocks rupture. As the rock ruptures, energy is released abruptly as seismic waves, which radiate in all directions. The rupture may result from plate tectonic movements (faulting, subduction), volcanism, or human causes. Earthquakes occur at all three types of plate boundaries (divergent, convergent, and transform fault) and in intraplate locations. Rupture may be incomplete during the main event, allowing subsequent releases of the residual stress as aftershocks. Segments of faults such as the San Andreas may creep slowly and release stress gradually, whereas other segments are locked and accumulating dangerous levels of stored stress before movement.

Earthquakes generate P, S, Love, and Rayleigh waves, each of which has a characteristic speed and movement pattern. An epicenter can be determined from the arrival time differences of the P and S waves recorded at three different seismic stations. Data from at least four stations are needed to determine the earthquake's depth (focus).

Earthquake intensity based on the effects on people and the damage caused is expressed on the Mercalli Scale. Magnitude is determined from the amplitude of seismic waves on a seismogram and is measured using the Richter scale. Each increase of 1 on the Richter Scale reflects a ten-fold increase in amplitude and a thirty-two-fold increase in released energy. The extent of earthquake damage depends on surface geology, proximity to the epicenter, structural stability, population and structure density, extent of utility line disruption, stability of evacuation routes, and time of day.

Earthquake damage results primarily from ground shaking. Shearing forces destroy foundations and structures and break gas and power lines, resulting in fires and explosions. Ground vibration results in subsidence due to packing changes, liquefaction of sands (quicksands), and formation of quickclays. Loosening of weathered, unconsolidated rock and soil on slopes produces landslides that cause structural damage and may close evacuation routes.

Tsunami are very low-amplitude, very long-wavelength seismic sea waves resulting from disturbance of the seafloor and the overlying water. They may be caused by submarine landslides, volcanic eruptions, and earthquake faulting. Tsunami are very fast in the deep ocean but upon encountering shallow coastal waters, they slow and increase amplitude greatly (up to 30 meters, or 100 feet). Tsunami damage depends on coastal exposure, tidal range, magnitude, and depth of focus of the generating earthquake, and the elevation, structure, and population density along the affected coast. The Pacific Tsunami Warning Center issues tsunami warnings shortly after an earthquake is detected.

Precursors of earthquakes include changes in rock properties, low-level seismic events, release of radon gas, changes in the ratio of velocities of P and S waves, changes in groundwater levels, bulging and tilting on the surface, and anomalous animal behavior—but to date none of these have enabled the successful prediction of an earthquake.

Seismic gaps are fault segments along which large strain-releasing seismic events have occurred in the past, but not recently. These are prime candidates as future earthquake epicenters. Recurrence rates are being calculated from historical records and studies of sediment layers disturbed by "paleo-earthquakes."

## KEY TERMS

aftershocks
amplitude
Benioff zone
body waves
cubic packing
dilation
earthquake
elastic rebound theory
epicenter
focus
foreshocks
intensity
intraplate earthquakes
lag time
liquefaction
Love waves
magnitude
Mercalli scale
packing
pancaking
period
precursors
P wave
quickclay
quicksand
Rayleigh waves
rhombohedral packing
Richter Scale
rift valleys
sand boils
sand dikes
sand ridges
seismic gaps
seismogram
seismograph
seismology
surface waves
S wave
tsunami

## REVIEW QUESTIONS

1. Describe the mechanism that causes each type of earthquake. Also, give a geographic example of each of the following: (a) divergent margin, (b) transform fault, and (c) subduction zone.
2. Describe differences in the origin of plate-margin earthquakes and intraplate earthquakes.

3. Name the four kinds of seismic waves. Show the motion caused by each.
4. What is the difference between earthquake *magnitude* and *intensity*? How is each expressed?
5. How does each of the following affect the degree of earthquake damage? Give examples. (a) surface geology, (b) structural stability, (c) time of day, and (d) aftershocks.
6. Describe the conditions in which shearing by earthquake waves would cause: (a) subsidence in a sandy fill, (b) collapse and flowing in an area underlain by former oceanic clays, (c) development of quicksand and sand boils on the surface, and (d) landslides.
7. Describe three different ways that tsunami may be generated.
8. Describe four factors that determine the degree of damage from a tsunami.
9. Earthquake aftershocks usually are weaker than the main shock, so why do they do so much damage?
10. For each of the following earthquake precursors, give a possible cause of these phenomenon: (a) bulging and tilting of the surface, (b) change in groundwater levels, and (c) release of radon gas.
11. What is a seismic gap? Why is it so important in predicting the epicenters of future earthquakes?
12. Describe how average earthquake recurrence intervals for a given location can be determined by analyzing the sediment layers exposed in a trench in that area.

## FURTHER READINGS


Atwater, B. F., and Yamaguchi, D. K., 1991, Sudden, probably co-seismic submergence of Holocene trees and grass in Coastal Washington State: Geology, v. 19, p. 706–709.

Bolt, B. A., Horn, W. L., MacDonald, G. A., and Scott, R. F., 1977, Geologic hazards (rev. second edition), Chapter 1 (Hazards from Earthquakes) and Chapter 3 (Hazards from Tsunamis): New York, Springer-Verlag Co., 330 p.

Coch, N. K., and Ludman, A., 1991, Physical geology: New York, Macmillan Publishing Co., 678 p., Chapter 19 (Earthquakes and Seismology).

Davidson, K., 1994, Learning from Los Angeles: Earth, v. 3, no. 5 (Sept.), p. 40–48.

Davidson, K., 1994, Predicting Earthquakes: Earth, v. 3, no. 3 (May), p. 56–63.

Fumal, T. E., Pezzopane, S. K., Weldon, R. J., and Schwartz, D., A 100-year average recurrence interval for the San Andreas fault at Wrightwood, California: Science, v. 259, p. 199–203.

Nicholas, D. R., and Buchanan-Banks, J. M., 1974, Seismic hazards and land-use planning, U.S. Geological Survey Circular 690, 33 p.

Nuttli, O. W., Bollinger, G. A., and Herrmann, R. B., 1986, The 1886 Charleston, South Carolina, earthquake: A 1986 perspective, U.S. Geological Survey circular 985, 52 p.

Petroski, H., 1994, Broken bridges: American Scientist, v. 82, no. 4 (July-August), p. 318–321.

Plafker, G., 1965, Tectonic deformation associated with the 1964 Alaska earthquake: Science, v. 148, no. 3678, p. 1675–1687.

Thenhaus, P. C., 1990, Perspectives on earthquake hazards in the New Madrid seismic zone, Missouri: Earthquakes and Volcanoes, v. 22, no. 1, p. 4–21.

Toussan, R. T., Bennett, J. H., Borchardt, G., Saul, R., Davis, J., Johnson, C., Lagorio, H., and Steinbrugge, 1989, Earthquake planning scenario for a major earthquake on the Newport-Englewood fault zone: California Geology, v. 42, no. 4, p. 75–84.

U.S. Geological Survey, 1992, Goals, opportunities, and priorities for the USGS Earthquake Hazards Reduction Program: U.S. Geological Survey Circular 1079, 60 p.

U.S. Geological Survey, 1990, Probabilities of large earthquakes in the San Francisco Bay region, California: U.S. Geological Survey Circular 1053, 51 p.

Wood, P. L., and Page, R. A., 1989, The Loma Prieta earthquake of October 17, 1989: Earthquakes and Volcanoes, v. 21, no. 6, p. 215–237.

# 6

# Soil Erosion and Sediment Pollution

Soil is one of our basic resources because it is necessary for growing food. However, usually it is neither very thick nor permanent. If protective vegetation is removed from agricultural topsoil, the soil's thinness enables it to be completely eroded by wind or water. Once soil is lost, all that remains is infertile sediment. Although new soil will eventually form, it requires hundreds or thousands of years. Consequently, agriculture cannot continue unless artificial fertilizers are used. The eroded soil and sediment lead in turn to another set of problems referred to collectively as **sediment pollution.**

## WHAT IS SOIL?

It is important to understand that the term *soil* has different meanings to farmers, engineers, and geologists:

1. *Farming:* The definition of soil introduced in Chapter 3 (regolith enriched in the nutrients necessary to support plant growth) is the *agricultural* definition. Thus, farming soil usually is no more than a meter thick and may extend only to the depth reached by a plow.
2. *Engineering:* Engineers view soil as the loose material that can be removed without blasting (this loose material is residual or transported regolith). By this definition, soil can be several tens of meters thick.
3. *Geology:* Geologists view soil as regolith formed from the weathering of the underlying materials (residual regolith). Thus, *geological* soils are not necessarily suitable for growing crops unless their upper portion has been enriched with the nutrients necessary to support plant growth.

Because of these different interpretations, we shall use the agricultural definition of soil in this chapter. For infertile loose materials, we shall use the term *regolith* (transported or residual) or sediment.

**Gully erosion near Half Moon Bay, California. (Courtesy of Raymond Pestrong, San Francisco State University.)**

## WHY ARE SOILS IMPORTANT?

The soil and underlying regolith have other uses beyond supporting vegetation growth. Wells driven into permeable and water-saturated regolith supply most of the drinking water in many areas (discusssed in Chapter 8). Sewage effluent from cesspool and septic tank overflow soaks through the soil, and as it does, oxygen and microorganisms in the soil decompose some of the organic and chemical waste (discussed in Chapter 12).

This chapter describes soil and regolith. It also explains what happens when this material is eroded by natural processes and is made more susceptible to erosion by the actions of people.

## FACTORS FAVORING SOIL DEVELOPMENT

Production of soil by physical and chemical weathering was introduced in Chapter 3. But what factors promote good soil development? They include:

- ☐ *Warm and wet climates,* which increase chemical weathering.
- ☐ *Parent material (rock or sediment) containing elements needed by plants* and released by weathering.
- ☐ *Vegetation with substantial root systems,* to aerate and stabilize the developing soil.
- ☐ *Soil drainage flow,* to allow chemical weathering processes to proceed.
- ☐ *Sufficient time,* allowing thick soils to develop.

The presence or absence of these factors explains why some areas are noted for excellent soils and agriculture, whereas others are barren, such as deserts.

Some of the most fertile soils develop where geologic processes have fragmented rock into smaller particles, thus increasing the surface area exposed to weathering. For example, across a large area of the north-central United States, glaciers have ground up bedrock in southern Canada and the north central states and transported the regolith southward, depositing it as glacial till and outwash as the ice melted (Figure 3–20). The surface of this finely ground parent material became weathered extensively in milder postglacial climates to produce the very fertile soils of the midwestern United States. These soils support the multibillion-dollar U.S. agriculture industry.

In a similar fashion, fertile soils develop on volcanic ash. Eruptions produce parent material that contains abundant fine mineral particles. These weather readily to form soil.

## SOIL HORIZONS

Chemical weathering produces aluminum oxides, iron oxides, and clay minerals in the soil. As weathering continues, these products become depleted in some parts of the regolith and enriched in others. This process leads to the formation of layers called **soil horizons.** They are distinguished from one another by organic content, color, particle size, and percentage of unweathered material. Soils that have developed over enough time (hundreds or thousands of years) to form distinct horizons are called **mature soils.**

Four major soil horizons are recognized in many mature soils. Starting with the uppermost layer, these are called the O, A, B, and C horizons. Figure 6–1 shows these horizons and illustrates the type of soil that develops in the northeastern United States.

The uppermost layer is the **O horizon,** which is characterized by the accumulation of organic material and the absence of mineral matter. The uppermost portion of this organic horizon is made up of fresh to partly decomposed plant matter, whereas the lower part contains a more decomposed plant matter. The organic accumulations in the O horizon give it the distinctive dark brown or black color that most people associate with fertile soil.

The next layer, the **A horizon,** is composed both of mineral particles and organic material. In humid climates, considerable water percolates downward through the soil, like water through coffee grounds. The water selectively removes calcium, aluminum, and iron as it passes through. This removal of soluble material is called **leaching.** For this reason, the A horizon sometimes is called the **leached horizon.**

The carbonates, along with clay minerals and iron oxides derived from weathering of minerals, are transported down through the soil by percolating water. They accumulate in the underlying **B horizon,** sometimes called the **accumulation horizon.** The concentrations of iron oxides in the B

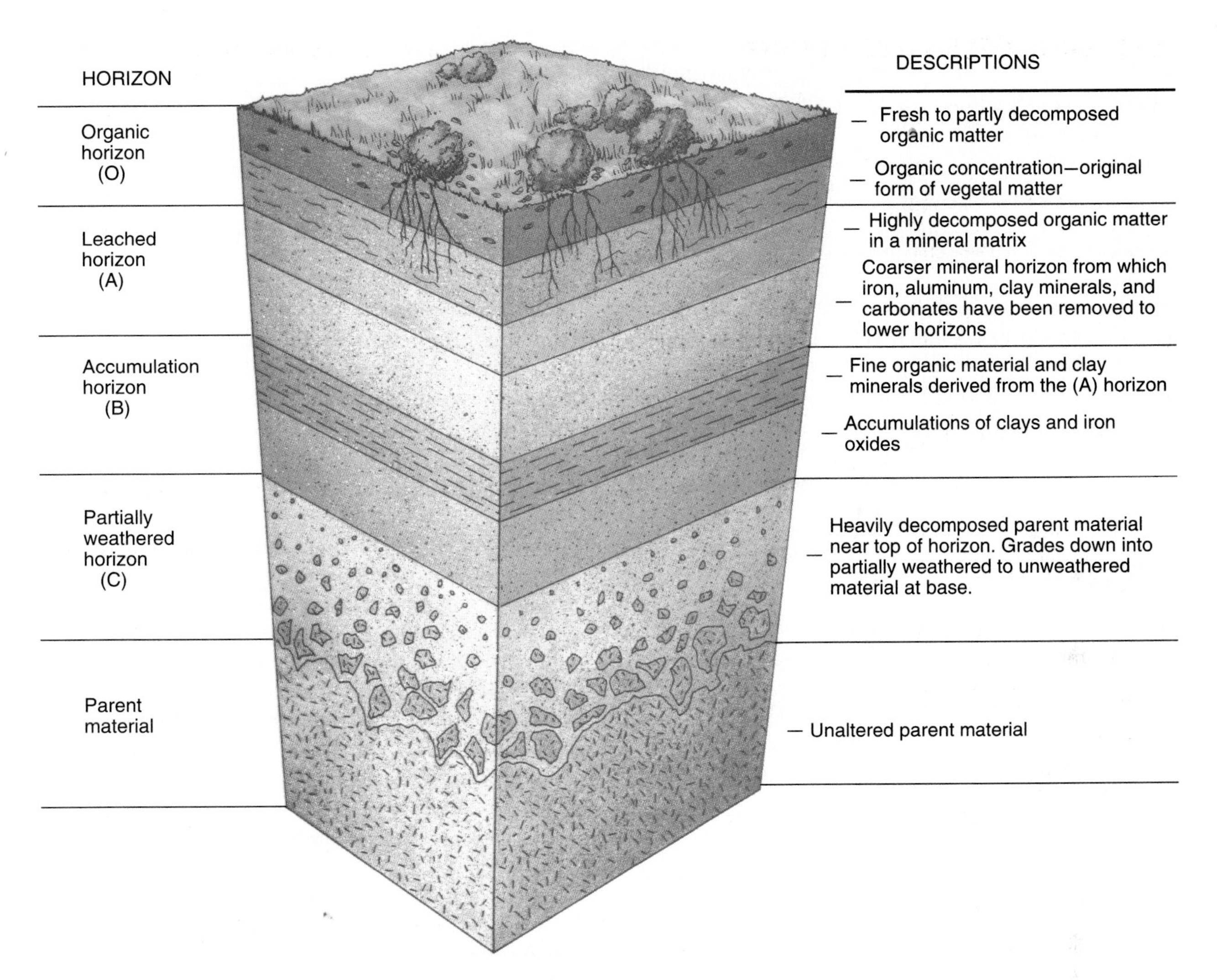


**FIGURE 6–1**
Cross section of a soil in the northeastern United States, showing the soil horizons. The boundaries between real soil layers are seldom as sharp as those shown here; they usually are more gradational. (Reprinted with the permission of Macmillan College Publishing Company from *Physical Geology*, by Nicholas K. Coch and Allan Ludman. Copyright © 1991 by Macmillan College Publishing Company, Inc.)

horizon commonly impart a yellowish to a red color, often visible at construction sites.

The **C horizon** is composed of partially weathered bedrock. At the top of the C horizon is highly weathered rock or sediment, which may preserve some characteristics of the parent material. The lower portion of the C horizon clearly contains pieces of the bedrock.

Where rainfall is greater than 50 to 60 centimeters (about 20 to 24 inches) per year, calcium carbonate and other soluble compounds are dissolved by carbonic acid and removed from the soil, whereas clay minerals and iron oxides accumulate in the B horizon. Such a soil is called a **pedalfer.** (The name comes from *ped* for soil, plus *al* for aluminum clays and *fer* for ferrum, or iron.) Pedalfers are common in the eastern United States (Figure 6–1).

Where rainfall is less than 50 to 60 centimeters (about 20 to 24 inches) per year, there is insufficient water to remove soluble compounds. The result is a soil rich in these compounds, called a **pedocal** (*ped* for soil, *cal* from the Latin *calc*, meaning lime). A type of pedocal in which the soluble minerals actually form crusts, layers, and pore fillings is called a *caliche*.

## SOIL CHARACTERISTICS

Soils vary from place to place because of differences in soil-forming factors such as the mineralogy of the parent material, climate, slope, vegetation, and time. The areal extent of soil types discussed in this chapter is shown in Figure 6–2. These differences not only affect soil fertility but may also lead to geohazards as well. Let us look at the properties soil scientists use to describe soils.

### Texture

Soils vary widely in their texture (particle size, particle roundness, and sorting). In general, these properties are inherited from the parent material on which the soil developed. Soils within an area generally are characterized by their texture, for example as "sandy soils" or "clayey soils." The terms for particle size (Wentworth Scale) are given in Table 3–1.

### Color

The color of a soil is determined by its composition. However, the color also gives clues to the *conditions* under which the soil formed. For example, poorly drained soils that are rich in organic matter tend to be dark. In contrast, soils with good drainage and some maturity tend to have tan, yellow, or even red "B" horizons. These colors reflect the oxidation of the iron in the minerals to produce rust-colored iron oxides.

### Cohesion

The degree to which a soil holds together is a measure of its **cohesion.** In general, sandier soils are non-cohesive (consider how easily a ball of wet sand falls apart). Finer-grained, clay-rich soils are cohesive (consider a ball of modeling clay). Cohesion is an important factor in **erodibility,** the ease with which a soil is eroded. In general, clay-rich soils resist erosion by water and air better than sandier soils. Coarser-grained non-cohesive soils tend to be

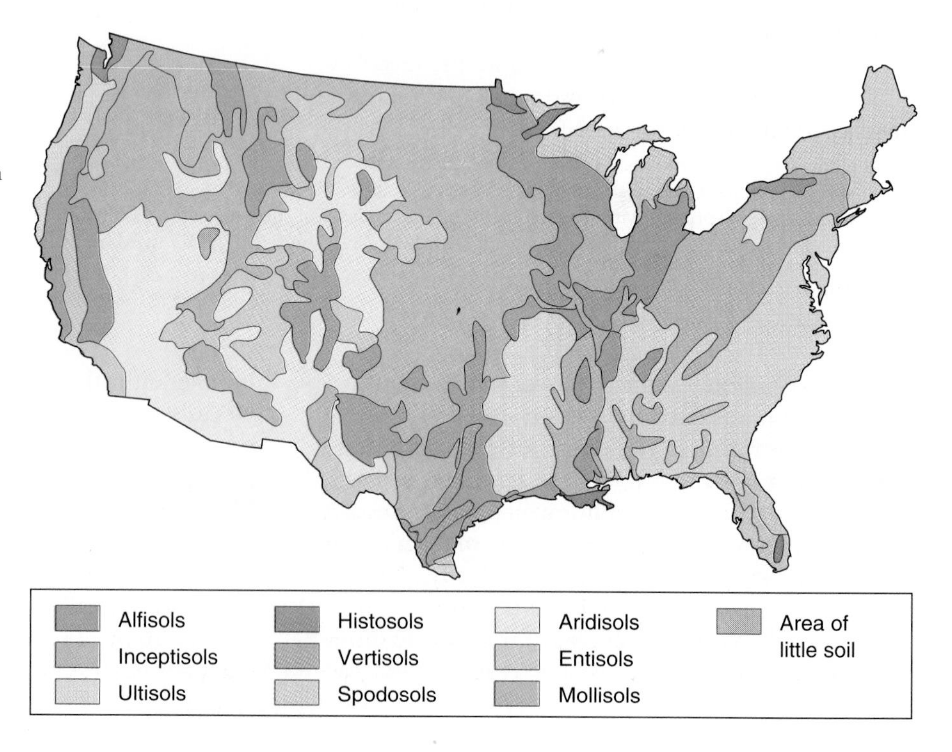


**FIGURE 6–2**
Map of U.S. soil types. Refer to Table 6–1 for definitions. (From T.L. McKnight, Essentials of physical geography, Prentice-Hall, 1992, Fig. 10-13, p. 222. Used with permission.)

**FIGURE 6–3**
Gullying in coarse-grained soils near a housing development in Franklin County, Ohio. (U.S. Department of Agriculture)

more easily gullied by water and removed by wind if protective vegetation is removed (Figure 6–3).

## Moisture Content and Engineering Properties

The moisture content of a soil can significantly affect its landslide or flow potential. As the amount of water in a soil increases, it becomes heavier and more fluid, so its potential for flowage increases (Chapter 9). The percent of water (by weight) in a soil is called its **water content.** As the water content increases, the engineering behavior of a soil changes between two limits, the plastic limit and the liquid limit. The **plastic limit** is the water content that causes a soil to become moldable, like modeling clay. Increasing the water content until the soil reaches its **liquid limit** causes the soil to flow.

Greater clay content in a soil increases its ability to retain moisture and therefore increases its liquid and plastic limits. The numerical difference between the liquid and plastic limits is the soil's **plasticity index.** A low plasticity index means that a soil will go from a plastic state to a liquid state with the addition of relatively little water. Therefore, a lower plasticity index for a given soil means that it will flow more easily. If the liquified soil is exposed on a slope, it may form a hazardous soil/water slurry called a **mudflow.** This geohazard is described more fully in Chapter 9.

## Compressibility

The degree to which the volume of a soil decreases when normal stress is applied perpendicular to its surface (Chapter 2) is a measure of the soil **compressibility.** The most compressible soils are fine-grained and/or organic-rich ones that hold a lot of water in their pores. When pressure is applied, the soil water is squeezed out and volume decreases. This compaction can create a construction problem: a building placed atop compressible soil will compress it, resulting in foundation cracks.

Another type of compressible soil is one with a high percentage of organic material and decomposing vegetation, such as peat, which can hold a lot of water. Building directly on this type of soil squeezes out the water and compresses the soil, sometimes causing damage to building foundations and walls. The geohazards associated with compressible surface materials are discussed more fully in Chapter 10.

## Fertility

The ability of a soil to supply the nutrients required for plant growth (nitrogen, phosphorous, and potassium) is called its **fertility.** Fertile soils are developed both in residual regolith and in transported regolith. Soils developed on transported regolith can be quite fertile for two reasons: they may already contain organic material, or they may be composed of finer mineral particles that weather easily to release elements needed by plants. For example, sediment deposited by river floods usually is quite fertile because it is composed of easily weathered fine material, including organic matter which settled out of the floodwater.

Fertile soils occur on the floodplains of many rivers. These soils remain fertile because periodic flooding replenishes the minerals and organic material in the soil. However, these favored farming sites put people and property at risk from the inevitable natural flooding that occurs. In addition, stream modification to minimize flooding in these areas only causes other problems somewhere else (Chapter 7) and cuts off local soil replenishment.

Similarly, volcanic soils usually are very fertile because the fine pyroclastic material (Chapter 4) weathers easily. For example, the devastated forests around Mount Saint Helens (Chapter 4) have shown remarkable recovery since the 1980 eruption. Some of the new trees had grown to heights of 5 or 6 meters (16 to 20 feet) by 1991. Rich volcanic soils draw people back to the slopes of active or dormant volcanoes time after time. Sooner or later, however, their luck may run out (Figure 6–4).

### Porosity and Permeability

The porosity and permeability of a soil (Chapter 2) determine the rate at which water moves through it. This flow is important (1) to prevent the soil from becoming waterlogged and (2) to remove dissolved ions so they do not build up in the soil. In general, the porosity increases with grain size and better sorting.

Sorting is a measure of particle-size uniformity in a soil or sediment (Chapter 3). Permeability increases with the size of the individual pore spaces and the degree to which they are interconnected (Figure 3–8b and c). As pore space diameters increase, the pore walls exert less frictional drag on fluids moving through the opening, thus increasing the permeability.

### Reactivity

Some soils are acidic, particularly wet, organic-rich ones. They can react chemically with steel pipelines and other metal objects placed within them. Such soils have a high **reactivity.** Remedial measures, such as coating a pipeline, must be taken to reduce corrosion of the metal. The addition of lime can temporarily reduce the acidity of reactive soils.

## SOIL CLASSIFICATION

The U.S. Department of Agriculture classifies soil into 10 categories (Table 6–1). The areal distribution of each category is shown in Figure 6–2. The first two soil classes (Entisols and Inceptisols) are soils that have little or no development of soil horizons because of their geography (desert, arctic) or because they are too young. The next six soil classes (Spodosols, Alfisols, Mollisols, Aridisols, Ultisols, and Oxisols) strongly depend on climate and are characteristic of various geographic regions.

**FIGURE 6–4**
Lahars from Mount Pinatubo inundate farms and villages on the Pasig-Patero River floodplain, Philippines. (Photo by Tom C. Pierson, Cascades Volcano Observatory, U.S. Geological Survey.)

TABLE 6–1
Soil classification (Seventh Approximation). Refer to Figure 6–2 for locations.

| Soil Order | Characteristics | Areas Where Common |
|---|---|---|
| Entisols | Soils without horizons (soil-forming processes have had insufficient time to produce horizons) | Wide geographic range, from desert sand dunes to frozen ground of arctic zones |
| Inceptisols | Weakly developed soil horizons; soil horizon A is developed | Wide geographic range wherever soils have just begun to develop on newly deposited or exposed parent materials, such as volcanic or glacial deposits |
| Spodosols | Humid forest soils with a gray leached horizon and a B horizon enriched in iron or organic material leached from above. Commonly beneath coniferous forests | New England, northern Minnesota, and Wisconsin |
| Alfisols | Soils with clay-enrichment in the B horizon. Lower organic content than Mollisols. Medium to high base supply. Commonly beneath deciduous forests | Western Ohio, Indiana, lower Wisconsin, northwestern New York, central Colorado, western Montana |
| Mollisols | Grassland soils, with a thick, dark organic-rich surface layer. High base supply (calcium, sodium, and potassium) | Widespread in central and northern Texas, Oklahoma, Kansas, Nebraska, North and South Dakota, and Iowa |
| Aridisols | Desert and semiarid soils; low organic content along with concentration of soluble salts within soil profile | Widespread in desert and semiarid areas of Nevada, California, Arizona, and New Mexico |
| Ultisols | Deeply weathered red and orange clay-enriched soils on surfaces that have been exposed for a long time | Humid temperate to tropical soils. Widespread in southeastern U.S. east of Mississippi Valley |
| Oxisols | Intensely weathered soils consisting largely of kaolin, hydrated iron, and aluminum oxide. Bauxite forms in these soils | Warm tropical areas with high rainfall |
| Histosols | Organic soils and peat | Mississippi Delta in Louisiana, Everglades in Florida, local bogs in many areas |
| Vertisols | Swelling soils with high clay content which swell when wet and crack deeply when dry | Southeast Texas, local areas |

SOURCE: U.S. Department of Agriculture, Soil Conservation Service, 1960.

The last two orders, Histosols and Vertisols, are described largely by their composition and are relatively independent of climate. For example, a Histosol is an organic-rich soil that can be found in Louisiana, California, New England, Kansas, or Canada. A Vertisol contains clays that shrink (when dry) and swell (when wet). These shrink-swell clays make up soils in a number of geographic areas, with consequent geohazards (Chapter 10).

## SOIL EROSION

Erosion of the land surface is a function of soil characteristics (as previously discussed), rainfall, vegetation, slope, wind, and land use. These factors do not act independently but combine, reinforcing or opposing each other to create the specific soil erosion pattern in a location. The areal variation in rates of soil erosion in the United States is shown in Figure 6–5.

### Rainfall and Vegetation

Soil erosion varies most with rainfall (Figure 6–6). The relation is not linear, however. In arid climates (annual rainfall less than 25 centimeters or 10 inches), little vegetation exists and sediment is readily eroded, so the sediment yield increases with rainfall. In addition, desert rainfall is intense (although infrequent and short-lived), increasing its erosional potential in the absence of vegetation. Erosion peaks at the precipitation level below where grasslands begin to develop.

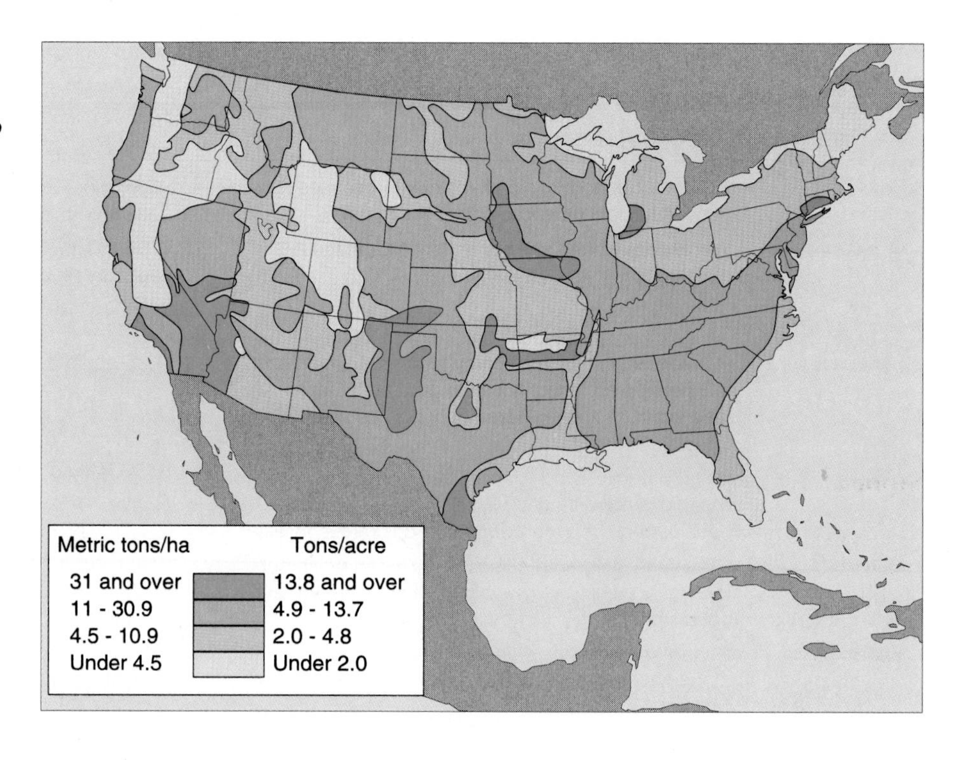


**FIGURE 6–5**
Soil erosion on cropland in the United States. Erosion rates greater than 11 metric tons per hectare (4.9 tons per acre) reduce productivity. Note the erosion severity in "America's Breadbasket," the Midwest. (Adapted from United Nations data, as presented in R. W. Christopherson, 1994, *Geosystems,* 2nd ed.: New York, Macmillan College Publishing Company, Inc.)

Erosion begins to decrease as rainfall reaches the level that is characteristic of grasslands (30 to 75 centimeters or 12 to 30 inches), because the grass roots anchor the soil and resist erosion. The roots also aid water infiltration. Further, the grass itself breaks the impact of the raindrops and thus decreases their erosion potential. Sediment yield remains the same above a rainfall value of 75 centime-

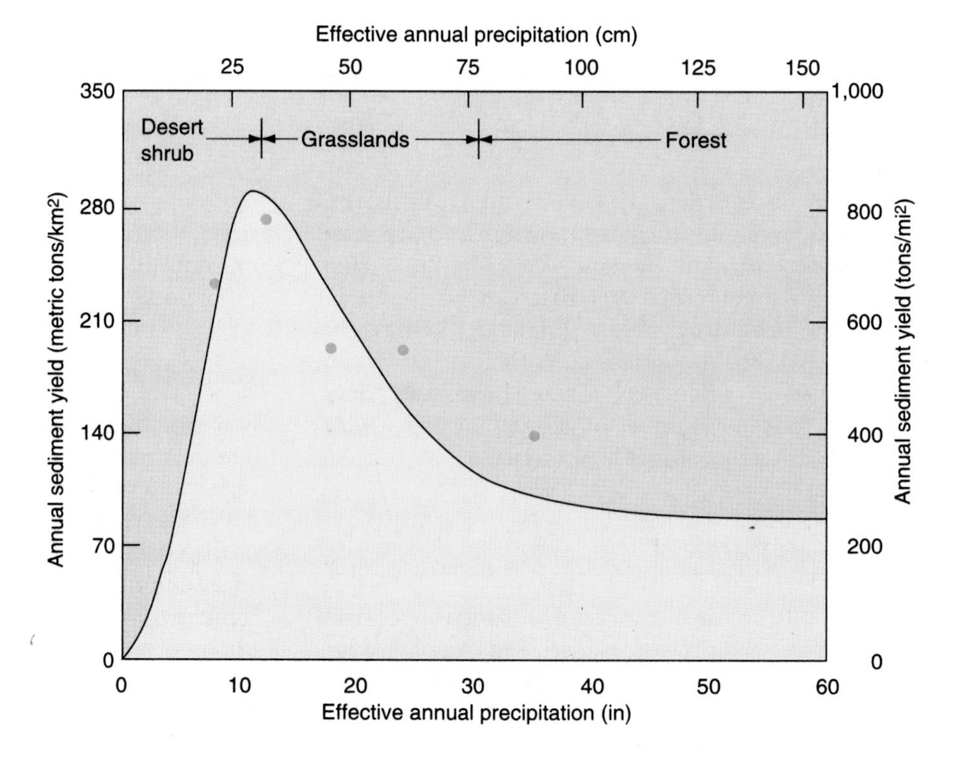


**FIGURE 6–6**
Sediment yield vs. total annual rainfall. (After Langbein and Schumm, 1958, American Geophysical Union Transactions v. 39, p. 1076–1084, Fig. 2, p. 1077.)

ters (30 inches) because trees, shrubs, and forest floor cover significantly retard erosion. Slope erosion has been limited in some areas by planting vegetation, such as kudzu vine, to stabilize slopes.

Heavy soil erosion can occur when drought thins the vegetation, as happened in the Dust Bowl during the 1930s in the U.S. Southwest, primarily Oklahoma and Texas. Major vegetation losses also occur from human activities. These include land clearing for construction or logging (Figure 6–7) and some forest fires. Air pollution, such as acid rain or industrial fumes, also may kill vegetation.

## Slope

Both the length of a slope and its steepness affect erosion. In general, the longer the length of exposed slope and the steeper its gradient, the greater the sediment yield. Longer slopes encounter more rainfall and accumulate more rainwater, creating a higher volume flow with greater erosion potential. Steeper slopes increase the velocity of runoff water and thus its erosion potential.

## Wind

Wind erosion is most prevalent in dry climates. Most of the world's dry climates are sparsely settled, and land use is adjusted to the dry conditions. Farmers either limit land use or artificially support agriculture by pumping groundwater or importing surface water through aqueducts.

Sometimes, changes in rainfall and temperature patterns cause desert areas to spread in a process called **desertification.** A good example is the expansion of the Sahara Desert southward into areas of Africa that used to be more humid (Figure 6–8). Not only are crops harder to grow, but natural vegetation is growing sparser. This is a natural change that has occurred before. However, the effect is worsened by poor land use, which leads to greater soil erosion.

Human activities often increase the effect of wind erosion of the soil. A combination of short-term natural dryness (drought) and destructive farming practices created a massive geohazard in the U.S. Southwest during the 1930s. The area became known as the "Dust Bowl" for the dangerous dust storms that deflated the dry, overworked soil.

## Land Use

Land use also strongly influences the soil erosion rate. A good example is the effect of changing land use in the Washington, D.C., metropolitan area. This region has been settled since early colonial times, so a good record exists of changes in land use and variations in sediment yield over a long period (Figure 6–9).

Prior to 1810, the area was forested and experienced minimal soil erosion. About 1820, land clearing for agriculture began, and sediment yield began to increase, peaking around 1900. A slow decline in sediment yield followed until the late

**FIGURE 6–7**
Gully and sheet erosion associated with housing development along I-5 north of San Diego, California. (Courtesy of Howard Wilshire, U.S. Geological Survey.)

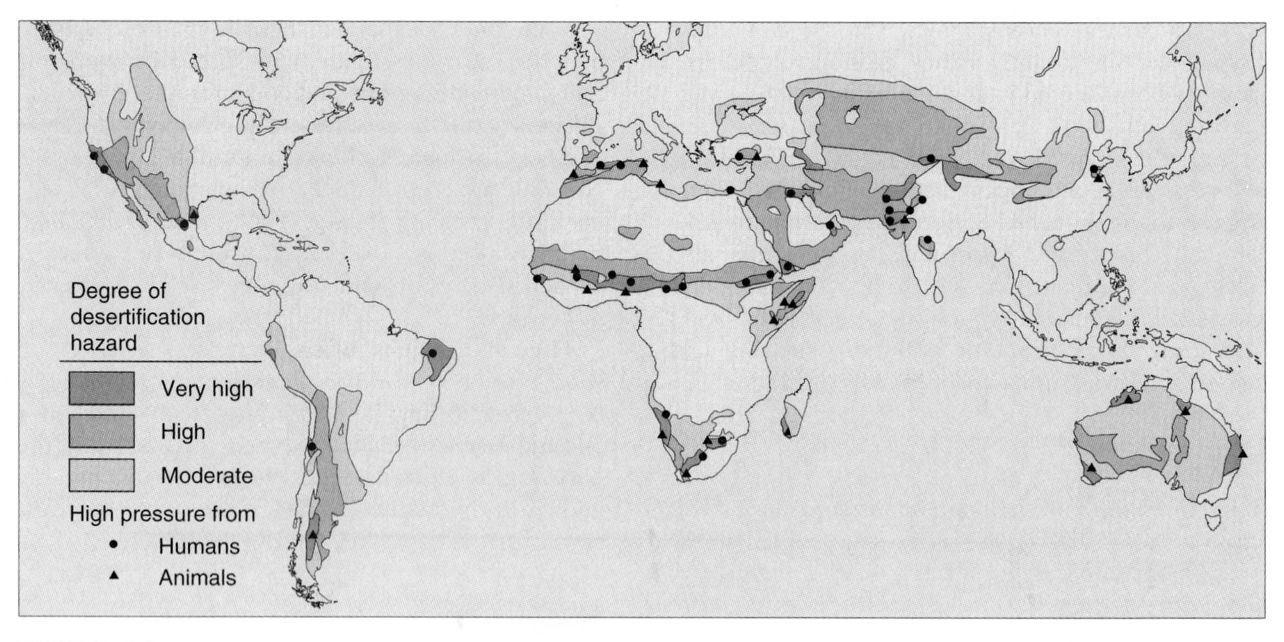


**FIGURE 6–8**
Map showing desertification hazards on a global scale. (Adapted from Council on Environmental Quality, 1978, Annual Report.)

1950s. This drop probably resulted from decreased farming in the area after the 1950s, which allowed ground cover to return once again and retard soil erosion.

Rapid urbanization began in the 1960s, with land clearing and construction of homes, offices, malls, factories, and new roadways to service them. This was accompanied by a dramatic increase in sediment yield (Figure 6–9). As urbanization peaked and began to wane, sediment yield began to decrease because most surfaces had been paved over or planted. Little bare ground remained to be eroded. While sediment erosion is less of a problem today, the present high rate of urbanization is causing new problems, such as increased flooding, as you will see in Chapter 7.

Many human activities promote soil erosion: surface mining, construction, and traffic. We will now examine each of these.

***Surface Mining.*** When valuable mineral resources are located deep underground, they are recovered by *deep mining*. When they are near the land surface, the resource is recovered by **surface mining** (either strip mining or open-pit mining). In surface mining, the overlying regolith is excavated and set aside. The exposed mineral resource then is mined (Figure 6–10). The final step is to put

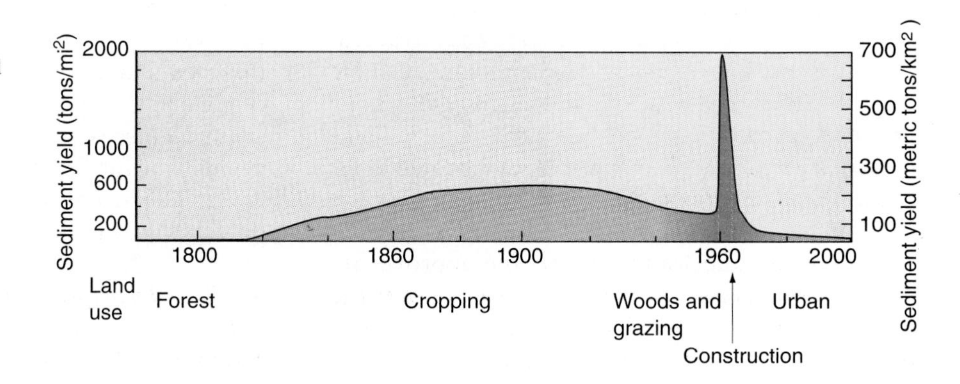


**FIGURE 6–9**
Changes in sediment yield with land use in the Washington, D.C., metropolitan area. (From M. G. Wolman, 1967, A cycle of sedimentation and erosion in urban river channels, Geografiska Annaler, v. 49A, p. 385–395, Fig. 1, p. 256. Used with permission.)

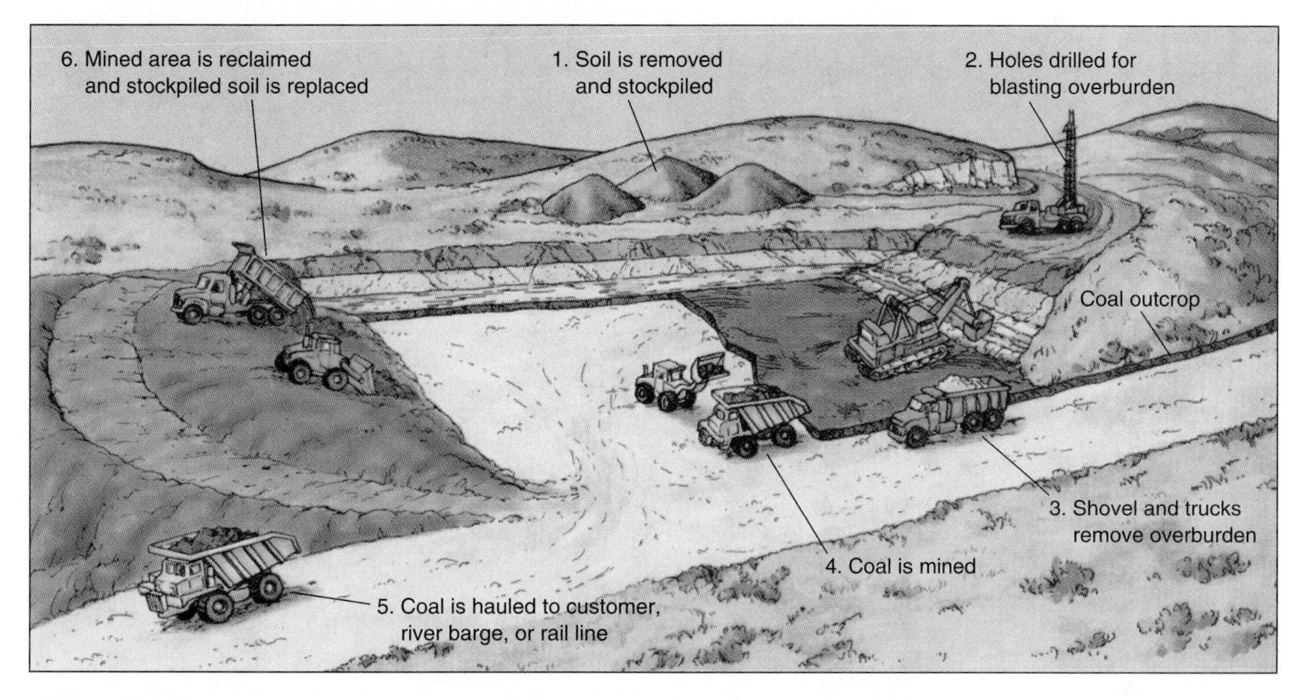


**FIGURE 6–10**
Surface mining process and products (stripped soil, regolith, and resource). This type of surface or "strip" mining is common in the Appalachian Mountains. The mining follows the contour of the coal where it crops out of a mountainside. (After Paul W. Queen, West Virginia Geological and Economic Survey.)

back the regolith and plant it, a process called *reclamation.*

However, prior to strong federal laws, soil and regolith were stripped off and indiscriminately mixed in piles. This left the land with a furrowed topography and large "waste piles" on which nothing grew well (Figure 6–11). Under such conditions of steep and bare slopes, sediment erosion reached a maximum. In addition, water infiltrating the piles could react with some minerals to produce toxic substances. (This problem is discussed in more detail in Chapter 12.)

Because of strong federal and state environmental laws, however, surface mining has changed. Most strip mining operations today minimize these problems (Figure 6–10). In a modern surface mine, the soil is carefully removed and stockpiled. The remaining infertile regolith and rock overlying the resource then are stripped off and stored in separate piles. After the mineral resource is removed, bulldozers smooth the waste piles to the approximate original topography. The soil then is spread over the restored area and planted with suitable vegetation. These procedures not only minimize sediment erosion but also restore the area to near its original condition (Figure 6–12).

***Construction.*** Construction activities are a major cause of soil erosion because large areas are laid bare and worked over at the same time. Slope changes may be significant. The removal of vegetation worsens the problem. Vegetation not only anchors the soil and helps shield it from raindrop impact, but it also slows the overland flow that causes soil erosion.

Erosion is most severe on hillside construction sites where excavations may significantly alter surface slopes (Figure 6–13). An excavation into a slope sometimes creates a steeper slope on the backside of the excavation. The steeper slope is prone to greater erosion and gullying unless it is planted promptly with fast-growing grass or other ground cover. The steepened slope also increases landslide potential, as will be discussed in Chapter 9.

***Surface Disruption by Pedestrians and ORVs.*** Even erosion-resistant fine-grained soils become readily erodible when the vegetation on their sur-

**FIGURE 6–11**
Abandoned surface mine in Ohio. This site was mined before the Surface Mining Control and Reclamation Act (SMCRA) was signed into law on August 3, 1977. (U.S. Department of the Interior, Office of Surface Mining, Reclamation and Enforcement)

**FIGURE 6–12**
Land reclamation after surface mining near St. Clairesville, Ohio. This former mine site is now productive farmland. (U.S. Department of the Interior, Office of Surface Mining, Reclamation and Enforcement.)

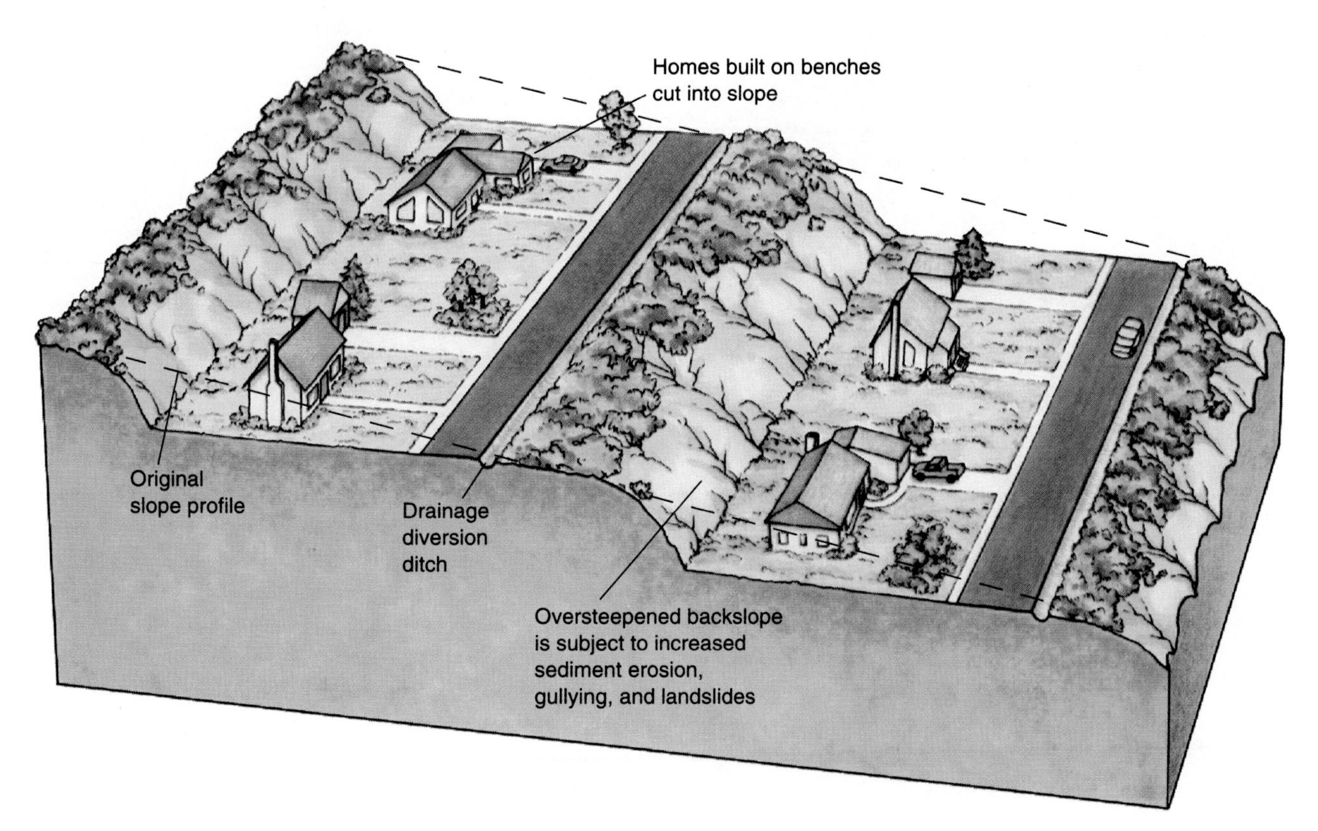


**FIGURE 6–13**
Diagram showing slope changes and potential erosion/landslide problems in upland building sites.

faces is worn away by pedestrians or the land surface is rutted by vehicles. A significant erosion agent is the off-road vehicle (ORV), which is becoming popular in coastal, mountain, and desert areas. These vehicles destroy vegetation and make ruts in the surface (Figure 6–14). Further, their weight compacts the soil, reducing rainfall infiltration. As infiltration decreases, runoff increases, and greater surface erosion and gullying is the result.

## SEDIMENT POLLUTION

The environmental damage to humans, plants, and animals resulting from sediment deposition is called **sediment pollution.** Sediment pollution causes varied aesthetic, health, and safety problems. It greatly increases the cost of maintaining such public services as pure drinking water, drainage channels, navigable waterways, and water supply systems.

### Aesthetic Problems

Soil erosion can result in surfaces stripped of vegetation and mud stains on structures. Overland runoff on denuded slopes first forms shallow **rills,** which resemble miniature gullies. They deepen and interconnect to form deeper **gullies** (Figure 6–3). The heads of the gullies extend themselves by eroding upslope over time, a process called **headward erosion.** If unchecked, the gully system can extend from the original bare surface to undermine areas having vegetative cover and even forests (Figure 6–15).

Once a gully system has developed, it expands rapidly unless remedial measures are taken. Gullying can be controlled by filling and vegetating the gullies and by placing diversion ditches to minimize overland runoff until the eroded slope is fully restored (Figure 6–16).

Chemically weathered soils (Chapter 3) commonly are enriched in iron oxides. When overland

FIGURE 6–14
Deep erosion channels cut by vehicles on a slope in Jawbone Canyon, California. In many places the channels reach down to the bedrock. (Howard Wilshire, U.S. Geological Survey.)

FIGURE 6–15
Gully erosion accelerated by farming practices near Half Moon Bay, California. Deep gullying extends up into vegetated slope. (Courtesy of Raymond Pestrong, San Francisco State University.)

waters erode these soils, the iron oxides give the stormwater a rusty color. This becomes a problem when rust-colored waters flood homes, for the muddy waters leave permanent rust-colored stains on all absorbing surfaces (Figure 6–17).

## Increase in Water Treatment Cost

People obtain water from wells in the ground (Chapter 8), streams and lakes, or reservoirs. Wells usually are sediment-free because the soil particles filter from the water as it flows through the ground toward the well. However, water from reservoirs, rivers, and lakes can be degraded seriously by sediment pollution.

Sediment pollution also decreases water storage capacity in reservoirs—in fact, it is the limiting factor in determining the life of a reservoir. Dams usually are built in sparsely populated areas to impound streamwater in reservoirs. Engineers predict the reservoir's life by determining the expected sediment volume supplied each year relative to the storage volume of the reservoir. These estimates are based on sediment yield from a drainage basin that is vegetated.

However, over time, and with increasing population, remote areas may become settled. Typically, they are first developed into recreation complexes and then evolve into new suburban areas as vacationers elect to live there year-round. The construction and change in land use can result in significant increases in sediment supply to the reservoir. This excess sediment reduces the storage volume of the reservoir significantly, unless remedial action is taken.

Engineers have tried trapping the sediment in **debris basins** upstream from the reservoir, dredging sediment from the reservoir, and utilizing controlled flushing to release sediment through the dam. These actions have mixed results and are expensive. Most scientists agree that the best solution is to control sediment at its source by on-site man-

FIGURE 6–16
Gully control by placement of stone riprap on a mine site in eastern Kentucky. Channel erosion decreased greatly after the channel was lined with rock. (Courtesy of Chuck Meyers, U.S. Department of the Interior, Office of Surface Mining, Reclamation and Enforcement.)

agement of runoff and trapping sediment in upland basins.

Sediment pollution also affects the *quality* of drinking water. Wherever water is taken from streams or lakes, the excess sediment must be removed before it is potable, or suitable for drinking. Chemicals such as alum (potassium aluminum sulfate) are used to coagulate the particles, causing them to settle from the water. Use of these chemicals and settling tanks increases water treatment cost.

FIGURE 6–17
Flood damage and wall staining in a machine shop. (Glade Walker, U.S. Bureau of Reclamation.)

## Health Hazards

Mosquitoes need standing pools of water to thrive, so one way to control mosquitoes is to lower the water table. This limits the water pools in which they can lay eggs. The water table is lowered by digging drainage ditches across marshes and swamps (Figure 6–18). Sediment-laden runoff from construction areas may clog these ditches and prevent adequate drainage.

Fine-grained clay-rich sediment particles have surfaces that can hold harmful substances extracted from the water, such as toxic chemicals or heavy metals. This process of attracting substances to particle surfaces is called **adsorption** (see Chapter 13). The particles concentrate the adsorbed materials. When moving water slows, the particles with their adsorbed substances are deposited. If they are subsequently eroded, the concentrated adsorbed pollutants may be released into water once again, creat-

FIGURE 6–18
Drainage ditch clogged with sediment on an Indiana mine site. (Courtesy of Chuck Meyers, U.S. Department of the Interior, Office of Surface Mining, Reclamation and Enforcement)

ing new pollution conditions and a potential health problem.

### Hazards to Plants and Animals

Aquatic plants depend on light penetrating through the water for photosynthesis. The diminished ability of light to pass through sediment-clouded water limits the growth of aquatic plants. Bottom-dwelling (benthic) organisms and fish also suffer because sediment pollution increases the sediment supply. The greater sediment volume literally can bury some organisms and interfere as fish try to lay their eggs on the bottom. The increased water turbidity also clogs the gills of fish and makes eating harder for filter-feeders, which are animals that filter water to obtain their food.

### Increased Flooding Risk

Clear gutters, clear drainage ditches, and clear, deep stream channels are essential for conducting stormwater from an area to prevent flooding. Increased sediment supply fills drainage structures and stream channels, forcing excess water to overflow onto the adjacent land. This problem is especially severe when the eroded sediment is poorly sorted and coarse-grained. Gravel and even boulders may be washed into drainageways and remain there because they are too large to be moved by the water. These particles serve as nuclei around which other coarse particles accumulate to form a plug of gravel, which further restricts flow through the channel.

Increased sediment supply in a navigable waterway may require continual dredging to maintain channel depth. This not only is expensive but increases the possibility of water pollution, because any buried contaminants may become resuspended in the water. This problem is discussed further in Chapter 13.

## REDUCING SEDIMENT POLLUTION

The most effective and inexpensive way to reduce sediment pollution is to do it at the source. Strategies used in site-management and waterway-management programs include siting, construction practices, and land usage.

### Siting

Sediment control begins with an intelligent choice of a site where sediment yield will be minimized during development. For example, gently sloping areas are better than steeply sloping ones. Areas with less erodible soils are preferred. Areas having good natural drainage are preferred over those that drain poorly.

### Construction Practices

Remedial measures must be in place *before* clearing the land and starting construction. Remedial measures include drainage culverts, diversion channels along the tops of slopes, and sediment traps (Figure 6–19) to hold eroded sediment on-site rather than allowing it to travel. This trapped sediment can be dredged from the debris basins later and used to fill rills and gullies formed during construction.

Good practice restricts land clearing to areas that will be built on immediately. Clearing a wide area risks extensive erosion and sediment pollution. If this cannot be avoided, the cleared areas should be seeded with fast-growing grass to give temporary soil-holding during construction. Covering the cleared construction surface with straw or mulch

**FIGURE 6–19**
Sediment trap utilizes filter fabric baffle on a Maryland site. The turbid, sediment laden water on the left contrasts markedly with the clearer water on the right at the outlet end of the settling pond. (Courtesy of Chuck Meyers, U.S. Department of the Interior, Office of Surface Mining Reclamation and Enforcement)

holds soil moisture and provides a protective mat from erosion by construction vehicles.

Removing trees, storing them, and replanting them after construction provides mature vegetation to the building site as soon as construction is completed. This cost is at least partially offset by the greater value of developed sites that have mature trees.

Special measures are required for excavation into slopes for structure sites (Figure 6–13). The flat **bench** excavated for structure sites also creates a steep backslope behind it, which must be protected promptly with vegetation or netting (Figure 6–20). **Diversion culverts** along the top of the backslope will protect it from runoff. **Paved drop chutes** along the slope and **storm drains and sewers** on the site complete the protection. These measures remove water immediately to prevent flooding. Multiple drainage channels across debris basins (Figure 6–20) trap sediment and reduce sediment pollution downslope.

## Land Usage

One way to limit soil erosion is to minimize the removal of natural vegetation. This is difficult because of the homestyle to which Americans are accustomed. In the typical **tract development** of many homes, each dwelling is at the center of its lot or tract. This has been standard in America for half a century.

In **cluster development** (Figure 6–21), many structures are grouped together in "villages," with the land in between kept in its natural state for walking, picnicking, and bicycling. This popular development requires a minimal road network. Electricity, power, phone, TV cable, trash removal, and sewage service are much cheaper to provide because each conduit and collection site serves multiple dwellings. Construction and heating costs are less because of common walls among the units. Many feel that cluster development is a less expensive, more efficient, environmentally safer way to house our growing population. Because smaller areas are cleared, there is far less soil erosion.

Cluster development exemplifies the compromises that a growing population may have to make. The personal space for each dwelling is greatly reduced in cluster development, but the gain is in the large common area available to all for recreation. It is one way in which a growing population can enjoy a pleasant life-style while maintaining an acceptable quality of life in the face of shrinking land availability and resource depletion we will face.

## LOOKING AHEAD

In the next chapter, we look at a major agent of sediment erosion and transportation: streams. We will consider how streams work, how they flood, how floods are predicted, and ways of mitigating

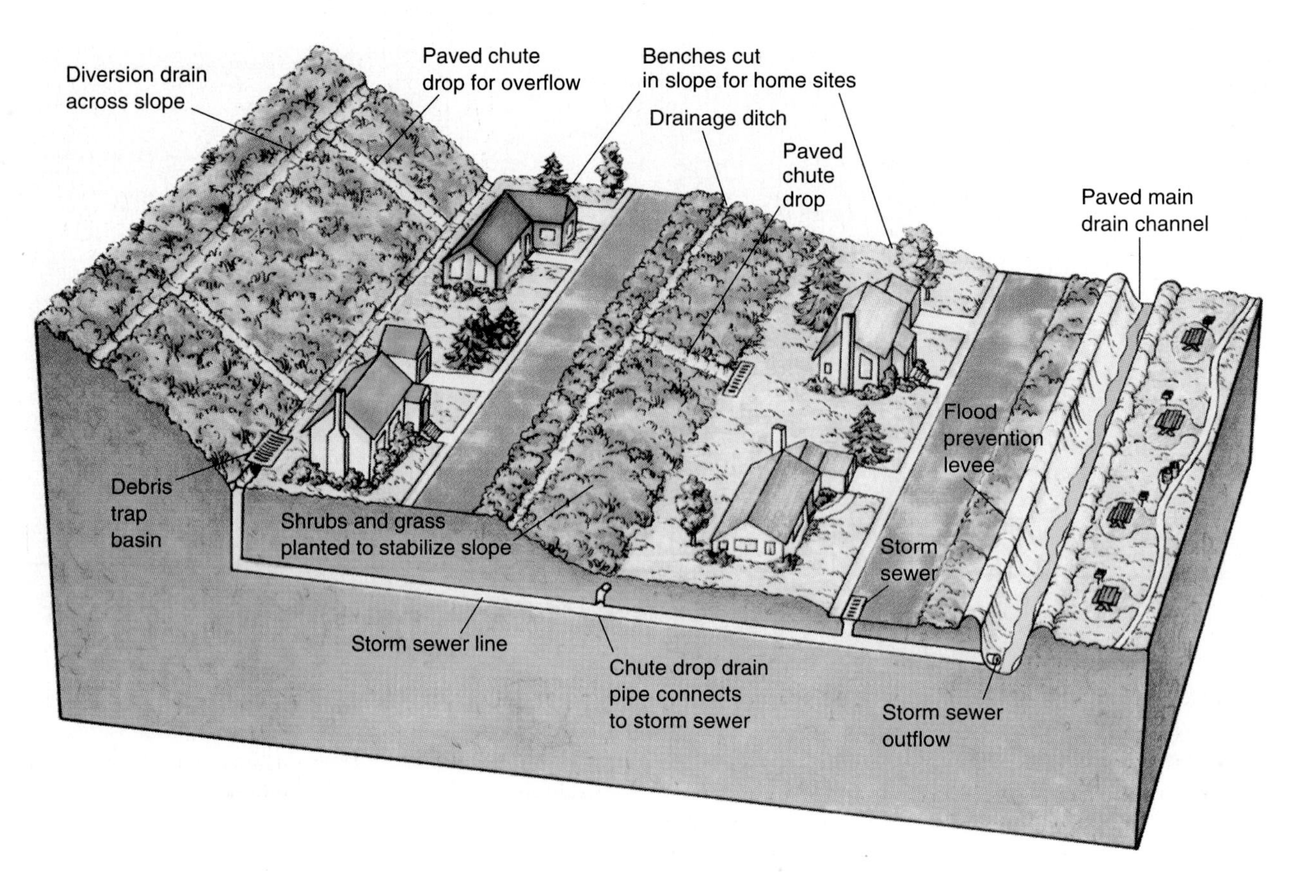


**FIGURE 6–20**
Proper drainage and erosion-control practices in development minimize sediment pollution and flooding.

**FIGURE 6–21**
Cluster development in Columbia, Maryland. The land is developed so that homes are concentrated in small areas (villages) separated by green space for common use. (Photo © Sky High Studios.)

this widespread geohazard. We will visit New Orleans and Rapid City, South Dakota, for case studies of flood control.

## SUMMARY

*Soil* has different meanings for farmers, engineers, and geologists. Agricultural soil is regolith derived from rock or sediment that has been sufficiently weathered to release nutrients needed for plant growth. Engineering soil is any loose material that can be excavated without blasting. Geologists consider soil to be residual regolith; only the upper part may qualify as an agricultural soil.

Once agricultural soil is washed away, only infertile sediment or rock remains and the land becomes useless for growing crops. Soil is useful not only for growing crops but also for retaining drinking water and purifying sewage waste. Certain "problem soils" have properties that can cause geohazards.

Soil profiles develop most fully in warm and humid climates. Continued weathering develops mature soils with four horizons: O (organic), A (leached), B (accumulation), and C (partially weathered parent material).

Soil properties vary widely. Soil color is related to its organic content, oxidation, and drainage conditions. Soils with greater cohesion are less erodible. As soil moisture increases, the soil first becomes plastic and then becomes a liquid that can flow downslope. Compressible soils, such as fine-grained or organic-rich soils, are readily reduced in volume when under pressure.

Fertile soils are rich in the nutrients nitrogen, phosphorous, and potassium. Soil porosity is the volume of open space within the soil. Permeability describes the ease with which water can move through the soil. Some soils are reactive and can corrode metal objects buried in them, such as pipelines. Soils are classified by their degree of horizon development, composition, and the type of material concentrated or depleted in each horizon.

Soil erosion is strongly related to rainfall and vegetative cover. Maximum erosion occurs in semiarid regions where annual rainfall is low, rains are infrequent but heavy, and vegetation is poorly developed. With greater rainfall, there is greater vegetation development and this limits soil erosion by slowing flow and aiding infiltration. Erosion also increases with slope length and the amount of slope exposed. The least erosion occurs in natural forests, more in cropland, and the most in cleared areas.

Sediment pollution creates multiple problems. Iron oxides in suspended soil particles color floodwater red, creating aesthetic problems as it colors every surface it encounters. Sediment-rich runoff fills mosquito-control drainage ditches, causing a health risk. When pollutants are adsorbed onto sediment surfaces and concentrated, another health risk is created.

Water cost increases as sediment fills reservoirs. Water treatment cost increases when suspended sediment must be removed from drinking water. Aquatic plants and animals are at risk when water turbidity and bottom sedimentation increase. Flooding increases when sediment clogs drainage ditches and streams, forcing the excess water onto the adjacent land.

Sediment pollution can be reduced by retaining natural vegetation. Prompt seeding with fast-growing plants can inhibit erosion on slopes denuded by lumbering or forest fires. Good construction practices are used to minimize soil erosion during construction by protecting the soil and diverting drainage around the construction area. Prompt replanting, protective berms, and effective drainage systems with debris traps minimize sediment erosion.

## KEY TERMS

A horizon (leached horizon)
adsorption
B horizon (accumulation horizon)
bench
C horizon (partially weathered horizon)
cluster development
cohesion
compressibility
debris basins
desertification
diversion culverts
erodibility
fertility
gullies
headward erosion
leaching
liquid limit
mature soils
mudflow
O horizon (organic horizon)
paved drop chutes
pedalfer
pedocal
plastic limit
plasticity index
reactivity
rills
sediment pollution
soil horizons
storm drains and sewers
surface mining
tract development
water content

## REVIEW QUESTIONS

1. How does the definition of soil differ according to farmers, engineers, and geologists?
2. Give four reasons why soils are important.
3. How do soil horizons differ? Account for the differences.
4. What information can be obtained from the color of a soil?

5. Define the liquid limit and plastic limit. What is the environmental significance of these parameters?
6. What makes some soils so compressible?
7. What is soil fertility? How do farmers enhance it?
8. Distinguish between soil porosity and permeability.
9. Use the soil classification in Table 6–1 to determine the types of soils in your area.
10. How is soil erosion affected by (a) rainfall, (b) vegetation, (c) land use, and (d) soil cohesiveness?
11. Describe how modern surface mining operations minimize soil erosion problems.
12. Compile a list of places in your area where significant soil erosion occurs. Try to determine the factors causing each.
13. How can soil erosion be minimized (a) before, (b) during, and (c) after construction?
14. Describe the environmental benefits of cluster development versus standard tract development.
15. What factors caused the Dust Bowl?

## FURTHER READINGS


Birkeland, P. W., 1984, Soils and geomorphology: New York, Oxford University Press, 285 p.

Hunt, C. B., 1972, Geology of soils: Their evolution, classification and uses: San Francisco, W. H. Freeman Co.

Jones, D. E., Jr. and Holtz, W. G., 1973, Expansive soils: The hidden disaster. Civil Engineering, August, p. 49–51.

Pestrong, R., 1974. Slope stability, American Geological Institute, New York, McGraw-Hill, 65 p.

Robinson, A. R., 1971, Sediment, Journal of Soil and Water Conservation, v. 26, p. 61–62.

Wilshire, H. G., and Nakata, J. K., 1976, Off-road vehicle effects on California's Mohave Desert, California Geology, v. 29, p. 123–132.

# 7

# *Streams*

Throughout human history, river valleys have been heavily settled, despite the geohazards of river erosion and flooding. These areas are heavily developed for several reasons. In mountainous and hilly regions, they often are the only flat land available for development, farming, highways, and railroads. Fertile soils have made stream floodplains favored agricultural sites. Streams provide abundant water for drinking, crop irrigation, watering livestock, domestic use, manufacturing, and waste disposal.

Access to water transportation is another reason for river valley development. River transport is the least expensive way to move some bulk goods between the continental interior and the coasts. Coal is an example: the National Coal Association calculates that a gallon of diesel fuel can move one ton of coal:

- almost 100 kilometers (60 miles) by truck
- about 320 kilometers (200 miles) by rail
- over 800 kilometers (500 miles) by river barge.

River transport had a significant role in developing many major U.S. cities, including New Orleans, St. Louis, and Pittsburgh.

The dramatic flooding of the Missouri-Mississippi River Basins in 1993 provides a powerful example of severe flooding, and it is cited throughout this chapter.

Urbanization of stream valleys is a growing problem, because these valleys flood periodically. Although most floods are geohazards of natural origin, our modification of stream channels and urbanization of valley floors has increased the frequency and severity of many floods, making them human-generated geohazards as well.

The massive flooding in the Missouri-Mississippi River valleys during the summer of 1993 was a natural event. However, its severity was increased by human intervention, and the damage was extreme in some places due to extensive development of former floodplains. These problems are discussed in Box 7–1.

**Flooding on the Mississippi River of St. Louis, Missouri, in the summer of 1993. (Allan Tannenbaum/Sygma.)**

## STREAMFLOW

Some basic aspects of streams were discussed in Chapter 3. To review, rainwater and meltwater from snow reach streams by traveling as runoff along the surface, or by infiltration and gravity flow underground.

Over time, the characteristics of flowing water can change considerably. These changes in flow influence the shape of the stream channel. Several variables describe streamflow and channel characteristics:

- **Discharge** is the volume of water passing a point over a period of time—usually expressed as cubic meters per second or cubic feet per second. Stream discharge can vary from a few cubic millimeters per second for a small rill to over 200,000 cubic meters per second (about 7 million cubic feet/second) for Earth's largest-volume river, the Amazon in Brazil.
- **Velocity** is the rate of water flow (for example, 10 meters per second or 33 feet/second).
- Stream **width,** the horizontal distance from bank to bank, is measured perpendicular to the flow (Figure 7–1).
- Stream **depth,** the vertical distance between the water surface and the stream bed, is averaged for several points across the stream (Figure 7–1).

### Discharge

Generally, natural stream discharge increases under three conditions: (1) during transition from a drought period to a flood period; (2) in times of heavier rainfall or snowmelt; and (3) where tributaries add water to the stream. Greater stream discharge erodes the streambed more and transports more eroded material from the area.

### Velocity

Water velocity varies with the gradient and discharge of a stream (Chapter 3). It also varies widely at different points across a stream. Maximum velocity occurs in the center of the channel, just below the surface (Figure 7–2). The lowest velocities occur near the bottom and on the channel sides, where friction is greatest.

*All other things being equal,* velocity is faster in streams having steeper gradients (Chapter 3), streams having smooth and straight channels, and streams having large discharges. The last point may not be as obvious as the other two. As stream dis-

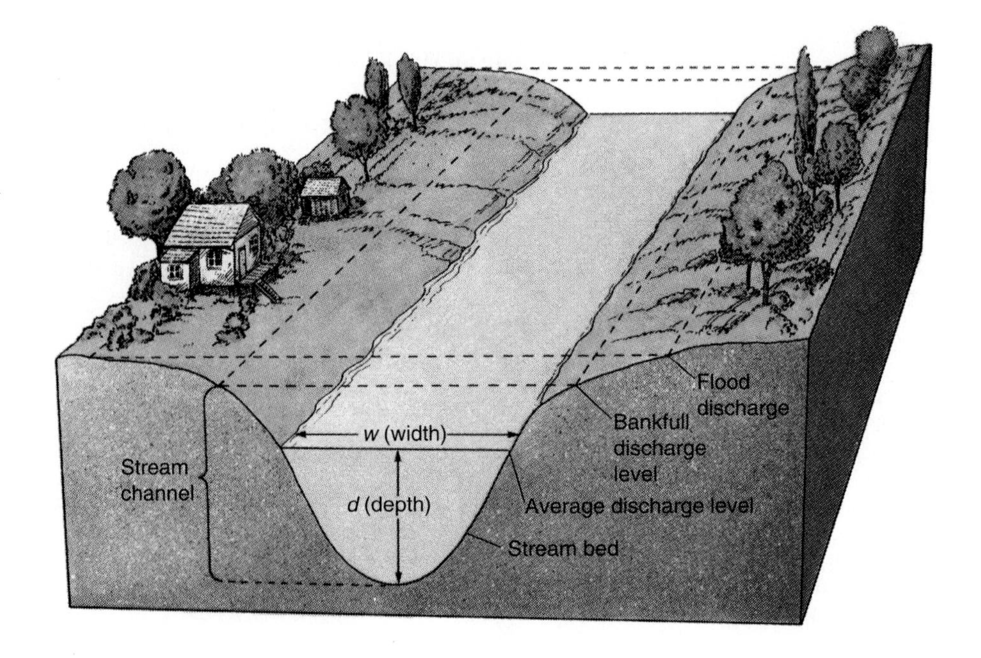


**FIGURE 7–1**
Variables used to describe streams. Note changes in the width *w* and depth *d* as discharge increases. (Reprinted with the permission of Macmillan College Publishing Company from *Physical Geology,* by Nicholas K. Coch and Allan Ludman. Copyright © 1991 by Macmillan College Publishing Company, Inc.)

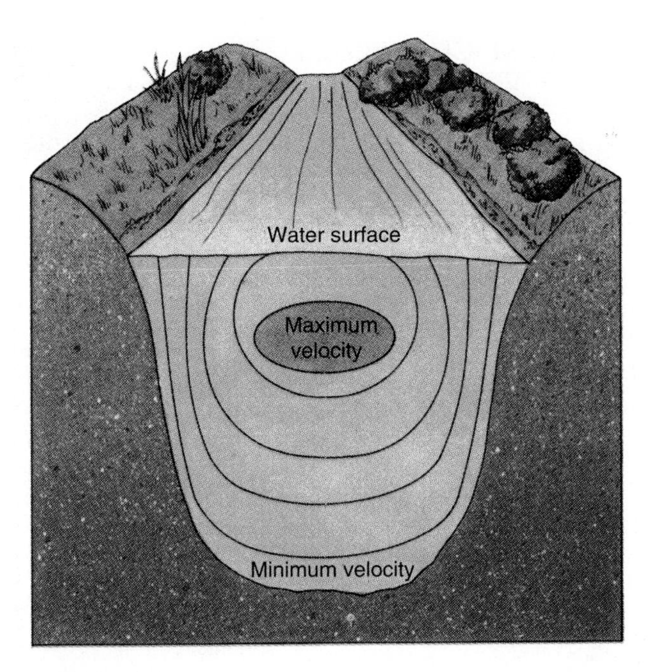


**FIGURE 7–2**
Water velocity varies across a stream. (Reprinted with the permission of Macmillan College Publishing Company from *Physical Geology*, by Nicholas K. Coch and Allan Ludman. Copyright © 1991 by Macmillan College Publishing Company, Inc.)

charge increases after a rainfall, the stream *must* flow faster to move the greater discharge through the existing channel.

## Relation Between Streamflow and Channel Variables

As stream discharge increases, predictable increases occur in stream width, depth, and velocity. These variables are systematically related:

$$\underset{(m^3/\text{sec})}{\text{Stream discharge}} = \underset{(m)}{\text{width}} \times \underset{(m)}{\text{depth}} \times \underset{(m/\text{sec})}{\text{velocity}}$$

You can see that an increase in any of the factors—width, depth, or velocity—increases the discharge. A decrease in any factor reduces the discharge. Such a system, in which a change in one part is *actively balanced* by a change in another part, is said to be in **dynamic equilibrium.**

To demonstrate how dynamic equilibrium operates in streams, consider the changes in flow that occurred in the San Juan River near Bluff, Utah, from September to October 1941 (Figure 7–3). Note the changes in the channel as discharge increased markedly from September to October. The

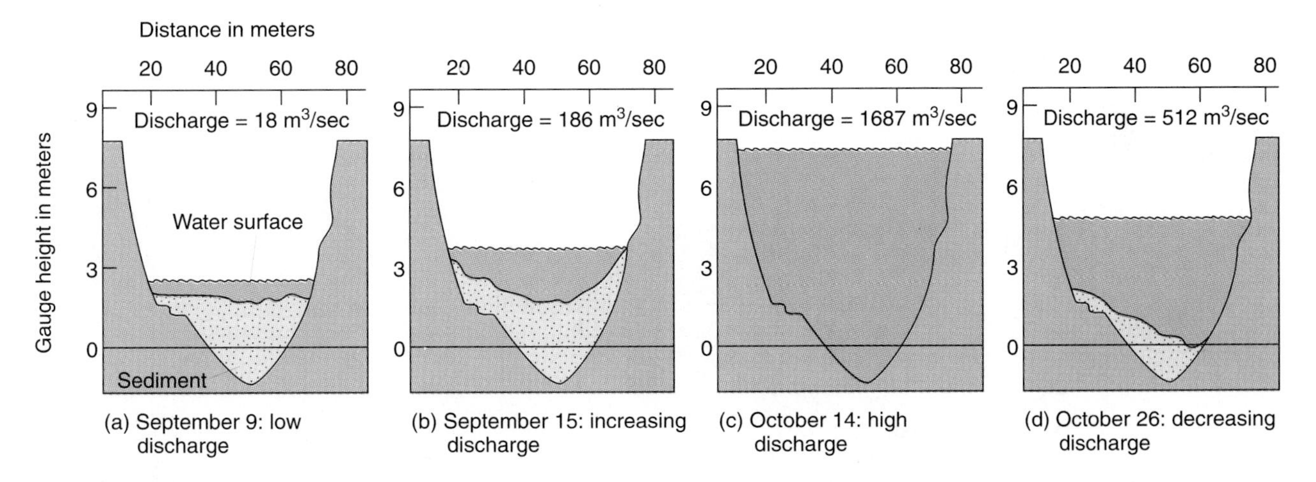


**FIGURE 7–3**
Cross sections showing the changes in water surface and stream bed elevations in the San Juan River, Utah. (a) During low discharge, the stream flows slowly over thick channel deposits. (b) A rise discharge is accompanied by increasing width, depth, and velocity. (c) High discharge results in erosion of the bed material so that the stream reaches its maximum cross-sectional area. (d) Decreasing discharge results in a deposition of stream deposits and lowering of water-surface elevation. (After Fluvial processes in geomorphology. By L. Leopold et al. Copyright © 1964 by W. H. Freeman and Co. Used with permission.)

increasing discharge was accompanied by faster velocities, a rise in elevation of the water surface (which also increased the stream width and depth), and greater scouring of material from the streambed (which increased the depth)—Figure 7–3.

As the stream discharge began to decrease (Figure 7–3d), the velocity slowed and the water level dropped as the channel bed was built up once again. These adjustments brought the stream channel back into equilibrium with the lower discharge.

It is important to remember that the base of the stream channel is being scoured and deepened at the same time. This increased erosion contributes much of the sediment that covers flooded land.

## STREAM CHANNELS

Most stream channels fit into one of three categories:

- □ A **straight channel** (Figure 7–4a).
- □ A **meandering channel** (Figure 7–4b) is a series of sinuous curves (a fine example is the lower Mississippi River).
- □ A **braided channel** resembles braided hair or a braided rope (Figure 7–4c). Braided channels break into numerous smaller channels separated by islands or sandbars.

Over the entire length of a stream, which can range from less than a kilometer to 6,693 kilometers

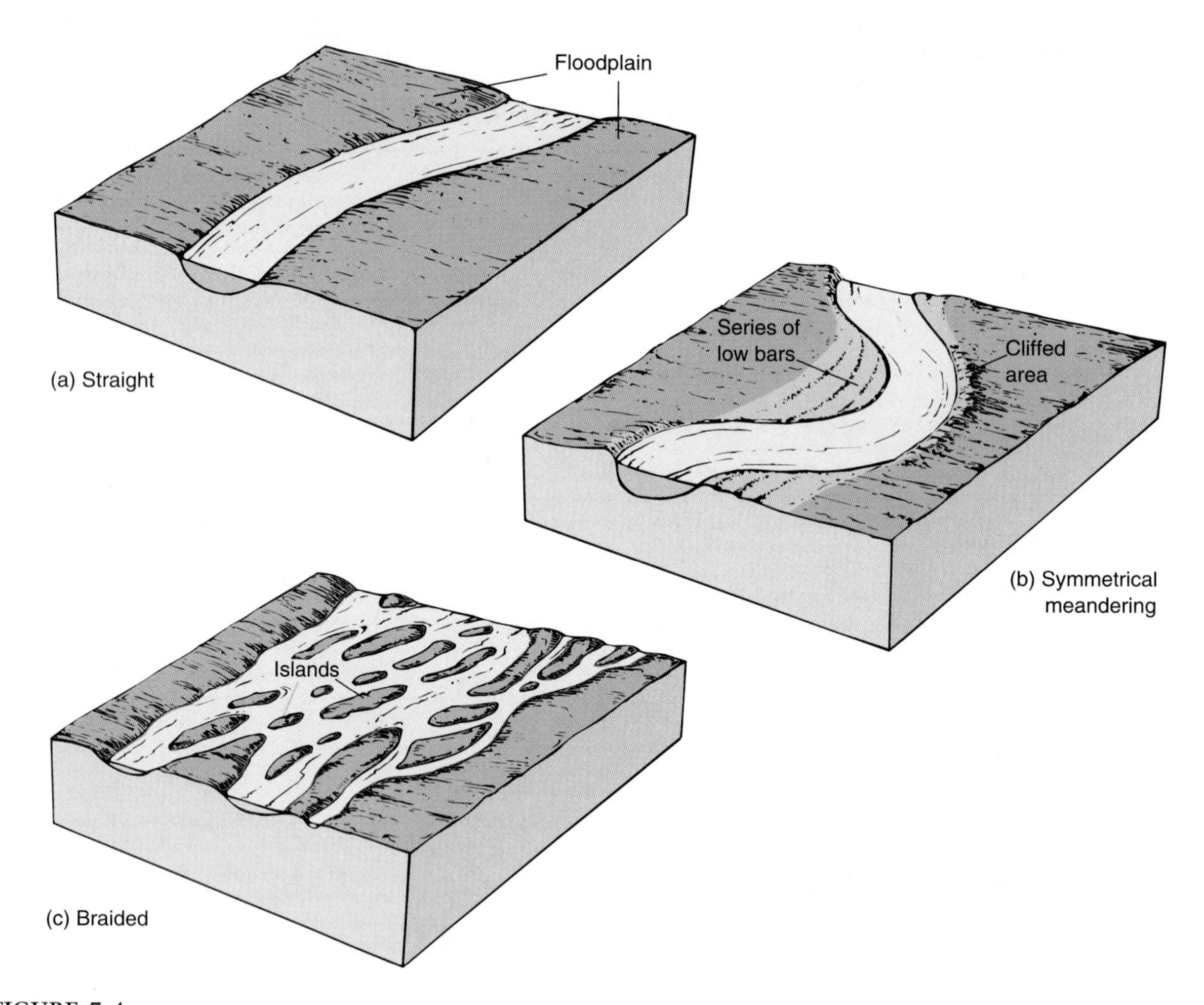


**FIGURE 7–4**
Three types of stream channel patterns: (a) straight, (b) meandering, and (c) braided. (Reprinted with the permission of Macmillan College Publishing Company from *Physical Geology,* by Nicholas K. Coch and Allan Ludman. Copyright © 1991 by Macmillan College Publishing Company, Inc.)

(4,160 miles) for Earth's longest river, the Nile, different segments may have different channel patterns. This results from varying local conditions of topography and geology. A stream segment also may change channel pattern during different seasons of the year, because discharge or sediment supply varies seasonally. The ability of a stream to change its channel pattern is another example of the dynamic equilibrium mentioned earlier in this chapter.

## Straight Channels

What makes some streams flow nearly straight? Factors include steep gradient, a channel of resistant rocks, and well-developed linear (straight) fractures such as joints in the rock. Steep gradients encourage water to move downslope in the most direct path—a straight line. Resistant rock inhibits the lateral (sideward) erosion of channel walls. Well-developed linear fractures in rocks provide a path along which erosion is likely to occur; thus the stream aligns itself with the fractures.

## Meandering Channels

Most streams have meandering channels, which reflect the most energy-efficient path the streams can take. Conditions favoring the development of a meandering channel include easily erodible bank materials and a gentle gradient. Streams with gentle gradients are close to base level (Chapter 3) and have less tendency to downcut vertically. They tend to erode the outside bends of their channels, a process known as **lateral erosion.** Lateral erosion is most marked when the bank materials are unconsolidated sediments.

As a meandering stream migrates laterally across its floodplain, the meanders may become so curving, or sinuous, that the stream eventually erodes through the narrow neck separating two meander curves (Figure 7–5). This cutting off and abandonment of a former meander curve leaves a crescent-shaped **oxbow lake** on the floodplain (c). Many oxbow lakes fill with vegetation and fine suspended sediments from subsequent floodwaters.

The continual slow migration of stream meanders can be a geohazard to structures on the outside of a meander (Figure 7–5). It can even have political effects, because streams often define boundaries between nations, states, counties, and other administrative units. In some cases, people

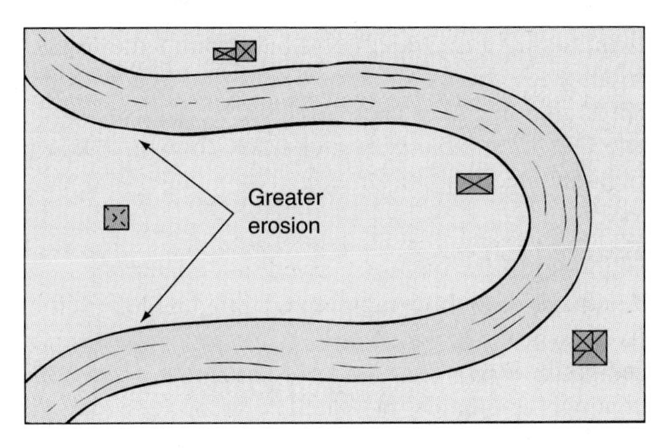


(a)

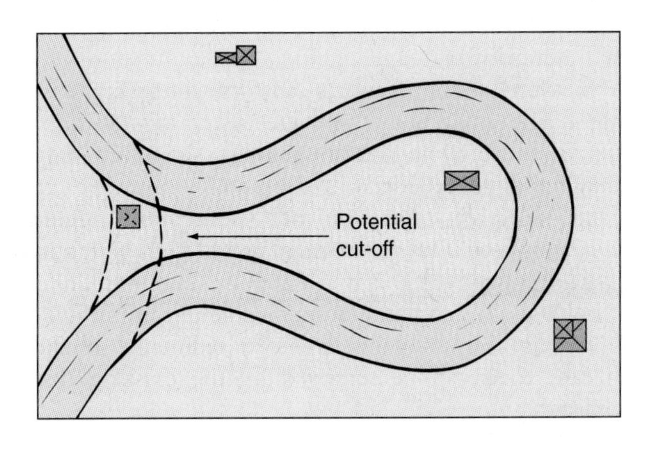


(b)

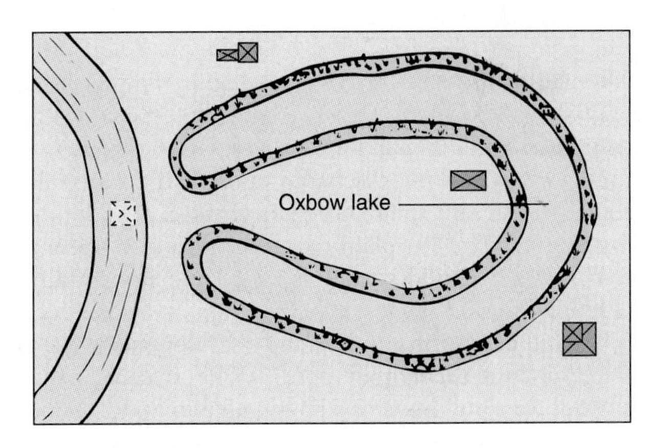


(c)

**FIGURE 7–5**
Formation of oxbow lakes by meander cutoff. Note the fate of one of the homes. (Reprinted with the permission of Macmillan College Publishing Company from *Physical Geology,* by Nicholas K. Coch and Allan Ludman. Copyright © 1991 by Macmillan College Publishing Company, Inc.)

living along a meander bend have found themselves in another county or state when the stream erodes through a meander neck! A classic example of this is the Omaha suburb of Carter Lake, Iowa—look at it on a detailed map of Omaha.

### Braided Channels

Braided channels develop where streams lose velocity to where they no longer can carry their coarser sediment load. This load is deposited within the channel as islands or bars. A braided stream is another example of dynamic equilibrium, because excess sediment is deposited temporarily until the stream again has sufficient discharge to move the sediment downstream.

Braided stream channels can form for several reasons. They may develop where a stream's gradient decreases abruptly and its velocity slows, or they may form where discharge decreases due to greater evaporation or infiltration into the ground. Braiding can occur if more sediment is added to a stream than it can move. This may happen where bank material is easily eroded or where a steep-gradient tributary or a landslide adds more sediment to the stream than can be removed by the existing discharge.

## STREAM GEOHAZARDS: FLOODS

The cause of a flood is of course quite simple; it is excessive stream discharge. Discharge that fills a stream channel is called a **bankfull discharge** (Figure 7–1). When the discharge exceeds the bankfull stage, the stream spills out of its channel and flooding begins. The flat plain on either side of a stream that is flooded periodically is called the **floodplain** (Figure 7–4).

It is important to understand that floods are *normal, natural events* that occur along streams over periods varying from several times a year to every few years. However, urbanization of floodplains and manipulation of stream channels has increased both the magnitude and frequency of floods in many areas.

### Conditions that Cause Floods

A flood occurs when a stream channel is incapable of containing the discharge. Contributing factors to this condition are heavy rains, rapid snowmelt, steep slopes, dam failure, storm surges, and human manipulation of the landscape, such as deforestation.

Heavy rains and rapid melting of snow cover are the most common causes of flooding. Although we normally think of flooding as a geohazard in humid climates, serious flooding occurs in very *dry* climates as well. In dry regions, the total annual precipitation may be scant, but it commonly is distributed in a few brief, intense weather events. This heavy rainfall lands on slopes that have little vegetation, so the water runs off immediately, rapidly filling stream channels. This runoff causes extensive lateral erosion and bank collapse, and the overflowing spreads across the land as sheetflow.

Flooding is even more serious if the area is bordered by steep slopes, because steepness increases the flow velocity. An example is Las Vegas, Nevada, which lies in a desert basin surrounded by mountains. Intense summer thunderstorms commonly flood parts of the city (Figure 7–6). As urbanization continues in the Las Vegas Basin, the problem will worsen unless diversion channels are built to conduct floodwater away from developed areas.

Floods also occur when dams fail, releasing large volumes of water that abruptly inundate areas downstream. Such a disaster happened in Johnstown, Pennsylvania, in 1889, killing 2,200 and causing massive destruction. Other notable dam failures include the 1963 Vaiont Dam collapse in

**FIGURE 7–6**
Warning sign in the parking lot of a hotel in Las Vegas, Nevada. (Courtesy of Marie Morisawa.)

Italy (see Chapter 9, Box 9–1), and the Buffalo Creek Dam break in West Virginia in 1972.

Rapid winter snowmelt causes flooding if the rate of melting exceeds the ground's capacity to absorb the water. If the ground is still partially frozen, making it essentially impermeable, flooding can be even more severe.

Flooding also occurs when tropical storms and hurricanes make landfall on a shoreline. The flooding (storm surge) at the coast is discussed in detail in Chapter 16. Here we consider what happens as the storm moves inland. Generally storms dissipate as they move inland and spread their precipitation over a wide area. However, if the storm is blocked by another weather system inland, it "stalls," and its precipitation saturates a smaller region, day after day, causing major flooding.

Massive flooding occurred after Tropical Storm Alberto hit the Florida Panhandle on the weekend of July 4, 1994. The storm moved inland from the Gulf Coast, heading over Florida and Georgia. A weather system to the east (the Bermuda High; see Chapter 16) prevented it from heading out and it stalled over the region for days. As much as 90 cm (36 in) of rain fell in the area, swelling the streams and flooding town after town downstream. Thirty-one counties in Georgia and 15 in Florida were declared federal disaster areas. The Governor of Georgia called it "the single most serious disaster in the history of the state." Federal estimates in mid-July were 31 people killed, 50,000 homeless, a million acres of crops and farmland flooded, and thousands of structures (homes, businesses, roads, bridges) suffering major water damage; damage was estimated at a minimum of 250 million dollars.

### Different Kinds of Flooding: Upstream and Downstream

Floods have different characteristics, depending on whether they occur upstream or downstream. *Upstream floods* result from locally intense rainfall in a portion of a drainage basin, as from a severe storm. Floodwaters rise and fall rapidly, and severe local flooding may result. However, the downstream portion of the stream may be little affected.

*Downstream floods* are related to larger-scale weather events, such as the movement of a weather front or storm system across a region. Water levels rise more slowly than in an upstream flood, but larger areas of the drainage basin may be inundated for days and property damage may be extensive (see Box 7–1).

## FLOOD SEVERITY

Flood severity depends on natural conditions within a drainage basin (rainfall, infiltration rate, slope, vegetation, climate, season) in combination with human activity (urbanization, agriculture, timbering, flood control).

### Rainfall and Infiltration Rates

The relation between rainfall rate and the infiltration capacity of the ground was discussed in Chapters 3 and 6. Any condition that decreases the infiltration capacity of soil or any condition in which rainfall rate exceeds the infiltration capacity will lead to an increase in overland runoff in the form of sheetflow. Under these conditions, streams quickly exceed bankfull discharge, and flooding results.

Infiltration is reduced naturally in some desert soils (the aridisols) when impermeable crusts and layers of salts develop in the upper part of the soil profile (Chapter 6). Human activity can reduce infiltration capacity by paving and where construction equipment and off-road vehicles compact the surface layers (Chapter 6). Brief, severe rainstorms flood the surface faster than infiltration can occur, and this excess water moves as sheetflow to cause flash flooding.

### Slope, Vegetation, Climate, and Time of Year

The steeper the slope, the faster water flows across its surface and the greater the flooding danger. Vegetation usually slows water flow, aiding infiltration and reducing flood risk. Conversely, lack of vegetation reduces infiltration and increases flood risk.

The climate and the time of year also are important in determining whether flooding will occur. In the winter, frozen soil has poor infiltration, a situation that favors runoff along the surface. In late winter, warm periods and rain may result in rapid melting of snow cover. This rapid increase in water supply to streams may exceed their bankfull discharge, resulting in flooding.

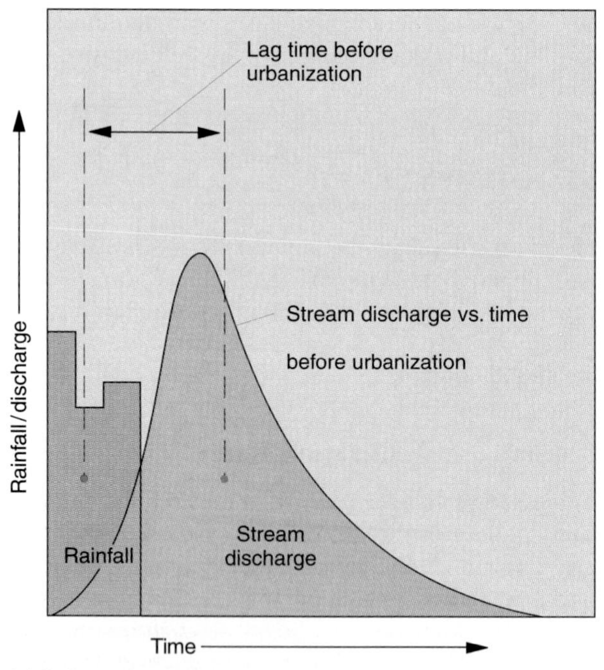


(a) Before urbanization

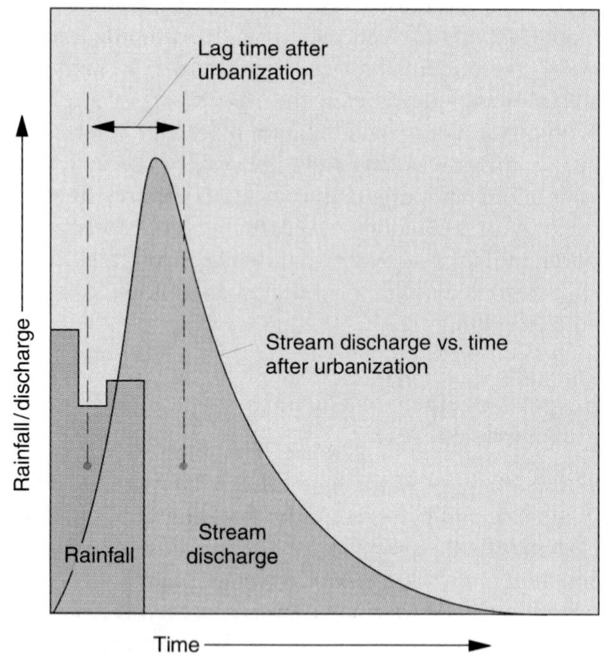


(b) After urbanization

**FIGURE 7–7**
Hydrographs showing lag times in natural and developed areas, (a) before and (b) after urbanization. (Reprinted with the permission of Macmillan College Publishing Company from *Physical Geology,* by Nicholas K. Coch and Allan Ludman. Copyright © 1991 by Macmillan College Publishing Company, Inc.)

## Urbanization

If an area is fully vegetated, a soaking rain increases stream discharge slowly, and the peak discharge will occur well after the rain begins (Figure 7–7a). The time difference between peak rainfall and peak stream discharge is called **lag time.** Lag time exists because it takes time for the rain to saturate soil and then create significant runoff. It also takes time for water that infiltrates soil to percolate through the ground to the stream.

Urbanization shortens lag time and increases peak discharge, thus increasing flooding. During urbanization, vegetation is removed, surfaces are paved, and stream channels are modified. Paving prevents water from sinking into the ground, so it drains faster over the surface into the stream. This reduces lag time and increases peak discharge of the stream. These changes shorten the lag time of streams (Figure 7–7b), which means that a greater volume of water runs off faster, increasing the risk of flooding. If the lag time becomes very short, especially dangerous "flash" floods may result.

The bar graphs in Figure 7–7 show the same rainfall. The curves show the increase and decrease of stream discharge over time. The gentler curve in (a) is for a stream flowing in a vegetated area (before urbanization). The steeper curve in (b) is the same stream after urbanization. You can see that an urbanized stream reaches peak discharge faster (shorter lag time). It also has a greater peak discharge than a similar stream in a natural area.

With urbanization also comes the construction of storm sewers and the paving over and construction on areas that border streams (Figure 7–8). Building sewers removes water from developed areas and delivers it faster to the nearest stream. The convergence of several sewer systems in a stream can greatly reduce lag time and increase peak discharge well above flood level.

Yet another aspect of urbanization that promotes flooding is building on floodplains. Floodplains are flat and easily developed, so available land is quickly occupied. To create even more building sites, developers commonly extend the floodplain by adding *fill* (rock and soil) to land that borders streams. The fill is confined behind wooden or steel walls called *bulkheads* to protect it from stream erosion. The drawback is that filling and bulkheading reduce the stream's cross section (Figure 7–8). To move the same volume of water, the stream must increase ve-

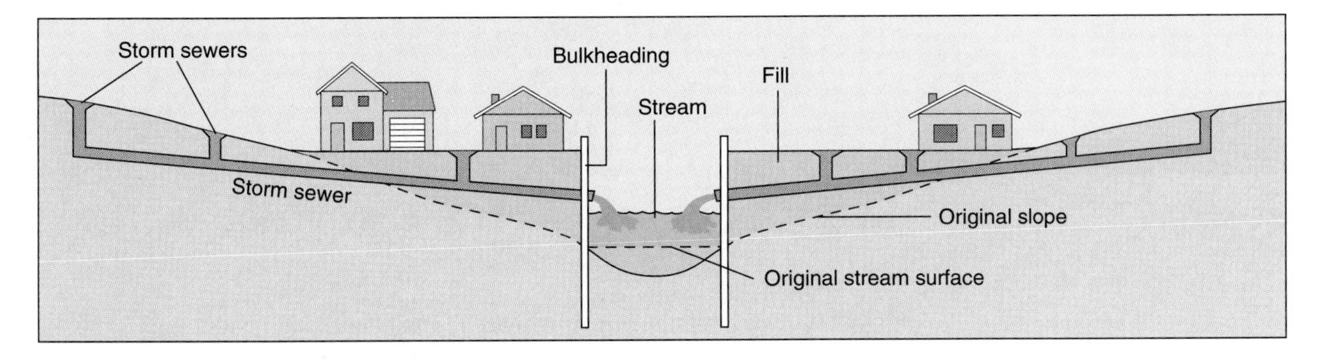


**FIGURE 7–8**
Cross section showing the effects of storm sewers and channel constriction in increasing the peak discharge of streams.

locity and depth, increasing erosion and causing local flooding where none existed previously.

## FLOODING EFFECTS

The 1993 inundation of the Missouri-Mississippi Basin provides an excellent example of the effects of flooding (see Box 7–1). If you live in the Midwest, you may have experienced the effects yourself. If not, you probably observed them on television. In this section, we will use this event—now the worst flooding on record in North America—as an example of the economic, health, and biological effects of flooding.

### Regional Disruption and Economic Loss

The great flood of 1993 was an economic disaster, manifest in many ways. For example, as the floodwaters rose, bridges closed, including the U.S. Route 24 bridge spanning the Mississippi River between Missouri and Illinois. The area's major population and commerce center, Quincy, Illinois, was cut off from many of its workers and shoppers who live in Missouri.

Along the river, businesses closed because of transportation problems or lack of water and power. In Iowa, the Des Moines River overtopped the levees and flooded the water treatment plant for the City of Des Moines, leaving the population of 200,000 waterless for over a week. The downtown area was unaffected by flooding, but businesses closed because water was unavailable in event of fire. A large meat packing plant laid off most employees due to a lack of potable (drinking) water.

The barge industry, which carries coal, grain, and other products up and down the Mississippi, lost $3 million to $4 million a day as shipments were stalled. A system of dams and locks on the upper Mississippi River is used to control floods and to allow the passage of ships. Although the system sustained little damage, it required weeks for the U.S. Army Corps of Engineers to remove the floating debris caught in the locks. Summer tourism and shopping were disrupted by flooded roads, closed bridges, and broken communications.

A major rail line was closed in Missouri when a bridge collapsed; shipments of coal, chemicals, and autos were disrupted; hundreds of businesses were wiped out in St. Louis suburbs and other river communities in Illinois, Iowa, Missouri, and Nebraska; and crop losses totaled in the billions.

### Health Problems

Flooded sewage treatment plants in Iowa—Cedar Rapids, Denison, Des Moines, Burlington, Davenport, and other locations—released untreated sewage into the floodwaters. In Missouri, 35 wastewater treatment plants failed. Flooded sewage plants spread untreated sewage wherever the water flowed. Because fecal bacteria in sewage can infect rescue workers, authorities suggested that they be immunized against tetanus during the 1993 flooding.

Pools of standing floodwater also make ideal mosquito-breeders. Mosquitoes are more than itchy nuisances—they are disease-carriers. Despite these widespread risks during the flooding, no increase in disease was noted.

In these river valleys, industrial, agricultural, and domestic chemicals were released when floodwaters

## BOX 7–1 THE MISSOURI-MISSISSIPPI VALLEY FLOOD OF 1993

The greatest North American flood in recorded history had its origins in a change in upper-atmospheric wind flow. The normal front between cold polar air and warmer air over the United States generates an eastward-flowing high-altitude wind called the *jet stream*. Normally, the jet stream flows in an arc high above the U.S.-Canadian border (Figure 1). In the summer of 1993, however, it moved southward, flowing along an arc over Oregon, down over Utah and Colorado, and up over Minnesota (Figure 1).

North of this polar front is cool, dry air. The unusual position of this air mass brought it in contact with warm, moist southerly winds over the

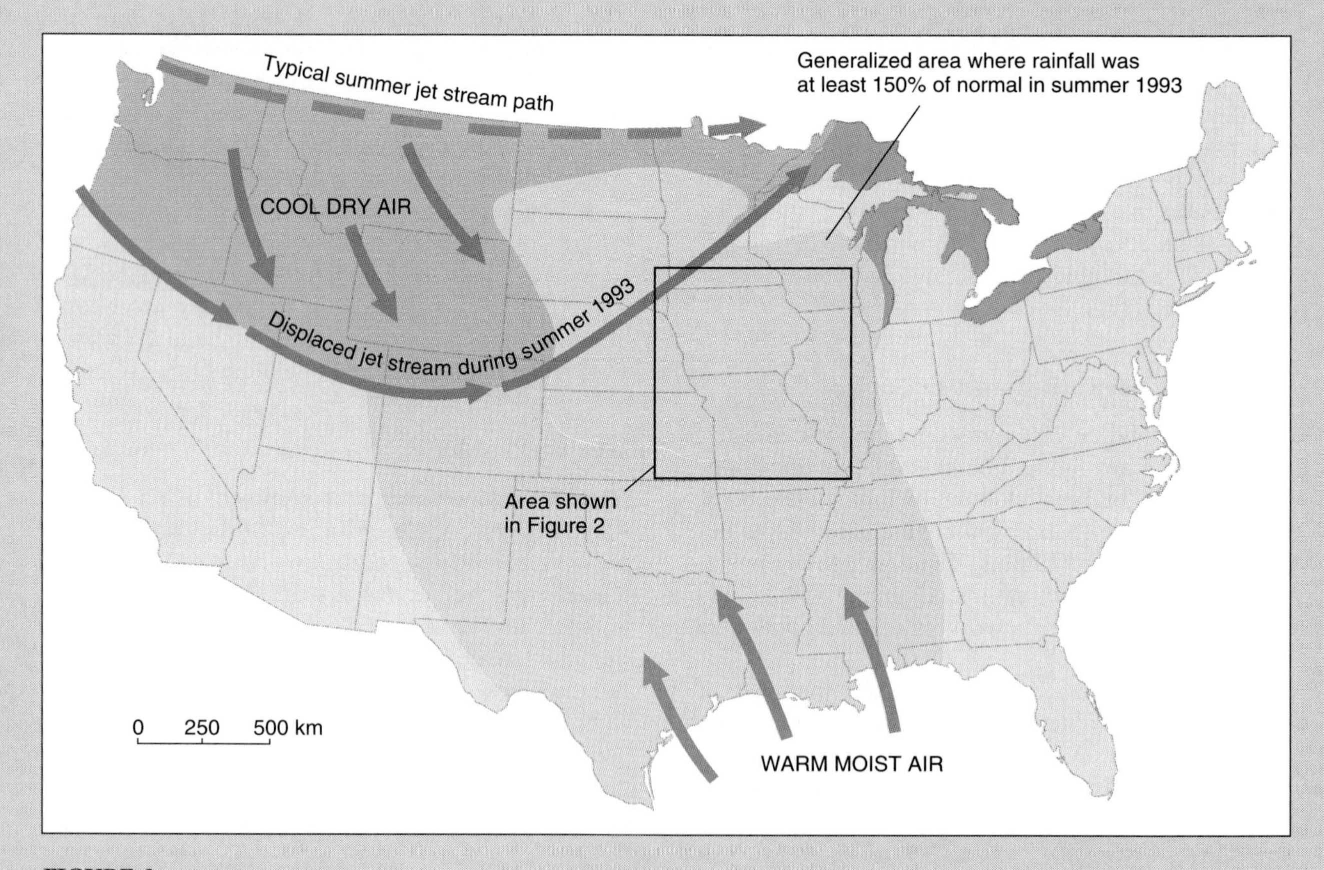


**FIGURE 1**
Generalized weather conditions in the Missouri-Mississippi River valleys during the flooding of 1993.

Midwest. Their interaction with the cold air produced rain over a large region including South Dakota, Nebraska, Kansas, Minnesota, Iowa, Missouri, Illinois, and Wisconsin. Had this pattern lasted briefly, there would have been no problem. But it persisted from early June to the end of July. The situation was complicated by the lack of a normal mid-winter thaw, which would have melted part of the snow cover earlier. Instead, when spring came, river levels rose rapidly before the heavy rains began in June.

Major streams draining into the Mississippi and Missouri Rivers have their headwaters where the rainfall was heaviest (Figure 2). This coincidence of an unusual weather pattern and geography set the stage. The confluence of the swollen Mississippi and Missouri Rivers led to a regional inundation of historic proportions (Figure 3).

**FIGURE 2**
Detailed map of area affected by flooding in 1993. Map shows rivers, dams and localities mentioned in text.

**FIGURE 3**
Composite satellite image of the confluence of the Illinois River (upper left) and Missouri River (lower left) with the Mississippi River (center left to right). The normal width of the rivers appears dark blue (1988 image, SPOT satellite). The width on July 14, 1993, is shown in light blue (ERS-1 image). The brownish red area at bottom right is St. Louis, Missouri. (ITD-SRSC/RSI/SPOT Image copyright ESA/CNES/SYGMA.)

## FLOOD LEVELS

The flooding posed special problems for St. Louis, which is just south of the confluence of the Mississippi and Missouri Rivers (Figure 2). The flood crests of both rivers arrived in tandem.

Downtown St. Louis is protected by a 15.8-meter (52-foot) concrete floodwall. The people of St. Louis watched nervously as floodwaters climbed the floodwall. On July 17, the river level at the nearby Gateway Arch (Figure 4) was 13.5 meters (44.3 feet). Engineers calculated that the discharge at 23.6 million liters per second (6.25 million gallons per second), *five times the normal flow* of 4.7 million liters per second (1.25 million gallons per second).

By August 1, the river level climbed to 15.1 meters (49.5 feet), only three-quarters of a meter (2.5 feet) below the top of the floodwall (Figure 4). The St. Louis floodwall is one of the strongest and highest on the Mississippi River, but week after week of high water levels exerted great strain on the structure. A leak of 150,000 liters per minute (40,000 gallons per minute) developed under a suburban section of the wall. Emergency workers

**FIGURE 4**
Mississippi River floodwall just north of St. Louis. Note the difference in height between the river (left) and the protected industrial area on the right. (U.S. Army Corps of Engineers, St. Louis District)

managed to stop the flow with thousands of sandbags and 5,400 kilograms (6 tons) of stone.

Fortunately, the floodwall held, and river level never reached the top. By August 3, river level at the Gateway Arch had dropped to 14.8 meters (48.6 feet). St. Louis was spared because of their massive floodwall. Will they be as fortunate the next time?

## WHY DIDN'T THE LOWER MISSISSIPPI FLOOD SEVERELY?

The U.S. Army Corps of Engineers' system on the Mississippi River is designed to contain the "project flood," a hypothetical inundation in which three of the worst storms on record occur over seven days. The Corps views the Mississippi River as upper and lower halves: the "upper Mississippi" is the drainage north of Cairo, Illinois, where the Ohio River joins the Mississippi (Figure 2). The "lower Mississippi" is the drainage south of Cairo. Normally, the upper Mississippi flow is small compared to that of the Ohio River. The project flood assumes that the Ohio River flow will be 10 times that of the upper Mississippi.

Fortunately, the 1993 rains had minimal effect on the Ohio River. In fact, it was flowing below normal at the same time that the Mississippi and Missouri Rivers were flooding. Thus, the lower Mississippi River was able to contain the massive upper Mississippi River flow without flooding. Had heavy rainfall also occurred in the Ohio River Basin (Figure 1), serious flooding would have extended into the lower Mississippi Basin as well.

## TO LEVEE OR NOT—DOLLARS AND SENSE

Liverpool and Kampsville are two towns on the Illinois River, 130 kilometers (80 miles) apart (Figure 2). Both were badly flooded in 1993. However, their approach to floodplain management after the flood was quite different. Liverpool has chosen a 1,300-meter (4,400-foot) levee, constructed by the U.S. Army Corps of Engineers for around $2.1 million.

In Kampsville, the Corps decided that a levee was not feasible. The government instead has spent $1.1 million buying buildings and demolishing them since 1986. However, some homeowners think the government is paying too little and have found other ways to be compensated. For example, one person offered his inn for $125,000 dollars, but the government refused. The owner already had collected over $250,000 from flood insurance from *past* floods and was expecting a settlement of $100,000 to $130,000 from the 1993 flood.

## CHANGING STRATEGIES IN FLOODPLAIN MANAGEMENT

In the past, floodplains were managed primarily by building structures to control streamflow. Today, floodplain management employs less control

damaged chemical storage areas. Although this created a potentially serious pollution problem, the huge volume of floodwater greatly diluted these toxic substances. However, certain chemicals may have bonded to sediment particles in the water. These particles were deposited in marshes and lakes by retreating floodwaters, possibly creating local concentrations that will adversely affect wildlife.

### Biologic Effects

Not all effects of flooding are bad; some very positive environmental effects occur. Overflowing waters carry nutrients that fertilize aquatic plants and wash fish into formerly isolated lakes. During the 1993 midwestern floods, commercial fisherman reported a significant population increase in catfish and other game fish in lakes and creeks on the floodplain. The abundant plant growth resulted in an explosion in fish populations, attracting ducks and waterfowl that feed on the plants, and eagles and hawks that feed on the fish. Flood sediments also add nutrients to existing soils on floodplains, although growing crops are destroyed.

## FLOOD CONTROL

Flooding is a geohazard along most streams, large and small. A variety of methods are used to reduce flood flows *(flood control)* and to minimize the effects of floods *(flood mitigation)*. These include floodways, diversions, floodwalls, dams, levees, and channelization. We will briefly examine each.

### Floodways and Diversions

Communities can reduce flood damage by creating **floodways.** These are areas on the floodplain of a stream where no new structures or homes are per-

through structural modification and more through a river's natural ability to contain floods. However, this strategy requires people to accept limits on development in hazardous areas.

Some feel that all levees should be eliminated, allowing rivers to naturally flood a bordering band of wetlands. This may be best environmentally, but it is unrealistic for major urban areas like St. Louis and New Orleans (Box 7–2). The method can be used in less populated areas by buying agricultural land beside the river and restoring the original wetlands. Another option is to use levees, but relocate them away from the river to allow wetlands to contain floodwater.

Building codes in some communities require homes to be built above the 100-year flood level. Other codes prohibit all buildings on floodways. Soldiers Grove, Wisconsin, solved the flooding problem by moving its entire business district to higher ground. The 900 inhabitants of Valmeyer, Illinois, about 35 kilometers (22 miles) south of St. Louis (Figure 2), voted to move their *entire town* 2.4 kilometers (1.5 miles) inland onto high bluffs above the east bank of the Mississippi. Townspeople noted ironically that in the days before flood control structures, farmers lived on the bluffs and came down to the floodplain to farm the rich land.

mitted. Existing homes and businesses are relocated from the floodplain, and those that must remain are suitably protected.

Features designed to slow, absorb, or store floodwater may be added to floodways. An example is the floodway created along Mingo Creek near Tulsa, Oklahoma, after disastrous flooding occurred in 1984. The city created a natural *greenway corridor* devoid of structures along the stream and excavated basins in the corridor to retain floodwaters. Between floods, these lakes are dry, and the corridor is used for athletics and recreation. A similar floodway was built in Rapid City, South Dakota, following disastrous flooding there.

## Floodwalls

Some stretches of rivers prone to flooding are protected by **floodwalls** (Figure 7–9). These reinforced concrete structures parallel river banks and

**FIGURE 7–9**
Floodwall along the Susquehanna River in Pennsylvania. (Photo by author.)

prevent floodwaters from inundating the settled areas behind them. Landings and stairways allow people access to the river for recreation. If the floodwalls are breached or overtopped, flooding conditions may persist for days in the area behind them, because the floodwall acts as a dam in reverse, preventing floodwater from returning to the riverbed.

## Dams

Dams impound streamwater for water supply, hydroelectric power generation, recreation, and flood control. Dams are not intended to retain all the water that backs up behind them; they simply *control* the rate of streamflow, like a large valve. Once water behind a dam reaches a certain level, a **spillway** allows the water to spill into the stream below the dam.

A dam whose sole purpose is flood protection is called a **flood impoundment dam** (Figure 7–10). Flood impoundment dams differ from others because they are kept *empty* as much as possible. They impound floodwater, which then is released promptly and gradually so the dam will be ready to contain the next flood.

Flood impoundment dams are especially important in areas prone to flash floods. Dry regions are good examples, such as the U.S. Southwest and parts of California. Extensive urbanization of dry mountain slopes in these areas increases runoff during rainstorms, which increases the potential for flash flooding (Figure 7–6). This operation often is in conflict with other needs of the community, such as recreation or hydropower generation, which require that the reservoirs be kept full.

Like all human modifications of nature, flood impoundment dams can create problems. These dams trap stream sediment in their basins, not a large problem in itself. But in California, trapping sediment creates a *displaced problem* (Chapter 1) along the seacoast. There, sediment carried to the coast by streams replenishes beach sand that is eroded by storms. However, flood impoundment dams have reduced the coastal sediment supply, so eroded beach sand is not being replaced. This problem is discussed further in Chapter 15.

Dams have proved very effective in controlling flooding. Perhaps the greatest regional flood prevention success has been in the valley of the Tennessee River, about 1,000 kilometers long (650 miles), which meanders across much of Tennessee and northern Alabama. The Tennessee Valley Authority (TVA) maintains a series of dams that slow floodwaters and generate hydroelectric power for the region.

In an extraordinary flood, such as the Missouri-Mississippi flood of 1993, the role of dams can become ambiguous. Consider the dam system in south central Iowa, shown in Figure 2 of Box 7–1. The Des Moines River backed up behind the Saylorville Dam northwest of Des Moines, protecting the city. However, during the height of the flooding, the overflow from Saylorville Dam added to

**FIGURE 7–10**
Flood impoundment dam, outlet of Little Dell Dam, Salt Lake City, Utah. (U.S. Army Corps of Engineers, Sacramento District.)

the flow from the undammed Racoon River, threatening Des Moines. The city asked that more water be impounded in Saylorville Lake. However, towns *above* the dam then were threatened by the expanding lake.

A similar problem occurred south of Des Moines. Inhabitants of Eddyville, below the Red Rock Dam, worried about high water levels in the Des Moines River. They felt that more water should have been released from Red Rock Reservoir earlier in the spring, to allow storage of more floodwater.

## Levees

Ridges of sediment called **levees** are deposited along stream banks each time a stream floods, whether the flood is great or small. (Refer to the explanation of levees in Chapter 3.)

These natural levees prevent flooding from small increases in streamflow. They are no barriers to large discharges, however. If an area has been urbanized, which increases runoff, natural levees will be overtopped more often. Thus, as an area is developed, natural levees must be built up artificially to contain the greater discharge.

Levees are constructed of earth embankments or reinforced concrete caps over a core of fill (Figure 7–11). Such reinforced levees are designed to prevent all but extreme floods from inundating the floodplain behind the structure.

Although levees minimize local flooding, they create other problems:

- □ Periodic flooding enriches soil fertility by depositing nutrient-rich mineral particles over the floodplain. By preventing most of this flooding, this soil enrichment is halted.
- □ Levees give people a false sense of security, tempting them to resist evacuation orders in case of an impending flood. Levee failures certainly occur, as happened hundreds of times during the 1993 Missouri-Mississippi flood.

## Levee Failure

Whether a levee is a simple bulldozed ridge of sand and clay or a carefully engineered concrete structure, it can fail. In a heavily urbanized riverside area such as New Orleans, levee failures can have catastrophic results, so a variety of flood control measures are utilized to protect the city (see Box 7–2). Three major types of levee failure are shown in Figure 7–12:

1. A river crest higher than the levee overtops the levee and flows onto the floodplain. Erosion during this overflow may breach the top of the levee.
2. Extensive saturation of a levee by weeks of high water levels may weaken it, resulting in collapse along the floodplain side. This may decrease the levee's thickness sufficiently for water to breach it.
3. Fluid pressure from the river can be transmitted through pore spaces (Chapter 3) in saturated sediment under the levee. Under increased pressure from the flooding river, the water rises through the floodplain near the levee, moving

**FIGURE 7–11**
An artificial levee prevents floodwaters (left) from inundating the developed area behind the levee (right). (U.S. Bureau of Reclamation.)

## BOX 7–2 NEW ORLEANS: A CITY AND ITS RIVER

No city in America is more dependent on a river than New Orleans, Louisiana. In fact, one of its nicknames is the "Crescent City" because it developed around a meander bend of the Mississippi River (the other side of the city borders Lake Ponchartrain). The river is important to the city for transportation and drinking water. However, New Orleans has a unique relationship with the Mississippi because *virtually all of the city is now below river level, and large portions are even below sea level* (Figure 1).

New Orleans' precarious location between the Mississippi River and Lake Ponchartrain compounds its potential for flooding (Figure 2). It sits atop the Mississippi Delta Plain, an area that is subsiding as fine-grained sediments beneath it are slowly being compacted. Accompanying this subsidence is the gradual post-glacial global rise in sea level that is accompanying the melting of continental glaciers.

The city's low elevation requires massive pumping stations to remove water runoff from the city during heavy rainfalls. The paths of many Gulf Coast hurricanes come close to New Orleans, and coastal flooding long has been a serious problem (Figure 3). Considering all this, you can see why flood protection is vital to New Orleans!

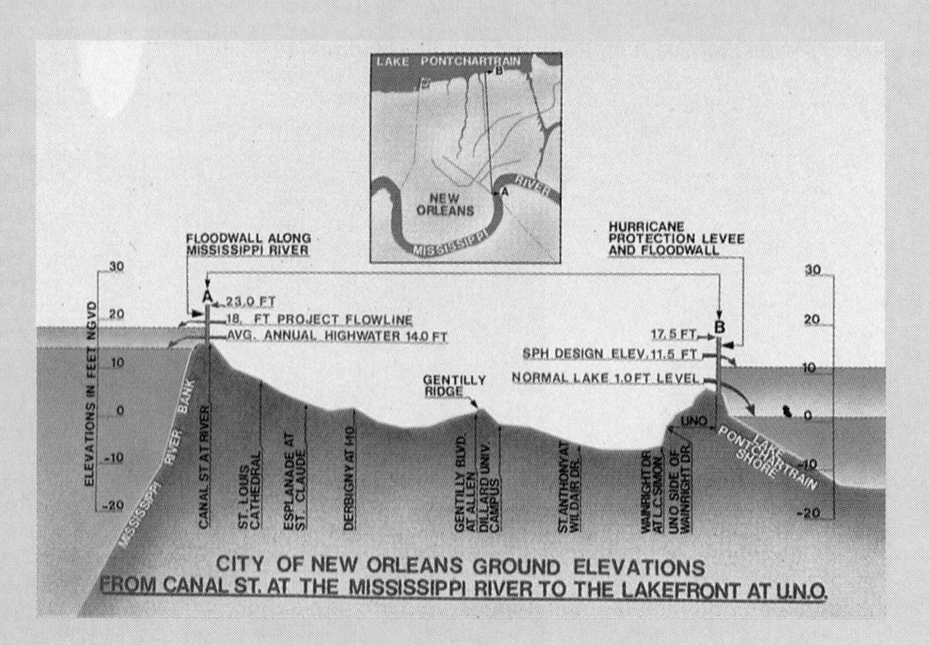


**FIGURE 1**
Cross section through New Orleans from the Mississippi River to Lake Ponchartrain. Most of the city is near or below sea level. Artificial levees along the Mississippi River and Lake Ponchartrain protect the city from flooding. (Courtesy of Brett Hess, U.S. Army Corps of Engineers, New Orleans District.)

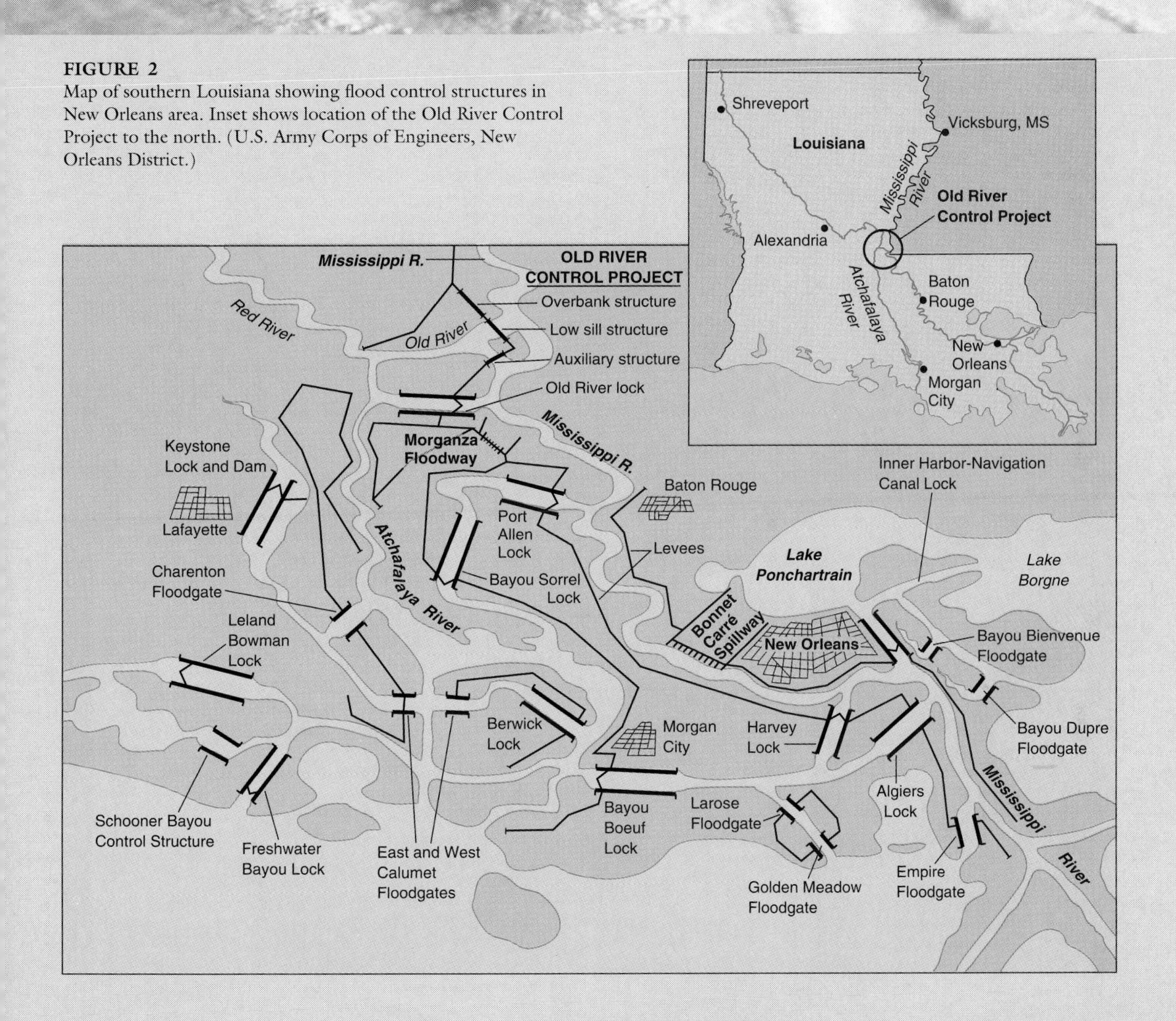


**FIGURE 2**
Map of southern Louisiana showing flood control structures in New Orleans area. Inset shows location of the Old River Control Project to the north. (U.S. Army Corps of Engineers, New Orleans District.)

## FLOOD CONTROLS

The fate of the city and the lives of its inhabitants depend on multiple flood-control devices built by the U.S. Army Corps of Engineers (Figure 2). The first level of protection is a series of artificial levees that parallel the river and its branches and canals (Figure 1). Massive gates in some of the levees normally remain open for vehicles to pass through, but they are closed when flooding threatens (Figure 4).

**FIGURE 3**
Flooding in New Orleans and its suburbs (top of photograph) resulting from Hurricane Betsy in 1965. (U.S. Army Corps of Engineers.)

Setback levees have been built further inland to provide additional protection. Visitors to the riverside tourist area known as the French Quarter may not realize that the massive levees there confine the Mississippi River *above* the level of the French Quarter. Perhaps people trust the levees too

**FIGURE 4**
A motorized flood gate protects New Orleans at Canal-Toulouse Street. The gate is kept open to enable tourists to see the river. It closes (note wheels on gate base at left) when a water level rise threatens the French Quarter to the left. (U.S. Army Corps of Engineers, New Orleans District.)

much, because several housing developments are located well below normal river level (Figure 5).

The second level of protection is provided by the Bonnet Carre Spillway and Floodway about 32 kilometers (20 miles) north of the city along the Mississippi (Figure 6). Its purpose is to provide a way to divert floodwater into a coastal bay called Lake Ponchartrain, reducing the threat to New Orleans. The 2.3-kilometer (1.4-mile) structure is divided into 350 bays (openings), each closed by twenty creosoted timbers. It takes about thirty hours to remove all the timbers, but this amount of warning time is available because New Orleans, at the mouth of a river that is 3,765 kilometers (2340 miles) long, knows well beforehand of any flood advancing downstream.

How many timbers to remove depends on the amount of excess discharge that is to be sent across the floodway into Lake Ponchartrain. The spillway has proved highly successful. During a disastrous earlier flood in 1973, opening the spillway lowered the river stage at New Orleans by about 1 meter (3.3 feet).

Use of the spillway is not without environmental problems. Each time it is opened, from 7 to 9 million cubic meters (9 to 12 million cubic yards) of sediment can be deposited on the floodway. This does not cause serious sediment pollution, however, because the area has been cleared of all structures, and it is *designed* to be flooded. Much of the sediment is used for fill in local construction. The massive infusion of freshwater and sediment into brackish Lake Ponchartrain seriously disrupts the lives of aquatic plants and animals there. However, as the salinity is restored, and the nutrient-rich silt settles onto the bottom, there is a bloom in life activity.

**FIGURE 5**
Levee in New Orleans protects housing developments and business on floodplain (left) at an altitude below the Mississippi River on the right. (Courtesy of Brett Hess, U.S. Army Corps of Engineers, New Orleans District.)

**FIGURE 6**
Aerial view of Bonnet Carre Spillway in open position. Floodwaters from the Mississippi River (at bottom) are channelled across the floodway into Lake Ponchartrain (distance) This decreases the water levels, protecting the city of New Orleans to the south. (U.S. Army Corps of Engineers, New Orleans District)

## THE ATCHAFALAYA DIVERSION

The *ultimate* protection for New Orleans lies north, halfway across the state, at the site of the Old River Control Structure (Figure 7). A series of human activities in the mid-1800s unexpectedly allowed the Atchafalaya River to divert more and more Mississippi River flow into the Atchafalaya Basin. This diversion was inevitable because the Atchafalaya has a steeper gradient and shorter path to the sea (230 kilometers or 140 miles) than the Mississippi (500 kilometers or 300 miles). Thus the Atchafalaya is tending to capture more and more of the Mississippi River drainage.

With each passing year, the Atchafalaya became deeper and wider. Without artificial controls, most of the Mississippi River would be diverted and New Orleans would face a disaster: the lower Mississippi would turn into a saltwater estuary, and the city would lose most of its drinking and industrial water. Serious channel shallowing (shallowing of the river channel through deposition) would limit transport into this major port city, with serious economic impact.

## THE OLD RIVER CONTROL STRUCTURE

The federal government responded to the increasing diversion of Mississippi River discharge into the Atchafalaya in 1963 with the $15,000,000 Old River Control Project. For the first time, dams and gates controlled the amount of water that was allowed to enter the Atchafalaya Basin. However, the great

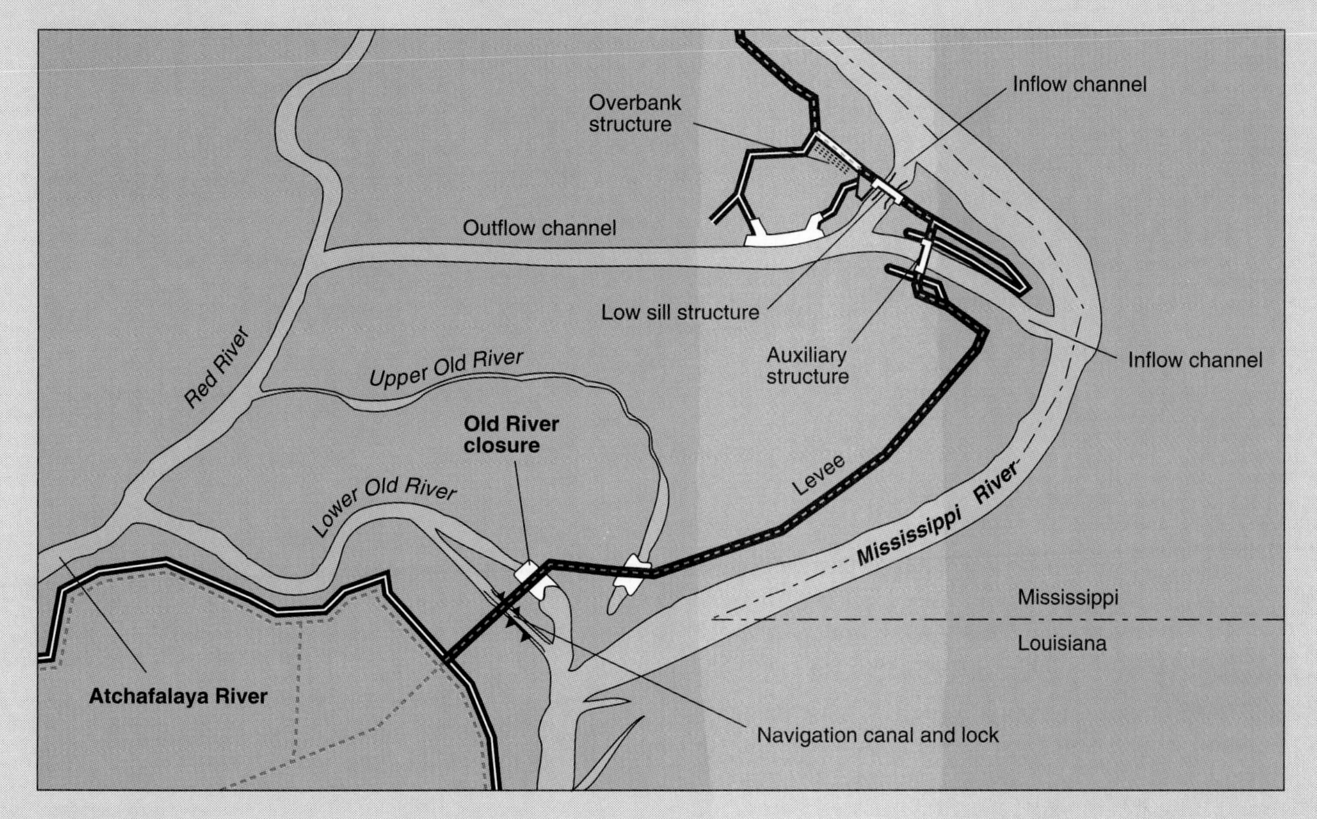


**FIGURE 7**
Diagram of Old River Control Project. This project controls the volume of water from the Mississippi River (right) that can enter the Atchafalaya River (left). (U.S. Army Corps of Engineers, New Orleans District.)

Mississippi River floods of 1973 almost destroyed parts of the structure. The damaged parts were repaired, and an Auxiliary Structure was completed in the mid-1980s at an estimated cost of $295,000,000. Engineers feel that this structure can prevent natural diversion well into the next century.

Another advantage of the Old River Control Structure: should a great flood threaten New Orleans, the option exists to divert large volumes of water into the Atchafalaya Basin. That basin already has been prepared as a floodway, and the cities on it have been protected by levees.

However, no one is really sure whether the diversion could be stopped after the flood or if the structure itself can resist the passage of such a volume of water. If not, the consequences for New Orleans would be disastrous. The city would be saved from the flood, but it might be destroyed economically without the mighty Mississippi running through it.

FIGURE 7–12
Types of levee failure. Levees can fail when (1) the river overtops the levee, (2) saturated earth slumps and flows, and (3) fluid pressure moves water through the levee and upward on the back side. The rising water forms sand boils that undermine the levee.

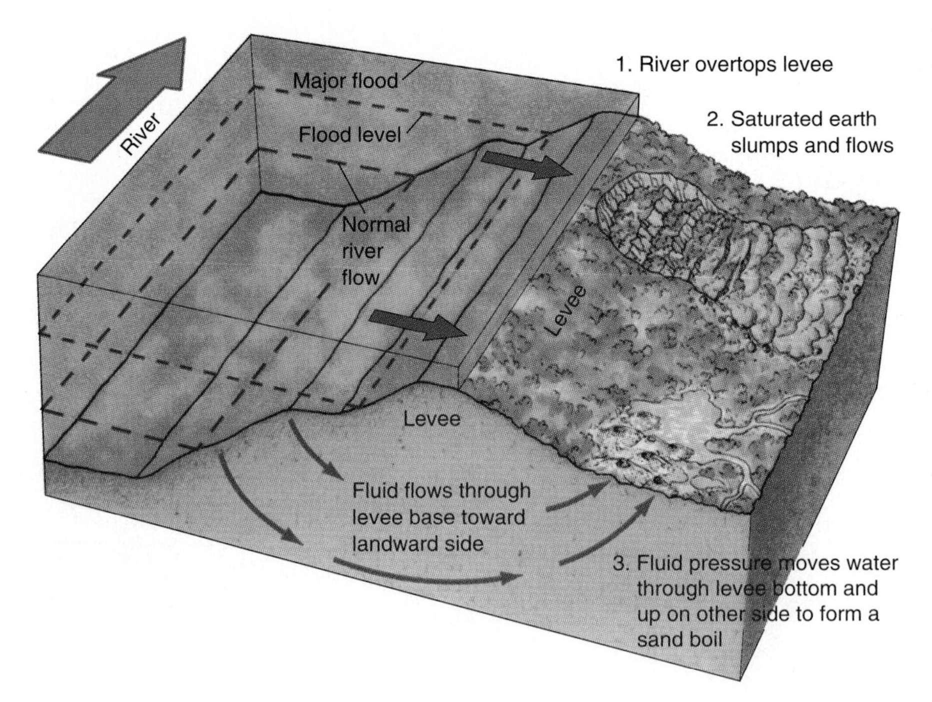


sediment upward to the land surface in a **sand boil** (Chapter 5). If the river level continues to rise or remains high, the sand boil will widen and may undermine the levee, causing a breach.

Water bursting through a narrow levee breach is moving much faster than the floodwaters in the main channel. The breaking out of this front of water and its fast flow can cause more destruction to structures behind the levee than floodwaters in the main channel would have caused.

A failed levee continues to cause damage long after it breaks. The breach allows large volumes of water to enter formerly dry areas, forming temporary lakes (Figure 7–13). Such lakes do not go away immediately, because the lake water is blocked from returning to the main channel by levee segments that were not destroyed. Consequently, the water level drops along the main river days before it drops behind breached levees. Not all of the water returns to the river after the flood. A portion of it is retained in lower elevations behind the levee.

Interestingly, levee failure in one area may prevent flooding in another area. The U.S. Army Corps of Engineers studied the effects of levee breaching along a stretch of the Mississippi north of St. Louis, between Rock Island and Quincy, Illinois (Figure 2 in Box 7–1). Every time a levee breach or overtopping occurred along this stretch, the river-level gauge at Quincy recorded a drop of 0.6 to 0.9 meters (2 to 3 feet). However, with the river flowing at 14,150 cubic meters per second (500,000 cubic feet per second), river height was restored within a few hours. Even so, this temporary discharge drop allowed numerous downstream communities more time to sandbag their levees.

## Levee Breaching During the 1993 Flood

The surge of floodwater through a breached levee can cause catastrophic damage. In 1993, millions watched on TV as Missouri farmhouses collapsed from powerful flows near a breached levee. The problems didn't end when the floodwaters subsided. As already explained, levees act as dams in reverse, impounding water on the floodplain for days after the main channel flow has subsided. The unbreached levees upstream also contribute to flooding of unprotected areas downstream by increasing the discharge that reaches them.

During the 1993 flood, a small Illinois town took the unusual step of breaching a levee to reduce the flood risk. Prairie du Rocher, a historic eighteenth-century town, is on the Mississippi River 48 kilometers (30 miles) south of St. Louis (see Figure 2 in Box 7–1). The river broke through a levee to the north at Columbia, forming a stream 3 to 5

**FIGURE 7–13**
Levee failure results in stream (left) inundating floodplain (right). (U.S. Bureau of Reclamation).

kilometers (2 to 3 miles) wide that flowed parallel to the Mississippi down the floodplain. The water inundated one small town and threatened to do the same to Prairie du Rocher.

Local officials and the U.S. Army Corps of Engineers decided to breach the levee just north of Prairie du Rocher. Their hope was that, by flooding adjacent farm land, they could slow the water's flow down the floodplain and relieve the pressure on the levee, protecting the town. Divers opened the levee floodgates, slowly releasing river water onto the floodplain. Later, an artificial breach was cut in the levee to release more water. The strategy succeeded, reducing the river level by 1.5 meters (5 feet) and saving the town—at the cost, however, of flooding some farms.

## How the Levees Held in 1993

An undeveloped river in its natural state reduces the force of flooding simply by overflowing its banks. Because of the concentration of industrial, commercial, agricultural, and residential development along the Mississippi, this great river has become one of the most controlled rivers on Earth. Following disastrous floods along the lower Mississippi in 1927, the U.S. Army Corps of Engineers began to construct flood control projects. Levees built by the Corps prevent flooding from normal high discharges and allow the floodplain to be fully developed.

Unfortunately, before the 1993 flood, many people did not clearly understand that these levees stand between them and death. The massive levees give a sense of protection that is justified for smaller floods, but not for the great flow that occurred. Further, not all levees along the rivers are built to Corps of Engineers standards. Some are simpler, less costly, less reliable structures.

The 1993 flood was so great that the limits of many levees were tested and sometimes exceeded. Of 275 Corps of Engineers levees, 85% did their job, but 31 were overtopped, 8 were eroded and ruptured, and 3 were breached. The performance of non-federal levees was much worse: only 43% withstood the trauma, and 800 of 1,400 failed.

In the historic town of St. Genevieve, Missouri, stand three dozen homes built by French settlers in the 1780s, making the town a National Historical Landmark. For protection from river flooding, this town had only the Mississippi's natural low levee during the 1993 flood. The town missed its chance to get a federal levee in the 1950s when the government was paying for construction, because some farmers refused to cede their land. The U.S. Congress recognized the historical value of the town and authorized a levee in 1986, but the town could not pay its 25% of the cost. As the 1993 floodwaters rose, hundreds of volunteers constructed a sandbag and rock barrier, saving this historic community.

**FIGURE 7–14**
Construction of a channelized stream segment. (U.S. Bureau of Reclamation)

## Channelization

The modification of a stream channel by straightening, clearing, deepening, widening, and lining with concrete or boulders is called **channelization** (Figure 7–14). Streams tend to overflow at river bends, so removing these bends by channelization reduces flooding locally. Channelization speeds water's passage through an area. It also increases the cross-sectional area of the stream so that more water can be carried.

The benefits of channelization are balanced by problems that it creates. Stream-widening means cutting trees along the bank, and removing this vegetation, with its soil-holding roots, increases sediment input into the stream (Chapter 6). Further, the reduction of shade increases water temperature. Both changes stress aquatic animals in the river.

Deepening the stream channel churns the bottom, disrupting the environment of bottom-dwelling animals. The greater water turbidity limits the depth to which light can penetrate, jeopardizing aquatic plants, which must have light to carry on photosynthesis. Channelization creates short, straight stream segments with greater cross-sectional areas. These stream segments can handle unusually large flows better, but tend to choke with debris during normal flows.

Although channelization can minimize flooding, it should include some curved sections to reduce velocity, should be done with minimal removal of bank vegetation, and requires periodic maintenance to remove debris that accumulates in channelized segments.

# CLUES TO PAST FLOODS

Anyone who builds anything near a river must be alert for potential flooding. When building along a stream bank, the flooding potential is obvious. But what about a site that is, for example, 15 meters (50 feet) higher than the floodplain? Has it ever been flooded? Flood hazard maps are available for many areas to answer this question (Figure 7–15). But if such a map is unavailable, you can look for several physical, biologic, and geologic clues that indicate past floods, and estimate their level.

## Physical Clues

Research at a site along the Pedernales River in Texas revealed physical evidence of past floods. The following types of evidence are keyed by letter to Figure 7–16.

**a.** *Stranded debris,* such as the log caught in the tree, indicates that water levels reached that height.
**b.** Large *bare, sandy tracts* indicate flood scouring above present stream level.
**c.** *Ripple marks in sand* above the present channel indicate that water flowed across that area.
**d.** *Lines of driftwood* may mark the former high-water mark.
**e.** *Trees bent downstream* suggest shearing by the current, further supported by *broken branches with new sprouts* arising from the limbs.
**f.** Floodwaters commonly cut into the river valley at their highest stages, forming cut surfaces called *terraces.*
**g.** Floodwaters cause *breaks in slope in sediment.*
**h.** *Erosional niches in bedrock* are eroded by debris carried by floodwaters.
**i.** High flooded areas also may have *scour holes* eroded by rock debris in the swirling, whirlpool-like flow of past floods.

In particular, features such as these should be noted by anyone interested in purchasing riverfront property. Many of these features remain visible for years.

Anyone interested in purchasing waterfront (or waterview) property should consult Federal Emergency Management Agency (FEMA) flood insurance rate maps, flood records on file in government offices, and talk to long-time residents of the area.

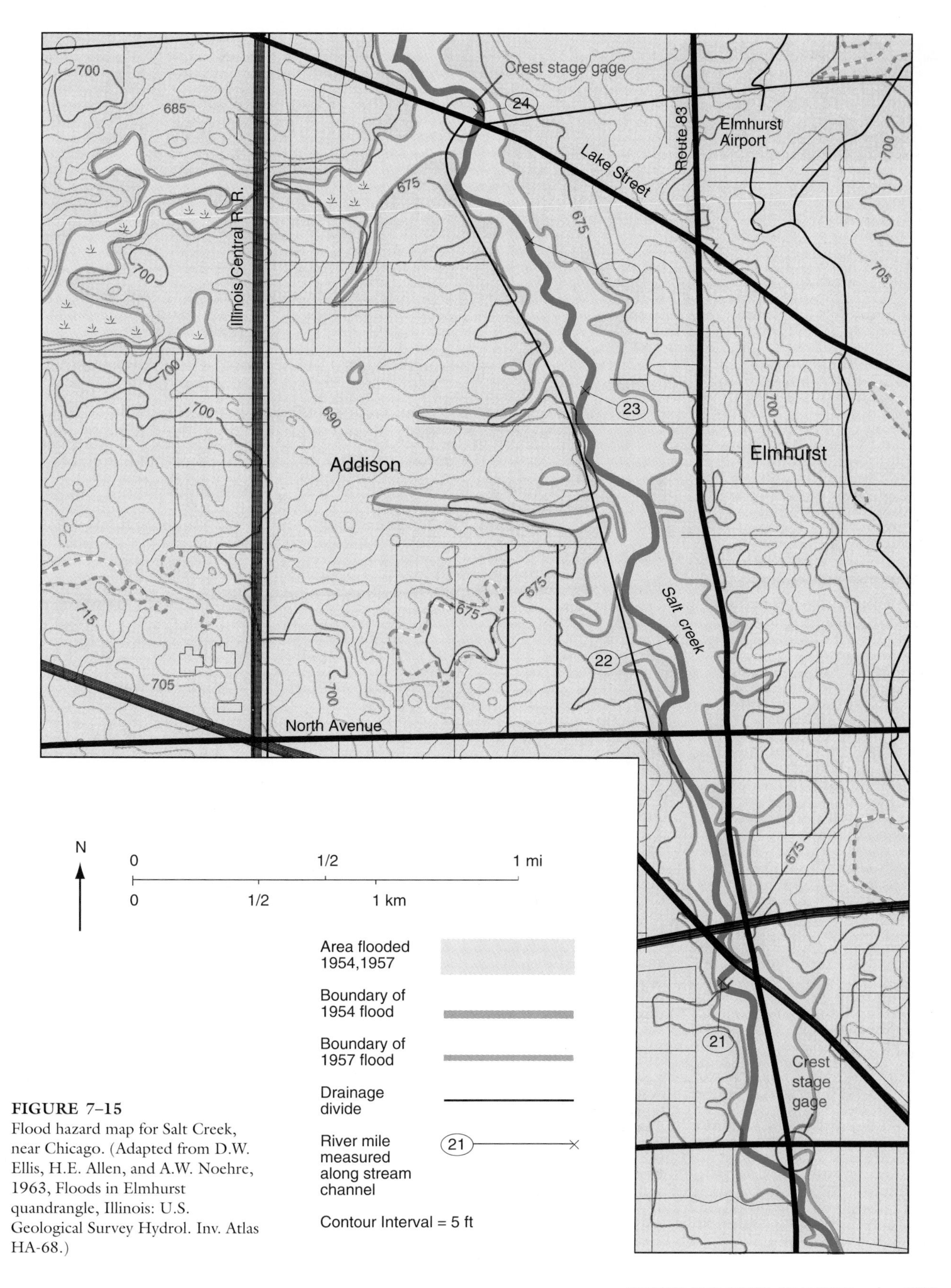


**FIGURE 7–15**
Flood hazard map for Salt Creek, near Chicago. (Adapted from D.W. Ellis, H.E. Allen, and A.W. Noehre, 1963, Floods in Elmhurst quandrangle, Illinois: U.S. Geological Survey Hydrol. Inv. Atlas HA-68.)

BOX 7–3

## WHAT IS A 50-YEAR FLOOD?

You probably have heard the terms "50-year flood," "100-year flood," and so on. What do they mean? Here is a look at how hydrologists and civil engineers determine flood frequency and likelihood.

Stream discharge is measured at **gauging stations** set up at several points along a stream. A gauging station consists of a well in the stream bank connected by a pipe to the river. As the river surface rises and falls in response to changes in discharge, so does the water level in the well. A float in the well continually measures the **stage height** of the river. Data on stage heights accumulated over time reveals the stream's range of stage heights.

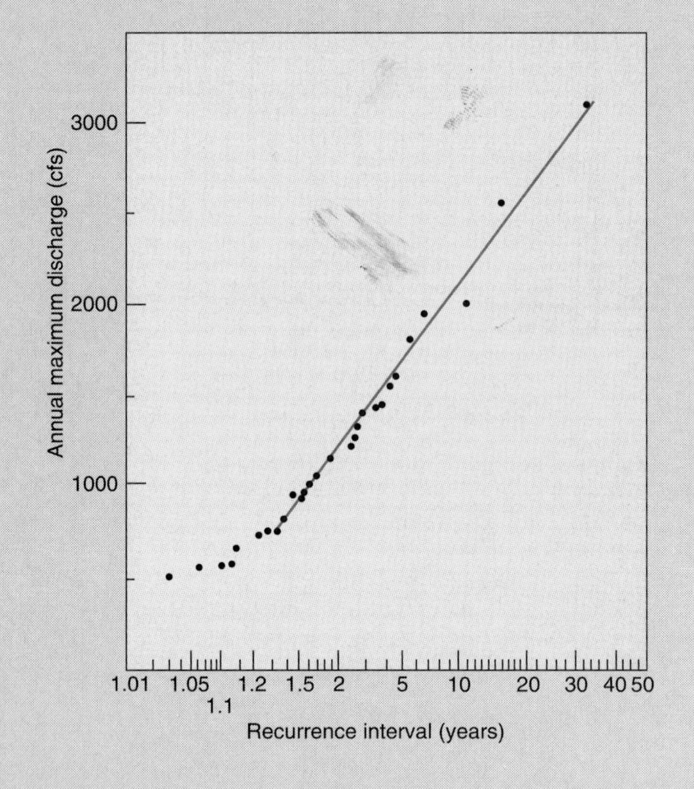


**FIGURE 1**
Flood frequency recurrence curve for Rock Creek near Red Lodge, Montana. (After D. F. Ritter, 1980, Process geomorphology (second edition): Wm. C. Brown, Fig. 5.27, p. 186.)

A stream's stage heights are correlated with its discharge volumes. Its discharge volume is calculated in this manner: velocity ($v$) of the current at the station is measured with *current meters.* The width ($w$) and depth ($d$) of the stream at the station are measured. Then, discharge ($Q$) is calculated with the *stream formula:*

$$\text{Discharge } (Q) = \text{width } (w) \times \text{depth } (d) \times \text{velocity } (v)$$

Once discharge measurements are correlated with stage heights, discharge is determined by reading the stage height alone. Discharge then is plotted over time as a **stream hydrograph** (see Figure 7–7).

Many stream discharge measurements are assembled to determine the *frequency* of each level of discharge. This indicates the *recurrence interval* for each level of discharge, or how often a given level of flooding occurs ("50-year flood," for example). The recurrence interval for each discharge then is plotted on a **flood frequency recurrence curve** (Figure 1).

The **recurrence interval** is the *average* period of years over which a discharge level is expected. For example, the following recurrences are expected around Rock Creek, Montana:

| **Recurrence interval** | **Stream discharge** |
|---|---|
| 5-year flood | 48.7 cubic meters per second (1,719 cubic feet per second, or cfs) every 5 years (average) |
| 10-year flood | 61 cubic meters per second (2,156 cfs) every 10 years (average) |
| 20-year flood | 74 cubic meters per second (2,625 cfs) every 20 years (average) |
| 30-year flood | 85 cubic meters per second (3,000 cfs) every 30 years (average) |

Note that if a 10-year flood level is reached one year, it does *not* mean that another 10 years must pass before a similar flood level is reached! Flood recurrence values are *averages* over a long period. In fact, floods of identical levels can occur in the same year, in two consecutive years, or not for decades, depending on weather conditions.

Also important is that these values refer only to the *conditions under which the original measurements were made*. If the stream is modified in any way, the discharge recurrence values can change considerably.

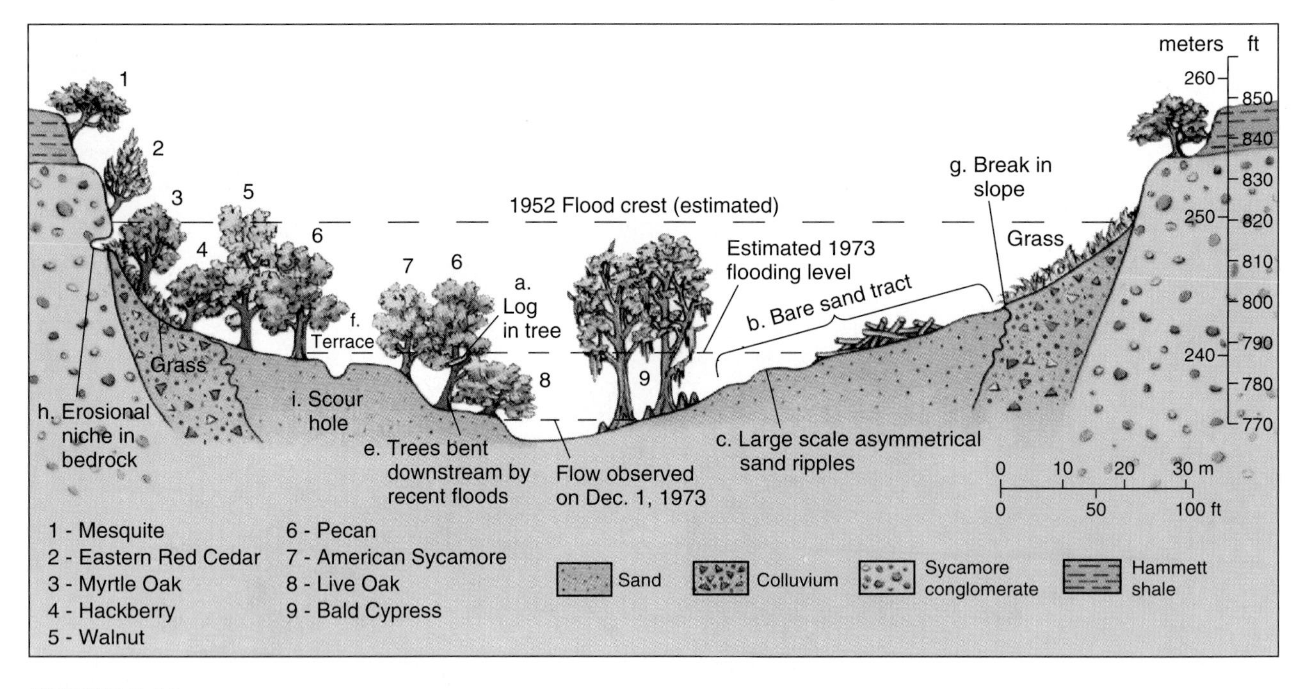


FIGURE 7–16
Features indicating successive flood levels of the Pedernales River at Trammel Crossing in Pedernales River State Park, Texas. See text for discussion. (After V. R. Baker, 1976, Hydrogeomorphic methods for the regional evaluation of flood hazards, Environmental Geology, v. 1, p. 261–281. Modified with permission of Springer-Verlag Co.).

### Biologic Clues

The types of vegetation shown in Figure 7–16 show that different species develop higher above the river channel than those along the river. These differences can be mapped by noting the vegetation patterns, which are indicated by different colors on color-infrared aerial photographs. For example, in the Pedernales River shown, a color-infrared photograph shows in pink frequently flooded lands adjacent to the river channel where sycamore, pecan, and water hickory are more common. Maroon characterizes hillslopes and areas between streams that are rarely flooded and are covered with eastern red cedar, juniper, and live oak.

### Geologic Clues

Coarse-grained stream deposits (gravel and sand) usually occur only within the stream channel, whereas finer sands, silt, and clay cover the floodplain. However, when a stream overtops or breaches its levees, the normal channel deposits (sand and gravel) may be deposited well above the low-water channel level out on the floodplain.

### Dating Former Floods by Geologic Methods

The physical observations of past floods (coarse sediment layers, scour holes, driftwood accumulations, and so on) provide visual proof of past floods. But *how old* are these floods? The carbon-14 dating method gives us answers.

The key is the ***organic material*** buried with the flood sediments. To recover material for dating, a trench is dug to expose sediment layers deposited by past floods. Pieces of downed trees, plant materials, and organic-rich soils are dated by geochemists (geologist-chemists) using the carbon-14 isotope dating method.

## PREDICTING FLOODS

To predict a flood for a stream or river, you first must know the stream's "normal" discharge. Then you need a record of discharge levels during past floods. This data can be put together to show how often different flood levels are expected to re-occur. Box 7–3 describes how this is done.

### Flood Hazard Maps

A **flood hazard map** shows areas that would be flooded by stream discharges of a given magnitude. For example, a "50-year flood" is an event that reaches a certain level on an average of about 50 years (see Box 7–3). The height that water will reach for a given discharge is available from stream gauging station records. Of course, this height refers only to the *immediate* area. Streamwater runs downslope, so its surface is never truly "level" but always has at least a slight slope, steeper upstream than downstream. When records from several gauging stations along the stream are assembled, you can determine how high a flood of a given magnitude will rise. These elevations are plotted on a topographic contour map of the area to produce a flood hazard map (Figure 7–15).

For example, if the 20-year flood discharge for an area rises to an elevation of 152 meters (500 feet), then you simply highlight the 152-meter contour line in that area. This indicates that all points in the stream valley below 152 meters will be flooded every twenty years, *on the average*. All points above 152 meters will not be flooded. This could be valuable information if you discovered that the beautiful home you plan to buy is enclosed by the 20-year contour on a flood hazard map of your area!

## LOOKING AHEAD

You have just looked at water on Earth's surface—streams—and its geohazards—flooding. In the next chapter, we will go underground—only a few meters—to look at groundwater. Groundwater permeates soil and rocks and is the prime source of water for plants and for dwellings that have wells. Groundwater is easily contaminated, and when it becomes polluted it constitutes a major geohazard.

## SUMMARY

The width, depth, and velocity of a stream changes to accommodate changes in discharge, maintaining dynamic equilibrium. A straight stream channel develops in rocks with prominent joints, on steeper slopes, and in harder rocks. A sinuous, meandering stream channel develops at lower gradients, in weak rocks, or in unconsolidated sediments. A braided stream, with multiple channels that flow around islands or bars, develops where discharge falls as sediment supply remains constant, or where sediment supply increases markedly at a stable discharge.

When water in a stream channel exceeds the bankfull discharge, the excess spills across the floodplain in a flood. Upstream flooding is local and intense within smaller drainage basins and results from local weather conditions. Downstream flooding rises slowly, covers wide areas for days, and results from regional weather fronts or storm systems.

Flooding is natural and results when rainfall intensity or rapid snowmelt exceeds the infiltration capacity of the land surface. Flood frequency in unprotected areas increases with human activity. Building of engineering structures helps to control streams and "prevent" flooding. Actions that increase the flooding frequency include removal of vegetation, paving and adding storm sewers to the upland surface, and filling and bulkheading the areas bordering the stream.

Data from past flooding events are used to calculate the flood recurrence interval (the average frequency when flooding of a given stage occurs in an area). Flood hazard maps outline areas that have been inundated by floods of different recurrence intervals (10 years, 20 years, and so on).

Flooding is controlled or mitigated by a wide variety of engineering structures. Flood impoundment dams retain excess water and release it slowly to the stream. Floodwalls and levees confine water within the stream channel and prevent flooding. Channelization straightens streams and allows floodwaters to pass through an area more quickly, although it also disrupts the environment of plants and animals in the stream and increases flooding downstream.

Areas of past flooding can be recognized by stranded debris, sandy or gravelly tracts above river level, trees bent or broken in a downstream direction, and erosional niches and scour holes above the present level of the stream. Flooded areas also may be detected by their distinctive vegetation cover. Flood deposits can be dated by using the carbon-14 method on wood or organic material.

## KEY TERMS

bankfull discharge
braided channel
channelization
depth
discharge
dynamic equilibrium
flood frequency recurrence curve
flood hazard map
flood impoundment dam
floodplain
floodwalls
floodways
gauging stations
lag time

lateral erosion
levee
meandering channel
oxbow lake
recurrence interval
sand boil
spillway
stage height
straight channel
stream hydrograph
velocity
width

## REVIEW QUESTIONS

1. If floodplains are sites of frequent flooding, why are they so heavily developed and populated?
2. Describe the interrelation among stream discharge, width, depth, and velocity.
3. What factors govern the development of (a) straight, (b) meandering, and (c) braided channels?
4. Distinguish between upstream and downstream floods.
5. Define lag time. How does lag time change as an area is urbanized?
6. Describe the factors that determine the relative rate of infiltration versus runoff after a rainfall.
7. Describe how a stream hydrograph is constructed for a gauging station.
8. Describe each of the following flood protection devices and discuss any environmental problems that accompany their implementation: (a) floodways, (b) levees, (c) flood impoundment dams, and (d) channelization.
9. Describe the factors that increased the severity of flooding in the Missouri-Mississippi River Valleys in the summer of 1993.
10. What features indicate that a given waterfront property was flooded in the past?

## FURTHER READINGS

Chen, A, 1983, Dammed if they do and dammed if they don't, Science News, v. 123, March 26, p. 204–206.
Federal Interagency Floodplain Management Task Force, 1992, Floodplain Management in the U.S.–An Assessment Report, v. 1, summary, 69 p.
Junger, Sebastian, 1992, The pumps of New Orleans, American Heritage of Invention and Technology, Fall, v. 8, no. 2, p. 42–48.
National Coal Association, 1992, Facts about coal.
Newsweek, 1990, Troubled waters, April 16, p. 66–80.
Tennessee Valley Authority, 1983, Floodplain management, 80 p.
U.S. Army Corps of Engineers, 1983, Streambank protection guidelines for landowners and local governments, U.S. ACE, Waterways Experimental Station, Vicksburg, Miss, 60 p.
U.S. Army Corps of Engineers, 1992, Floodproofing regulations, Washington D.C., U.S. Govt. Printing Office, 1992-621-474/41697.
U.S. Army Corps of Engineers, 1976, Flood control in the lower Mississippi Valley, U.S. ACE, Vicksburg, Miss. 43 p.

# 8

# *Groundwater*

Surface water visibly flows and accumulates on Earth's surface (Chapter 7). Groundwater is far less visible, but it exists in much greater volume. All of the water in rivers, streams, and lakes totals only about 1.5% (estimated) of the water volume that exists beneath the land surface. Groundwater generally remains out of sight, but it can surface as a spring. Groundwater springs also feed lakes and streams, usually below the water level where they are not readily seen. Some groundwater even flows as streams in underground caverns. In dry regions, groundwater may be the major source, or *only* source, of water.

Surface water may be clouded with sediment, but groundwater is relatively clear, making it a good drinking water source for communities and individual dwellings. Surface water temperature varies seasonally, whereas groundwater flows at a constant cool temperature year-round, making it valuable for industrial cooling.

Increasing groundwater use is creating shortages and water-quality problems in many areas. To understand these problems, we first must look at how water circulates and is naturally stored underground.

## GROUNDWATER ACCUMULATION AND MIGRATION

Water that infiltrates the ground moves downward by the force of gravity. It flows through any openings it finds, ranging from joints and fractures in rock to tiny pore spaces. Pores exist between particles of soil, sediment, and in "solid" rock, between the rock grains (see Figure 2–17).

**Groundwater emerges as springs at Vesey's Paradise (mile 32) on the Colorado River, Grand Canyon National Park. (Photo by author.)**

### Permeability of Rock, Soil, and Sediment

The capacity of a rock or sediment to transmit a fluid is its **permeability.** For groundwater to migrate, it must be in permeable sediment or rock. It is not enough for the material to be porous. The pore spaces also must be *intercon-*

*nected* to create a pathway that permeates the material. Well-sorted gravel, with connected pore spaces, is quite permeable. Granite lacks pore spaces and thus is impermeable. However, granite with well developed joints (Chapter 2) can be fairly permeable, depending on the degree of fracturing and how well connected the fractures are.

Even porous materials can be impermeable if the pores are small and not connected. Clays are a good example. They may have porosities of over 50% (that is, half the volume of a piece of clay actually may be air spaces). While they may hold large volumes of water within the pores when saturated, few of the tiny pore spaces in clays are connected, resulting in a low permeability. The small pore spaces exert a greater frictional effect on any fluid passing through, slowing the flow and decreasing permeability.

### Aquifers and Aquicludes

A *permeable* rock or sediment capable of both *storing* and *transmitting* groundwater is called an **aquifer.** Examples of aquifers are sand and gravel beds, sandstones, and limestones that have solution cavities. To obtain adequate water, people who drill water wells always seek to drill into an aquifer. Aquifers range in size from very local (a permeable rock layer within a hill) to massive (the Ogallala aquifer, which underlies eight states from South Dakota to Texas, supplying water to millions of people and thousands of farms).

An **aquiclude** is an *impermeable* rock or sediment that transmits groundwater too slowly to be useful to people. Layers of clay or shale rock usually are aquicludes. The movement of groundwater is controlled by the presence of aquifers and aquicludes. (Another name for aquiclude is *aquitard,* but we use *aquiclude* here because this term is better known.)

## GROUNDWATER AND THE WATER TABLE

When rainfall infiltrates the ground, it percolates downward under the influence of gravity until it is stopped by an impermeable layer—an aquiclude. The groundwater then accumulates above this barrier, gradually filling the pores upward toward the surface. If the rainfall is sufficient and spread out over a long period, all the pore spaces in the ground may become saturated. However, permeable rocks and sediments usually do not become saturated all the way to the surface, and the groundwater becomes distributed in distinct zones (Figure 8–1).

A small portion of the rainfall that infiltrates the ground is retained within the root system of plants just below the surface. This thin belt of partially saturated pores is the **soil water zone** (Figure 8–1). Below this zone is the thicker **unsaturated zone,** in which pores are partially filled with air and partially with water that is traveling between the surface and the layer below. The next lower zone is the **saturated zone,** where all pores are filled with water. The saturated zone extends downward to impermeable layers below.

The contact between the unsaturated zone and the saturated zone is the **water table.** The water table in any area naturally fluctuates up and down with precipitation, typically changing from season to season. The replenishing of the groundwater in the water table increases in the spring and fall, when rainfall is greater, resulting in a rise in the water table. In summer and winter, rainfall is generally lower, and the water table lowers.

People also influence the elevation of the water table. The water table rises and falls, keeping a balance between groundwater removal (from pumping) and addition (from infiltration of precipitation into the ground). If groundwater is pumped out of the ground faster than it can be resupplied, the elevation of the water table drops.

Water table elevations also reflect longer-term climatic changes. During a time of extreme drought, the water table can drop below the bottom of wells, making the wells dry up (Figure 8–1).

### Groundwater Flow Rate

Water can flow rapidly across the land surface or in channels. In contrast, the flow of groundwater in most situations is much slower due to the frictional effects of pores. Consequently, minute "threads" of water seep their way *slowly* through the pore spaces of sediments and of rocks and fractures in sedimentary, metamorphic, and igneous rocks (Figure 2–17a, b). Groundwater moves much faster in fractured or dissolved rock, where the openings are large (Figure 2–17c, d).

Groundwater flow rates generally are only a few centimeters per day. For perspective, consider that

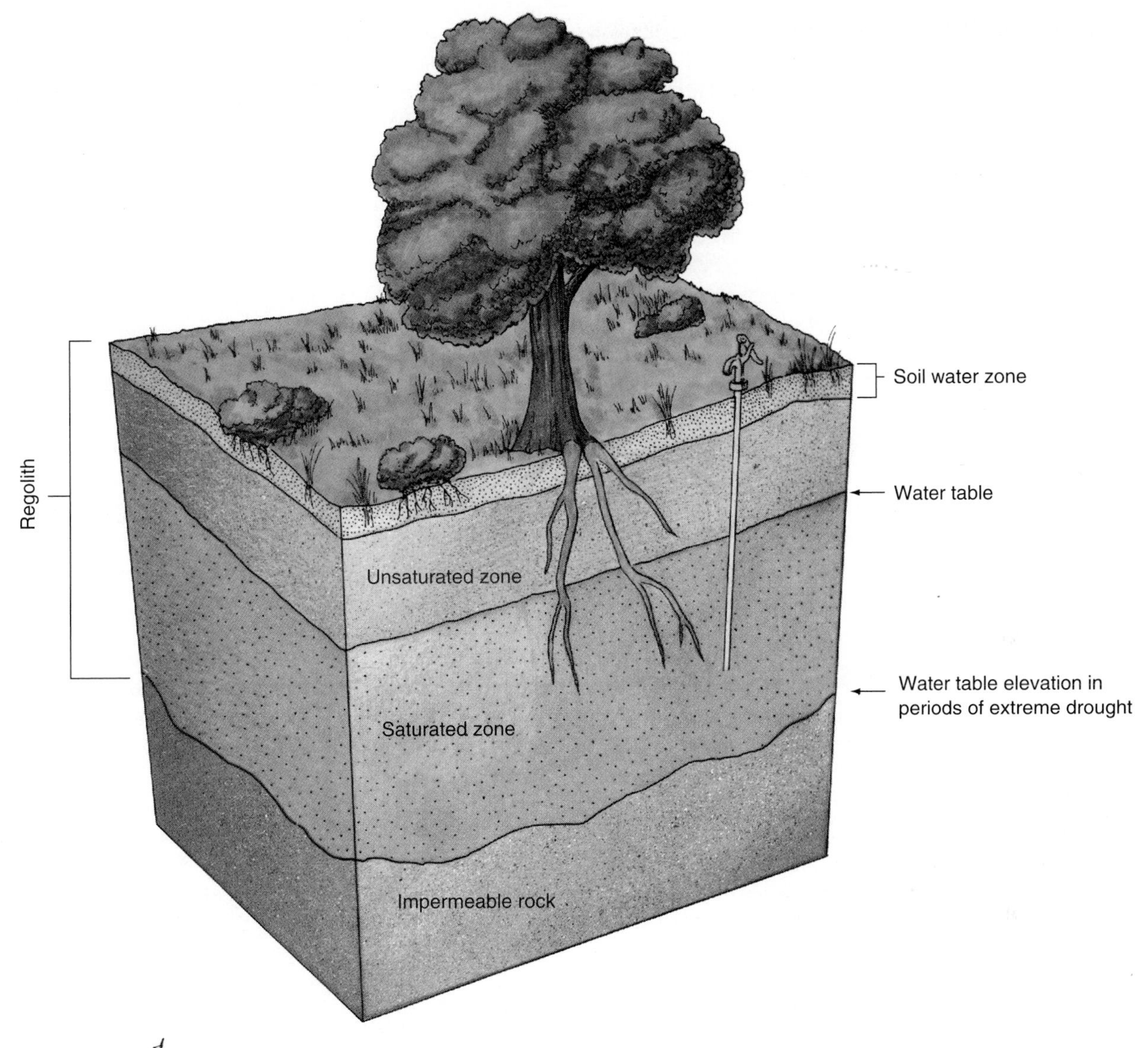


**FIGURE 8–1**
The vertical distribution of groundwater zones. (Reprinted with the permission of Macmillan College Publishing Company from *Physical Geology*, by Nicholas K. Coch and Allan Ludman. Copyright © 1991 by Macmillan College Publishing Company, Inc.)

with a flow rate of 1 meter (3.3 feet) a day, groundwater would move only 365 meters (1,200 feet) a year. With a flow rate of 2 centimeters (0.8 inch) a day, groundwater would move only 700 centimeters (275 inches) a year. Faster and slower rates occur under special conditions. For example, water flowing through a limestone cavern may move at normal streamflow speeds.

These slow rates of groundwater flow have great environmental significance. It is this slow flow through interconnected pore spaces that allows oxygen, soil microorganisms, and bacteria sufficient time to break down some kinds of pollutants, such as sewage, that may have been introduced into the groundwater.

The slow rate of groundwater flow also controls the availability of water from wells. Large volumes of water are pumped from aquifers every day. This removal is called **discharge.** Replenishment of groundwater by infiltration of surface water is called

**recharge.** If groundwater is pumped out of the ground faster than it can be recharged (that is, if discharge is greater than recharge), wells go dry.

The rate of groundwater movement was computed by French engineer Henry Darcy in 1856. Darcy found that water discharge through an aquifer increases under three conditions: as the aquifer's permeability increases, as the aquifer's cross-sectional area increases, and as the difference in elevation increases between where water enters the aquifer and where it leaves. This relationship between these three conditions and the distance of transport is called **Darcy's Law:**

Discharge = (cross-sectional area) × (permeability) × (vertical distance)\(distance of transport)

Darcy's Law is illustrated in Figure 8–2. Thus, for a given cross-sectional area and length of aquifer, the groundwater flow rate increases with both the permeability and the inclination of the aquifer.

## Shape of the Water Table

The water table is a surface that generally conforms to the shape of the land surface above. It is highest in elevation under hills and lowest under valleys (Figure 8–3).

You have seen how groundwater moves through uniformly permeable material. What happens when infiltrating water encounters an impermeable layer, or aquiclude? The water accumulates above this layer to form a *local water table*, perched above the level of the *regional water table* below (Figure 8–3). This small, local water table therefore is called a **perched water table** (Figure 8–3). The larger water table underlying the perched water table is fed by water that sinks through the sides of the hill beneath the aquiclude.

## Springs

If the perched water table is inside a hill, the groundwater above may flow along the top of the aquiclude and emerge from the side of the hill as a line of springs (Figure 8–3). Springs also form where groundwater flows up through the bottoms of water bodies such as lakes, rivers, or even the ocean. A **spring** is any natural discharge of groundwater at the surface. Because groundwater is insulated from surface temperature variations, springwater may be cooler in summer and warmer in winter than local surface waters.

A common misconception is that water from springs is "purer" than water from wells or adjacent streams. This has no scientific basis and is a romantic notion propagated by companies that bottle springwater. Springwater and well water are the same groundwater, reaching the surface by two different routes. If a harmful liquid, such as gasoline, is spilled on the surface in Figure 8–3, it could contaminate springwater, wells dug into the regional water table, or water from the stream. Springwater

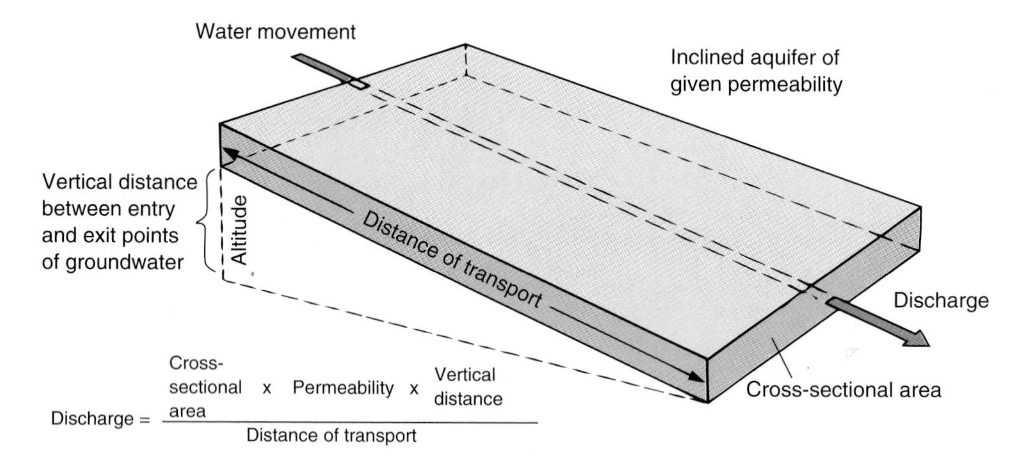


**FIGURE 8–2**
The rate of groundwater flow can be predicted by Darcy's Law. (Reprinted with the permission of Macmillan College Publishing Company from *Physical Geology*, by Nicholas K. Coch and Allan Ludman. Copyright © 1991 by Macmillan College Publishing Company, Inc.)

**FIGURE 8–3**
Elevation of the water table varies with changes in surface topography. An impermeable bed (aquiclude) can cause a perched water table (center); groundwater may migrate across the top of the aquiclude and exit the hillsides as springs. (Reprinted with the permission of Macmillan College Publishing Company from *Physical Geology,* by Nicholas K. Coch and Allan Ludman. Copyright © 1991 by Macmillan College Publishing Company, Inc.)

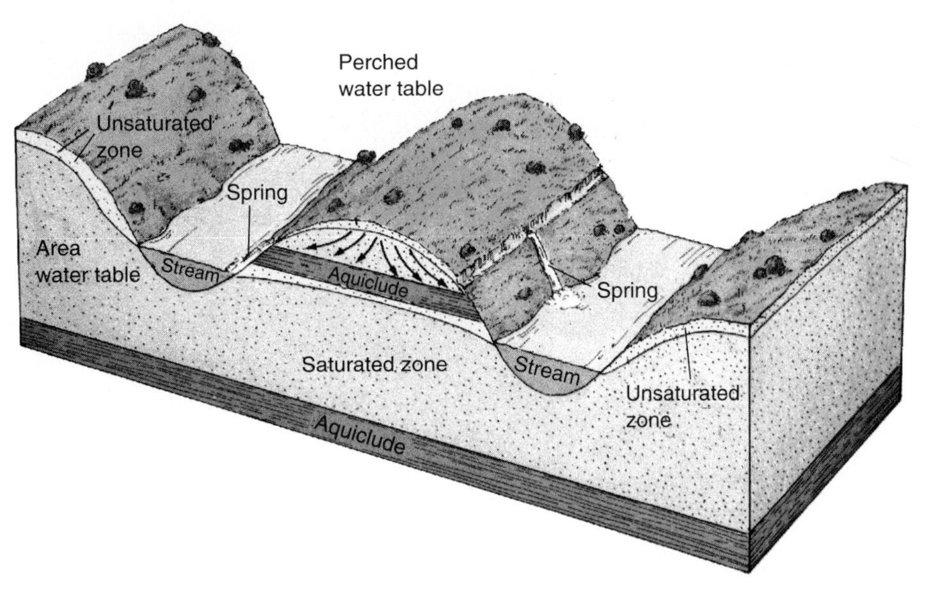


differs only in the geologic conditions that bring it to the surface, not in its purity.

## Depressions in the Water Table

The elevation of the water table reflects a balance between recharge (from precipitation and streams) and discharge (from springs, streams, and pumping wells). It is important to remember that pumping usually depletes water from rock fractures or sediment pores at a far greater rate than they can be refilled by slow groundwater flow. This excess withdrawal creates a **cone of depression** in the water table (Figure 8–4).

The cone of depression shown in Figure 8–4 is a local one, only a few tens of meters in diameter at the top. However, pumping for public water supply, agricultural, and industrial uses can produce much larger *regional* depressions in the water table. They may be hundreds of kilometers across (Figure 8–5).

**FIGURE 8–4**
A cone of depression forms in an unconfined aquifer when discharge exceeds recharge. The greater the water consumption, the deeper is the cone of depression. (Reprinted with the permission of Macmillan College Publishing Company from *Physical Geology,* by Nicholas K. Coch and Allan Ludman. Copyright © 1991 by Macmillan College Publishing Company, Inc.)

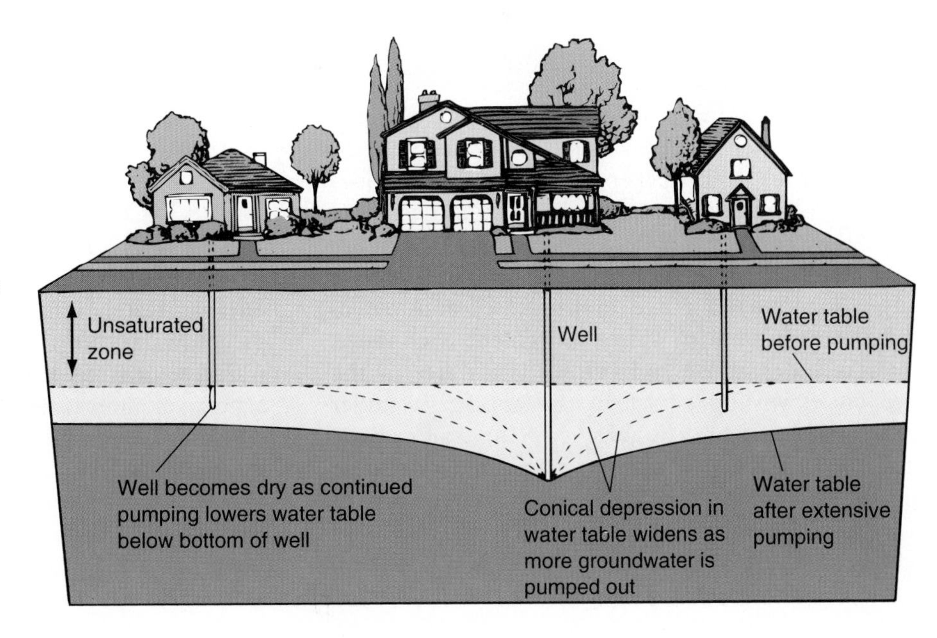

FIGURE 8–5
Areas of regional depression in the water table. (U.S. Geological Survey, as adapted in E. A. Keller, 1992, Environmental geology (sixth edition): New York, Macmillan, Fig. 11.14a, p. 257.)

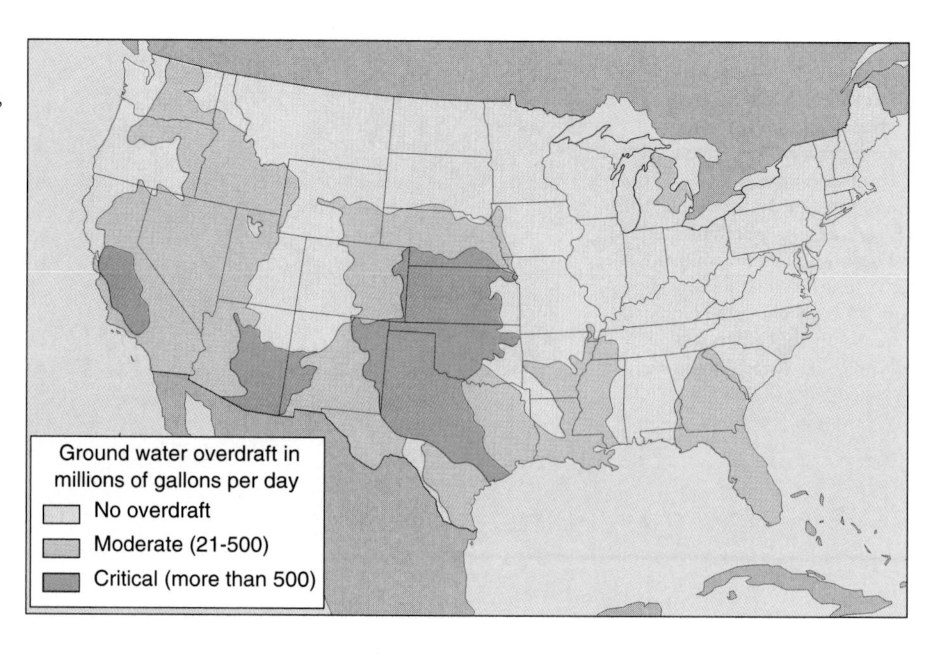


## Confined and Unconfined Aquifers and Artesian Wells

An **unconfined aquifer** has no aquiclude above it (the upper blue aquifer in Figure 8–6). Its upper surface is exposed to atmospheric pressure through the interconnected pores in the aquifer. Most shallow wells are drilled into unconfined aquifers.

A **confined aquifer** is confined above and below by impermeable beds, or aquicludes (the lower blue aquifer in Figure 8–6). It contains groundwater under pressure that is significantly greater than atmospheric pressure. The conditions that create a confined aquifer include a body of permeable sediment or rock (the thicker the better) that slopes away from a recharge area that has a humid climate.

In the "confined aquifer recharge area" in Figure 8–6, water infiltrates at the aquifer's exposed upper edge (a mountain range in this case) and moves downward by gravity. Water movement is confined by aquicludes above and below. The water in the deeper (downslope) part of the aquifer is under great pressure from the weight of the water above it. In some cases, the confining aquicludes become fractured, and at these points, the aquifer receives additional recharge from above. Or, it may lose some water into the stratum below or above if there is sufficient confining pressure.

When a well is drilled into a confined aquifer, the pressure is relieved and the water rises upward toward the ground surface. A well in which water rises above the aquifer without pumping is called an **artesian well.** It may even rise above the ground surface and flow freely from the ground, in which case it is called a free-flowing well. In the early stages of water withdrawal from artesian aquifers, pumping may not be required. However, as more and more wells tap the aquifer, the water pressure drops and eventually pumps must be used to bring the water to the surface.

# RELATION BETWEEN SURFACE WATER AND GROUNDWATER

Surface water infiltrates the ground to become groundwater, but the reverse also happens: groundwater returns to the surface through springs and artesian aquifers. In any location, the direction in which the water moves depends on the slope of the water table. As a result, some streams gain water from the ground, whereas others lose water into the ground. Add the variable of climate, and a wide variety of surface and groundwater relations exists worldwide.

**FIGURE 8–6**
Geologic section showing confined and unconfined aquifer systems. (Reprinted with the permission of Macmillan College Publishing Company from *Physical Geology*, by Nicholas K. Coch and Allan Ludman. Copyright © 1991 by Macmillan College Publishing Company, Inc.)

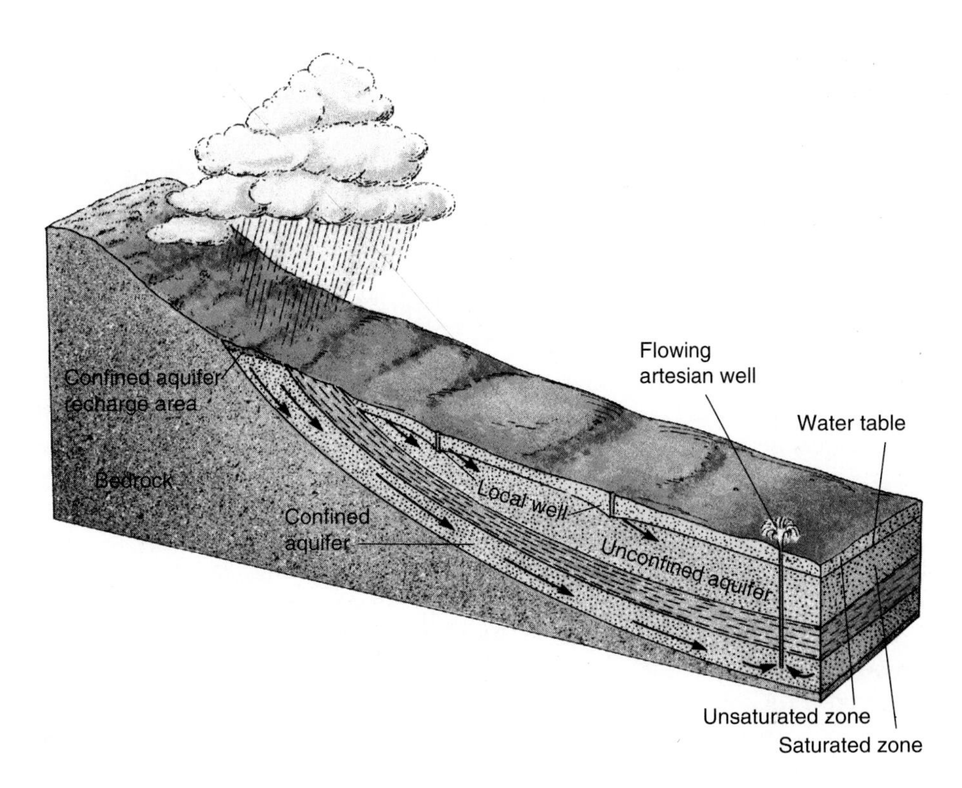


## Effluent and Influent Streams

In humid areas, the water table is high, and groundwater recharges the stream through springs in the stream channel. The amount of groundwater entering a stream is more consistent than surface water added by variable precipitation, so the entering groundwater is called the stream's **base flow.** Surface flow adds to the stream's discharge. Streams that receive most of their discharge from groundwater are called **effluent streams** because the water flows *from* the groundwater system into the stream.

In dry climates, the water table is low, and the reverse happens: instead of groundwater recharging a stream, the stream recharges the groundwater system. Streams that supply water to the ground are called **influent streams** because they lose water *into* the ground.

The Colorado River originates in the Rocky Mountains and flows southward through the deserts of the American Southwest until it empties into the Gulf of California. How can the Colorado maintain its flow as it crosses one of the driest areas in America? It loses large volumes of water, not only through leakage into the ground, but through evaporation into the air. The Colorado continues to flow because its discharge—from mountain rain, snowmelt, thousands of springs, and tributaries—still is greater than its water loss to the desert.

## Perennial, Intermittent, and Ephemeral Streams

Streams can be classified by their continuity of flow. The word *perennial* means "through the year," and thus we call streams that flow year-round **perennial streams.** They are supplied largely by base flow from high water tables. They are more common in wet regions, such as the eastern United States. However, in drier parts of the western United States, perennial streams flow wherever the amount of water in the stream exceeds the losses from infiltration into the ground or evaporation.

Many streams, especially in the drier western United States, do not flow year-round, but are seasonal. These **intermittent streams** flow only part of the year, receiving most of their water from precipitation or snowmelt. Some of their water comes from the water table when it is high enough.

The word *ephemeral* means "short-lived," so we call streams that are dry most of the year **ephemeral streams.** They are mostly dry because the water table is far beneath their channels. Ephemeral streams can form a geohazard because they transport water through an area very quickly, usually as a flash flood (Chapter 7). The only vegetation bordering an ephemeral stream is deep-rooted plants that can reach the water table, or cactus, which can absorb moisture quickly from rain and retain it through the dry intervals.

## WATER RESOURCE DEVELOPMENT

The water supply system in an area evolves as population, industry, and agriculture increase. Population centers near major rivers (like New Orleans) or lakes (like Chicago) generally use these sources throughout their development.

Sparsely settled areas commonly are served by individual shallow wells drilled into local unconfined aquifers (Figure 8–6). Some cities also tap water from unconfined aquifers. However, by the time an area grows into a suburb or a small city, most unconfined aquifers have become polluted. At this point, these areas often construct municipal systems that tap confined aquifers at depth (Figure 8–6). Many areas in the Midwest tap confined aquifers where relatively impermeable till is at the surface, overlying a deeper aquifer in glacial outwash. The water then is piped into homes and factories.

Great metropolitan regions, such as Boston, New York, Los Angeles, and San Francisco, long ago contaminated their unconfined aquifers. Today even their confined aquifers are either polluted or are insufficient. These cities thus must build massive systems to collect surface water or groundwater in distant areas and transfer it by *tunnels or open aqueducts* into the metropolitan region.

Let us now take a look at different sources of groundwater.

### Individual Water Supplies

For individual homes, farms, and communities, most groundwater is pumped from unconfined aquifers in permeable sediments or rocks (Figure 8–6). The most commonly tapped aquifers are those that developed in former stream deposits, because they are so abundant and usually quite permeable. For example, in the Midwest and New England, large volumes of groundwater are obtained from sand and gravel originally deposited by streams from melting glaciers (Chapter 3). Other sources of groundwater are from wells drilled into floodplains or into islands within the channels of modern rivers.

### Aquifers of Fractured Rock

Local groundwater supplies also can be obtained from impermeable rocks that have interconnected fractures (Figure 8–7). Groundwater can move faster through the large spaces in fractured rock than through the tiny pore spaces of sands and gravels. Consequently, water moving through fractured rock rather than sediment pores undergoes less natural filtration of pollutants. This is because adsorption contributes to the filtration process. This means that pollutants introduced into a fractured rock aquifer are filtered to a lesser degree and can travel faster, polluting other groundwater supplies more rapidly.

### Limestone Aquifers

Solid, unfractured limestone is not very permeable and is a poor aquifer. Also, in pure water, limestone is not very soluble. However, limestone is easily fractured, and it dissolves readily in acidified water. Limestone with fractures enlarged by acid groundwater forms an excellent aquifer.

Rainwater is slightly acidic (due to its dissolving carbon dioxide from the atmosphere). In some areas that are subject to air pollution (see Chapter 11), rainwater is significantly acidic. As the rainwater infiltrates the ground, it picks up more carbon dioxide from decaying vegetation in the uppermost part of the soil. As the acidified groundwater migrates through fractures in the limestone, it dissolves the rock, gradually widening the fractures.

The chemical reactions increase the concentration of calcium ions in the groundwater. This is called **hard water,** which requires more soap when cleaning and leaves a hard calcium deposit on plumbing fixtures. (Water that lacks such a concentration of calcium ions is called “soft water.”)

Limestone aquifers differ from the others we have discussed because the groundwater dissolves the limestone in some places and then deposits some of this dissolved calcium carbonate in other places. This is the origin of limestone caverns, with their massive underground chambers (dissolved by acidic groundwater) and their spectacular deposits extending down from the ceiling (stalactites) and up from the floor (stalagmites).

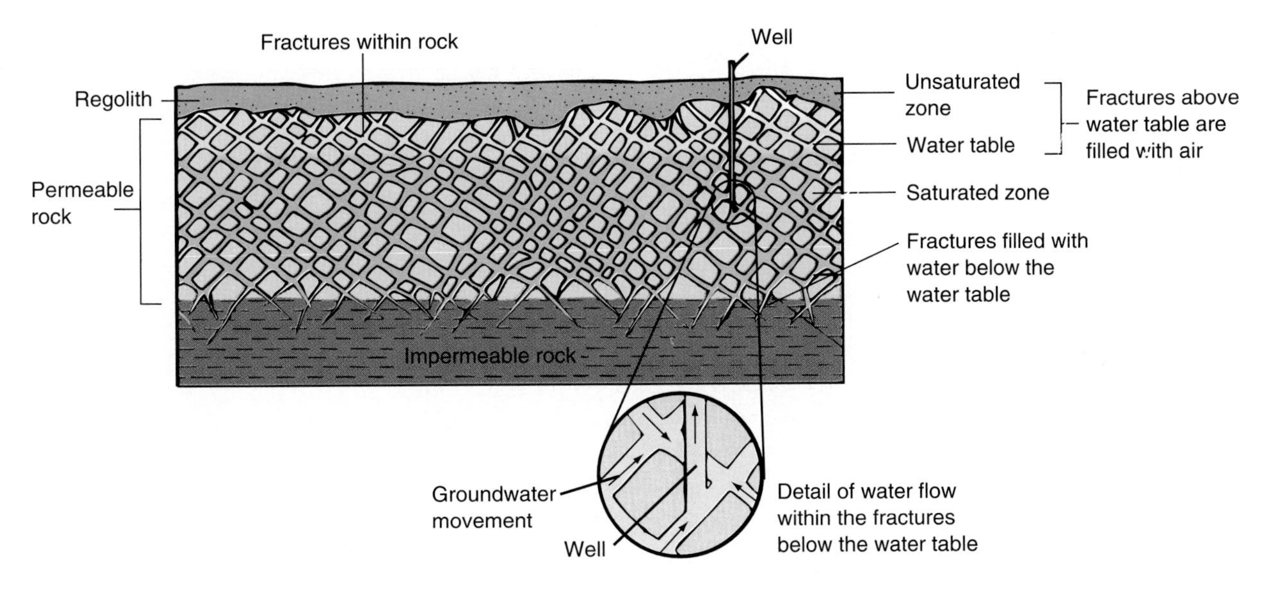


**FIGURE 8–7**
Groundwater distribution in fractured rock. (Reprinted with the permission of Macmillan College Publishing Company from *Physical Geology*, by Nicholas K. Coch and Allan Ludman. Copyright © 1991 by Macmillan College Publishing Company, Inc.)

The continual underground dissolution of a limestone aquifer can create serious surface problems, such as sinkholes (Chapter 10). The enlarged fractures and solution cavities (caverns) allow groundwater to move much faster in limestone aquifers than in other unconfined aquifers. This makes areas underlain by extensive limestone especially susceptible to groundwater contamination from surface pollutants (Chapter 12). Such areas in the United States include Florida, portions of Kentucky and Indiana, the "Great Valley" of West Virginia, Virginia, and Pennsylvania, and the limestone plains of Missouri.

## Island Aquifers

Islands in the ocean are surrounded by saltwater, yet have abundant natural freshwater. How can this be? The answer lies in the different densities of freshwater and saltwater. Freshwater is less dense than saltwater. Where freshwater and saltwater meet in the ground beneath an ocean island, they tend to remain separated, with the less dense freshwater remaining above the denser saltwater.

As long as an island has a permeable surface and rainfall occurs, freshwater will infiltrate the soil and rock of the island. The weight of this freshwater acts like a piston, displacing saltwater from the pores under the island and forcing it outward (Figure 8–8). Over time, this mechanism builds a lens-shaped body of freshwater that overlies the denser saltwater in the pores beneath the island. Thus, a well drilled on such an island will provide fresh (not salty) groundwater.

Due to the density difference, a column of freshwater 41 meters (134.5 feet) high equals the mass of a column of saltwater only 40 meters (131 feet) high. This relationship allows us to determine the thickness of the freshwater at any point on the island. For every meter the freshwater table is elevated above sea level, there are 40 meters (131 feet) of freshwater below.

This relationship gives the freshwater accumulation area beneath a permeable island the lens shape shown in Figure 8–8. Dry periods or overpumping may lower the water table elevation, and the base of the lens rises as the freshwater lens shrinks. The saltwater encroaches inward and upward to replace the withdrawn freshwater, and consequently saltwater eventually may be drawn into wells, contaminating drinking water.

Island aquifers provide water on sandy barrier islands along the Atlantic and Gulf Coasts, on limestone islands of the Florida Keys, and on the many coral islands of the South Pacific. The Hawaiian Islands are volcanic island aquifers. Rainwater infil-

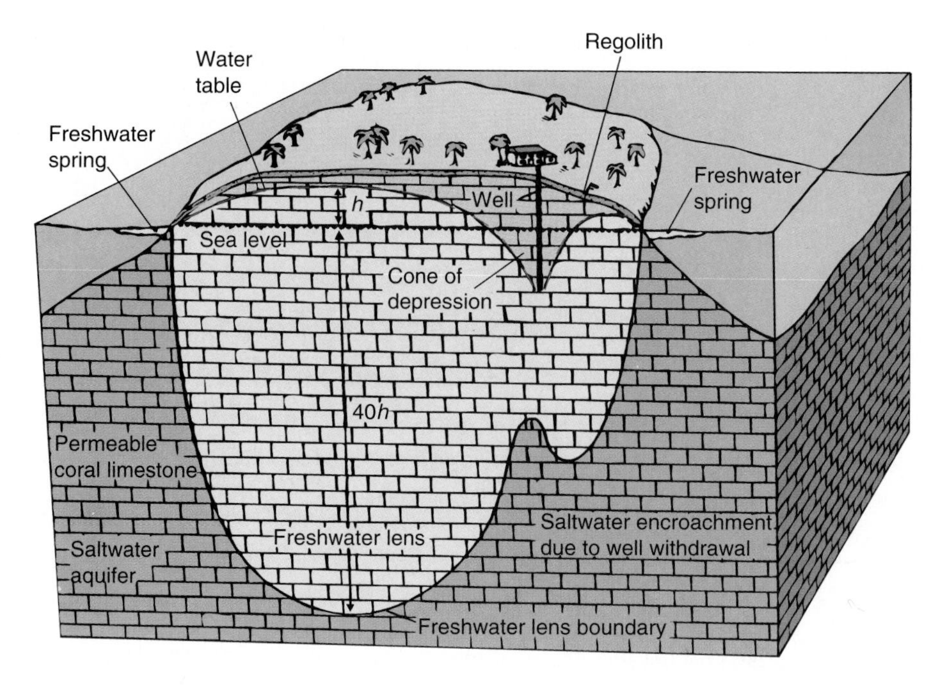


**FIGURE 8–8**
Geologic section showing the development of an island aquifer. The lens-shaped freshwater body sits atop the saltwater below. For every meter the freshwater table is above sea level *(h)*, there are 40 meters (130 feet) of freshwater below that point. (Reprinted with the permission of Macmillan College Publishing Company from *Physical Geology*, by Nicholas K. Coch and Allan Ludman. Copyright © 1991 by Macmillan College Publishing Company, Inc.)

trates joints and fractures in basalt flows and accumulates as a freshwater lens that overlies the saltwater.

## Confined Aquifer Systems

When wells drilled in shallow unconfined aquifers are inadequate, deeper wells are drilled into confined aquifers. However, there is a limit to how deep a well can be drilled. Below a certain depth, pressure (compaction) and cementation reduce the pore spaces, and greatly decrease permeability. These aquifers may be vast, extending under plains in the continental interior, spanning several states. They may develop in basins between mountain ranges, or extend under coastal plains and the ocean. We will consider these aquifers in both settings: continental aquifers and coastal aquifers.

## Continental Aquifers

Extensive confined aquifers provide groundwater to major regions of the United States. The basic structure of these confined aquifers is shown in Figure 8–6. A good example is the Ogallala Aquifer, a sand and gravel aquifer that underlies parts of eight states, including Nebraska, Kansas, and Texas. The Ogallala Aquifer is used to irrigate an area of about 5.7 million hectares (14 million acres). Intensive withdrawal of water from this aquifer has significantly lowered the regional water table, a problem considered later in this chapter.

Valleys may fill with stream deposits derived from nearby mountains, creating alluvial basins. The stream deposits may be porous and permeable enough to form thick aquifers, storing great volumes of water. These aquifers are recharged by rainfall entering the valley at the edges of the basin. In some basins, additional recharge is provided by influent streams. Groundwater is obtained from these basin aquifers in many areas of Nevada, western Utah, western Colorado, southern Arizona, and southwestern New Mexico.

However, the most intensive use of basin aquifers in the United States is in California. About 40% of the state is underlain by basins that contain groundwater. This resource is especially important to populations and agriculture during droughts.

The combined storage capacity of California's groundwater basins is estimated at nearly 30 times the total surface water storage of the state's reservoirs. However, only about 11% of basin aquifer water is of acceptable quality (very low salt content) or can be withdrawn economically. Further, excessive groundwater withdrawal from basin aquifers in California has resulted in subsidence of the land surface (Chapter 10).

## Coastal Aquifers

Many cities on the Atlantic and Gulf Coastal Plains of the United States obtain their drinking water from aquifers that are recharged inland and extend under the coastal plains, draining into the ocean. These are called **coastal aquifers** (Figure 8–9). Coastal aquifers are similar to unconfined aquifers that exist inland, except that the pores in their oceanward ends are filled with saltwater. This is important, because wells may draw either freshwater or saltwater, depending on conditions.

In a coastal aquifer, the freshwater tends to overlie the saltwater where the two are in contact, because freshwater is less dense than saltwater (see detail in Figure 8–9). The saltwater boundary shown

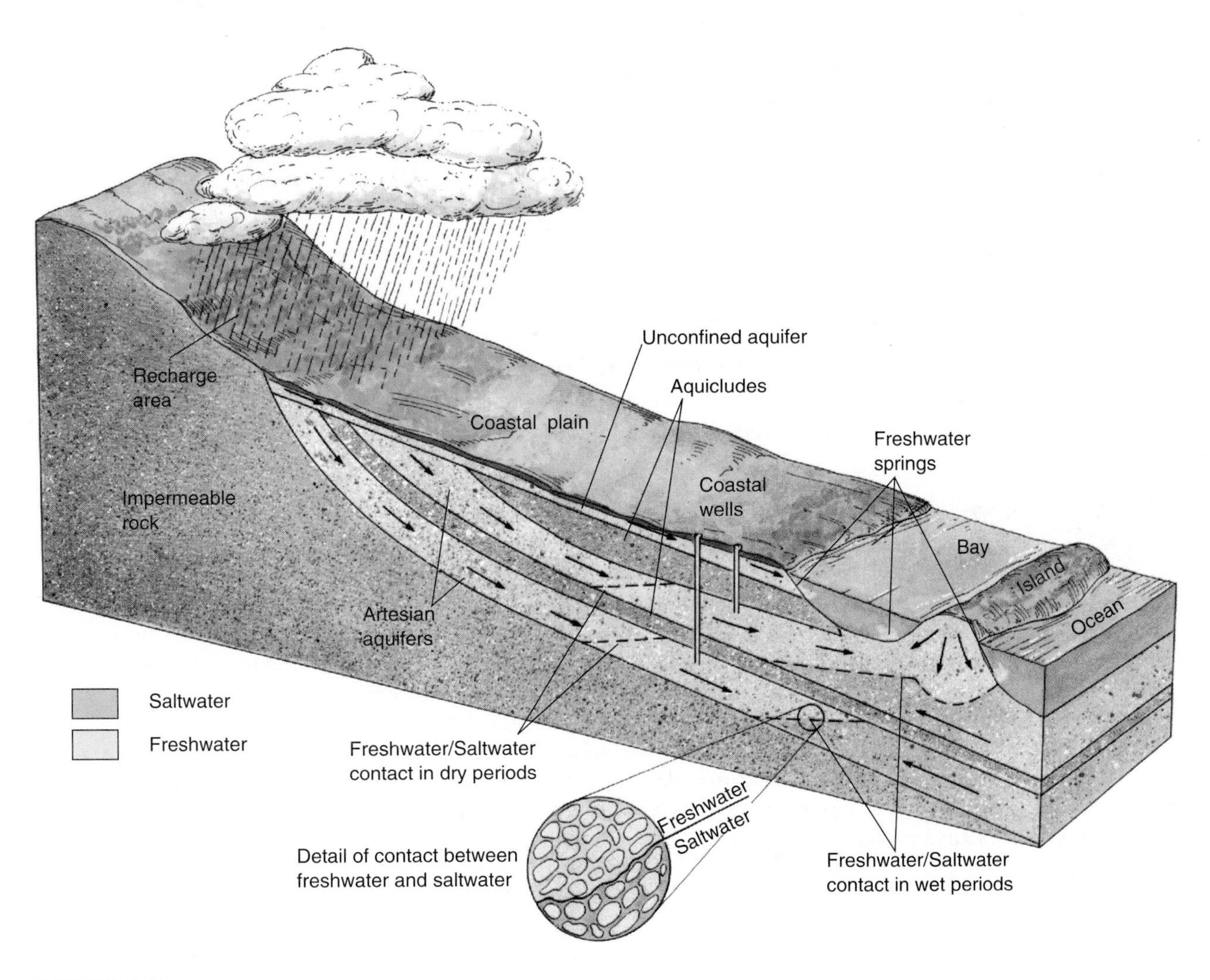


**FIGURE 8–9**
Freshwater-saltwater relationships in coastal aquifers. During periods of high rainfall, freshwater displaces saltwater in the aquifers, and freshwater springs occur in the bays. During periods of drought, saltwater moves up the aquifers (dashed lines show freshwater/saltwater boundaries) and contaminates coastal wells. (Reprinted with the permission of Macmillan College Publishing Company from *Physical Geology,* by Nicholas K. Coch and Allan Ludman. Copyright © 1991 by Macmillan College Publishing Company, Inc.

in the figure represents the position of the freshwater/saltwater boundary during high rainfall. The pressure from the freshwater keeps the saltwater from rising in the aquifer. During this high-rainfall condition, coastal wells draw freshwater from both the upper and lower artesian aquifers, and freshwater artesian springs flow into the saltwater bay.

During a drought, the freshwater/saltwater balance in coastal aquifers changes markedly. The reduced rainfall brings less freshwater recharge into the aquifers. This allows the saltwater to rise in the aquifer, to the dashed lines shown in Figure 8–9. The same thing happens if wells are overpumped. Such saltwater penetration into aquifers is called **saltwater encroachment.** During a drought, coastal wells would be pumping saltwater from the aquifers, and freshwater springs would cease flowing into the bay. The island in Figure 8–9 preserves its own freshwater supply through the mechanism discussed for Figure 8–8.

### Importing Water by Aqueduct

Groundwater supplies seldom are adequate for large metropolitan areas. The next option is to use water from a nearby river or lake. If this too is inadequate, water must be imported.

The Florida Keys import water over 200 kilometers (124 miles) from wells on the Florida mainland (Box 8–1). New York, Boston, San Francisco, and Norfolk import water from distant reservoirs, created many years ago by damming streams and rivers in sparsely settled areas. Los Angeles and San Diego import much of their water through an aqueduct and canal system that extends eastward to the Colorado River. It is important to anticipate the need for importing water long before it becomes a problem, because aqueducts are expensive, take a long time to build, need extensive engineering, and require acquisition of land for pipeline access.

## GROUNDWATER GEOHAZARDS

Now that we have shown how groundwater travels and is affected by human use, we turn to its associated geohazards. Groundwater geohazards include changes in the stability of the land surface (discussed in Chapter 10), contamination from surface pollutants and leaking underground storage tanks, temperature increase, and depletion. In this section, we look at problems of quantity, temperature, and how pollutants migrate.

### Water Table Depression

Groundwater abundance depends on the balance between discharge and recharge. In many areas, widespread water table depression occurs because this balance has been disturbed (Figure 8–5). This problem stems from three major causes:

- □ Pumping rates far exceed recharge rates (Figure 8–10a).
- □ Recharge is reduced by paving in rapidly urbanizing areas (Figure 8–10b).
- □ Well water is being used but not returned to the ground near its source (instead, it is sent to a stream or the ocean—Figure 8–10c).

Rapidly growing areas have such great demand for groundwater that natural infiltration cannot recharge the groundwater reservoir sufficiently. This leads to a steady drop in the water table (Figure 8–10a). Increased development results in more paved roads and parking lots, drastically reducing the recharge area (Figure 8–10b). The problem is especially severe because water demand is increasing at the same time recharge is decreasing.

Groundwater for domestic, sanitary, and industrial use commonly is treated and returned via pipelines to nearby water bodies for disposal. This is cheaper than pumping it back into the ground. However, it decreases the recharge, and the water table becomes depressed (Figure 8–10c). This short-sighted approach reduces pollution, but decreases groundwater resources in that area.

The transition of a rural area into an urban environment dramatically lowers the water table. Conservation and environmental control will help ease the problem, but a water crisis probably is inevitable. It is important to act early before it becomes too late to import water by aqueduct in time to meet the rising need.

### Increasing Groundwater Temperature

Groundwater has a relatively constant year-round temperature in any given region. In the broad area of the United States east of the Rockies, groundwater temperature at about 30 meters (98 feet) depth is about the same as the average annual air temperature at that locality:

## BOX 8.1 WATER FOR THE FLORIDA KEYS—PARADISE LOST?

The Florida Keys are a 240-kilometer-long (150 mile) chain of limestone islands that stretch from Miami southwestward to Key West (Figure 1). The islands were isolated units until they were connected by the Atlantic Coast Railroad in 1913. A disastrous hurricane destroyed the railroad in 1935 (Chapter 16). The railroad was replaced in the late 1930s by a highway-and-bridge system (Figure 1). Key West developed into a major Gulf Coast port and naval base. The town

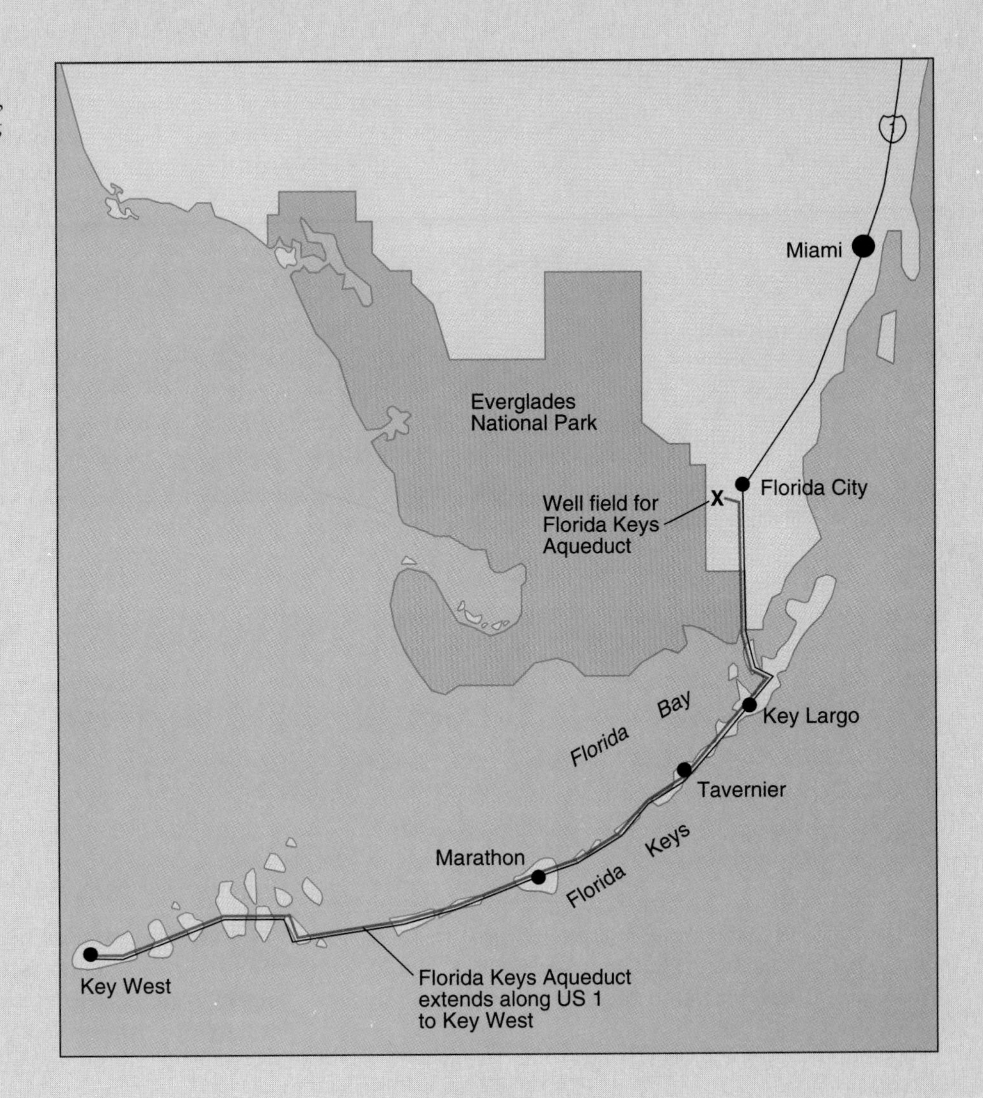


**FIGURE 1**
Map showing Florida Keys, Miami, Key West, and well fields supplying the Keys Aqueduct System.

obtains water from the limestone aquifer that makes up the island (Figure 8–8) and from rainwater, stored in cisterns by individual homeowners.

Visitors to Key West cannot fail to be impressed with the architectural detail in older houses. Their overlapping roof systems not only provide shade in this subtropical climate but also catch as much rainwater as possible. The rainwater flows into gutters and vertical piping systems at several sides of the roof (Figure 2), and then moves downward into an underground cistern. This mode of water supply is called a **catchment system.**

This island aquifer/catchment system served Key West and the sparsely settled Florida Keys until World War II. Then, larger and more dependable water supplies were required for naval facilities.

Wells were drilled into confined limestone aquifers on the Florida mainland, near Florida City (Figure 1). The groundwater was pumped into an aqueduct constructed for 209 kilometers (130 miles) along U.S. 1 southward to Key West (Figure 3). Years later, a larger-diameter pipeline replaced the original to provide virtually all the water for the Florida Keys.

**FIGURE 2**
Water collection pipes on a home in Key West, Florida. The pipes collect water from the roof and transport it to a storage cistern under the home. (Photo by author.)

- 6–11°C (43–52°F) in the northern states
- 11–20°C (52–68°F) in the middle states
- 20–25°C (68–77°F) in the southern states

Groundwater temperatures are slowly increasing in many areas that are developed or are rapidly developing. This is a geohazard because warmer water retains less oxygen (Chapter 12). Less oxygen in groundwater makes it more hospitable to pathogenic bacteria from spills or leaking sewage lines and tanks. The temperature increase occurs because groundwater is used for industrial cooling (the

**FIGURE 3**
Old Florida Keys aqueduct suspended under former U.S. 1 bridge (at right). The new, larger aqueduct is carried under the new bridge visible at left. (Photo by author.)

Today the Florida Keys Aqueduct Authority pumps nearly 42 million liters (11 million gallons) per day of water throughout the Keys. The water's journey takes about four days from the Florida City wells to Key West.

Bringing abundant water into the Florida Keys has been a mixed blessing. Prior to construction of the aqueduct, the area was relatively uninhabited and a pristine tropical paradise. Since the aqueduct and new road system were constructed, the Florida Keys have rapidly urbanized. Development is severely straining the only reef system in the continental United States, eliminating many vital wetland habitats (see Chapter 14). Development is bringing many more people to live in one of the most hurricane-prone areas in the United States (Chapter 16).

Although hurricanes are not a groundwater hazard, this is a good time to look at the 1935 hurricane that devastated the central Florida Keys. It was the most powerful in U.S. history. Emergency managers are concerned with how to evacuate tens of thousands of people when the next "big one" approaches the Florida Keys. Evacuees must travel the Overseas Highway (a mostly two-lane roadway, with three or four lanes in places), cross numerous bridges, and cross causeways that are easily flooded.

water absorbs the heat) and because paved surfaces heat runoff before it infiltrates the ground.

The constant, relatively cool temperature of groundwater makes it useful for industrial air conditioning in the summertime. Unfortunately, the groundwater becomes heated by this process (Figure 8–11). When pumped back into the ground by injection or diffusion wells, this warmer water increases the temperature of the groundwater reservoir over time. Also, dark paved areas absorb solar energy more quickly than natural surfaces and heat the rainwater that runs across them (Figure 8–11).

FIGURE 8–10
Three reasons that depressions form in the water table: (a) more water is pumped out than can be recharged by infiltration; (b) structures and paved surfaces prevent infiltration to recharge a groundwater reservoir depleted by pumping; (c) discharge is high but recharge is minimal because treated water is discharged into nearby water bodies rather than being returned to the ground.

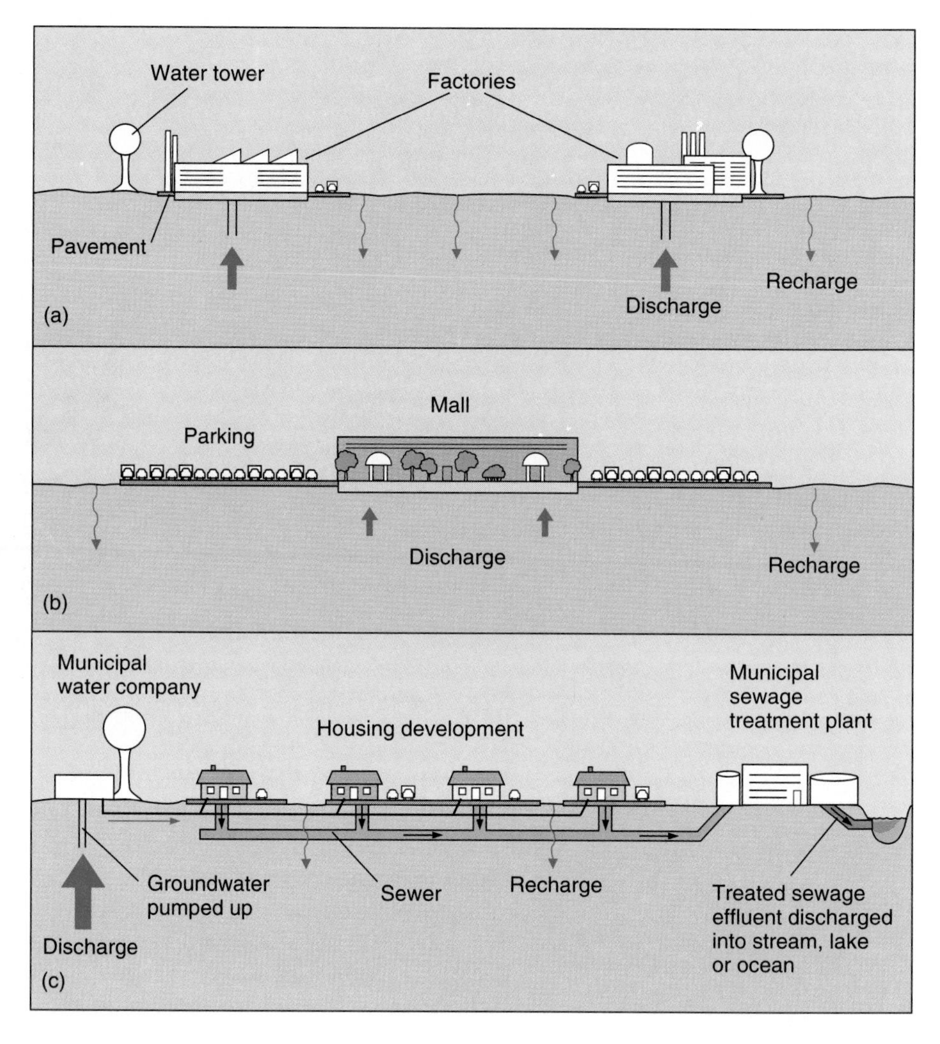


When it infiltrates, it raises groundwater temperature.

## Groundwater Pollution

Groundwater supplies are highly vulnerable to surface spills or to ruptures of underground tanks and pipelines. These spills and leaks often are of toxic chemicals. When these pollutants enter the groundwater, they move within the groundwater flow as a **contaminant plume** (Figure 8–12). The shape of this three-dimensional plume varies with the aquifer's geology (rock type, opening size, layering), its flow rate, and the density and chemical reactivity of the pollutant. A plume may extend hundreds of meters or even several kilometers.

Contaminants in the plume are much more concentrated near the source, and the concentration gradually decreases with distance. The outer rim of the plume is much less concentrated because it has been diluted with large volumes of groundwater. Thus, when a contaminant is first detected in low concentrations in a well, you can expect the concentration at that locality to become greater as the plume progresses. However, this sometimes is not the case, because the contaminant may be diluted, adsorbed onto sediment particles, or may chemically decompose.

Another important factor governing pollutant movement is its density compared to water. If a pollutant is more dense than water, such as a salt solution, it will sink through the groundwater until it

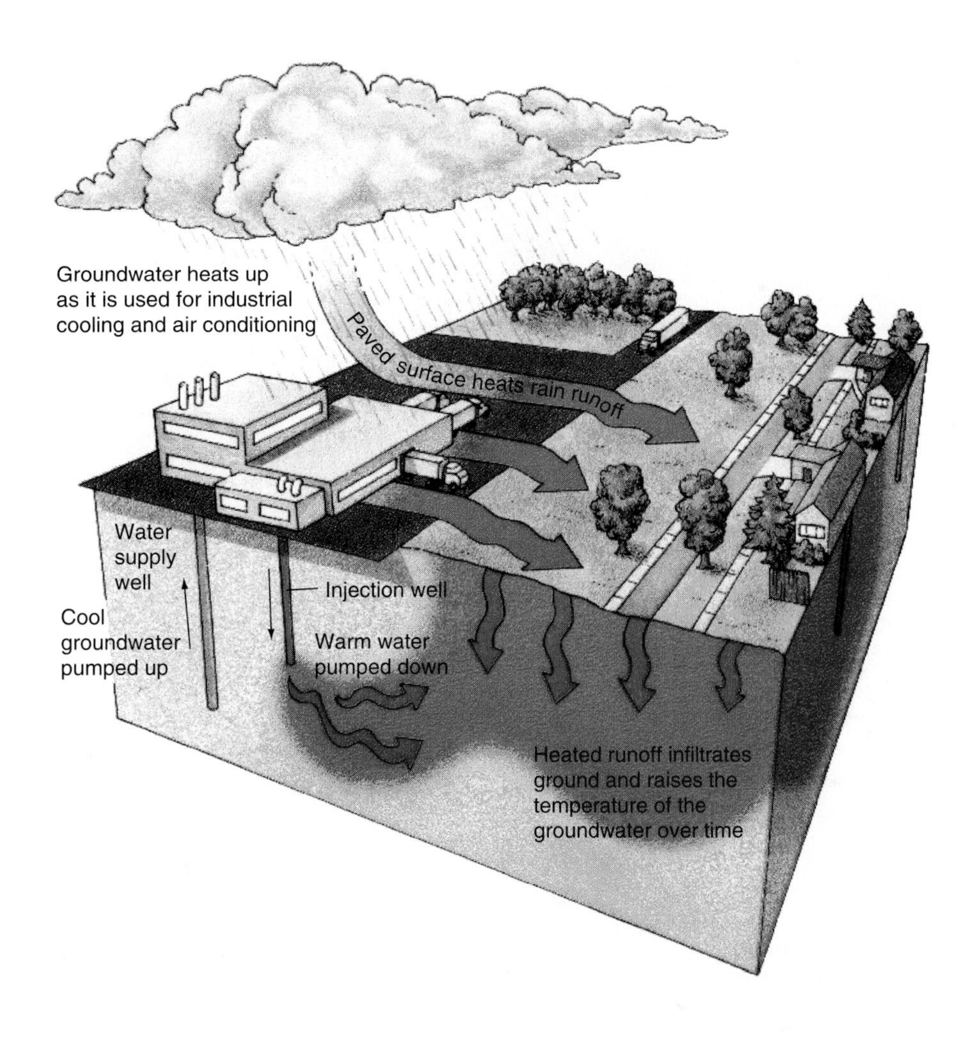


**FIGURE 8–11**
Geologic section showing how groundwater temperature can increase in an urbanized area.

reaches a lower bounding aquiclude (Figure 8–13). At this point, the plume may flow along the top of the aquiclude, as shown in the figure. However, dilute solutions of the pollutant will form contaminant plumes and be carried by the prevailing groundwater flow, as shown. Pollutants that are less dense than water tend to sit atop the water table, with dilute plumes moving with the groundwater flow, away from the pollution source (Figure 8–13).

Additional problems arise when a contaminant has a vapor (gas) phase as well as a liquid phase. Examples are chemicals from landfills (see Chapter 12) or petroleum products. The gases generated at or just below the water table can move upward into the unsaturated zone, contaminating it (Figure 8–13). They also can migrate into buildings through cracks, permeable foundations, and floor drains, making people ill and possibly building to explosive concentrations.

A special problem exists with flammable or explosive pollutants in vapor form. This can happen from gasoline vapors from a leaking underground tank or from methane produced by anaerobic decomposition of garbage in a landfill (see Chapter 12). If these gases migrate into the basements of adjacent homes and are ignited by furnaces, water heaters, or electrical sparks, fire and explosion may result.

## PROTECTING AND MONITORING GROUNDWATER QUALITY

The best way to protect groundwater quality is to prevent pollutants from entering the groundwater

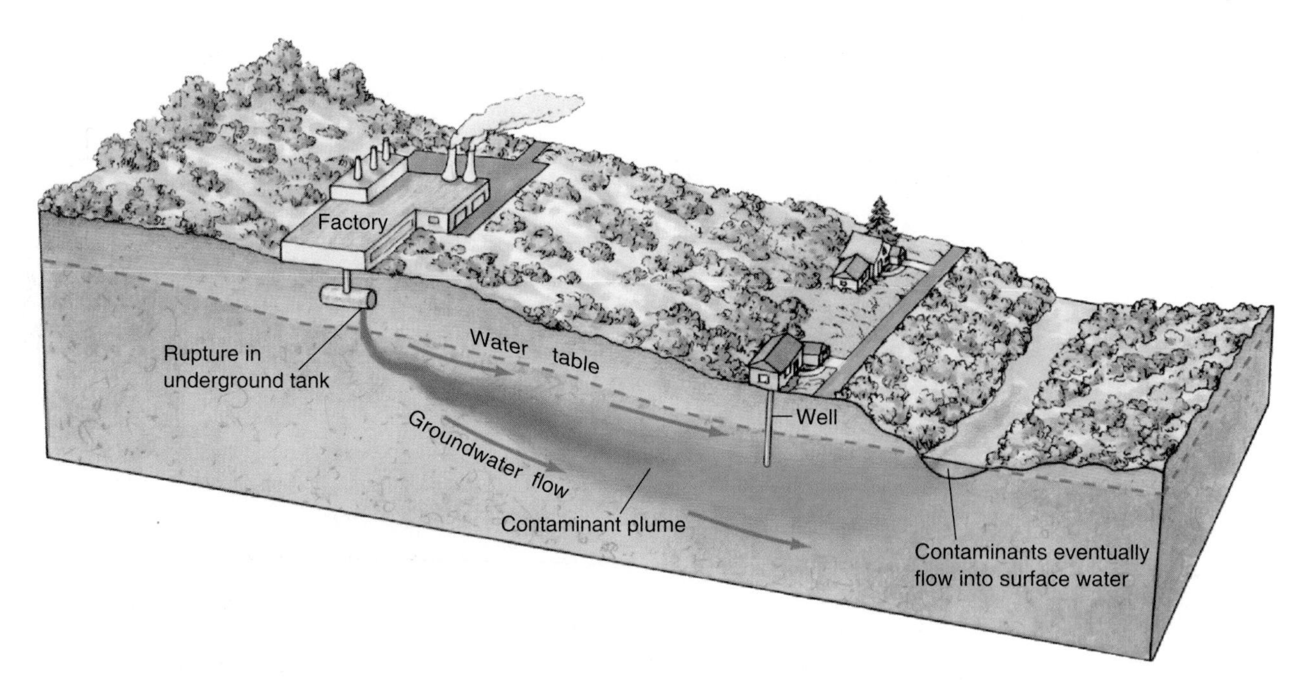


**FIGURE 8–12**
Geologic section showing movement of a contaminant plume and its effect on surface and ground water supplies.

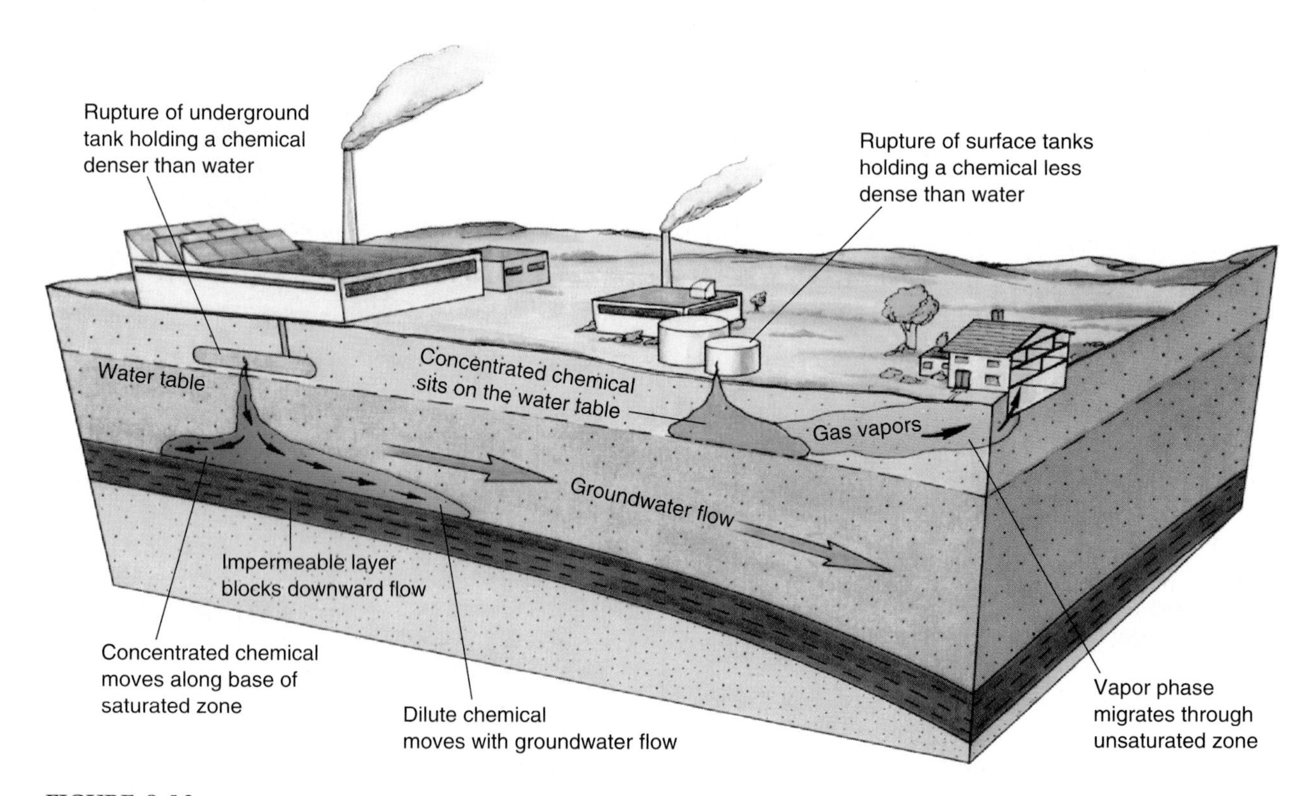


**FIGURE 8–13**
Geologic section showing movement of a vapor phase and of pollutants that are heavier and lighter than water.

system. Spills and tank ruptures are inevitable, however, and various methods have been devised to monitor groundwater quality.

### Preventing Contaminant Spills, Ruptures, and Leaks

The effects of spills can be minimized by erecting **containment structures** around surface storage areas. These devices keep the spill from leaving the site and limit infiltration into the ground. Common types of containment structures are walls around surface tanks to contain spills. Another type uses floating barriers to prevent oil from a ruptured tanker or a spill from dispersing in the surrounding water.

Pipelines should be constructed to assure that joints between pipe segments are tightly sealed and won't leak. Underground steel tanks are especially subject to corrosion and should be scheduled for periodic replacement or replaced with a noncorroding type, such as fiberglass. In some urban areas, underground tanks are being banned and replaced by above-ground tanks, the condition of which can be closely monitored.

### Monitoring Groundwater Quality

Groundwater quality can be monitored with water samples from **monitoring wells.** These wells are drilled between potential pollution sources and domestic water supplies. Water samples taken from various levels in the wells can indicate contaminant migration and determine its concentration.

## MAINTAINING GROUNDWATER SUPPLY

Several methods are used to maintain the volume of groundwater available to a community. These include decreasing discharge (conservation), increasing recharge (preserving infiltration areas), and augmenting infiltration by injecting water from another area into the local aquifers.

### Water Conservation

We can conserve water by learning to use less. Americans use more water per capita than any other population in the world. Many water companies suggest shorter showers, inflatable bags in toilet tanks to make them hold less water, fixing leaky plumbing, restricting lawn-watering, car-washing, and pool-filling, and so on.

In watering livestock and irrigating crops, agriculture uses more water than any other human activity. It might appear that irrigation just borrows water and doesn't really *consume* it, because it goes back into the ground and streams. However, so much of the water evaporates that irrigation constitutes a major loss to the groundwater system. The amount of water lost to the atmosphere in this case is called a *consumptive use*. Modern irrigation methods minimize **consumption** by delivering water in smaller quantities directly to the plants or trees.

### Preserving Infiltration Areas

Effective infiltration is essential to provide adequate aquifer recharge. This can be accomplished by restrictive zoning—fewer houses per area. Another method is the cluster development, described in Chapter 6. Cluster development (Figure 6–21) leaves maximum open space around developed areas, facilitating infiltration of water to recharge the aquifers below.

### Other Recharge Methods

Natural recharge by infiltration often is supplemented by other methods. Many rapidly growing areas try to forestall a groundwater crisis by actively increasing recharge. In suburban Nassau and Suffolk Counties on Long Island, New York, laws require construction of surface basins as a water conservation measure. These **recharge basins** are fed by culverts that channel rainwater so that it infiltrates the ground and recharges the aquifers below. Aquifers beneath the Los Angeles area are recharged by controlled flooding of specially prepared permeable ground.

Another recharging method is to pump water from elsewhere through injection wells and into the ground. In Los Angeles, aqueduct water obtained from the Colorado River is pumped down injection wells to displace saltwater in coastal aquifers.

## LOOKING AHEAD

Our study of groundwater leads naturally to what water does to the soil and rock that it permeates. In simple terms, water lubricates and weakens soil and rock. This makes groundwater an important factor in triggering landslides, the subject of the next chapter.

## SUMMARY

Groundwater is an estimated 68 times more abundant than surface water. It is an important resource because of its widespread geographic occurrence. In dry regions of few streams, groundwater is commonly the only source of drinking, industrial, and agricultural water.

Groundwater occurs beneath Earth's surface in pore spaces between sediment particles, in rock fractures, and in solution cavities. The water moves slowly through the ground under the influence of gravity and confining pressure. The boundary between the unsaturated zone, where the pores are only partially filled with water, and the saturated zone, where the pores are completely filled, is the water table. Permeable materials are porous and have plentiful connections between pores. Impermeable materials may or may not be porous, but they lack connections between pores.

Aquifers are rock or sediment layers that can store and transmit water. An unconfined aquifer has no aquiclude above it and is open to the atmosphere. A confined aquifer is bounded above and below by impermeable aquicludes. The water in confined aquifers is under pressure and may rise to the surface without pumping to form an artesian well. The discharge in aquifers is predicted from Darcy's Law.

Maintaining groundwater supplies requires that the water withdrawn by pumping is balanced by aquifer recharge. When discharge greatly exceeds recharge, a cone of depression forms in the water table and wells may run dry until the aquifer is adequately recharged.

Limestone aquifers pose special problems because acidic groundwater reacts with limestone to dissolve it. This process enlarges rock openings so that there is less support for the surface. When the unsupported surface collapses, it forms a sinkhole. Coastal aquifers pose problems because saltwater may intrude their lower ends if recharge is insufficient.

Groundwater that enters a stream forms its base flow, enabling the stream to flow even during drought. Effluent streams are supplied by groundwater and are common in humid regions. Influent streams supply water to the ground and are more common in dry areas. Streams can be classified by their continuity of flow as perennial, intermittent, and ephemeral.

In a developing area, water first is obtained from individual wells into unconfined aquifers. If these become inadequate, it may be necessary to drill wells into confined aquifers at greater depth. Large metropolitan areas may need aqueducts to import water from distant surface or underground sources.

A wide variety of groundwater geohazards occur. In some areas, groundwater withdrawal has caused subsidence of the land surface. Elsewhere, heavy discharge has resulted in deep depressions in the water table. Also, contaminants are entering the groundwater reservoirs in many areas.

Depressions in the water table occur when discharge locally exceeds recharge, when infiltration is reduced, or when groundwater is used and not returned to the ground. People increase recharge by constructing recharge basins, maintaining open areas for infiltration, or by injecting used water back into the ground. Conservation practices are increasingly used to preserve groundwater resources.

Groundwater pollution forms contaminant plumes that travel within the groundwater flow. The path of the pollutant depends on the aquifer's geology, flow rate, and the density and chemical reactivity of the pollutant.

## KEY TERMS

aquiclude
aquifer
artesian wells
base flow
catchment system
coastal aquifers
cone of depression
confined aquifer
consumption
containment structures
contaminant plume
Darcy's Law
discharge
effluent streams
ephemeral streams
hard water
influent streams
intermittent streams
monitoring wells
perched water table
perennial streams
permeability
recharge
recharge basins
saltwater encroachment
saturated zone
soil water zone
spring
unconfined aquifer
unsaturated zone
water table

## REVIEW QUESTIONS

1. Recalling information from Chapter 2, describe how groundwater is stored in the following types of rocks: (a) jointed granite, (b) limestone, (c) sandstone, (d) shale.
2. Describe the difference between the movement of a pollutant plume in sandstone and in limestone.
3. How do unconfined and confined aquifers differ from one another? Describe five conditions necessary for the development of a confined aquifer.

4. What are the differences between influent and effluent streams?
5. Describe a coastal aquifer. How is the groundwater quantity and quality of this type of aquifer affected by a period of drought?
6. Describe how freshwater supplies exist under an island. How can the thickness of the freshwater reservoir be estimated?
7. Describe the water supply for the area where you live. Is your water supplied by individual wells, deep wells into confined aquifers, lake, stream, rainwater (cistern), or imported from elsewhere by aqueduct?
8. Describe several different causes for large-scale water table depressions that are occurring in many developing areas.
9. Explain two different ways in which aquifers can be artificially discharged.
10. Describe a pollutant plume and how it moves. How can the approach of a pollutant plume be detected before it reaches domestic water wells?
11. Contrast the movement in an aquifer of two different pollutants, one lighter and the other heavier than groundwater.
12. Describe the special problems associated with oil spills that infiltrate the water table. Can you think of ways to remove this contaminant?

## FURTHER READINGS

American Institute of Professional Geologists 1985, Ground water: Issues and answers: Arvada, Colorado, 24 p.

Easterbrook, D. J., 1993, Surface processes and landforms: New York, Macmillan, 520 p., Chapter 7 (Groundwater).

Ford, R. S., 1978, Groundwater—California's precious resource: California Geology, v. 31, no. 2, p. 27–32.

Manning, J. C., 1992, Applied principles of hydrology (second edition): New York, Macmillan, 294 p.

National Geographic, 1993, Water—The power promise and turmoil of North Americas fresh water, v. 84, no. 5A, 120 p.

U.S. Geological Survey, 1984, Groundwater regions of the United States, Water Supply Paper 2242, 78 p.

U.S. Geological Survey, 1988, Groundwater and the rural homeowner, 36 p.

U.S. Geological Survey, 1970, Urbanization and its effect on the temperature of the streams on Long Island, N.Y., Prof. Paper 627-D, 110 p.

# 9

# *Landslides*

- ☐ *California*—heavy rains saturate masses of sediment and rock that flow onto a coastal highway.
- ☐ *Great Lakes*—high water levels erode cliff bases, resulting in mass movements that damage homes atop the cliffs.
- ☐ *New York*—frost-loosened rock falls onto an interstate highway, killing a motorist.
- ☐ *Illinois*—flooded rivers erode bluffs and levees, causing them to collapse into the raging water.

Each of these events is a *downslope displacement of regolith and rock.* Such events popularly are called **landslides.** However, many geologists prefer the term *mass movement* because these displacements do not always occur on "land" but on the seafloor as well, and many move by creeping or falling, rather than "sliding." In this text, we will use both terms.

Mass movement can be a serious hazard where expanding populations build homes on steep slopes. The hazard is especially severe when hillside development is done in a seismic zone (Chapter 5), where ground shaking is a constant threat. As Figure 9–1 shows, downslope movements of surface material are a hazard in many parts of the United States.

Why do some mass movements travel downslope very fast, whereas others migrate so slowly that people notice the displacement only over a long period? Why are many areas free of major mass movements, whereas others are continually at risk? How can we tell if mass movements are a potential risk when buying a home or seeking a building site? We will answer these questions by comparing factors that trigger mass movement with factors that prevent such movement.

**Vehicles scattered by May 30, 1983, debris flow in Ophis Creek, near Reno, Nevada. (Photo by Patrick Glancey, U.S. Geological Survey.)**

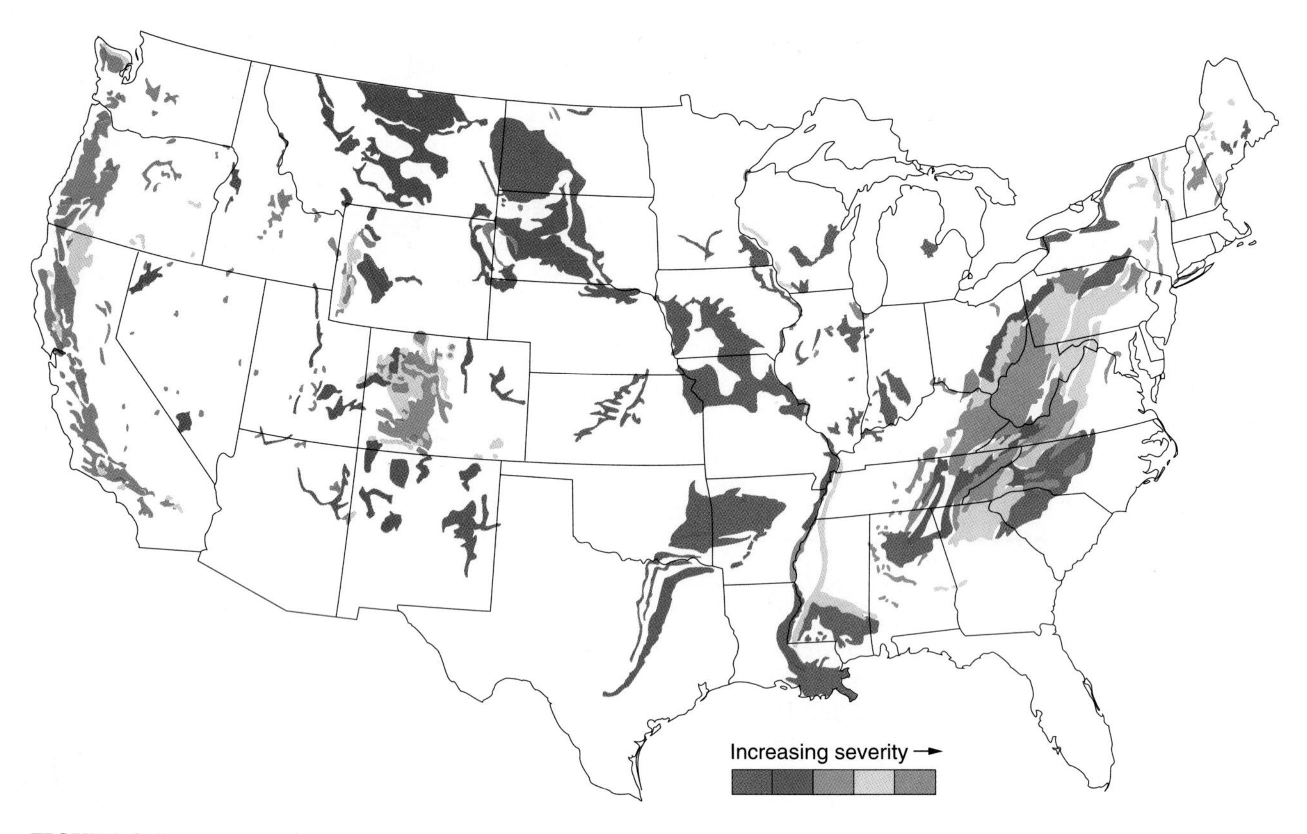


**FIGURE 9–1**
Map of mass movement problems in the United States. Areas that may contain landslides or be susceptible to landsliding on a scale too small to be shown are not colored. (Modified from American Institute of Professional Geologists (AIPG), 1993, The citizens' guide to geologic hazards: Arvada, Colo., AIPG, 79 p. Used with permission.)

## FACTORS AFFECTING MASS MOVEMENT

Two classes of forces are involved in mass movements: **driving forces,** which promote them, and **resisting forces,** which deter movement. Mass movement is far more frequent on steeper slopes, indicating that gravity is a major driving force. Although water is an agent that can cause mass movement under most conditions, water also can resist movement under other conditions. Other factors are more passive, weakening material on a slope so that gravity can move it downslope more easily. We first will look at the role of gravity in triggering mass movement because it is the major driving force.

### Gravity as a Driving Force

We know from experience that the steeper the slope, the greater is the tendency of materials to move downward. When driving down a steep hill, you use brakes (resisting force) to act against the strong pull of gravity (driving force) that tugs the car downhill. Similarly, rock and regolith on steep slopes are affected by strong forces that tend to move the material downslope. This tendency is resisted by the strength or cohesiveness of the material and the force of friction.

If the resisting forces exceed the driving forces, the material will stay in place. However, if the driving forces are greater than the resisting ones, the material will break loose and move downslope, a process called **mass movement.**

Driving forces can be increased by natural processes and human activity. For example, water erosion or human excavation can remove a portion of the lower slope, thus steepening the slope. People may build structures on the upper slope, increasing its weight. For example, waves cutting into the base of a cliff remove support for the upper part

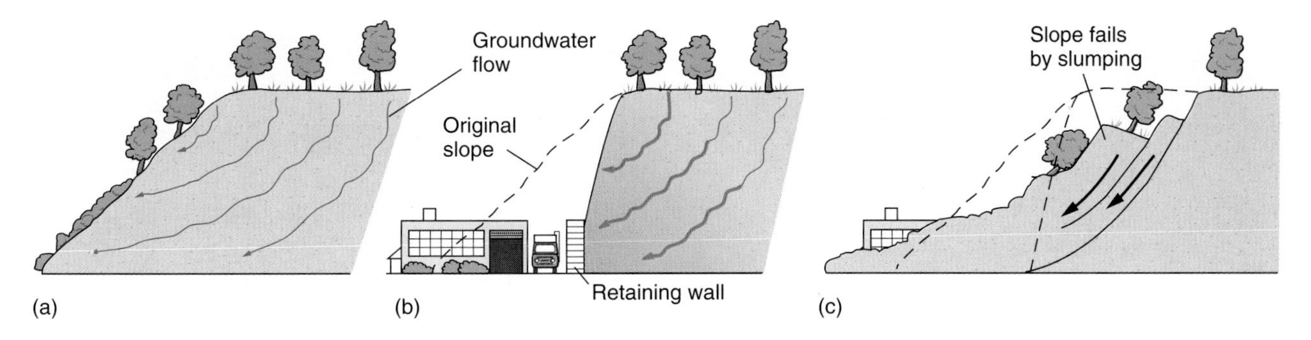


FIGURE 9–2
Human activities affecting slope stability. (a) Natural slope conditions. (b) Excavation into the slope creates a flat area for construction and steepens the slope above it. (c) Slope failure results in a landside that buries part of the construction site.

and steepen the slope. When a portion of a slope moves downward, this process is referred to as **slope failure** (Figure 3–8). In a similar fashion, excavating into a slope to create a level building area can also affect slope stability and cause a landslide (Figure 9–2).

## Role of Rock Structures

The phrase "solid as a rock" is a misnomer because most rocks contain a variety of openings that can affect their strength:

- ☐ Pore spaces between particles allow fluids to pass through, dissolving the cementing material and weakening the rock.
- ☐ Intersecting sets of parallel joints allow rocks to break into smaller masses that move more easily downslope (Figure 2–18).
- ☐ Contact surfaces between beds of rock that have different characteristics are points of weakness along which rocks can break. (Figure 2–18).

Plate tectonic pressures can reorient rocks after their formation. Such dipping layers (Figure 2–22) facilitate mass movement when they are inclined in the same direction as the slope of the land.

Even steep slopes can be quite stable as long as the rock layers are inclined *away* from the slope—see Figure 9–3a, "stable slope," where the layers slope downward away from the roadcut. However, rock layers that are tilted downslope decrease slope stability (the opposite side of the roadcut in Figure 9–3a). Thus, slope stability on opposite sides of a highway cut may be quite different.

Sometimes rock layers are inclined away from the slopes but *fractures* within the rock are inclined downslope (Figure 9–3b). In this instance, such a slope is quite unstable. If driving forces eventually exceed resisting forces, the rock masses can slide over each other along the bedding or fracture surfaces.

To determine slope failure potential, detailed geologic studies are necessary before excavating a slope. This is extremely important to transportation and civil engineers, who are concerned with the stability of slopes along highways, railways, canals, streams, and construction sites.

## Water as a Driving Force

Many mass movements occur after rain has saturated slopes. Water promotes movement in two ways. As an active agent, it increases the loading (weight) of sediment or rock by filling previously empty pores and fractures. Water also decreases the strength of the rock or sediment by reducing cohesion among particles. Clearly, the role of water is complex. Small quantities of water in a sediment actually may enable it to resist mass movement, as we shall explain later.

***Loading.*** Water becomes a driving force by increasing the weight of slope material. A stable mass of dry sand exposed on a slope can have up to 35% of its volume composed of dry pores. After a prolonged period of rain, these pores may be completely filled. The water increases the weight of the sediment and thus increases the force driving the

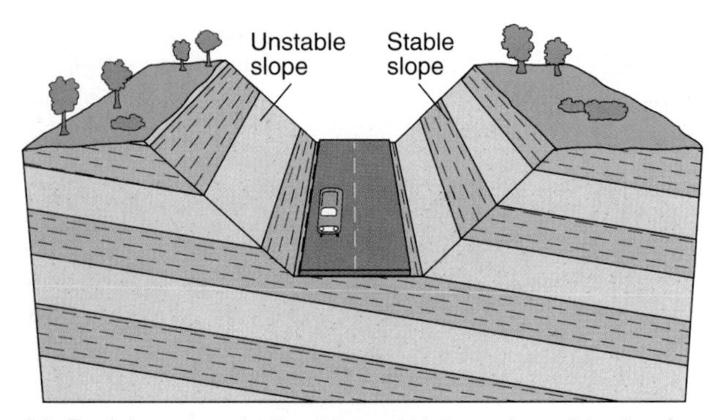


(a) Rock layers are inclined toward highway (unstable slope) and away from highway (stable slope) on opposite sides of the roadcut.

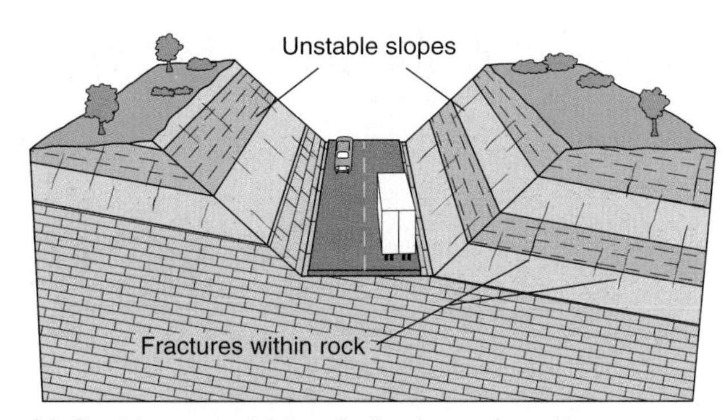


(b) Rock layers on right are inclined away from the roadcut, but fractures within the rocks slope toward the roadcut.

FIGURE 9–3
Slope stability depends partly on the orientation of rock layers. Slopes with rock layers or fractures dipping *into* a valley are potentially unstable.

sediment downslope. Loading causes many mass movements to occur during or shortly after prolonged rainfall.

***Reducing Rock Strength.*** Water can reduce rock strength in several ways. Water circulating through the pores of some rocks dissolves soluble cementing materials, such as calcium carbonate. This action reduces cohesion, allowing the grains to move past one another downslope more easily. Water can soften layers of shale and even cause some types of clay minerals to expand, reducing the frictional forces acting between rock layers. The force exerted by the growth of ice crystals in rock crevices can break apart rock (frost wedging) along joints and bedding surfaces (see Chapter 3). The loosened rock subsequently can move downslope.

***Expanding Clays (Shrink-Swell).*** Some clay-rich sediments can absorb large quantities of water and can swell up to many times their original volume if unconfined. **Bentonite** is perhaps the best-known example of swelling clays. Bentonite is composed of clay minerals formed by the chemical alteration of some volcanic igneous rocks, such as tuff.

When water infiltrates the ground from snowmelt, rainfall, lawn watering, or leaking pipes, any layers of bentonite that are present swell. They exert great pressure on overlying rocks, soil, and structures. Between rains, the layers dry and contract, forming large surface cracks that can damage overlying structures. Sediments of this type commonly are called "shrink-swell" clays. They are a common geohazard to foundations of buildings in many parts of the United States (Figure 9–4).

If water-saturated bentonite is on a slope, the material forms a slick surface that reduces friction and facilitates downslope movement of any overlying layers.

***Liquefaction of Clays.*** A **quickclay** is a type of clay that is transformed rapidly from a solid to a liquid under certain conditions. One type forms from clay minerals that originally accumulated in salty seawater. Ions from the saltwater remain in the pore water, holding the clay minerals together by the attraction of their electrical charges. The result is a "house of cards" structure (Figure 5–26).

When such clays are subjected to freshwater flowing through their pores, saltwater ions are flushed out. With these ions gone, the structure is greatly weakened, making the formerly solid clay unstable (Figure 5–26). If the slope is subjected to vibration, the clay structure collapses and transforms quickly into a viscous fluid, which flows downslope. Quickclay is a serious geohazard in many areas, such as the St. Lawrence Valley in Canada.

## Water as a Resisting Force

In some cases, water becomes a resisting force. In sediment pores that are not filled *completely* with water, the thin water film actually makes the parti-

FIGURE 9–4
Map of expansive soils in the United States. Expansive soils are most abundant in areas colored red and decrease in order of blue and purple. Areas that may contain expansive soils on a scale too small to be shown are not colored. (American Institute of Professional Geologists)

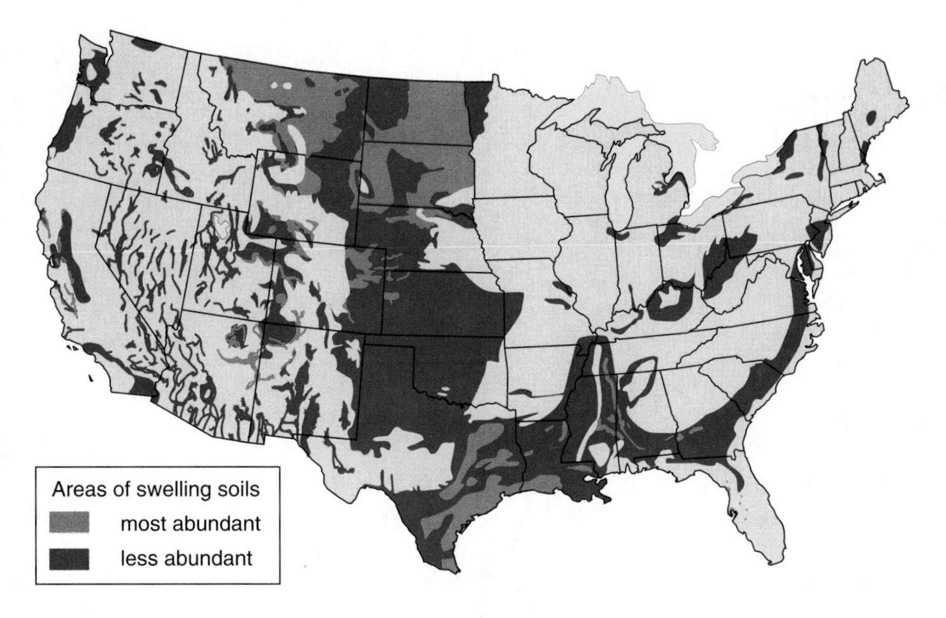


cles cohesive. **Cohesion** is the ability of particles to attract and hold each other together. This happens because the water molecules in the thin water films that line the pores attract one another. This attraction is called **surface tension,** a force that tends to "hold water together." (Surface tension gives a water droplet its rounded shape and thickness; without surface tension, a droplet would spread over a much larger area and be an extremely thin film.)

Because the water films are also attracted to the particle surfaces, the effect of surface tension is to pull the particles together (Figure 9–5). Surface tension enables you to build a castle of wet sand at the beach even though sand normally (dry) is a non-cohesive material. The castle remains standing, even though it has vertical walls, because the surface tension of water in the partially filled pores holds the sand grains together. As the water evaporates, the surface tension weakens, the sand loses cohesion, and the castle begins to crumble. When the tide rises and submerges the base of the castle, the pores become saturated, the surface tension effects are eliminated, and the castle crumbles into the water.

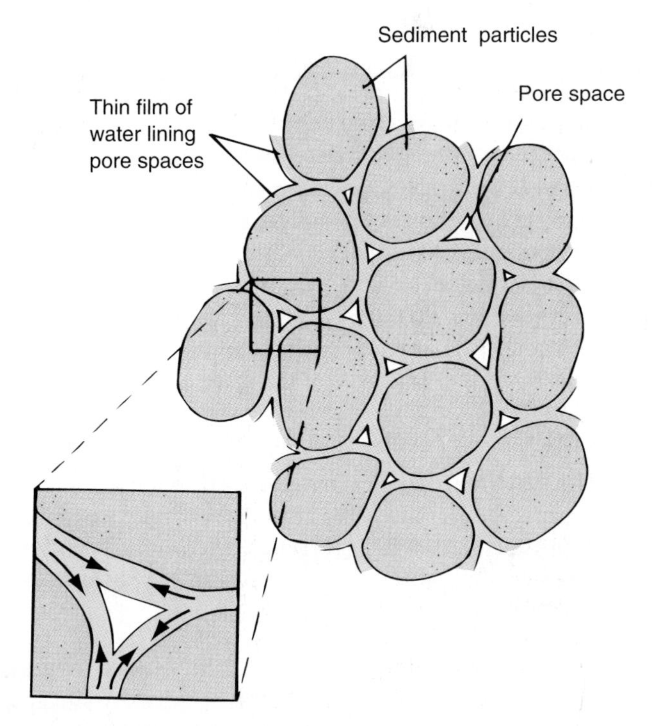


FIGURE 9–5
Surface tension forces. Surface tension, in the thin films of water that line the pores, holds together partially saturated sediments. Inset shows how surface tension pulls on both the water surface and the sediment particles, pulling the particles together. (Reprinted with the permission of Macmillan College Publishing Company from *Physical Geology,* by Nicholas K. Coch and Allan Ludman. Copyright © 1991 by Macmillan College Publishing Company, Inc.)

## Angle of Repose

Did you ever pour sugar, salt, sand, or some other granular substance into a cone-shaped pile and watch the sides build and collapse as more material

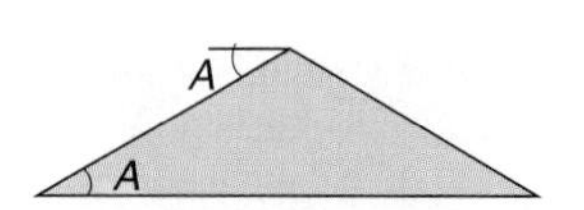


(a) Stable pile of sand with side slope angle (*A*) equal to the critical angle of repose for that size of sand grain.

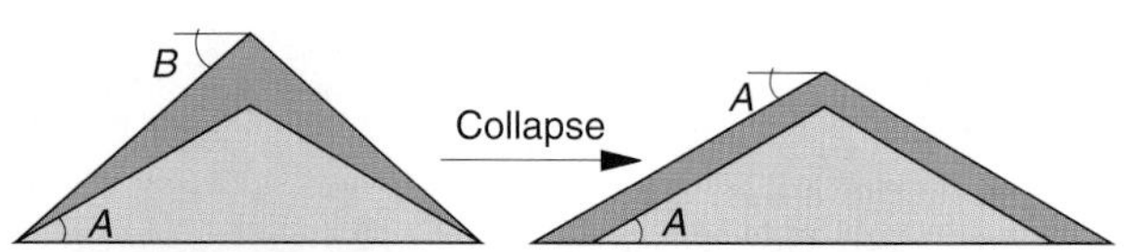


(b) More sand is added, forming a steeper side slope angle (*B*). This is greater than the critical angle of repose for that size of sand grain. The pile is now unstable and the sides collapse.

(c) A higher stable pile forms with side slope angles equal to the critical angle of repose (*A*) for that size of sand grain.

**FIGURE 9–6**
Angle of repose diagram. Effect of adding more sand to a stable sand pile. The new and higher pile has the same side slope angle as the original. (Reprinted with the permission of Macmillan College Publishing Company from *Physical Geology*, by Nicholas K. Coch and Allan Ludman. Copyright © 1991 by Macmillan College Publishing Company, Inc.)

is added? No matter how high the pile becomes, a few centimeters or a hundred meters, the sides maintain the same slope angle (Figure 9–6). The maximum angle at which granular materials can be piled is called the **angle of repose.**

Particle size and shape are the dominant factors in angle of repose, but other factors are also important. Larger particles maintain a steeper slope than smaller ones. Angular particles can interlock along their rougher edges and maintain steeper slopes than more rounded particles of the same size. Poorly sorted sediments have a steeper angle of repose because the smaller particles fit between the larger ones, giving the overall collection of particles stability at a steeper angle. Partially saturated sands, such as those in the beach castle just mentioned, have a steeper angle of repose because water surface tension holds the grains together (Figure 9–5).

## Particle Packing

The way that particles are arranged in a deposit can affect slope stability. The arrangement of particles in a deposit is called **packing.** We discussed this in Chapter 5 (Figure 5–22), but will amplify it here.

There are two extremes of packing:

- ☐ **Cubic packing** (Figure 5–22a) occurs when grains are positioned so their *centers* are directly above those of grains below. This is the loosest type of packing and has the most pore space. Cubic packing is most closely approximated where material has been dropped or bulldozed into space with little reworking.
- ☐ **Rhombohedral packing** (Figure 5–22b) occurs when the centers of grains are located over the *spaces* between grains below. This is the tightest form of packing and has the least pore space. Rhombohedral packing occurs in nature where material has been vibrated into place or deposited by water or wind currents.

Slope stability can be affected by a change in packing, because a change from loose (cubic) to tight (rhombohedral) packing decreases the volume and lowers the surface. Any type of ground movement, such as an earthquake, construction activities, blasting, or highway traffic, can be sufficient to change the packing of material, affecting slope stability. Structures built on such material may be damaged because the support for their foundation changes.

Another effect of a packing change is a reduction in pore space. This can expel pore fluids, most commonly water but sometimes oil, and can cause liquefaction in sand-sized and silt-sized sediments. This process commonly occurs in some sediments as a result of seismic shearing forces (Figure 5–23). The resulting deformed sediment patterns provide evidence of past earthquake vibrations (Chapter 5).

# TYPES OF MASS MOVEMENT

The three major types of mass movement are distinguished by the type of movement of rock or sediment:

- ☐ **Falls** move through the air and land at the base of a slope.
- ☐ **Slides** move in contact with the underlying surface.
- ☐ **Flows** are plastic or liquid movements, in water and more rarely in air, in which the mass breaks up and flows during movement.

Further subdivisions of these three major types of mass movement are based on the material, moisture content, and speed of movement. The major mass movement processes are illustrated in Table 9–1.

Mass movements occur at a wide range of speeds. Some occur so slowly that it may take years before the movement is noticeable from the downslope displacement of telephone poles, fences, and trees. Other mass movements occur so fast that the initiation of slope failure is followed within minutes by the covering of the slope base with debris.

## Rockfalls

The fall of rock particles through the air from a cliff is called **rockfall** (Table 9–1a and Figure 9–7). Rockfall can be a dry process, triggered by vibration or occasionally by root wedging (in which plant roots expand existing cracks in rocks). Rockfall often is initiated when water freezes in rock crevices. As the water freezes, it expands and exerts pressure on the crevice walls, breaking the rock apart, in a process called **frost wedging.**

**FIGURE 9–7**
Rockfall from the slope above formed this talus slope in the Sierra Mountains of California. (Courtesy of Raymond Pestrong, San Francisco State University.)

Rockfall is an extremely rapid process and thus is a serious geohazard. It can demolish structures near the bases of cliffs and is a danger to motorists where a highway cuts through rocks. The fan-shaped pile of rock fragments deposited by rockfall at the base of a cliff is called a **talus slope.**

## Slides

Movement along a plane surface is called a **slide** (Table 9–1b). Sliding movements are differentiated primarily by the character of the surface along which slope failure occurs, but also by type of material (rock or sediment), by water content, and by speed of movement.

***Rockslides.*** The downslope movement of a rock mass along a plane surface is a **rockslide.** The sliding surface is commonly a bedding plane, but rockslides can develop on other surfaces such as fractures that cut across layered rocks (Figure 9–3b). Any area where such rock surfaces are inclined toward an open space, such as a road cut or valley, has the potential for rockslide development. (See the unstable slopes in Figure 9–3 and Box 9–1).

***Slumps.*** The sliding of material along a *curved* surface is a **slump** (Table 9–1c). Slumping is common in unconsolidated sediments and in some weaker rocks. A common cause of slumping is erosion at the base of a slope, which removes support for the material above. This erosion may be natural, such as undermining of a river bank by streamflow, or cutting away of the base of a coastal cliff by storm waves (Figure 3–8). When the slope fails, the slump block rotates downward and a *scarp* (cliff) is formed at the top of the slope (Figure 9–8).

Slumping is an especially serious geohazard where structures are built on bluffs above the shoreline. Wave energy erodes the bases of the cliffs, causing them to fail. In many cases, homes built along the bluffs are lost and the scarp progresses landward, posing a geohazard for the next line of houses on the bluff.

The 1964 Alaska Earthquake (Chapter 4) caused serious land displacement in the city of Anchorage. Much of Anchorage is underlain by the Bootlegger Cove Clay, a marine layer that originally had saltwater in its pores. In some areas, this saltwater was replaced by freshwater, making those areas a potentially unstable quickclay (Figure 9–9). The ground

**TABLE 9–1**
Characteristics of major mass movement processes.

| Mass Wasting Type | Character of Movement | Subdivision | Speed and Type of Material |
| --- | --- | --- | --- |
| Falls | Particles fall from cliff and accumulate at base | Rockfall – see (a) below | Extremely rapid; develops in rocks |
| | | Soilfall | Extremely rapid; develops in sediments |
| Slides | Masses of rock or sediment slide downslope along planar surface | Rockslide – see (b) below | Rapid-to-very rapid sliding of rock mass along a flat inclined surface |
| | | Slump – see (c) below | Extremely slow-to-moderate sliding of sediment or rock mass along a curved surface |

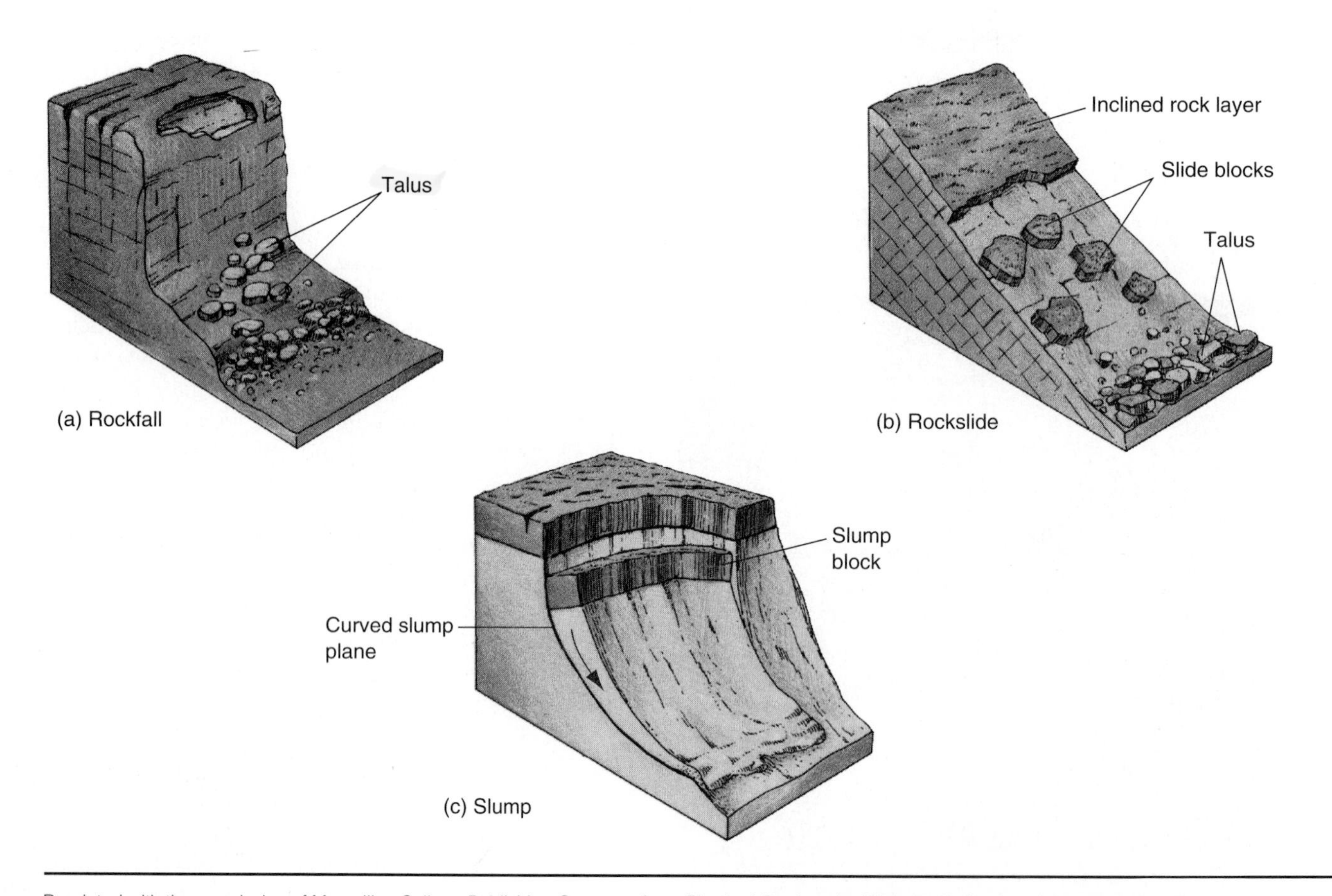

| Mass Wasting Type | Character of Movement | Subdivision | Speed and Type of Material |
|---|---|---|---|
| Flows | Displaced mass flows as a plastic or viscous liquid | Creep<br>– see (d) below | Extremely rapid; develops in rocks |
| | | Solifluction<br>– see (e) below | Very slow-to-slow movement of water-saturated regolith as lobate grows |
| | | Mudflow<br>– see (f) below | Very slow-to-rapid movement of fine-grained particles wtih up to 30% water |
| | | Debris flow<br>– see (g) below | Very rapid flow of debris; commonly starts as a slump in the upslope area |
| | | Debris Avalanche<br>– see (g) below | Extremely rapid flow; fall and sliding of rock debris |

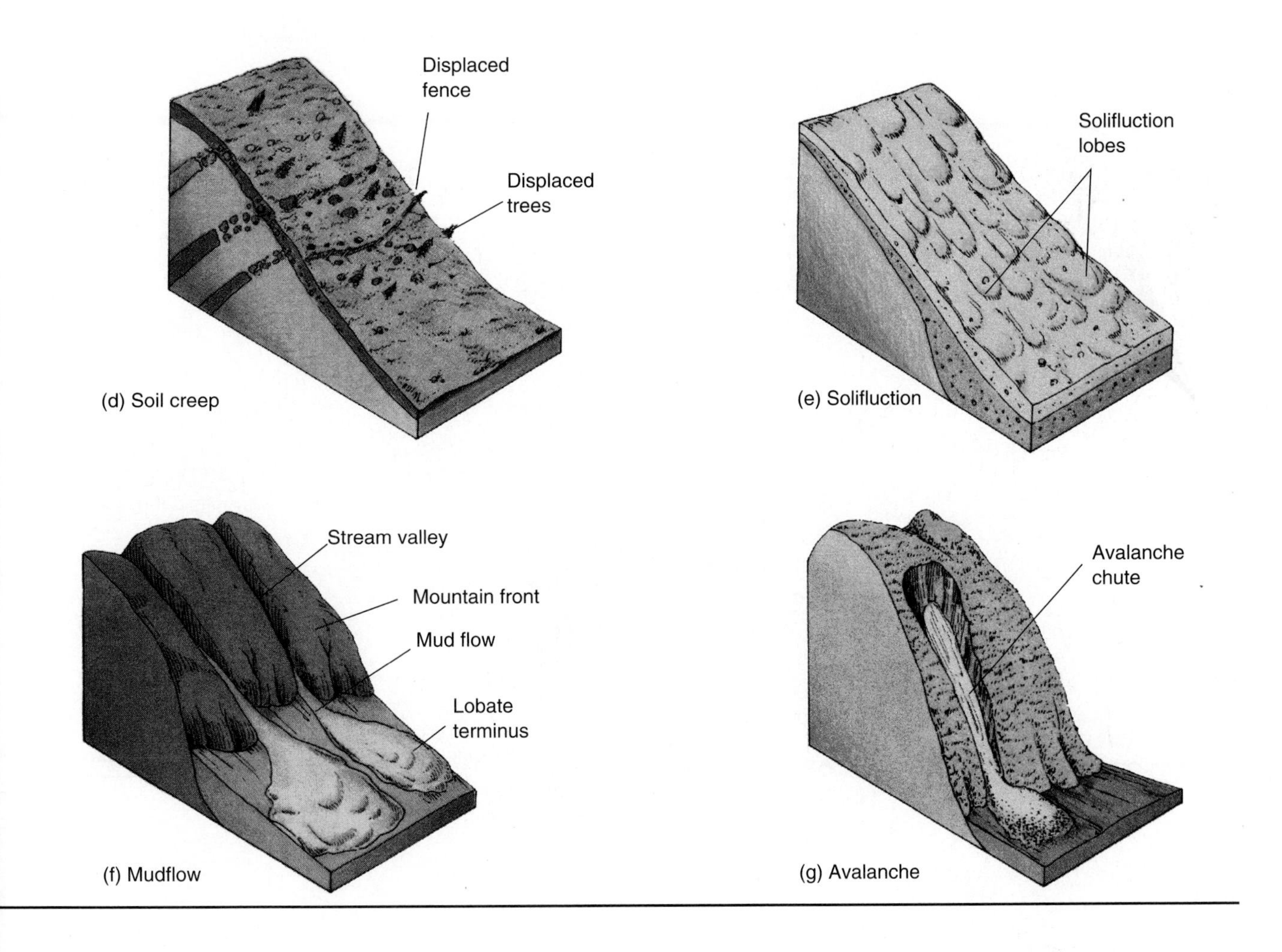

## BOX 9–1 THE VAIONT DAM DISASTER

One night in 1963, a geologic catastrophe—a completely unnecessary one—claimed 1,900 lives in northern Italy. It began seven years earlier, with construction of a massive concrete dam. At 261 meters (856 feet) high, the dam was built across the Vaiont River Valley to impound a large lake for generating hydroelectric power (Figure 1).

This valley's walls are underlain by limestone interbedded with shale (Figure 2). Strata on both sides are inclined toward the valley's center, creating potential instability. A large, ancient scar on the northern side of the valley testified to large-scale rockslides in the past. As workers poured concrete for the dam foundation, they watched it disappear into the rocks, forewarning that fractures or solution cavities existed below.

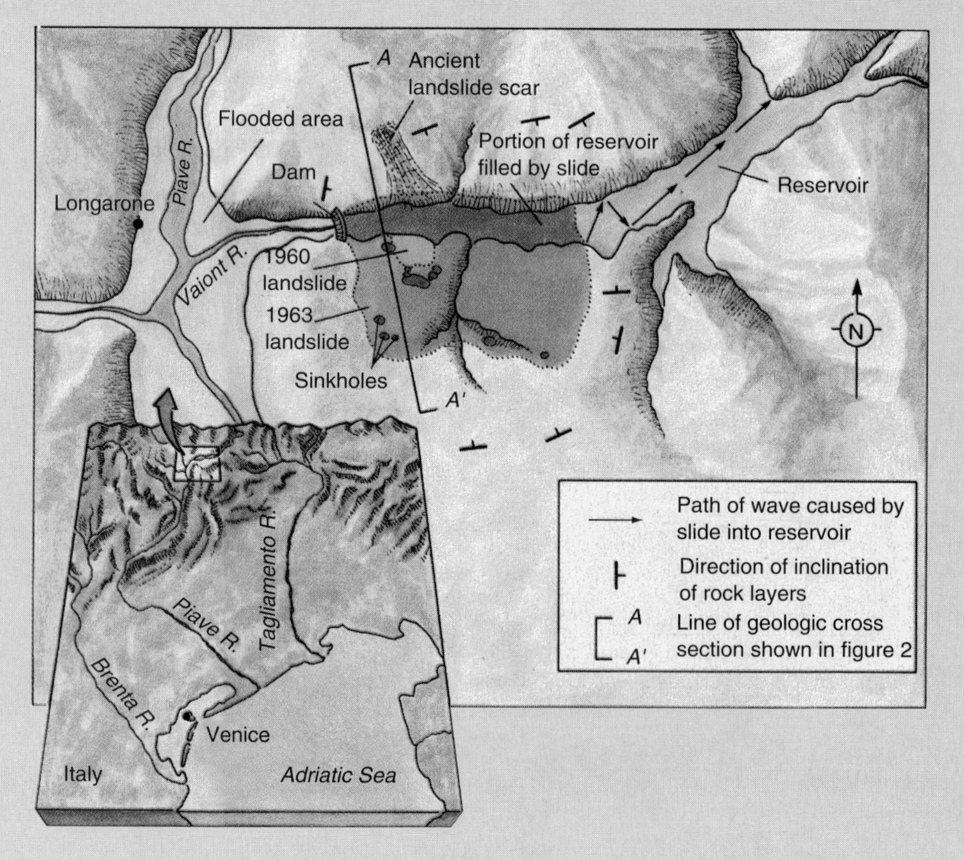


**FIGURE 1**
The Vaiont Dam area and Piave River valley, showing surface features mentioned in text. Section line *A–A′* is detailed in Figure 2. (After G. A. Kiersch, 1964, The Vaiont Reservoir Disaster: Civil Engineering, v. 34, p. 3. Reprinted with permission of American Society of Civil Engineers.)

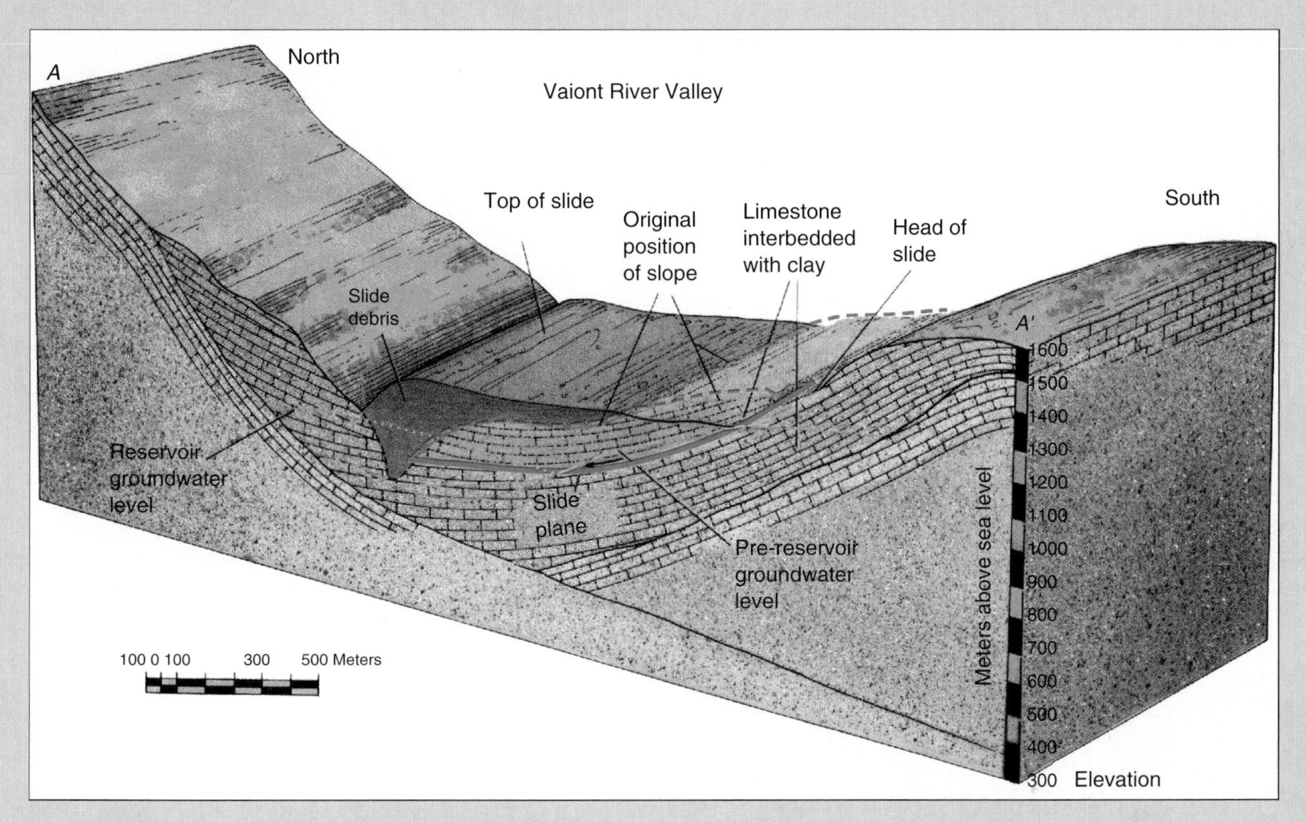


**FIGURE 2**
Section of Vaiont valley along line *A–A′* in Figure 1. All layers except the base are limestones. The mass that slid (shaded) was limestone interbedded with shale. (After G. A. Kiersch, 1964, The Vaiont Reservoir Disaster: Civil Engineering, v. 34, p. 3. Reprinted with permission of American Society of Civil Engineers.)

The completed reservoir was allowed to fill in stages. The rising water created pressure against the valley walls, profoundly affecting stability. In 1960, 700,000 cubic meters (916,000 cubic yards) of rock and soil slid off the south slope into the reservoir (Figure 1). This generated a wave 2 meters (6.5 feet) high that spread quickly across the lake surface. However, engineering studies and model analyses concluded that the reservoir could be safely filled.

As the reservoir level rose, surface creep accelerated dramatically. On September 4, 1963, creep rate was 3 to 6.5 millimeters (0.12 to 0.25 inches) a day. By October 6, it had accelerated to 40 millimeters (1.5 inches) a day. New cracks began to form on the slope, which was becoming saturated with extraordinary rainfall.

**FIGURE 3**
Upper half of Vaiont Dam, showing a portion of the mass movement which filled the reservoir. (Photo courtesy of F. W. Fletcher.)

Engineers realized that mass movement was inevitable, but they were confident that the slope would come down in relatively harmless blocks, and not in a mass. Such had been the behavior of previous slides in the region. Also, scale-model experiments forecast no danger from a wave formed in the lake by a slide.

Clearly, natural and human-induced changes were combining to create a gigantic rockslide. The river had carved a steep-walled valley and eroded lateral support from the steep, fractured rocks in the valley walls (Figure 2). The exceptional rainfall increased the weight of the rocks as it filled fractures and solution cavities. Cohesion was reduced within the shale as water saturated its clay minerals.

The evacuation order came shortly before 10:00 P.M. on October 9, 1963. But, 39 minutes later, 300 million cubic meters (400 million cubic yards) of fractured rock and water-laden limestone tore loose from the top of the south slope and roared downward toward the reservoir. The mass slid, not in a series of blocks as predicted, but as one huge intact block on slick, clayey limestone beds.

When it reached the water, the mass displaced about 50 million cubic meters (67 million cubic yards) of water, forming a wave 200 meters (650 feet) high. The wave spread across the lake, wiping out lakeshore villages (Figure 1). The well-designed dam held, but the giant wave rose over it and plunged into the narrow river gorge below. The confined floodwater surged across the Piave River Valley, killing 1,450 in the town of Longarone. Hundreds more died as the wave surged upstream and downstream. Today, the former reservoir is largely filled by slide debris (Figure 3) and no electricity is produced as originally planned.

This account describes the scene a few years later:

> . . . one comes across a crumbling stone wall, all that remains of someone's home, or some simple memorial to the victims—a crude wooden cross in a meadow or, in a shallow rock hollow, a small, carefully placed cluster of fresh, native alpine flowers or, sometimes, just a short list of names, often with the same last names, on a plaque. At the west end of the valley stands the dam, still the highest double-arch dam in the world; but no electricity is produced and exported to Italy's prosperous industries. Three hundred million cubic meters of rock and soil fill the reservoir and 1,899 people are dead.
>
> On November 24, 1968, as the trial began [the officials involved were prosecuted] . . . the [number of victims] came to nineteen hundred: Engineer Mario Pancini . . . taped the cracks around the doors of his Venetian room and turned on the jets of his gas range.*

*From Frank Fletcher, 1970, A terrifying equality: The story of the Vaiont Dam disaster: Susquehanna University Studies, vol. 8, no. 4, p. 300. Information in this box is based on his report.

**FIGURE 9–8**
Slumping (right) and arcuate slump scars (center) on a slope near La Honda, California. (Courtesy of Raymond Pestrong, San Francisco State University.)

shaking that accompanied the earthquake caused the clay to liquefy (Figure 5–26), and ground failure occurred.

Some of the most serious damage was in the suburb of Turnagain Heights. It is on a flat-topped bluff some 30 meters (100 feet) above the level of Knik Arm, a branch of Cook Inlet. Slump scars began to form in the overlying glacial deposits as soon as the underlying clay lost its cohesion. Within minutes, the area was changed from a flat bluff into a mass of downwardly rotating slump blocks, covered with tilted trees and twisted homes. Seventy homes were destroyed and three people were killed.

## Flows

**Flows** are mass movements in which the material behaves like a viscous fluid (Table 9–1f). In many cases, mass movements that start as falls, slides, or slumps are transformed into flows further downslope. Flows have broad characteristics. They encompass the wettest, driest, slowest, and fastest types of

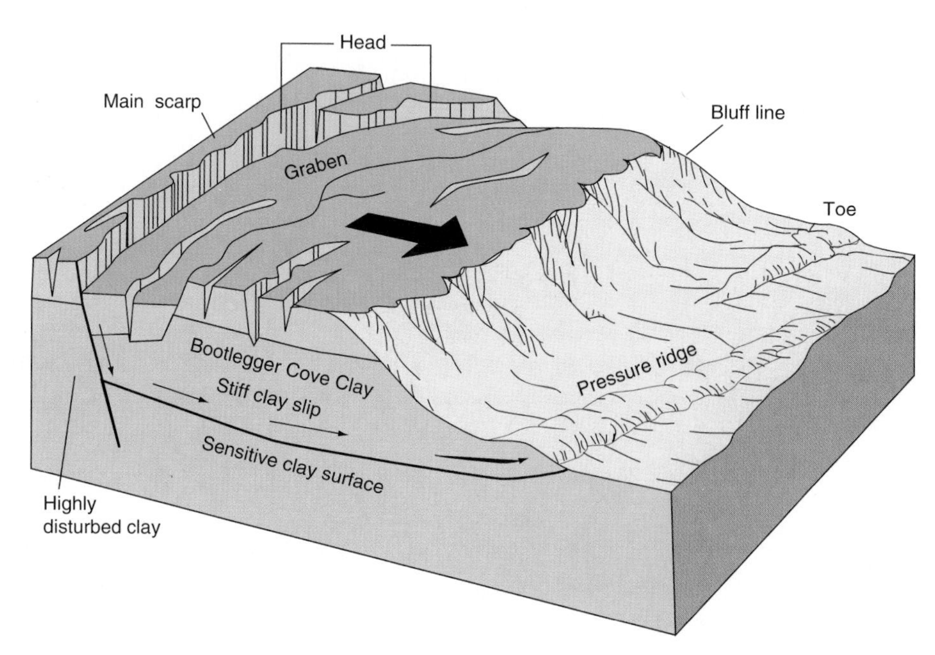


**FIGURE 9–9**
Diagram showing how mass movements developed in response to the 1964 Alaskan earthquake. Shown are the outcrop of the Bootlegger Cove Clay and mass movment areas. (Reprinted with the permission of Macmillan College Publishing Company from *Environmental Geology,* sixth edition, by Edward A. Keller. Copyright © 1992 by Macmillan College Publishing Company, Inc.)

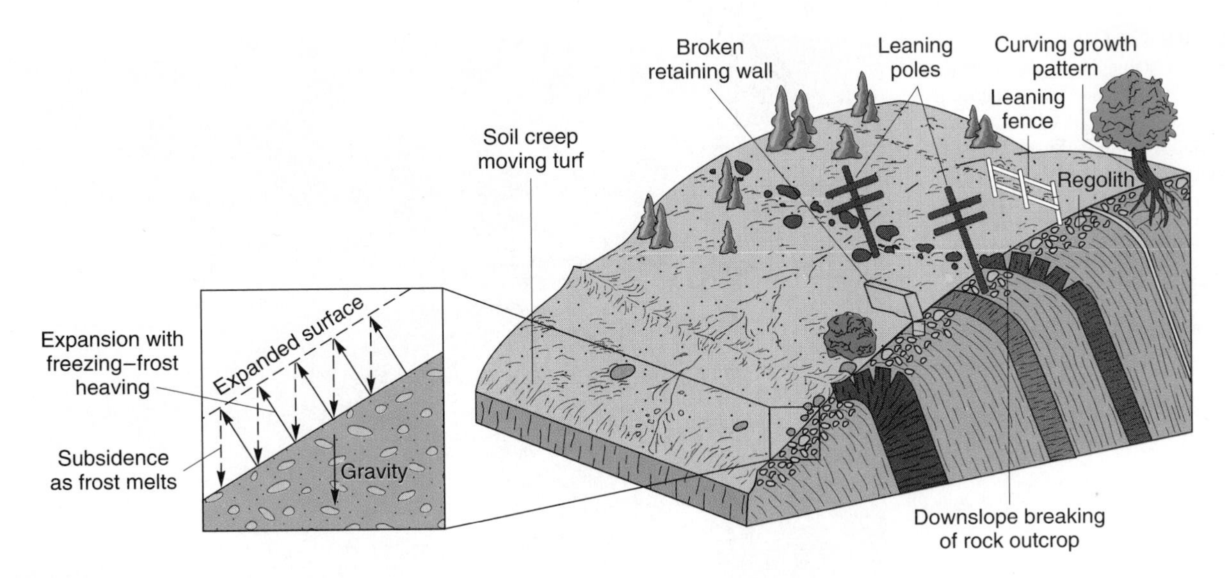


**FIGURE 9–10**
Results of creep on a slope. The enlargement shows the freeze-thaw cycle that occurs anywhere ground becomes even slightly frozen. In this cycle, freezing soil moisture expands, lifting particles. When the ice melts, gravity pulls the particles straight down, but at a point slightly farther down the slope. Repeated many times, this action moves each particle downslope, so soil gradually creeps downslope. (Reprinted with the permission of Macmillan College Publishing Company from *Geosystems,* second edition, by Robert W. Christopherson. Copyright © 1994 by Macmillan College Publishing Company, Inc.)

mass movement. We will discuss flows by considering them in order of increasing speed.

***Creep.*** Depending on where you live, you may have observed a downslope displacement of trees, fences, or utility poles (Figure 9–10). This slow displacement is the result of **creep,** an extremely slow movement of regolith, soil, and rock under the influence of gravity (Table 9–1d).

Creep is the least spectacular of all mass movements. However, its continuity of operation and its widespread action over Earth make it the most important mass movement process in terms of the *total volume* of material moved downslope each year. Abnormally high rates of creep may indicate that a potentially deep failure is imminent (see Box 9–1). Creep acceleration is a mass movement indicator and is discussed later in this chapter.

***Solifluction.*** The downslope movement of water-saturated regolith is called **solifluction** (Table 9-1e). Movement rates can be faster than creep and may attain a few centimeters per year. Solifluction may occur in any climate in which regolith becomes saturated with water. However, it is most common in cold climates where the upper part of the regolith freezes and thaws periodically.

Many cold areas are underlain by permanently frozen ground called **permafrost.** During warmer parts of the year, the uppermost part of this ground thaws and releases water within the sediment. This soggy mass of soil then can flow downslope over the permafrost beneath. Solifluction creates a topography characterized by curved lobate flows on the surface.

***Mudflows.*** **Mudflows** contain significant water (up to 30%) and a large proportion of fine-grained material (Table 9–1f). Mudflows are common on slopes in semiarid areas, where rare but intense, short-lived rainstorms rapidly convert regolith into a mass of viscous mud and rock that moves downslope at high velocity. Mudflows are a serious geohazard in urbanized dry areas such as in the Los Angeles Basin, where development has spread onto steep mountain fronts (Figure 9–11). Mudflows also are major geohazards on the slopes of volcanoes, where they are called *lahars* (Chapter 4).

The hazard becomes especially serious when heavy rains follow a drought or forest fires remove

**FIGURE 9–11**
Homes destroyed by a debris flow in Pacifica, California. Torrential rains during the winter of 1983 triggered slope failures throughout the San Francisco area. (Courtesy of Raymond Pestrong, San Francisco State University.)

slope vegetation. The disastrous forest fires in the Malibu area of Los Angeles in November 1993 set the stage for mudflows to begin with the winter rains. The destruction of vegetation and drainage control structures, such as runoff pipes and catch basins, permitted mud to cascade into dwellings and across the Pacific Coast Highway with the first major rain that fell on the denuded slopes. As expected, heavy rains on February 8, 1994, resulted in up to a meter of mud covering sections of the coastal road between Malibu and Topanga Canyon. Flooding and mudflows in Malibu trapped several residents, who were rescued by crews using earth-moving equipment.

***Debris Flows.*** Very rapid downslope movement of rock debris and regolith is a **debris flow.** Many start as slumps or slides, but are transformed into debris flows downslope as the mass breaks up and mixes with air and water. Many debris flows enter drainage systems and then follow them downslope.

The Alaska earthquake of 1964 triggered a massive debris flow from a mountain bordering the Sherman Glacier. The debris flow traveled 5 kilometers (3 miles) from its source and deposited a layer of debris 1.5 meters (5 feet) thick as it swept across the glacier without disturbing the fresh snow on its surface. The debris showed little sorting with distance, and the slope across which it moved was inclined only a few degrees.

This raises an interesting question: How do unsorted masses of debris travel so far over gentle slopes? It is now believed that at least a few debris flows are able to travel great distances over gentle slopes because they ride *on a cushion of air.* The air cushion is formed when tumbling rock debris traps air beneath itself. The presence of this low-friction air cushion enables the flow to move gently over land surfaces in much the same way that a "hovercraft" skims at high speeds over water.

***Debris Avalanches.*** The general term **avalanche** is used for the most rapidly flowing, sliding, and falling mass movement. If you have seen video of a snow avalanche, you know the material takes only a few minutes to travel from high on a mountain down the slope and across the valley below.

Rapid to extremely rapid movements of rock and sediment are **debris avalanches.** Many debris avalanches are characterized by semicircular heads, with elongate tongues of debris extending downslope (Figure 9–12). Chutes cut down the slope, terminating in piles of debris at the base of the slope.

A strong 1970 earthquake in Peru triggered a massive debris avalanche that killed nearly 30,000 people in the towns of Yungay and Ranrahirca, in the Andes Mountains. A mixture of vibration-loosened rock and sediment, glacier ice, and meltwater flowed rapidly 11 kilometers (7 miles) down the

**FIGURE 9–12**
Avalanche head, chute, and debris fan (base of slope) in a coastal valley near Skagway, Alaska. (Courtesy of Raymond Pestrong, San Francisco State University.)

mountain and into the towns in less than three minutes. The flow was split by a rock ridge, which diverted the debris toward the two towns.

Survivors described the front of the advancing debris as a rolling wave as high as 80 meters (260 feet). The rumbling and roar of the approaching debris was preceded by a strong surge of wind. Those few who managed to reach safety on Cemetery Ridge watched in horror as the rapidly moving debris flattened the towns and buried most of the inhabitants.

***Snow Avalanches.*** Rapidly moving fluidized masses of snow, called **snow avalanches,** are a serious threat to skiers and structures in mountainous areas (Figure 9–13). As more mountainous areas are developed for winter sports and housing, snow avalanches are causing more deaths, structural damage, and blocked highways.

The main event that prepares a snow-covered mountain slope for an avalanche is the development of an intermediate snow layer with a special characteristic: lower resistance to shearing than the other layers. This "slip surface" permits the snow layers above it to move quickly downslope.

This shearing layer is produced when snow covers a rock surface that has been warmed above freezing. Heated air rises from the rock through the snow, melting the edges of snowflakes, producing aggregates of granular snow crystals. This granular layer has less resistance to shearing than the layer of flat snowflakes from which it developed.

Similar surfaces also can develop during warm periods, when the surface snow melts and refreezes to form an icy crust. Fresh snow accumulating above this slick surface is potentially unstable.

An avalanche can occur spontaneously when the snow thickness accumulating above a slip surface results in a driving force that exceeds the shear resistance of the snow mass. Avalanches also are triggered by vibrations (seismic waves, noise) or by the disturbance of the snow surface by skiers or vehicles. The avalanche begins with deep cracks developing before masses of snow break loose along the slip surface. These snow masses mix with air and quickly break up into a churning fluidized mass of snow that rolls in turbulent waves downslope (Figure 9–13).

Avalanches commonly move preferentially through steeply sloping linear lows called **avalanche chutes** that channel the flow into a valley. Many of these chutes are excavated by debris avalanches at other times of the year (Figure 9–12).

Avalanches are very dangerous because of their great force and their high speed, which can reach up to 80 kilometers (50 miles) per hour. Thousands or millions of tons of snow, moving down slopes of 30 to 45 degrees, exert forces sufficient to destroy nonreinforced structures in their path. The potential for sliding can be determined by monitoring temperature, snow thickness, snow surface slope, and by identifying potential slip planes within the snow through coring or excavation.

To reduce avalanche severity, smaller avalanches may be triggered with controlled explosions or artillery bombardment. These small avalanches preempt a larger, more dangerous one that would be triggered naturally or by human activity.

FIGURE 9–13
Snow avalanche moving downslope. (U.S. Department of Agriculture.)

## PREDICTING MASS MOVEMENT

Several factors indicate areas that have mass movement potential. Some factors reveal potentially unstable surfaces; others indicate mass movement in its earliest stage.

### Slopes and Seismic Activity

The steeper a natural slope, the greater is its potential for downslope movement. Construction commonly oversteepens natural slopes. If these artificial slopes are not properly protected, they will be subject to mass movement (Figure 9–2). The material underlying a steep slope should be examined before construction, because such slopes are most subject to landsliding. This is especially true if there is seismic activity in the region.

### Geology and Structure

Slopes underlain by soluble rocks, or rocks easily weakened by water, have a greater potential for sliding. The potential is further increased if either bedding planes or joints in the rock are inclined toward valleys (Figure 9–3). These weaknesses are described in Box 9–1.

### Surface Water Buildup

The buildup of water within slope materials can destabilize them, as noted earlier. The problem is how to determine the water conditions within the slope without directly drilling into it. Valuable clues are springs along the slope, areas of continually wet ground, and pools of standing water, especially if they are oriented parallel to the edge of a cliff. These features suggest a high degree of water saturation within the slope.

### Topographic and Vegetational Features

A low area delimited by a semicircular scarp (low cliff) may represent an old landslide (Figure 9–14). This is especially true if the material in the low area has a hummocky topography (irregular or knobby) caused by flow masses or slide blocks, and if the area's vegetation is distinctly younger than that on either side. Such features indicate a past landslide and warn that the area requires detailed examination before construction.

### Accelerated Creep and Associated Features

Natural creep is very slow. However, creep rate can increase markedly if the mass of material is on the

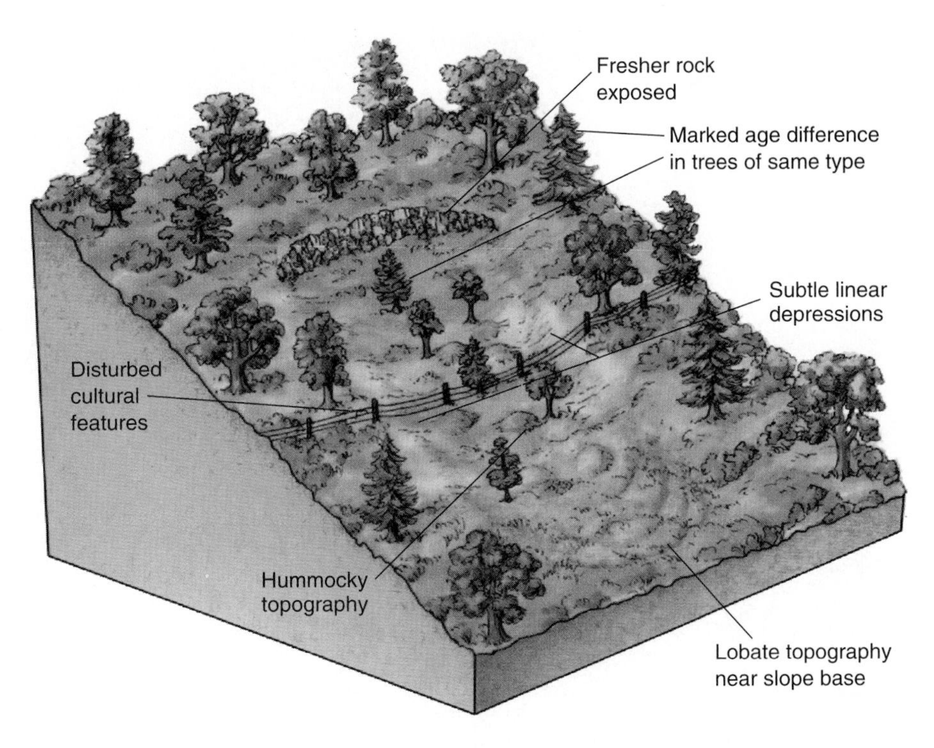


**FIGURE 9–14**
Topographic evidence of an old slump-flow event on a slope. The semicircular head marks the top of the former slump. The rolling topography in the foreground represents the area where the slump was transformed into a flow. (U.S. Geological Survey, Menlo Park)

verge of failing. Such accelerated creep was observed before the catastrophic Vaiont Valley rockslide in Italy, described in Box 9–1. Creep rates can be measured by devices implanted within the slope.

Creep also can be detected by studying surface features (Figure 9–15). Downslope tilting of recently planted vegetation or recently constructed fences and utility poles indicates high rates of creep. Deformed and fractured road surfaces cut into the slope also may be indicators.

**FIGURE 9–15**
Deformed birch trees on a slope near Duluth, Minnesota. The active soil creep on this slope moves the base of the trees downslope while the tree tops tend to grow vertically toward the sun. (Courtesy of Raymond Pestrong, San Francisco State University.)

## Landslide Potential Maps

Examination of past landslides and geologic studies of potentially unstable slopes can be used to construct *landslide potential maps* (Figure 9–16). To make such a map, areas underlain by slide-prone soils or rock are identified. These areas then are superimposed on a topographic map. Where steep slopes (shown by the topographic map contours) and landslide-prone soils or rocks coincide, these areas are mapped as *slide-prone.* The map in Figure 9–16 highlights areas of recent landslides, older ones, slide-prone areas, and stable areas.

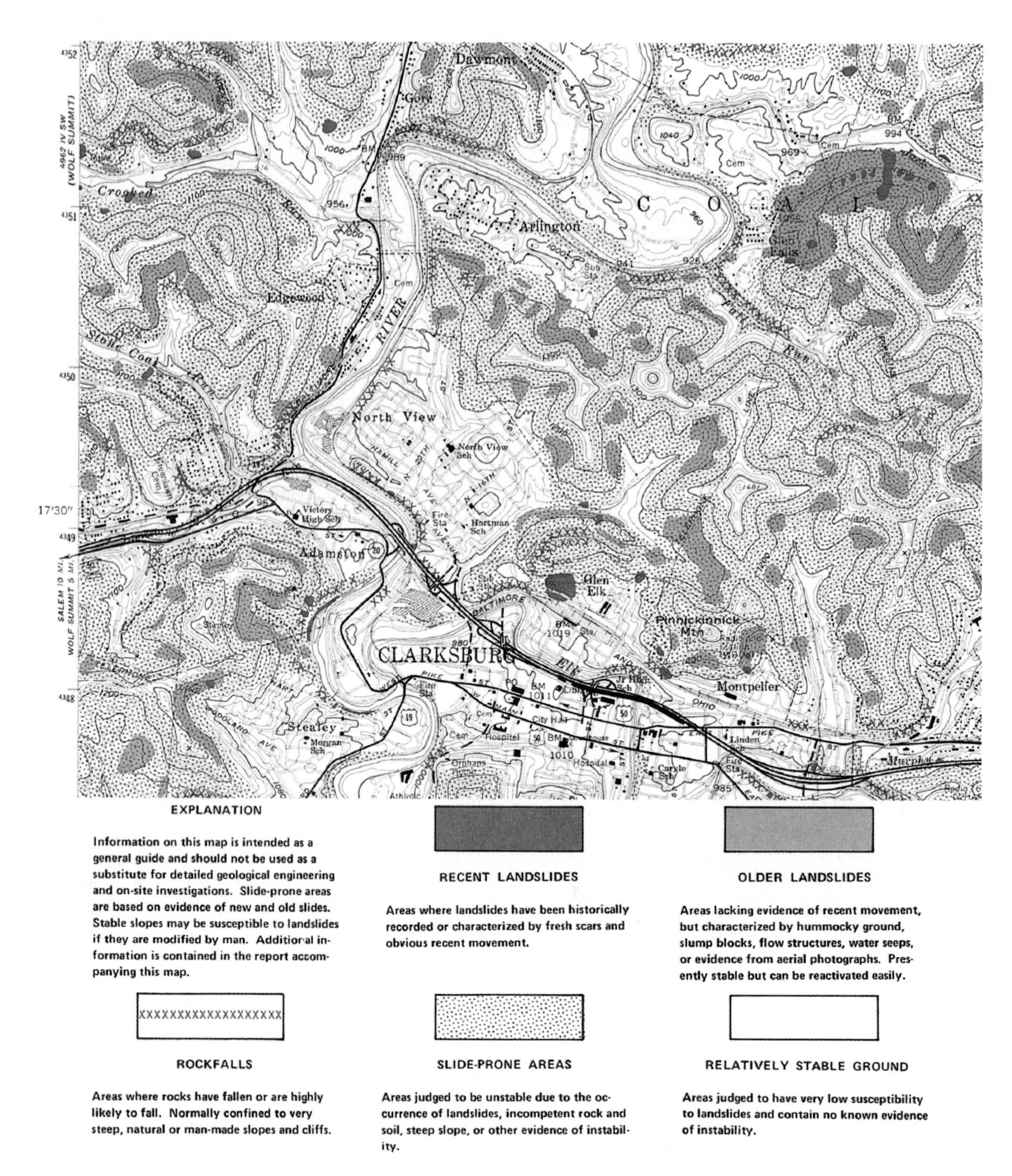


**FIGURE 9–16**
Portion of a mass movement potential map of the Clarksburg, West Virginia area, created by the West Virginia Geological and Economic Survey. Various government agencies publish similar surveys for many areas. People seeking home sites should study such maps before building. (West Virginia Geological and Economic Survey.)

## MASS MOVEMENT PREVENTION AND MITIGATION

Obviously, it is best to avoid building in any area subject to mass movement. However, railroads and highways often must traverse hazardous areas to connect population centers. Growing populations and limited land often require building on areas having potential landslide hazards.

How do engineering geologists and civil engineers prevent mass movement or mitigate its consequences when it does happen? We present here some methods they use, including slope drainage, slope reduction, and engineering structures.

### Slope Drainage

Water buildup within slopes can be reduced by several engineering techniques. *Interceptor drains* can be excavated along the top of the slope. These concrete-lined drains capture runoff and transport it away from the slope. *Perforated pipe* can be driven into the slope to collect water and drain it by gravity away from the slope. *Wells driven into the slope* can be pumped to remove water rapidly in especially critical areas.

### Slope Reduction

Steep slopes can be graded into gentler ones to reduce the landslide danger. If there is not enough room for such extensive grading, *benches or terraces* may be excavated into the slope. Breaking the slope into terraces or "stair steps" not only improves stability, but it also stops falling material before it can reach the protected area at the bottom of the slope and prevents water erosion by interrupting the flow of rills or channels that drain the slope.

### Engineering Methods to Resist Mass Movement

One way to discourage slope disintegration is to protect the surface from rain and snow. All crevices are sealed and layers of concrete and crushed rock are sprayed onto the surface in layers 8 to 10 centimeters (3 to 4 inches) thick. This coating of *shotcrete* (concrete "shot" onto a surface) prevents water from entering the rock to cause frost wedging (Figure 9–17).

**FIGURE 9–17**
Application of shotcrete to a surface. (Courtesy of Johnson Western Gunite Company.)

Slopes underlain by sediment or loosely consolidated rock are protected from sliding by *retaining walls* (Figure 9–18). Retaining walls may be built of a variety of materials. They usually include drains so groundwater can exit the slope before fluid pressure builds against the wall.

**FIGURE 9–18**
Concrete retaining wall near Tsukuba City, Japan. The holes in the wall are for drainage, to prevent water pressure from building up behind the wall. (Courtesy of Raymond Pestrong, San Francisco State University.)

**FIGURE 9–19**
Rock bolts stabilize a highway slope in British Columbia. (Courtesy of Raymond Pestrong, San Francisco State University.)

Inclined rock layers are stabilized with *rock bolts* (Figure 9–19). Holes are drilled into the rock below the level of any possible slip surface. Steel rods are inserted and cemented in place within the holes. Plates are placed over the ends of the rods and tightened against the rock face.

Overhanging rock is stabilized by supporting it with *buttresses*. These concrete and steel structures shore up the overhang and prevent movement.

## Engineering Structures to Mitigate Damage

Where mass movement is inevitable, engineering structures can minimize damage. They often are backups for upslope structures built to prevent landslides, but which fail to do so.

Slopes subject to rockfalls are protected by *cable nets and wire fences* that catch rock blocks before they can do damage (Figure 9–20). Commonly these structures are associated with *intercept ditches* excavated along the slope base or by *berms* (ridges) built along the base. These structures catch any debris that gets past upslope protection devices.

Sometimes a structure is intended to allow a landslide to pass over an area without disturbing whatever it is designed to protect. *Rock sheds and tunnels* are reinforced structures built over highway and railroad segments that are subject to landslide (Figure 9–21). They allow the mass to pass over without collapsing. Similar structures are used in snow avalanche areas to keep transportation routes open all winter.

**FIGURE 9–20**
Cable nets and fencing protect the highway from falling rock in San Francisco, California. (Photo by author.)

**FIGURE 9–21**
Rock shed protecting vehicular traffic from rock falls and rock slides from the slope above. (Federal Highway Administration.)

## LOOKING AHEAD

In this chapter, you have seen how lateral mass movement occurs on Earth's surface. In the next chapter, our attention shifts underground to the problem of vertical movement: land subsidence and surface collapse. Whether caused by natural phenomena, such as groundwater, or human activity, such as mining and drilling, surface subsidence is a major geohazard.

## SUMMARY

Mass movement is triggered when driving forces exceed resisting forces. The major driving force is gravity, assisted by water. Resisting forces include friction and the shear strength of the material. Rock structures such as bedding and joints are important in facilitating mass movement. Whenever these features are inclined toward a valley or excavation, greater danger of slope failure exists.

Expanding clays and quickclays create conditions favorable to mass movement if the surface is disturbed by vibrations. Granular materials piled at an angle exceeding their angle of repose are prone to slope failure upon vibration, precipitation, excavation, or loading.

Mass movement is characterized as falls, slides, or flows, depending on the nature of debris movement. In falls, the debris moves through the air to land in a pile on the ground. Slides are movements along a plane. Rockslides occur when rock slides along a flat plane in the rock. Slumps occur when sediment or poorly consolidated rock moves down along a curved plane.

Flows can be mudflows, debris flows, or avalanches, depending on the amount of solid material being moved and its speed. Mudflows are mixtures of fine-grained sediment and water, whereas debris flows contain a larger percentage of rock material. Avalanches are fluidized masses of debris that move very rapidly.

Potential landslide areas can be identified by steep slopes, indications of past landslide, water buildup, rock structures inclined toward valleys, younger trees covering an irregular topography, and accelerated creep of surface sediments.

Mass movement danger is reduced by controlling drainage and reducing the slope angle. Engineering structures used include shotcrete coating, retaining walls, rock bolting, and buttresses. Landslide impact can be mitigated with cable nets and wire fences, berms, intercept ditches, rock sheds, and tunnels. The best way to minimize damage from landslide is to avoid building in such areas, not a likely solution on a planet with a burgeoning population.

## KEY TERMS

angle of repose
avalanche
avalanche chutes
bentonite
cohesion
creep
cubic packing
debris avalanches
debris flows
driving forces
falls
flows
frost wedging
landslide
mass movement
mudflows
packing
permafrost
quickclays
resisting forces
rhombohedral packing
rockfall
rockslide
slide
slope failure
slump
snow avalanches
solifluction
surface tension
talus slope

## REVIEW QUESTIONS

1. Water can be both a driving and resisting force in mass movement. Explain.
2. Give several examples of how human activities can trigger mass movement.
3. How does the orientation of rock features such as bedding and joints affect the development of mass movement?
4. Describe the formation of quickclays and their role in mass movements.
5. How do the slope angle and the type of grain packing in a sediment affect slope stability?
6. Distinguish among the types of mass movement we call falls, slides, and flows.
7. How do rockslides and slumps differ?
8. Why are mudflows such a geohazard in dry regions?
9. Describe the conditions that result in a snow avalanche.
10. List all of the precursors that indicated there would be a rockslide in the Vaiont River Valley (Box 9–1).
11. What observations would you make to determine whether a slope had experienced landsliding in the past?
12. Describe five different engineering structures designed to reduce mass movement damage.
13. What features would you look for in the field for constructing a landslide potential map?
14. Many mass movements occur after rainfalls. Why?

## FURTHER READINGS


American Institute of Professional Geologists (AIPG), 1993, The citizens' guide to geologic hazards, Subsidence: Arvada, Colo., AIPG, 79 p.

Coates, D. R. (ed.), 1977, Landslides: Geological Society of America Reviews in Engineering Geology, v. 3, p. 278.

Cupp, D., 1982, Avalanche!: National Geographic, v. 162, p. 280–305.

Fleming, R. W., and Taylor, F. A., 1980, Estimating the cost of landslide damage in the United States: U.S. Geological Survey Circular 832.

Kenny, N. T., 1968, Southern California's trial by mud and water: National Geographic, v. 136, p. 552–73.

Kiersch, G. A., 1964, Vaiont Reservoir disaster: Civil Engineering, v. 34, p. 32–39.

Radbruch-Hall, D., et al., Landslide overview map of the coterminous United States: U.S. Geological Survey, Prof. Paper 1183, 25 p.

U.S. Geological Survey, 1981, Facing geologic and hydrclogic hazards, Chapter 4 (Hazards from ground failures) *in* Hays, W. W., ed., U.S. Geological Survey Prof. Paper 1240-B, 28 p.

U.S. Geological Survey, 1987, Landslides of eastern North America, *in* Schultz, A. P., and Southworth, C. S., eds.: U.S. Geological Survey Circular 1008, 43 p.

# 10

# *Subsidence and Collapse*

For centuries, people have exploited Earth's underground resources: groundwater, oil, coal, rock, and metal ores. Each year, we withdraw increasing volumes of these fluids and solids from the subsurface to meet the demands of industry and a growing population. But removal of these materials creates problems underground. When solids are removed (coal, ores), large voids are created. When fluids are removed (oil, water), the pores in which they existed become empty. In both cases, support is decreased for Earth materials above.

What happens at the surface depends on the depth of the void and the strength of the remaining material. Under the steady pull of gravity, unsupported surface material will subside or collapse. If this happens at shallow depth, the surface will depress in a process called **subsidence** (Figure 10–1a). This usually happens slowly, over periods of weeks to years.

If a void is near the surface, down to roughly 15 or 20 meters (50 to 66 feet) (but varying greatly with the rock at any location), surface material may break through and tumble into the space. This process is called **collapse** (Figure 10–1b). Where collapse into a void occurs at greater depth, overlying materials may deform downward, causing surface subsidence (Figure 10–1c).

Subsidence and collapse are serious geohazards. They are growing more common because land development is expanding into areas that are more subsidence-sensitive. The groundwater demand in growing communities lowers the water table and reduces artesian pressure (Chapter 8), thus aggravating subsidence. In some areas, old underground mines may collapse, causing subsidence damage to new developments on the surface.

**Collapse sinkhole and structural damage resulting from surface failure. (J. Watson/Sygma.)**

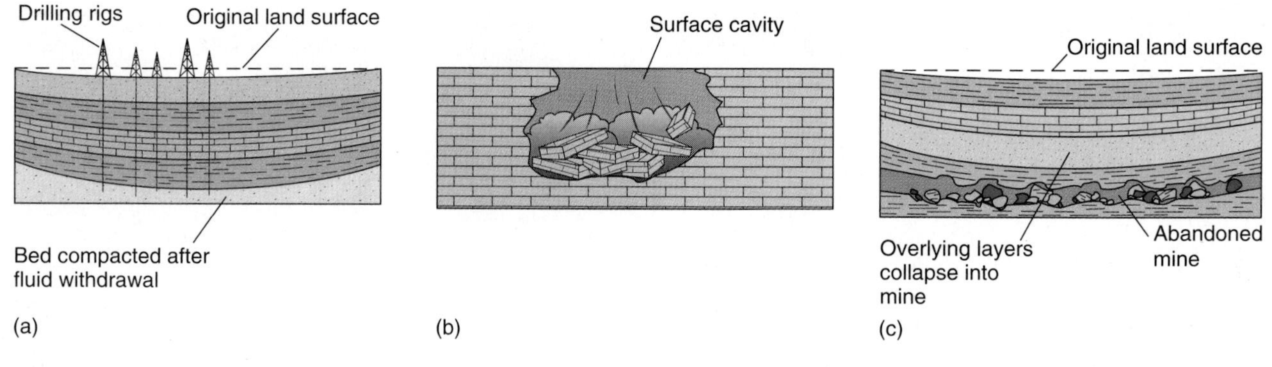


**FIGURE 10–1**
Relation between subsidence and collapse. (a) Fluid withdrawal results in bed compaction and surface subsidence. (b) Enlarging solution cavity eventually causes surface collapse. (c) Collapse of deep mine can cause overlying rocks to deform downward and create subsidence at the surface.

## SUBSIDENCE AT OR NEAR THE SURFACE

Three types of subsidence occur at or near the surface:

1. Subsidence above compressible fine-grained sediments.
2. Subsidence where clayey soils shrink and swell with changes in water content.
3. Subsidence where draining and subsequent decomposition of organic-rich deposits lower the surface.

We will now consider each of these.

### Subsidence Above Compressible Deposits

As building sites become harder to find, it has become commonplace to develop new land by adding sandy fill on top of fine-grained water-saturated sediments. Unfortunately, the weight of structures built on such land presses the water from the fine-grained sediments. The sediments then compact, and the surface subsides.

To avoid this problem for high-rise buildings, which concentrate great weight in a small area, their weight must rest upon something solid, not on the compressible sediments. Flat-footed pilings are driven into non-compressible materials, such as sand, or they may be driven through sediments to rest directly on bedrock below (Figure 10–2). However, smaller structures, roads, and utility lines rarely are supported in this way.

Figure 10–2 shows the different performance over time of supported high-rise structures and unsupported smaller structures. The surface between the supported high-rises will subside, damaging the smaller structures. Parking lots subside and crack, roadways and outdoor stairways shift, utility pipes raise bulges in the road as the road surface subsides over them, and surface depressions result when strained gas and water lines burst under the pressure from subsidence movements.

Many of these problems are exemplified in New York's Co-op City, the largest apartment complex in the United States, completed in 1971. It includes 35 apartment towers, each 24 to 33 stories high, and several townhouse complexes. Co-op City is home to more than 50,000 people. The complex is built on 300 acres of former wetlands and garbage dumps that were filled. The high-rise buildings were anchored into bedrock or built on pilings driven into firm sand. However, smaller structures on the adjacent land were not supported this way.

The result has been extensive structural damage. Cracks have opened on unsupported roads, athletic fields, and parking lots. Sunken garages (Figure 10–3) and burst utility lines have resulted from the land subsidence. Measurements comparing the original 1971 land surface with elevations in 1989 reveal that subsidence rates at twenty locations within Co-op City averaged about 2.4 centimeters per year (almost an inch), varying between 1.25 and 4.25 centimeters (0.5 to 1.7 inch). Total subsidence over the 18-year period was 22.5 to 76.5 centimeters (9 to 30 inches).

Remedial measures have been underway for several years and have been effective but expensive. One lesson to be learned from this experience is that, in an area where compaction is likely, it is important that

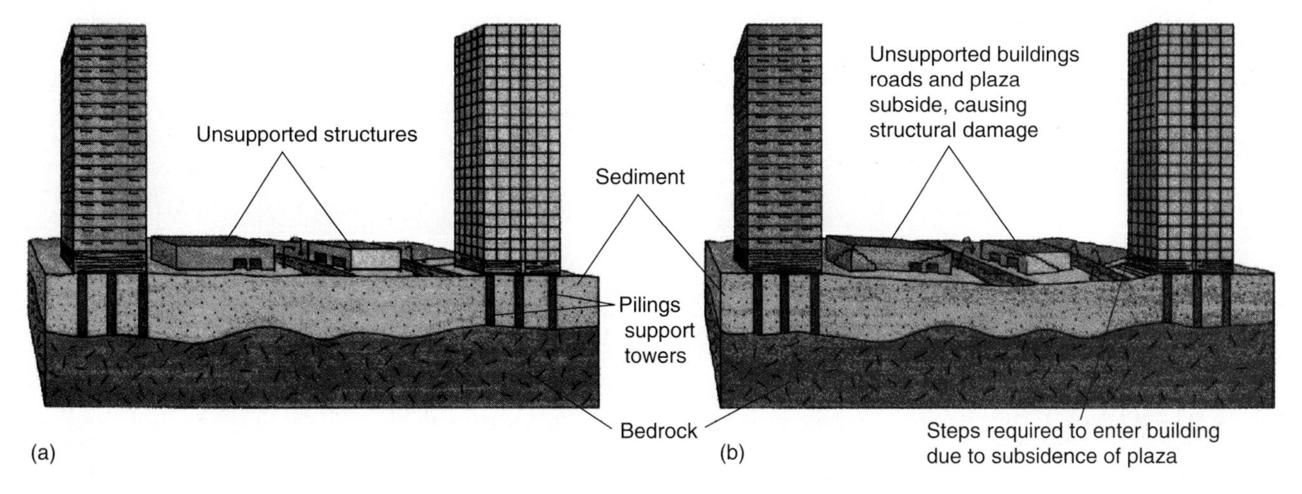


**FIGURE 10–2**
Differential subsidence. (a) The high rise buildings are built on piers set on bedrock; smaller buildings are not. (b) After a period of time the unsupported buildings have subsided, resulting in structural damage.

**FIGURE 10–3**
Ground subsidence resulting from compaction of underlying fill. The high-rise buildings in the distance are supported on solid foundations. The foundation of the parking garage on the right side of the photo was anchored into fill and has subsequently subsided below the land surface on the left; this made the lower level of the garage unusable. (Courtesy of Antonia Lioudakis.)

*both* the major structures and smaller ones be supported on pilings that rest on a strong layer below.

## Subsidence in Areas Underlain by Expandable Clays

A clay-rich rock called bentonite can absorb large quantities of water, swelling to many times its original volume (bentonite was discussed in Chapter 9). Clay-bearing sediments with this property are called "shrink-swell" clays. They are a widespread and expensive problem in many areas (for example, Denver and Boulder in Colorado, and Austin, Texas—Figure 9–4).

Shrink-swell clays cause problems in this manner: When water enters the ground, these layers swell and exert great pressure on overlying building foundations. This upward pressure can greatly exceed the downward pull of gravity on small structures. As a result, these structures may be differentially lifted, resulting in foundation and structural damage (Figure 10–4). Between rains, the clay layers dry out and shrink, forming large surface cracks up to 1.5 meters (5 feet) deep. These cracks can seriously damage overlying structures, utility lines, highways, and rail lines.

The problem is especially severe where part of the ground is covered with concrete or asphalt. When it rains, the unpaved ground bordering the pavement swells and exerts compressive stress on

**FIGURE 10–4**
Structural damage caused by shrink-swell clays.

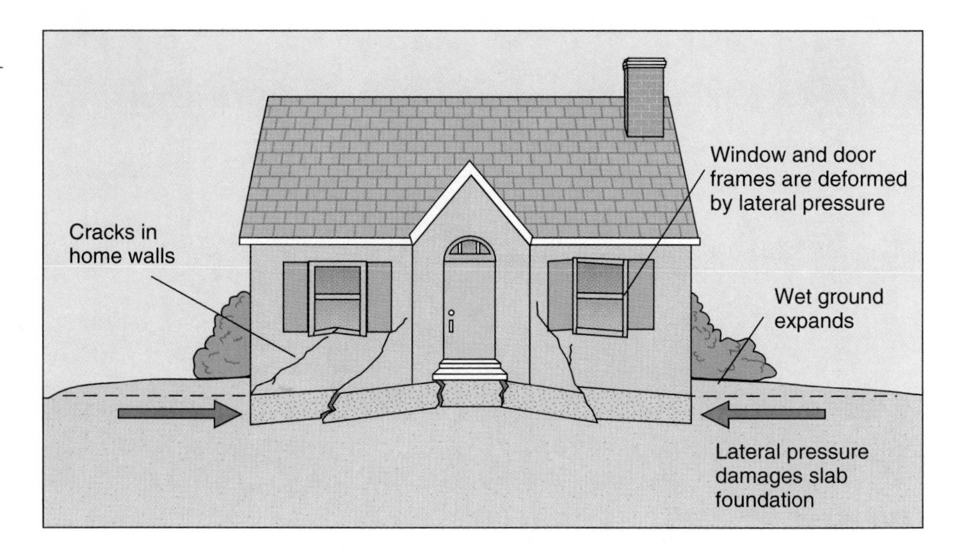


the dry paved areas. The result is that cracks, folds, and potholes formed in the road asphalt require constant repair. Shrink-swell clays also cause walls to tilt inward in homes built on concrete slabs.

When the ground dries out, the surface around the pavement shrinks and cracks, and water trapped under the covered portion causes the clays to swell and exert strong upward pressure. This can dome up the central part of a concrete slab, damaging the foundation (Figure 10–4). It also may rupture water or gas pipes in the foundation.

Damage can, however, be minimized in shrink-swell clay areas. Concrete slab foundations can be strengthened with steel reinforcing bars ("rebar") so that they are less easily cracked. Structures can be placed on pilings that rest on firm beds at depth. The expansive soil can be excavated and replaced with porous material, such as gravel, before the concrete slab is poured.

Water control is important to prevent damage. Strategies include draining roof water as far as possible from the structure, repairing leaky water pipes, and avoiding overwatering of lawns. Another important preventive measure is to avoid planting trees near homes, because they remove groundwater from around and under the foundation, resulting in shrinkage pressures that can crack the outer walls.

## Subsidence in Areas Having Organic Soils

In Chapter 6, we looked at a soil type called a *Histosol.* This organic-rich soil type occurs in a wide range of climates. Areas underlain by water-saturated Histosols have been drained for agriculture or building sites. The removal of water allows the soil to compact, creating significant subsidence.

The predrainage condition of a Histosol is shown in Figure 10–5a. A canal is dug through the Histosol, peat in this case, to promote gravity drainage (Figure 10–5b). The water flows from the soil into the canal, and the water table drops steadily over time. This water loss results in aeration, oxidation, shrinkage, and compaction of the Histosol. As biochemical decomposition breaks down the peat, additional subsidence occurs.

Major areas underlain by peat soils exist in Alaska and in formerly glaciated areas in the northern continental United States. However, geohazards from the draining of organic soils are most serious in California and Florida. The following are two examples.

***Sacramento River Delta, California.*** The Sacramento River delta is a mass of channels, wetlands, and islands. In the delta, streams that drain the Sierra Nevada Mountains meet saltwater moved by tides up the Sacramento River from San Francisco Bay (Figure 10–6). The organic soils of the delta were drained for agriculture many years ago. The land surface has since subsided an average of about 8 centimeters (3 inches) per year.

Many parts now are below river level and are protected from flooding by dike systems. Humans, structures, and wildlife are jeopardized by flooding during high river stages that overtop or breach the dikes. Also, the dikes could be damaged by an earthquake in this seismic risk zone (Chapter 5).

***Bordering the Everglades, Southern Florida.*** This area (Figure 10–7) contains over 12% of the organic

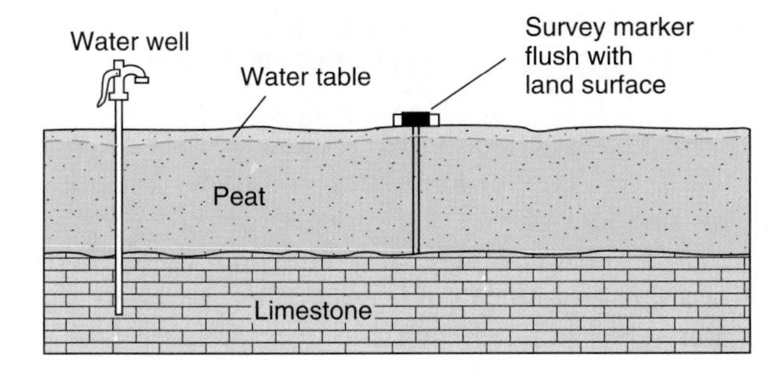

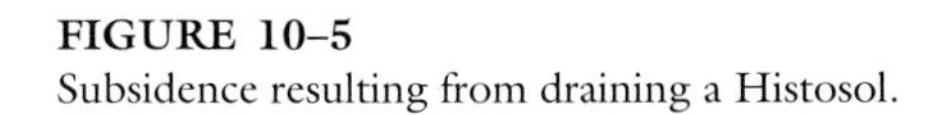

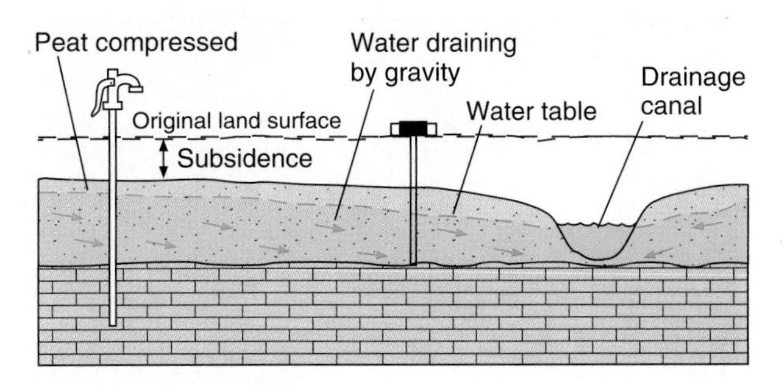


**FIGURE 10–5**
Subsidence resulting from draining a Histosol.

soils in the 48 contiguous states. Cleared for agriculture early in the twentieth century, it now is the major U.S. sugar cane growing area, as well as a major fruit and vegetable production area.

Subsidence began in the 1920s as drainage canals were excavated through the peaty soil. Drainage-induced subsidence has attained rates up to 2.5 centimeters (1 inch) per year. The subsidence results from organic decomposition of the peat, loss of buoyancy formerly provided by pore waters, and compression by agricultural machinery. Over time, gravity drainage becomes less effective as the slope of the water table decreases and subsidence rates become slower. Additional drainage is obtained by pumping the water from the ground, which is accompanied by renewed subsidence (Figure 10–8).

The effects of subsidence are visible in communities such as Belle Glade, in the Everglades agricultural area south of Lake Okeechobee. Concrete elevation markers that were installed flush with the ground seventy years ago now protrude more than 1.8 meters (6 feet) above the land surface (Figure 10–5). Homes built on pilings in this area, so that they do not subside, may require an additional doorstep periodically as the surrounding ground subsides!

Organic-rich soils cannot be drained without producing some subsidence. However, subsidence effects can be minimized by proper drainage man-

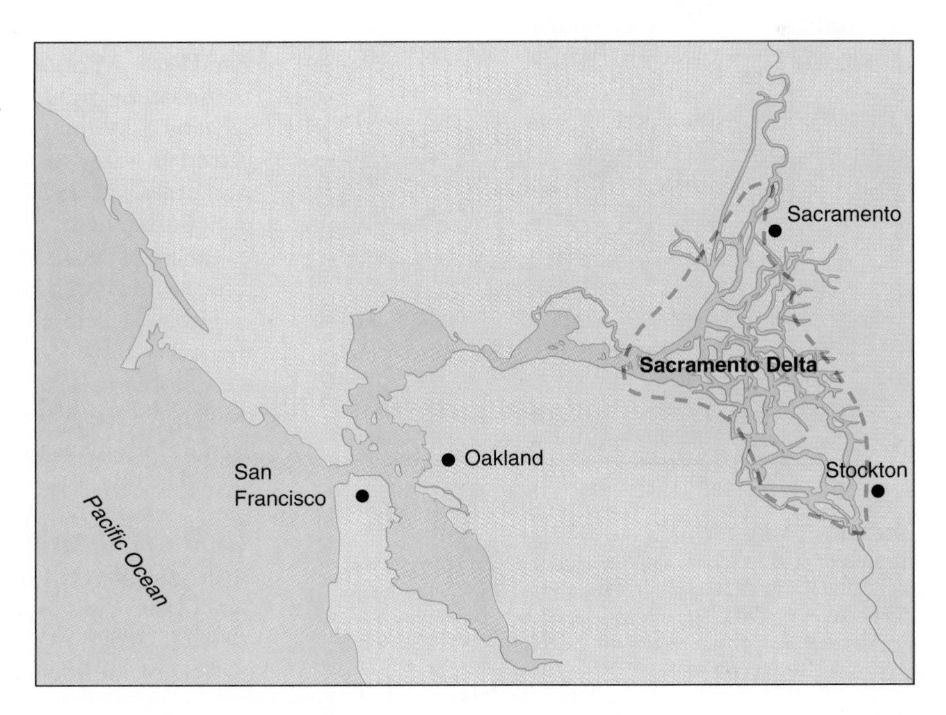


**FIGURE 10–6**
Map showing location of Sacramento River delta in California.

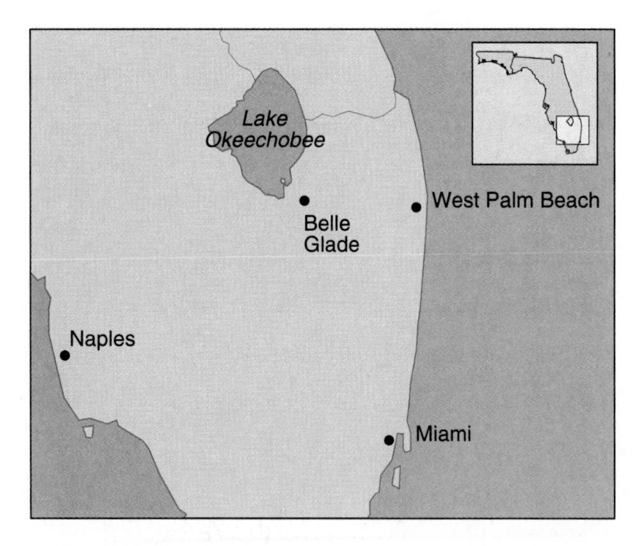


**FIGURE 10–7**
Map of South Florida showing location of Lake Okechobee and Belle Glade in the agricultural lands between Lake Okechobee and the Everglades.

agement, reinforcement of structures, and placement of structures on pilings that rest on a firm substrate. Protective dikes are very important where subsidence lowers areas to below river or sea level, thus increasing the threat of flooding.

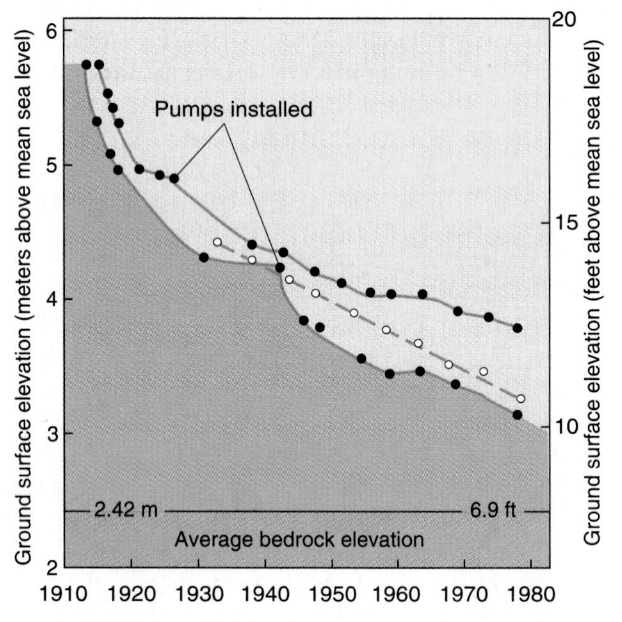


**FIGURE 10–8**
Graphs of ground surface subsidence in the Florida Everglades agricultural area after initial draining around 1912. (From J. C. Stephens et al, 1984, Organic soil subsidence: Geological Society of American Reviews in Engineering Geology v. VI, Fig. 3, p. 113. Used with permission.)

## SUBSIDENCE CAUSED BY WITHDRAWING OIL OR WATER

In the pore spaces of buried sediment or rock, fluids such as oil and groundwater are under pressure. This provides part of the support for the overlying material. However, we are removing these fluids from the ground at increasing rates, lowering the water table and reducing the pore pressure within the material. This permits compaction of the subsurface layer. The overlying material gradually subsides to produce a *subsidence basin.* Subsidence caused by fluid withdrawal at depth has become a serious problem in Houston-Galveston, Texas; Las Vegas, Nevada; Long Beach, California; Mexico City, Mexico; Venice, Italy; and other places.

As fluid is withdrawn, tensional forces (Chapter 2) pull the surface apart into **subsidence fissures** (Figure 10–9). Surface erosion then widens the fissures into gullies. In agricultural areas, these fissures can be plowed over and cause little harm. In urban areas, however, subsidence fissures can be serious, because most structures have brittle supports such as concrete slabs. When these structures are subjected to tensional stress, they crack and may fail.

In many places, the ground not only cracks, but portions of it are lowered by **subsidence faults** that cut across the surface. Where these faults pass under structures and transport systems, they can cause serious structural damage (see Box 10–1).

***Las Vegas, Nevada.*** Las Vegas lies in a desert basin bordered by mountains. Aquifers in the basin are recharged by mountain rainfall, and they long have provided water to Las Vegas. The first artesian well was drilled in 1907 for agricultural and municipal use. Subsequent development has removed great volumes of groundwater and lowered the water table as much as *55 meters* (180 feet).

Urban development accelerated after 1955, and the water table dropped faster. Subsidence fissures were noted in 1957. The progressive subsidence of downtown Las Vegas since 1935 is shown in Figure 10–10. Subsidence up to 1.5 meters (5 feet) has occurred in this area.

***Santa Clara Basin, California.*** One of the most severe subsidence areas in the United States is in the Santa Clara Basin, south of San Francisco Bay (Figure 10–11). Groundwater withdrawal caused 1.2 meters (4 feet) of subsidence between 1960 and 1967, with

**FIGURE 10–9**
Subsidence fissure, Picacho fault, Arizona. (Courtesy of Thomas Holzer, U.S. Geological Survey.)

the greatest movement just south of San Jose. In downtown San Jose, total subsidence between 1912 and 1967 was nearly 4 meters (13 feet). At Long Beach, California, pumping of oil from underlying strata caused severe subsidence. Subsidence has been so severe in some areas that seawalls have been built to prevent flooding from the Pacific Ocean.

## SUBSIDENCE CAUSED BY MINE COLLAPSE

The underground mining of coal and ores removes so much material that it seriously reduces the support for overlying rocks. If mine workings are relatively near the surface (within perhaps a few dozen meters), the overlying rocks may be too weak to support their own weight. Eventually, they will collapse into the mine below. This creates a series of *collapse pits* at the surface (Figure 10–12).

In many mines, the mined area is sufficiently deep that no significant surface collapse occurs. The overlying rocks are strong enough to permit slow downward deformation. The eventual collapse of the lowermost rocks into the mining cavity is followed by more gradual subsidence, which progresses toward the surface. Even if no collapse takes place at the surface, this gradual subsidence can warp the surface downward, seriously damaging buildings, roads, and pipelines. When surface subsidence is caused by mining, therefore, its pattern depends on mine depth, overlying rock strength, the acreage mined, and the mining method.

### Underground Mining Methods and Surface Subsidence Patterns

Some underground mining in the United States is done by removing most of the coal or ore, but leaving pillars of the resource to support the overlying strata, somewhat like using jackposts in a basement to support the upper floors (Figure 10–13a). This is called the **room-and-pillar method** (Figure 10–13a). Maintaining this support for the mine roof reduces surface subsidence if the mine is deep. However, if the mine is near the surface, this method may only delay surface subsidence or collapse. In shallower underground mines, when the surface subsides or collapses, it forms a gridwork of surface basins or pits that reflect the unequal support underneath (Figure 10–12).

In **longwall mining,** the resource is completely removed as mining advances (Figure 10–13b). This method is safer, faster, cheaper, and recovers more coal than any other underground mining method. A rotating cutter wheel travels 100 meters (330 feet) or more along a "long wall" of coal, shearing off many tons in minutes. In the active mining area, hydraulic shields support overlying rocks. Longwall mining should result in a more uniform subsidence of the surface as mining proceeds, but this system still causes considerable subsidence damage.

### Mining and Overlying Rocks

In underground mining, the overlying rocks deform downward and may crack and separate (Figure

## BOX 10–1 SUBSIDENCE IN SPACETOWN

In the Houston-Galveston region of Texas, Earth yields two valuable commodities: groundwater and petroleum. Figure 1 shows the concentration of urban areas and the Goose Creek Oil Field. Numerous industries line Galveston Bay and the Houston Ship Channel, a river segment that connects the Bay with Houston. Large volumes of groundwater are withdrawn by these industries, by the rapidly growing residential areas of Clear Lake City and Pasadena, and by the Johnson Space Center.

As water and petroleum are withdrawn, however, the surface is lowered, creating a serious subsidence geohazard. Withdrawing groundwater from coastal aquifers has caused subsidence fissures, cracks, and significant structural damage, and has increased flooding, especially from coastal storms and hurricanes. The cumulative subsidence from 1906 to 1978 is, at its greatest extent, 2.5 meters (over 8 feet), as shown in Figure 1.

Ellington Air Force Base has some of the oldest structures in the area, built in the early 1940s. From 1942 to 1973, subsidence at the base reached up to 1.5 meters (5 feet). Surface cracks sliced parking lots and paved surfaces, and offset buildings.

In 1975, the author accompanied NASA personnel to document a visible subsidence fault under the building shown in Figure 2. The fault was 8 to 10 centimeters wide (3 to 4 inches), with a displacement of about 30 centimeters (12 inches) (Figure 2).

Subsidence jeopardized the extensive collection of lunar rocks in a Johnson Space Center building in Clear Lake (Figure 1). Here, repeated elevation checks showed that subsidence slightly over 0.6 meters (2 feet) had occurred at two places at the Center between 1964 and 1973. This is a serious concern, because this area near Galveston Bay is less than 6.5 meters (21 feet) above sea level.

Concerned that a major hurricane could create a storm surge (Chapter 16) sufficient to damage the building holding the moon rocks, NASA constructed a hurricane-resistant facility that now protects these valuable samples for study by future generations.

Subsidence also affects communities near Galveston Bay and its tributary streams. As the land surface is lowered, these areas become more easily flooded by storms each year. In a major hurricane, Galveston Bay could easily rise 5 to 6 meters (16 to 20 feet), and flooding would be extensive. To ensure access to these coastal areas despite the subsidence, many roads and their protective seawalls are built up periodically (Figure 3).

Structural damage from subsidence is common in Clear Lake City. Walls are cracked, window and door frames sag, and differential movement of the

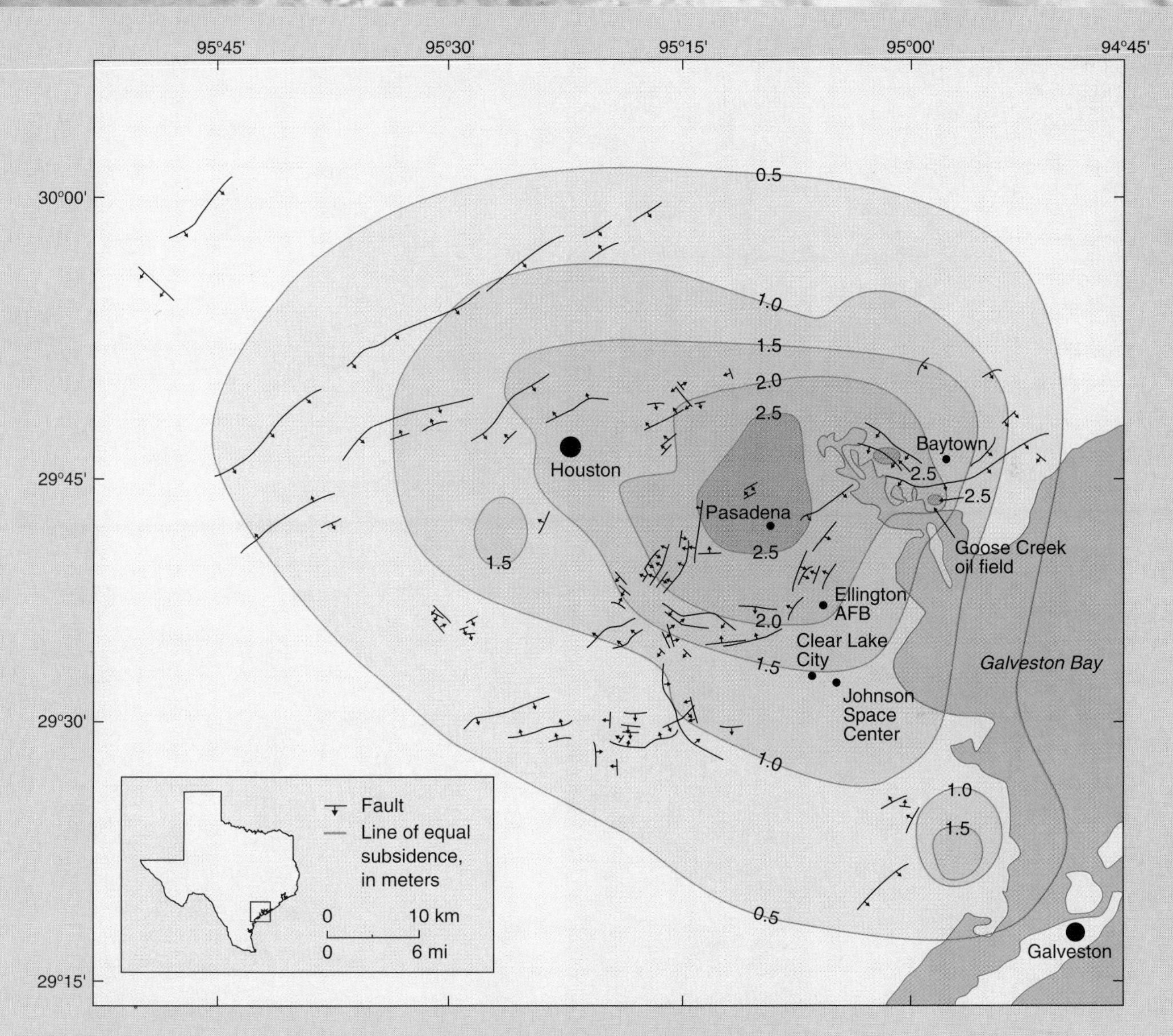


**FIGURE 1**
Map of central Texas coastal area showing localities mentioned in the text and contours showing land subsidence from 1906 to 1978 and location of subsidence faults in the Houston-Galveston, Texas, area. (Modified from T. H. Holzer, 1988, Ground Failure induced by Groundwater Withdrawal, Fig. 13., p. 80 *in* Geological Society of America Reviews in Engineering Geology, v. VI. Used with permission.)

**FIGURE 2**
Subsidence fault preserved under a building at Ellington Air Force Base near Clear Lake City, Texas. The surface in the foreground has dropped down relative to the surface in the background. The vertical displacement on the fault has been about 30 cm (12 in.) since the construction of the overlying building in 1942. (Photo by author.)

**FIGURE 3**
Road in Seabrook, Texas, raised to prevent flooding as a result of subsidence of surrounding surface. (Photo by author.)

land is visible in structures (Figure 4). Subsidence in one subdivision of Baytown caused such severe flooding that all the trees died from saltwater intrusion and all homes were abandoned due to structural damage and flooding (Figure 5). In other areas, flooded homes were razed, the land built up, and new homes were built on the same site, even though subsidence is continuing (Figure 6).

**FIGURE 4**
Subsidence fault passing through a building in Clear Lake City. Note that the water level in the left side of the pool is near the land surface, but it is below the patio surface on the right. The crack system on the right side of the building in the foreground results from tensional forces across the fault. (Photo by author.)

**FIGURE 5**
Homes submerged and trees killed by saltwater encroachment as a result of subsidence at Bayport, Texas. (Photo by author.)

Subsidence can be stabilized and some recovery in elevation may be obtained by pumping water back into the coastal aquifers. Subsidence also can be slowed by reducing groundwater withdrawal and relying more on surface water imported by aqueduct (Chapter 8).

**FIGURE 6**
(a) Homes at Seabrook, Texas, in 1975, which had been abandoned as a result of flooding from land subsidence. (b) Same area in 1990. The land was raised by filling and new homes were built on the same area. (Photo by author.)

(a)

(b)

10–14). The area affected by subsidence at the surface is broader than the mined area at depth. The area affected can be determined by the **angle of draw,** the vertical angle between the edge of the mined-out area and the edge of surface subsidence (Figure 10–14). The angle of draw is between 20 and 35 degrees in most U.S. mines. It increases where a thin section of unconsolidated sediments exists over a mine and decreases where thick sections of strong rock overlie the mined-out void.

Because subsidence varies over an area, different features appear. The outer part of the subsiding zone is being pulled apart (subsidence cracks) by tensional forces as the surface material is displaced toward the center of subsidence. However, in the central subsidence area, the ground is under compression (Figure 10–14).

Leaving pillars to support the surface does not always work well. Some mines continuously pump out groundwater that seeps in, but as the mines are closed, pumping stops, and the mines flood. This saturates and weakens the pillars (coal is quite porous; in fact, water wells often are drilled into coal aquifers). The weight of the overlying rocks,

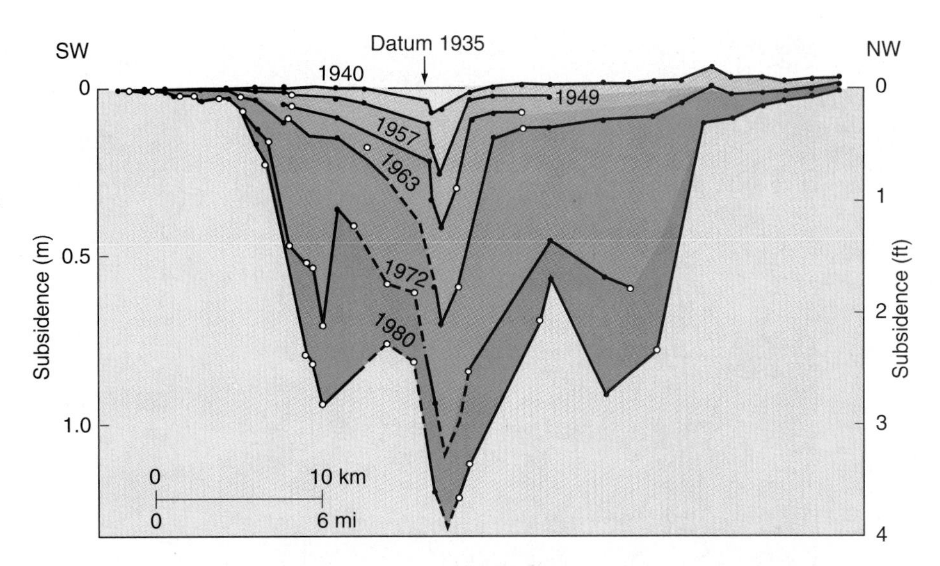


**FIGURE 10–10**
Subsidence profiles through downtown Las Vegas from 1935 to 1980. (After T. H. Holzer, 1988, Ground failure induced by groundwater withdrawal, Fig. 20. p. 86, *in* Geological Society of America Reviews in Engineering Geology, v. VI. Used with permission.)

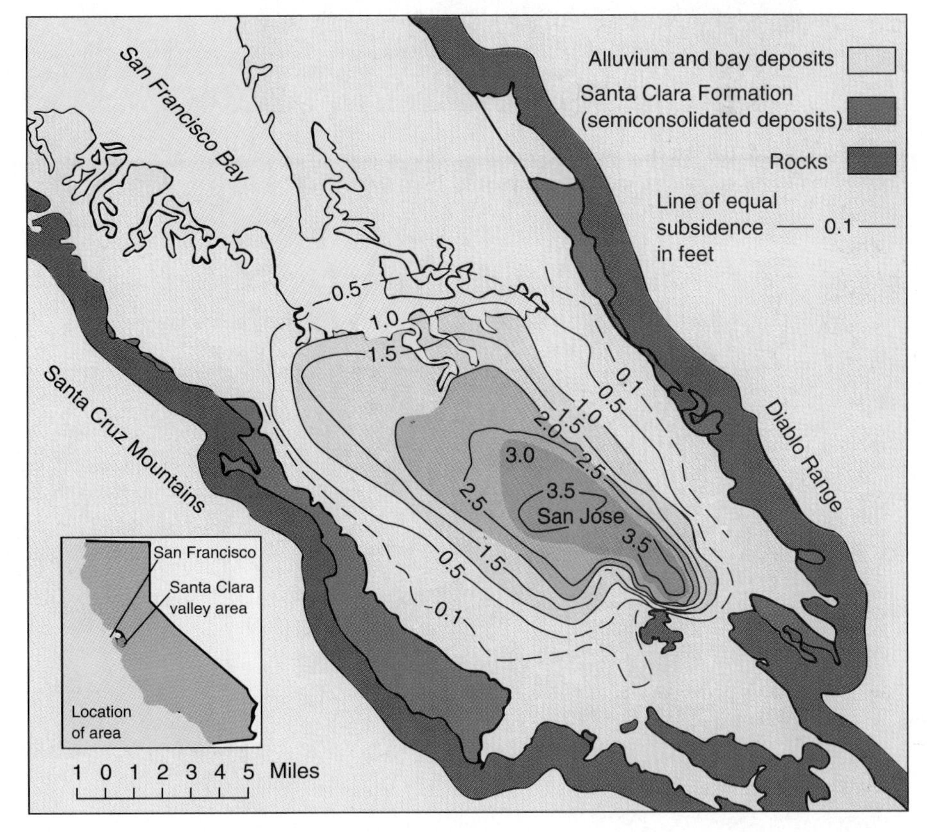


**FIGURE 10–11**
Contour lines showing points of equal surface subsidence resulting from excessive groundwater withdrawal in the Santa Clara Basin, California. (After Poland, 1969, Land subsidence in western U.S., *in* Olsen and Wallace, eds., Geological hazards and public problems, May 27–28 Conference Proceedings, U.S. Geological Survey, p. 77–96. Modified with permission.)

now concentrated on the weakened pillars, may split or disintegrate the pillars, resulting in collapse (Figure 10–15).

Seeping groundwater may weaken the rock below the coal so that the pillars may slowly drive down into it. This is most common where the underlying rock is shale. As a result, the mine floor becomes domed and the mine roof subsides (Figure 10–15). Surface problems thus may appear decades after mining has been completed, sometimes too late for surface owners to seek compensation.

Differences in strength of the overlying rocks may cause differential failure of the mine roof. Generally, sandstone is stronger and provides support

**FIGURE 10–12**
Subsidence pits at the surface resulting from collapse of the mine workings at depth. (U.S. Bureau of Mines.)

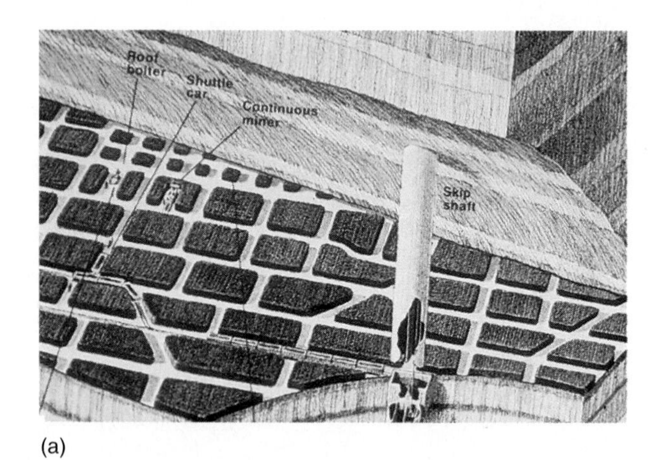


(a)

(b)

**FIGURE 10–13**
The most common types of underground mining today. (a) Room and pillar mining operation (U.S. Bureau of Mines.) (b) Longwall mining method. Over 6 kilometers (3.7 miles) into this eastern Kentucky mine, the longwall machine shears about 30 metric tons of bituminous coal each minute from the seam. It falls onto the conveyor for transport from the mine. The overhead shields support rock strata some 750 meters (half a mile) thick. Behind the shields to the left, the roof of the mine is allowed to cave in as the mining operation moves forward. (Grubb Photo Services, Arch of Kentucky #37 Mine, Arch Mineral Corporation)

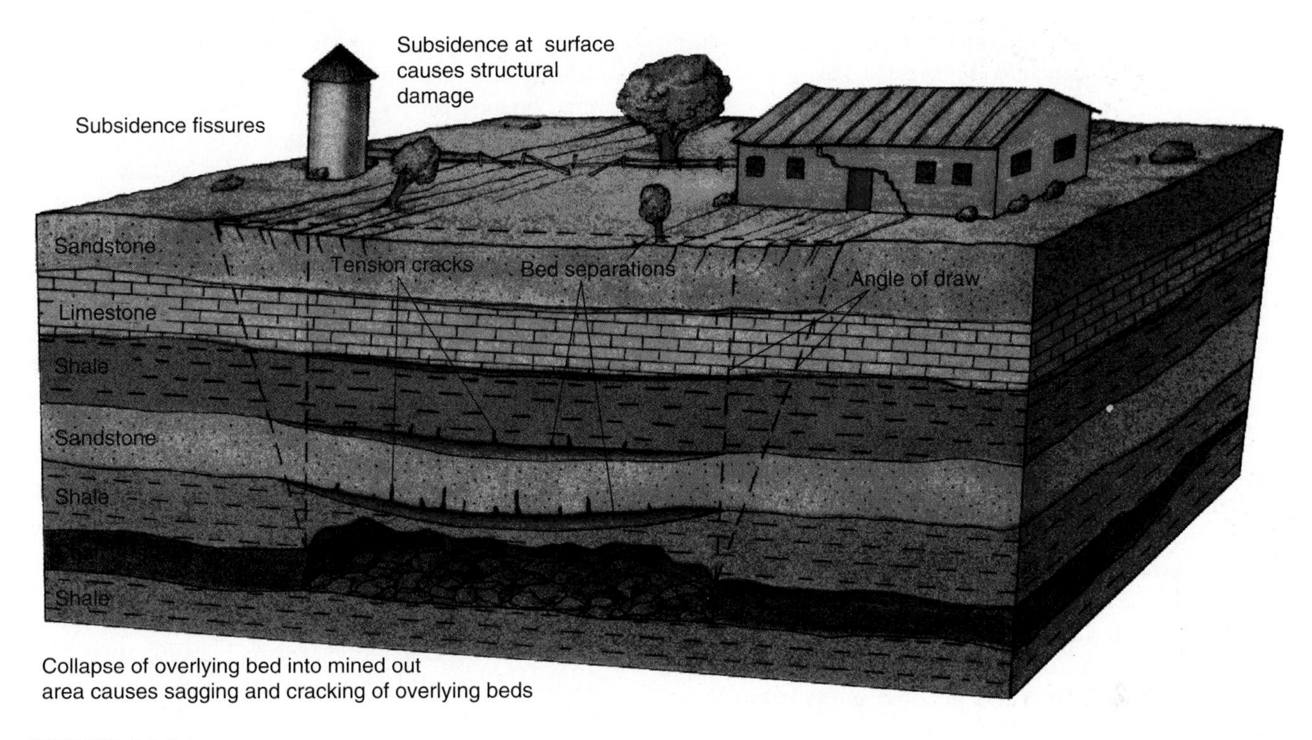


**FIGURE 10–14**
Vertical section showing subsidence deformation features in the underlying rocks and on the surface.

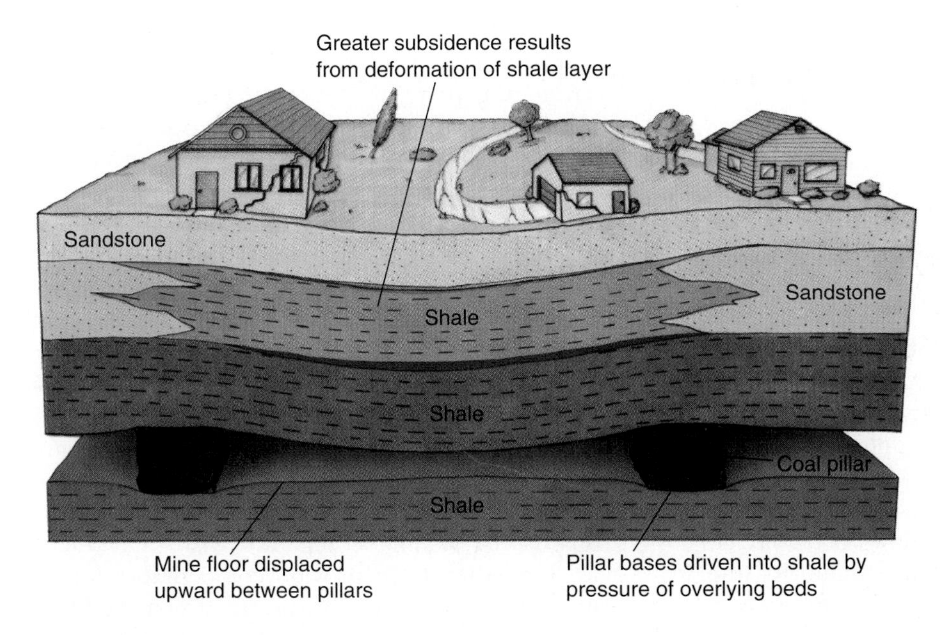


**FIGURE 10–15**
Vertical section showing changes that can occur to mine roof, floor, and land surface after the mine is abandoned. (Scale is exaggerated considerably for clarity.)

between pillars, whereas shale layers are weaker and may collapse completely (Figure 10–15).

### Subsidence Damage

Tensional forces from subsidence may crack brittle structures. If the displacement is sufficient, structures may fail. Such damage is common over abandoned coal mines in Pennsylvania, West Virginia, and Kentucky.

Surface drains, culverts, and underground pipes are easily damaged by subsidence. When constructed, they are graded (sloped) to ensure a steady flow under the force of gravity. As the surface subsides, drainage may become impeded due to a change in slope or bending or breaking of pipes. Ponding and flooding may result.

Subsurface sagging and cracking that accompany subsidence (Figure 10–14) may disrupt groundwater flow in aquifers, draining them and causing springs and wells to dry up. Fractured aquifers may lose artesian pressure and become more vulnerable to pollution from surface spills. The water freed from the aquifer may drain through cracks into the mine, contributing to the development of acidic mine waters (See Chapter 12).

The most typical signs of subsidence in a home are several windows and doors that continually stick despite efforts to correct them. As the subsidence worsens, other effects appear—plumbing leaks and cracked plaster. The plumbing leaks can be especially hazardous in a home with gas pipes.

### Coal Mine Fires

Most underground mine fires start where coal is exposed at the surface, whether a natural outcrop or exposed by mining. Ignition commonly begins from a brush fire or burning of trash near the exposed coal. The fire then spreads from the outcrop into underground areas.

Underground, you might expect these fires to extinguish when the oxygen supply is consumed. However, subsidence can form fissures in the overlying rocks, and these fissures become vents that supply surface oxygen to the fire below. As the fire continues, more coal is burned out, leading to more collapse and subsidence, which creates more fissures, and more oxygen from the surface feeds the fire.

Underground coal fires are extremely difficult to extinguish because of their depth and inaccessibility, noxious fumes, and great heat (coal, after all, is a fossil fuel burned as a heat source). For example, an underground coal mine fire in Carbondale, Pennsylvania, burned for *33 years* before being extinguished in 1965. This fire finally was put out by excavating down to the fire and smothering it with water and mud.

Cracking at the surface not only permits oxygen to enter the mine and feed the fire; it also permits smoke and hot gases to escape into the atmosphere. This causes air pollution, ground heating, and melting of snow cover, all with potentially adverse effects on plants, animals, and humans (Figure 10–16).

Under Centralia, Pennsylvania, a coal mine fire burned for over two decades. Carbon monoxide and carbon dioxide began seeping into homes. The carbon monoxide is poisonous, of course. The carbon dioxide is not poisonous, but it is hazardous because it displaces oxygen. One homeowner found that his furnace would not run because there was too little oxygen in his basement. When the oxygen level was measured, it revealed only a 9% concentration, compared to the normal atmospheric oxygen level of about 21%. This homeowner converted to electric heat.

### A Matter of Rights

It is very difficult for a mining company to obtain a permit to mine beneath an urban area today. However, in the past, simply owning the coal gave a company the right to mine it, regardless of what buildings, industry, or homes existed on the surface above. Such historic mining activity now makes urban subsidence a serious problem in cities such as Pittsburgh, Pennsylvania, and Fairmont, West Virginia. For many homeowners and businesses, the damage is done and the mining company has been out of business for decades.

Today, mining beneath a populated area is done in ways that minimize surface subsidence. The mining company can provide support for overlying structures and utilities. If too many structures require support, however, the cost may cancel mining plans. Some mining companies write agreements with homeowners to provide compensation or repair in case of damage. But practices vary, and legal problems often result when an area is undermined.

Many property owners simply do not know whether mining has occurred, or might occur, under their property. Anyone owning property in a region

**FIGURE 10–16**
Anthracite coal seam fire under an eastern Pennsylvania town. The fire periodically burns to the surface, emitting smoke and noxious fumes that kill the local vegetation. (U.S. Department of the Interior.)

of underground mining can ask their state's Department of Mines or Geological Survey whether their property has been undermined, or might be some day. Many states maintain maps of underground mining activity. If mining is planned, the homeowner can purchase mine-subsidence insurance.

One more problem is that a coal company's deed may allow them to remove the coal without regard to the current surface owner's concerns. Mining may be unpreventable and no compensation may be available for certain categories of damage.

### Mitigating Mine Subsidence

Subsidence can be mitigated by filling the void below. *Mine waste* (rock) can be moved into mined areas to support the roof. Where these areas are less accessible, they can be filled by pumping sand or cement through holes drilled from the surface.

Another approach is to delay surface development until mining and filling operations are completed. Sometimes building foundations are placed on pilings which pass through the rock and rest on the mine floor below.

## SUBSIDENCE AND COLLAPSE CAUSED BY DISSOLVING OF LIMESTONE

In areas underlain by limestone that is near the surface, subsidence and collapse can result naturally, without human involvement. Groundwater action can dissolve limestone to form cavities, creating a surface of collapsing sinkholes called *Karst topography,* named after a region in Yugoslavia where it is well developed. The groundwater action weakens the overlying rock, letting the surface subside or collapse into voids below.

This process forms circular depressions, called **sinkholes,** at the surface. Florida provides excellent conditions for sinkhole formation because of its warm, humid climate and because it is the only state completely underlain by limestone on the surface or at shallow depths. Limestone solution and sinkhole formation also is a problem in states that are partially underlain by shallow limestone, such as Kentucky, Indiana, Tennessee, West Virginia, Virginia, and Pennsylvania.

## Factors Controlling Limestone Solution

Limestone is poorly soluble in pure water, but dissolves readily in water that has been acidified by the addition of carbon dioxide ($CO_2$), forming carbonic acid ($H_2CO_3$). Rain dissolves some $CO_2$ as it falls through the atmosphere, but most $CO_2$ becomes dissolved as groundwater infiltrates through decaying vegetation and root systems at the top of the soil profile (Figure 10–17).

Dissolving of limestone increases greatly where fractures increase the surface area exposed to groundwater (Figure 10–17). The solution rate increases even more where fracture zones intersect. Limestone solution is believed to be greatest at the water table and just below it. Lowering of the water table results in downward solution of the limestone and the formation of large voids within it.

Many limestones are not pure calcium carbonate. They also contain poorly soluble components, such as quartz sand and clay. As the limestone is dissolved, these essentially non-dissolvable components accumulate as a fine-grained **insoluble residue.** This residue can be washed into solution cavities or accumulate at the top of the limestone, where it acts as an aquiclude (Chapter 8). It also

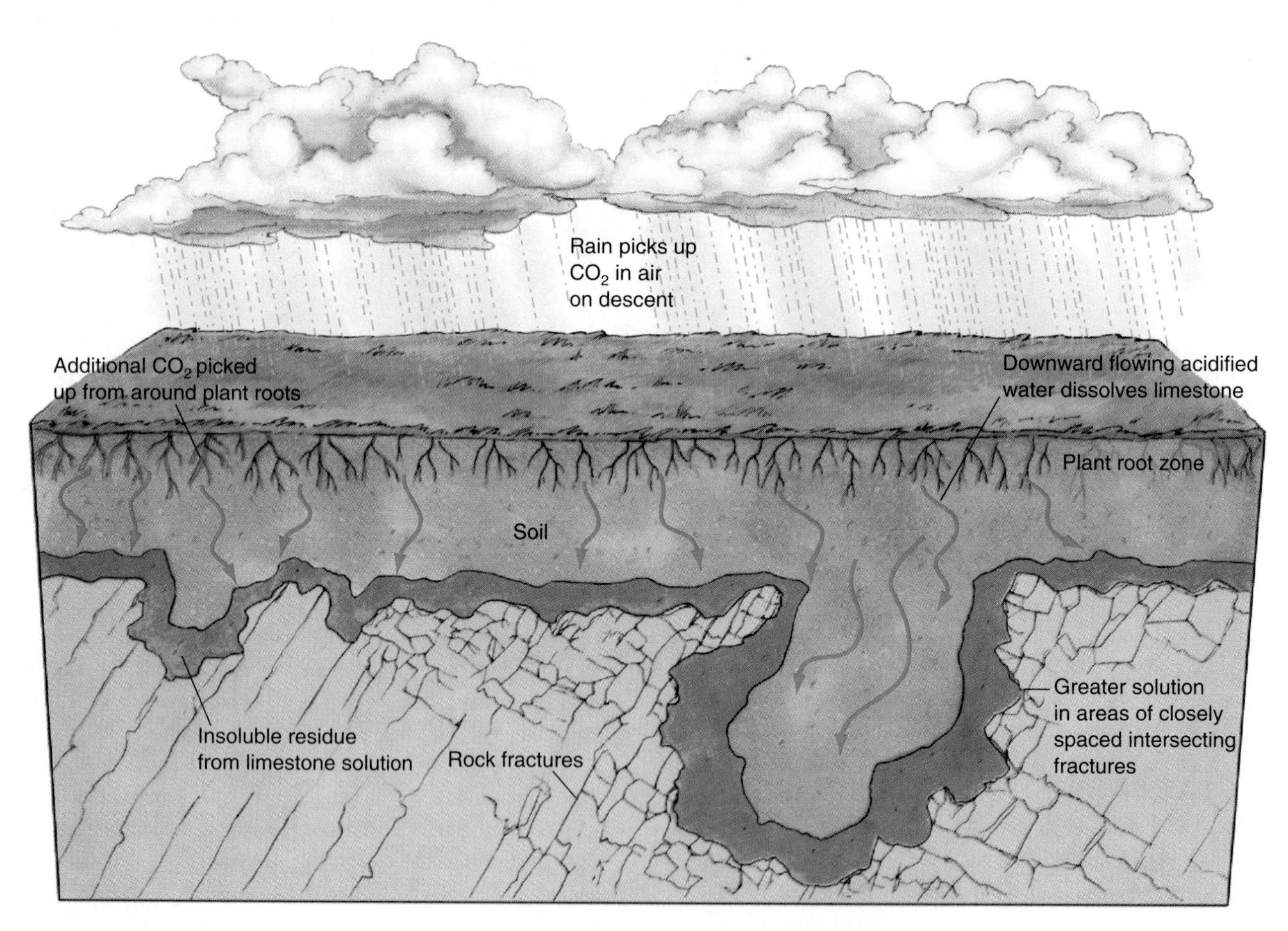


**FIGURE 10–17**
Factors in the solution of limestone. Downward-percolating groundwater dissolves carbon dioxide as it passes through root systems of plants. The acidified water dissolves limestone along fracture systems where water flows. Greater solution occurs where fracture systems intersect because the surface area exposed to solution is greater at these points.

helps support overlying material by filling solution cavities.

## Factors Promoting Subsidence and Collapse

Underground limestone solution creates cavities and reduces support for the overlying materials. If these materials are thick and strong, they may deform, causing the surface to subside gradually. However, if the cavity is near the surface, or if the overlying materials are weak, the surface may collapse suddenly.

Human activities have greatly increased subsidence and collapse in limestone areas such as Florida. The increasing demand for groundwater has lowered the water table, thus accelerating limestone solution, removing more of the surface support, and possibly washing some overlying material into the cavities. As development increases, homes, roads, stores, and factories increase the weight on the surface. The likelihood of collapse and subsidence clearly becomes greater.

## Sinkhole Development from Limestone Solution

Florida is one of our fastest growing states. Increasing population, industry, agriculture, and rapid urbanization have accelerated sinkhole development. Work by the Florida Sinkhole Research Institute showed that different types of sinkholes develop in response to differences in surface cover and the position of groundwater level. However, sinkholes are of two basic types—those that form where limestone is at or near the surface, and those that form where thick sections of sand or clayey sand cover the surface. Four varieties exist in Florida and are discussed next.

**Solution sinkholes** are saucer-shaped depressions formed by solution of surface limestone that has only a thin cover of soil or sediment (Figure 10–18). The water table usually is well below the surface. These sinkholes are best developed where surface water infiltrates fractured limestone. They form very slowly and pose few problems because they subside slowly, rarely collapse, and are generally small-scale features.

**Cover-subsidence sinkholes** form when overlying loose sand moves continually and slowly downward to fill a developing solution cavity (Figure 10–19). As the sand migrates downward, a subsidence sinkhole forms in the surface sand, and subsequent subsidence occurs very slowly. Downward movement is not dramatic, but it can damage nearby structures. Figure 10–20 shows how tensional forces from a cover-subsidence sinkhole have caused structural damage to a Florida home.

**Cover-collapse sinkholes** form where a thick section of sand overlies a clay layer on top of limestone (Figure 10–21). The clay layer is an aquiclude that separates two distinct groundwater systems (the sand layer and the limestone). The clay confines water under artesian conditions in the limestone below it, and forms a perched water table in the sand above it (Chapter 8).

Cover-collapse sinkholes form in this manner: over a long period, a solution cavity develops in the limestone and enlarges upward (Figure 10–21a). For a time, the intact clay layer has sufficient strength to span the underlying solution cavity and not collapse into it. However, as the cavity enlarges, more of the clay falls into it until the clay layer finally is breached. The water and sand then move downward into the cavity, like the flow of sand in an hourglass (Figure 10–21b). This causes a rapid drop in the sand cover and the formation of a cover-collapse sinkhole at the surface.

Cover-collapse sinkholes are especially serious because they develop so quickly. They can be triggered easily by a rapid withdrawal of water, resulting in a drop in the elevation of the water table. Cover-collapse sinkholes also can be triggered when

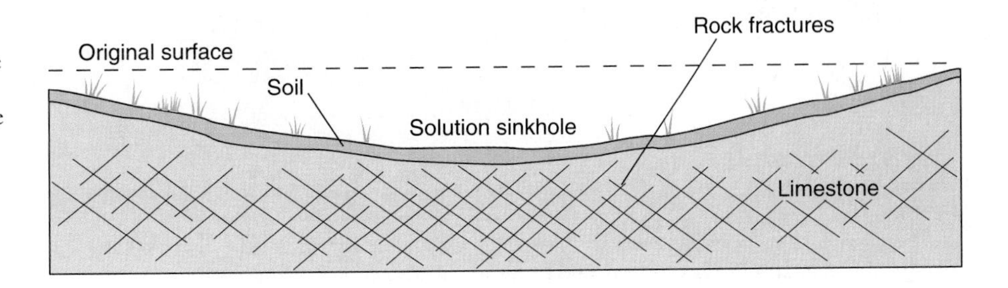


**FIGURE 10–18**
Development of a solution sinkhole in fractured limestone. Solution sinkholes form very slowly and pose few problems.

**FIGURE 10–19**
Formation of a cover-subsidence sinkhole.

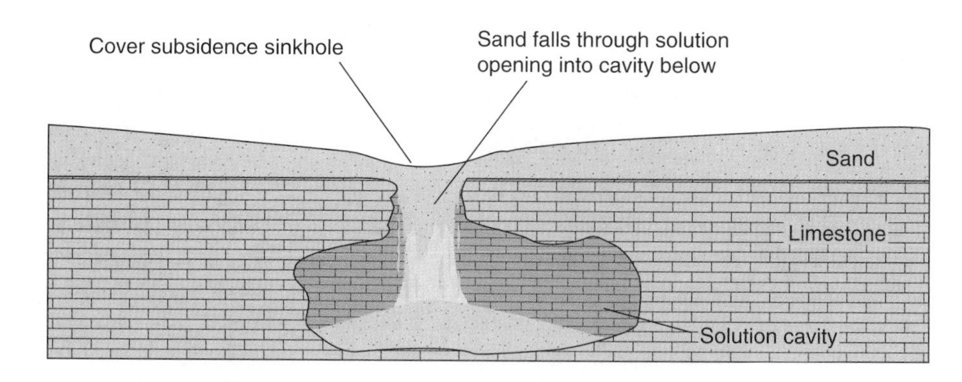


**FIGURE 10–20**
A cover-subsidence sinkhole is forming and structural damage to this home is occurring as the sand layer slowly settles into a limestone cavity below. (Modified from Beck, 1988, Environmental and engineering effects of sinkholes- the processes behind the problems: Environmental Geology and Water Science, v. 12, no.2, Fig. 5, p. 75. Used with permission.)

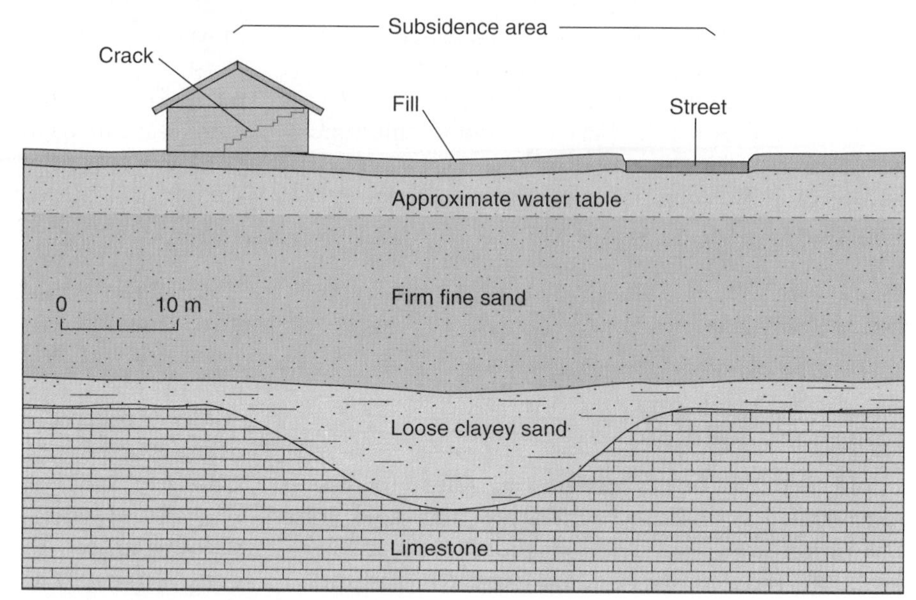


**FIGURE 10–21**
Formation of a cover-collapse sinkhole. (a) Solution cavity develops in underlying limestone and expands upward into capping clay layer. (b) Solution cavity development breaches clay seal. Groundwater and sand rush into cavity below, producing a cover collapse sinkhole on the surface.

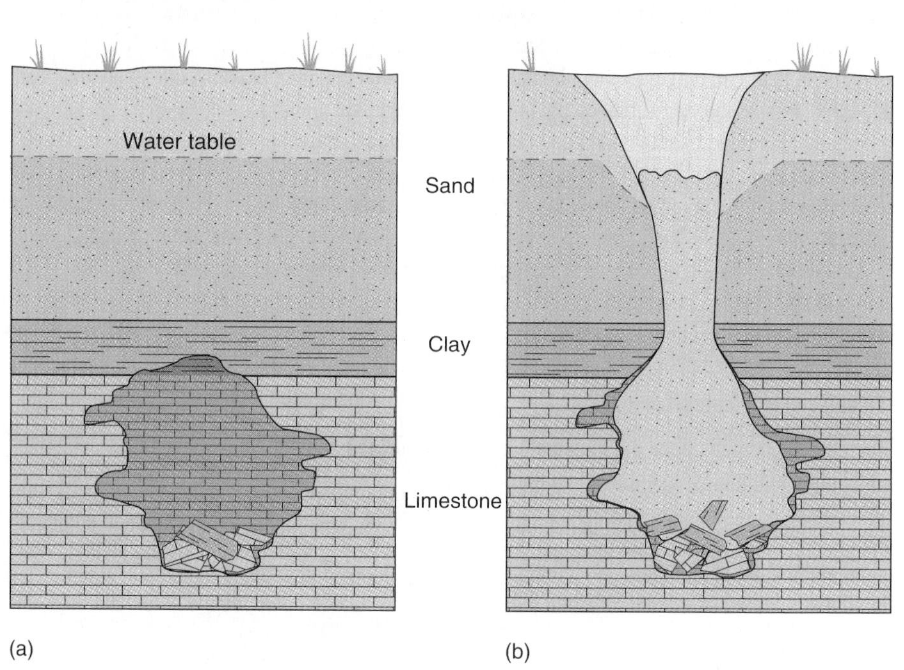

unusually heavy rains cause rapid infiltration and sediment is washed down into solution cavities below. On a spring day when about one-third meter (1 foot) of rain fell on Ocala, Florida, over an eight-hour period, a dozen sinkholes developed, including one beneath a state highway.

**Collapse sinkholes** are steep and rock-walled, forming quickly due to the sudden collapse of the roof of a solution cavity. Limestone solution at depth steadily works upward until the roof is too thin to support itself (Figure 10–22a) and it collapses into the cavity below (Figure 10–22b), leaving a sinkhole at the surface. Soil and sediment may subsequently fill the cavity. Subsidence of the fill leaves a depression on the surface over the buried cavity below (Figure 10–22c). This is the least common type of sinkhole.

## HUMAN ACTIVITIES THAT TRIGGER SINKHOLES

Another cause of sinkholes is a rapid drawdown of the water table due to excessive pumping. As rapid pumping lowers the water table, reducing the water pressure within the limestone, increased seepage from the sand layer above flushes sediment downward into the limestone, resulting in cover-collapse sinkholes (Figure 10–23).

Studies made in two Florida counties concluded that 70% of recent sinkhole development occurred during the dry months of April and May. Also, heavy groundwater withdrawal occurs during shorter periods in winter when large volumes of water are sprayed onto trees and plants to prevent freezing. During one January, 27 sinkholes formed in a strawberry-growing area northwest of Tampa within two days of water withdrawal for freeze-prevention.

Some sinkholes have been triggered by drilling water wells (Figure 10–24). The drill may inadvertently breach a clay layer on top of the limestone or create a new pathway, permitting rapid downward movement of water-saturated sand into solution cavities below. In fact, the only sinkhole-related fatality reported in Florida occurred in 1959, when a driller's helper was buried in such a collapse.

The construction of more and more housing developments, factories, shopping malls, and highways is increasing Florida's surface loading. This may increase collapse as new solution cavities extend upward to the surface.

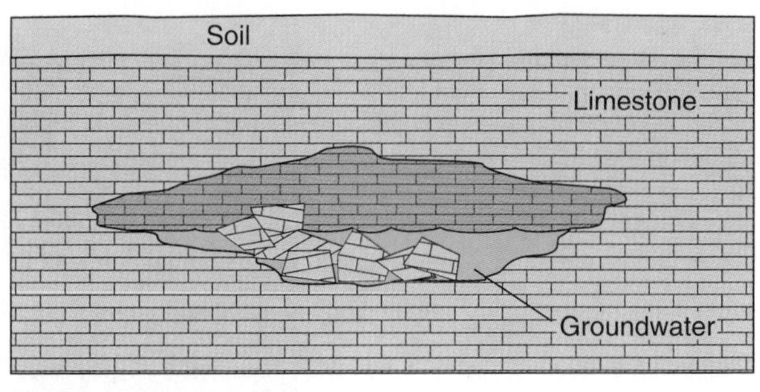


(a)

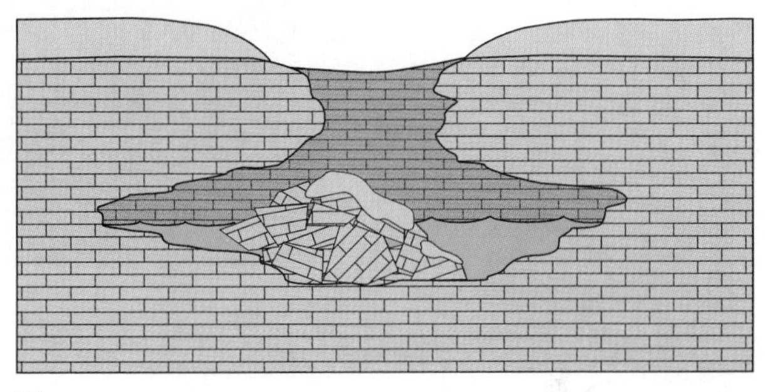

(b)

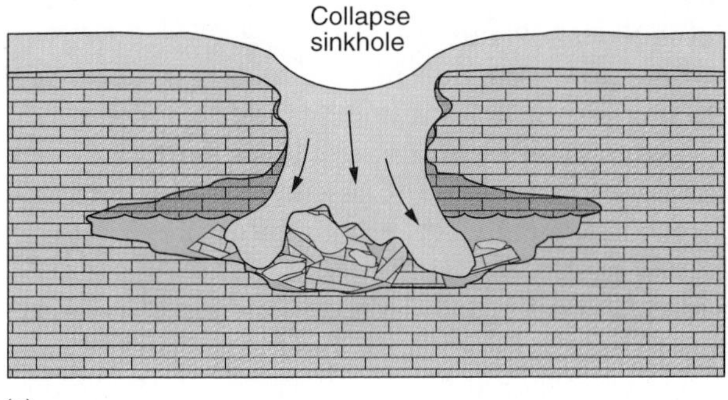


(c)

**FIGURE 10–22**
Formation of a collapse sinkhole. (a) Solution cavity develops and expands over time, undermining support for overlying rocks. (b) Overlying rocks and soil collapse into solution cavity below. (c) More soil and sediment wash in and fill the cavity. The fill subsides, causing a collapse sinkhole to form at surface.

### Recognition and Mitigation of Sinkholes

Where sinkholes might develop can be predicted by noting land subsidence and surface cracking patterns and by determining where voids exist in the subsurface rocks. *When* a sinkhole might develop, however, is not precisely predictable. Subsidence in the land surface can be detected and monitored through altitude surveys. Structural cracking may indicate that ground subsidence is in progress (Fig-

**FIGURE 10–23**
Cover-collapse sinkhole formed by water table drop due to discharge for freeze protection and irrigation in Pierson, Florida. (P. E. La Moreaux & Associates, Inc.)

ure 10–20). Ground cracks also may indicate imminent collapse. Crack patterns that form closed ellipses or circles are especially suggestive of sinkhole development (Figure 10–25).

But how can subsurface voids be detected? Drilling of test holes is one method, but it is time-consuming and expensive, and holes must be drilled close enough together to assemble an accurate underground picture. A promising improvement is *ground-penetrating radar (GPR)*. The GPR antenna is placed directly on the surface and microwave energy pulses are radiated into the ground. These pulses are reflected back from layers. Layers of different composition reflect different characteristics. The reflections are recorded graphically (Figure 10–26a).

GPR recordings reveal not only subsurface layering, but cavities and shafts developed in the bedrock. Use of GPR reduces the need for drilling, which can be used more cost-effectively to determine the character of the sediments that fill the cavities and shafts (Figure 10–26b) and their potential for causing cover-collapse.

The possibility of collapse always must be considered and evaluated in an area underlain by limestone. Failure to do so could result in structural cracking, destruction, or leaks from reservoirs or lakes. Leaks are especially serious under landfills and sewage lagoons because the leaking liquid may also contaminate groundwater below.

For example, in 1978 in West Plains, Missouri, three large sinkholes developed below a sewage treatment lagoon, allowing *25 million gallons* of sewage to leak into underground caverns and aquifers in southern Missouri and northern Arkansas. This contaminated the underground water supply and at least 700 people became ill. Twenty-five National Guard tank trucks were required to supply pure water to isolated residents in the West Plains area.

A sinkhole that developed on June 29, 1994, near Mulberry in central Florida posed both a collapse hazard and a water pollution hazard. The sinkhole was 40 meters (130 feet) wide and 90 meters (295 feet) deep with steep walls. Unfortunately it developed under a pile of gypsum (calcium sulfate) at a fertilizer plant. A portion of the pile collapsed into the sinkhole, leading to concern that it would contaminate the aquifer (Chapter 8) that provides drinking water for the area.

## LOOKING AHEAD

From the subsidence and collapse of Earth's surface in this chapter, we look upward, to the geohazards in the atmosphere. Acid precipitation, the greenhouse effect, ozone depletion, air pollution, radon gas—each is a serious problem that affects many of us, or promises to do so.

**FIGURE 10–24**
Cover-collapse sinkhole triggered by water well drilling, Spring Hill, Florida. (P. E. La Moreaux & Associates, Inc.)

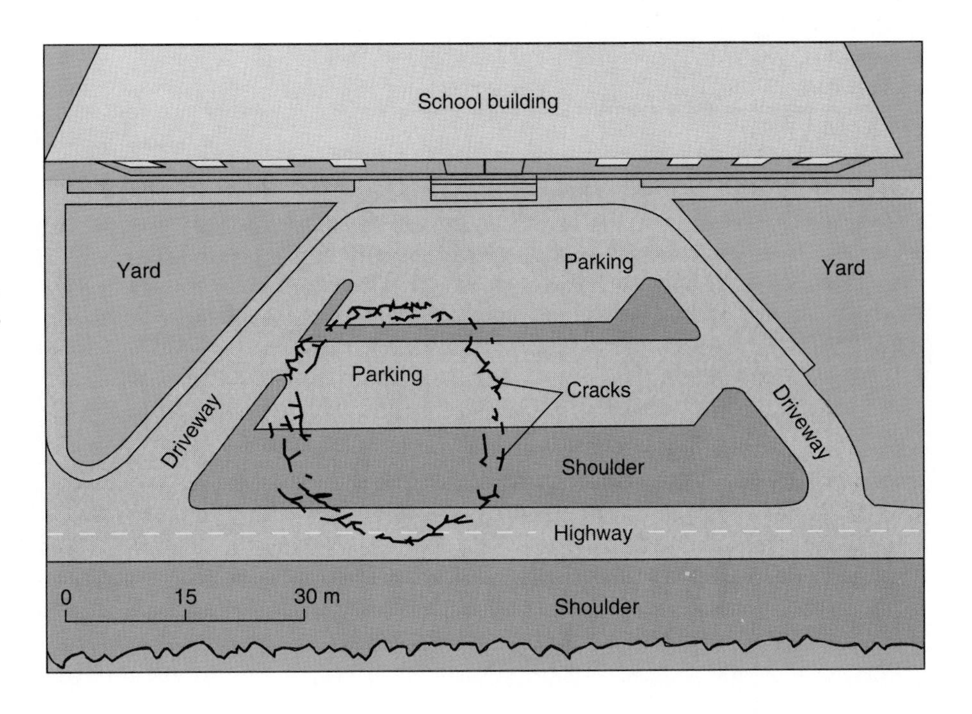


**FIGURE 10–25**
Map of incipient sinkhole that developed in the parking lot of a Florida high school. The discontinuous circular crack pattern outlines the sinkhole. (Modified from Wilson and Beck, 1988, Evaluating sinkhole hazards in Mantled Karst Terrane, Proc. Geotechn. Aspects of Karst Terrains, American Society of Civil Engineers, Fig. 5, p. 9. Used with permission.)

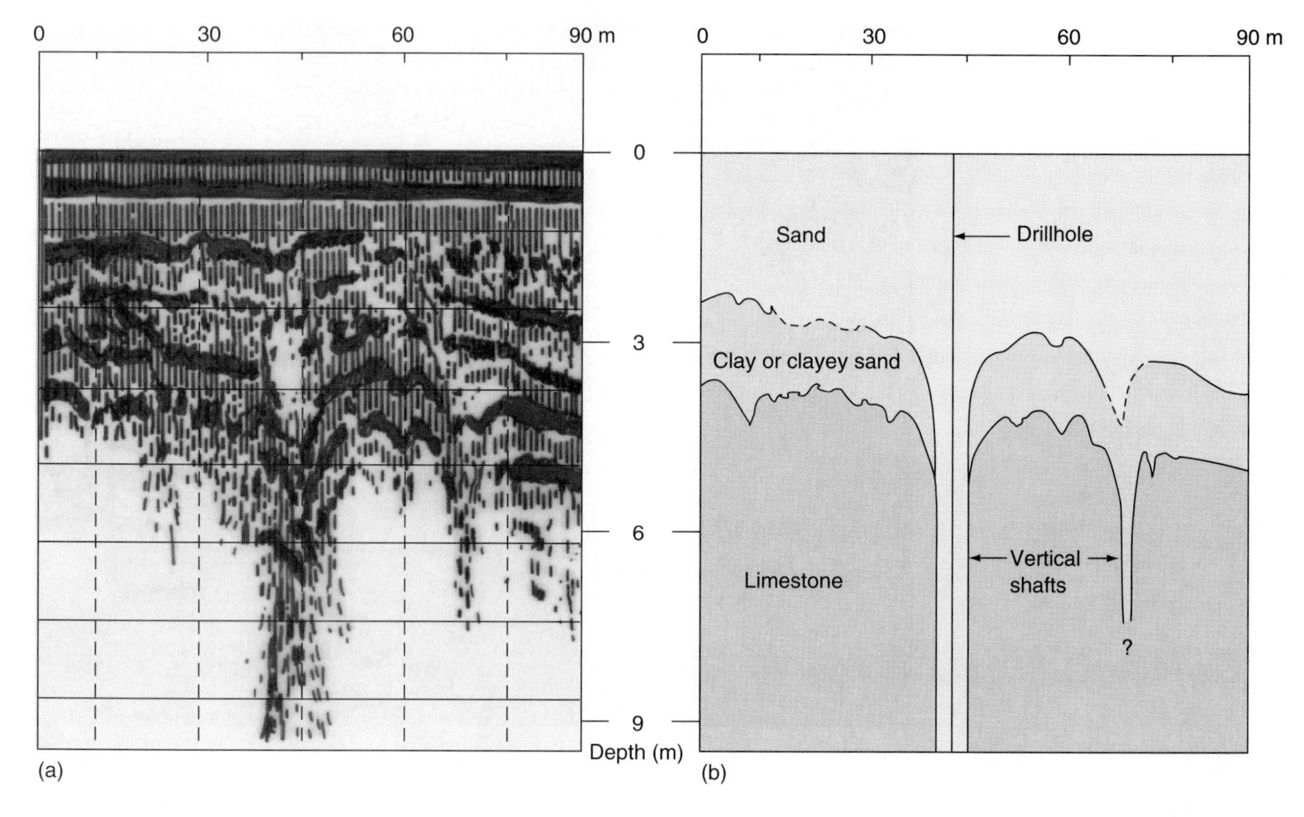


**FIGURE 10–26**
Cross section showing underground cavity on ground-penetrating radar (GPR). The profile shows a portion of a newly excavated landfill cell near Gainesville, Florida. (Modified from Wilson and Beck, 1988, Evaluating sinkhole hazards in Mantled Karst Terrane, Proc. Geotechn. Aspects of Karst Terrains, American Society of Civil Engineers, Fig. 9, p. 17. Used with permission.)

## SUMMARY

Subsidence is a slow lowering of the land surface. Collapse is a relatively fast opening of the land surface and movement of surface materials into cavities below. Subsidence may be caused by compaction of compressible material such as water-saturated clays, drying of shrink-swell clays, or dewatering, oxidation, and compression of organic materials such as peat.

The most areally extensive subsidence occurs when fluids such as petroleum or groundwater are removed from the pore spaces of sediments and sedimentary rocks. This is a serious problem in many rapidly growing areas such as Las Vegas, Houston-Galveston, San Jose, and southern Florida.

Subsidence also results from collapse into cavities at depth. Such subsidence is most common where rocks that overlie an underground mine collapse into the mine. Thick, strong overlying rocks may limit the problem to subsidence and surface cracks. Weaker overlying rocks or a mine nearer to the surface may experience collapse or dramatic subsidence.

In room-and-pillar mining, surface effects may be delayed for years, especially if large pillars are left throughout the mine. In longwall mining, surface subsidence occurs in a progressive wave as mining proceeds underground.

Surface subsidence results in structural cracking, fissures and faults in the surface, changing land slope, interference with water flow through pipes and culverts, ponding of surface water, and changes in groundwater availability. Subsidence greatly increases the danger from coal mine fires, by cracking the overlying rocks to provide a pathway for oxygen to reach the fire.

Collapse is a natural process that becomes a serious geohazard in developed areas underlain by limestone. Acidic groundwater moves through joints in the rock and enlarges them into fissures. Where fissures intersect, large cavities may develop. Differences in the character and thickness of overlying sediments cause four different types of sinkholes. Where limestone is near the surface, solution sinkholes or collapse sinkholes may form. Where thick layers of sand overlie the limestone cover, cover-subsidence sinkholes form where sand moves slowly into cavities below. If the sand moves down faster into cavities, far more serious cover-collapse sinkholes form.

Human activities can promote collapse in limestone areas, particularly Florida. As more surface structures are built, they create greater pressure on the rocks weakened by cavity development below. Channeling of surface drainage below may wash sand into cavities, also resulting in rapid cover-collapse.

Increasing groundwater demand draws down the water table, lowering the pressure in confined limestone aquifers. This allows the water pressure in unconfined sand aquifers above to drive sand downward through fissures in the limestone. Such drawdown usually occurs in drier months but also occurs in winter months when water is sprayed onto crops to prevent frost damage.

Collapse in limestone areas can be predicted by noting surface warping, decreases in elevation of surface features, circular ground cracking patterns, or by using drilling or ground penetrating radar (GPR) to determine the degree to which voids have developed in the limestone below.

## KEY TERMS

angle of draw
collapse
collapse sinkholes
cover-collapse sinkholes
cover-subsidence sinkholes
insoluble residue
longwall mining
room-and-pillar method
sinkholes
solution sinkholes
subsidence
subsidence faults
subsidence fissures

## REVIEW QUESTIONS

1. A high-rise development is to be built on reclaimed marsh land underlain by fine-grained sediments that overlay bedrock at depth. What subsidence problems could develop? How could they be avoided?
2. When peat bogs are drained for agriculture or building, subsidence develops. Give three reasons why.
3. How can subsidence (and land bulging) be minimized when developing an area underlain by shrink-swell clays?
4. Explain why subsidence is an especially serious problem in coastal areas and give an example.
5. Describe the difference in surface subsidence that results from (a) room-and-pillar mining and (b) longwall mining.
6. What factors determine the kind, degree, and area of surface disruption that occurs from an underground mine? (Assume that you know the area of the mine and its depth below the surface.)
7. Describe five different problems resulting from surface subsidence.
8. (a) How do coal mine fires start? (b) How does subsequent subsidence affect them?
9. How can surface subsidence effects above an old coal mine be minimized by (a) actions on the surface and (b) changes made to the void below?
10. Describe ways in which human activities have resulted in sinkhole development, subsidence, and collapse in limestone areas such as Florida.
11. Describe the differences in geologic conditions that result in the development of (a) solution sinkholes, (b) collapse sinkholes, (c) cover-subsidence sinkholes, and (d) cover-collapse sinkholes.
12. What three techniques can be used to analyze the potential for development of collapse sinkholes in a limestone area?

## FURTHER READINGS


American Institute of Professional Geologists (AIPG), 1993, The citizens guide to geologic hazards, landslides and avalanches: Arvada, Colo., 134 p.

Beck, B. F., 1988, Environmental and engineering effects of sinkholes, the processes behind the problem: Environmental Geology and Water Science, v. 12, no. 2, p. 71–78.

Dolan, R., and Goodell, H. G., 1986, Sinking cities: American Scientist, v. 74, p. 38–47.

Dougherty, P. H., and Perlow, M., Jr., 1987, The Macungie Sinkhole, Lehigh Valley, Pennsylvania, cause and repair: Environmental Geology and Water Science, v. 12, no. 12, p. 89–98.

Dunrud, C. R., and Osterwald, F. W., 1986, Effects of coal mining subsidence in the Sheridan, Wyoming, area: U.S. Geological Survey, Prof. Paper 1164.

Holzer, T. L. (ed.), 1984, Man induced land subsidence: Geological Society of America Reviews in Engineering Geology, v. VI, 232 p.

Marsden, S. S., and Davis, S. N., 1967, Geological subsidence: Scientific American, v. 216, no. 6, p. 93–108.

Newton, J. G., 1987, Development of sinkholes from man's activities in eastern U.S.: U.S. Geological Survey Circular 968.

Poland, J. F., and Davis, G. H., 1969, Land subsidence due to withdrawal of fluids, *in* Varnes, D. J., and Kiersch, G., eds.: Reviews in Engineering Geology, Geological Society of America, p. 187–269.

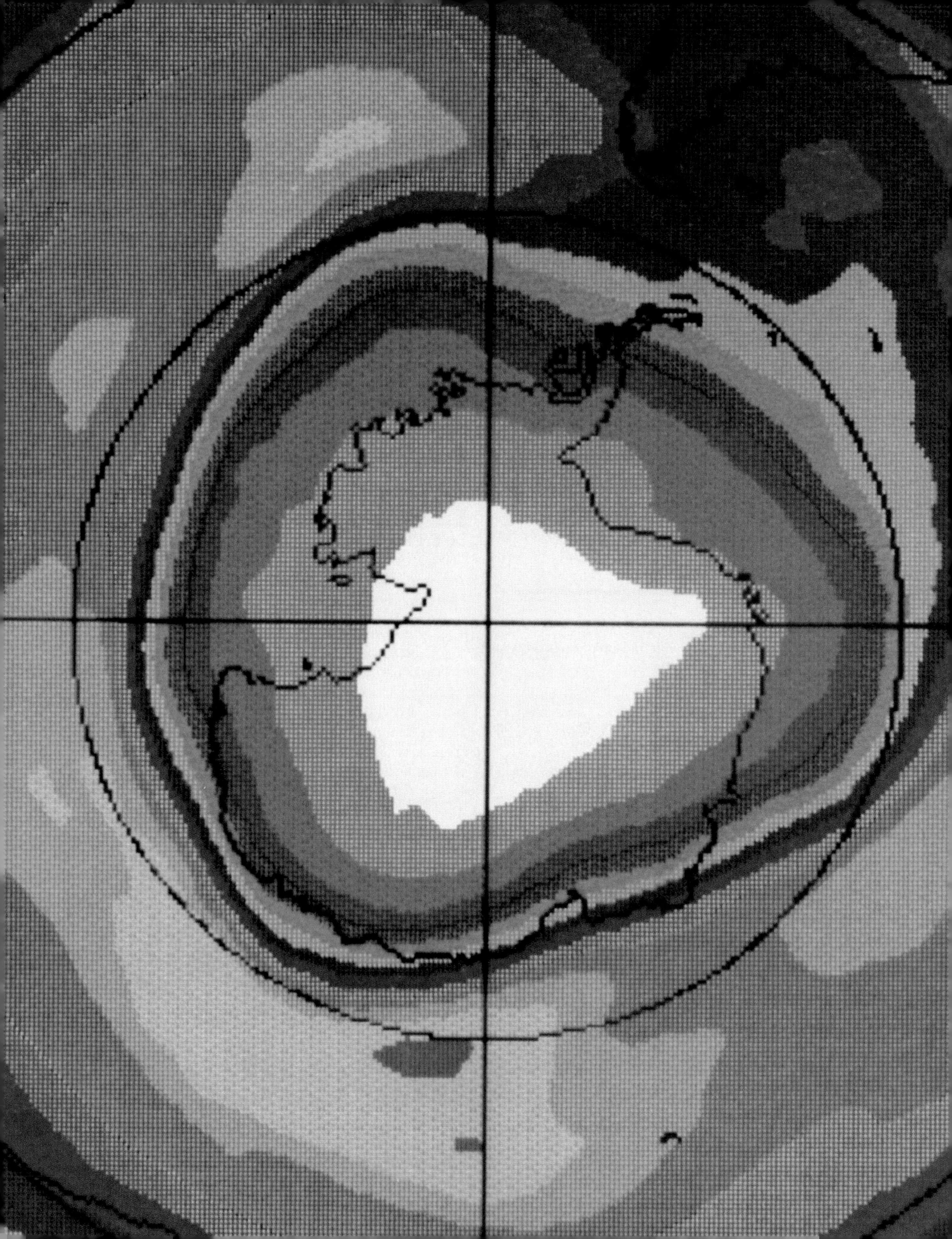

# 11

# *Atmospheric Geohazards*

- ☐ Fish are dying at unusually high rates in remote mountain lakes in the northeastern U.S.
- ☐ Some marble statues and limestone facings on buildings are pitting and crumbling faster than they used to.
- ☐ Climate patterns appear to be changing.
- ☐ Sea level is rising faster than expected.
- ☐ Ultraviolet radiation is increasing as ozone levels in the upper atmosphere are decreasing.

These are just a few consequences of the increasing degradation of our atmosphere. People typically think of air pollution as causing breathing difficulty, and it can. But air pollution has far more serious consequences for the atmosphere, lithosphere, hydrosphere, and the biosphere—all living things. Air pollution is especially serious because regional and global wind patterns disperse pollutants far from their sources.

For example, acid rain, caused by sulfur and nitrogen gases in the air, was recognized as a geohazard in Scandinavia in the mid-1940s. More than 70% of the sulfur in Sweden's air is anthropogenic, or resulting from the influence of human beings on nature. Of that 70%, about 77% comes from *outside* Sweden. Suspected sources (burning coal, ore smelting) are hundreds of kilometers away, in industrialized regions of England and in Germany's Ruhr Valley.

Similar problems exist in North America. Many industries now discharge smoke through taller stacks (up to 300 meters or 1,000 feet) to place pollutants higher in the atmosphere so wind can better disperse them. This reduces local pollution, but the smoke is dispersed by prevailing westerly winds, which blow eastward in the northern United States. This movement simply shifts the air pollution hazard eastward (Figure 11–1), a good example of a *displaced problem,* as discussed in Chapter 1.

The United States and Canada have negotiated intensively to reduce air pollution in the Canadian Great Lakes area. Its source is believed to be in-

**Concentration of ozone over the South Pole on October 6, 1993. The ozone "hole" (area of low ozone concentration) is the white area in the photo center. (NASA)**

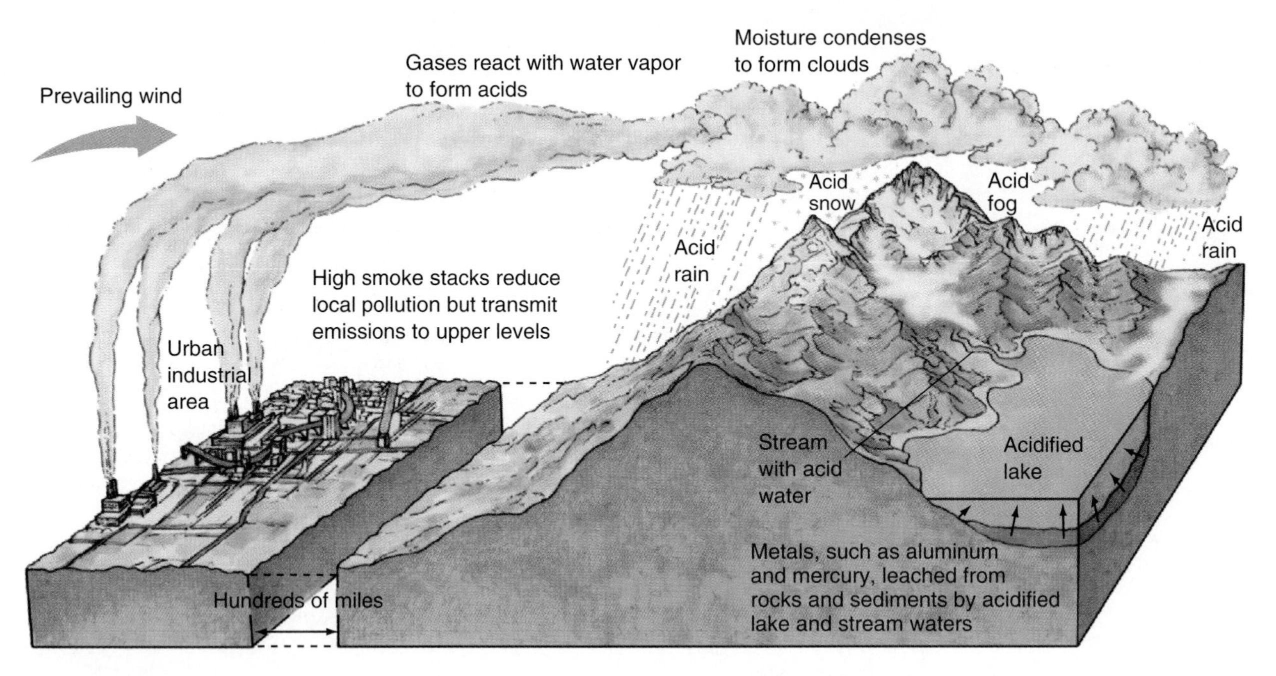


**FIGURE 11–1**
Atmospheric cross section showing the origin and dispersal of acid precipitation.

dustrial emissions from the north-central United States, primarily Missouri, Illinois, Kentucky, Indiana, Ohio, West Virginia, and Pennsylvania.

In this chapter, we will examine five of the most compelling atmospheric geohazards: acid precipitation, the greenhouse effect, ozone depletion, smog, and radon gas.

## ACID PRECIPITATION

By burning fossil fuels such as oil, natural gas, and coal, the United States annually discharges approximately *50 million metric tons of sulfur and nitrogen oxides* into the atmosphere, according to the U.S. Environmental Protection Agency (EPA). These *primary pollutants* include mainly sulfur dioxide and nitrogen oxides. In clouds, chemical reactions between water and these gases convert the oxides into *secondary pollutants,* such as sulfuric and nitric acids.

As moisture in these clouds condenses, **acid precipitation** occurs. Depending on the temperature, it falls either as **acid rain** or **acid snow,** or it may mantle mountaintops in *acid fog*. The acids also may settle from the atmosphere as solids or on dust particles in the atmosphere. This dry fallout is called *acid deposition*.

### Rain and Snow Acidity

How *acidic* is acid precipitation? Acidity varies with conditions, from rainfall to rainfall and from place to place. The acidity or alkalinity of rain or any other liquid reflects the concentration of hydrogen ions (**$H^+$**) in the solution (see Chapter 2). This concentration is expressed on the **pH scale** with a number between 0 and 14, as shown in Figure 11–2. A pH approaching 0 is increasingly acidic. A pH approaching 14 is increasingly *basic* or *alkaline*. In the middle of the scale, a pH of 7 is *neutral*—neither acidic nor alkaline.

A very strong acid, such as the sulfuric acid in an automobile battery, has a pH of 1.0. A very strong alkali, such as lye, has a pH of 13.0. The pH scale is not linear but changes tenfold with each unit. For example, a liquid with a pH of 3 is ten times more acidic than a liquid with a pH of 4, a hundred times more acidic than one with a pH of 5, a thousand times more acidic than one with a pH of 6, and so

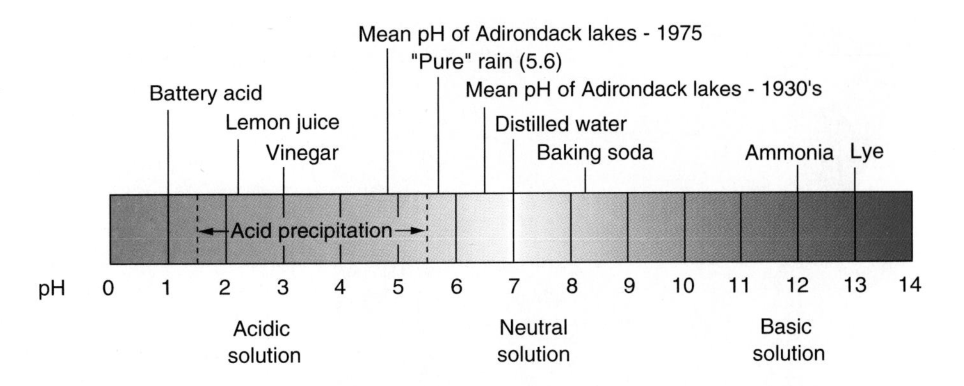


**FIGURE 11–2**
Scale showing variations in pH and the acidity and alkalinity of common fluids. (The term *pH* abbreviates "potential of hydrogen.")

on. Viewed another way, a pH of 3 is one-tenth as acidic as a pH of 2, and so on.

Normal rain and snow are seldom at a neutral pH of 7. They are mildly acidic with a natural pH around 5.6, resulting when $CO_2$ from the air is dissolved in the rain to form a weak acid (carbonic acid, or $H_2CO_3$). The U.S. Environmental Protection Agency defines **acid precipitation** as that having a pH less (smaller number) than 5.6.

The Adirondack Lakes in upper New York State illustrate how the steady input of acidic precipitation can alter a lake's chemistry. In 1930, the mean pH of several Adirondack lakes was 6.5 (almost neutral). By 1975, the value had changed to 4.8, significantly more acidic. Over a 45-year period, the lake waters had become about 100 times more acidic. These pH changes had a profoundly negative effect on aquatic animals and plants in the lakes.

If acid rain is harmful, acid snow is worse. As water freezes, the crystallizing ice excludes everything but the water. Loosely speaking, the pollutants are "squeezed out" and become concentrated on the outside of the ice crystals. The first melting of a snow pack removes a large part of these pollutants, forming a concentrated "slug" of acid water. Panther Lake, in the western Adirondack Mountains of New York State, normally has a neutral pH of about 7. But when the snow pack melted in the spring of 1979, the pH dropped to 5.0; thus the acidity had increased 100 times.

## Effects of Acid Precipitation

***Lakes, Plants, and Animals.*** Acid precipitation is causing tree loss, leaf and plant mottling, and reduced growth rates in the Appalachian Mountains (Georgia to New England) and Adirondack Mountains. Europe's forests have been affected similarly, with an estimated productivity decrease of 16% and damage of $30 million per year.

Conifers such as pine and spruce are most heavily affected because their evergreen needles do not fall each autumn like the leaves of deciduous trees. Conifer needles remain exposed all year. With continued exposure to acid precipitation, the needles turn yellow or brown and fall off. Acid precipitation breaks down the oily coating on leaves and needles that helps to hold in moisture and protect trees from insects and disease.

Acid precipitation also has serious effects on aquatic life. The pH of lake waters affects many life processes in aquatic animals. Many aquatic organisms tolerate acidified water (pH of 5.5 to 4.5), but few can tolerate water more acidic than a pH below 4.5. For example, fish cannot live in water that has a pH below 4.5.

Acidic water can leach harmful elements from some rocks. Mercury is one example; its level has increased in the tissue of fish that dwell in acidified lakes in the Adirondack Mountains of New York State. Aluminum is another example. Acid waters dissolve aluminum from soil minerals. The metal travels in acidified soil waters into lakes, increasing their aluminum levels (Figure 11–1). Aluminum concentrations usually are less than 0.1 part per million (ppm). However, acidified waters in the Adirondacks have a concentration of 0.2 to 1.0 ppm, and 0.2 ppm is toxic to fish. Plants also are affected adversely by aluminum.

Acidified lakes may look clear and blue from the air, but a closer look shows them to be barren of most life except algae. Adults of many aquatic species are more tolerant of increasing acidification, but their eggs and developing young are not. Even-

**FIGURE 11–3**
This 150-year-old Carrara marble column in Philadelphia has been damaged by acid rain on its exposed side. Note that on the sheltered side the detail is preserved though darkened by soot. (Photo by L. H. Nelson, U.S. Park Service.)

tually, species after species is extinguished because its reproduction is hampered.

***Buildings, Textiles, Paper, and Metals.*** Acid precipitation can decompose popular building materials such as concrete, marble, and limestone (Figure 11–3). The acid water dissolves calcium carbonate and etches the stone's surface. Sometimes the calcium carbonate that composes marble and limestone reacts with sulfuric acid in the rain. This forms a calcium sulfate mineral called gypsum ($CaSO_4 \bullet 2H_2O$), which is weaker. The outer part of the stone may crumble.

Acid precipitation accelerates weathering of surfaces. Damage may be less dramatic than the crumbling of a famous statue or the destruction of a landmark building, but its aggregate cost is far greater. For example, car finishes and house paints erode and fade, textiles deteriorate and fade, paper becomes discolored and brittle, rubber cracks and loses strength, and metals become pitted and tarnished when exposed continually to acid precipitation.

## Sensitivity to Acid Precipitation

The geologic characteristics of some areas naturally enhance or moderate the effect of acid precipitation. Consequently, some areas are more sensitive to acid precipitation than others. The EPA has mapped the varying sensitivity to acid precipitation across the United States (Figure 11–4). More sensitive areas tend to be underlain by silica-rich rocks such as granite, or to have pine forests with acidic soils. Areas underlain by limestone tend to become less acidified because the limestone is alkaline and reacts with the acid precipitation. This tends to neutralize the acidity and to raise its pH (Figure 11–2).

This differential acidification can be seen in northeastern New York State. Here, both naturally acidic rocks (silicates) and naturally alkaline rocks (carbonates) underlie different lake basins in an area

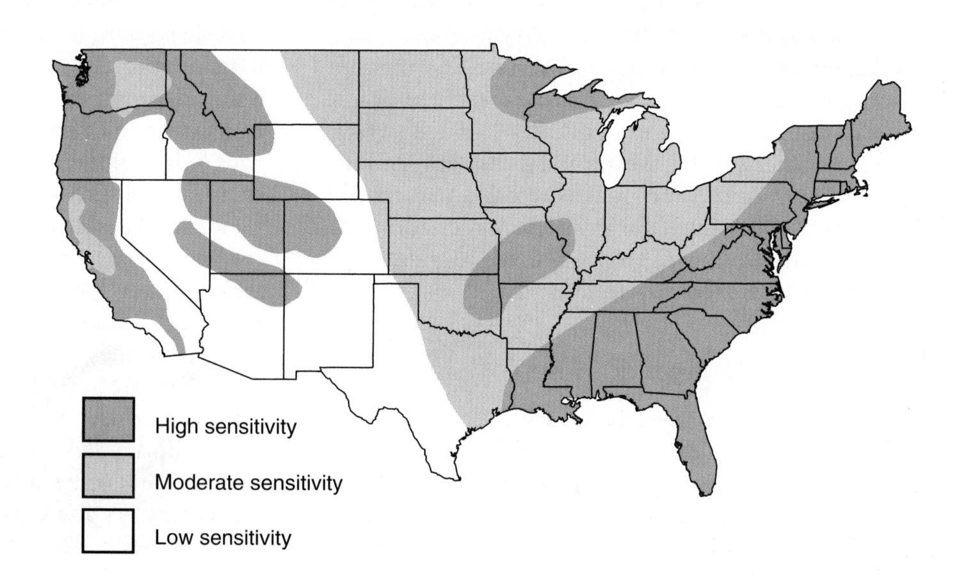


**FIGURE 11–4**
Map showing areas of different sensitivities to acid precipitation. The sensitivity varies with the soil composition, climate, and types of vegetation. (From U.S. Environmental Protection Agency, 1980.)

that experiences intensive acid precipitation. In the lakes underlain by limestone, fish survive, whereas their populations may be severely restricted or even eradicated in lakes underlain by granite.

### Mitigating Acid Precipitation

The effect of acid precipitation can be lessened by certain practices, both at its source and at sites affected by the acidity. Emissions can be reduced at the source by burning lower-sulfur fuels in power plants and industrial complexes. The United States has abundant supplies of "low-sulfur" coal (less than 1% sulfur by weight), concentrated in Montana, Wyoming, Colorado, and Utah, and some in West Virginia and Kentucky. Unfortunately, the western coal has a lower heating value, and the cost of transporting it to eastern power plants limits its use.

Emissions can be reduced at the source by cleaning waste gases with "scrubbers" that remove harmful substances. In a scrubber, gases are passed through a spray of water and lime before being emitted from smokestacks. Sulfur oxides ($SO_x$) in the gases react with the lime ($CaO$) to produce calcium sulfate ($CaSO_4$), a harmless solid.

Reducing harmful emissions with scrubbers is expensive. For example, Northern Indiana Public Service Company installed scrubbers that remove 90% of sulfur pollutants from its coal-fired electric generating plant, but the cost exceeded $100 million. They added scrubbers so they could continue to buy coal from mines in nearby Illinois, where the coal contains higher levels of sulfur (around 3%). But many other utilities have elected to import cleaner-burning low-sulfur coal from other states. Consequently, large mines in Illinois are closing, costing hundreds of mining jobs and millions in lost tax revenue.

At an affected site, damage can be reduced by adding crushed limestone to acid water bodies. Limestone is alkaline, so it *buffers* or reduces the acidity of the water. Liming of acidified lakes is only a temporary solution, however, and it is very expensive, considering the number of affected lakes.

## THE GREENHOUSE EFFECT

Even in cold climates, plants grow during winter in glass-walled greenhouses. The sun's short-wavelength radiant energy passes through the greenhouse glass and is converted to heat energy when it strikes surfaces inside. This heat energy has longer wavelengths (infrared) that cannot pass through the glass, and the energy becomes trapped inside the greenhouse. This trapped heat energy warms the air inside, creating an artificially warm climate (Figure 11–5).

Gases in the atmosphere can act in a similar fashion to the glass in a greenhouse by preventing heat from being reflected back into space. Increases in natural and anthropogenic carbon dioxide and other atmospheric gases are trapping more and more of Earth's reflected heat. This process is causing a gradual increase in atmospheric temperature that we call the **greenhouse effect.**

The greenhouse effect keeps Earth habitable. However, anthropogenic activities have exacerbated the greenhouse effect, and we now face a potentially serious problem.

The extent to which the greenhouse effect is a problem is quite controversial. So are the possible consequences if we fail to start controlling the emission of greenhouse gases. Some scientists see no problem, whereas others forecast catastrophic changes to human health, food sources, climate, and water supplies.

### Origin of Greenhouse Gases

Gases involved in the greenhouse effect are primarily water vapor ($H_2O$), carbon dioxide ($CO_2$), methane ($CH_4$), nitrogen oxides ($NO_x$), and chlorofluorocarbons (CFCs). The CFCs are strictly anthropogenic. The other gases are both natural and human-generated. The amount of $CO_2$ in the atmosphere is of particular concern. It has increased markedly because of the burning of fossil fuels such as coal, petroleum-based fuels (gasoline, kerosene, diesel), and natural gas. In tropical forests, significant $CO_2$ also is released when trees are cut and the debris is burned or left to rot. Another source is organic material in exposed soil.

These anthropogenic changes have markedly increased the $CO_2$ concentration in the atmosphere from a 1958 average of 315 parts per million (0.0315%) to a 1980 average of 338 parts per million (0.0338%). The rate of $CO_2$ released into the atmosphere will only increase as more fossil fuels and tropical rain forests are burned.

The other gases play a lesser role. Although their emission rates are a fraction of the $CO_2$ rate, they are a problem because they absorb infrared radia-

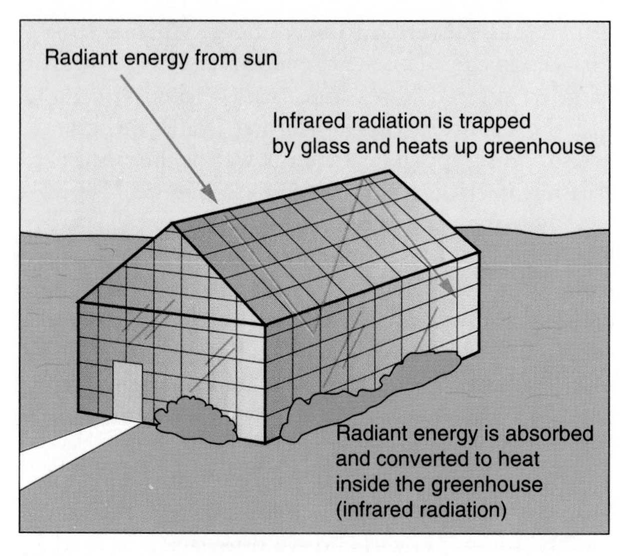


(a) Greenhouse

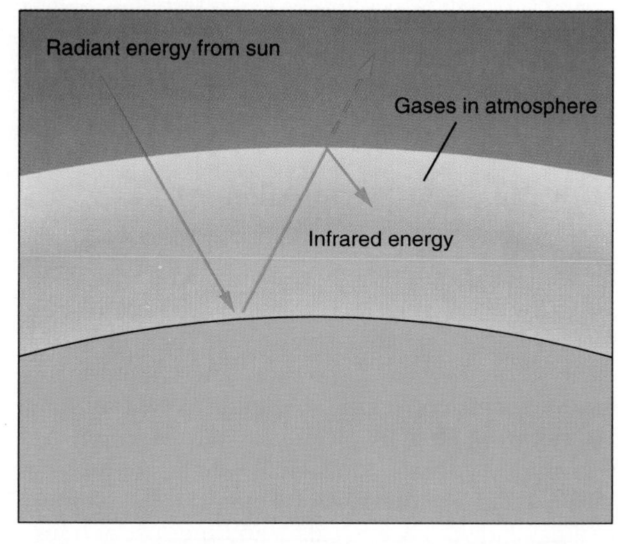


(b) Greenhouse effect in atmosphere

**FIGURE 11–5**
The greenhouse effect mechanism. (a) In a greenhouse, the glass admits solar radiation, which strikes surfaces and is converted to heat. It is reradiated as infrared (heat) energy, which cannot penetrate the glass readily. Thus it prevents much of the energy from returning to space and heats the air inside. (b) In a polluted atmosphere, the gases absorb some solar radiation and heat the atmosphere. Radiation that passes through to Earth's surface is converted to heat, which further raises the temperature of the atmosphere.

tion 50 to 100 times more effectively than $CO_2$. The contribution that major groups of natural and anthropogenic atmospheric gases have made to the greenhouse effect is shown in Figure 11–6.

The greenhouse effect is nothing new or unnatural. It has functioned since Earth developed its present atmosphere. Long before humans evolved onto the scene, decaying vegetation, volcanoes, and animal waste were producing gases that contribute to the greenhouse effect.

What worries many scientists today is the *marked increase* in the greenhouse effect noted in recent years. The growth of human population and our need for food, fuel, and products have greatly increased greenhouse gas emissions. For example, increased agriculture, livestock production, and sewage treatment all produce more methane ($CH_4$). The burning of carbon-rich fossil fuels continues to increase levels of carbon dioxide ($CO_2$).

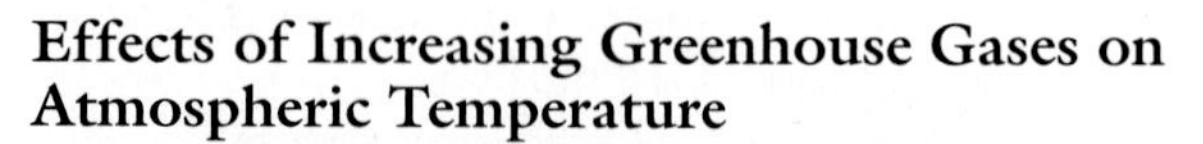

## Effects of Increasing Greenhouse Gases on Atmospheric Temperature

At Mauna Loa in Hawaii, atmospheric $CO_2$ levels have been monitored since 1958, and the data re-

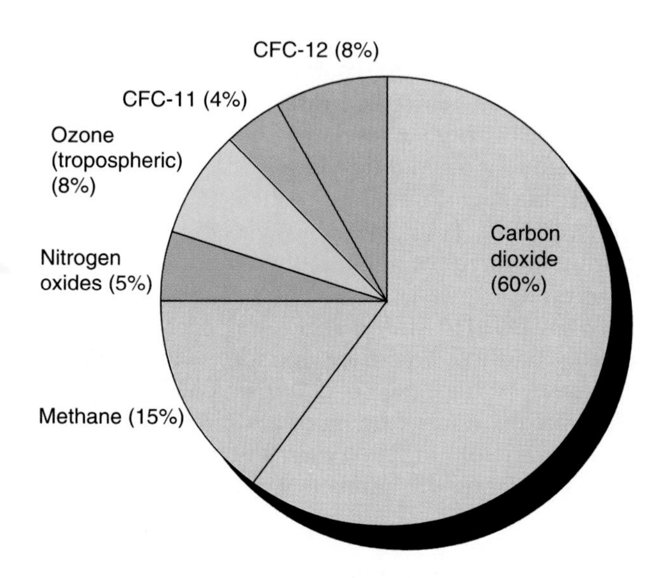


**FIGURE 11–6**
Contributors to the greenhouse effect during the 1970s. (Data from H. Rodhe, 1990, A comparison of the contribution of various gases to the greenhouse effect: Science, v. 248, p. 1218–1219.)

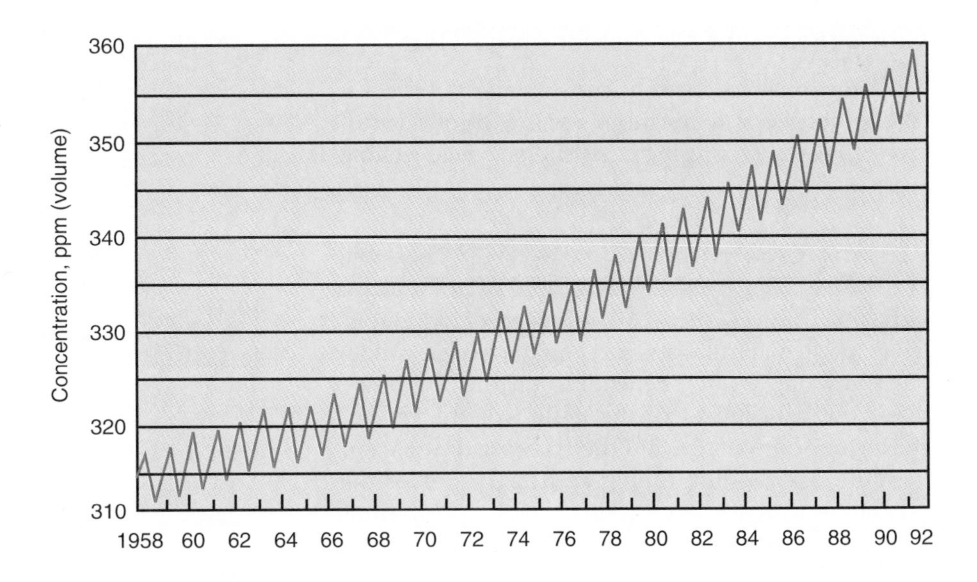


**FIGURE 11–7**
Graph showing rising $CO_2$ levels measured at Mauna Loa Volcano, Hawaii. (From C. D. Keeling, R. B. Bacastow, A. F. Carter, S. C. Piper, T. P. Whorf, M. Hermann, W. G. Mook, D. J. Moss, and H. Roeloffzen, Scripps Institution of Oceanography, National Oceanic and Atmospheric Administration)

veal steadily increasing values (Figure 11–7). Such measurements are significant because the observation point is on a mid-Pacific island that is not industrialized and thus should represent average atmospheric conditions in the Northern Hemisphere.

Recent research reveals a close relation between increasing $CO_2$ levels and atmospheric temperature. The average worldwide surface temperature in 1990 was nearly 16°C (60°F). This was the highest recorded since record-keeping began in the nineteenth century. In addition, seven of the warmest years since 1880 have occurred in the last eleven years.

Some scientists suggest that this process is self-perpetuating. As $CO_2$ levels cause temperatures to rise, plant activity and the decay of organic matter increase, releasing greenhouse gases into the atmosphere and increasing the greenhouse effect (Figure 11–8).

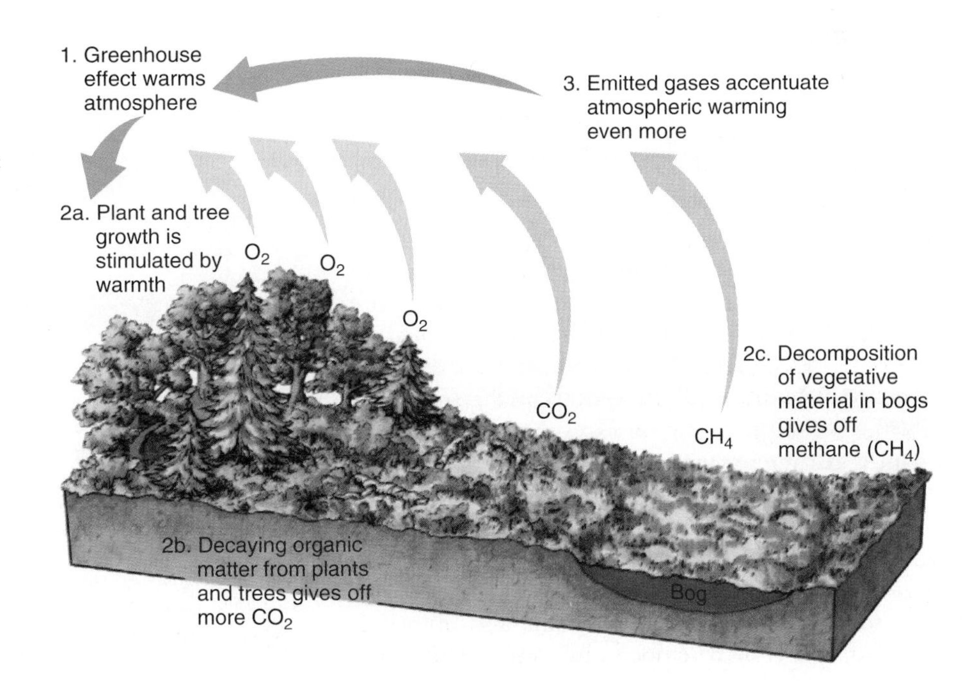


**FIGURE 11–8**
The greenhouse effect may feed itself. Increasing temperature may lead to more gas generation, which further elevates atmospheric temperature. (Based on the research of Dr. George Woodwell, Woods Hole Oceanographic Institution, Woods Hole, Mass.)

## Consequences of the Greenhouse Effect

The most obvious result of the greenhouse effect is that atmospheric temperatures will continue to increase. However, this increase will have far-reaching consequences.

***Climatic Change.*** The *actual* changes that will occur as a result of the greenhouse effect are not known. However, they have been modeled (simulated mathematically) on a computer. Each model has some uncertainty, depending on the technique used, the data included, the design of the model, the assumptions made, and the accuracy of the data.

Although absolute timing and the extent of the greenhouse effect are controversial, projections from different models do show similar trends. Assuming that current $CO_2$ increases continue, $CO_2$ levels will double by the year 2050. As a result, models predict an overall warming of 1.5°C to 4.5°C (3 to 8°F). Temperatures will rise faster at middle and lower latitudes and less in equatorial regions. These models suggest that the effect will be greater in the Northern Hemisphere than in the Southern Hemisphere (Figure 11–9).

Major shifts in rainfall patterns are predicted. For example, northern Africa will become wetter, ameliorating its present desert conditions. The central United States and south-central Canada will become drier. Unfortunately, this area is one of the most productive agricultural regions on Earth. Currently, parts of this region are affected periodically by drought, and rainfall is marginal now. Predicted higher temperatures and lower rainfall in the future will dry the surface. Wind erosion will remove the soil, converting "America's breadbasket" into a new Dust Bowl (see Chapter 6).

***Effect on Plants and Animals.*** A temperature increase in agricultural areas where rainfall is adequate may cause faster growth, but it also may create a problem: plant disease organisms and insect pests that normally are killed by winter cold may survive warmer winters. These diseases and insects may claim a greater toll on the harvest, increasing year-round plant damage.

Geologists have established that repeated climatic changes in Earth's history were accompanied by migration of plants and animals. This reflects their effort to survive, because each plant and animal is adapted to a certain temperature range. These past

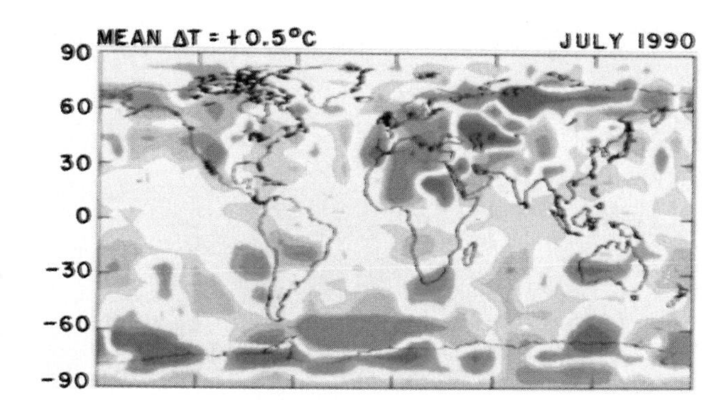


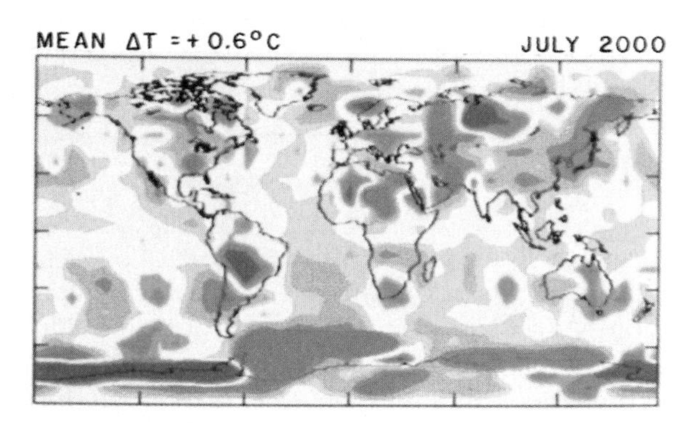


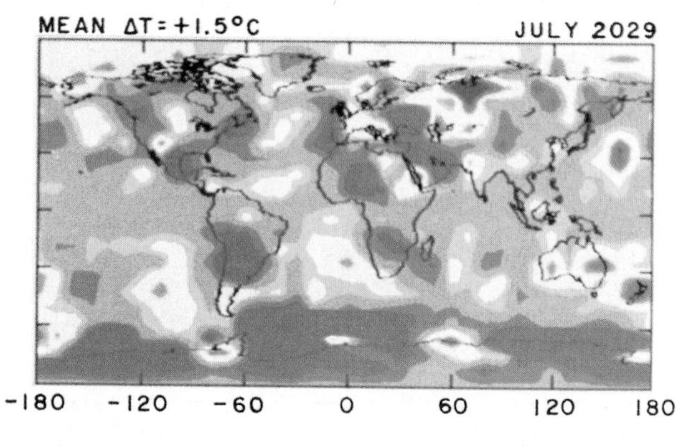


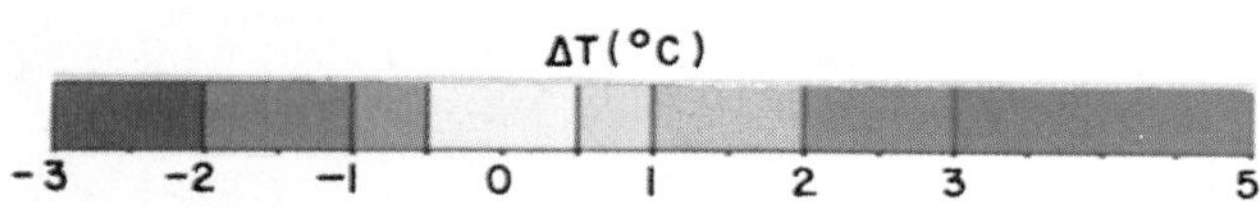


**FIGURE 11–9**
Temperature changes caused by global warming projected for the month of July in the years 1990, 2000, and 2029. This model is based on a scenario in which human activity is modified to freeze emissions at current levels. This merely delays the doubling of $CO_2$ until the middle of the next century. (From J. Hansen, I. Fung, A. Lacis, S. Lebedeff, D. Rind, R. Reudy, and G. Russ, 1988, Global climate changes as forecast by Goddard Institute for Space Studies Three-Dimensional Model: Journal of Geophysical Research, v. 43, no. 8 (August), plate 6. Published by the American Geophysical Union.)

natural temperature changes occurred slowly, over ten of thousands or hundreds of thousands of years. This gradual change allowed sufficient time for plants and animals to migrate.

However, the model-predicted greenhouse temperature increases will occur in *less than a hundred years.* Many animals are mobile enough to move with the shifting climate. But plants may not be able to migrate fast enough. (They "migrate" by the spread of their seeds and spores. Those that land where conditions are unsuitable will perish; those that land in a hospitable environment will survive. In this manner, plant species slowly shift their location, generation by generation.) Many animals eat plants, so their food may not be able to migrate fast enough with them. This could severely dislocate ecosystems.

Researchers at the University of Vienna reported (*Nature,* June 9, 1994) that plants in the Alps are seeking higher ground in response to Earth's warming temperatures. They compared 1992 temperatures and abundances of primarily flowering plants at altitudes above 3,000 meters (10,000 feet) with data from the early and mid 1900s. Mean annual temperature has increased 0.7°C (1.26°F) since the early 1900s. Nine common plant species have ascended the mountains at a rate of 1 to 4 meters (3.3 to 13 feet) per decade. The researchers concluded that if global temperatures continue to increase, some mountain plants may have nowhere to go.

***Rise in Sea Level.*** The greatest geohazard from the greenhouse effect will occur along coastlines. Warmer temperatures following the last glaciation are causing the slow melting of continental ice masses over Antarctica and Greenland. This meltwater drains into the ocean, resulting in a general sea-level rise. The greenhouse effect may accelerate sea-level rise, because global warming will accelerate ice sheet melting. Further, as ocean water heats, it expands, raising the ocean surface yet higher. This thermal expansion is predicted to contribute more to sea-level rise than the melting of ice. Very few scientists question that sea-level rise has accelerated. The only controversy is over *how fast* it is rising.

Rising sea-level will extend saltwater further onto the land. Freshwater marshes and swamps will become covered with saltier water. Animals and plants that are sensitive to salinity will find their habitats destroyed or displaced.

Serious disruption will occur to drinking water supplies as sea level rises. Coastal cities like New Orleans rely on rivers for part of their water supply. As saltwater moves up these rivers, their supplies will become less potable. Many areas rely on coastal aquifers (Chapter 8) for their water supply. As sea level rises, saltwater will encroach into coastal aquifers, contaminating wells that are the principal water supply for many communities.

Low-lying nations, such as Bangladesh, will be in serious trouble. Low-lying U.S. coastlines from New York to Texas will undergo major dislocation as sea level rises into the next century. For example, a sea-level rise between 0.9 and 1.6 meters (3 and 5 feet) is predicted for Charleston, South Carolina, into the next century. The city is on a low peninsula (elevation below 6.3 meters, or 20 feet), bounded by rivers on two sides, with a bay between it and the ocean.

Complicating this is another geohazard for Charleston: this city and many other low-lying coastal cities on the Atlantic and Gulf coasts are in hurricane-prone areas. In 1989, the weaker side of Hurricane Hugo caused a storm surge over 3 meters (10 feet) above mean sea level in Charleston. Imagine the future damage if the overall sea level rises another meter or more, and if the Charleston area is struck directly by a major hurricane. The problem is even more serious because the warming of the ocean may provide more energy to generate more frequent, more intense hurricanes! (This problem is discussed further in Chapters 15 and 16.)

## Mitigating the Greenhouse Effect

It may be possible to significantly reduce geohazards posed by the greenhouse effect changes. The effect may be slowed by burning less fossil fuel, reducing deforestation, limiting use of CFC aerosol propellants (discussed later in this chapter), and reducing the methane released from landfills (Chapter 12).

A natural phenomenon could delay an increase in global temperatures: volcanic eruptions. The 1991 explosion of Mount Pinatubo volcano in the Philippines produced a haze of sulfur dioxide vapor and a dust cloud. They reflected incoming sunlight and may have lowered average global temperatures up to 0.6°C (1°F) between 1992 and 1995. The Mount Pinatubo explosion, described in Chapter 4, produced an estimated 15 million tons of sulfur dioxide ($SO_2$) gas—twice the amount from Mexico's El Chichón volcano. The Pinatubo cloud also spanned 10° of latitude around Earth, a wider band than El Chichón's 5°.

Volcanic eruptions provide only temporary moderation, of course. Global warming probably will continue, because volcanic emissions are episodic, whereas the greenhouse effect is continuous and cumulative.

## OZONE DEPLETION

In 1985, British scientists published long-term observations, made since 1957, of a curious pattern in the ozone gas layer of the stratosphere over Antarctica. Ozone gas concentration in the layer diminished during the Antarctic spring and early summer (September to December). This phenomenon was creating a thinning of the ozone layer called the **ozone "hole"** (Figure 11–10). Since 1985, the "hole" has expanded in most years. In 1993, it set a record: about 70% of the layer, an area almost the size of North America, vanished over Antarctica.

The ozone layer forms an absorbing shield against hazardous ultraviolet radiation from the sun. Thus, the ozone layer is extremely important to life on Earth, and any thinning of it could spell medical problems, including cancer and genetic damage. You might think that the increasing radiation from the Antarctic hole would affect only penguins and a few hardy plants near the South Pole. However, more recent data suggest that the ozone "hole" is larger than expected and is widening. This could have far-reaching consequences for life at other latitudes, including ours.

### The Ozone Layer and Its Significance

Ordinary oxygen exists as a 2-atom molecule, $O_2$, and makes up about 21% of Earth's atmosphere. Ozone is a variant form, a 3-atom molecule, or $O_3$. Ozone makes up a tiny fraction of Earth's atmosphere and is concentrated in the ozone layer in the stratosphere. (To understand how little ozone exists in the atmosphere: if it all could be gathered at

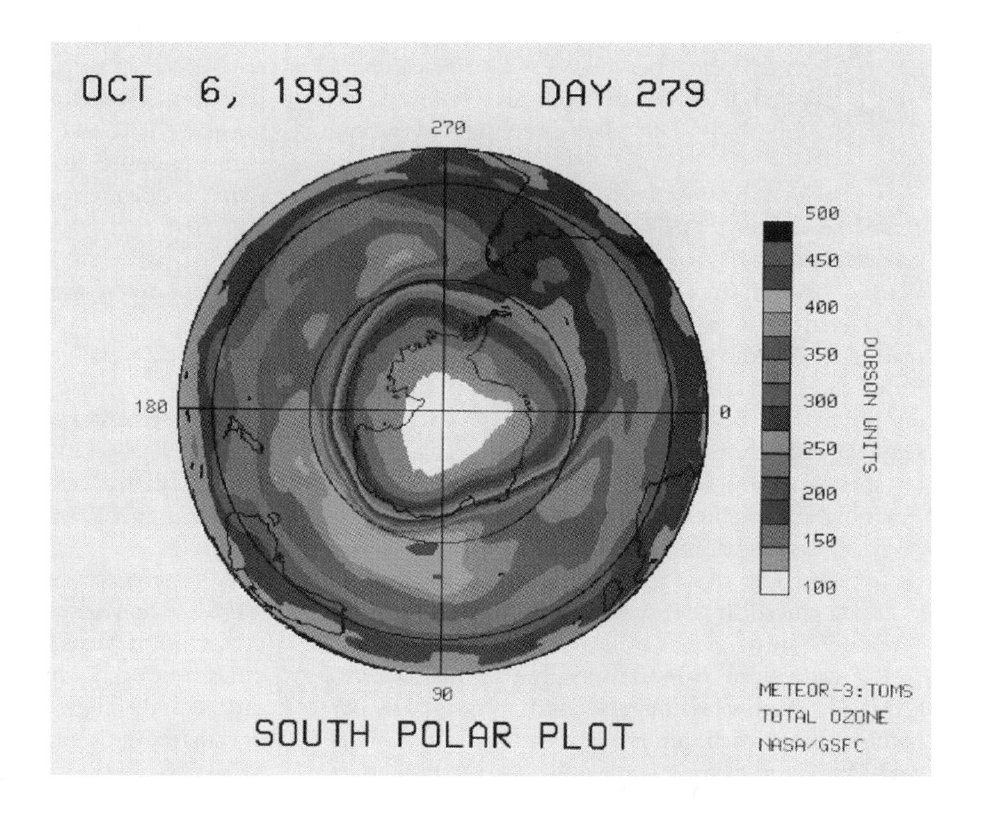


**FIGURE 11–10**
Nimbus-7 satellite view of ozone depletion (central area) over Antarctica on October 6, 1993. (NASA)

Earth's surface, it would form a layer about 3 millimeters or ⅛ inch thick.)

Upper-atmosphere ozone is created when ultraviolet radiation (UV), a component of sunlight, bombards $O_2$ oxygen molecules, separating the two oxygen atoms. Some of these free oxygen atoms then combine with oxygen molecules ($O_2$) to form ozone ($O_3$). As long as the atmosphere contains a consistent percentage of $O_2$ and Earth receives steady UV radiation from the sun, the ozone layer is maintained.

Ozone absorbs harmful ultraviolet (UV) radiation. Solar ultraviolet radiation arrives in two forms, UV-A and UV-B. UV-A radiation reaches Earth's surface and is tolerated by organisms (although overexposure causes temporary "sunburn"). In contrast, UV-B radiation can cause genetic damage. Fortunately, most UV-B is absorbed by the ozone in the ozone layer. What little UV-B reaches Earth's surface is believed to cause some skin cancers in people.

The ozone balance is maintained by the natural ozone-creating mechanism described. However, a serious geohazard exists today because humans have created artificial chemicals that rise into the stratosphere and react with ozone, decreasing the stratospheric abundance of this important gas.

## Destruction of the Ozone Layer by CFC Compounds

Our ability to manipulate molecules and design chemicals is wonderful—but often there are consequences. Such is the case with chlorine-containing compounds called **chlorofluorocarbons (CFCs).** CFCs originally were developed in the 1940s as substitutes for toxic ammonia used in refrigeration. Their characteristics also made them ideal as foaming agents for plastic packaging (Styrofoam), insulation, air conditioning, and especially as aerosol propellants in spray cans.

CFCs are chemically inert, nontoxic, and easily liquified. Problems began when they were released into the atmosphere during use. Their widespread use after 1970 marked the beginning of the serious destruction of the ozone layer.

In contrast to other atmospheric components, CFCs are not degraded by sunlight, oxidation, or precipitation. They slowly rise into the stratosphere, where they are bombarded with UV radiation. This energy breaks apart the CFC molecules, freeing chlorine atoms. The CFCs react with ozone molecules to convert them into ordinary $O_2$ molecules. In a subsequent reaction, liberated chlorine atoms break apart more ozone. This process continues at rates dependent on the supply of CFCs and the amount of UV radiation.

According to the EPA, the atmospheric lifetimes for the most commonly used CFC compounds, known as CFC-11 and CFC-12, are 60 and 120 years respectively. During that time, each chlorine atom may destroy almost 100,000 ozone molecules. With increasing degradation of the ozone layer, more UV-B radiation reaches Earth's surface, constituting a serious biological threat.

## Geohazards of Ozone Depletion

In 1991, the EPA announced that the ozone layer loss over the United States since 1978 has been 4 to 5%. During the winter of 1992-1993, over North America, Asia, and Europe, ozone levels dropped 10 to 15% below the average for 1979 through 1990. This greatly increases the possibility of radiation damage to humans, animals, and vegetation. The ozone loss increases poleward. EPA computer models estimate an ozone loss of 10 to 12% over the next 20 years. Each 1% reduction in ozone allows 2% more UV radiation to reach Earth, resulting in a calculated 5 to 7% increase in skin cancer.

EPA estimated an additional 200,000 deaths from skin cancer in the United States over the next 50 years, beyond the 400,000 already expected skin cancer deaths for that period. UV-B also can cause cataracts of the eye and weaken the immune system.

## Extent of Ozone Depletion

At first, scientists thought that ozone depletion occurred largely in the area of the Antarctic ozone "hole" (Figure 11–10). However, skin cancer incidence is increasing in the Southern Hemisphere countries of Argentina, Chile, South Africa, Australia, and New Zealand. The problem is especially serious in Chile. It is the only developed country directly under the ozone "hole," and Chile's incidence of the virulent skin cancer, malignant melanoma, has increased fourfold since 1980.

Subsequent studies in 1991 show the problem to be far more extensive. One NASA scientist reported a significant ozone decrease over both the Northern and Southern Hemispheres, not only in winter, but also during spring and summer—when most people sun bathe.

### Mitigating Ozone Depletion

Ozone depletion can be alleviated by eliminating the use of CFC compounds. The United States has committed to eliminating CFC use by 1995. Other nations, however, vary in their willingness to stop CFC manufacture and use. About 100 countries have agreed to reduce their CFC production 50% by 1999.

Although significant reduction of CFCs helps, it may be too late to avoid some serious radiation damage. Even if CFC use were to stop right now, CFC compounds that already have been released are presently rising from the lower atmosphere and will continue to do so for about 10 years. Even if CFC production is banned completely, CFC releases will continue for years through breakdown and leakage of older air conditioning and refrigeration units, discarded aerosol cans, and deteriorating foam packaging. Conservative estimates predict at least an additional 3% ozone loss by the end of this century. Use of an appropriate sunscreen to reduce skin cancer risks may become essential for anyone venturing outdoors.

## SMOG—ORIGINS AND TYPES

Up to this point, we have looked at regional and global pollution. However, in many urban areas, local weather conditions and topographic features confine pollution over a city or metropolitan region. The problem is especially severe in Los Angeles, New York, Denver, Mexico City, and Beijing. In some cases, pollution levels are so great that people become sick if they spend any time outdoors, particularly senior citizens. Because this urban pollution is often smoky and foglike, it is known as **smog** (smoke plus fog).

There are two types of smog, reflecting different origins. One is *industrial smog,* more common in the past when utilities in major cities burned large amounts of coal and heavy fuel oil containing sulfur impurities. Industrial smog consists largely of sulfur dioxide gas, droplets of sulfuric acid, and suspended particles. Industrial smog is rare in the United States today because of controls on sulfur levels in fuels. However, it still exists in central Europe (Poland and Ukraine particularly) and in China, where coal still is burned with inadequate pollution controls.

The other type is *photochemical smog,* so-called because sunlight is involved in its formation (Figure 11–11). Photochemical smog is most common in cities with warm, dry climates and great numbers of motor vehicles. It is exacerbated where mountains ring a city, such as Mexico City, inhibiting wind flow and allowing pollutants to accumulate.

In photochemical smog, the basic culprits are *nitrogen oxides,* which are generated by vehicle exhausts, and *volatile hydrocarbon compounds* released by oil refineries, dry cleaners, and evaporation of gasoline and oil from fuel tanks and engines. Together, the nitrogen oxides, water, and hydrocarbons are affected by ultraviolet radiation in sunlight, generating a mixture of ozone, nitric acid, and other chemicals that make up the brownish mist known as photochemical smog. It is ironic that *reduced* ozone in the upper atmosphere is a hazard, whereas *excessive* ozone is a hazard in the lower atmosphere!

### Inversion Layers

Under normal conditions, warm air near Earth's surface cools upward at a steady rate. As surface air is heated, its density becomes reduced and it rises, carrying pollutants with it. They then are transported from the area by upper-level winds. However, local or regional weather conditions sometimes create an inverted condition, with colder air near the surface and warmer air above. This is an **inversion** (Figure 11–12).

During an inversion, the lower and often more polluted air cannot rise and be dispersed, because it is blocked by the warmer air above. The pollution level in the lower layer can build for hours or even days. It can become severe enough to affect people's respiratory systems, visibility, and air traffic. Historically, severe inversions have been quite dangerous, notably in London (4,000 people perished in 1952) and Donora, Pennsylvania (20 died and thousands were sickened in 1948). Eventually, stronger winds or the passage of a weather front stirs the lower atmosphere and disperses the polluted air.

FIGURE 11–11
Reactions producing photochemical smog.

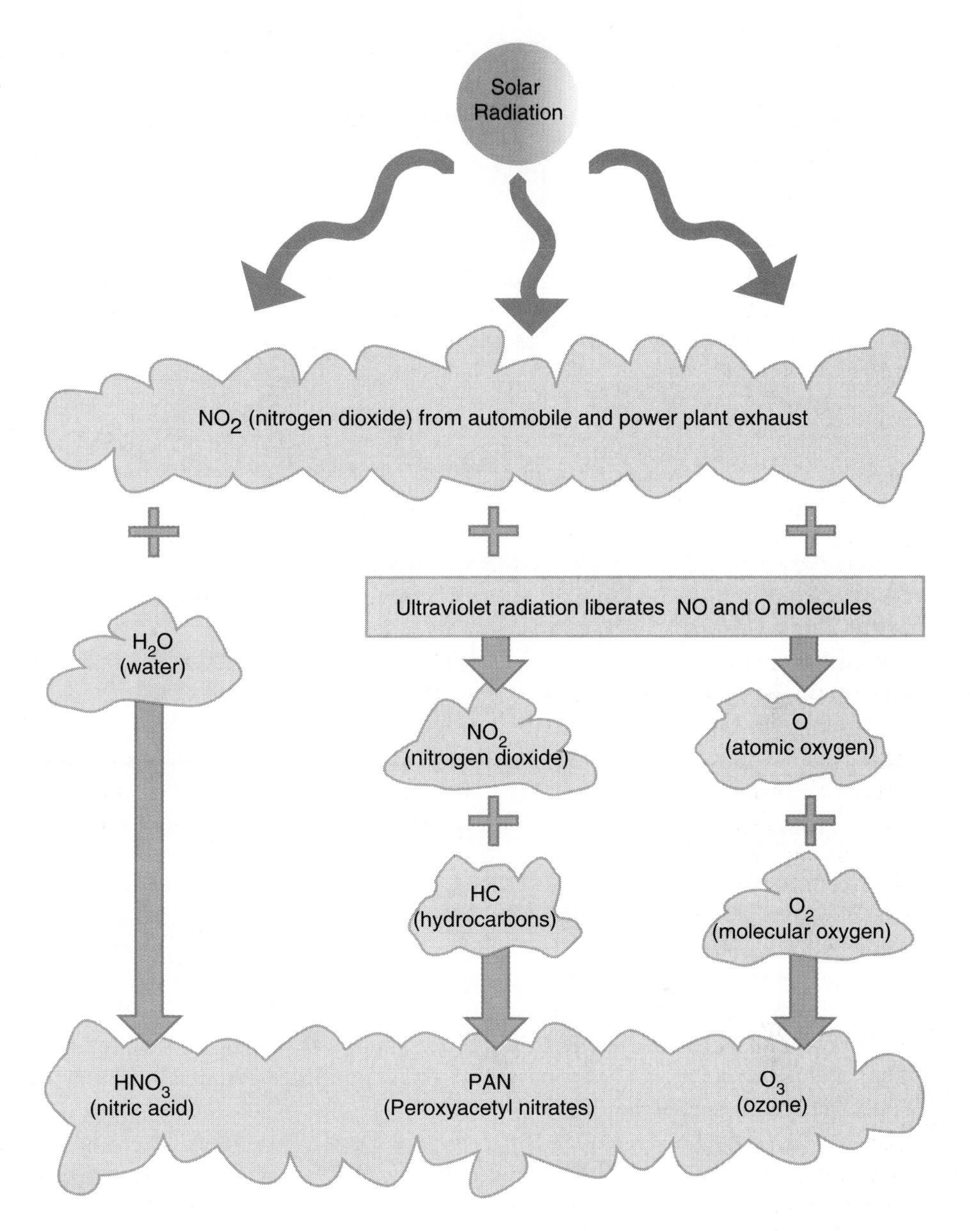


## Industrial Smog

A dangerous inversion occurred over New York City during the Thanksgiving holiday weekend in 1966. Just prior to this event, New York City's air had a sulfur dioxide concentration of 0.16 ppm, the highest in the nation at that time. Local levels to 3.5 ppm were recorded (a level of only 1.2 ppm for eight hours is fatal).

On Thanksgiving Day, low, dark clouds obscured the city skyline from the suburbs. By mid-morning the *pollution index* (a measure of the combined sulfur dioxide, carbon dioxide, haze, and smoke in the air) read 60.6, ten points above the "health danger" mark. The city was put on an "air emergency alert."

Coal-burning incinerators were shut down, and power companies that were capable of burning multiple fuels were asked to switch from coal to other fuels, such as oil and gas. The city asked people to reduce indoor temperatures to 16°C (60°F) to reduce heating needs and to cut emissions. People were asked not to drive, and truck deliveries were curtailed. Despite these measures, 200 deaths were attributed to this pollution event.

**FIGURE 11–12**
Cross section of the atmosphere showing conditions during an inversion.

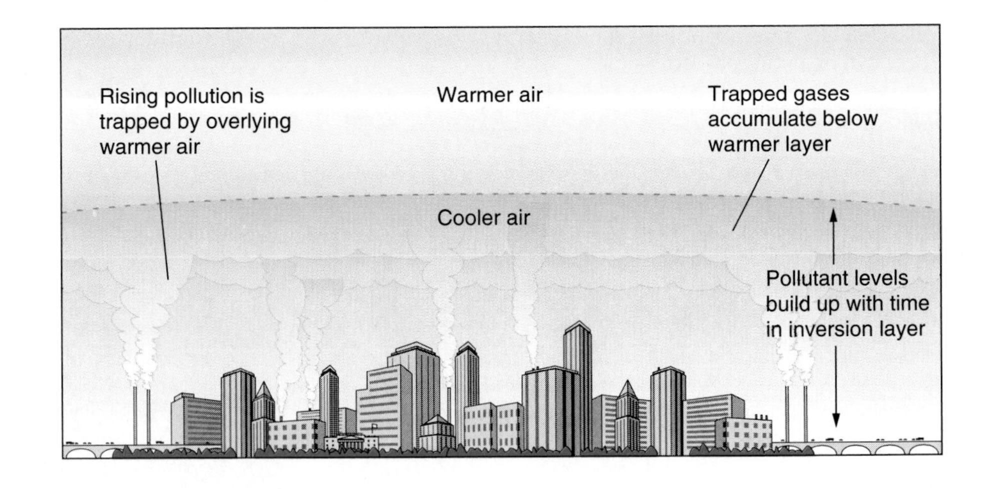


In 1971, New York City enacted its Air Pollution Control Act, one of the strictest, most comprehensive pollution-control laws in America. As a result, no industrial smogs have occurred in the New York metropolitan region since that time. Smog remains a problem at times, but it is the photochemical type.

## Photochemical Smog

Another serious smog situation occurs in cities confined by mountains and having warm, dry climates. When cool winds blow into the city, they displace warm air upward, creating an inversion. Pollution generated in the city then becomes trapped beneath the warm air layer and confined by the mountains. The pollution concentration increases with time. Such problems are common in Los Angeles (see Box 11–1). They also occur in Salt Lake City and Denver.

***Denver.*** Denver is rimmed by mountains, forming the Denver Basin, and is subjected periodically to photochemical smog. When an inversion occurs, pollution builds in the lower layer. It is especially bad in the morning, when surface air is colder and denser, inhibiting pollution dispersal upward. Only when the surface air warms sufficiently later in the day can the polluted air rise, allowing the pollutants to disperse.

Weak winds push the dirty brown air back and forth across the basin during a smog event, and high mountains block the winds that can stir the lower atmosphere. A particularly severe smog event in 1979 ended only when strong winds and a westerly snowstorm dispersed the polluted air from the Denver Basin.

***Mexico City.*** Smog is especially severe in Mexico City, one of the largest metropolitan areas on Earth. Its physical setting, population growth, and numerous pollution sources combine to make it one of the world's most polluted cities. Mexico City lies in a valley about 2,100 meters (7,000 feet) above sea level. It is surrounded by mountains, some with elevations to 5,000 meters (16,500 feet), except to the north. Prevailing winds from the north back polluted air against the surrounding mountains.

Over 16 million people live in this basin. More than 2 million vehicles (autos, taxis, buses, and trucks) supply 75% of the pollutants. The other 25% comes from natural phenomena (such as fires) and industrial sources. The valley is filled with small, energy-intensive industries. Many larger industries are concentrated to the north, so their emissions are blown across the city by the prevailing winds.

Air quality in Mexico City is very poor. Ozone levels in 1990 exceeded hazardous levels on many days. In part, this is because the city is at a high altitude, where fossil fuels burn less completely due to the lesser amount of oxygen available. Consequently, the burning of fossil fuels produces more carbon monoxide and ozone. The worst year for air pollution in Mexico City's history was 1991, with a significant increase in respiratory ailments reported among its inhabitants.

To reduce emissions, Mexico City is closing some industries, upgrading vehicle engines to burn fuel more efficiently, curbing the use of autos, making mandatory the use of low-sulfur fuels, and closing some major air-polluting industries. For example, Mexico City's giant PEMEX oil refinery, northwest of the city, was closed because it alone generated 7% of the city's atmospheric pollution. The closing was painful economically, because the refinery provided a third of the city's gasoline and most of its diesel fuel, and Mexico temporarily had to import some refined fuels.

The people of Mexico City are making a serious effort to reduce air pollution, but their attempts may be only partially effective. Rapid population growth and more vehicles will increase pollution, and the topographic and wind conditions won't change—they still will trap pollution in the basin.

### Air Pollution Legislation

In the United States, the Clean Air Act Amendments of 1990 provide stringent controls on air pollutants. These include controls on sulfur dioxide emission from power plants ( which cause industrial smog and acid precipitation), automobile emissions (which cause photochemical smog), and chemicals that deplete ozone levels in the atmosphere.

## RADON

In 1985, an engineer at a Pennsylvania nuclear power plant underwent a routine radiation check at the plant entrance and set off a radiation alarm. The remarkable thing was that he did so upon *entering*, not as he was leaving! This indicated that he was bringing radiation *into* the plant. Where could he be picking up this radioactivity? The EPA tested the air in his home and recorded unusually high levels of the radioactive gas called **radon.**

Radon accounts for about 40% of the radiation dose received by the average American (Figure 11–13). Although most homes do not have hazardous levels of radon, awareness of this geohazard is growing. For years, uranium miners in the western states displayed an unusually high frequency of lung cancer. The cause was discovered to be radioactive radon gas, which they inhaled daily in the mines.

Experts believe that long-term exposure to similar levels in homes would have the same result. But can homes attain such high levels? Where strong radon generation occurs beneath the surface, and where well-sealed homes prevent the escape of radon once it is inside, 5 to 10% of homes may reach dangerous radon concentrations. This may seem a small percentage, but it could mean thousands of homes (Figure 11–14).

### Radon Levels and Risks

Radon concentrations in the United States are measured in *picocuries per liter of air (pCi/L)*. One pCi is equal to the decay of about two radioactive atoms per minute. Radon levels can vary greatly in outdoor air, indoor air, air between soil particles, and groundwater (Figure 11–14). Ranges of radon levels reported by the U.S. Geological Survey are:

| | |
|---|---|
| Outdoor air | 0.1 to 30 pCi/L (average 0.2) |
| Indoor air | 1 to 3,000 pCi/L (average 1 to 2) |
| Soil air | 20 to 100,000 pCi/L (most between 200 and 2,000) |
| Groundwater | 100 to 3,000,000 pCi/L. |

What do given radon levels mean in terms of human risk? The radon risk and a comparison with the risk of lung cancer death is given in Table 11–1.

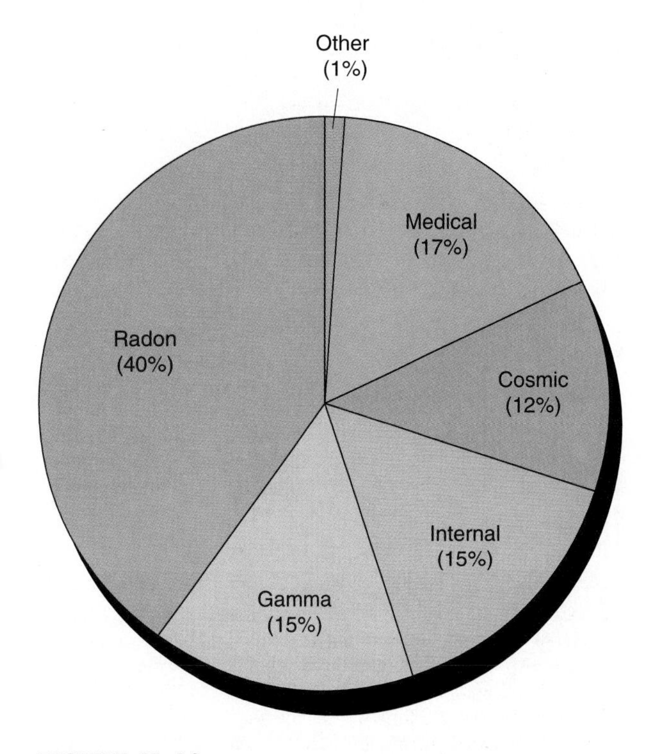


**FIGURE 11–13**
Sources of radiation. (Adapted from Edward B. Nuhfer, Richard J. Proctor, Paul H. Moser, and others, 1993, The citizens' guide to geologic hazards: Arvada, Colorado, American Institute of Professional Geologists, p. 29)

BOX 11–1

## THE LOS ANGELES BASIN: THREE ANNUAL GEOHAZARDS

Los Angeles sits in a basin, flanked on three sides by mountains and open to the Pacific Ocean on the west (Figure 1a). Prevailing westerly winds blow from the ocean onto the land. When cold ocean air moves in, creating an inversion, it traps pollutants against the surrounding mountains. From late winter through early summer, the great volume of pollution produced in the Los Angeles Basin is retained as smog in the area. The smog level can increase to hazardous levels for several

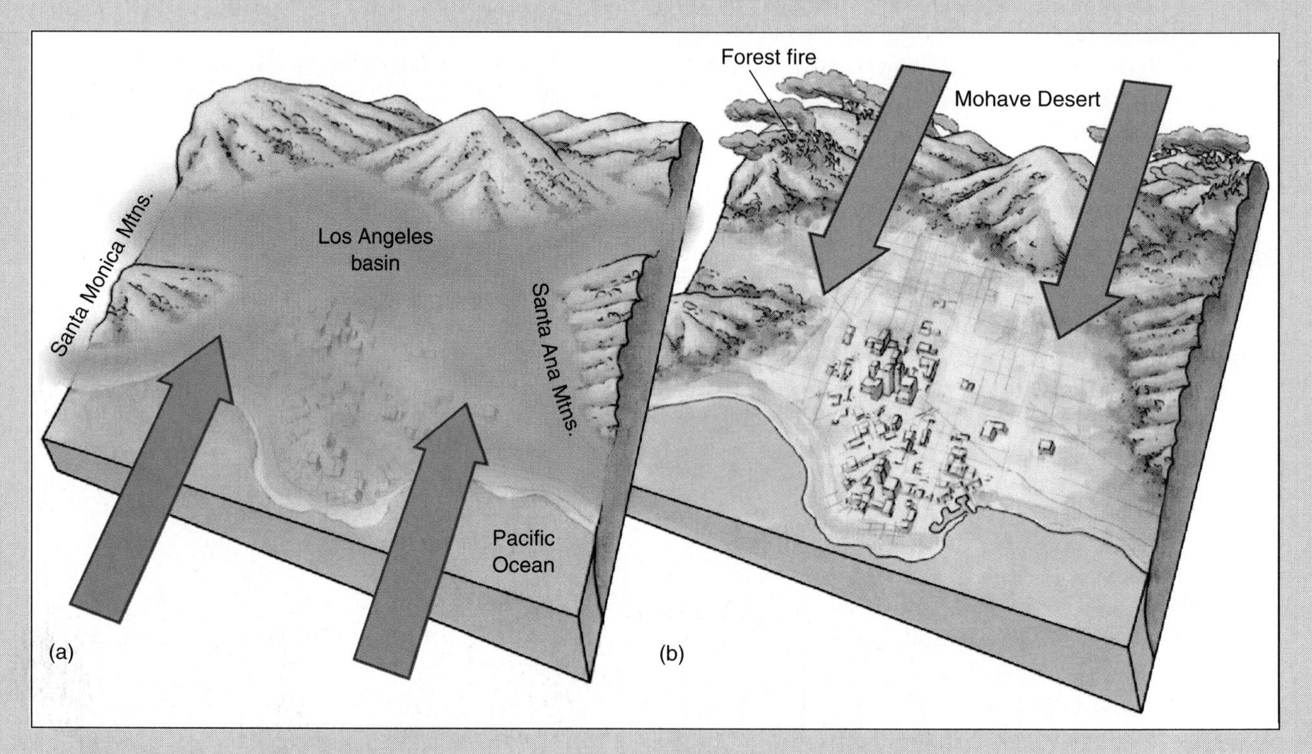


**FIGURE 1**
Common wind conditions in the Los Angeles Basin. (a) *Late winter into summer:* Pollution generated within Los Angeles Basin is backed up against the surrounding mountains by onshore winds. Smog levels increase. (b) *Fall into early winter:* Offshore winds blow pollution out of the basin and smog levels drop. These hot, dry Santa Ana winds result in numerous forest fires. Later, in the spring, rains falling on fire-denuded slopes trigger mudflows.

weeks before changing weather conditions finally disperse it. This is the first of three major geohazards that afflict our nation's second-largest metropolitan area.

Relief from the smog comes in the fall, when the wind direction reverses and warm air from the interior blows the smog out to sea, clearing the air in the basin (Figure 1b). These hot, dry desert winds are called the *Santa Ana Winds,* named for a local canyon.

The Santa Ana winds disperse the smog, but they bring the second geohazard: forest fires. Day after day of hot, dry winds dries the vegetation and fans any minor flame into forest fires, which are common during the fall and early winter (Figure 1b). The area vegetation is chaparral, oil-rich shrubs that ignite readily. This sequence of events triggered the disastrous 1993 fires in the Los Angeles suburbs (Figure 2), and similar fires in 1991 to the north in the San Francisco area.

In late winter, the wind reverses, blowing onshore once again and bringing the third geohazard: the spring rains, which fall on fire-denuded steep mountains slopes, triggering massive mudflows in parts of the basin. At the same time, smog levels resume building up in the basin as winds blow from the ocean (Figure 1a).

The smog episodes continue until the dry Santa Ana winds blow again, and the environmental cycle of forest fires and mudflows continues.

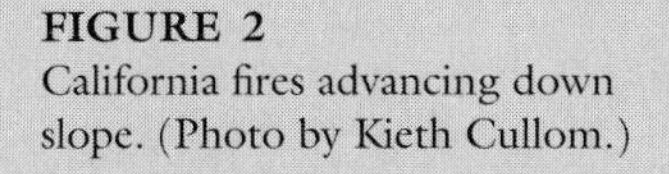

**FIGURE 2**
California fires advancing down slope. (Photo by Kieth Cullom.)

**FIGURE 11–14**
Radon levels in the air and in homes can vary greatly. (U.S. Geological Survey, 1992, p. 4.)

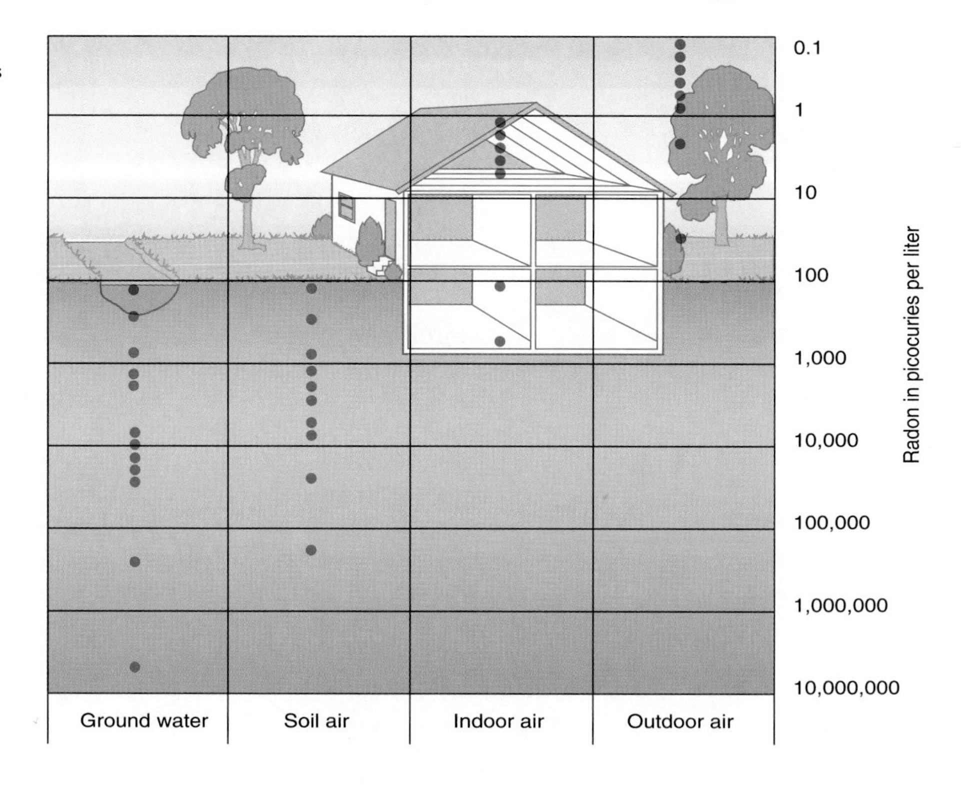


**TABLE 11–1**
Comparative risk of different radon levels.

| Radon Level pCi/L | Estimated Fatal Lung Cancers/1000 | Comparable Exposure Levels | Comparable Risk Estimate |
|---|---|---|---|
| 200 | 440–770 | 1,000 times average outdoor level | More than 60 times nonsmoker risk |
| 100 | 270–630 | 100 times average indoor level | Four pack/day smoker Or 20,000 chest X-rays/yr |
| 40 | 120–380 | 100 times average outdoor level | Two pack/day smoker |
| 20 | 60–210 | | |
| 10 | 30–120 | 10 times average indoor level | One pack/day smoker |
| 4 | 13–50 | 10 times average outdoor level | Five times non-smoker risk |
| 2 | 7–30 | | |
| 1 | 3–13 | Average indoor level | Non-smoker risk of fatal lung cancer |
| 0.2 | 1–3 | Average outdoor level | 20 chest X-rays/yr |

Source: Environmental Protection Agency.
EPA recommends action be taken if indoor radon levels exceed 4 pCi/L, which is 10 times the average outdoor level. Some EPA representatives believe the action level should be lowered to 2 pCi/L; other scientists dissent and claim the risks estimated in this chart are already much too high for low levels of radon. The action level in European countries is set at 10 pCi/L. Note that this chart is only one estimate; it is not based upon any scientific result derived from a study of a large population meeting the listed radiation criteria.

## How Radon Gas Enters the Environment

Radon gas comes from the decay of uranium. Uranium atoms are unstable, so uranium is radioactive, slowly releasing particles and bursts of energy from its nucleus (see Nuclear Reactions in Chapter 2). One uranium isotope, uranium-238, is widespread and occurs in the minerals of rocks and sediments worldwide. As it decays, it forms isotopes of other elements. One is a metal, radium-226, which then decomposes into a gas, radon-222.

As a gas, radon is readily inhaled. Like all radioactive elements, unstable radon begins to decay as soon as it is formed, giving off particles and becoming radioactive polonium-218, the actual agent that damages tissue. The solid polonium-218 lodges in the lungs. As it breaks down, it releases alpha and beta particles that damage the tissue lining and may cause lung cancer.

Smokers are at elevated risk, because smoking and radon inhalation increase the risk of developing lung cancer. Radon has a very brief half-life, only 3.8 days. Thus, half of a given amount of radon has decayed after 3.8 days, and half of the remainder after another 3.8 days, and so on.

As radon is produced, it accumulates in fractures or pore spaces in the rock or sediment. Radon moves more readily through permeable soils such as coarse sand and gravel than through denser clays. It moves more slowly through water-filled pores and fractures. For example, in water-saturated rock or soil, radon decays within 2 centimeters (0.8 inch) of travel. Through dry rocks and soils, however, radon can travel up to 180 centimeters (70 inches) before most of it decays. If these openings are connected to the surface, the radon may move into the atmosphere and be dispersed by wind, eliminating the hazard by dilution.

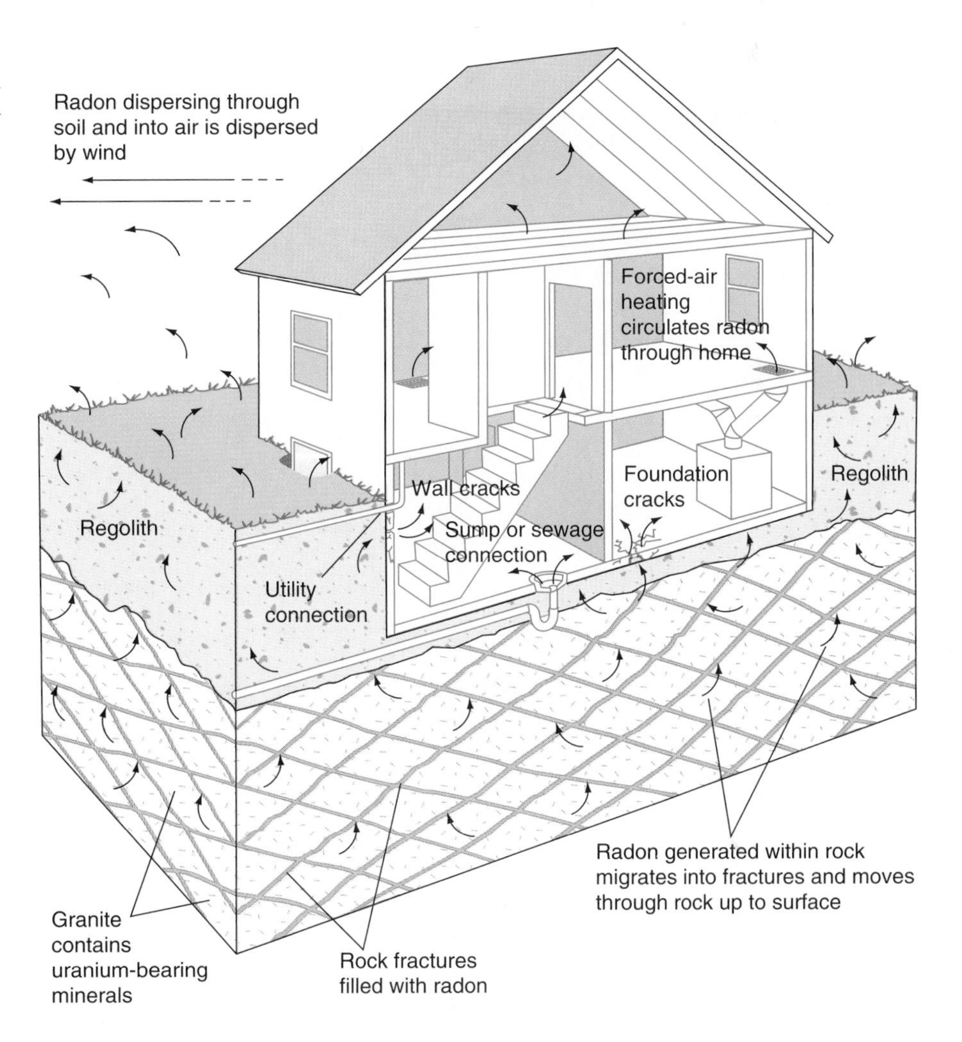


**FIGURE 11–15**
Production of radon and its dispersal into and through a home overlying a radon-emitting deposit.

Problems occur when a home's foundations and basements are constructed within radon-generating sediment or rock. Radon moves through soil and rock fractures toward homes, because air pressure in the ground is greater than the air pressure within the home. The radon enters cracks or other openings in the foundation and moves through the house (Figure 11–15).

In recent years, people have made their homes more energy-efficient to reduce fuel use during cold weather. The filling of holes, use of thick insulation, tight seals around windows, and installation of double-glazed storm windows certainly prevents cold outside air from getting in. However, it also reduces ventilation and increases the risk of indoor pollution.

Where radon seeps into such an energy-efficient house, it frequently remains there, usually in the basement, posing health problems as its concentration increases. Subsequent inhalation can produce radiation damage to the delicate linings of the lungs.

## Variations in Radon Emission

The amount of radon emitted by a sediment or rock depends on its uranium content, the degree to which its openings are developed (fractures or pore spaces), and whether it is covered by soil or other sedimentary deposits.

Radon emission varies with rock and sediment type. Igneous rocks rich in silica and alumina, such as granites and rhyolites, generally contain more uranium than igneous rocks rich in iron and magnesium, such as gabbro and basalt (see Chapter 2 to review these rock types). Sedimentary rocks generally are low in uranium, except for black shales and phosphate-rich rocks. Metamorphic rocks, such as gneiss and schist, are richer in uranium than marble, slate, and quartzite.

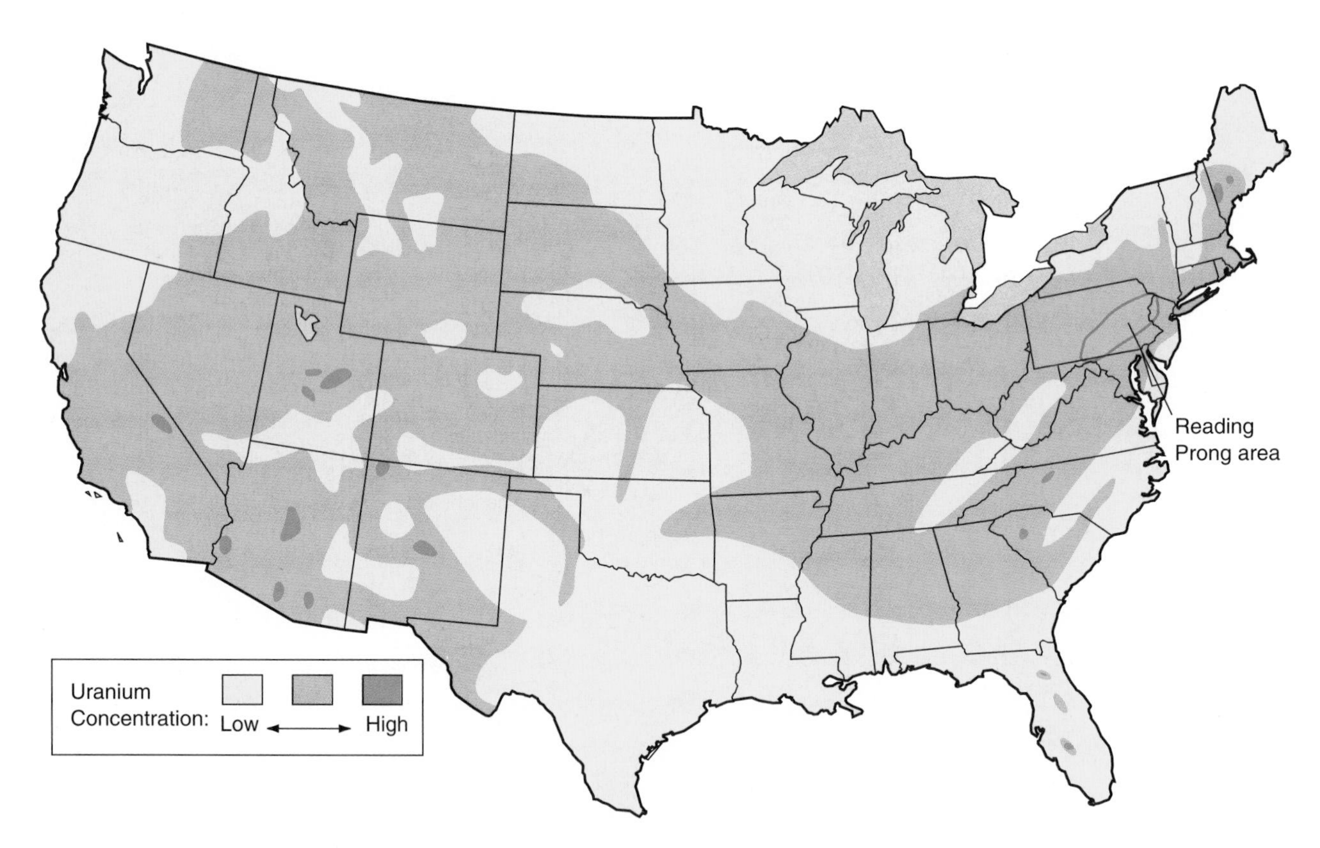


**FIGURE 11–16**
Variations in radioactivity in the United States. This map indicates the uranium level in surface materials. (U.S. Geological Survey, 1992, p. 18–19.)

Thus, radon concentration varies with the type of rock exposed at the surface in an area. In general, the more fractures in a uranium-bearing rock, the greater is its radon potential. The variation in uranium levels of U.S. surface materials is shown in Figure 11–16.

## Radon Levels Within Homes

Radon concentration in a home is greater if cracks and openings in the foundation (Figure 11–15) allow more radon to enter. In general, the more airtight the home, the greater the radon level, because the gas cannot escape. Basement concentrations are greatest because they are closest to the source, whereas upper floors have smaller and more uniform concentrations because air moves more freely than through a basement.

Radon measurement in a home also varies with time of year. Readings are lower in summer because windows are open or air conditioners are exhausting air from homes. Accumulating radon thus is dissipated to the outside. In winter, more radon is trapped inside the homes. These seasonal differences are important in determining when to make radon pollution readings to determine if there is a risk.

Some homes have been inadvertently *constructed* of radon-emitting materials. Radon contamination was discovered in several homes around a factory in Landsdowne, Pennsylvania, where radium-226 was extracted from ore in the 1920s—ironically, for treating cancer. Sandy waste was used by local contractors to make building materials. All of these homes have been emitting radon directly from their walls since they were built. Of 25 contaminated homes, the government offered to relocate families from the 13 most hazardous. These had a concentration 23 times the acceptable value.

## Mitigating Radon Problems

Significant portions of the United States contain surface rocks having uranium concentrations and, consequently, higher radon emissions. Some of these areas are densely populated and are of special concern. The *Reading Prong* is a strip of metamorphic and igneous rocks extending from central Pennsylvania across New Jersey into New York (Figure 11–16). Parts of the Reading Prong are heavily urbanized, with over 100,000 people living on the New Jersey section alone. Home sites in these areas are safe as long as homes are tested for radon and any necessary remedial measures are taken.

In risk areas, indoor radon concentrations can be reduced. In a "slab" home, which is built on a concrete slab without a basement, an impermeable material placed between the slab and the ground or a ventilated crawl space can reduce the risk of radon entry. Sealing all cracks and openings can reduce the entrance of radon into basements. Exhaust ventilation in basements can remove any radon that does enter. Ventilation is wise even in winter months. While this may raise fuel costs slightly, it will also lower health risks from the inhalation of indoor radon. The radon level within these homes should be measured periodically.

## Assessing Radon Risk

Radon levels can be checked in homes. Both passive and active devices are used. Long-term measurements are made by burying a device in the soil.

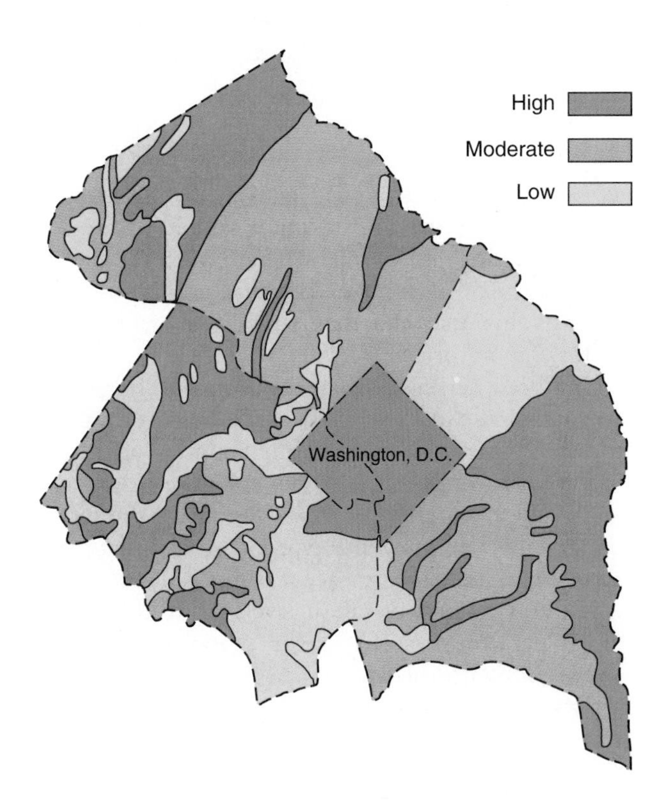


**FIGURE 11–17**
Radon potential map of counties surrounding Washington, D.C. (U.S. Geological Survey, 1992, p. 25)

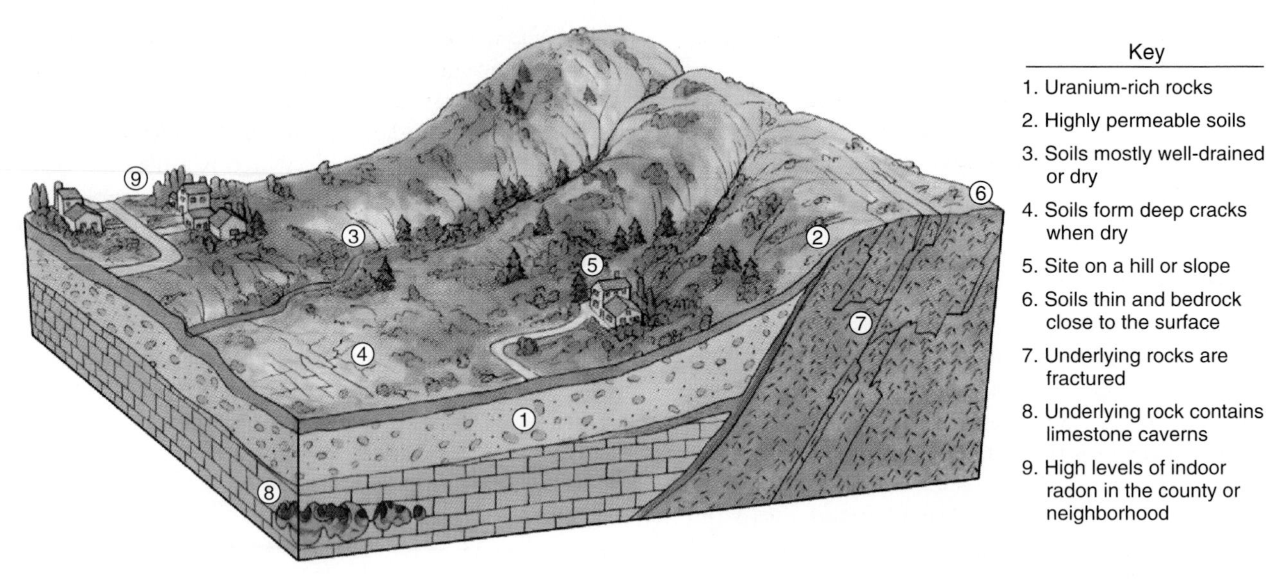


**FIGURE 11–18**
Scientists who understand the geology and soils of an area can evaluate the radon potential of housing sites. The factors shown can increase the odds of above-average radon levels: (1) uranium-rich rocks, (2) highly permeable soils, (3) well-drained soils that are dry most of the time, (4) soils that form deep cracks during dry spells, (5) hillside or sloping location, (6) thin soils with bedrock near the surface, (7) fractured underlying rocks, (8) limestone caverns in underlying rocks, (9) counties or neighborhoods with previous reports of high radon levels. (U.S. Geological Survey, 1992, p. 26.)

It can also be determined actively by inserting a probe into the ground, taking a sample of the soil and analyzing it. Homeowners can test the air themselves by using test kits that are exposed to the building's air and then sealed and mailed to a testing laboratory.

Homeowners also can assess the radon hazard by consulting *radon potential* maps if they exist for one's location. One such map, for the counties surrounding Washington, D.C., is shown in Figure 11–17.

In any given area, scientists can assess the radon danger by examining the type of surface materials, land slope, drainage, surface features, soil characteristics, and reports of radon levels from homeowners. Characteristics that favor radon buildup are shown in Figure 11–18.

## LOOKING AHEAD

From the gaseous geohazards examined in this chapter—acid precipitation, the greenhouse effect, ozone depletion, smog, and radon gas—we shift our focus to solid and liquid geohazards in Chapter 12. Solid waste, landfills, incineration, recycling, sewage, acid mine drainage, and PCBs are topics we generally would rather not think about. However, we have made them a major geohazard on our planet, so now we must deal with them.

## SUMMARY

Air pollution is a serious geohazard to respiration in humans and animals, and it damages vegetation. The problem is not a local one because regional wind systems disperse emissions far from their source.

Acid precipitation is produced when emissions of sulfur dioxide and nitrogen dioxide, produced by burning fossil fuels, react with water vapor in the atmosphere to produce acid rain and snow. The acidity of precipitation is described by the pH scale, a measure of the concentration of hydrogen ions in the solution. A pH of 7.0 is neutral; alkaline solutions have a higher pH between 7.0

and 14, and acidic solutions have a lower pH between 7.0 and 0.

The acidity of surface water is increased in areas of acid precipitation that are underlain by acidic (silicic) rocks, such as granite, or acidic soils, such as under pine forests. By contrast, areas underlain by alkaline (carbonate) rocks, such as limestone, react with the acid precipitation to reduce acidity (raise the pH).

Acid precipitation is a geohazard. Acidified waters leach harmful elements, such as mercury and aluminum, from surface rocks and sediments. Fish die in lakes that become too acidic. Leaves become mottled and plant growth is stunted. Marble statues and limestone ornamental facings become pitted and dissolved. Acid precipitation can be mitigated by the burning of low-sulfur fossil fuels or by expensive "scrubbing" of harmful gases before they are vented from industrial smokestacks. The pH of acidified lakes can be raised (made less acidic) by adding crushed limestone to the water.

The greenhouse effect results when carbon dioxide and other gases released by the burning of fossil fuels accumulates in the atmosphere and traps heat radiated from the ground. This may be causing a gradual global warming, with widespread consequences. The warmth melts continental ice sheets, which causes water to expand; both effects raise sea level. Rising sea level will slowly inundate coastal communities and cause saltwater encroachment into surface water and groundwater supplies in coastal areas. Increasing global temperatures will change temperature and rainfall patterns and therefore influence agricultural productivity.

The marked increase in use of chlorofluorocarbon (CFC) compounds since 1970 is decomposing ozone in the stratospheric ozone layer. This ozone depletion has already created a hole in the ozone layer over Antarctica and, to a lesser degree, over the North Polar region. The ozone layer is Earth's shield against the harmful effects of ultraviolet radiation. As ozone levels become depleted in the atmosphere, the risk of skin cancer and cataracts in humans and animals increases.

Smog contains a mixture of water vapor and harmful gases that accumulate wherever conditions prevent dispersal of pollutants. In winter, cold air may become overlain by warmer air during an inversion. Pollutant levels in urban areas build in the lower atmosphere until weather changes destroy the inversion structure. Another type of smog buildup occurs where surrounding mountains prevent winds from blowing pollution from an urban basin, such as Los Angeles or Mexico City.

Radon is a radioactive gas produced by the natural breakdown of radioactive uranium-bearing minerals in sediments and rocks. Normally, radon is dispersed by winds as it is emitted from the ground. However, when it leaks into homes and is inhaled over a long period, it greatly increases the risk of lung cancer. The radon hazard can be mitigated by sealing foundations from the ground so that no leakage can occur into the home, by maintaining adequate ventilation, and by frequent monitoring of indoor radon levels.

## KEY TERMS

acid precipitation
acid rain
acid snow
chlorofluorocarbons (CFCs)
greenhouse effect
inversion
ozone "hole"
pH scale
radon
smog

## REVIEW QUESTIONS

1. Why is it more difficult to develop laws that deal with air pollution than for other kinds of pollution?
2. Describe the geologic controls that affect the acidity of surface water in an area affected by acid precipitation.
3. Distinguish between the environmental effects of acid rain and acid snow.
4. **a.** What is the greenhouse effect?
   **b.** Describe five different negative effects that may result from this phenomenon.
5. Describe another geohazard that may decrease, if only for a short time, the effects of global warming.
6. What are two ways in which global warming increases sea level?
7. Describe the two mechanisms that permit smog to reach harmful concentrations over urban areas.
8. **a.** Describe the origin of radon.
   **b.** What allows dangerous radon levels to build in homes?
   **c.** What geologic conditions favor greater radon emissions?
   **d.** How can indoor radon levels be reduced in areas that have high radon emission from the ground?
9. a. How has the hole in the ozone layer formed?
   **b.** What are the consequences of reducing the ozone concentration in the stratosphere?
10. Describe the air pollution problems in your own area.
    **a.** What are their origins?
    **b.** What is being done to mitigate their effects?

## FURTHER READINGS


Barnett, R., 1986, Ozone protection: The need for a global solution: EPA Journal, v. 12, no. 10 (Dec. 1986), p. 10–11.

Federal Emergency Management Administration, 1991, Projected impact of relative sea-level rise on the National Flood Insurance Program, 72 p.

Hendry, G. R., 1981, Acid rain and gray snow: Natural History, v. 90, n. 2, p. 58–64.

Mintzis, M. M., 1986, Skin cancer: The price for the ozone layer: EPA Journal, v. 12, no. 10 (Dec. 1986), p. 7–9.

Mohnen, V. A., 1988, The challenge of acid rain: Scientific American, v. 259 (August) p. 30–38.

Popkin, R., 1986, Two killer smogs the headlines missed: EPA Journal, v. 12, no. 10 (Dec. 1986), p. 27–29.

Rampino, M. R., and Self, S., 1984, The atmospheric effects of El Chichón: Scientific American, v. 250 (Jan.), p. 48–57.

Rind, D., 1986, The greenhouse effect: An explanation: EPA Journal, v. 12, no. 10 (Dec. 1986), p. 12–14.

Rodhe, H., 1990, A comparison of the contribution of various gases to the greenhouse effect, Science, v. 248, no. 6, p. 1217–1219.

Rowland, F. S., 1986, A threat to Earth's protective shield: EPA Journal, v. 12, no. 10 (Dec. 1986), p. 4–6.

Rowland, F. S., 1989, Chlorofluorocarbons and the depletion of stratospheric ozone, American Scientist, v. 77 (Jan.–Feb.) p. 36–45.

Titus, J. G., 1986, Rising sea levels: The impact they pose: EPA Journal, v. 12, no. 10 (Dec. 1986), p. 17–20.

U.S. Geological Survey, 1992, The geology of radon, General Interest publication 19920-326-248, 28 p.

# 12

# Waste Disposal

The marked growth of human population (Chapter 1) has been accompanied by a dramatic increase in resource use. Processing resources to produce consumer items and packaging has generated a great volume of solid and liquid waste.

As urban centers have grown, less space remains for storing solid waste. It is piled higher and higher on remaining open land or moved by truck and barge, at great expense, to disposal sites far away. Solid waste is not inert, and much of it breaks down over time into noxious gases and liquids. The gases include "greenhouse gases" (carbon dioxide and methane) and carcinogenic gases (benzene and vinyl chloride), and they contaminate the air around solid waste disposal sites. The liquid (leachate) from solid waste can contaminate groundwater supplies.

Liquid waste is a particularly dangerous problem because of its mobility. Once it enters surface water bodies or underground aquifers, liquid waste can be dispersed far from its source. Two general categories of liquid waste are natural and anthropogenic. Natural liquid waste decomposes from bacterial degradation and oxidation. But anthropogenic liquid waste often results from chemical manufacturing, and normal waste breakdown processes have little effect on these "unnatural" liquids. Consequently, some anthropogenic liquid waste remains unchanged—and hazardous—as it migrates away from its source.

In this chapter, we will look at solid waste and its disposal in landfills, by incineration, and through recycling. We will examine liquid waste—sewage, acid drainage, and toxic liquids—and its disposal in secure landfills, by incineration, and by deep injection into rock strata. We also will study a waste category that is a unique product of the Nuclear Age: radioactive waste.

## THE GARBAGE EXPLOSION

**Operations at a North Carolina landfill. (Ron Sanford/Black Star.)**

Garbage has long been used to fill low-lying areas that are later capped with clean fill and used as building sites. But land has become more scarce and the

## BOX 12–1 SOLID WASTE—NOT JUST A PROBLEM ON LAND

The problems of solid waste on land are commonly known. But the problem in lakes and oceans has attracted attention only recently. Debris washed into the ocean from land and dumped into the ocean from vessels creates aesthetic problems and a hazard to marine life. Even remote beaches now are littered with floatable debris, such as plastic bottles, foam cups, and plastic grocery bags.

A number of components make up the debris washed into the ocean (Figure 1), primarily plastic (62%), metal (mostly cans, 11%), paper (12%), and glass (10%). This debris is deposited in the water from land and ocean/lake sources (Figure 2). This oceanic littering requires periodic "beach sweeps" by government workers and volunteer groups to keep the beaches clean. The problem usually recurs with the next tide following a beach cleanup.

**FIGURE 1**
Components of solid waste collected during 1988 beach cleanups. (Center for Marine Conservation, 1989)

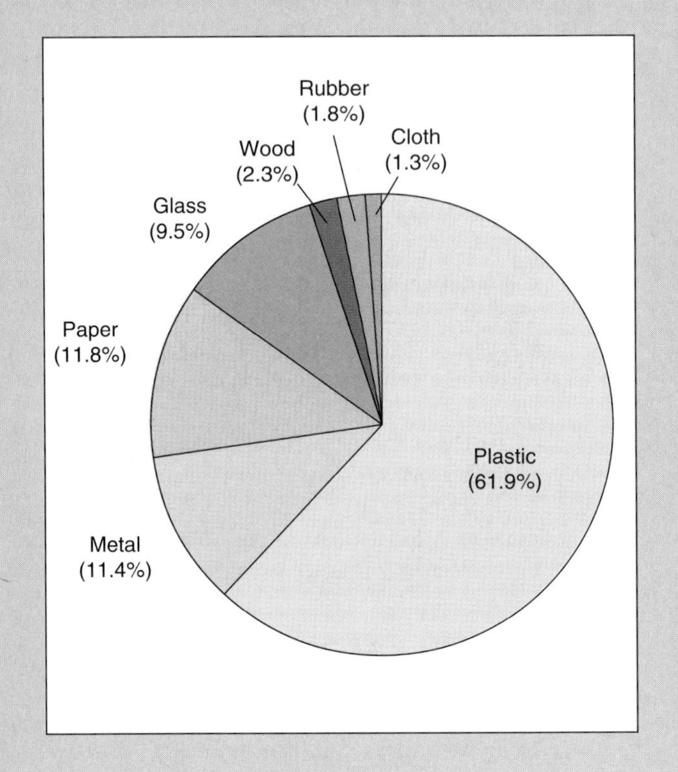

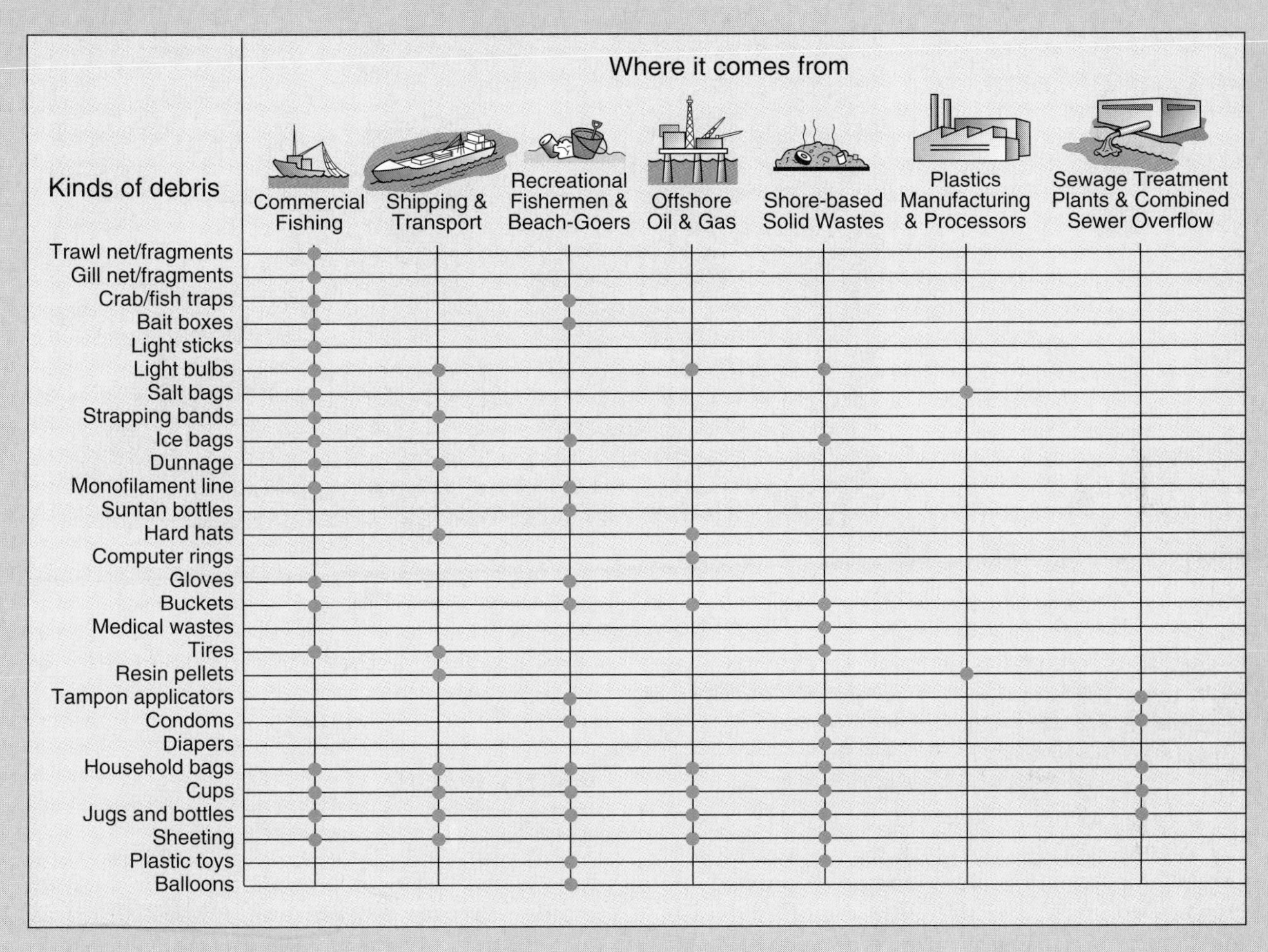

Where it comes from

| Kinds of debris | Commercial Fishing | Shipping & Transport | Recreational Fishermen & Beach-Goers | Offshore Oil & Gas | Shore-based Solid Wastes | Plastics Manufacturing & Processors | Sewage Treatment Plants & Combined Sewer Overflow |
|---|---|---|---|---|---|---|---|
| Trawl net/fragments | ● | | | | | | |
| Gill net/fragments | ● | | | | | | |
| Crab/fish traps | ● | | ● | | | | |
| Bait boxes | ● | | ● | | | | |
| Light sticks | ● | | | | | | |
| Light bulbs | ● | ● | | ● | ● | | |
| Salt bags | ● | | | | | ● | |
| Strapping bands | ● | ● | | | | | |
| Ice bags | ● | | ● | | ● | | |
| Dunnage | ● | ● | | | | | |
| Monofilament line | ● | | ● | | | | |
| Suntan bottles | | | ● | | | | |
| Hard hats | | ● | | ● | | | |
| Computer rings | | | | ● | | | |
| Gloves | ● | | ● | | | | |
| Buckets | ● | | ● | ● | ● | | |
| Medical wastes | | | | | ● | | |
| Tires | ● | ● | | | ● | | |
| Resin pellets | | ● | | | | ● | |
| Tampon applicators | | | ● | | | | ● |
| Condoms | | | ● | | ● | | ● |
| Diapers | | | | | ● | | |
| Household bags | ● | ● | ● | ● | ● | | ● |
| Cups | ● | ● | ● | ● | ● | | ● |
| Jugs and bottles | ● | ● | ● | ● | ● | | ● |
| Sheeting | ● | ● | | ● | ● | | |
| Plastic toys | | | ● | | ● | | |
| Balloons | | | ● | | | | |

**FIGURE 2**
Kinds and sources of marine debris. (State of Alaska, Sea Grant)

Floating debris poses special problems for marine life, especially mammals, birds, sea turtles, fish, and crustaceans. For example, Alaskan fur seals become entangled in fishing net fragments (Figure 3). Hawaiian Monk seals become snarled in gill nets and in the buoy lines of lobster traps. Sea birds also become entangled in nets, plastic beverage container rings, and fishing line.

Large leatherback turtles, which subsist largely on jellyfish, and young marine turtles sometimes mistake plastic debris for jellyfish, and dead turtles have been found with plastic bags and sheets in their stomachs. Fish and turtles choke on the plastic rings that hold six-packs together.

FIGURE 3
Fur seal pup trapped in a fishing net. (Courtesy NMFS/NOAA)

solid waste volume has increased steadily. As a result, many solid waste sites have grown vertically to the point that they can became uneconomical to operate and even an aircraft hazard. An interesting example is New York's Fresh Kills Landfill.

It began as a dump in a wetland in 1948. Today, its volume is about 25 times that of the greatest Egyptian pyramid! By the time it is abandoned and landscaped in the year 2005, this pile of solid waste and sediment cover will attain 168 meters height (550 feet), which is higher than the Statue of Liberty by 61 meters (200 feet). When completed, it will contain at least 60 million cubic meters of trash (78 million cubic yards). The U.S. Geological Survey says it will be the highest feature on the Atlantic coast between Florida and the hills along the coast of Maine. It is being capped and abandoned because of its height and because the scavenging sea gulls it attracts are a hazard to aircraft.

Clearly, trash is a serious problem both on land and in aquatic environments (Box 12–1). Little space remains for storing garbage in most urban and suburban areas. Fortunately, solutions exist for significantly reducing waste volume, providing beneficial products, and reducing consumption of natural resources.

## SOLID WASTE

The solid portion of our waste is composed of varied material. Each component (Figure 12–1) differs greatly in **reactivity**—the degree to which it reacts with oxygen, water, and soil bacteria. For example, glass will remain unchanged for centuries, whereas organic waste ultimately decomposes completely. In general, the greater the reactivity of a solid waste, the greater will be its volume reduction over time, especially in landfills.

The portion of this waste that is organic—agricultural waste, manure, food processing by-products, household food waste, paper, yard waste—will decompose and lose volume over time. This fraction of solid waste is said to be **biodegradable,** because it is broken down by biological agents. Other solid waste—concrete, metals, glass, whole automobiles, appliances—is **nonbiodegradable** and undergoes minimal breakdown with time. Nonbiodegrad-

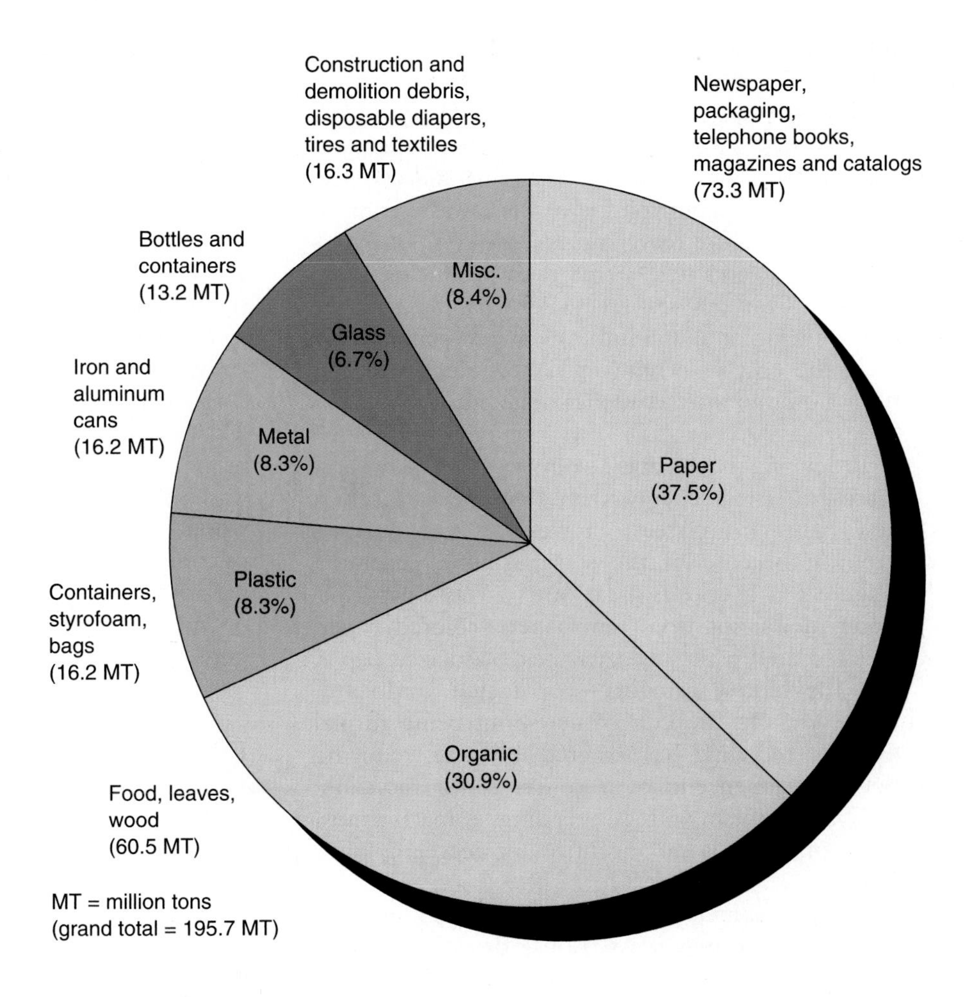


**FIGURE 12–1**
Components of U.S. municipal solid waste, the total of which is 178 million metric tons (196 million tons). MT = million tons. (EPA, July 1992, Characterization of municipal solid waste in the U.S.)

able solid waste preserves much of its initial volume. (Plastics are a special case. They are organic, but their structure makes their decomposition occur very slowly, requiring centuries for some types.)

## Reducing Solid Waste Bulk

The major obstacle to reducing the volume of solid waste is that *all of its components are mixed together* in a landfill. These components include *compostables* (food and yard waste), *combustibles* (wood, paper), and *recyclables* (metal, glass, paper, plastics). When these are separated at their source or at waste processing centers, each component can be separated and only certain components are buried, greatly reducing the volume of solid waste.

In waste management today, the goals are to prevent pollution, to reduce consumption of new materials, and to recycle. Environmental managers follow a distinct priority of disposal methods to achieve these goals:

1. Reduce waste at the source (for example, use less paper by making double-sided photocopies).
2. Use reusable materials (cloth shopping bags instead of paper or plastic, ceramic mugs instead of plastic foam cups, remanufactured appliances, reusable cloth diapers).
3. Recycle materials (melt and reuse glass, plastics, and metals; turn paper to pulp and manufacture new paper).
4. Compost organic materials (food and yard waste).
5. Convert waste to energy (incinerate waste, but use the heat to generate electricity).
6. Incinerate (no energy recovery; it just reduces the bulk).
7. Landfill what remains.

## RECYCLING

Recyclable products—paper, glass, metals, and plastics—average about 60% of the material in municipal landfills (Figure 12–1). Recycling makes sense because of the large volume of these waste components and because recycling of some components is cheaper and uses less energy than producing these materials from raw resources. For example, the cost of producing an aluminum can by recycling is only 1/10 of the cost of producing a new aluminum can from aluminum ore (bauxite). The energy saving is 95%.

Producing new paper from recycled paper requires 60% less energy than does manufacturing new paper from trees. However, paper poses a problem because of the sheer volume generated (Figure 12–1) and its great variety. Waste paper includes newsprint from newspapers, colored paper, glossy coated paper in magazines, plasticized paper, and the largest category—corrugated cardboard. Unfortunately, the glut of newsprint being turned over for recycling has lowered its value, reducing the financial incentive to collect it for recycling. This is a problem with all recycling. A market must exist, or there is no incentive for recycling businesses.

Glass containers can be sorted by type and melted to produce new glass. Bottle deposits in some states not only have increased glass recycling efforts but also have reduced roadside litter. Plastics are a special problem because of their durability, the great volume produced, and the widely varied types (and mixtures of types) presently in use. Their bulkiness makes them less economical to collect, and a lack of market is a frequent problem.

## COMPOSTING

Gardeners long have composted organic waste. **Composting** is an aerobic process (meaning that it uses oxygen) in which bacterial degradation and oxidation break down organic material. In commercial composting plants, the organic material is placed in a vented chamber, which allows air to permeate the pile and drive the aerobic decomposition. This produces an organic rich residue that is useful as a fertilizer.

Composting is supposed to happen in a landfill. However, mechanical compaction of waste prior to burial, plus daily covering, plus rapid burial under new waste seriously reduce the oxygen and water supply, which inhibits decay. You probably have heard about researchers digging in landfills and finding decades-old newspapers that still are readable and hot dogs from the 1950s that still are recognizable (if no longer appetizing). Organic waste—food scraps, leaves, wood chips, paper—is a significant part of landfill volume (Box 12–2).

Many communities no longer accept yard waste (leaves, grass clippings, branches) in their landfills. They send homeowners to municipal composting facilities or recommend processing yard waste at home. Composting can produce noxious smells if waste is poorly mixed, but stirring and adding more wood chips ameliorates the problem. The product of composting is a useful fertilizer and its sale covers part of the processing cost. In some cities, "leave it on the lawn" programs are catching on for letting grass clippings decompose where they fall.

## INCINERATING

In the past, incineration of solid waste was quite common, although it was criticized for only partially combusting the waste and polluting the air in the process. The practice of incineration declined over the past two decades as air pollution laws grew more stringent. Now the severe need to reduce solid waste volume has forced communities to reconsider incineration. Fortunately, the incineration process, emissions control, and monitoring systems have improved greatly.

The first step in incineration is to reduce the waste volume, which involves separation of materials, typically into recyclables, combustibles, and noncombustibles. In some communities, separation is done by consumers. They place paper, glass, and metal in separate containers, and this waste is collected and sent to a recycling facility. In other areas, mixed waste is collected and the separation/recycling is done at a waste management facility. The nonrecyclable waste then is buried, composted, or burned.

In a modern incinerator, a temperature of 900 to 1,000°C (1,650 to 1,830°F) is sufficient to burn all combustibles, leaving only ash and noncombustibles such as metal cans (if they were not separated beforehand). After burning, the hot residue is quenched with water, drained, and trucked to a landfill. The volume reduction by incineration is considerable, averaging about 50%, and can be very high (75 to 90%) if all recyclable materials are separated beforehand.

Waste combustion generates hot gases and fine ash. The ash is filtered and trucked to a landfill. The gases are cooled and scrubbed of toxins, using various methods. For example, some use ammonia to render harmless the nitrogen oxides ($NO_X$). Activated carbon captures most mercury and dioxins. The cleaned gas travels up tall stacks from which it is dispersed by the wind.

Incineration not only reduces the waste volume to be placed in landfills but also can generate electric power. In such a facility, the hot gases produced by incineration heat water in pipes, converting it to steam, which turns turbines, generating electricity.

Incineration has problems, of course. Emissions may cause air pollution if scrubbing is inadequate. Incinerator ash and emissions may contain high levels of heavy metals such as lead, cadmium, and mercury. These plants are expensive to build, especially with proper pollution controls. And no one wants to live next door to an incinerator, which has lead to protests of incinerator construction. (This is an example of the NIMBY phenomenon—Not In My Back Yard!)

## LANDFILLS

Earlier in this century, most solid waste was piled in town dumps. Some were old sand or gravel pits, or sometimes the waste was dumped in a ravine or over a hill, or piled on flat land in remote locations. The piling of waste in the open air led to pollution and other geohazards (Figure 12–2). Wind blew trash onto nearby areas. Garbage attracted disease-carrying rodents, insects, and birds. Gases generated by the biologic breakdown of organic waste, foul-smelling and even flammable, were emitted from the landfill.

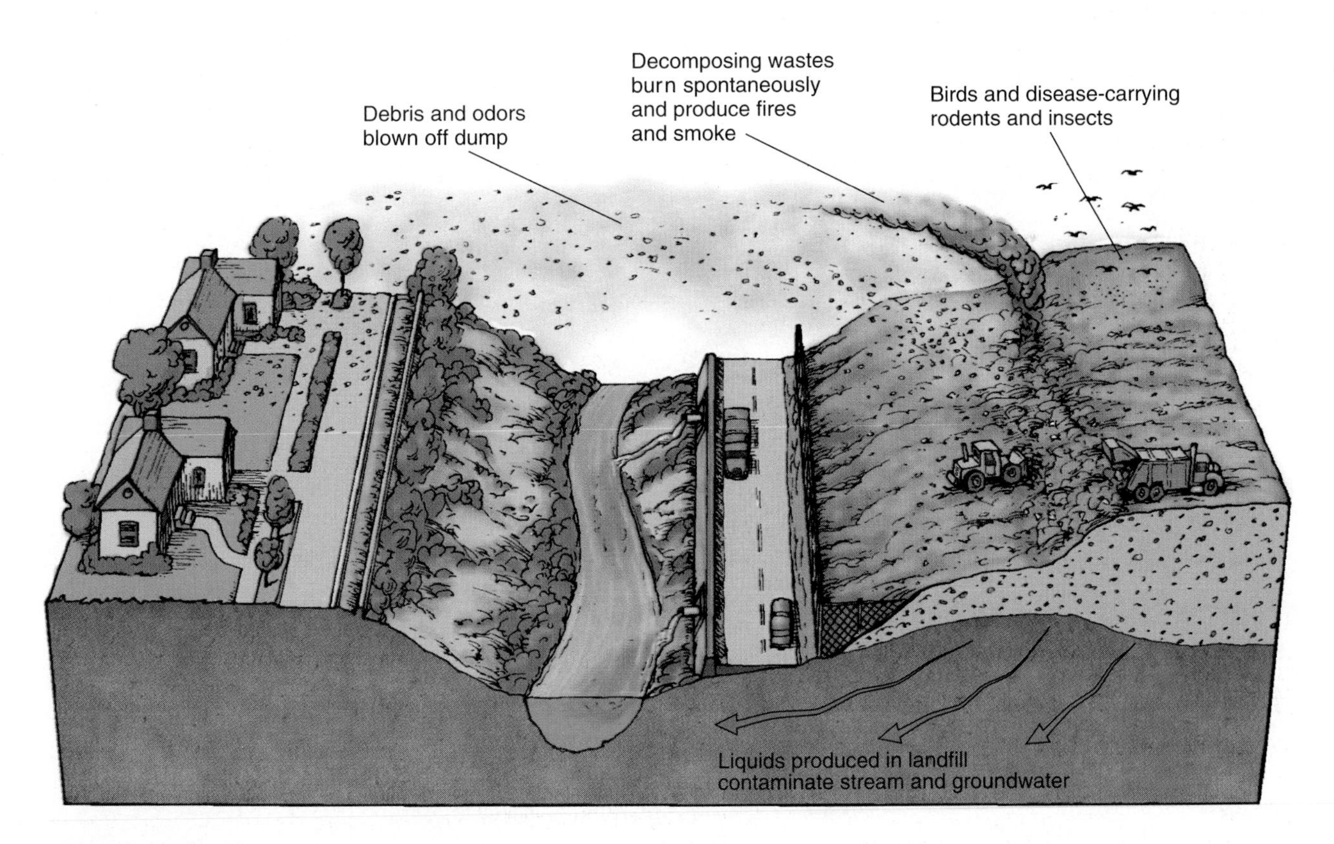


**FIGURE 12–2**
Hazards associated with open dumps.

## BOX 12–2 GARBOLOGY—A SCIENTIFIC LOOK AT SOLID WASTE

The tools of archeological research are being applied to landfills—excavating to expose layers, dating them, and sorting, describing, and determining the frequency of materials in each layer. Anthropologist William Rathje of the University of Arizona in Tucson and his students have operated the Garbage Project, excavating landfills since 1972.

Using a truck-mounted bucket auger (Figure 1), they obtain a core of garbage. They determine its age from readable newspapers, some as much as 40 years old (Figure 2) and from the distinctive style of containers or openers (such as "pop-top" rings).

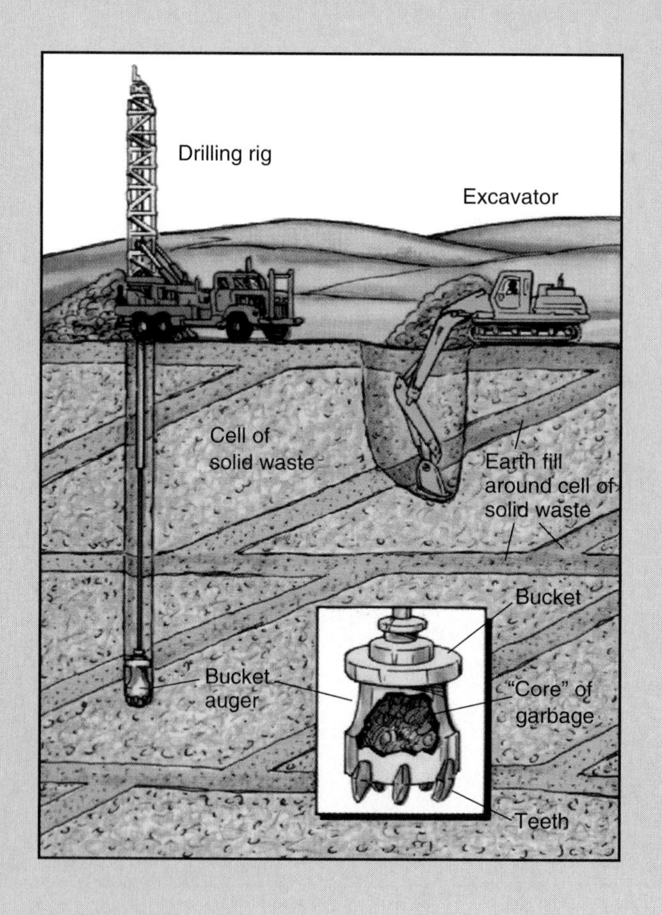


**FIGURE 1**
Methods of excavating solid waste for scientific analysis.

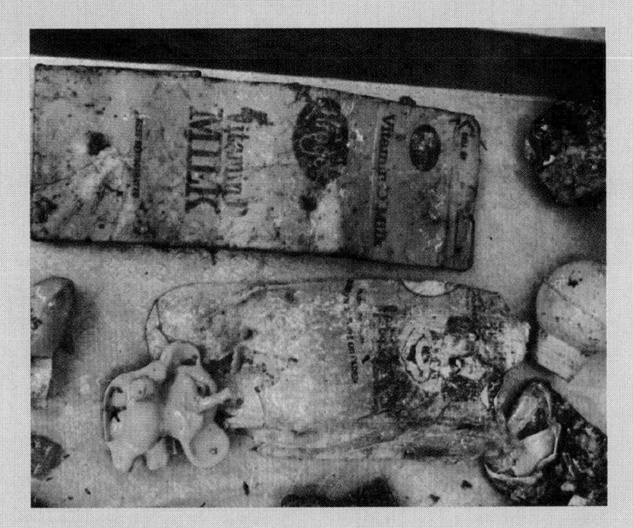


**FIGURE 2**
Readable printed material after 10 years in a sanitary landfill. (Courtesy of American Plastics Council/Riley N. Kinman, University of Cincinnati.)

**FIGURE 3**
Food waste after 10 years in a sanitary landfill. (Courtesy of American Plastics Council/Riley N. Kinman, University of Cincinnati.)

The results of the Garbage Project have been surprising. Materials that decompose in the atmosphere do not break down in a landfill because the environment is largely anaerobic. Newspapers from years ago are easily read. Some food even remains recognizable (Figure 3).

The researchers found that the component occupying the largest volume in landfills—50%—was paper (Figure 12-1). Plastics occupy only 10% by volume because their large initial volumes in household trash are reduced by compaction. Plastic foam occupies a surprisingly low 1% of landfill volume. Disposable diapers occupy only 1.4%.

The researchers also were surprised at how much hazardous waste can enter a landfill legally. For example, fingernail polish can contain four or five chemicals that EPA classifies as hazardous. Under the Resource Conservation and Recovery Act (RCRA), an industry would have to dispose of these chemicals at a regulated hazardous-waste facility. But household waste is exempt from RCRA.

The Tucson landfill is estimated to receive 350,000 nail polish bottles each year. Although the amount in each is small, the total hazardous waste from this item alone is considerable. For example, if 10% of the nail polish remains in a discarded bottle, nearly 257 liters (68 gallons) accumulate in the Tucson landfill per year. This is another example of the urban concentration factor discussed in Chapter 1.

The breakdown of organic material in the presence of oxygen is called an **aerobic reaction.** Aerobic reactions involving organic material form carbon dioxide. The $CO_2$ combines with rainwater to produce carbonic acid, which flows from the dump into groundwater. Decomposing organic material also generates heat, which can ignite combustible waste materials in a process known as **spontaneous combustion.** Smoldering fires were common in old open dumps and proved a great nuisance. Windblown debris, noxious smells, fires, and smoke provided the impetus to develop a better way to store solid waste. Thus the sanitary landfill was invented.

## Sanitary Landfills

A solid waste disposal site in which each day's accumulation is covered with a blanket of sediment is called a **sanitary landfill.** Solid waste is trucked to the landfill and placed in the active working area for that day (Figure 12–3). Bulldozers pile the waste against the side of the area. At the end of the day, *fill* (soil, sand, clay) is deposited over the day's accumulation to form a **cell** of solid waste. This makes the landfill "sanitary" because the sediment cover prevents debris from blowing off the landfill, decreases the presence of vermin, and reduces noxious odors and the risk of fire. However, this basic design corrects only a few of the problems associated with landfills.

## Sanitary Landfill Hazards

Although sanitary landfills alleviate the problems of open dumps, they generate new hazards. Covered material continues to decay, but in the absence of oxygen. Thus it decays **anaerobically,** a process that results in the generation of methane gas ($CH_4$). Methane is invisible, odorless, and combustible. It can infiltrate a landfill and any permeable adjacent ground. In one notable incident on New York's Long Island, methane from landfills moved beneath frozen ground during winter, entering the basements of nearby homes. The gas subsequently was

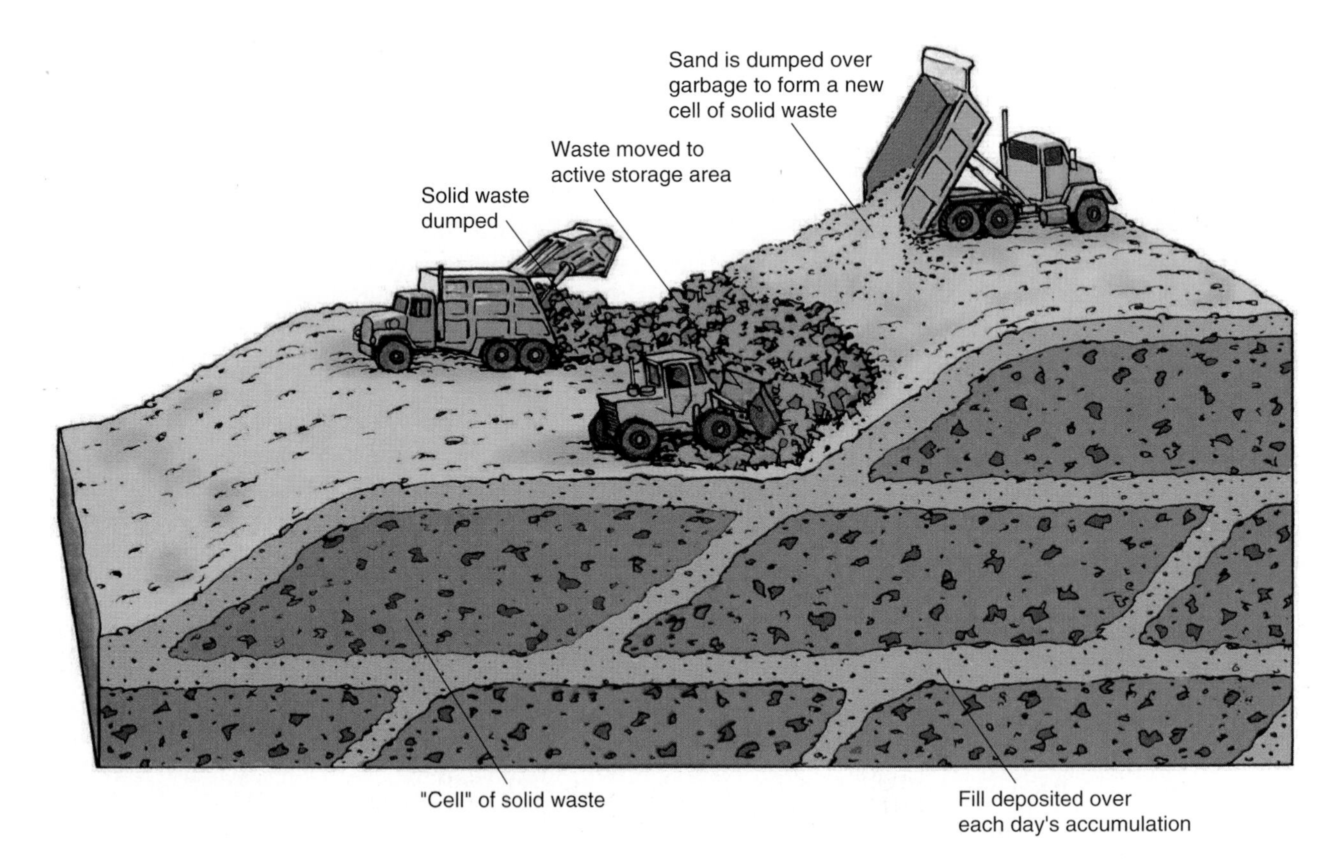


**FIGURE 12–3**
Cross section showing the operation of a sanitary landfill.

ignited from flames in oil furnaces, with explosive results.

Rainwater and groundwater that enter a landfill combine with reactive material to form a foul liquid called **leachate.** The chemical composition of a leachate varies with the waste, its reactivity, the climate, depth to the water table, and the degree to which aerobic or anaerobic conditions prevail.

Leachates commonly contain metals (chiefly iron, manganese, and zinc), nitrates, and phosphates. They often contain levels of lead and cadmium metals that exceed drinking water standards. Leachates usually include organics, too. At a minimum, leachate increases the salt content of groundwater, concentrates solid particles, and may introduce harmful metals and pathogens (bacteria) that degrade drinking water.

## Sanitary Landfill Design to Eliminate Hazards

A well-designed sanitary landfill (Figure 12–4) is properly sited geologically and minimizes methane and leachate released into the environment. To reduce leachate volume, an *impermeable cap* of plastic or clay is placed over the inactive part of the landfill to limit rainfall infiltration. Leachate within the landfill is collected by internal *piping systems* and removed by pumps for treatment.

Leachate not removed by the piping system percolates downward through the landfill. At the bottom, a properly designed landfill employs one or more heavy plastic liners to prevent leachate from entering aquifers beneath the landfill. Such liners are both impermeable and unreactive, and they have a thickness of 20 to 60 mils, or thousandths of an inch. (Compare this to a common trash bag, which is only 1 or 2 mils thick.)

A liner makes a piping system essential, because leachate accumulates above a liner, like water filling a bathtub. The piping system drains the leachate to a treatment point. Some landfill leaks can be detected by the oily appearance of leachate and the death of vegetation around the landfill's base.

Methane gas produced within a capped sanitary landfill can build to dangerous levels unless it is removed periodically. Various gas-handling methods are used, depending on the landfill's size, gas volume, and proximity to housing. In any case, methane gas control requires a system of wells with-

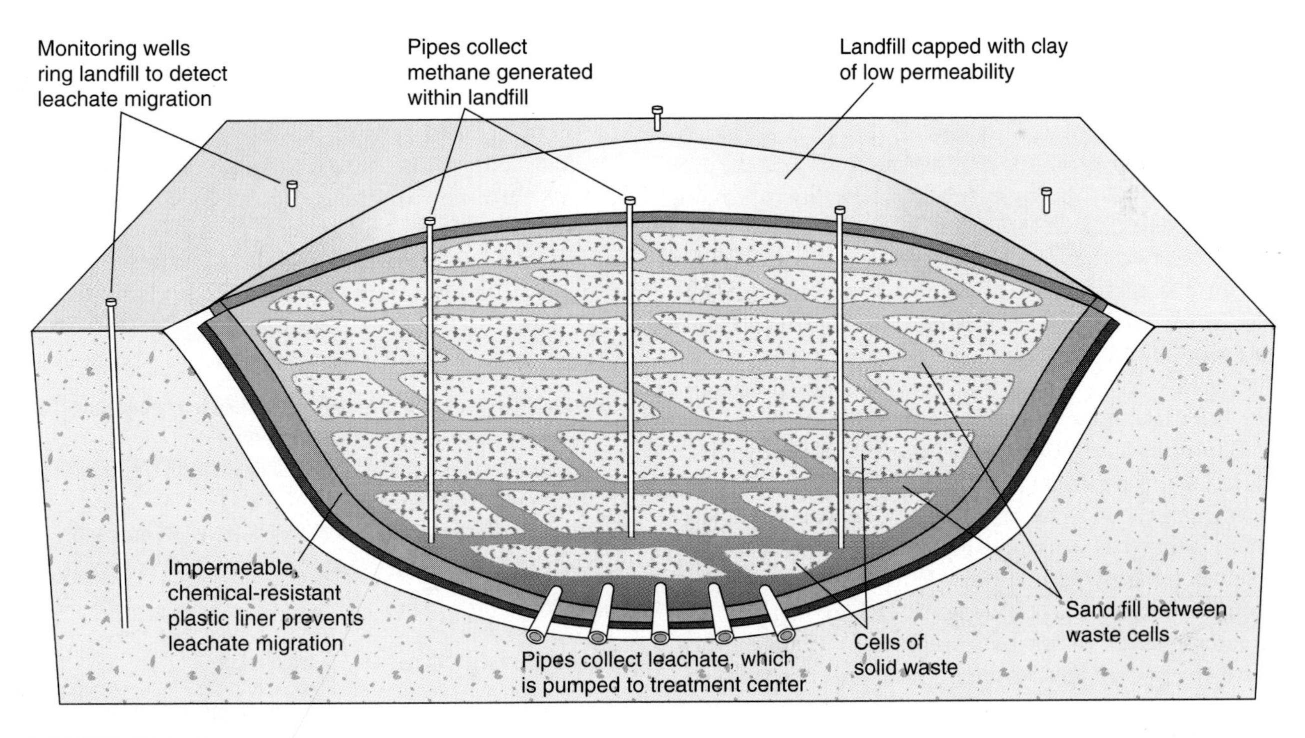


FIGURE 12–4
Proper design of a sanitary landfill to eliminate leachate and gas production and migration.

in the landfill to collect the gas and pump it to the surface.

If the gas volume is small, it can be vented or burned at the surface. In larger landfills, the gas is collected and sold to gas utilities as a substitute for natural gas. Earth's largest sanitary landfill, the Fresh Kills Landfill in New York City, sells landfill-generated methane to a local gas company. This reduces the risk of explosion and helps pay for landfill operation.

## Geologic Siting of Sanitary Landfills

Factors to consider in siting a landfill include the nature of the underlying rock or sediment, water table elevation, groundwater flow direction, and local climate. Most important is the kind of geologic material underlying the site. If it is sediment (gravel, sand, silt, clay), it needs to be impermeable to limit the migration of leachate. The only sediment that meets that criterion is clay, which makes an ideal landfill substrate because it has low permeability.

Rock strata also can form a good substrate, provided there are no extensive fractures, joints, or faults and that the rocks will not react with leachate. For example, limestone may be problematic because acidic leachate passing through it will react with and dissolve the limestone. To avoid erosion and release of the landfill contents during a flood, potential landfill sites should also not be in flood-prone areas.

Because a basic goal in landfill design is to avoid contaminating groundwater, the most important aspects of landfill siting are subsurface sediment/rock permeability and elevation of the water table. Sediment or rock that has low permeability is desirable because it discourages migration of leachate. Also, the lower the water table is beneath the bottom of the landfill, the better the site because this makes it harder for any leachate that gets through breaks in the liner to contaminate local water supplies.

The best landfill site usually is underlain by silts, clays, or unfractured shales (Figure 12–5a). The low permeability of these materials (Chapter 3) makes them behave as a natural liner, preventing leachate from percolating downward. Where silts and sands overlie fractured rock (Figure 12–5b), there is greater chance of leachate migration. It may infiltrate the sediment and enter the fracture system in the rocks below.

Surface sands overlying an impermeable layer at depth (Figure 12–5c) pose a greater problem because leachate may contaminate shallow wells in the area. An especially bad situation occurs when sandy surface material overlies limestone (Figure 12–5d), which may have solution cavities (Chapter 10). Leachate may contaminate the shallow sandy aquifer and then seep down through solution cavities to contaminate the limestone aquifer as well.

Careful geologic investigation must be conducted when siting a new landfill, but this was not done for many older ones. They are unlined, and little or no geologic thought was given to site selection. Many older landfills now are being closed and capped with impermeable material to minimize rain infiltration. However, a cap leak, a water table fluctuation, and continuing anaerobic decomposition will generate leachate far into the future. Such older landfills are "time bombs" that threaten water supplies with contamination. Despite such problems with older landfills, a properly designed modern landfill can be a model of responsible land use.

All of these factors are important to consider even if the landfill has a liner, because no liner is perfect or permanent. It can crack or be punctured or attacked by chemicals. It is best to acknowledge that leachate will migrate sooner or later, and plan to minimize it.

Because a landfill liner is a water barrier, leachate inevitably will accumulate above it. In a properly designed landfill, the leachate is collected by a system of perforated pipes (pipes with holes punched in them) placed in the base of the landfill. The leachate enters the pipes through the holes and drains through the pipes to a collection point, where it is treated. This system, plus proper venting of gases released in the landfill, alleviates most of the problems commonly associated with landfills.

In summary, there are three key elements in a proper sanitary landfill operation:

- ☐ Choosing the site geologically (low-permeability substrate; not a flood-prone area).
- ☐ Properly designing the landfill (liner, gas vents, leachate collection).
- ☐ Properly maintaining all systems and monitoring peripheral areas for air and water pollutants from the landfill.

## Monitoring Sanitary Landfills

The potential for geohazards requires that even properly designed landfills be monitored through their designed life span and beyond. **Monitoring**

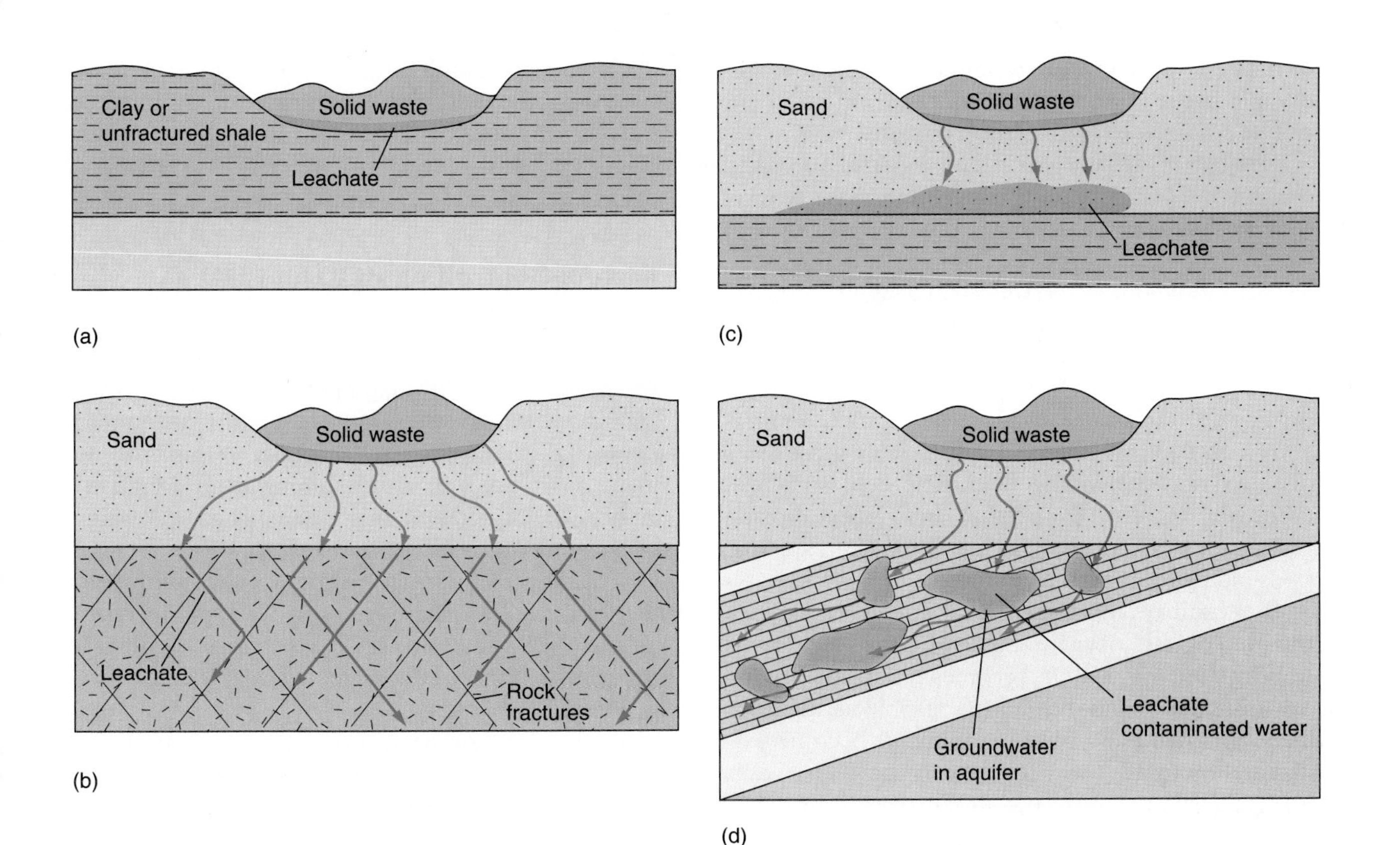


**FIGURE 12–5**
Possible geologic conditions that must be considered in the siting of sanitary landfills. (a) Leachate is trapped at the top of an aquiclude. (b) Lechate enters the rock below and moves along the fractures. (c) Lechate migrates in the upper aquifer. An aquiclude prevents downward migration. (d) Leachate enters solution cavities in the limestone below and migrates down the slope of the aquifer. (Adapted from Schneider, U.S. Geological Survey, Circular 601-F, Fig. 2, p. F8.)

**wells** are drilled into a landfill and at various distances around it. Water samples from various depths in each monitoring well are tested regularly. If water quality deviates significantly from normal, the changes recorded in different wells can help locate the source (such as a liner leak) and indicate the direction and speed of leachate movement. Remedial measures then can be taken, such as drilling wells around the landfill to withdraw and treat contaminated water.

## NONTOXIC ORGANIC LIQUID WASTE

Nontoxic, organic liquid waste such as sewage, effluent from garbage disposals in kitchen sinks, and animal manure is not hazardous, although it can carry dangerous bacteria. Nor is it as dramatic as the toxic waste discussed later in the chapter. However, the sheer volume of nontoxic organic waste exerts a profound effect on surface water quality worldwide. To understand how this organic material affects water quality, it is necessary to understand what happens in the breakdown of organic particles in water.

### Biological Oxygen Demand (BOD) and Dissolved Oxygen (DO) Level

The aerobic breakdown of organic material (sewage waste, plant and animal tissue) requires oxygen. Animals also require oxygen for respiration. The total oxygen demand for respiration and organic breakdown is the **biological oxygen demand (BOD)** of a volume of water or bottom sediment. If sufficient oxygen is available, aquatic animals can breathe and organic particles in the water and bottom sediment can be oxidized to form carbon dioxide.

Oxygen in aquatic environments is dissolved in the water. A measure of its concentration is the level of **dissolved oxygen (DO).** The DO varies with water temperature. As the temperature increases, less gas can be dissolved in a fluid and vice-versa. You can experience this physical law by opening warm soda and cold soda. What happens? The warm soda bubbles from the bottle (less carbon dioxide gas can be dissolved at the elevated temperature). In contrast, the cold soda retains most of its dissolved gas within the fluid. The cold soda will eventually "go flat" as it warms and carbon dioxide gas escapes. This principle has important applications in environmental situations.

If the dissolved oxygen level equals or exceeds the biological oxygen demand, the system is healthy. Organic matter becomes oxidized, aquatic organisms have no respiration problems, bottom life is abundant, and bottom sediments become oxidized (Figure 12–6a). However, if the DO level is not sufficient to satisfy the BOD, oxygen becomes depleted, swimming organisms die, most bottom organisms die, and organic particles accumulate on the bottom as a dark-colored, anaerobic sediment (Figure 12–6b). A body of water thus deprived of oxygen is undergoing **eutrophication.**

## Factors Controlling BOD and DO

The major source of DO is oxygen from the atmosphere. Atmospheric oxygen enters the water by diffusion through the water surface. The more turbulent the water surface, the greater the amount of oxygen that is added to the water (think of the many air bubbles in water coming from a tap with an aerator, or aeration that occurs as coastal waters crash onto the coast in the surf zone). In general, DO decreases downward in the water column from the surface. Seasonal changes in water temperature affect water's ability to hold oxygen in solution. In the summertime, the warmer water holds less oxygen than during winter.

Water also can be heated anthropogenically by industrial processes, including electrical power generation. When heated water is added to a water

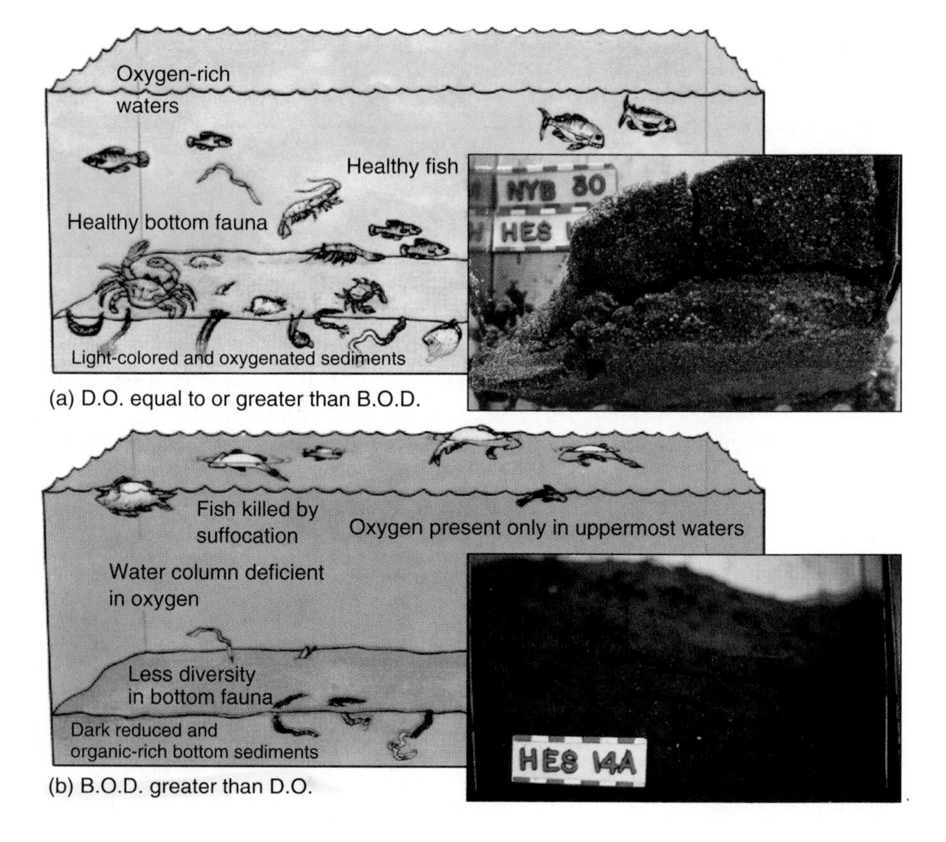


**FIGURE 12–6**
Cross sections showing the effects of changes in the Biological Oxygen Demand (BOD) versus the Dissolved Oxygen (DO) in a water body. (a) If DO equals or exceeds BOD, organisms thrive. (b) If BOD exceeds DO, the water body eutrophies, and organisms perish. The photographs show the differences in sediment characteristics under the two conditions. (Photos by author.)

body, it locally raises the water body's temperature and decreases the DO level. This is called **thermal pollution.** Its destructive effect on organisms is shown in Figure 12–7.

In the figure, the industrial plant discharge is free of toxic pollutants and differs from the river water only in its elevated temperature. It is summertime, and the stream has a high BOD that is just balanced by the DO in the portion upstream of the industrial complex (A-B). The heated industrial discharge water raises the stream temperature downstream of the plant (B-C). This warmed water can hold less oxygen, so aquatic animals develop respiration problems, resulting in a *fish kill,* essentially from suffocation.

BOD is controlled not only by respiration but also by the addition of natural organic debris (plant and animal tissue) and anthropogenic debris (organic waste and sewage). The presence of anthropogenic organic material is far more significant than the natural debris. A high anthropogenic input to aquatic systems occurs from livestock feedlot runoff, food processing by-products, and especially from sewage, raw or treated, that is discharged into nearby waterways.

Any liquid discharge from an industrial plant is called **effluent.** Organic effluent can be treated by infusing it with large volumes of air through submersible pumps, a technique called aeration. Another method is to treat waste by filtering it through gravel. This allows oxygen and bacteria time to break down organic material. Bacteria reside on the surface of the gravel and feed on the organic material.

## Sewage

Sewage is the largest anthropogenic organic component supplied to the environment. It is a *slurry* (mixed liquid-solid), rich in nitrogen and phosphorus. In many nations, raw sewage is released into nearby water bodies. The extremely high BOD of untreated sewage places a great oxygen demand on the water into which it is discharged. Fortunately, most sewage is treated to some degree before it is released.

***Sewage Treatment—Private Septic Tanks.*** In rural farms and homes and many suburban areas, sewage usually is disposed of on-site by organic digestion in an underground **septic tank** (Figure 12–8). Waste flows into the tank by gravity and is digested by anaerobic bacteria. This produces a sludge that settles to the bottom and must be pumped out after a year or two, depending on usage. The liquid effluent (partially purified water) flows from the tank into a system of perforated pipes in a leaching field. The leachate is absorbed into the permeable soil. Here the effluent is further purified by the action of soil bacteria.

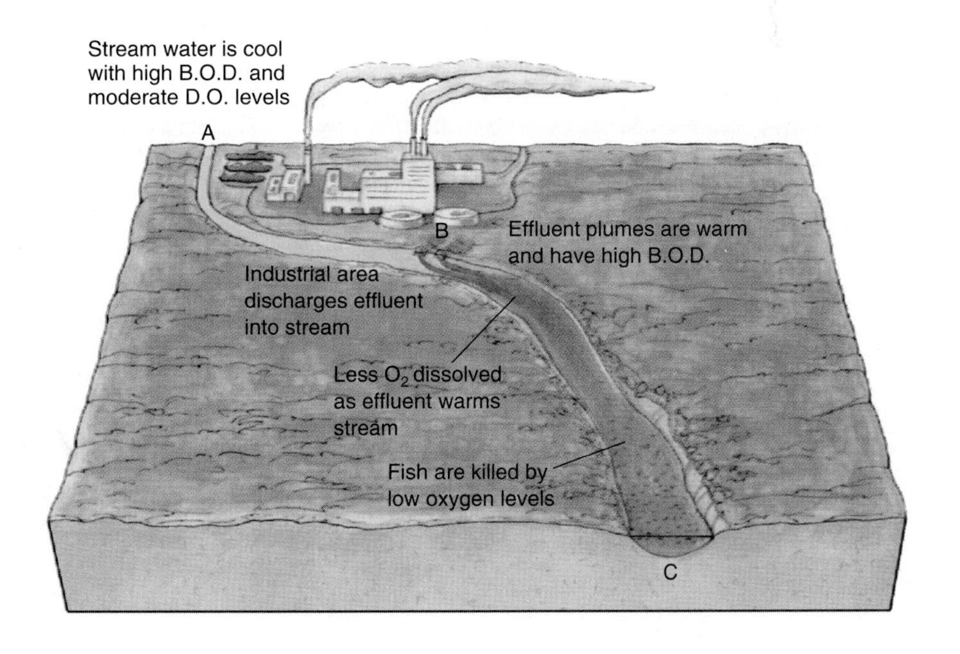


**FIGURE 12–7**
Diagram (map view) showing how thermal pollution from industrial water emissions causes fish kills in a stream with a high BOD.

FIGURE 12–8
Individual treatment of sewage by a septic tank-drainage field system on a home lot.

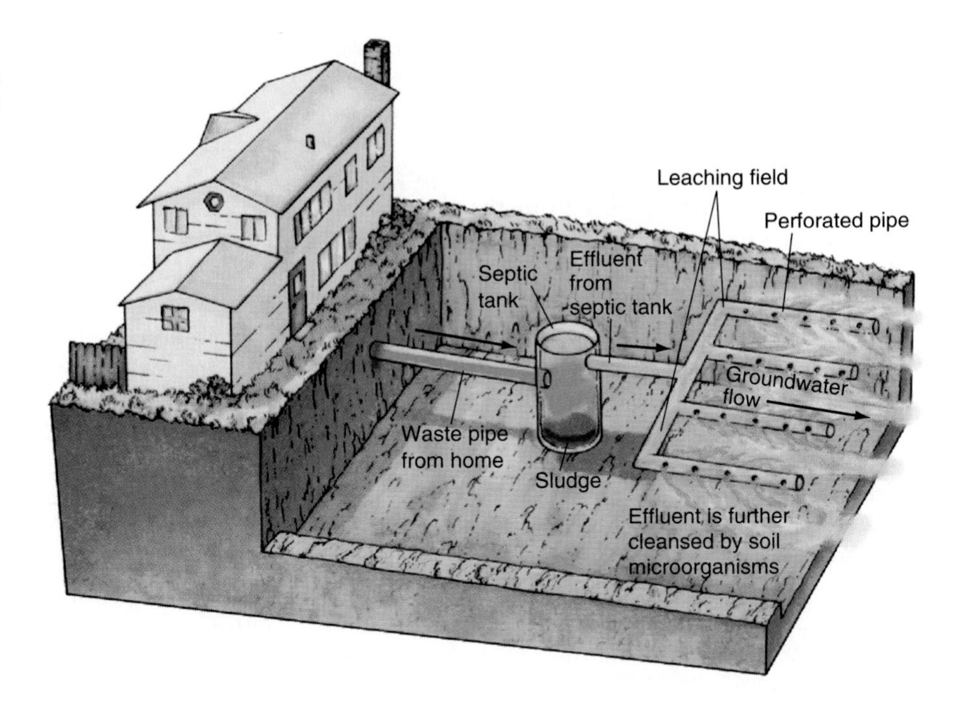


The location of a home's septic tank is very important, especially if the drinking water source is a well. Contamination of drinking water by septic tank effluent occurs easily, so a home septic system must be installed with care and according to regulations.

*Sewage Treatment—Urban Sewers and Plants.* In urban areas, domestic waste is collected by an underground system of **sanitary sewers** that deliver it directly to sewage treatment facilities. A separate system of **storm sewers** collects storm runoff from the land and streets and delivers it directly into nearby streams, lakes, or the ocean. In many older cities and towns, these sewer systems are joined to form **combined sewers** that collect all the flow and deliver it to a sewage treatment plant.

Most plants can process more waste water volume than their planned capacity. However, sustained heavy rainfall adds so much water to storm sewers that treatment plants must divert part of the flow to the nearest body of water—untreated. Thus, in a heavy storm, an area with combined sewers might be forced to contaminate nearby surface water with raw sewage and bacteria. Increasingly, these areas are building **storm water waste storage facilities** to store the excess flow temporarily until it can be sent to a sewage treatment plant for proper treatment.

## Urban Sewage Treatment

Sewage treatment (1) screens floating material (wood, cloth, plastic objects), (2) removes solids by settling, (3) digests organic material to varying degrees, (4) settles the remaining particles to clear the water, and (5) chlorinates the effluent to kill any remaining pathogenic bacteria.

Sewage can be treated at three different levels:

- □ *Primary treatment plants* screen material, digest some of the organic material, and chlorinate the effluent.
- □ *Secondary treatment plants,* currently the most widely used, employ all processes of primary treatment, plus extensive digestion of organic material to reduce BOD significantly (Figure 12–9).
- □ *Tertiary treatment plants* eliminate virtually all BOD and produce a potable effluent (one of drinking quality). Very few tertiary plants exist in the United States because they are expensive and require considerable land for settling ponds and filtration facilities. One example is at Lake Tahoe, Nevada.

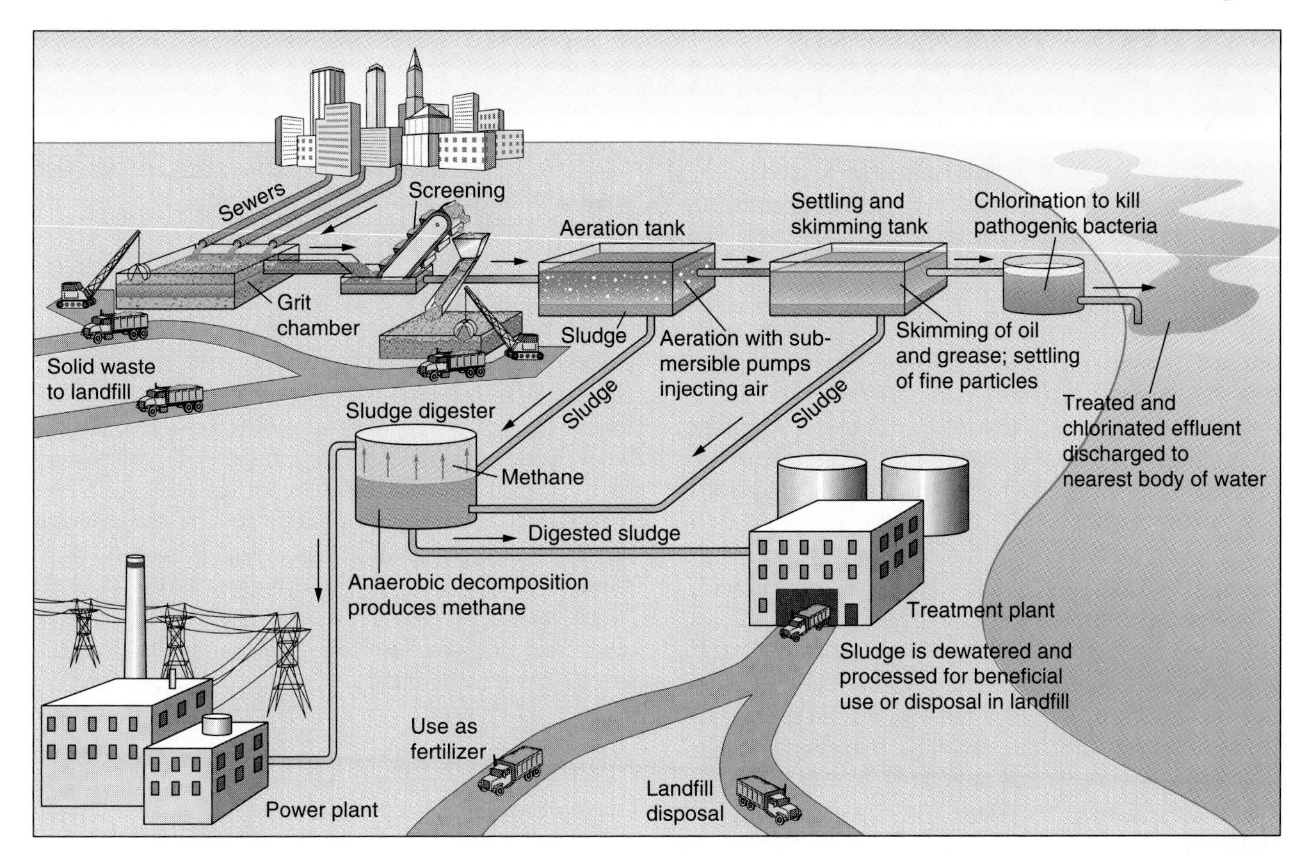


**FIGURE 12-9**
Generalized operation and products of a secondary treatment system.

Secondary sewage treatment involves several processes that yield solid, liquid, and gaseous waste (Figure 12–9). Sewage entering the plant first passes through a *grit chamber,* where sand and gravel are settled. It then moves through a series of screens that remove floatable debris. These solids are trucked to sanitary landfills.

The sewage then travels to large *aeration tanks* where pumps aerate the mixture. This provides more oxygen to oxidize organic particles. At the same time, bacteria are breaking down or digesting sewage particles. The solid residue, **sewage sludge,** settles to the bottom and is sent to a *sludge digester.*

Sludge digestion employs intense anaerobic bacterial action, a by-product of which is methane gas. In some plants, this gas is used to generate electricity. (The methane is burned; the heat generates steam; the steam drives turbines, which turn generators to produce electricity.) Some large plants run virtually all of their operations with electricity produced from their sewage-generated methane. These self-powered plants not only reflect good resource management but also have the advantage of continuing to operate during power failures at utilities.

The digested effluent from the aeration tank moves to the *settling tank,* where oil and grease are skimmed from the surface, and fine suspended particles settle from the effluent, decreasing its turbidity. The effluent is chlorinated to kill pathogenic bacteria and is discharged into a nearby water body.

Sludge on the settling tank floor is pumped out for disposal off site. In coastal areas, sewage sludge once was pumped into tankers and deposited in the ocean, but federal law now prohibits ocean dumping. If its metal content is low, sludge can be treated and processed into fertilizer. Otherwise it can be incinerated and the ash deposited in lined landfills.

## ACID DRAINAGE FROM MINES

Some regions are afflicted with acidic surface water, and the source is not acid precipitation. Rather, it is produced when iron sulfide minerals in the rocks react with water and oxygen to produce sulfuric acid. When the acidic water drains into streams, it is called **acid drainage.** In sufficient quantities, it can acidify a stream, reducing its pH. A plentiful iron sulfide mineral involved in forming such acid drainage is *pyrite,* commonly called fool's gold because of its resemblance to real gold.

Acidic drainage was first recognized as a problem in underground coal mines. Many coal seams are bounded above and below by dark shales that contain pyrite. Groundwater enters the mine through rock fractures and mined-out areas and reacts with the pyrite and oxygen in the air to form sulfuric acid.

When the resulting acidic water accumulates within a mine and drains into groundwater and streams, it is called *acid mine drainage.* The increasing acidity of groundwater enables it to leach metals from rocks, so the acid drainage may become enriched in metals as it migrates (Figure 12–10). When the drainage works its way into streams, dissolved iron becomes exposed to oxygen and turns formerly clear streamwater an ugly color—rusty or bright orange. The decrease in stream pH can result in fish kills.

Acid drainage is best developed where coal mining has exposed iron sulfide minerals to air and water. Pollution controls (See Chapter 11) now are driving utilities to use low-sulfur coal. Consequently, less high-sulfur coal is being mined, which reduces the incidence of acid mine drainage.

Acid drainage is not limited to underground rocks, however. It also develops at the surface, wherever pyrite-rich rocks are exposed to weathering. This can occur at natural outcrops, at a surface coal mine rich in pyrite (Chapter 8), or where pyrite-bearing rocks have been used as fill. This geohazard is long-lived because the weathering process and percolation of acidified waters takes time, and the acid drainage may not be observed until years after mining or construction has ended. Its effect can continue for decades, until all of the pyrite has weathered.

### Mitigating Acid Drainage

In mine waste and construction areas, acid drainage is mitigated by covering reactive rock with imper-

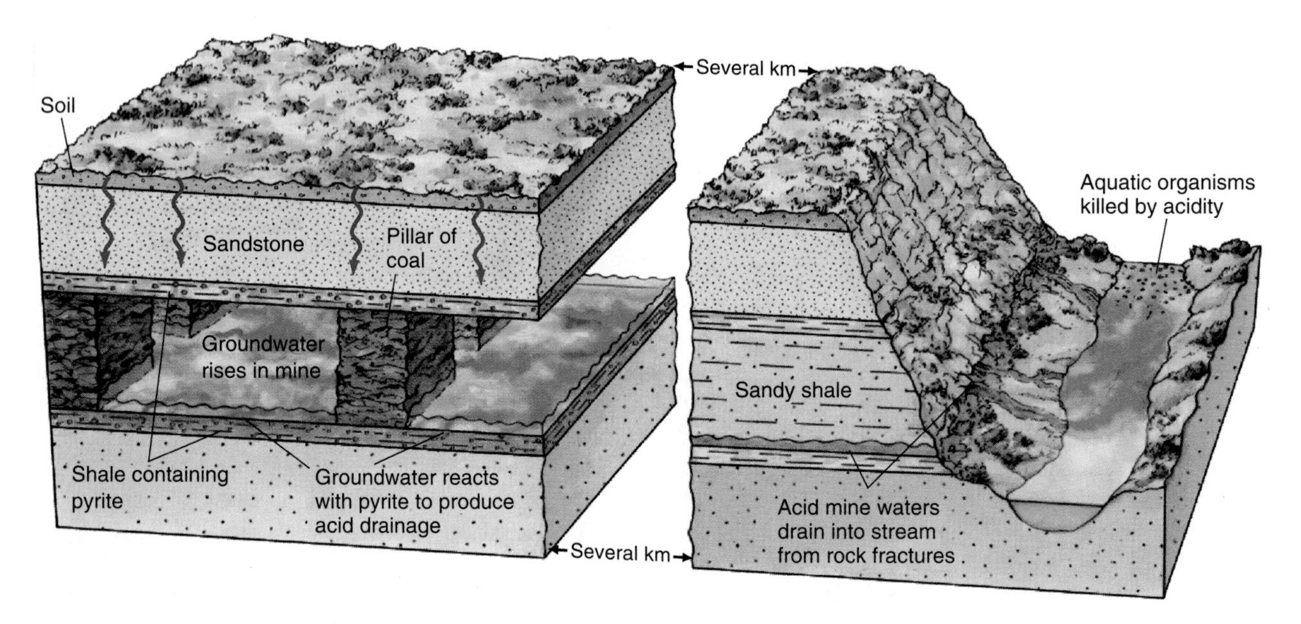


**FIGURE 12–10**
Geologic section through an abandoned coal mine showing how acid water forms and migrates laterally into surface water bodies.

meable clay to prevent rain infiltration. Seepage from the site is drained into treatment ponds, where lime or crushed limestone is added to neutralize the acidity.

Such treatment is expensive and was not required of mining companies until about two decades ago. As a result, hundreds of Appalachian and Rocky Mountain streams became polluted with acidic drainage. Many remain polluted, although acid production has declined. The Surface Mining Control and Reclamation Act of 1977 now requires companies to address this geohazard. We gain improved water quality, safer water supplies, and preserved habitats, but the cost is passed on through higher coal prices and electric bills.

In the case of acid drainage from construction sites, the best way to avoid it is to choose a less reactive rock for fill. An interesting example is the Halifax, Nova Scotia International Airport, built in 1982, which inadvertently created an acid drainage hazard. Airport runways were dug in the Meguma Slate, a pyritic rock. Part of the excavated material was used as runway fill and the rest was placed in a waste pile near the runways. Soon after completion, runway underdrains and the waste pile began to discharge acid drainage, including high concentrations of trace metals that threatened a prime salmon fishery on a nearby river.

To correct the problem, the waste pile was capped with 0.7 meters (2.3 feet) of clay at a cost of $800,000. A $500,000 lime treatment facility was also built. Treating the drainage continues at an annual cost of $240,000. Because of this incident, all future construction involving the Meguma Slate requires environmental evaluation by Environment Canada, the equivalent of the U.S. EPA.

## HAZARDOUS LIQUID WASTE

A liquid waste is considered hazardous by the U.S. Environmental Protection Agency if it has one or more of the following characteristics (Figure 12–11):

- ☐ *Ignitability:* the waste can catch fire. Examples are solvents, paints, and petroleum products.
- ☐ *Corrosivity:* the liquid is capable of corroding metal, such as "eating through" a storage barrel. Acids are an example.
- ☐ *Reactivity:* these liquids are unstable under normal conditions and can explode if ignited or, in some cases, if mixed with water.

**FIGURE 12–11**
Characteristics of hazardous waste. (Environmental Protection Agency.)

**TABLE 12–1**
Examples of hazardous waste generated by businesses and industries

| Waste Generators | Waste Type |
|---|---|
| Chemical manufactures | Strong acids and bases<br>Spent solvents<br>Reactive wastes |
| Vehicle maintenance shops | Heavy metal paint wastes<br>Ignitable wastes<br>Used lead acid batteries<br>Spent solvents |
| Printing industry | Heavy metal solutions<br>Waste inks<br>Spent solvents<br>Spent electroplating wastes<br>Ink sludges containing heavy metals |
| Leather products manufacturing | Waste toluene and benzene |
| Paper industry | Paint wastes containing heavy metals<br>Ignitable solvents<br>Strong acids and bases |
| Construction industry | Ignitable paint wastes<br>Spent solvents<br>Strong acids and bases |
| Cleaning agents and cosmetics manufacturing | Heavy metal dusts<br>Ignitable wastes<br>Flammable solvents<br>Strong acids and bases |
| Furniture and wood manufacturing and refinishing | Ignitable wastes<br>Spent solvents |
| Metal manufacturing | Paint wastes containing heavy metals<br>Strong acids and bases<br>Cyanide wastes<br>Sludges containing heavy metals |

*Source:* Environmental Protection Agency.

- *Toxicity:* these liquids are harmful if absorbed (on the skin) or ingested.

Examples of hazardous wastes are shown in Table 12–1.

In an approved facility (Figure 12–12), hazardous wastes are tracked from their generation to final disposal ("cradle to grave"). Transporters pick up properly packaged and labeled hazardous waste from the industries that generate them (generators). Each shipment is accompanied by a manifest (a list of items) and moves in vehicles clearly labeled with the type of waste. This enables emergency workers to determine quickly any potential danger in case of leakage, fire, explosion, or submergence.

After the hazardous liquids are used, residue is collected and transported to a treatment facility operating with an EPA permit. After treatment, these chemicals are disposed of in an EPA-approved disposal site, such as a secure landfill, surface impoundment, or deep-well injection site (described later in this chapter).

## TOXIC LIQUID WASTE

Prior to the passing of the 1970 Clean Air Act (amended in 1990) and the 1972 Clean Water Act (amended in 1991), liquid waste routinely was burned off, buried, poured into pits, or discharged into the nearest body of water (Box 12–3). Much of this liquid waste was toxic. Some was explosive or inflammable (petrochemicals), corrosive (acids used in industrial processing), radioactive (mining waste), toxic (PCBs), or infectious (medical waste).

Fortunately, we have come a long way. Ever-stricter laws regulate the disposal of these pollutants each year (Box 12–4). Unfortunately, some companies have found it more profitable to delay environ-

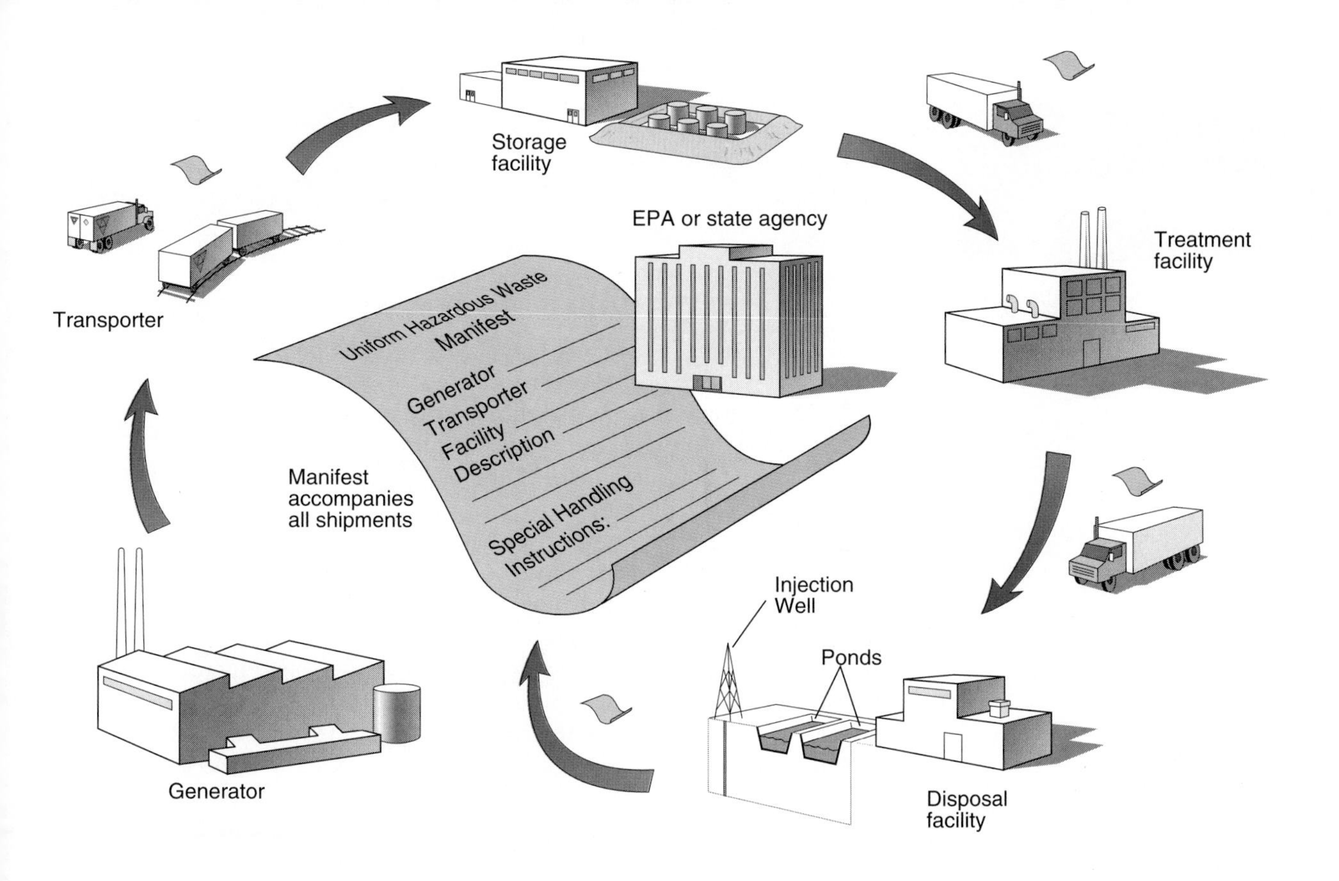


**FIGURE 12–12**
Tracking hazardous wastes from generation to disposal. (Environmental Protection Agency.)

mental action in court as long as possible rather than to pay for far more expensive remedial action. Environmental agencies have spent considerable money and time bringing polluters to justice.

## A Legacy, Past and Present

Despite legal progress in the United States, serious liquid waste disposal problems remain. Accidents, uninformed actions, and illegal dumping cause the continuing release of toxic material.

***The PCBs of Hudson River.*** PCBs, or polychlorinated biphenyls, are oils used as insulators in electrical power transformers. The problem is that they are highly toxic. For years, the General Electric Company legally released PCBs into the Hudson River behind a dam at Fort Edward, New York (Figure 12–13). In 1973 the Niagara Mohawk Power Company removed the dam. Some of the massive build-up of PCB-contaminated sediment behind it began to wash into the Hudson. Serious river flooding three years later eroded more of these deposits, and PCB levels in sediment and fish rose steadily downstream each year.

***Love Canal.*** Much toxic waste remains dangerous for long periods, even hundreds of years. Before environmental protection laws were passed, much toxic liquid waste was poured into steel drums and either placed in surface dumps or buried. In both settings, these containers corroded within decades. Few knew the location of these dumps or even what type of material was buried there. As containers corroded and leaked into surface water and groundwater, health problems emerged.

A notorious example is the Love Canal area near Buffalo, New York. Toxic waste in containers was dumped—legally—by the Hooker Chemical Company in the abandoned Love Canal during the 1940s and 1950s. The canal subsequently was filled and the land sold for $1.00 to developers and the local school board. A school, playground, and many homes were built on the site (Figure 12–14).

**BOX 12–3**

## AMERICA'S ONLY RIVER FIRE HAZARD—A TURNING POINT IN THE ENVIRONMENTAL MOVEMENT

The fire lasted only one-half hour. Damage was modest, about $50,000. However, the event was momentous in the environmental movement. On June 22, 1969, the surface of the polluted Cuyahoga River, just southeast of Cleveland, Ohio, *caught fire* (Figure 1). Fires on the river had occurred before, in 1936 and 1952, but this unusual fire sparked the consciousness of an emerging environmental movement. America had a *river* so badly polluted with flammable chemicals that it was declared a fire hazard!

The fire came at the time of growing recognition of other environmental problems such as smog, lake eutrophication, pollution, and major oil spills. It became a symbol of the scope of environmental degradation, and it was important in stimulating the passage of major environmental laws of the 1970s.

The story has a happy ending. Today, water quality of the Cuyahoga between Cleveland and Akron, Ohio, has greatly improved and it has been made useful for recreation. Its banks are lined with shops and restaurants. The Cuyahoga episode is an excellent example of how a badly deteriorated area can be restored through environmental regulation.

**FIGURE 1**
Fire on the Cuyahoga River, near Cleveland, Ohio, in 1969. (*Plain Dealer,* Cleveland, Ohio.)

## BOX 12–4 WHO'S WATCHING THE STORE?

A variety of federal, state, and local laws now control the use and disposal of waste. Some control the generation of waste components. Some control manufacturing and processing of products. Others regulate transportation, and still others control consumer and industrial use and disposal.

The more important federal laws aimed at reducing air, land, and water pollution are shown in Figure 1.

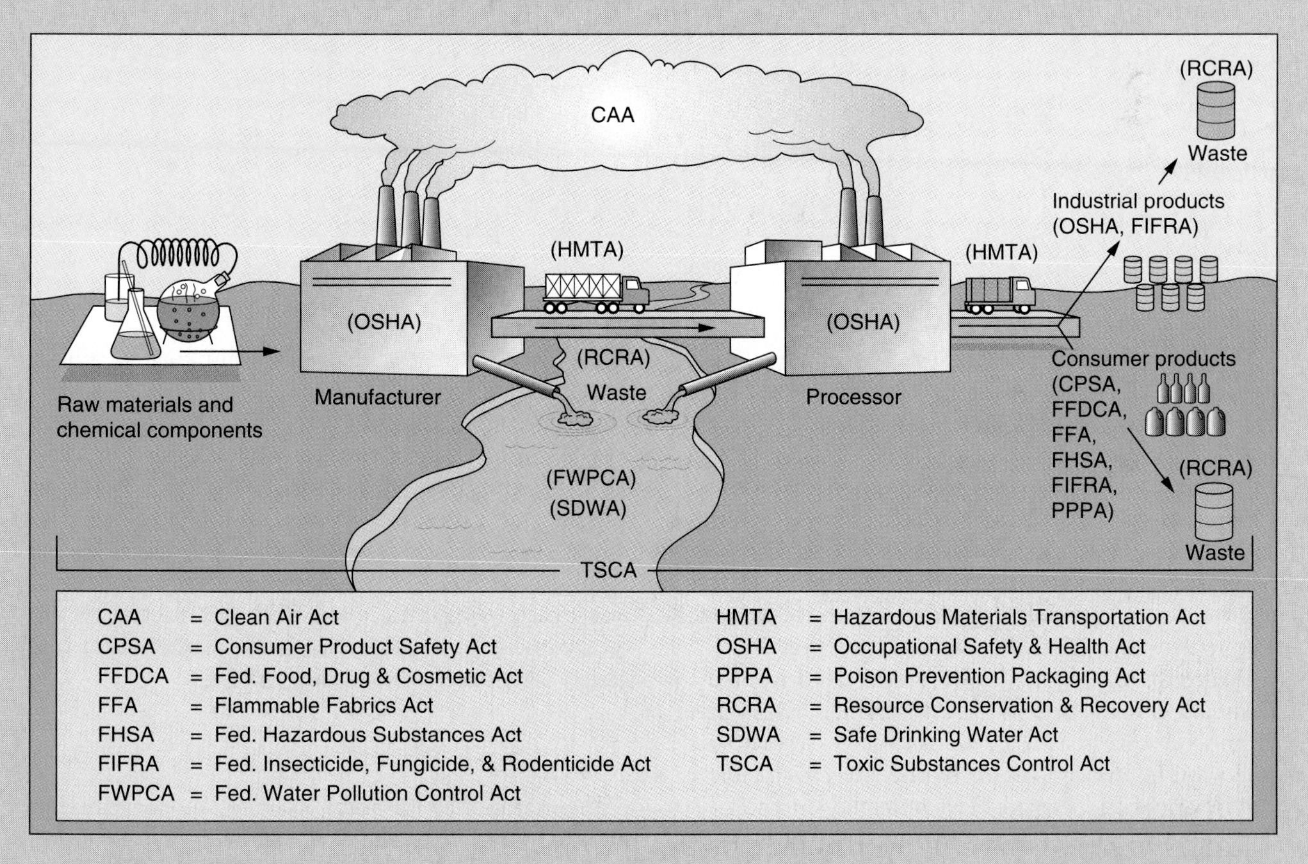


**FIGURE 1**
Hazardous waste control, cradle to grave. (EPA)

**FIGURE 12–13**
Diagram showing result of removing dam at Fort Edwards that had empounded PCB waste from General Electric facilities. (Modified from K. E. Limburg, 1985, PCBs in the Hudson, *in* Limburg, et al. (eds.), The Hudson River Ecosystem: New York, Springer Verlag Co., p. 87. Diagram is modified from Toffelmire & Quinn, 1979, NYSDEC Tech Paper 55.)

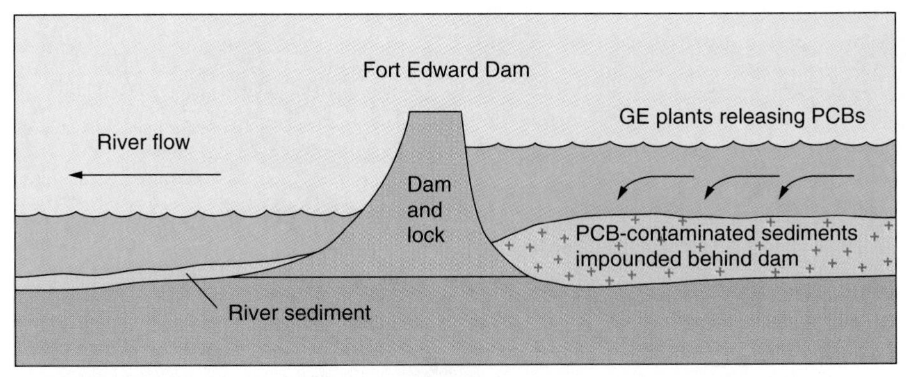


(a) Initial conditions

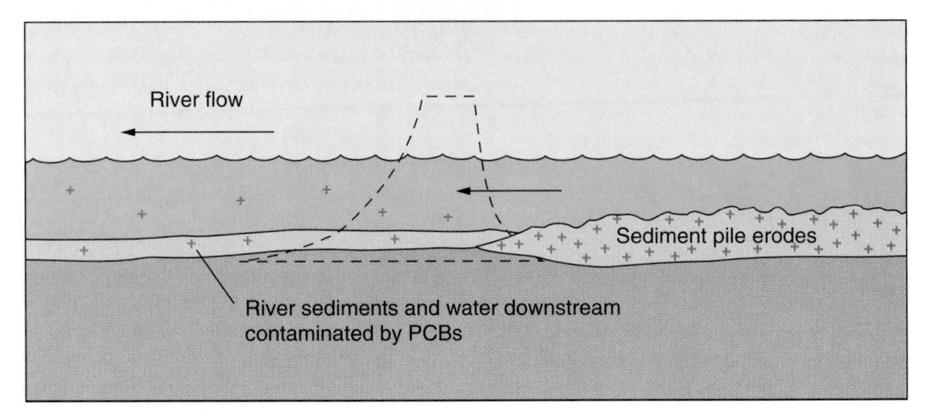


(b) After dam removal

As the buried chemical drums corroded, their contents leaked, the barrels collapsed, and the ground above them subsided. Their contents reached the surface, both as liquid and vapor. Toxic substances such as benzene, dioxin, and chloroform were thus released. Tests conducted in 1977 and 1978 revealed the presence of toxic materials in basements. Skin rashes developed on children playing outdoors. Residents experienced above-average frequencies of birth defects and miscarriages.

More than 100 families were evacuated from the immediate area beginning late in 1978. As the cleanup progressed in 1979, the full extent of the pollution became apparent. By the late 1970s, many carcinogenic agents had been detected on the site, and Love Canal became a national symbol of the problem of improperly stored toxic waste. In 1980, the federal government declared an emergency and committed $15 million in loans and grants for remediation and home purchases. By the early 1980s, remedial measures to neutralize and isolate the waste began.

***Midnight Dumpers.*** Illegal dumping of toxic liquid waste is a continuing problem, known as "midnight dumping." Toxic liquid waste has been sprayed on roads, dumped into isolated rural streams, drainage ditches, and mountain ravines, buried illegally in landfills, and pumped down abandoned mines. The high cost of proper disposal makes these illegal methods attractive to unscrupulous operators. In most cases, illegal dumping is carried out not by the generator of the waste, but by "treatment and disposal" companies that evade the law.

## Liquid Toxic Waste Disposal

Four methods of disposal, storage, and destruction of toxic liquid waste are incineration, injection wells, lined surface ponds, and secure landfills. Although currently in use, these techniques are still undergoing study. They seem to work quite well, so far, but will they continue to be effective through your lifetime and your children's lifetime? Earth movements, weathering, groundwater flow, and cli-

**FIGURE 12–14**
Aerial view of Love Canal disposal site (open area). (J Goera, N.Y. State Department of Environmental conservation.)

mate change may have long-term effects that we cannot predict.

## Incinerating Liquid Waste

Liquid waste can be burned at very high temperatures. In 1993, the United States disposed of nearly 4.5 million metric tons (5 million tons) of hazardous liquid waste a year by burning it in 184 incinerators and 171 industrial furnaces, including 34 cement kilns. For perspective, this volume is equivalent to a line of tank trucks holding 23,000 liters (6,000 gallons) stretching from the nation's capital to Los Angeles (3,800 kilometers or 2,400 miles).

The incineration process is shown in Figure 12–15. Waste is burned at 1,000°C (1,830°F). Some gases are burned in a secondary combustion chamber at 1,300°C (about 2,400°F). Particulates from the combustion are removed and stored in a landfill. The gas passes through a scrubber to remove pollutants that can cause acidic air pollution (see Chapter 11). The gas then is emitted from the plant smokestacks.

Incineration of hazardous waste is highly controversial. Such facilities seem to provide a magic solution (the liquid waste harmlessly disappears). However, they are expensive to build. They produce solid waste that may contain heavy metals. The scrubbed gases emitted into the atmosphere still contain $CO_2$, a major contributor to the greenhouse effect.

Proponents of waste incineration insist that, *if properly operated and maintained,* incinerators burn most pollutants or reduce them to ash, and that scrubbers remove most potential air pollutants from the gases. Others feel that the process is less efficient, that incinerators may not be carefully maintained, or that even low levels of pollutants may cause long-term health problems.

## Injecting Liquid Waste into Wells

Some liquid waste is pumped down injection wells into deep, porous strata that are believed to be isolated from groundwater sources (Figure 12–16). Hazards of deep-well injection include spills, which can occur during surface operations. Other risks are leaks from corroding pipes or pipe joints and leaks from the isolated bed the waste is pumped into. Existing fracture systems or new ones activated by earthquakes or the fluid pressure of injection can

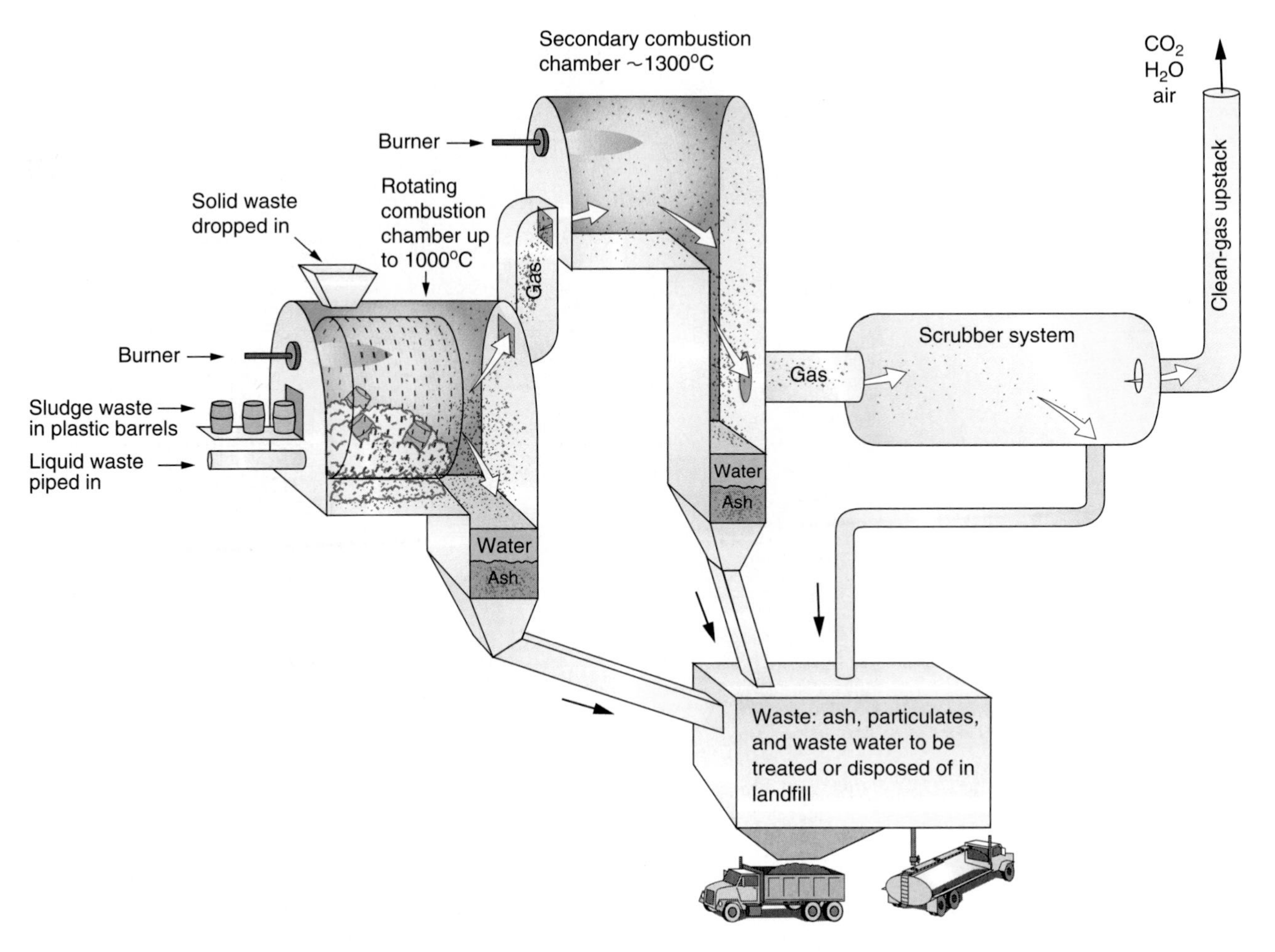


**FIGURE 12–15**
Operation of a high-temperature incinerator to burn toxic waste. (Adapted with the permission of Macmillan College Publishing Company from *Environmental Geology,* 6th ed., by Edward A. Keller. Copyright © 1992 by Macmillan College Publishing Company.)

become conduits along which the toxic liquid moves upward into aquifers.

A rock layer might appear to be a good storage reservoir in a particular place, but what happens as the liquid waste migrates far from the injection point over time? The waste may encounter fractured rock, allowing it to contaminate overlying aquifers or even to enter deep water wells.

## Evaporating Liquid Waste in Lined Surface Ponds

Relatively weak concentrations of toxic chemicals can contaminate large volumes of water, which happens in industrial washing processes. To avoid this problem, the contaminated water may be disposed of in lined surface ponds (Figure 12–17). The waste is pumped into a large pit lined with chemically resistant plastic. The water evaporates, leaving a toxic residue, which may be stored in a secure landfill.

Potential problems with this system include spills, leaks in surface piping, and failure of the liner due to corrosion or stress from temperature extremes. In addition, evaporation of volatile components may jeopardize local air quality. When working correctly, this method helps to reduce the volume of toxic liquids.

## Landfilling Liquid Waste

A secure landfill is similar to the sanitary landfill described earlier in this chapter (Figure 12–4). How-

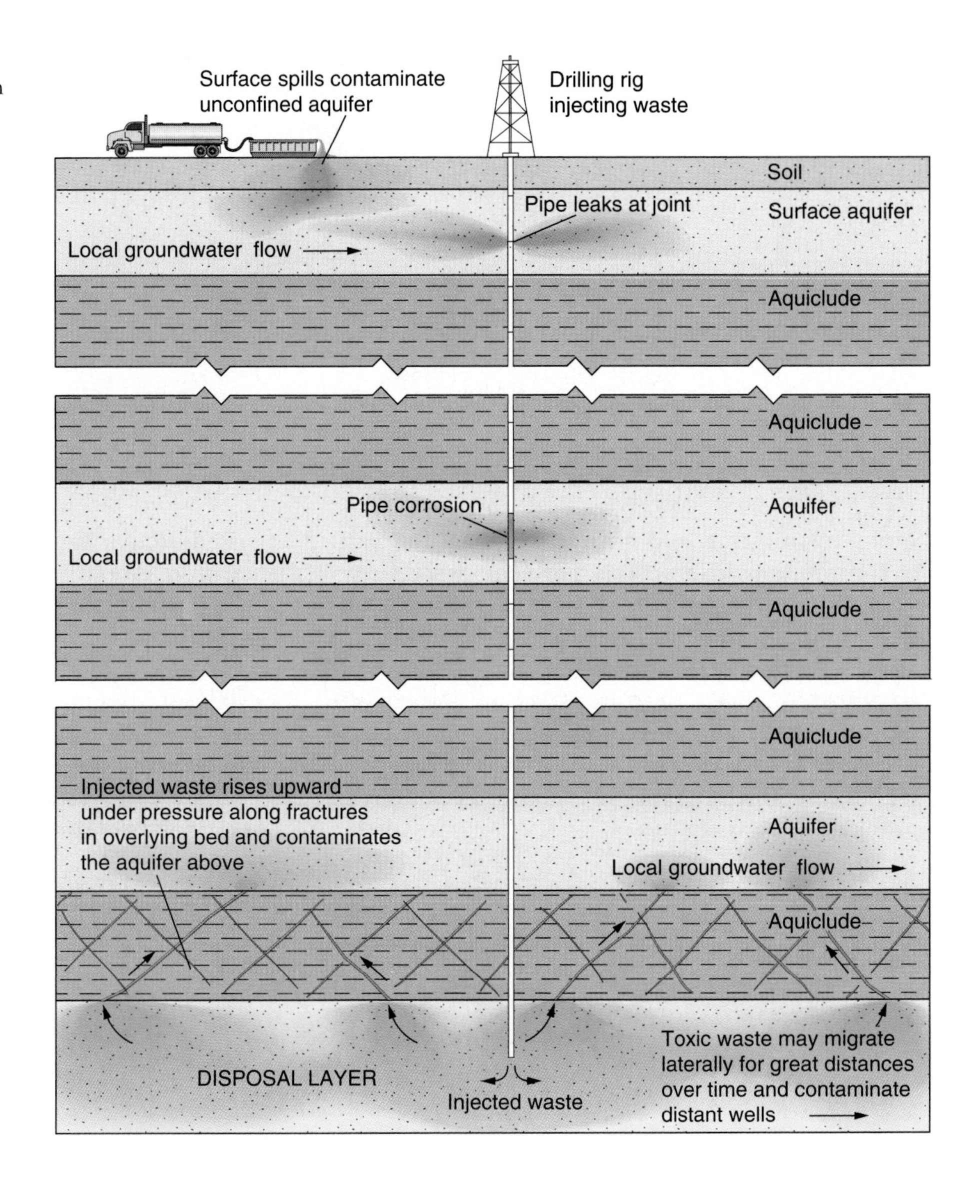


**FIGURE 12–16**
Geologic section showing operation and potential problems with deep-well injection of liquid waste.

ever, a secure landfill is made *secure* because it is specifically designed to resist leakage. Secure landfills are needed for toxic liquid waste because the consequences of leakage are far more serious than for ordinary leachate.

A secure landfill pit is lined on the bottom and sides with 2 meters (6.5 feet) or more of clay (Figure 12–18). A chemically resistant plastic liner is placed against the clay. Chemical waste then is added in clusters of stacked containers, separated from each other by fill. Within the landfill, pipes collect waste from any leak and from any water entering the landfill. The top is capped by a layer of clay.

Even a secure landfill is not 100% secure, of course. The clay seal can be breached by cracking from settling of the fill, or it can be punctured by animal burrows. The underlying liner, although chemically very resistant, may be eaten through by corrosive chemicals. It is essential to separate wastes into compartments so that accidental leakage underground cannot create explosive mixtures.

## RADIOACTIVE WASTE

The Nuclear Age has brought many wonders—nuclear power plants, nuclear medicine, precision mea-

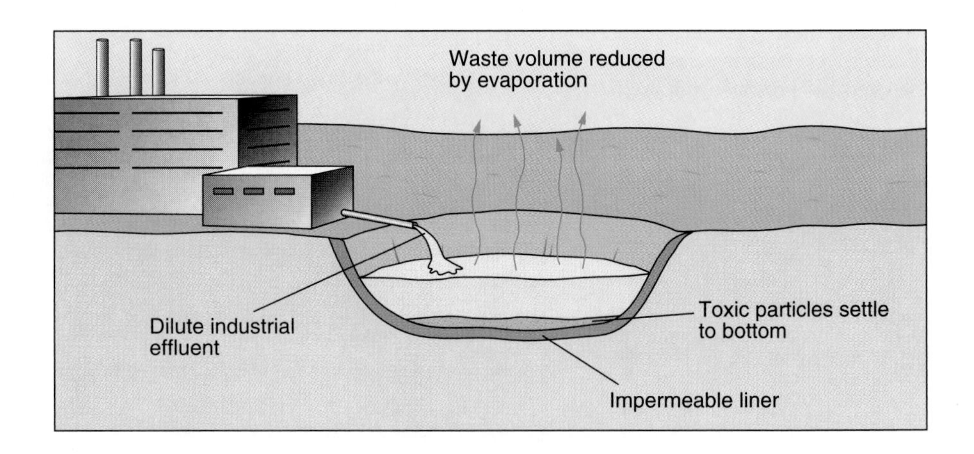


**FIGURE 12–17**
Operation of a lined surface pond for evaporative concentration of relatively dilute toxic waste.

suring instruments, and nuclear weapons. But each is contributing to a serious problem: what to do with the growing volume of nuclear waste. As nuclear power plants continue to operate (Figure 12–19), they generate "spent" fuel rods that must be disposed of. As they are decommissioned (removed from service and dismantled), their radioactive parts must be safely discarded. The same is true for disassembled atomic weapons. The disposal problem is growing.

Radioactive waste is classified as either *low-level* or *high-level*. About 90% of nuclear waste is low-level, including contaminated materials from nuclear medical laboratories, materials used to clean up spills of radioactive liquids, and clothing. Generally, this low-level waste requires no special disposal precautions. Low-level waste is either held in storage until its radioactivity decays, or it is disposed of in landfills.

High-level waste, including spent fuel rods from nuclear power plants and atomic weapon parts,

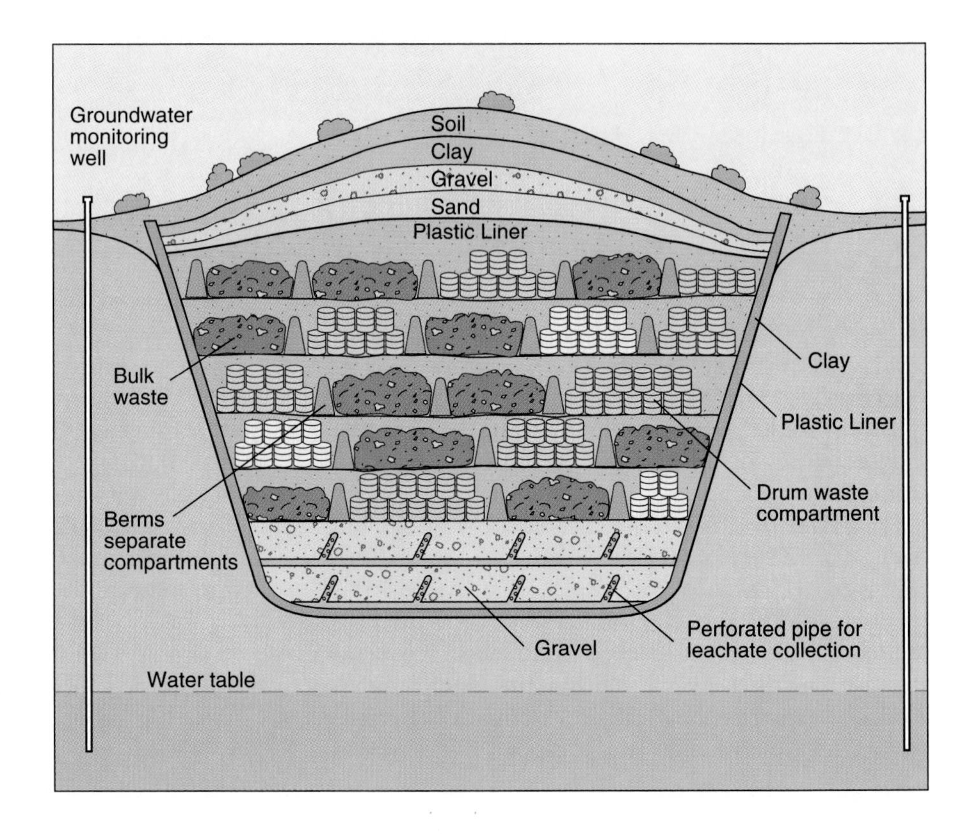


**FIGURE 12–18**
Cross section of a secure landfill.

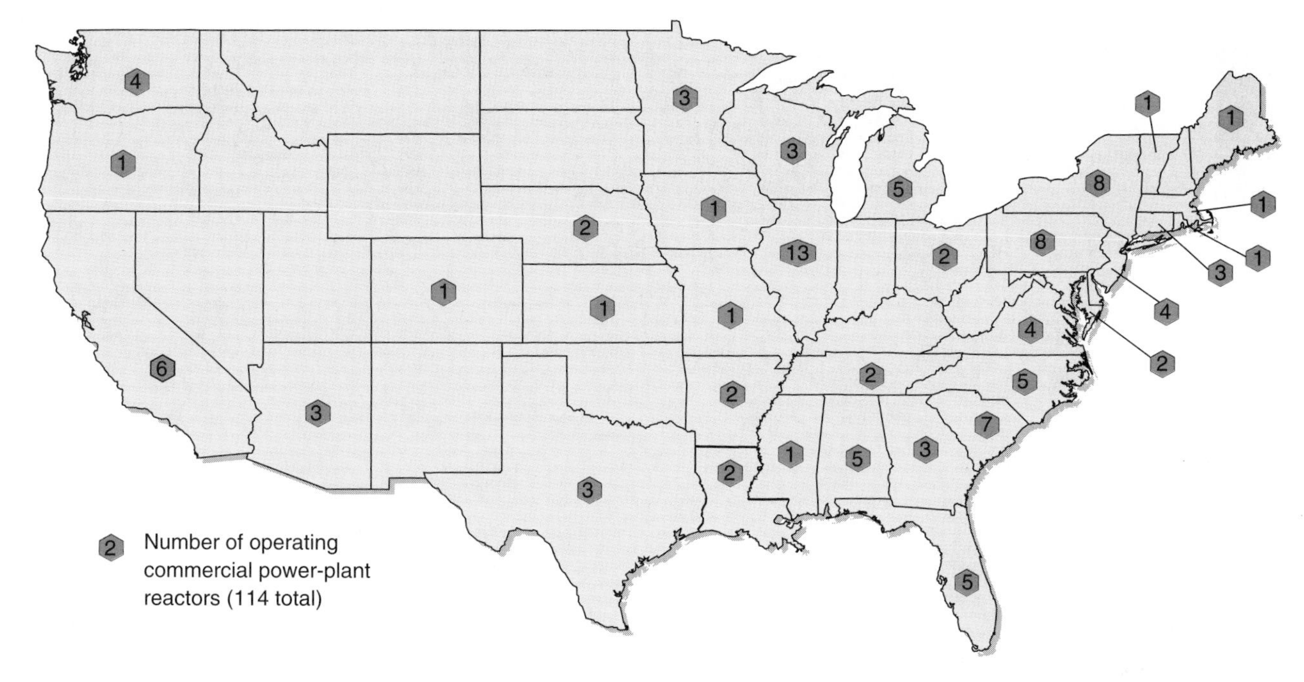


**FIGURE 12–19**
Power plant reactors in the United States as of December 1989. (U.S. Department of Energy.)

poses a much greater problem. High-level waste remains highly radioactive and very hazardous for millennia. At present, even after half a century of generating nuclear waste, no permanent high-level waste disposal site exists in the United States. The serious hazard of nuclear waste and the technical complexity of safe disposal require careful analysis and time to develop politically acceptable choices.

In the United States, high-level waste presently is stored "temporarily" as a liquid in underground tanks, as a granular solid, or as a liquid that has been solidified into "salt cake." The four major U.S. high-level storage sites and their waste volumes as of 1984 are shown in Figure 12–20.

### Radioactive Waste Disposal

The problems of nuclear waste disposal can be summarized in a few questions:

- ☐ What is the long-term stability of a given storage site?
- ☐ Are there unforeseeable leakage conduits (faults, fractures, or rock porosities)?
- ☐ Nuclear waste generates considerable heat as it decays; what effect will this heat have on the repository over time?
- ☐ What community would welcome a high-level radioactive waste depository as a neighbor? (This is the NIMBY phenomenon again.)

At present, the only plan for high-level waste disposal under formal consideration is storage in a suitable rock formation deep underground. The problem is to find such a formation. It must be strong, impermeable, stable, and fracture-free. The only site selected by Congress for study is in the Nevada desert, 100 miles north of Las Vegas: Yucca Mountain. The U.S. Department of Energy is considering a facility like that shown in Figure 12–21.

Because a majority of Nevadans oppose the Yucca Mountain site, a legal battle will be in progress for some time. If the site is rejected, it will take years for another site to be researched and approved.

## THE WASTE PROBLEM AHEAD

The United States leads the world—by far—in per capita production of most types of waste. Evolving legislation is making progress toward better waste disposal. However, we are running out of space for solid waste in existing landfills. New sites meet in-

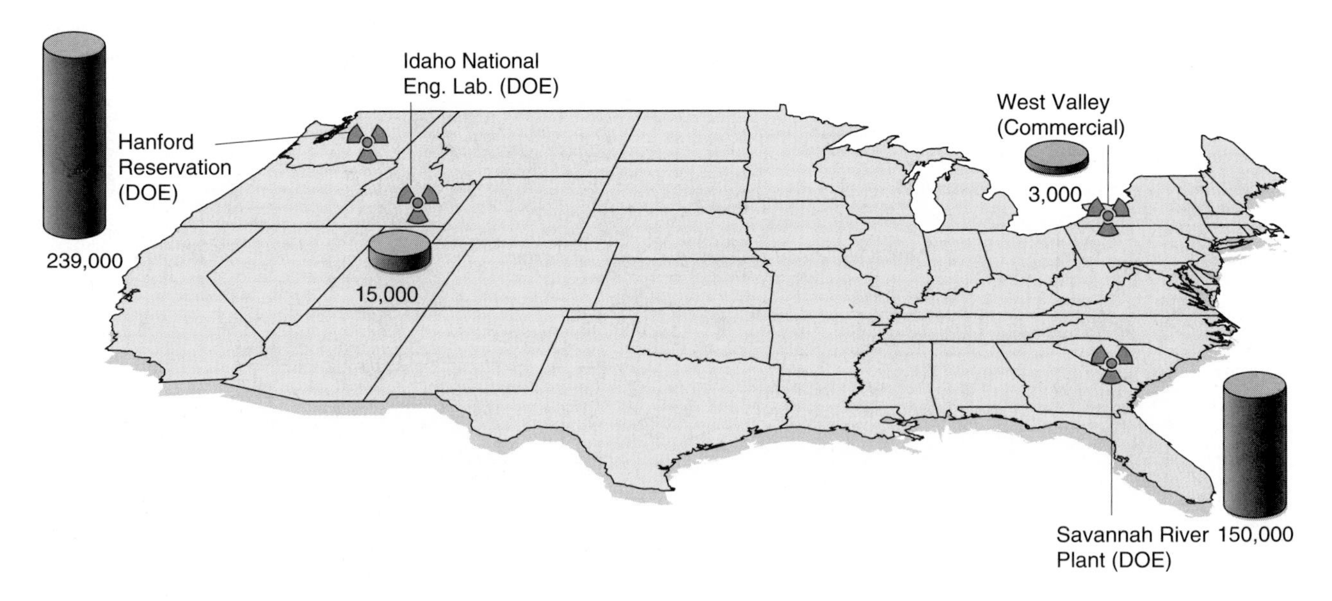


**FIGURE 12–20**
High-level radioactive waste storage in the United States; volume in cubic yards. (AIPG, 1984, Issues and answers: Radioactive waste: Arvada, Colorado, AIPG.)

creasing opposition, largely from the NIMBY phenomenon. As for liquid waste, even if concentrations released into water bodies are drastically reduced, the large *volumes* produced annually create a serious geohazard for the future.

Four strategies can help to sustain the quality of life for our rapidly growing population:

1. Because so much of the waste problem is its sheer volume, we must learn to do with less, called *conservation*. For example, we could eliminate elaborate packaging, eliminate throwaway products, prohibit planned obsolescence, and maintain things we own for longer periods.
2. Wide-scale *recycling* will save energy, decrease the solid waste volume, and slow the consumption of nonrenewable resources.
3. Wider use of *alternative materials* that are environmentally sound will reduce solid waste.
4. Continuing evolution of *regulations* and their *enforcement* will improve control over waste (see Box 12–4).

## LOOKING AHEAD

From the unpleasant, worrisome world of solid waste, sewage, acid drainage, and toxic liquids, we turn in the next chapter to some of Earth's most beautiful places: estuaries and wetlands. From estuaries like Chesapeake Bay in Maryland to wetlands like the Tule Lake marshes in California, these moist places nourish abundant life—and pollution problems abound.

## SUMMARY

Waste disposal is reaching a crisis in some areas. We are running out of space to store solid waste. A long legacy of discharging liquid waste onto the ground, into surface water, and into groundwater is creating a geohazard today. Also, the natural resources that go into making waste are being depleted.

Solid waste is stored in sanitary landfills located in geologic formations of very low permeability (commonly, clay) or protected from leaking by a plastic liner. Water percolating through the landfill plus heat generated by biodecomposition of the waste produces leachate and methane gas. Under proper landfill management, leachate is drawn off and treated and methane is processed for sale to utility companies. Modern landfills are monitored for leachate movement by regularly analyzing water samples from wells surrounding the landfill.

Other methods of dealing with solid waste are composting, incineration, recycling, source reduction, and reuse. Innovative methods are helping to reduce problems of solid waste disposal. However, we must reduce

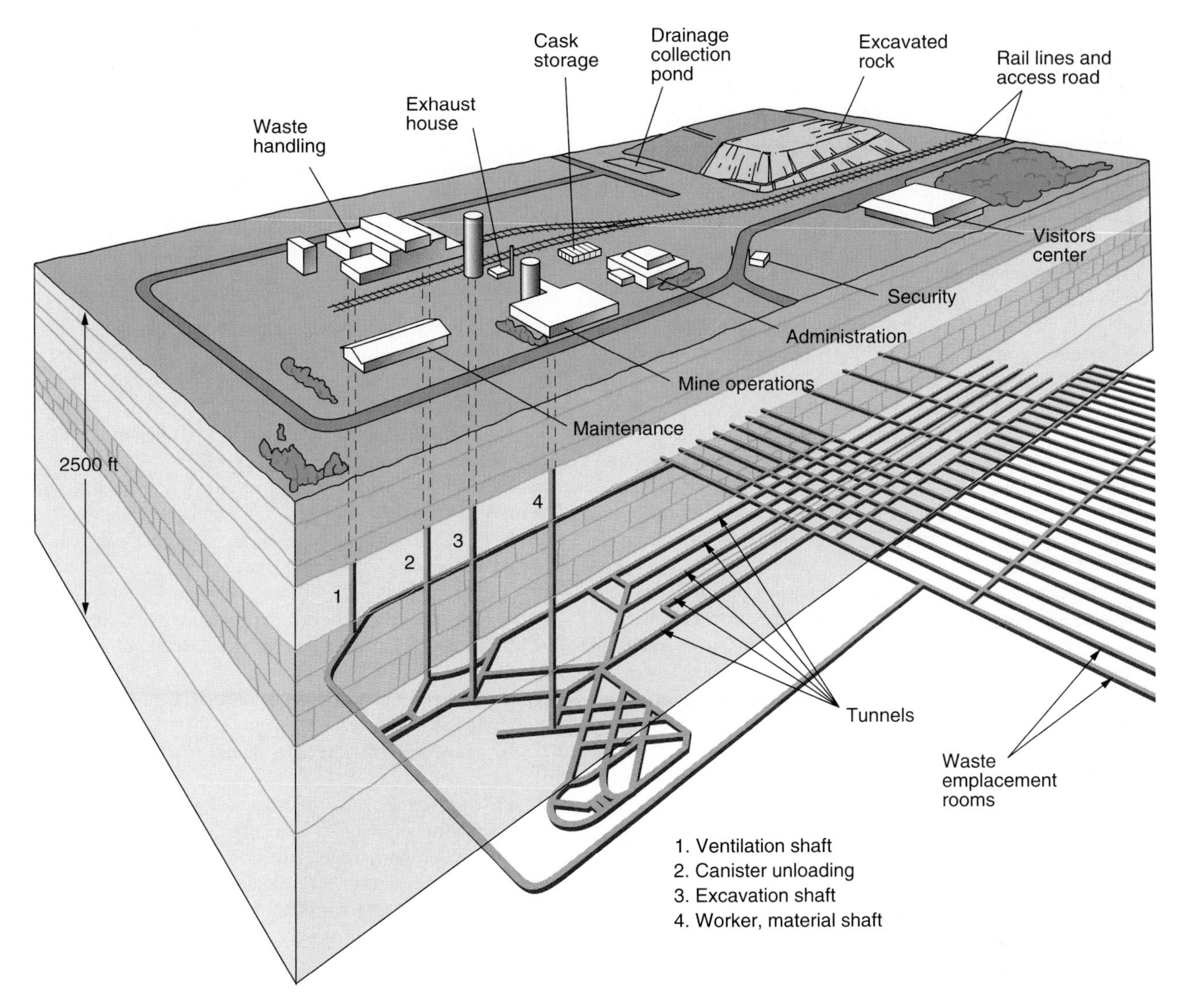


**FIGURE 12–21**
Operation of a radioactive waste repository. (AIPG, 1984, Issues and answers: Radioactive waste: Arvada, Colorado, AIPG.)

waste volume. This can best be accomplished by consuming less, sorting waste by type at its source (homes, factories) and recycling, using biodecomposible alternative products, and using less packaging.

Organic waste is a serious problem because of its volume and high biological oxygen demand (BOD). Organic effluent discharged into surface water consumes dissolved oxygen. As oxygen is depleted, fish and other aquatic animals develop respiratory problems and may die.

Sewage is treated by screening floatable debris, collecting sand and gravel, and depositing these solid components in landfills. The liquid slurry is treated by aeration, bacterial action, skimming, and settling. Waste digestion generates methane that can be burned to generate electricity to power the treatment plant. Sludge has been deposited at sea (no longer acceptable) or in secure landfills, or processed into fertilizer. Liquid effluent is chlorinated to reduce pathogenic bacteria and discharged to the nearest water body.

Acidic drainage results from water and oxygen reacting with iron sulfide minerals (commonly pyrite, or fool's gold) in rock. The reaction produces sulfuric acid, which can kill aquatic plants and animals. The low pH of acid drainage also leaches toxic metals from rocks. These metals may become concentrated in the sediment and in aquatic animals.

Acid mine water develops in coal mines, where pyrite in rock reacts with groundwater and air to form sulfuric acid. The acid water may flow from the mine and cause fish kills in nearby streams. Acid drainage problems also develop on the surface where pyrite-rich rocks are exposed or used for fill. Rain percolating through the fill combines with the pyrite and oxygen to produce acidic water that destroys vegetation and adversely affects aquatic life in streams.

Liquid waste is a special concern because of its mobility. It can travel considerable distances in surface water and groundwater. Liquid waste can be toxic (PCBs), flammable (petrochemical spills), corrosive (industrial acid washes), or infectious (medical or sanitary waste). Improperly stored liquid waste containers corrode. The release of their contents causes serious health problems.

Liquid waste is being disposed of by incineration at high temperatures, by injection into deep rock layers, by deposition into lined surface ponds where evaporation reduces the mixture to a residue, or by burial in containers in a secure landfill.

High-level nuclear waste remains highly radioactive and very hazardous for millennia. At present, even after half a century of generating nuclear waste, no permanent high-level waste disposal site exists in the United States. The serious hazard of nuclear waste and the technical complexity of safe disposal require careful analysis and time to develop politically acceptable choices.

## KEY TERMS

acid drainage
aeration
aerobic reaction
anaerobically
biodegradable
Biological Oxygen Demand (BOD)
cell
combined sewers
composting
Dissolved Oxygen (DO)
effluent
eutrophication
leachate
monitoring wells
nonbiodegradable
reactivity
sanitary landfill
sanitary sewers
septic tank
sewage sludge
spontaneous combustion
storm sewers
storm water waste storage facilities
thermal pollution

## REVIEW QUESTIONS

1. Describe the geologic factors that should be considered in properly siting a landfill.
2. **a.** Describe factors that favor the breakdown of solid waste.
   **b.** How well is this accomplished in a landfill? What is your evidence?
3. **a.** Describe how solid waste landfills produce gases and liquids as by-products.
   **b.** How are these waste by-products managed?
4. **a.** What methods reduce the pollution potential of a landfill?
   **b.** How can effluent from a landfill be monitored?
5. **a.** What steps are involved in sewage treatment?
   **b.** What by-products are produced by sewage treatment?
6. **a.** What is BOD? DO?
   **b.** What factors govern the DO level in water?
7. What is thermal pollution and how does it effect aquatic animals?
8. How do human activities affect BOD and DO of local water bodies?
9. **a.** Compare the functions of storm sewers and sanitary sewers.
   **b.** What problems develop in sewage treatment in areas utilizing a combined (storm and sanitary) sewer system?
10. **a.** Describe the process that produces acid drainage.
    **b.** How can the effects of acid drainage be minimized?
11. **a.** Describe an example of a toxic liquid waste discharge in your area.
    **b.** How were its effects reduced—or why were they not?
12. First describe each of the following liquid waste treatment processes, and then list what problems may accompany each treatment method: (a) incineration, (b) deep-well injection, (c) lined surface ponds, (d) secure landfills

## FURTHER READINGS


American Institute of Professional Geologists, 1985, Hazardous waste—Issues and answers: Arvada, Col., 24 p.

American Institute of Professional Geologists, 1985, Radioactive waste—Issues and answers: Arvada, Col., 28 p.

Byerly, D. W., 1992, Acid drainage vs. construction: Professional Geologist (Nov.), p. 10–11.

Clarke, M. J., and others, 1991, Burning garbage in the U.S.: Practice vs. state of the art: New York, INFORM, 275 pp.

Clarke, M. J., 1992, Waste characterization studies and the solid waste hierarchy: Resource Recycling (Feb.), p. 75–84.

Cottingham, D., 1988, Persistent marine debris: Challenge and response: The federal perspective, Alaska Sea Grant College Program public, no. 1, 41 p.

Fishbeir, B. K., and Gelb, C., 1992, Making less garbage: A planning guide for communities: New York, INFORM, 180 pp.

Rathje, W. D., and Murphy, Cullen, 1992. Rubbish! The archeology of garbage: New York, HarperCollins, 250 pp.

Rathje, W. D., and Psihoyos, L., 1991, Once and future landfills: National Geographic (May), p. 115–134.

U.S. Geological Survey, 1989, Geohydrologic aspects for siting and design of low-level radioactive waste disposal, circular no. 1034, 36 p.

White, P. T., 1983, The fascinating world of trash: National Geographic (April).

# 13

# *Estuarine and Wetland Problems*

Most rivers do not abruptly empty into the sea. Instead, they merge with the sea in a transition area called an estuary (Figure 13–1). An estuary is a valley carved by streams when sea level was lower, during a glacial period or "ice age." It subsequently became flooded as glaciers melted and sea level rose. Estuaries are bordered by saltwater wetlands—areas flushed by the tide—with characteristic vegetation and fauna.

In a similar fashion, rivers and streams are associated with freshwater wetlands—marshes and swamps—that become inundated whenever the rivers flood. Freshwater wetlands also can occur as isolated bogs or as marshes around lakes (Figure 13–1).

All of these components—freshwater wetlands, saltwater wetlands, and estuaries—are very productive biologically. They are host and nursery to widely varied plants, mammals, birds, amphibians, reptiles, fish, and invertebrates. They also are subject to a unique set of geohazards.

## ESTUARIES

Many of America's great cities are located where the freshwater of a river meets the saltwater of the sea—Seattle, San Francisco, Norfolk, New York, Baltimore, and many others. These cities are located on estuaries. An **estuary** is both a river's mouth and a branch of the ocean, a place where saltwater and freshwater mix, and where tides flow.

There are important differences between a river and an estuary (Table 13–1). A river flows in one direction (downstream) and contains only freshwater (no saltwater). An estuary contains brackish water (mixed saltwater and freshwater) because the river's freshwater discharge mixes with ocean saltwater that is carried into the estuary by tides.

**Satellite view of a portion of Chesapeake Bay. (NASA)**

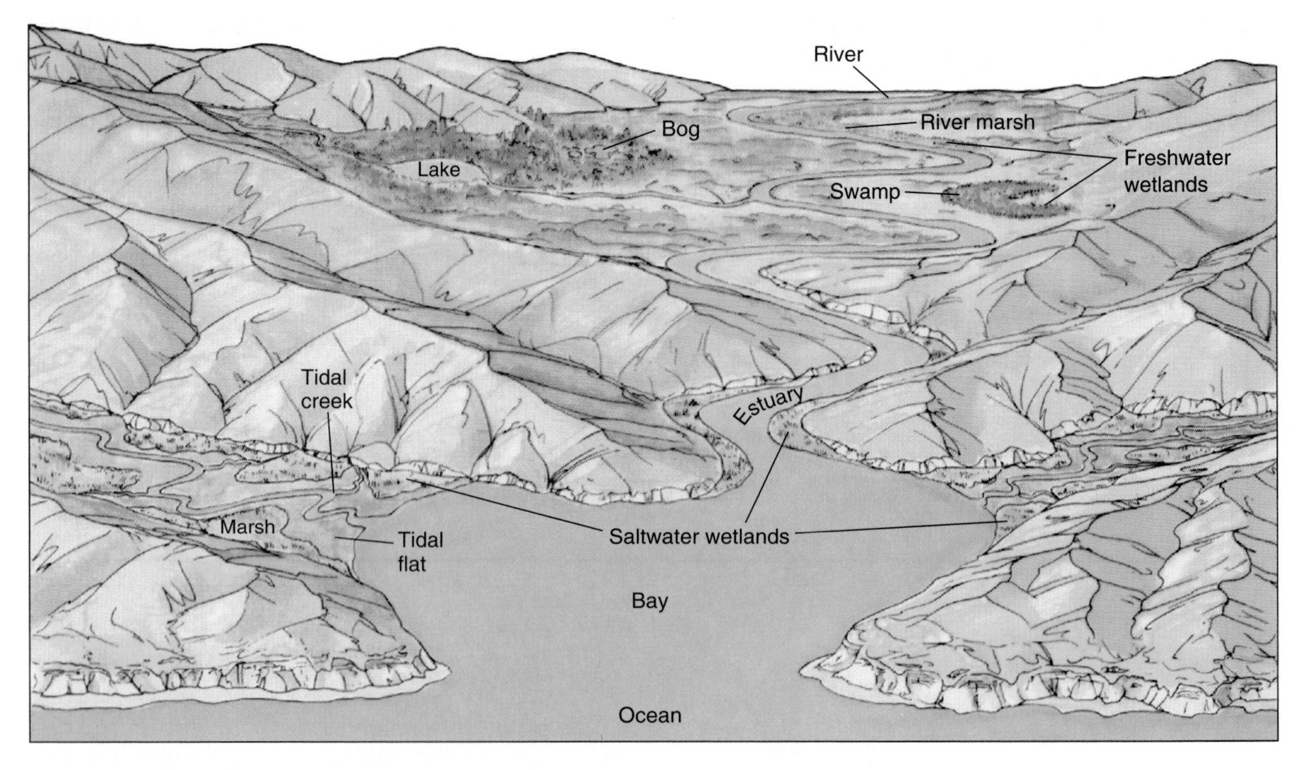


**FIGURE 13–1**
Estuaries, rivers, and wetlands.

Because estuary water level rises and falls with the tides, the water can flow either up-estuary or down-estuary. Estuarine water flows up-estuary with the rise of the incoming tide, or **flood tide.** Estuarine water flows down-estuary with the fall of the outgoing tide, or **ebb tide.** This back-and-forth movement of estuarine water can cause a progressive concentration of pollutants along an estuary, much more so than along a river. In a river, the pollutants are washed downstream.

The differences between a river and an estuary are very significant environmentally. Waste discharged into a river can pollute a city downstream, but not upstream (Figure 13–2a). However, waste discharged into an estuary at ebb tide (Figure 13–2b) may return up-estuary at the next flood tide! The great population growth of urban centers built around estuaries has caused increasing pollution, reducing the quality of estuarine water.

## How Do Estuaries Form?

Earth's climate has warmed slowly during the last 15,000 years, following the last major advance of continental glaciers during the Pleistocene Epoch. The melting of glaciers during this period has raised sea level worldwide by about 130 meters (425 feet). One result was the formation of estuaries, a process

**TABLE 13–1**
Differences between rivers and estuaries.

| Factor | River | Estuary |
|---|---|---|
| Salt Content | Freshwater | Saline, brackish, mixed |
| Flow Direction | Downstream only | Down-estuary or up-estuary, depending on tidal stage |
| Tidal Effects | None | Strong |

(a) River

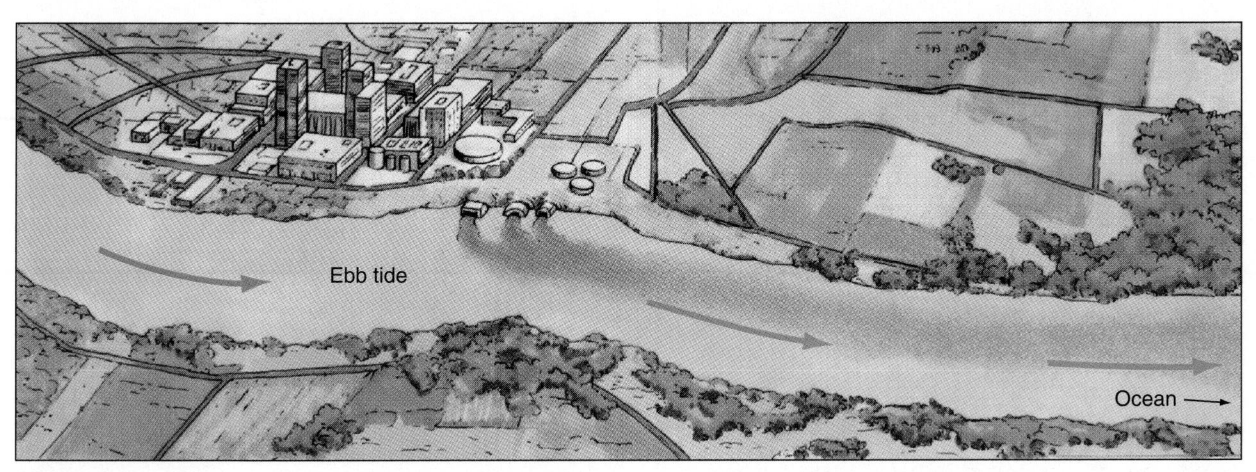


(b) Estuary - ebb tide

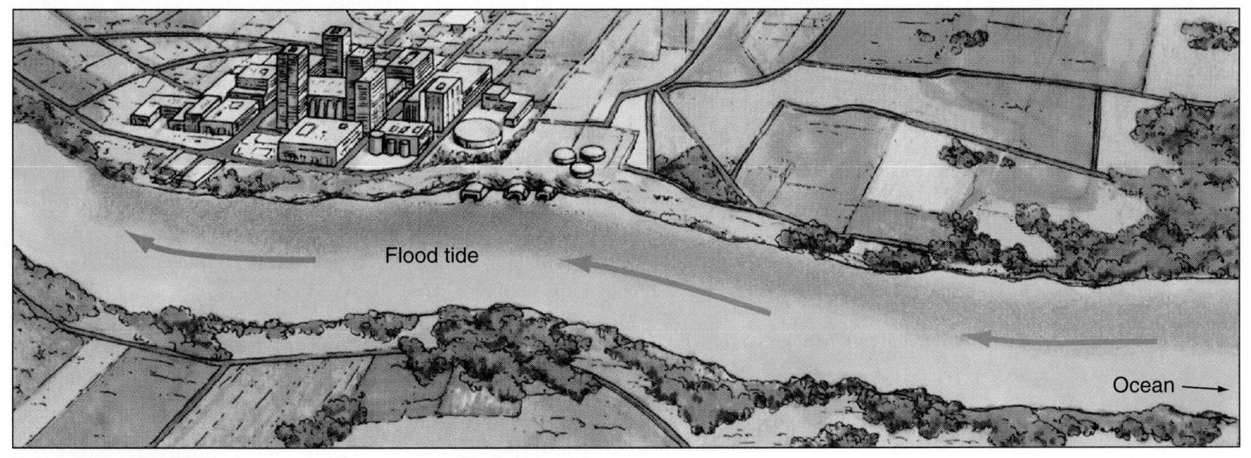


(c) Estuary - flood tide

**FIGURE 13–2**
Difference in water flow between rivers and estuaries, illustrated by the paths of pollutants released by a city. (a) Wastes discharged into a river are carried downstream. (b) Wastes discharged into an estuary on ebb tide are carried seaward. (c) Waste-contaminated estuary waters may return on the next flood flow.

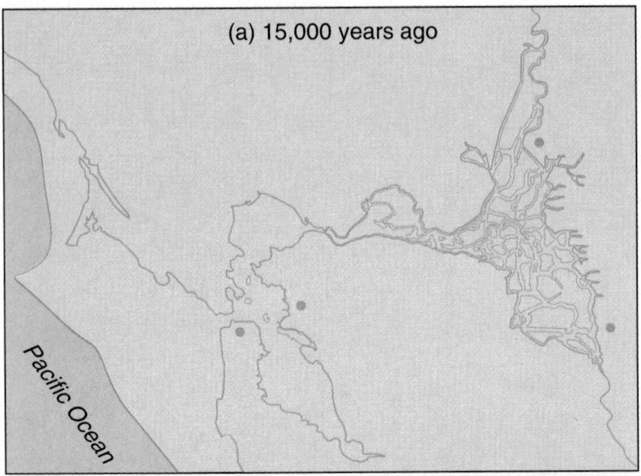


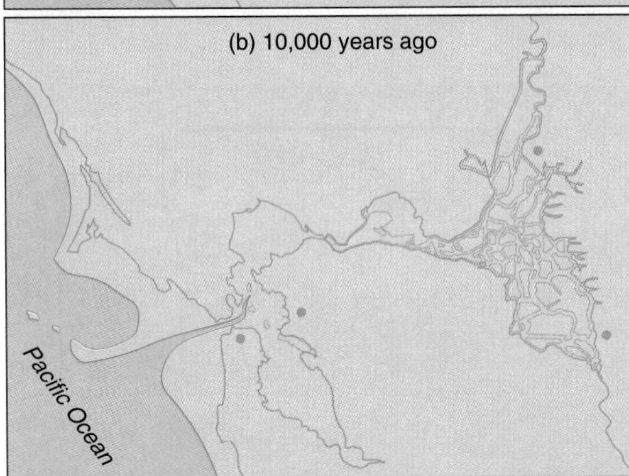


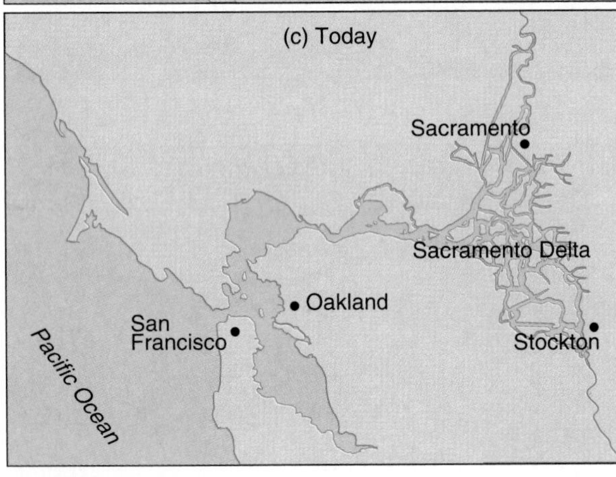


**FIGURE 13–3**
Sea-level rise created the San Francisco estuary in California. (a) 15,000 years ago. (b) 10,000 years ago. (c) today. (Adapted from San Francisco Estuary Project p. 31.)

illustrated by two fine examples, the San Francisco estuary (Figure 13–3) and the Chesapeake Bay (Figure 13–4).

Post-glacial sea-level rise created three basic types of estuaries, depending on local geologic conditions (Figure 13–5). In some cases, the sea moved into existing river valleys to form **river estuaries** (Figure 13–5a) at right angles to the coast. In other places, the rising sea filled existing low areas along the shoreline to form **bays** (Figure 13–5b).

The third estuary type formed along much of the Atlantic and Gulf Coasts. As sea level rose, linear sand islands developed near the shore. These long islands, called *barrier islands,* isolated water in linear estuaries that run parallel to the shore, giving rise to **back barrier bays.** Along the northeastern Atlantic, they are called *barrier bays,* and along the southeastern Atlantic they are called *sounds* (Figure 13–5c).

## How Salty Is an Estuary?

Again, estuaries are regions where freshwater from a river flows into saltwater from the ocean. The salt content of water is measured by its **salinity,** which is its concentration of dissolved salts. You are familiar with concentrations expressed as a *percentage,* or so many parts per hundred. Average seawater is about 3.5 parts per hundred, or 3.5% salt by weight. But because salt concentrations are so small, it is more convenient to express them in parts of salt *per thousand* parts of water. Thus, ocean water has an average salinity of about 35 parts per thousand, or 35 ppt. **Brackish water** (mixed saltwater and freshwater, as in an estuary) has a salinity between about 7 and 15 ppt.

The actual boundary between freshwater flowing from land and saltwater that intrudes inland from the ocean in an estuary is not a sharp divide. The mixing grades over a zone extending tens of kilometers along the channel. The thickness and position of this zone is constantly changing, varying with the rate of streamflow and the tidal range. When streamflow is strong or the tidal range is small, saline water cannot enter the estuary very far. When river flow is weaker, as in drier seasons, or when tides are higher, saltwater may intrude far upstream.

A common type of estuary is the **salt wedge estuary,** also called a *partially mixed estuary* (Figure 13–6). It develops where the tidal range is moderate and stream discharge dominates over tidal intrusion. In this type of estuary, less-dense freshwater rests on top of denser saltwater in a wedge that thins toward the ocean as more of the fresh and saltwater mix.

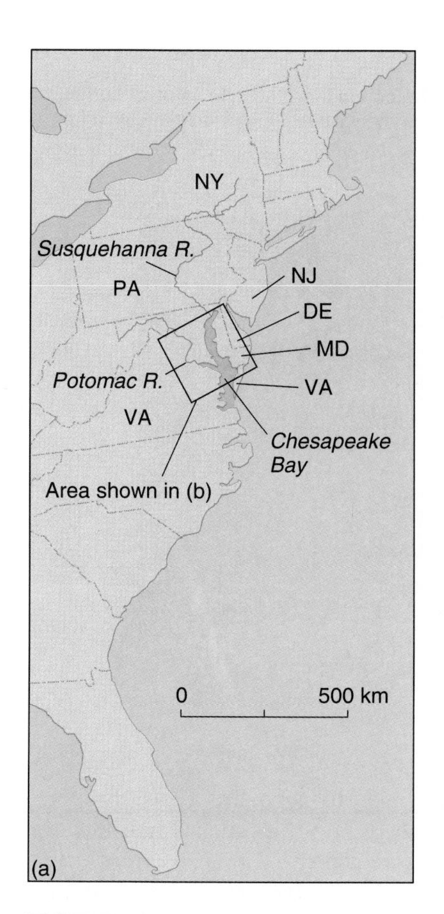

(b)

**FIGURE 13–4**
Chesapeake Bay—the largest estuary in America. (a) Index map. (b) Satellite view. (NASA.)

The movement of water in such a partially mixed estuary is complex. It can be illustrated by watching what happens if floats are suspended in the water (Figure 13–6). Consider two floats, one suspended in surface water (S) and the other suspended in bottom water (B). The arrows on the floats show that, during flood tide (up-estuary), both floats move inland, whereas on ebb tide (down-estuary), both move oceanward. This is called **tidal circulation.**

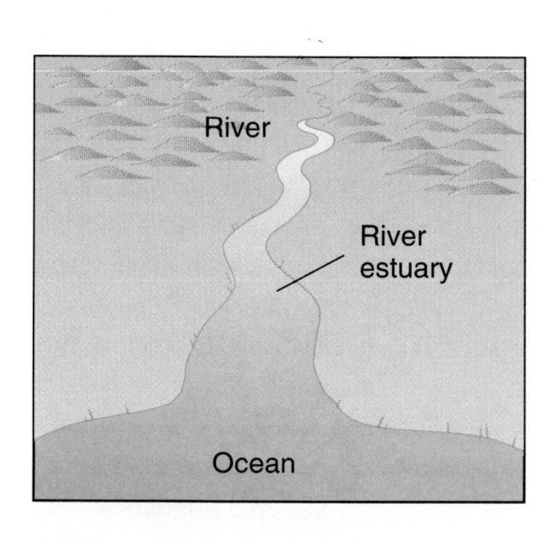


(a) River estuary (Hudson River, NY)

Bay
Ocean

(b) Bay (San Francisco, CA)

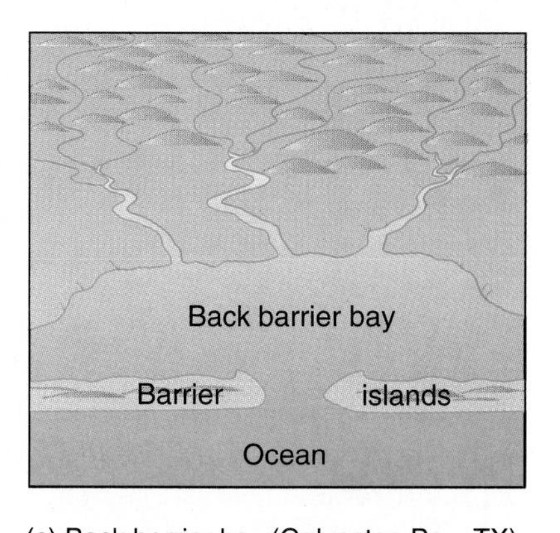


(c) Back barrier bay (Galveston Bay, TX)

**FIGURE 13–5**
Three basic types of estuaries: river estuaries, bays, and back barrier bays.

FIGURE 13–6
Salinity relations and circulation in a salt wedge estuary. The surface float (S) and bottom float (B) illustrate the net nontidal flow described in the text.

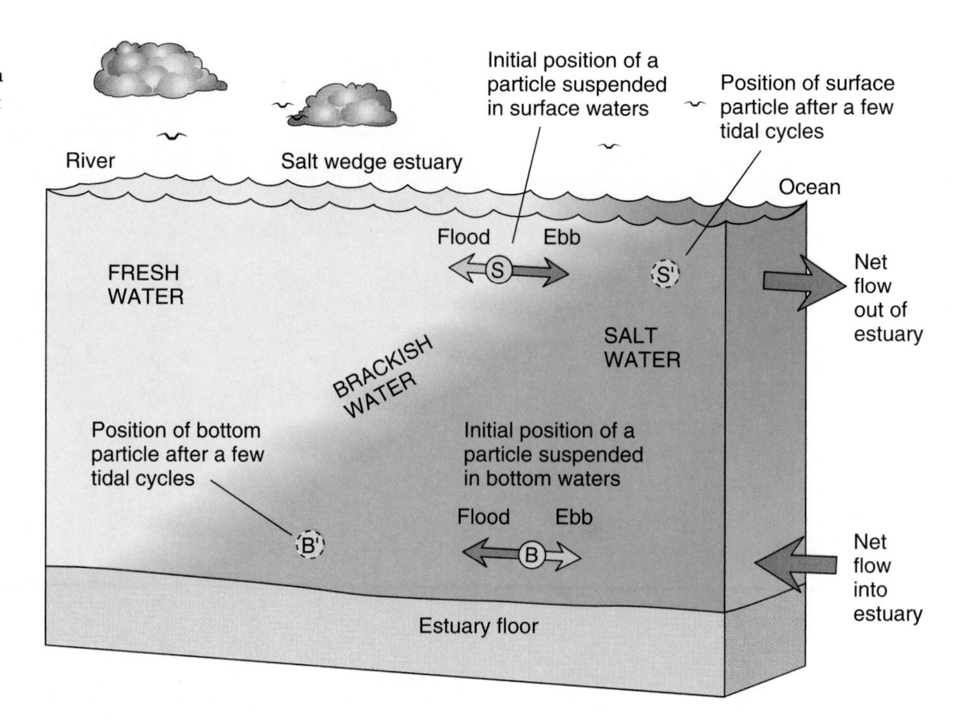


However, the two floats do not remain in the same area. They have a *net movement over several tidal cycles.* Freshwater flow into the estuary adds to the movement of the surface float, creating a net movement toward the ocean. Saline ocean water moving along the bottom adds to the movement of the bottom float, creating a net movement inland. The position of each float, after several days of tidal cycles, is indicated by S′ and B′.

The *non*tidal flow also results in a *net seaward movement of surface water and a net inland movement of bottom water.* This net nontidal flow has considerable environmental significance. Consider the case of a city discharging sewage waste into surface water during an ebbing tide (Figure 13–2). Contaminated particles settle from the water that flows toward the ocean. They fall into the bottom water, where net flow can return some of the particles back *toward* the city!

## Life in an Estuary

The daily shifting of the tides circulates the water in an estuary, constantly bringing fresh food and oxygen and removing waste. Thus, estuaries teem with life. The brackish water brings a diversity of highly adapted—and therefore fragile—organisms. Nutrients from decaying vegetation and animal tissue nourish plants, which in turn are food for many species. This is the base of complex estuarine food webs of invertebrates and vertebrates.

Microscopic plankton carried by the tides provide food to filter-feeding organisms such as clams and oysters. Worms, crabs, and snails extract food from the nutrient-rich mud. Grass and algae provide a sheltered environment for juvenile fish and food for waterfowl such as geese and ducks.

## Flocculation and Bioagglomeration of Sediments in Estuaries

When salinity changes occur in estuaries, they trigger chemical and biological processes. These cause sediment particles carried in the water to be deposited on the bottom. Many suspended sediment particles are clay-size, and clay particles have electrical charges on their edges. In freshwater, these charges make the particles *repel* each other, which keeps particles suspended in freshwater and makes the water cloudy.

As salinity increases, however, ions in the saltwater *attract* the electrically charged clay particles, making them aggregate or clump together into larger particles called *flocs.* The process is called **flocculation** (Figure 13–7a). Flocculation of clays at the boundary between saltwater and freshwater

**FIGURE 13–7**
Processes resulting in estuarine sediment deposition: (a) flocculation; (b) bioagglomeration.

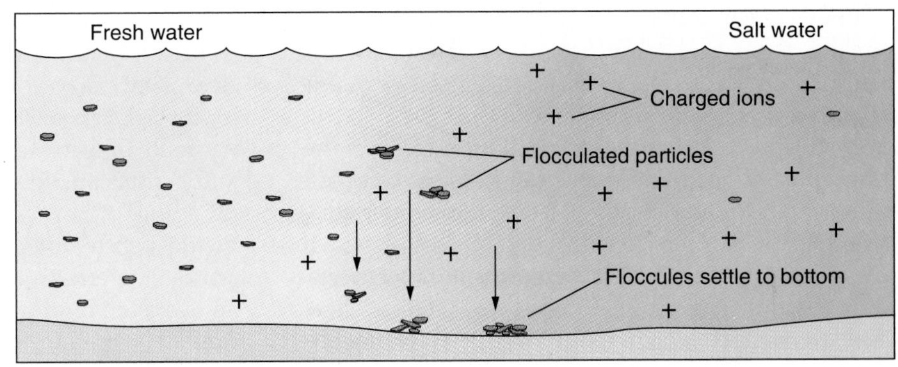


(a) Flocculation

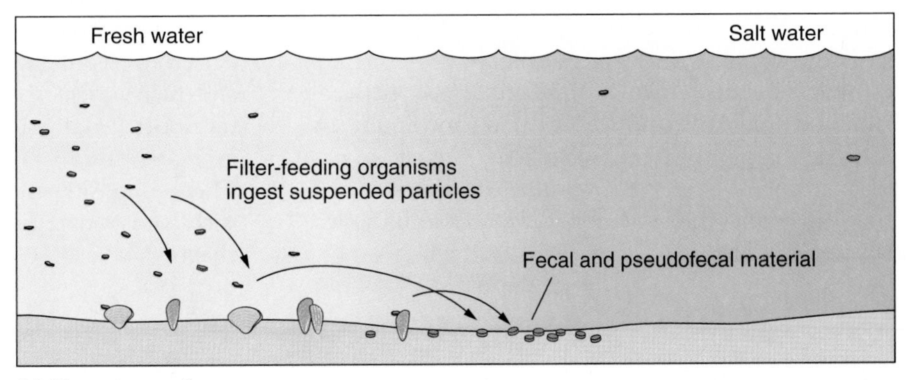


(b) Bioagglomeration

causes the flocs to fall to the bottom, resulting in a clearing of the water and a buildup of sediment on the bottom. Over time, this buildup on the bottom can be considerable. The resulting decrease in water depth is called *shoaling,* and it can cause passing ships to run aground.

Biological processes also affect sediment action. Aquatic organisms called *filter feeders,* such as clams and mussels, obtain their food by drawing in large volumes of water and filtering out the solid portions. Part of this solid material is metabolized for food and excreted as fecal pellets; the rest is ejected as pseudofecal material. This organic concentration of suspended sediment particles is called **bioagglomeration** (Figure 13–7b).

The greater abundance of animals in saline water increases bioagglomeration and therefore the rate of sedimentation. Any change in freshwater discharge in estuaries—natural or human-caused—can change the position of the saltwater/freshwater boundary. This shifts the location of maximum sedimentation and shoaling of the bottom. These changes can create a navigational hazard.

## ESTUARINE PROBLEMS

The increasing density of human settlement around estuaries places greater stress on them and at the same time decreases estuary quality. This decline in quality results not only from increasing use but also from attempts to change natural estuary conditions.

### Fertility, Cooling, Waste, and Transportation

Estuaries and the marshes and tidal flats that border them are among the most fertile places on Earth. They are fertile because continually decaying plant and animal material releases nutrients, such as phosphorus and nitrogen. The nutrients support a rich variety of aquatic life. Fish lay their eggs in the shelter of estuaries where their offspring can grow to maturity before returning to the more rigorous ocean environment. When people eliminate these nursery areas by filling them or reducing their water quality, the offshore fish population eventually decreases.

One use of water in estuaries is for industrial cooling. After use, the water, now much warmer, is returned to the estuary. However, when water is heated, the risk of thermal pollution exists. Recall from Chapter 12 that thermal pollution reduces the amount of oxygen that water can hold and can kill aquatic organisms (Figure 12–7). Thermal pollution can be minimized with cooling canals, which are open, lined ditches through which water circulates to cool (Figure 13–8). Another method uses heat exchangers, which cool industrial water by absorbing heat from it and releasing the heat into the atmosphere (heat exchange). The cooled water then is released into the estuary.

Large volumes of waste inevitably pass through estuaries because they are the interface between land and sea. Waste must be treated thoroughly before discharge to avoid disturbing the ecological balance of estuaries and because estuarine circulation can return the contaminated water to its source (Figure 13–2).

Transportation is a very important function of estuaries. Estuaries provide sheltered harbors for river craft from the continental interior and for seacraft. Deep navigation channels within estuaries naturally fill with sediment that is delivered by rivers and carried inland by tides. This requires continual dredging of the estuary bottom to maintain adequate depth for vessels. In the Port of New York–New Jersey (New York Harbor), for example, an average volume of nearly 7.6 million cubic meters (10 million cubic yards) of sediment was dredged each year from 1978 to 1988.

Shoreline development in large ports, such as Baltimore and San Francisco, disturbs the natural pattern of sedimentation. The construction of docks and piers interferes with water flow and sediment transport (Figure 13–9). As the water slows down, it deposits more of its sediment load between piers. These deposits create further need for dredging with its associated problems (discussed later in this chapter).

**FIGURE 13–8**
Turkey Point nuclear power plant at Homestead, Florida. The facility has an extensive system of canals (shown in background) to cool the effluent from the plant. (Photo by author.)

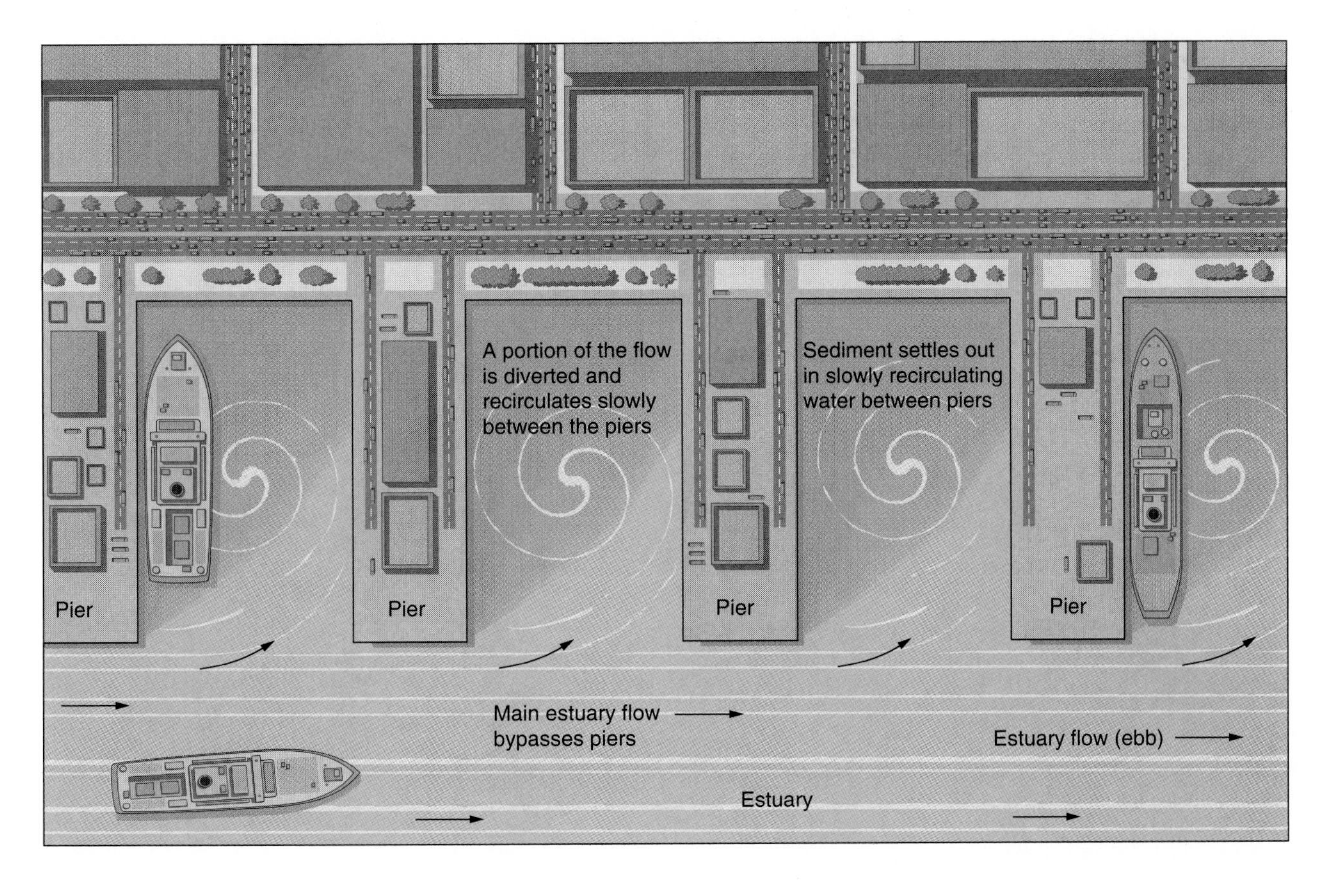


**FIGURE 13–9**
Structures built out into estuaries interfere with tidal flow and cause sediment to be deposited.

## Organic Loading

One of the most serious estuarine problems is the increased input of organic waste, which is rich in nitrogen and phosphorus. The sources are sewage and industry. In Figure 12–6, you saw how this waste increases BOD (biological oxygen demand), which leads to a drop in DO (dissolved oxygen). Wherever raw sewage, treated sewage, or other organic waste is discharged into estuaries or tributary streams, the potential exists for serious reduction in DO and the consequent destruction of aquatic animals.

Long-standing discharge of heavy organic waste in isolated locations along a flowing river-estuarine system can cause permanent stretches of alternating clean water and polluted water. This is **segmental pollution** (Figure 13–10). The polluted segments maintain themselves, despite the river-estuarine flow because the BOD always exceeds the DO in the industrial-urban areas. In the natural segments between those stretches, the DO level is much higher, and the greater available oxygen reduces the pollution level.

Problems of organic loading in estuaries can be reduced by secondary treatment of sewage waste (Figure 12–9) and by aeration of industrial organic waste.

## Toxic Pollution

The pollution of streamwater from spills was discussed in Chapter 12 ("The PCBs of Hudson River"). However, pollution also can attain high levels when modest day-to-day releases accumulate. Pollutants may become concentrated in both the water and the bottom sediment. Contaminants in solution may be flushed out to sea, but some will remain in the estuary if they are held to the surface of fine grained sediment particles in a process known as **adsorption** (Figure 13–11). The particles and adsorbed pollutants settle and form a sediment layer at the bottom. By this mechanism, both bot-

**FIGURE 13–10**
Segmental pollution in an estuary.

tom sediments and the pore water in them may become contaminated in pollutants.

After the pollution plume is fully dispersed, the adsorbed pollutants in the bottom sediment may become covered with "clean" sediment. The contaminants thus may remain stored for years. However, if the bottom is disturbed by dredging or if a major flood scours the bottom of the estuary, the pollutants may become remobilized and become a hazard once again.

## Dredging Problems

Estuaries are being filled continually with sediment brought downstream by rivers and carried inland by tides from the ocean. To maintain the depth necessary for navigation, excess sediment must be dredged. Unfortunately, dredging releases pollutants on the bottom, increases turbidity, and creates the problem of moving the dredged material to some new location.

Dredging must be done properly to avoid releasing buried organic material or pollutants trapped earlier in the sediments (Figure 13–12). It is essential to test the bottom material for pollutants before dredging. To do this, a long, cylindrical sample is removed from the bottom with a coring apparatus. Chemical and biological analysis of the sediment sample within the core can reveal buried pollutants.

A primary effect of dredging is to temporarily cloud the water, creating turbidity, which decreases penetration of light that is needed by aquatic plants for photosynthesis. Dredging also causes deposition of suspended sediments away from the sides of the channel being dredged, contributing to shoaling on the banks on either side of the channel. The increased sedimentation and turbidity also may adversely affect bottom-dwelling organisms locally.

The greatest problem with dredging is what to do with the material. It varies from clean sand (dredged from ocean-approach channels) to contaminated fine-grained sediment (dredged from heavily industrialized harbor shipping channels). The majority is clean, with only traces of contaminants. What to do with it depends on the type of sediment, the contamination level, volume, and site

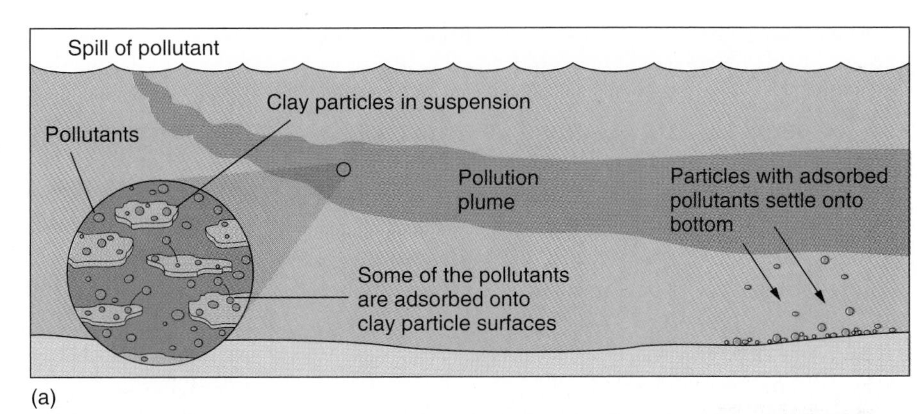


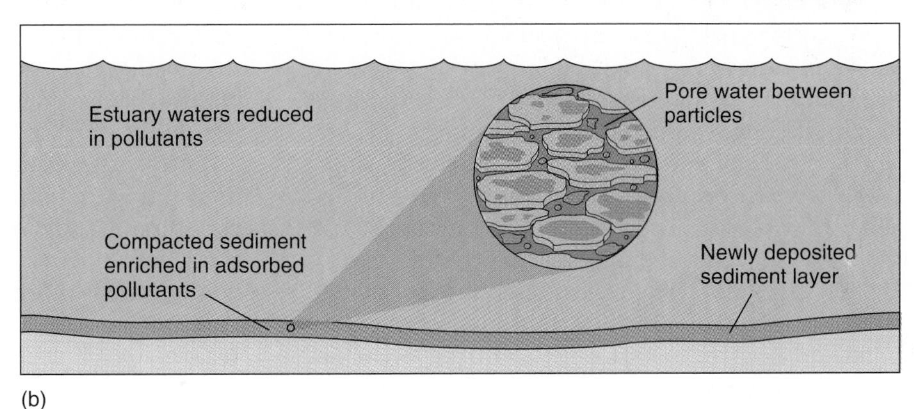


**FIGURE 13–11**
(a) Adsorption of toxic substances onto the surfaces of suspended particles. (b) When these particles settle and accumulate on the bottom, they may concentrate pollutants in the bottom sediment.

availability. About 80% is stored separately in a contained disposal facility, which is separated from the marine environment by dikes, or is disposed of on land in containment basins or along the estuary's banks. The remaining 20% is dumped in the ocean.

*Contained land disposal* involves pumping dredged material into diked areas on land. One problem with this method is getting permission from landowners and government to install such facilities. This is another example of the NIMBY effect.

*Containment area disposal* involves building a dike from shore into the estuary to enclose a basin, into which dredged material is pumped. Complete filling of the diked area can form new land. It may be covered with soil, vegetated, and made available for recreation, industrial parks, or other uses. An example of a large containment facility is Craney Island in the harbor of Hampton Roads, Virginia.

*Containment islands* are another disposal technique, creating islands by building a diked enclosure in the estuary itself. This technique has been used to store dredged material in the harbors of Baltimore, Maryland; Charleston, South Carolina; and Mobile, Alabama.

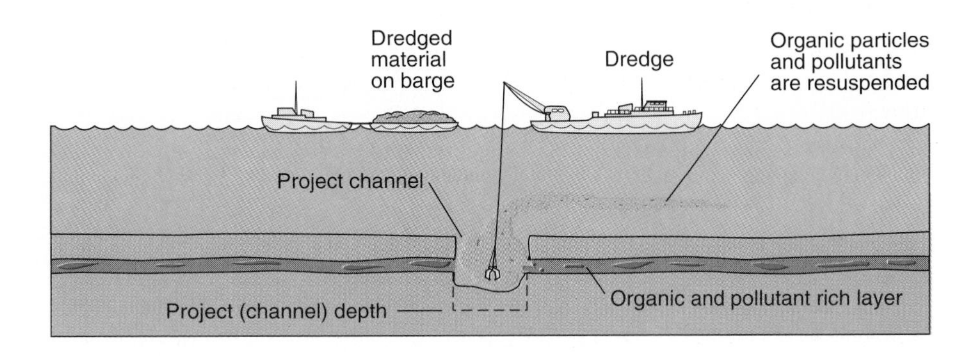


**FIGURE 13–12**
Potential problems of estuary dredging.

*Ocean disposal* sometimes is required for the great volume of material dredged from some harbors. Dredged material is deposited in mounds on the ocean floor. If contaminated sediment is dumped, it is covered by a layer of clean dredged material. The material falls very quickly to the bottom, with minimal lateral dispersal in the overlying water. One such disposal area, 12 miles south of Coney Island, New York, in 24 to 27 meters (80 to 90 feet) of water, showed no disturbance in the surface of the dredged material even after the passage of Hurricane Gloria in 1985.

*Borrow pits* are another method for disposal of dredged material. These are seafloor pits, either existing from previous sand mining or dug solely for the purpose of disposal (Figure 13–13a). Barges dump dredged sediments above the pit (Figure 13–13b). The sediments then are capped with a layer of clean dredged material to prevent erosion (Figure 13–13c). This method has been used in Portland, Oregon, and is being studied elsewhere.

Some dredged material is used beneficially to create new wetlands or to stabilize old ones. Clean sand, dredged from tidal inlets and the ocean-approach channels to harbors, is used to replenish eroded beaches (Chapter 15) and to build offshore berms (submerged sand bars) that absorb the energy of incoming waves that cause beach erosion.

## Anthropogenic Changes in Stream Flow

The freshwater/saltwater balance within a given estuary varies with natural fluctuations in stream discharge into the estuary. This balance governs the habitats of different organisms as well as the zones of most rapid sedimentation, which are near the freshwater/saltwater contact.

***Charleston Harbor.*** What happens when human activity markedly increases freshwater discharge into an estuary? This occurred in Charleston, South Carolina, in the 1940s and created serious shoaling problems in the harbor (see Simmons and Hermann, 1992). One of the major estuaries draining into Charleston Harbor is the Cooper River (Figure 13–14). In late 1941, construction was completed on a dam to divert water from the Santee River (to the north), through the Pinopolis Dam and hydroelectric plant, and down the Cooper River into Charleston Harbor.

Prior to the Santee-Cooper diversion, freshwater input into Charleston Harbor was modest, averaging about 4 cubic meters per second (5 cubic yards per second). Tidal forces and saltwater movement dominated the harbor. Dredging of about 105,000 cubic meters (140,000 cubic yards) of sediment per year maintained the 11-meter (35-foot) channel across the harbor.

In 1942, the Santee-Cooper Hydroelectric Project went into operation, and the average freshwater flow down the Cooper River into Charleston Harbor increased to 425 cubic meters per second (550 cubic yards per second). This *hundredfold* increase significantly affected sedimentation. The annual dredging requirement grew *55-fold* to 7,650,000 cubic meters (10,000,000 cubic yards).

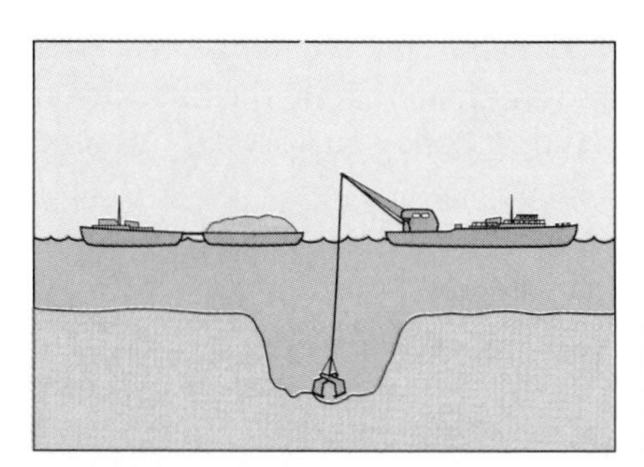

(a) Depression (borrow pit) made in ocean floor from sand mining

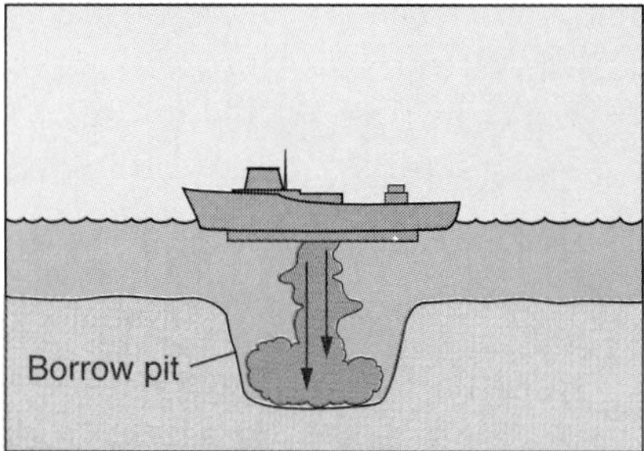


(b) Hopper dredge deposits dredged material into pit

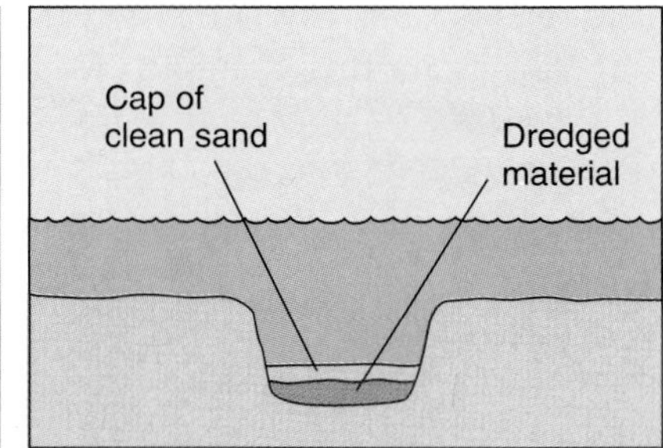


(c) A layer of sand is deposited over the dredged material

**FIGURE 13–13**
Creation of an ocean borrow-pit disposal area for dredged sediment. (a) Borrow pit is created on the seafloor by sand mining. (b) Dredged material is deposited within the pit. (c) A cap of clean dredged material is placed on top of the contaminated dredged material to render it harmless to the marine environment.

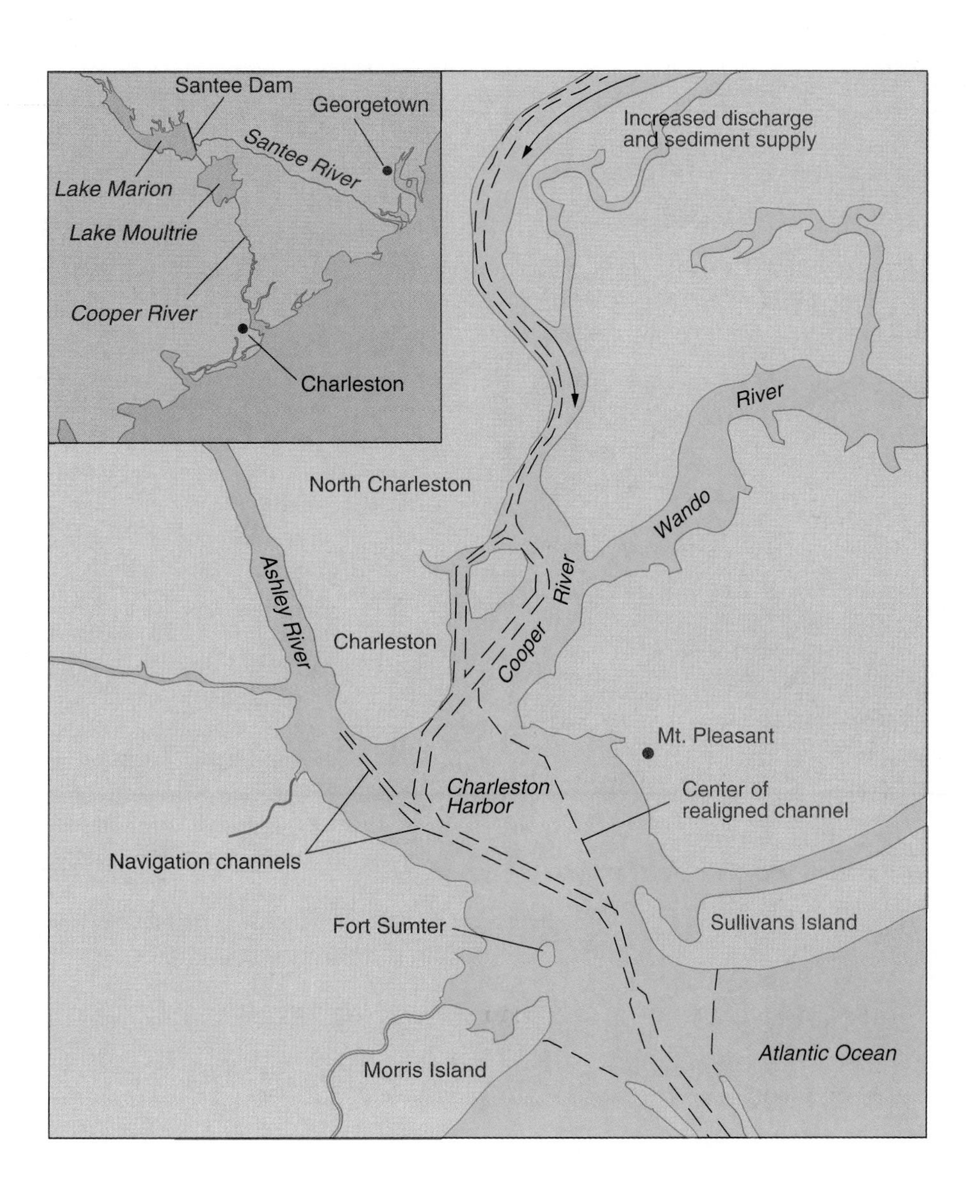


**FIGURE 13–14**
The Santee-Cooper River water-diversion scheme near Charleston, South Carolina.

The Army Corps of Engineers attributed this drastic increase to two changes caused by the Santee-Cooper diversion. The greater flow in the Cooper River increased erosion of its drainage basin and the transport of material. The expanded freshwater supply resulting from the diversion also modified circulation within the estuary to greatly increase sedimentation within the harbor.

***San Francisco Estuary.*** Changes in freshwater discharge also have affected the San Francisco estuary in California (Box 13–1). However, in this case, problems have resulted from a *decrease* in freshwater inflow into the estuary. Roughly 50% of the average annual freshwater flow into the estuary has been diverted to San Francisco Bay-area cities, to farmers in the Sacramento Delta (see Chapter 10), and to southern California (Figure 13–15). Water diverted to southern California moves southward via the California Aqueduct to Los Angeles.

In the San Francisco estuary, the natural westward flow of freshwater toward the Pacific prevented saltwater from moving very far up into the freshwater delta. However, the large freshwater diversion has permitted up-estuary intrusion of saltwater, causing environmental damage in the delta.

Low flows now limit the growth of phytoplankton, shrimp, and the fish that feed on them. The less-

**BOX 13–1** **SAN FRANCISCO'S FORMER WETLANDS**

The history of wetlands elimination in the San Francisco-Sacramento area has been documented by the San Francisco Estuary Project. From 1850 to 1914, hydraulic gold mining was extensive in the Sierra Nevada Mountains. This method, which flushes gold-bearing sand from the riverbanks, washed about 2 billion cubic meters (2.6 billion cubic yards) of sediment into the bay. This reduced the bay's volume and expanded the tidal wetlands around its periphery.

However, between 1860 and 1930, 97% of the Sacramento–San Joaquin River Delta's 140,000 hectares (350,000 acres) of freshwater wetlands were diked and planted. Following this, tidal wetlands around San Francisco Bay were converted to agricultural and salt production.

As the bay area grew in population, thousands of additional acres were used for housing. Another burst of urbanization followed World War II, when more housing, roads, airports, and landfills were created by filling and building upon former wetlands. At midcentury, massive flood control and water-diversion projects altered wetlands and greatly reduced natural freshwater flow into San Francisco Bay, increasing the bay's salinity. The greatest effect was on the tidal marshes bordering the bay.

According to San Francisco Bay Estuary Project, in 1850 there were 220,000 hectares (550,000 acres) of tidal wetlands. By 1985, only about 27,000 hectares (66,000 acres) remained, a reduction of 88%!

er flow also reduces the frequency with which the southern part of the San Francisco Bay is flushed of pollutants. The resulting stagnant water could adversely affect bottom-dwelling organisms and thus cut off a food source for migratory shorebirds. Another problem is that fish eggs and juvenile fish are sucked into the pumps of water diversion projects. Clearly, displacement of the saline/freshwater contact affects the distribution and abundance of many organisms in the river-estuarine system.

The problem soon may be ameliorated. In 1993, four federal agencies used the Federal Clean Water Act and the Endangered Species Act to propose new water-quality standards for the estuary. They would require reducing the volume of freshwater that is diverted southward by about 7%. This would reduce the environmental damage in the delta.

## WETLANDS, FRESH AND TIDAL

Many rivers, lakes, estuaries, and bays are bordered by areas where the ground is constantly wet, generally called *wetlands*. Several federal agencies regulate these biologically important lands, including the U.S. Army Corps of Engineers, the U.S. Environmental Protection Agency, the U.S. Fish and Wildlife Service, and the U.S. Department of Agriculture's Soil Conservation Service. They define a **wetland** as an area that is inundated or saturated with water for at least 10 continuous days, that contains vegetation adapted to saturated soil conditions, and that has highly organic (histosolic) soils.

Most wetlands are **freshwater wetlands** that develop where water covers the ground. **Tidal wetlands** occur where salty estuarine water covers the

**FIGURE 13–15**
Saltwater intrudes upstream as freshwater in the Sacramento Delta is diverted southward by the California Aqueduct, resulting in a net loss of freshwater input to San Francisco Bay.

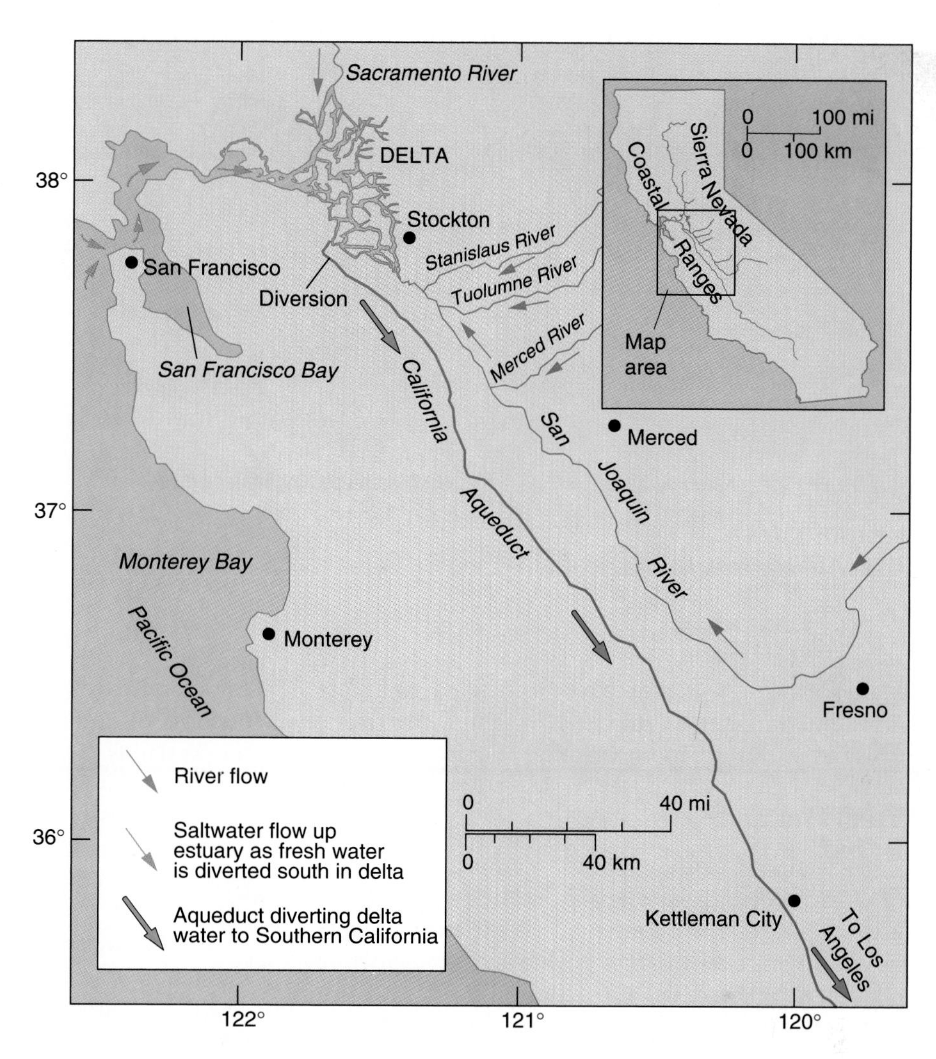


land. Tidal wetlands are just as important as freshwater wetlands, although far less abundant. This chapter deals only with wetlands in temperate latitudes (above 30°, which is the latitude of central Florida, for example). Special problems of subtropical wetlands in southern Florida will be considered in Chapter 14.

Wetlands are fertile, flat, and easily filled with sediment, so they have been increasingly drained and converted for farming, domestic, and urban use. About 50% of our wetlands have been lost since the early 1600s, when colonization of America began in earnest. Based on the U.S. Fish and Wildlife Service National Wetlands Inventory, the United States will lose an additional 1,700,000 hectares (4,250,000 acres) of wetlands by the year 2000. This will be a tragic loss, for wetlands are one of the most important parts of Earth's ecosystem.

## Freshwater Wetlands

Freshwater wetlands make up over 90% of total wetland acreage and consist of swamps, marshes, and bogs (Figure 13–16).

- ☐ **Swamps** contain woody plants and trees and sometimes are covered with standing water.
- ☐ Freshwater **marshes** often are covered by shallow water, are treeless, and are characterized by soft-stemmed plants such as cattails, sedges, and grasses. They usually border lakes and streams.
- ☐ Bogs are local depressions, generally peat-filled, that form where standing water and vegetation accumulate. Mosses are common in bogs.

In Chapter 7 you saw how wetlands that border streams help reduce flooding because the spongy soils absorb floodwater and release them slowly.

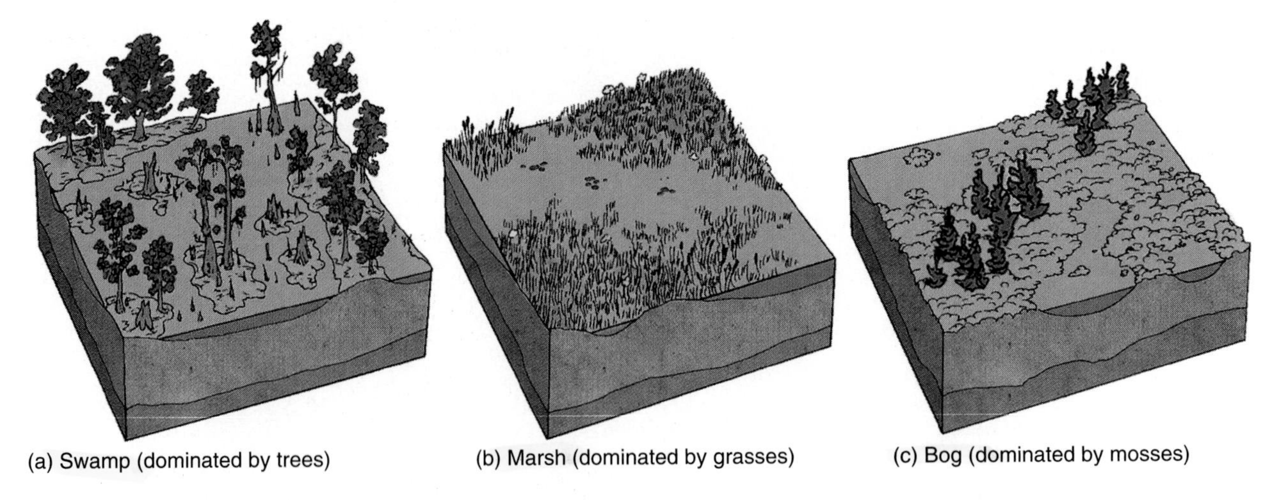


**FIGURE 13–16**
Types of freshwater wetlands: (a) swamps, (b) marshes, and (c) bogs.

Many of these wetlands have been filled to create more building sites as towns grew along the rivers. This has reduced flood protection (Figure 13–17).

## Tidal Wetlands

Tidal wetlands (Figure 13–18) comprise only 10% of U.S. wetland acreage, but their importance greatly exceeds their abundance. Nutrients supplied by tidal wetlands are vital to life in the nearshore portion of the ocean. Destruction of wetlands eliminates the vital exchange of nutrients and oxygenated water between wetlands and the adjacent ocean. Both tidal and freshwater wetlands have been extensively eliminated for agriculture, industry, housing, marinas, and transportation (Figure 13–19). These developments frequently are at great risk during severe storms and floods.

Tidal wetland components are defined by their vegetation and elevation relative to low tide. At low tide, water remains only in the *tidal channel*. The tidal channel system provides water exchange between the wetlands and the ocean. The *tidal flat* is the unvegetated portion between low-tide level and high-tide level. The *tidal marsh* is a vegetated area near and above higher tide (Figure 13–18).

## Exchanges Between Tidal Wetlands and the Ocean

As mentioned, tidal wetlands are among the most fertile environments on Earth. This results from nutrients released by plant and animal decay and the deposits of organic-rich wastewater that washes into the wetlands. The biological oxygen demand of this organic decay, plus the needs of animal respiration, lower the dissolved oxygen in the wetlands and estuarine water. Unless oxygen is periodically resupplied by tidal circulation, aquatic life perishes (Figure 12–6).

Tidal exchange provides the mechanism for maintaining healthy aquatic life in both the wetlands and the ocean (Figure 13–20). As the tide ebbs, organic-rich water moves from the wetlands into the ocean (Figure 13–20a). This ebb flow provides the nutrients necessary to support the abundant life characteristic beyond the mouth of an estuary.

Nearshore ocean water has greater DO than inland wetlands and bays because waves and currents mix oxygen from the atmosphere into the near-surface of the water. Flood-tide movement of oxygenated nearshore ocean water into the wetlands (Figure 13–20b) provides the oxygen needed for animal and plant respiration and organic breakdown.

This system explains the most vital function of tidal wetlands—maintaining the ocean environment in a condition that can support abundant life. When we interfere with the water-nutrient exchange between wetland/estuaries and the ocean, we create problems. On the one hand, when wetlands are filled, they no longer supply nutrients to the ocean. On the other hand, if we inhibit tidal flow inland, oxygenated water no longer can reach the wetlands

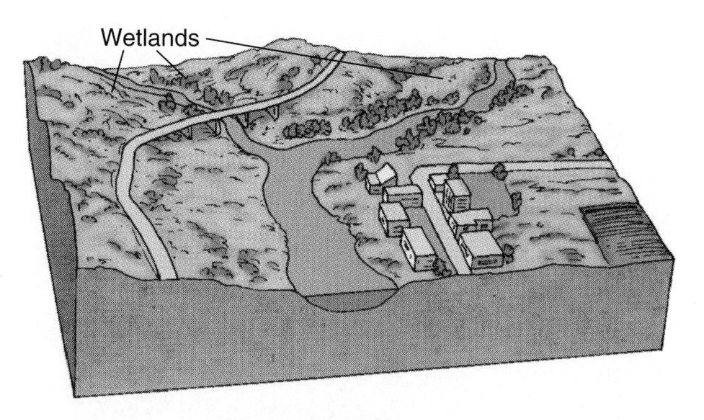


(a) Stream bordered by freshwater wetlands

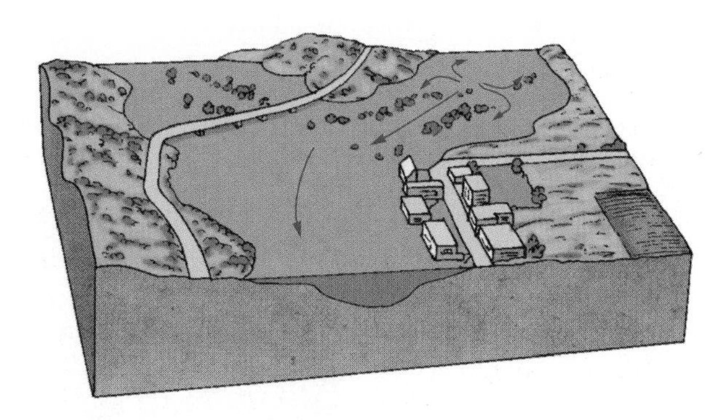

(b) Wetland during submergence river flooding

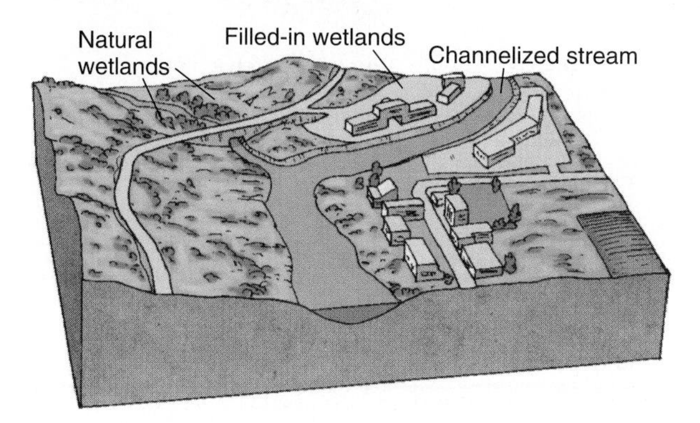


(c) Wetland areas filled and stream channelized

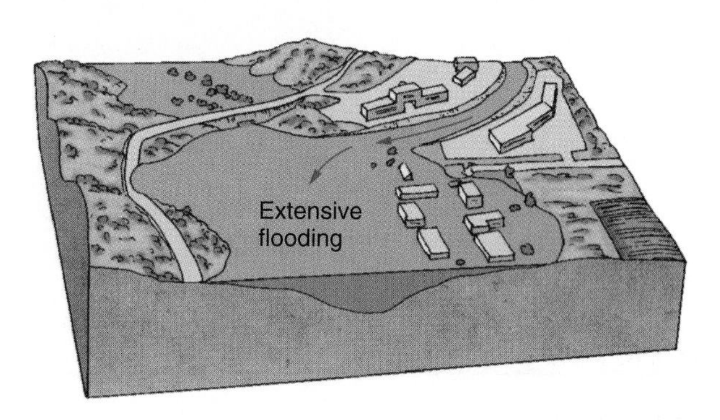


(d) Severe flooding damage occurs in developed areas

**FIGURE 13–17**
Role of wetlands in flood protection. (a) Natural river system and bordering wetlands. (b) Flood in a natural system. Wetlands absorb spreading floodwaters, which increases lag time and decreases peak discharge downstream. (c) Urbanized floodplain. Wetlands have been partially filled and built upon. The absorbent wetlands have been replaced by impermeable pavement and building sites. (d) Flooding in an urbanized basin where wetlands have been partially eliminated. Rainfall drains quickly into streams, decreasing lag time and increasing peak discharge downstream. (Adapted from diagram in Interagency Task Force Report, Our nation's wetlands, 1978, p. 26.)

to prevent their stagnation. Structures such as docks and marinas near ocean inlets also may restrict tidal flow and further reduce oxygen supply to the wetlands.

## IMPORTANCE OF WETLANDS

Wetlands are valuable for their plant life and wildlife, flood mitigation, pollution control, and simple scenic beauty.

### Nutrient Productivity and Wildlife Habitats

Wetlands are among the most fertile ecosystems on Earth. Ecologist Eugene Odum studied Georgia salt marshes and found that they produce 7.3 metric tons (8 tons) of organic material per acre per year. For comparison, a fertile dryland hay field produces only 3.6 metric tons (4 tons) per acre per year. Water flowing into wetlands provides all the minerals needed to support abundant vegetation. As the vegetation dies and decomposes, it releases large amounts of nitrates and phosphates, which are natural fertilizers.

The growing vegetation supports widely varied fish, animals, and birds, providing them with food and shelter. When the plants die, they decompose. The organic nitrogen and compounds they release either become trapped in sediment or become food for fungi, bacteria, or debris-feeding shellfish such as crabs.

FIGURE 13–18
Aerial view of wetlands complex on the South Carolina coast. Features shown are marsh (vegetated areas), tidal flats (bare areas), and a tidal inlet through which the wetland drains into the Atlantic Ocean. (Photo by author.)

## Flood Mitigation

Wetlands provide natural flood protection by providing broad areas where floodwater can spread out, by storing and slowly releasing this water, and by slowing the velocity of flood flow. The spongy, organic soil of wetlands, characterized by intertwining root systems and vegetation, acts as a natural baffle, reducing floodwater velocity. The absorbent mass soaks up excess water. After a storm, the wetlands

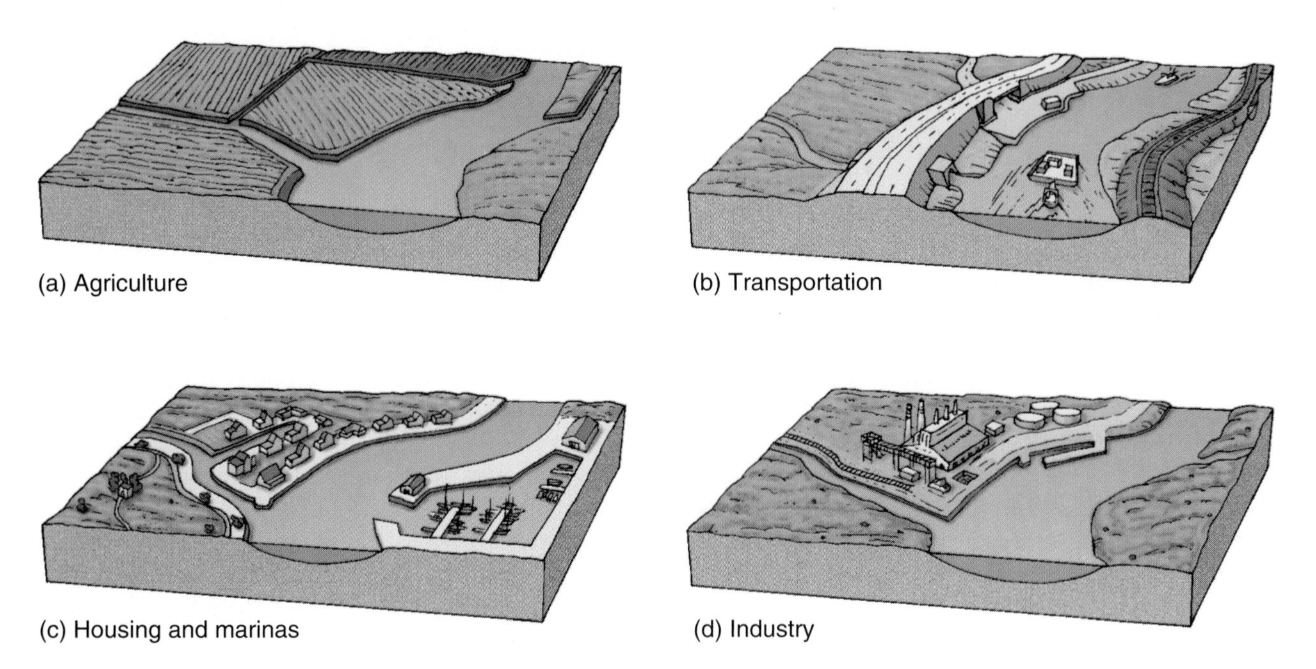


FIGURE 13–19
Wetlands (freshwater and saltwater) have been eliminated for development of (a) agriculture, (b) transportation, (c) housing and marinas, and (d) industry.

**FIGURE 13–20**
Tidal exchange of nutrient-rich and oxygen-rich waters between wetlands and the ocean. (a) *Ebbing tide.* Nutrient-rich waters move from wetlands into the ocean on the ebbing tide. These nutrients support near-shore ocean life. (b) *Flooding tide.* Oxygen-rich waters move from the ocean into bays and wetlands on the flood tide. This oxygen is required for animal respiration and organic decomposition in wetlands.

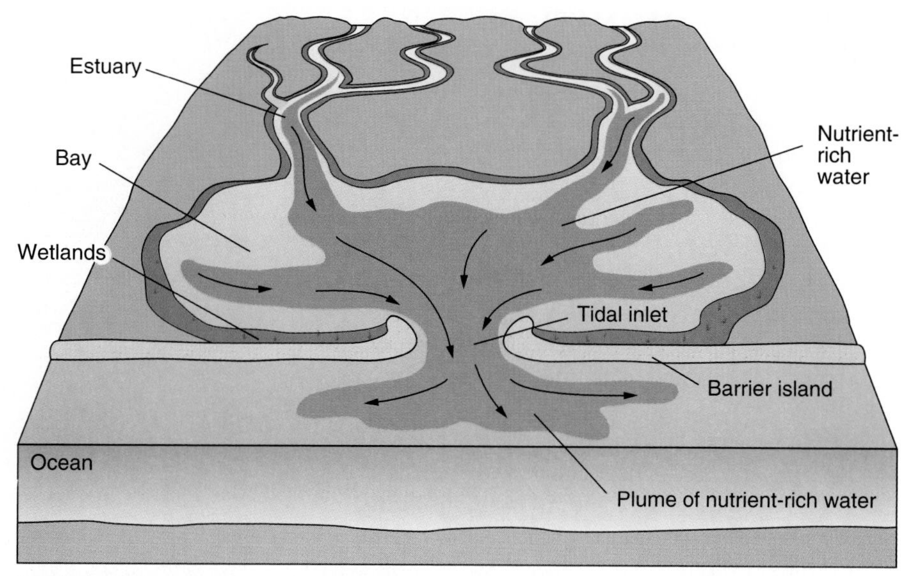


(a) Ebbing tide

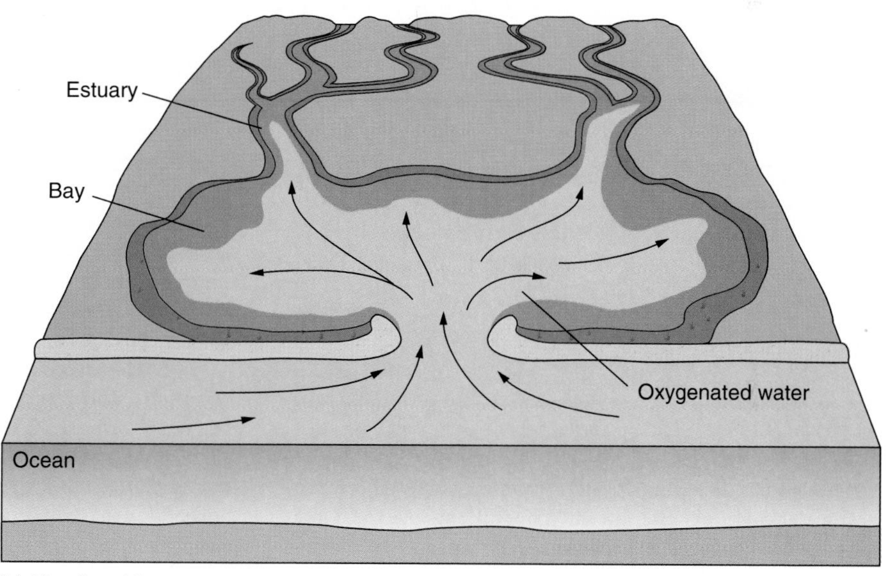


(b) Flooding tide

release the trapped water slowly. The effect is to increase lag time in streams and estuaries bordered by wetlands (Chapter 7).

## Pollution Control

Wetlands remove pollutants from the runoff they receive. Organic effluent, such as wastewater from sewage treatment plants, is rich in nutrients such as phosphates and nitrates. These are taken up and used by wetlands plants. As polluted water moves through the wetlands, metals and other toxic substances can be adsorbed onto clay-sized sediment particles and organic material. Some toxic material is even used by certain plants that have a greater tolerance for many toxic substances. This filtering action of wetlands vegetation improves the quality of runoff.

## WETLANDS ELIMINATION

Wetlands are being eliminated rapidly. Because wetlands have long been regarded as useless land that breeds robust, disease-carrying insects, communities have voiced little opposition to filling and converting them to agricultural, domestic, or urban use. All major coastal cities and many inland communities have expanded by filling wetlands and building on them (Figure 13–21). Only recently have people realized the tremendous environmental importance of wetlands and the resources they provide.

It is important to remember that wetlands are, by their nature, at or near local water level. When they are paved over and built upon, flooding increases in severity (Figure 13–17c and d). Structures built on them often are the first to be flooded (Figure 13–17).

**FIGURE 13–21**
Foster City, California, was built on former marsh land bordering San Francisco Bay. (Courtesy of Raymond Pestrong, San Francisco State University.)

## LOOKING AHEAD

While this chapter discussed the wetlands of temperate latitudes, the next chapter looks at subtropical wetlands and coral reefs. Both are fragile and important ecosystems. Both are threatened by increasing pressure from development, tourism, and fishing.

## SUMMARY

Estuaries are the transitional water passages between freshwater streams and the saltwater of the ocean. They were formed as post-glacial sea level rose and filled depressions along the shoreline or invaded river valleys inland to form river-estuaries, bays, or back barrier bays. They are affected by the tides. Flood tide occurs as tide level rises and oxygenated water moves from the ocean inland. Ebb tide occurs as the tide level drops and the nutrient-rich water from the wetlands and estuaries moves oceanward.

Salinity is the concentration of salt per thousand parts of water. Ocean water has an average 35 ppt of salt. Estuaries are branches of the ocean that have brackish (7 to 15 ppt) to salty (35 ppt) water. The boundary where saltwater meets freshwater in an estuary shifts position, depending on freshwater input into the estuary. In an estuary, there is a net flow of fresh surface water *out* of the estuary and a net flow of saline bottom water *into* the estuary.

In freshwater, suspended fine particles repel each other due to their electrical charges, causing the particles to remain in suspension and the water to be cloudy. In saltier water, ions draw the particles together (flocculation). This increases sedimentation and clarifies the water. Filter-feeding organisms ingest water and sediment particles, expel the water, and excrete the sediment as part of fecal pellets (bioagglomeration).

Estuaries are nurseries for juvenile fish, provide ship channels, afford cooling water for industry, and are used for waste disposal. The waste disposal has an environmental cost. Organic-rich wastewater increases BOD and thermal pollution decreases DO. Contaminants can be adsorbed onto sediment particles. The deposited particles may become resuspended later during dredging operations or when the estuary bottom is scoured in a storm or flood.

The dredging necessary to maintain ship routes can disturb and resuspend bottom sediments, which may cloud the water, locally reduce light penetration needed by plants, and cover bottom-dwelling organisms with sediment. The largest problem with dredging is what to do with the dredged material. It can be pumped to confined disposal areas on land, used to create islands in the estuary, or deposited on the ocean floor to fill borrow pits where sand has been removed for construction. Anthropogenic changes in freshwater discharge cause sedimentation and habitat loss within the estuary and even in ocean areas near the estuary's mouth.

Wetlands are inland or shore areas that are saturated or covered with shallow water for periods sufficient to support vegetation adapted to wet conditions. Wetlands are important habitats, providing shelter for animals and fish, nutrients for plants, flood protection, and adsorption of pollutants. Tidal wetlands provide nutrients to ocean water during ebb tide. During flood tide, oxygenated ocean water moves into the wetland to prevent stagnation.

Great areas of wetlands have been drained, filled, and built upon over the years, resulting in the loss of the benefits of wetlands to many areas. This practice is a potential geohazard because filled-in wetlands form an unstable base in an earthquake (Chapter 5), subside over time (Chapter 10), or may be rapidly flooded in a severe storm (Chapter 16).

## KEY TERMS

adsorption
back barrier bays
bays
bioagglomeration
bogs
brackish bater
ebb tide
estuary
flocculation
flood tide
freshwater wetlands
marshes
river estuaries
salinity
salt wedge estuary
segmental pollution
swamps
tidal circulation
tidal wetlands
wetlands

## REVIEW QUESTIONS

1. **a.** Describe water circulation within a river estuary.
   **b.** Give an example showing the environmental impact of this water circulation.
2. **a.** How do flocculation and bioagglomeration influence sedimentation in estuaries?
   **b.** How do anthropogenic changes influence sedimentation in an estuary?
3. Describe what happens when freshwater discharge to an estuary is reduced.
4. Estuaries are used for waste disposal and transportation. Describe any problems that result from these activities.
5. **a.** What problems can occur in dredging?
   **b.** Describe the methods for disposing of dredged material.
6. **a.** What problems can develop when estuary water is used for cooling?
   **b.** How can these problems be mitigated?
7. Some sections of rivers and estuaries maintain low oxygen levels (DO) even though the water is flowing. How can this be?
8. What role can sediments play in spreading toxic pollution?
9. Describe four different functions of wetlands.
10. What problems result when wetlands are eliminated?
11. **a.** Describe the components of a tidal wetland.
    **b.** Although tidal wetlands comprise only 10% of all wetlands, they are disproportionately important. Explain.
12. Describe what happens in the tidal exchange between wetlands and the ocean. Why is this so important?

## FURTHER READINGS


Horton, T., 1993, Hanging in the balance—Chesapeake Bay: National Geographic, v. 183, no. 6 (June), p. 2–35.

Interagency Task Force, 1978, Our nations wetlands: USGS Printing Office Stock # 041-011-00045-9, 70 p.

Mitchell, J. G., 1992, Our disappearing wetlands: National Geographic, v. 182, no. 4 (Oct.), p. 3–45.

New York Times, 1990, Efforts to halt wetland loss are shifting to inland areas, (Mar. 13), p. Cl and C 12.

San Francisco Estuary Project, 1990 (a series of 4-page pamphlets) on Agricultural drainage; Dredging and Waterway Modification; How we use the estuary's water; Pollution; Sacramento-San Joaquin Delta; and Wetlands.

Shabman, L., Riexinger, P., and Brown, T., 1993, Clarifying classification: National Wetlands Newsletter, v. 15, no. 1 (Jan.-Feb.), p. 4–23.

Simmons, J. R., and Hermann, F. A., 1972, Effects of man-made works on the hydraulic, salinity, and shoaling regimens of estuaries, p. 555–570 *in* Nelson, B., ed., Environmental framework of coastal plain estuaries: Geological Society of America Memoir 133, 619 p.

South Carolina Coastal Council, 1990, Understanding our coastal environment: Charleston, 40 p.

South Carolina Sea Giant, 1989, Wetlands: Coastal Heritage, v. 4, no. 1 (Winter) 12 p.

U.S. Army Coastal Engineering Research Center, 1972, Marsh building with dredge spoil in North Carolina, CERC R-2-72, 28 p.

# 14

# Problems of Mangrove Wetlands and Coral Reefs

Some of our finest water recreation and most valuable commercial sea animals exist along the Gulf Coast and around islands of the Caribbean. Offshore portions of these areas contain wetlands populated with mangroves, and some parts have coral reefs.

The wetlands provide nutrients that support fish, crayfish, crabs, and shrimp. Magnificent coral reefs occur along some coasts, sheltering the coastline from storms and affording an unparalleled underwater vista for snorkelers and scuba divers. Such areas attract tourists to these paradises, with more and more visitors each year.

Some visitors never leave. They decide to stay, building homes along these ecologically fragile coasts. As development and population increase, wetlands are being eliminated and coral reefs are dying. Many are concerned that we may be enjoying these resources to death.

## MANGROVE WETLANDS

Mangrove wetlands provide the same benefits afforded by temperate wetlands to the north (Chapter 13). The key difference lies in the trees called **mangroves,** which are salt-tolerant and thus can root in either saltwater or freshwater. Mangroves flourish along coasts that lie between the Equator and approximately latitude 30°, in both the Northern and Southern Hemispheres (Figure 14–1). In the United States, they are best developed in Florida and Louisiana. Their root systems provide sanctuary for marine organisms and offer very effective protection from storm waves attacking the coast (Figure 14–2).

**Coral reef assemblage, Florida Keys. (Photo by Steven C. Cook.)**

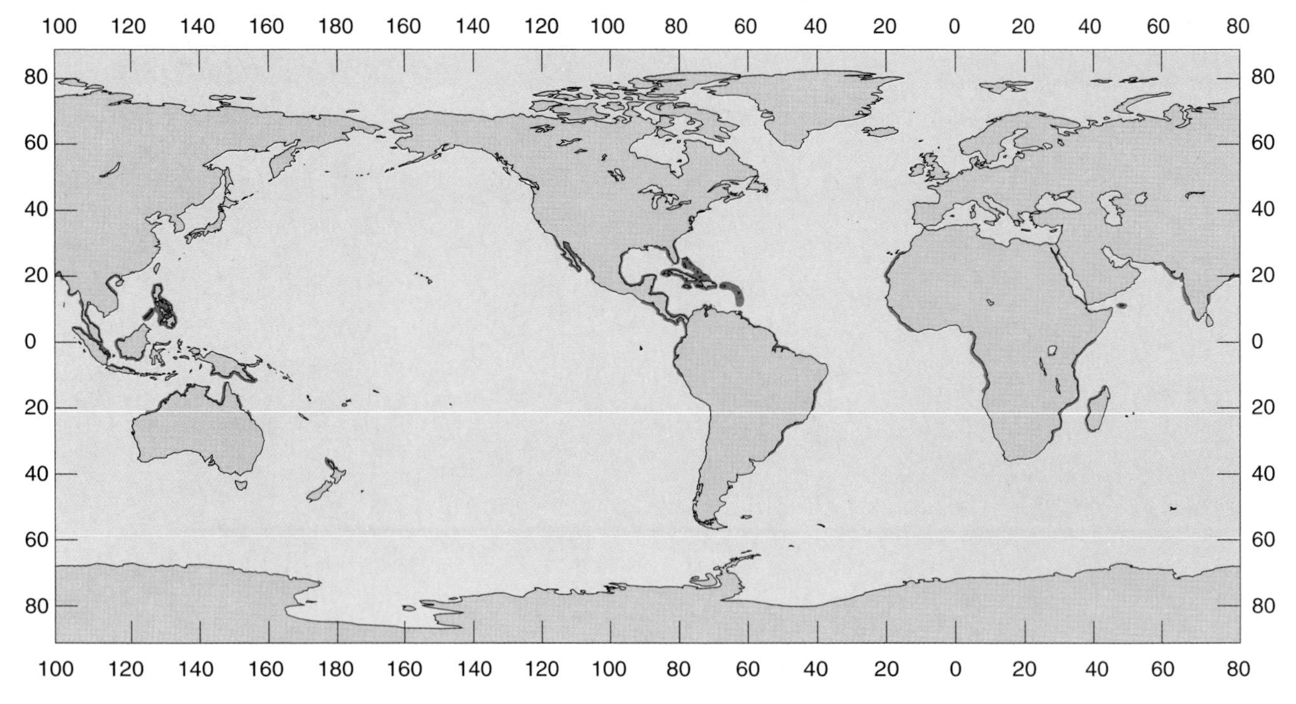


**FIGURE 14–1**
Distribution of mangrove forests in the world. (Modified with permission from James W. Nybakken, 1993, Marine biology (third edition): New York, HarperCollins, Fig. 9.25, p. 375.)

The salt tolerance of mangroves is a remarkable adaptation. Their specialized roots "filter" the salt in water, so that only salt-free water is taken up. Some species also have evolved salt-collecting glands so they can excrete salt through their leaves. A common misconception is that mangroves *require* saltwater. In fact, they thrive in freshwater, but some species have evolved a way to *tolerate* saltwater. This adaptability enables them to dominate some areas of the coastal zone.

**FIGURE 14–2**
Mangrove wetlands in the Dry Tortugas, Florida. (Photo by author.)

(a)

(b)

(c)

**FIGURE 14–3**
Types of mangroves: (a) red mangroves, (b) black mangroves, and (c) white mangroves.

Three mangrove species are common in the southeastern United States (Figure 14–3). They live in different zones, and their growth is governed by elevation relative to high tide. The *red mangrove* is the most abundant species along the seaward fringe of the lowest coastline—along sheltered subtropical and tropical shorelines, along tidal channels, on tidal flats, and on small islands within bays (Figure 14–3a). When people think of mangroves, it is the red mangrove they usually picture, distinguished by its arching system of supporting roots, anchored in mud. Red mangroves are rooted in seawater or are bathed in it by the daily tides.

The *black mangrove* occupies areas in which saltwater inundation occurs only during the highest monthly tides (spring tides), rather than daily. Black mangroves are distinguished by *pneumatophores* (breathing organs) that extend from its roots above the ground surface (Figure 14–3b).

The *white mangrove* usually occupies areas where saltwater inundation occurs only a few times a year (Figure 14–3c). Regardless of their position relative to the shoreline and the tides, all mangrove species are vital to the ecosystem in which they live.

## Mangrove Reproduction

Red mangroves produce a flower from which a single elongate seedling or *propagule* develops. These seedlings grow down from the branches (Figure 14–4) and eventually drop to the ground or into the water. Currents transport the propagules, and when they reach water shallow enough for the tip to become imbedded in the bottom, the mangrove seedling begins to grow.

In areas where sediment is being deposited to form shoals, mangrove colonization can stabilize the sediment against moderate erosion. Mangroves anchor the land with their extensive root systems

**FIGURE 14–4**
Mangrove stand with propagules extending down from the branches (center of photograph). (Photo by author.)

**FIGURE 14–5**
Mangrove trees colonizing a tidal flat in the Florida Keys. (Photo by author.)

and colonize former water-covered areas (Figure 14–5).

Because mangrove stands are dense, they are effective barriers to storm winds, waves, and storm surge. Early settlers in Florida and the Mississippi Delta learned that, when evacuation before a storm was impossible, their salvation lay in riding out the storm in boats, tied deep within the mangroves. We should not forget this natural storm-buffering capacity of mangroves later in the chapter when we consider the fate of housing developments built on former mangrove swamps.

## Mangroves and the Food Chain

Mangroves are important, not only for providing significant nutrients to support estuarine and coastal marine life but also for providing a sheltered habitat where certain fish lay their eggs and juvenile fish can develop. In Florida, it is estimated that 90% of all commercially important fish spend at least part of their lives in the mangrove wetlands and mangrove-lined estuaries. Mangrove wetlands also are extremely important as sheltered areas in which shrimp and other shellfish develop, a major economic resource in Florida and the Gulf of Mexico.

Red mangroves, which dominate the fringe forests, have leaves that are an important source of enriching nutrients in tropical estuaries. Their leaves fall into the water, where they are consumed by various animals and bacteria. Flushed by daily tides, the leaves are also washed into bays, spreading their nutrients. One hectare (about 2.5 acres) of red mangroves can drop 8 to 11 metric tons (9 to 12 tons) of leaves per year into estuary waters. This decomposition enrichment provides food for higher organisms.

Red mangroves also contribute *particulate organic matter (POM)* to the water. However, in the more protected inland areas that are characteristic of black mangroves, fallen leaves decompose on the forest floor and little particulate organic matter is exported to adjacent bays. However, the decomposed organic material is dissolved and then is exported in tidal water that flushes these areas during the highest monthly and seasonal tides. This *dissolved organic material (DOM)* forms flakes that are colonized by microbial organisms. This ultimately becomes food for crabs, shrimp, and other aquatic invertebrates. Thus, mangroves are a very important part of the estuarine and coastal food chain.

## Mangrove Wetland Destruction

As is the case with temperate wetlands to the north, increasing development is eliminating mangrove wetlands, particularly in Florida, through filling and development of the land. Wetland elimination in some areas has been extensive.

Florida's Tampa Bay is one of the fastest growing regions in the nation. The Florida Department of Natural Resources estimates that this area has lost 44% of its coastal wetland acreage, including both mangroves and salt marshes. Similar changes have occurred near Palm Beach on the Atlantic Coast, where mangrove acreage has decreased 87% in the last 40 years. It has been replaced with non-native trees, such as Australian pine, or by building on the sites.

The degree to which mangrove wetlands have been obliterated is demonstrated in the changes to Marco Island on Florida's Gulf Coast. In 1952, prior to development, a wide range of ecosystems thrived in the area, including red, black, and white mangroves (Figure 14–6a). By 1984, only 32 years later, most of the natural environment had been urbanized (Figure 14–6b) and replaced with extensive finger-fill canal development (Figure 14–6c).

The beneficial role of mangroves to nearshore fisheries became established in the 1970s, which lead to the enactment of laws to conserve mangrove areas against rampant destruction. However, because the original research was carried out on *red* mangroves along the coastal fringe, the public and regulatory agencies were led to believe that black and white mangroves, which grow further inland, have less importance in the estuarine food web.

Consequently, Florida developers are sometimes permitted to destroy black and white mangrove forests, as long as they do not destroy the coastal fringe of red mangrove. More recent studies have disclosed the high output of dissolved organic matter from black and white mangrove areas inland. Thus, *all* types of mangroves, regardless of color (species), are important in sustaining the estuarine food web.

***Finger-Fill Development.*** Figure 14–7a shows the pattern of water movement in an undeveloped mangrove wetland. Figure 14–7b shows the same land following elimination of the mangroves and dredging of channels across the former wetland. The dredged material is piled behind bulkheads constructed around the undredged sections. This forms elevated building sites that are separated by water-filled canals. The resulting pattern of land is called **finger-fill development.**

Finger-fill development creates popular living sites. Homes can be sold for premium prices because they are "waterfront property" with boat docks and fishing—at least initially—on canals that are only meters from the door. However, the environmental damage from such development is great. For this reason, finger-fill development is no longer permitted in many areas.

Such development is especially dangerous in the hurricane-prone southeastern United States. The building surfaces produced in finger-fill development usually are no higher than 2 to 3 meters (6.5 to 10 feet) above sea level, which is well below the storm surge of a moderate hurricane (Chapter 16). People with homes in such developments should be aware of this potential danger and the problem of obtaining insurance.

Finger-fill development has eliminated mangrove wetlands, destroyed habitats, cut off freshwater supply, and replaced a naturally self-cleansing system with one highly prone to pollution. Flow from overland runoff, septic systems, and spills eventually enters the canals. The "fingers" interfere with the tidal flushing that disperses pollutants. The consequence is stagnant water and floating debris, which accumulate at the landward end of these dead-end canals.

A causeway is an elevated highway built across low or wet ground (Figure 14–7b). A causeway can be built on pilings to elevate the roadway above the water and allow tidal currents to flush the bay. However, it is cheaper to build causeways by raising the land by fill and placing a roadway on top. Conduits are placed under the road fill so that tidal currents can flow through, but they allow only moderate tidal exchange. Thus, when a causeway is built across a bay, water stagnation can become even more serious. This type of development is common in places such as Miami, Tampa, and Sarasota, Florida.

Box 14–1 provides a look at some consequences of wetland loss along the Gulf Coast.

## CORAL REEFS

A *reef* is a ridgelike or moundlike wave-resistant structure occurring in the shallow waters along some coastlines. Reefs have different compositions, but the best known, and the ones of interest here, are built of the cemented skeletons of tiny animals called corals. These form **coral reefs** (Figure 14–8).

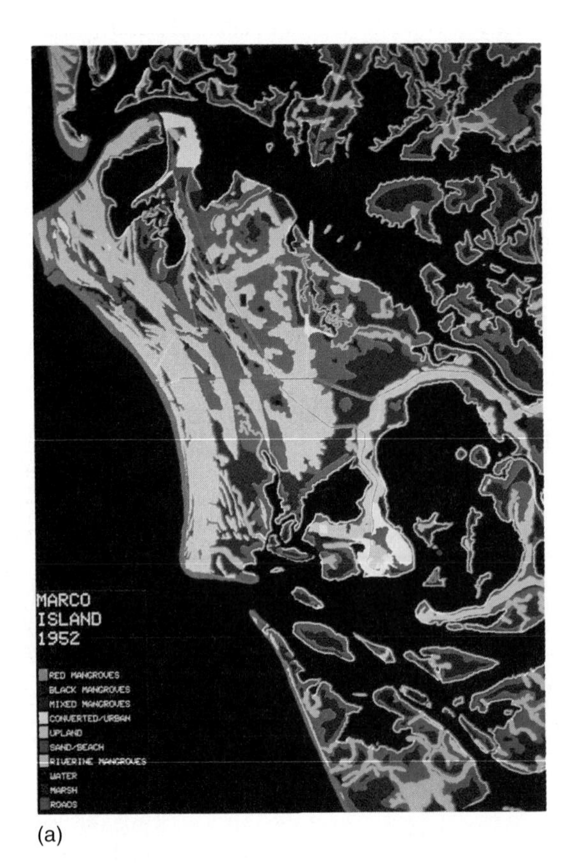


(a)

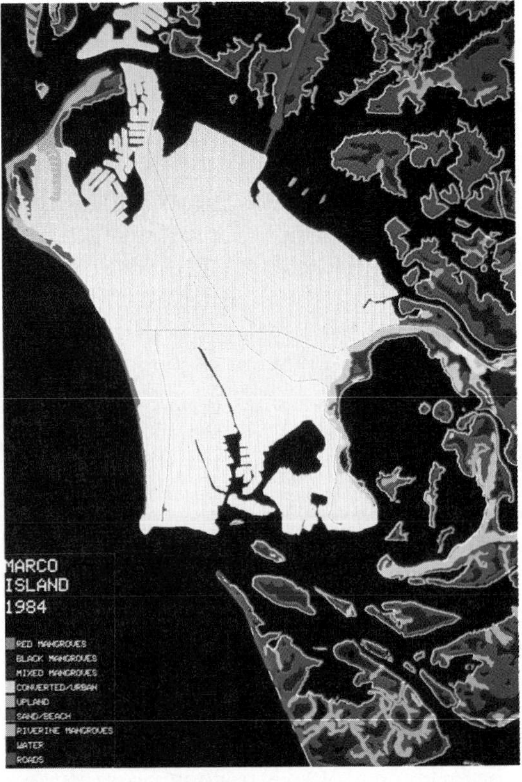


(b)

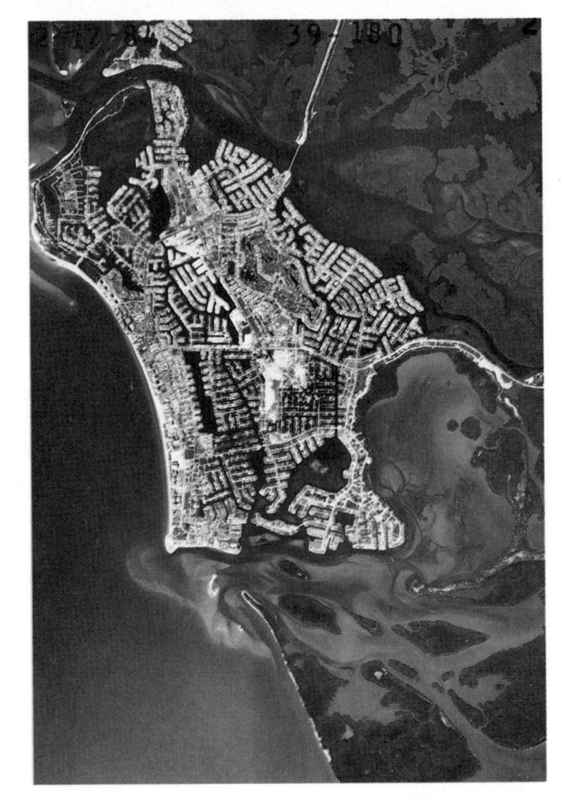

(c)

**FIGURE 14–6**
(a) 1952 and (b) 1984 digitized inventories of Marco Island, Florida, showing dramatic development; (c) 1984 color-infrared photograph of the island. (From Samuel Patterson, 1986, Mangrove community boundary interpretation and detection of areal changes on Marco Island, Florida: Biological Report 86(10), for National Wetlands Research Center, U.S. Fish and Wildlife Service (Aug.), p. 23, 46, 49.)

**FIGURE 14–7**
Elimination of mangrove wetlands and creation of a finger-fill development along a shoreline. (a) *Natural conditions:* barely emergent land is covered by vegetation. (b) *Urbanized conditions:* channels have been dredged through the wetland. The edges of the fingers (and dredged areas) have been bulkheaded and filled with dredged material to form building sites.

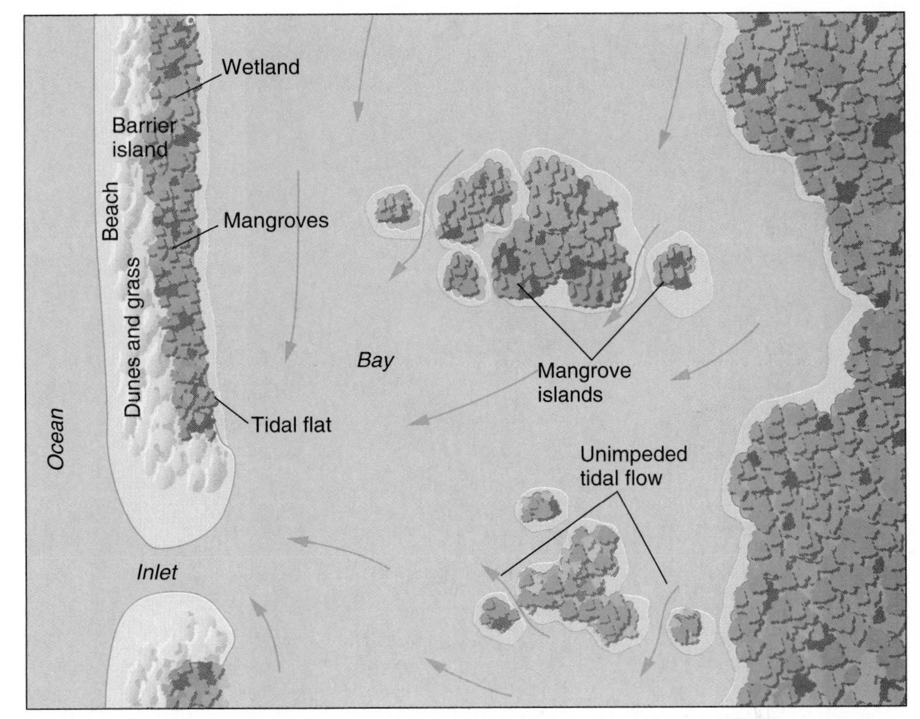


(a) Undeveloped bay and wetlands

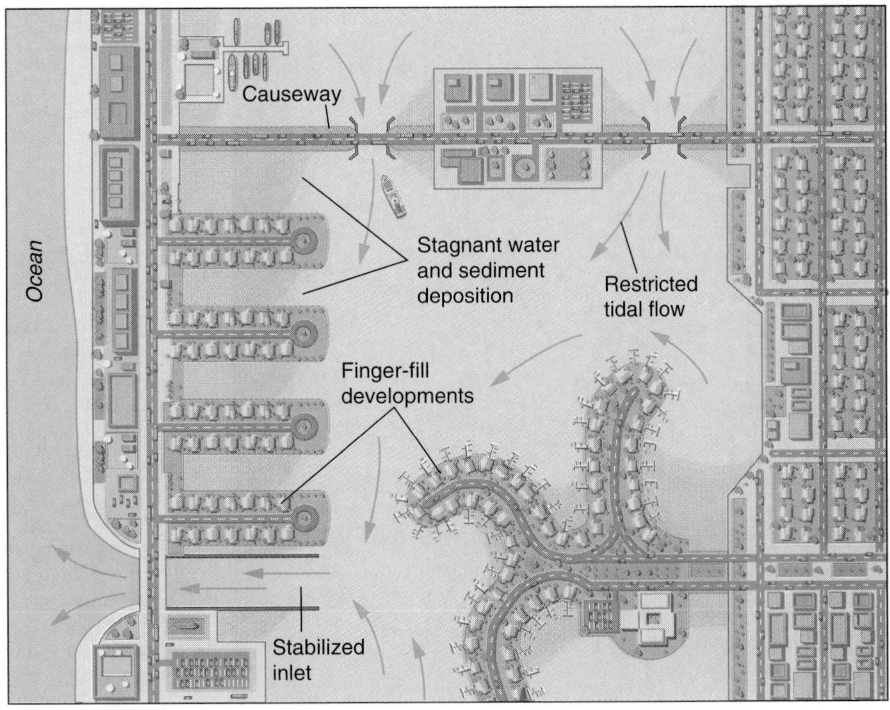


(b) Urbanized barrier island-bay-wetlands system

## BOX 14–1 LOUISIANA'S VANISHING COASTAL WETLANDS

Louisiana has 40% of the coastal wetlands in the United States. Unfortunately, Louisiana is losing these wetlands rapidly. According to the U.S. Army Corps of Engineers, the recession rate is 5 meters (16 feet) or more per year along 70% of the Louisiana shoreline. Along the remaining 30%, it is 3 to 5 meters (10 to 16 feet) per year (Figure 1). Although the cause of this wetland loss is natural, it is under the influence of anthropogenic activities.

Louisiana's shoreline is shaped naturally by the continual interplay of two processes: sediment deposition on the Mississippi Delta, and shoreline

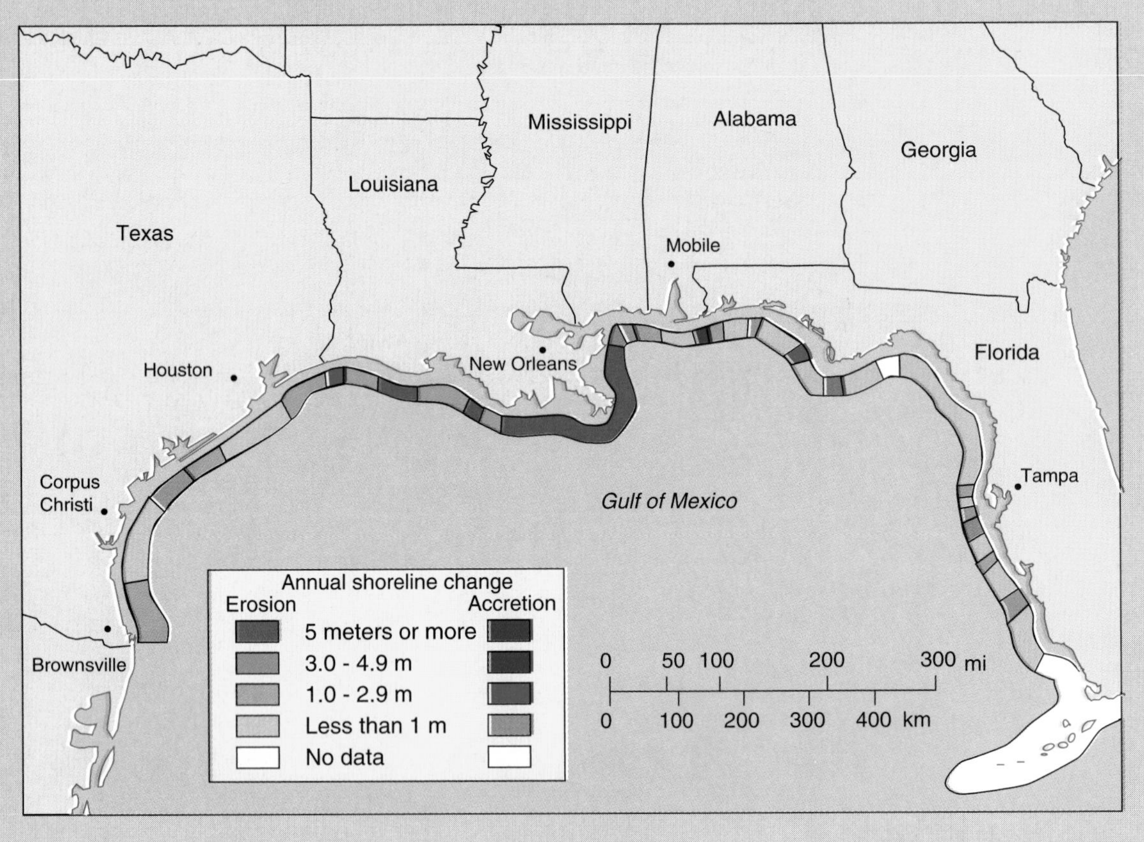


**FIGURE 1**
Shoreline erosion rates along the Gulf Coast. (U.S. Geological Survey, 1988)

erosion from waves and storms. Prior to human intrusion, this system was in equilibrium. Sediment freely overflowed onto the delta plain during floods, building up the delta, and less sediment went directly into the Gulf of Mexico.

When these processes were in balance, the shoreline position remained essentially stable, except for occasional catastrophic storms. However, the system now has tilted sharply out of balance, toward the erosion side. Louisiana now has the greatest rate of erosion of any state in the United States.

## SUBSIDENCE, SEA-LEVEL RISE, AND UPSTREAM CHANGES

Three major processes contribute to this extraordinary rate of coastal wetland loss:

1. Gradual subsidence of the Mississippi River Delta region (Figure 2a and b).
2. Rising postglacial sea level (Figure 2b).
3. Upstream changes in river systems that enter the Gulf of Mexico along the coast, especially the Mississippi River.

In Chapter 7, you saw that many flood-control alterations have been made to the Mississippi River's channel, floodplain, and tributaries. Prior to these controls, overflowing streams spread sediment laterally from their channels and built floodplains and wetlands (Figure 2a).

As more levees were built along the river for flood protection, flooding provided less sediment supply to the adjacent floodplains and wetlands were reduced. Rather than being deposited on floodplains and in wetlands along the streams, the sediment was carried down the river and deposited in the Gulf of Mexico (Figure 2a). Thus, the depositional/erosional balance along the coast was disturbed by levee construction because the material removed by wave and storm erosion was not being replaced.

Post-glacial sea-level rise is causing the shoreline to migrate landward each year (Figure 2b). This causes steady landward erosion and the intrusion of saltwater into water bodies and coastal aquifers.

The most significant cause of this increased erosion, however, is subsidence of the Mississippi River Delta. The continually accumulating mass of sediment from the river compacts older sediment underlying the delta. As the delta surface subsides, the shoreline migrates landward, eroding the shore farther and farther inland (Figure 2b). The net erosion from delta subsidence, sea-level rise, and anthropogenic stream changes accounts for remarkable loss of land in coastal Louisiana (Figure 1).

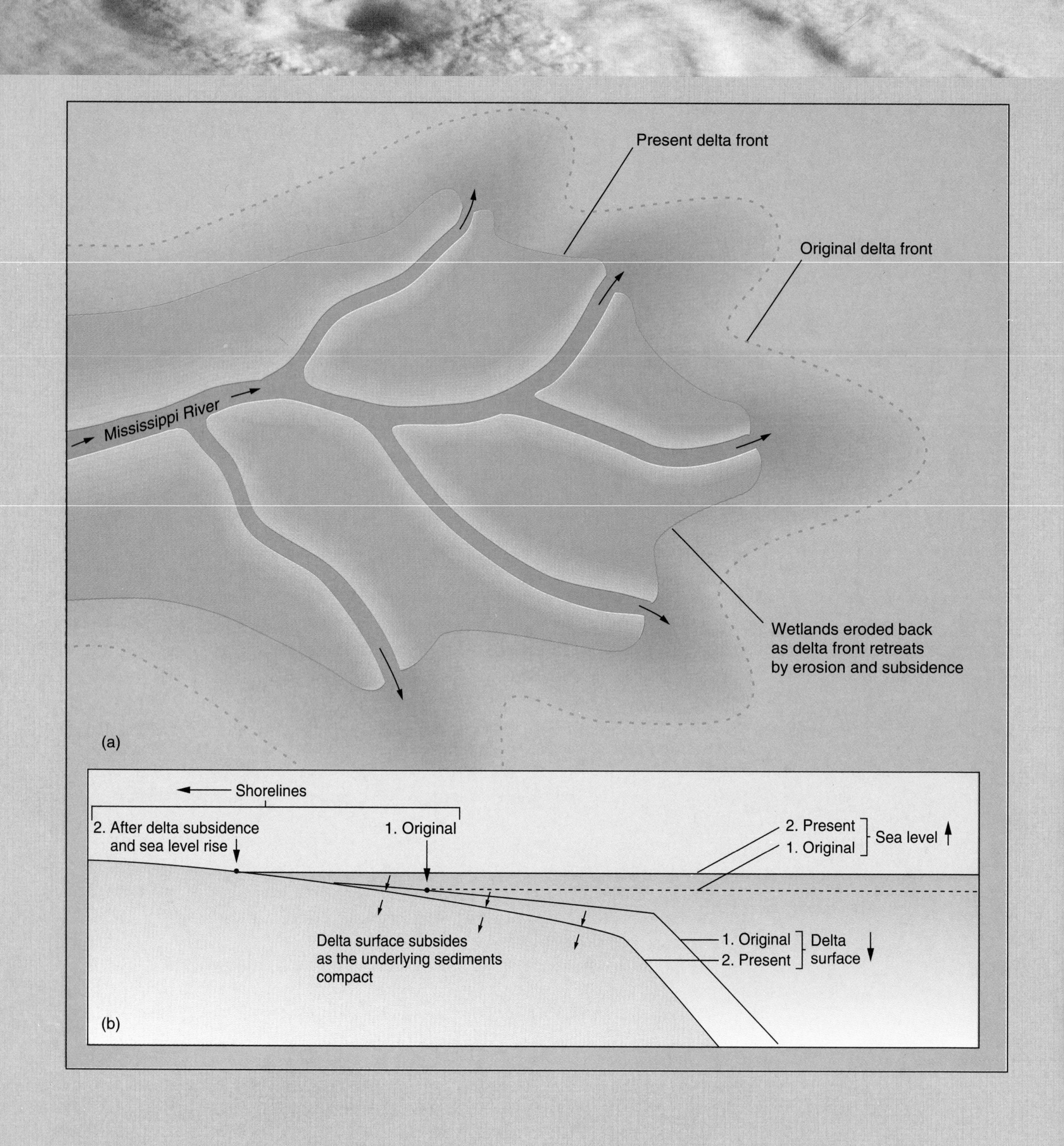

Present delta front
Original delta front
Mississippi River
Wetlands eroded back
as delta front retreats
by erosion and subsidence
(a)
Shorelines
2. After delta subsidence
and sea level rise
1. Original
2. Present
1. Original
Sea level
Delta surface subsides
as the underlying sediments
compact
1. Original
2. Present
Delta
surface
(b)

**FIGURE 2**
Coastal erosion and wetland loss in the Mississippi River Delta is caused by sediment compaction and subsidence during the present sea-level rise. (a) With the construction of more levees, flooding streams no longer supply water for wetlands or deposit sediment on floodplains. (b) Rising sea level erodes the shoreline and pushes saltwater into coastal water supplies. (After S. Penland, D. Boyd, D. Nummedal, and H. M. Roberts, 1981, Deltaic barrier development on Louisiana Coast: Gulf Coast Association of Geological Society Transactions, v. 39, Fig. 4, p. 690.)

## THE COST

Problems abound from this loss of real estate. As the delta surface subsides and the shoreline retreats landward, saltwater intrusion kills freshwater vegetation. Offshore islands such as Isles Dernieres have become severely eroded (Figure 3). With the passing years, southern Louisiana faces potential saltwater intrusion and greater storm damage.

Shoreline retreat has left former inland structures at or within the surf zone (Figure 3). It also has seriously dislocated the fur, fish, and waterfowl industries—a $1 billion-a-year business in Louisiana. Finally, the state's claims to offshore oil deposits, according to federal law, depend on the *distance from the shoreline out to the oil deposit.* As the shoreline retreats landward, Louisiana loses control over more and more potential offshore oil drilling sites!

Information in this box is based on work by S. Penland, H. H. Roberts, S. J. Williams, A. H. Sallenger, Jr., D. R. Cahoon, D. W. Davis, and C. G. Groat, 1990, Coastal land loss in Louisiana: Transactions Gulf Coast Association of Geological Societies, v. XL, p. 685–699.)

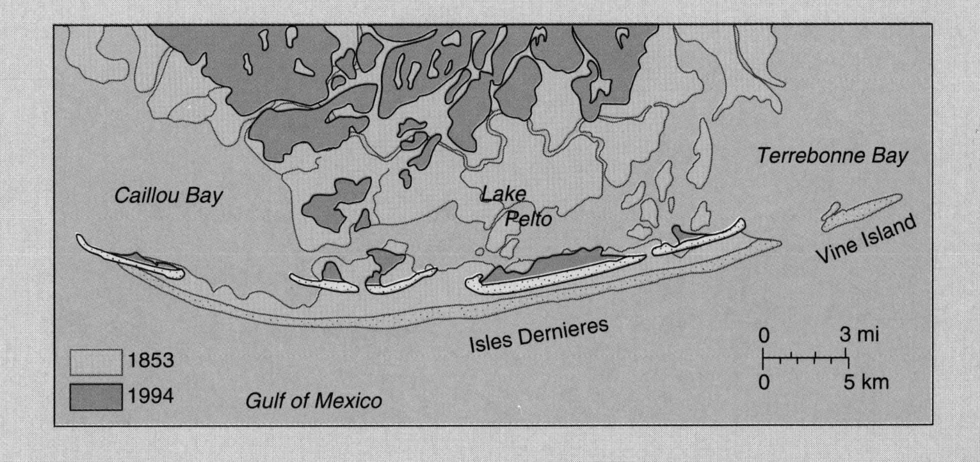


**FIGURE 3**
Isles Dernieres in 1853 and 1978. (After S. Penland, D. Boyd, D. Nummedal, and H. M. Roberts, 1981, Deltaic barrier development on Louisiana Coast: Gulf Coast Association of Geological Society Transactions, v. 39, Fig. 4, p. 471–480)

Coral animals grow rigid calcium carbonate skeletons, which collectively form the reef. Reefs incorporate in their structure many other organisms as well—algae, mollusks, and foraminifera. Reefs also form a habitat for perhaps as many as 3,000 other marine organisms, including fish, eels, octopus, clams, and crabs. Coral reefs are effective storm barriers, shielding the mainland behind them. Unfortunately, coral reefs are being destroyed by a number of agents.

## Corals Are Symbiotic Animals

Coral animals are pinhead-to-dime size. Individual animals are called **polyps.** The reef-building corals are called **hermatypic.** Hermatypic corals are a remarkable example of cooperation in nature, in this case between the corals and a type of algae. Such cooperation is called *symbiosis.* Single-celled algae called *zooxanthellae* (zoo-oh-ZAN-thell-eye) live within the tissue of individual coral polyps. The zooxanthellae produce oxygen and sugars by photosynthesis and these benefit the coral polyps. In turn, the polyps shelter and nourish the algae with their waste products, such as carbon dioxide, phosphates, and ammonia.

The important thing to know is that coral polyps and their symbiotic algae cannot live without each other. Anything that kills the algae, such as a reduction of the light needed for photosynthesis, also kills the coral. Anything that kills the coral also kills the algae.

Thus, *both algae and coral polyps must be present for a healthy reef.* When algae remove carbon dioxide from the coral polyps, this favors the deposition of calcium carbonate that lines the polyp chamber. Thus a head of coral is composed of a great many chambers, each occupied by individual coral polyps and their zooxanthellae.

Visitors who enjoy the beauty of a reef in the daytime never see the coral polyps because they retract into their chambers for safety from daytime predators (Figure 14–9a). However, at night the polyps extend their tentacles to catch plankton carried past them by currents (Figure 14–9b).

## Conditions Necessary for Coral Growth

Coral reefs require specific environmental conditions to thrive, including shallow depth, warmth, and clear, agitated water. This set of favorable conditions exists over a limited area, explaining why coral reefs are found only in certain parts of the ocean. Reefs are well developed in the United States in southern Florida, Hawaii, Puerto Rico, and the U.S. Virgin Islands (Figure 14–10).

***Depth.*** The algae within hermatypic corals require sunlight to conduct photosynthesis. Consequently, they can survive only within the top 50 meters (160 feet) of the ocean. Within this narrow range, deeper corals are quite delicate. In shallower waters, corals are more massive in structure and size and far more abundant because conditions are ideal for their growth.

**FIGURE 14–8**
Coral reef assemblage, Florida Keys. (Steven C. Cook.)

**FIGURE 14–9**
Coral polyps (a) closed and (b) open with mouth (center) and tentacles visible. (From Stephen Frink.)

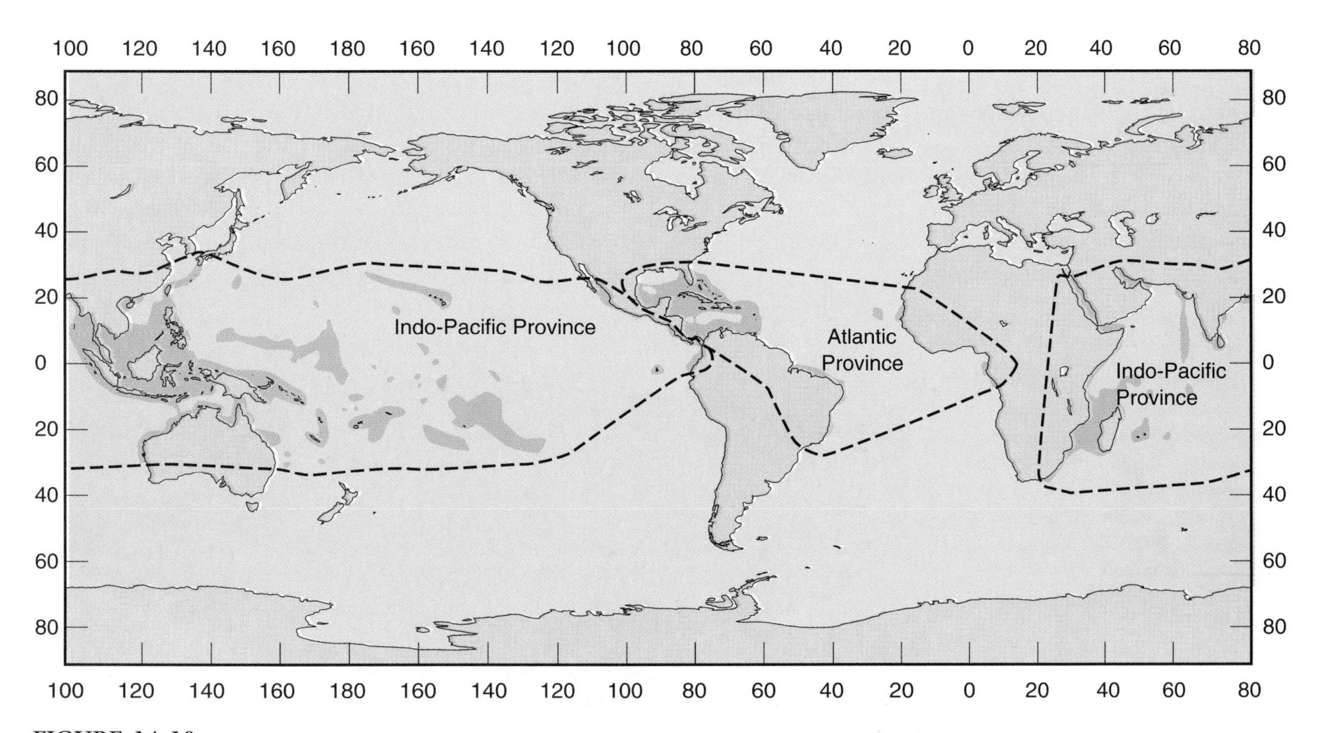


**FIGURE 14–10**
Areal distribution of coral reefs.

***Temperature and Salinity.*** Hermatypic corals require warm water, ranging from 18° to 30°C (64° to 86°F). Optimum coral development occurs between 23° and 27°C (73° to 80°F), a temperature range that exists in a wide band straddling the Equator between approximately 30° N and 30° S. Although hermatypic corals cannot grow at colder temperatures, their development also is retarded at higher temperatures. They may be killed by natural warming (an El Niño current) or thermal pollution from human activity.

Corals develop best within normal ocean salinity, which ranges from 27 to 40 parts per thousand (averaging 35 ppt—see Chapter 13). Influxes of freshwater may lower the salinity to a level at which corals cannot survive.

***Water Clarity.*** Corals thrive in clear water but die in water that is continuously or frequently clouded with suspended sediment. Turbid water reduces light penetration, which limits photosynthesis in the symbiotic algae. Sediment suspended in the water also smothers the coral polyps within their chambers.

***Water Movement and Salinity.*** Coral reefs thrive in rough water. This might seem surprising, because one usually thinks of harsh environments as being less hospitable to life. However, because they are attached to the reef, corals must have food delivered to them by the water. Waves and tidal currents drive large volumes of water with suspended organic particles across a reef and through openings in it, providing food for the coral polyps. Consequently, reefs usually are wider and better developed on coasts that face prevailing winds.

## Types of Coral Reefs

There are three basic types of coral reefs—fringing reefs, barrier reefs, and atolls.

A **fringing reef** is built against the coast. They are more common along coasts with steep offshore slopes (Figure 14–11A). Fringing reefs are common on volcanic islands such as Hawaii and in the American Virgin Islands.

A **barrier reef** is a linear feature parallel to a coast and separated from it by a wide lagoon or bay. Barrier reefs are most common where the offshore slope is very gentle (Figure 14–11B). Natural channels between coral reef segments permit water exchange between the bay and the ocean. Major parts of the reefs in southern Florida are of this type. Earth's longest barrier reefs are in the Gulf of Mexico along the eastern coast of Yucatán (Mexico) and Belize, and the Great Barrier Reef along the eastern coast of Australia.

An **atoll** is a circular reef that surrounds a lagoon. In 1842, Charles Darwin first proposed an origin for these structures, based on his field observations on several islands. Other hypotheses for atoll formation have been proposed, but Darwin's theory remains the most plausible. He proposed that atolls form from the subsidence of volcanic islands and their associated fringing reefs.

The fringing reef continues to grow upward as the volcano subsides (Figure 14–12a). When most of the island has subsided, a barrier reef forms, encircling the island (Figure 14–12b). After all of the original volcanic island has subsided, a lagoon takes its place as the circular reef pattern of the atoll forms. Storms breaking on the reefs of the atoll pile material above sea level as islands on the reef (Figure 14–12c).

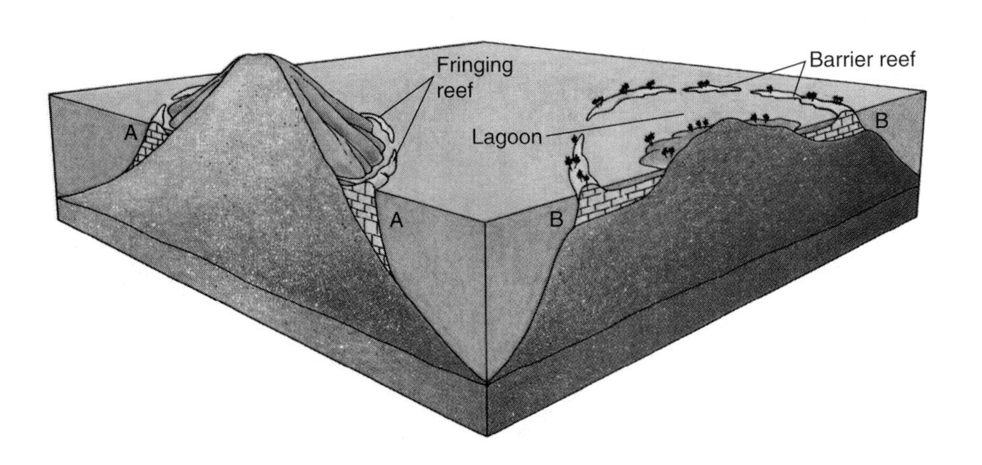


**FIGURE 14–11**
A fringing reef forms on steep slopes, and a barrier reef forms on gentle slopes. (Reprinted with the permission of Macmillan College Publishing Company from *Physical Geology* by Nicholas K. Coch and Allan Ludman. Copyright © 1991 by Macmillan College Publishing Company.)

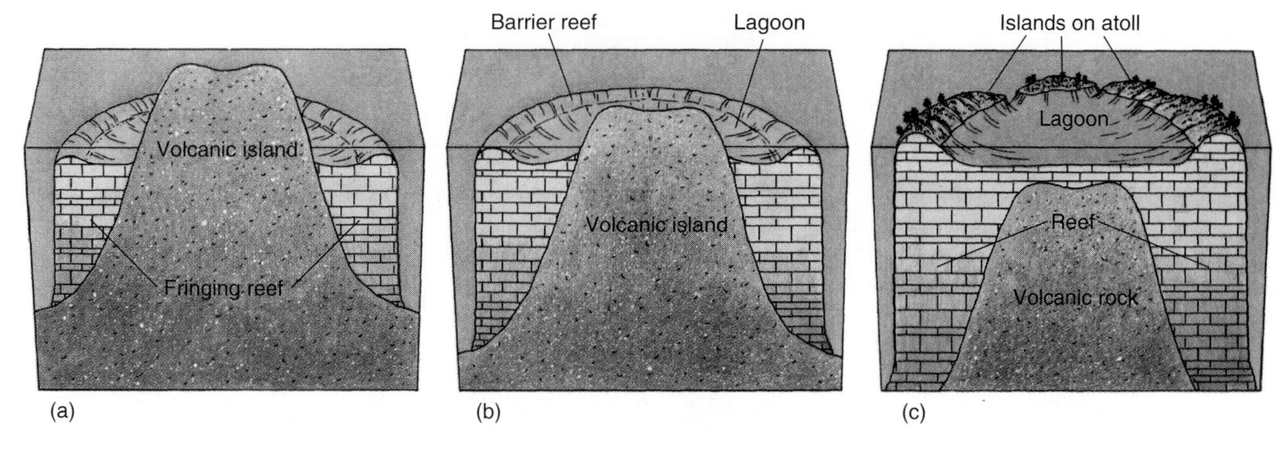


**FIGURE 14–12**
Diagram showing the formation of an atoll. (a) A fringing reef develops on the edges of a subsiding volcanic island. (b) The reef continues to grow upward, keeping pace with the island's subsidence. The fringing reef evolves into a barrier reef, with development of a lagoon. (c) After the volcanic island subsides below sea level, storms pile material upon the remnants. The barrier reef forms the circular atoll around a central lagoon. (Reprinted with the permission of Macmillan College Publishing Company from *Physical Geology* by Nicholas K. Coch and Allan Ludman. Copyright © 1991 by Macmillan College Publishing Company.)

Although many coral atolls are uninhabited, some larger ones can support human populations. For Darwin's subsidence theory to be correct, there should be volcanic rock beneath the carbonate rocks of an atoll. Drilling on atolls in the western Pacific revealed volcanic rock under the coral limestone.

## Zonation of Coral Reefs

A "coral reef" is not all coral. Corals make up at most 30% of the volume of the living reef. The intervening spaces are occupied by algae, mollusks, calcareous sediment, and debris ripped from the reef during storms. Further, the types of coral in a reef vary in zones from the open-ocean side to the shoreward side. In general, the type of coral found at any one place on a reef depends on wave intensity. This reef zonation is shown in Figure 14–13. The section shown is a *generalized* summary of coral distribution in the Florida Keys, the only coral reef system in the continental United States.

The intermediate-to-deep reef extends from the oceanic front of the reef into deeper water. In this area are massive corals, sponges, and coral debris broken from the reef front and moved down the fore reef slope. The front part of the reef is the *fore reef,* an area of extensive coral growth populated mainly by elkhorn coral (Figure 14–14). The fore reef is the highest and shallowest portion, frequently exposed to harsh conditions of low tide and crashing waves. Staghorn coral (Figure 14–15) is the dominant coral on the back reef side.

The *bay* environment stretches from the back reef to the shoreline. The character of the bay floor varies, with patches of sand, scoured bedrock, extensive areas of sea grasses, and isolated patch reefs. *Patch reefs* (Figure 14–16) are moundlike structures composed of massive corals, sponges, and other sea plants and animals.

Bordering the bay are intertidal areas. These are fringed with red mangroves, which provide shelter, food, and nurseries for many reef-dwelling animals that spend their juvenile stages near the shore.

## Reef Development in the United States

U.S. reefs are of two types, fringing and barrier (Figure 14–11). The steep-sided U.S. Virgin Islands have numerous fringing reefs. But the greatest of American reefs is the barrier reef that begins near Miami and spreads southward intermittently over 240 kilometers (150 miles) to Key West (Figure 14–17). The active reef is at the easternmost edge of the submerged Florida platform (continental shelf) where the bottom drops gradually into the Florida Strait, through which the Gulf Stream flows northward.

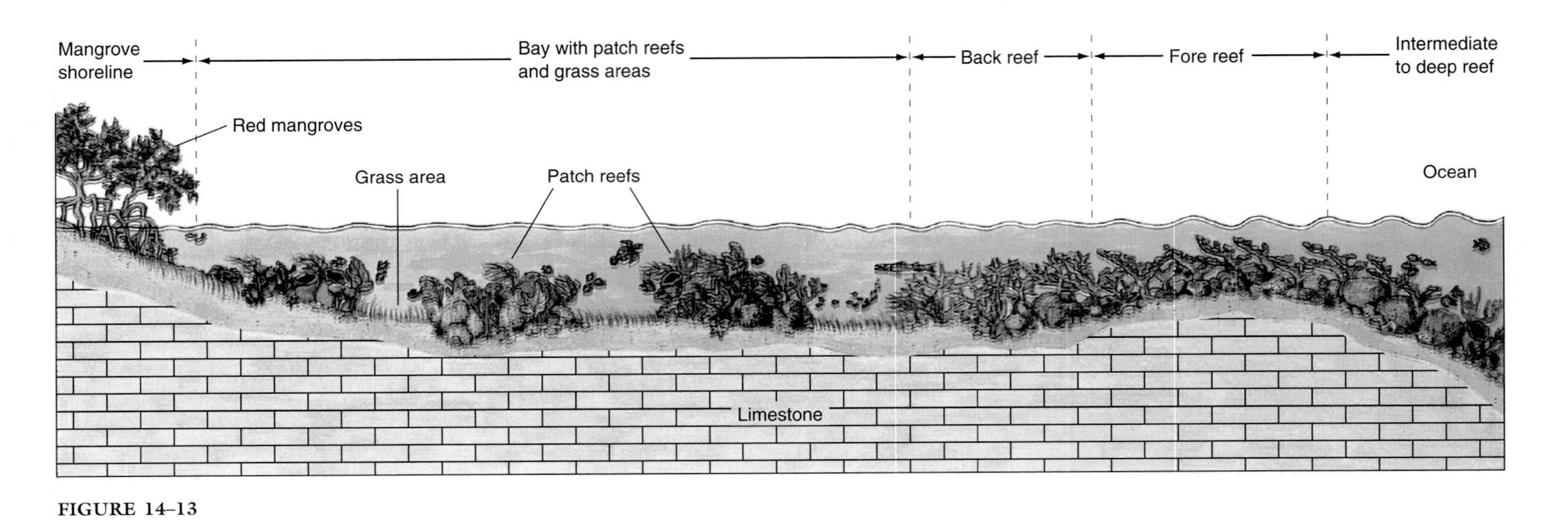


**FIGURE 14–13**
Generalized cross section of the Florida Keys reef-bay system, the only coral reef system in the continental United States. The types of coral in a reef vary in zones from the open-ocean side (right) to the shoreward side (left). In general, the type of coral found at any one place on a reef depends on wave intensity. (Based in part on NOAA-National Marine Sanctuary Program/Florida Audubon Society map, 1991.)

**FIGURE 14–14**
Elkhorn coral *Acropora palmata* in the Florida Keys. (Photo by author.)

**FIGURE 14–15**
Staghorn coral *Acropora cervicornis* in the Florida Keys. (Photo by author.)

A several-kilometer-wide shelf area separates the offshore barrier reef from the Florida Keys. The Keys and the barrier reefs are part of an older Pleistocene (Ice Age) reef system. Canals cut through the Keys reveal Pleistocene coral limestone. These rocky outcrops show that corals make up only a small part of the limestone. Other calcareous organisms, coral debris, calcium carbonate sediment, and pore spaces compose the rest of the rock. Many massive, intact coral heads are visible in the cross-section of limestone. Westward from the Keys is Florida Bay, a mixture of open bay and mangrove islands extending toward peninsular Florida (Figure 14–17).

**FIGURE 14–16**
Patch reef in the Dry Tortugas, Florida. (Photo by author.)

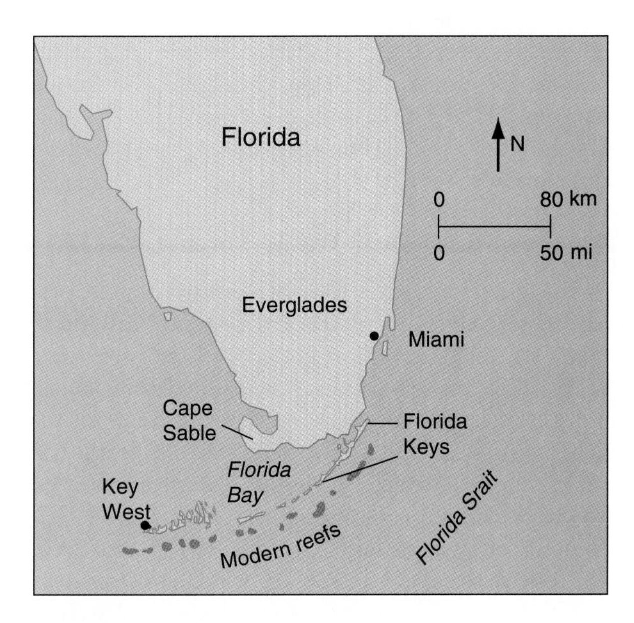


**FIGURE 14–17**
Map of southern Florida showing peninsular Florida, Florida Bay, Florida Keys, and the living reef off-shore bordering the Florida Strait.

## CORAL REEF DESTRUCTION

Increasing urbanization and human alteration of natural coastal systems pose a serious threat to coral reefs. The perils for coral reefs are shown clearly in southern Florida, from south of Miami to Key West. Although much of the following discussion uses this area as an example, the same effects can be observed in reefs throughout the Caribbean, the Pacific Ocean, and the Indian Ocean.

## Visitors Threaten Reefs

The beauty of coral reefs and their colorful dwellers have made them prime tourist attractions. This means that more people are anchoring on, boating across, swimming through, and collecting souvenirs from reefs, as well as developing land near them. The cumulative effects are destroying reefs.

The only living part of a reef is its outermost layer, where individual coral polyps live within their chambers. Humans are a threat to coral reefs. For example, dropping boat anchors on coral heads can cause serious damage. To reduce this damage in reef areas that are regulated by state and federal government, boats can be tied only at specified sites where permanent moorings have been set up.

Although the majority of snorkelers, fishermen, lobster gatherers, and scuba divers are respectful and careful not to damage the coral, inevitably some visitors are careless or even malicious. As visitors increase, the *cumulative* effect can seriously damage a coral reef.

## Sewage and Organic Waste Threaten Reefs

Anything that increases the nutrient input to a coral reef fosters algae growth. In extreme instances, algae can blanket coral heads and smother the polyps. The mechanism is simple: algae block the sunlight necessary for photosynthesis in the zooxanthellae, the algae that live symbiotically with the coral polyps. The algae covering also promotes the growth of oxygen-consuming bacteria.

Increased use of land that borders living coral reefs has multiplied the sewage waste and fertilizer runoff that enters coastal waters. Individual homes treat their sewage in *septic tanks*, where it undergoes bacterial decomposition. Like all sewage-treatment systems, the quality and thoroughness of treatment varies, so effluent released into the environment ranges from benign to barely treated. Consequently, this waste generally is rich in nutrients and bacteria.

The septic effluent infiltrates adjacent ground. In southern Florida, limestone underlies the soil, and it often is riddled with solution cavities and even cavern systems (Chapter 10). Some effluent may enter cavernous rock underground and move laterally under a bay to reach a reef. Also, a portion of the effluent can enter canals (especially in a finger fill development) or the bay near the shore. This addition of nutrients from sewage can lower the dissolved oxygen level and promote algae growth.

How significant is this addition of nutrients? Greater nutrient concentrations have been observed in the nearshore area of bays. However, these nutrient concentrations have not been observed in the bays themselves, perhaps because sea grasses in the bay take up most of the nutrients. Nevertheless, algae blankets are occurring on offshore reefs (Figure 14–18). These algae growths could result from *additional* nutrient input traveling seaward through the cavernous limestone under the bay and then flowing upward under the barrier reef.

The problem can be just as serious in an urban area served by municipal sewage treatment plants, such as Miami to the north and Key West to the south. Recall from Chapter 12 that secondary sewage treatment eliminates pathogenic bacteria and a *portion*—not all—of the organic load. The secondary treatment plant at Key West discharges approximately 317 metric tons (350 tons) of nutrients a year, including phosphates, nitrates, and ammonia. Individual septic tanks along the Keys may supply an equal mass.

This is a good example of how *correlation* does not establish *causation*. We observe more sewage and more algae growth, but they are not *necessarily* related. Many feel that the connection between sewage-supplied nutrients and algae growth on a reef is proven cause-and-effect, but the connection is only speculation at present. Continuing research may present alternative answers. However, if the

**FIGURE 14–18**
Algae blanketing a coral reef. (John Halas, Florida Keys National Marine Sanctuary.)

suspected relation is correct, it promises serious degradation of patch and barrier reefs.

Unquestionably, serious degradation of water quality has occurred in coastal development "canals." The once-clear waters, grass beds, and abundant aquatic life of the 1950s have been replaced by mostly murky, stagnant, fishless waters having great algae concentrations.

Organic pollution can be greatly alleviated with sewage treatment plants and by better organic digestion in those facilities. Although local businesses and inhabitants will pay more, thorough sewage treatment will help save the reef that provides many people, directly or indirectly, with their livelihood.

## Changes in Water Temperature and Salinity Threaten Reefs

As previously stated, corals live best within narrow limits of water temperature and salinity. Increasing changes in both factors threaten reefs worldwide.

Coastal water temperatures may be increasing from urban runoff, thermal pollution from industry and power plants, and possibly from global warming brought about by the greenhouse effect (Chapter 11). Increasing water temperatures decrease dissolved oxygen levels and accelerate algae growth, which are detrimental to a coral reef.

Increasing salinity is becoming apparent in Florida Bay. In the past, terrestrial water flow was primarily southward, through the Everglades and into Florida Bay (Figure 14–17). This freshwater flow diluted the saltwater entering Florida Bay from the Atlantic and the Gulf of Mexico, making the water brackish. Brackish water is an ideal nursery for marine shellfish and shrimp but is not favorable for coral growth. Thus, in the past, no coral reef development existed where Florida Bay waters flow out onto the reef.

Human intervention has changed all this. Massive water diversions for agriculture since the 1920s have eliminated much of the southward freshwater flow. Canal systems now divert much of the freshwater laterally to serve Florida's rapidly growing coastal communities (see Box 14–2). Consequently, the waters of Florida Bay have steadily increased in salinity, eliminating many shrimp nurseries. In times of greater water withdrawal and lesser rainfall, Florida Bay salinity now equals or even exceeds that of the nearby ocean.

In addition, the shallower waters of Florida Bay are warmer than the ocean. Some of this warmer water now reaches the reef through inlets between the Florida Keys (Figure 14–17), although its effects on the reef are not known at present.

## Sediment Pollution and Toxic Pollution Threaten Reefs

In Chapter 6, we described the problems caused by excessive sediment that might be produced by surface mining or by clearing land for agriculture, urban development, or highway construction. Rain then washes the sediment into coastal waters where it reduces normal light penetration. This process severely limits photosynthesis in the zooxanthellae algae in the coral. At the same time, the sediment literally smothers the coral polyps. Thus, the impact of sediment-laden water moving across a reef is that the coral are killed by smothering and starvation.

Urbanization, industry, and boat traffic release toxic liquids into coastal waters. These substances include household or industrial chemicals flushed into septic systems or washed from roads and parking lots, and gasoline and oil from marine engine exhausts, leaks, and spills. Rarely do these pollutants attain "killer" concentrations, but the cumulative effect of small doses over time can seriously degrade coastal ecology, including reefs.

## Coral Mining Threatens Reefs

In some areas, particularly in the Indian Ocean, coral reefs are being mined as a source of limestone for cement, road surfaces, and building. This not only destroys the reef but also produces turbid water, which threatens nearby reefs. Mining also decreases the storm protection for nearby coasts.

One study estimates that, at India's southern tip alone, nearly 36,000 metric tons (40,000 tons) of coral are removed each year. The government of the island of Sri Lanka (formerly Ceylon), just south of India, has been trying to prevent divers from removing as much as 9,000 metric tons (10,000 tons) of coral a year. Similar mining is occurring in the Maldive Islands south of India and along parts of Africa's eastern coast. Both India and Sri Lanka have experienced greater coastal erosion on their western coasts, which receive the brunt of monsoon storms. Sri Lankan biologists also have noted reduced coastal fish stocks as a result of the mining.

## BOX 14–2 CAN WE SAVE THE EVERGLADES?

The Everglades in southern Florida is an extraordinary ecosystem. Nowhere else do both freshwater and saltwater wetlands exist in such variety (Figure 1). The central part includes freshwater wetlands and environments such as cypress swamps, prairie, pinelands, and stands of hardwood. Bordering this central area are estuaries and saltwater wetlands with mangroves.

Continuously moving through the Everglades is a shallow, extremely broad flow of fresh water (Figure 2) that creeps seaward over a gently sloping bed. It is only 15 centimeters (6 inches) deep and 80 kilometers (50 miles) wide, a most unusual river. Over its course, the water level drops 4.6 meters (15 feet), eventually emptying into Florida Bay.

The Everglades is best known for its great diversity of birds, alligators in its freshwater swamps, and crocodiles in the bordering estuaries. It contains many other animals as well, a diverse flora, and some endangered species such as the Florida panther.

The constant flow of water through these ecosystems sustains their life. However, in recent years threats to the vitality of this great ecological system have appeared.

In the past 30 years, the volume of southward-moving water has decreased. Its quality also has declined. Under the natural system (Figure 3a), prior to extensive human presence, surface water spread to the coast, reducing the salinity of Florida Bay. But as southern Florida developed from 1950 onward, more and more water was diverted by canals to supply developing cities on the east and west coasts and to agricultural areas bordering Lake Okeechobee to the north. At the same time, agricultural interests "reclaimed" land for growing sugar cane, vegetables, and fruit. Housing and industrial developments spread over the land reclaimed from the Everglades.

The reduced freshwater flow also has poorer quality because of high nutrient levels (phosphorus) from fertilizers used in the "reclaimed" agricultural lands bordering Lake Okeechobee. As the freshwater flow decreased southward, saltwater encroached inland.

The problem became especially serious in times of drought, when critical water habitats dried up. Dried, organic-rich soils are flammable. They began to oxidize, giving off heat and igniting. These "peat fires" had devastating consequences.

The seriousness of this problem has been recognized and remedial measures are underway. In 1993, negotiations were in progress between farmers and the government to help increase water flow and quality. Growers agreed to take some land out of production, but were seeking guarantees against losing more land later.

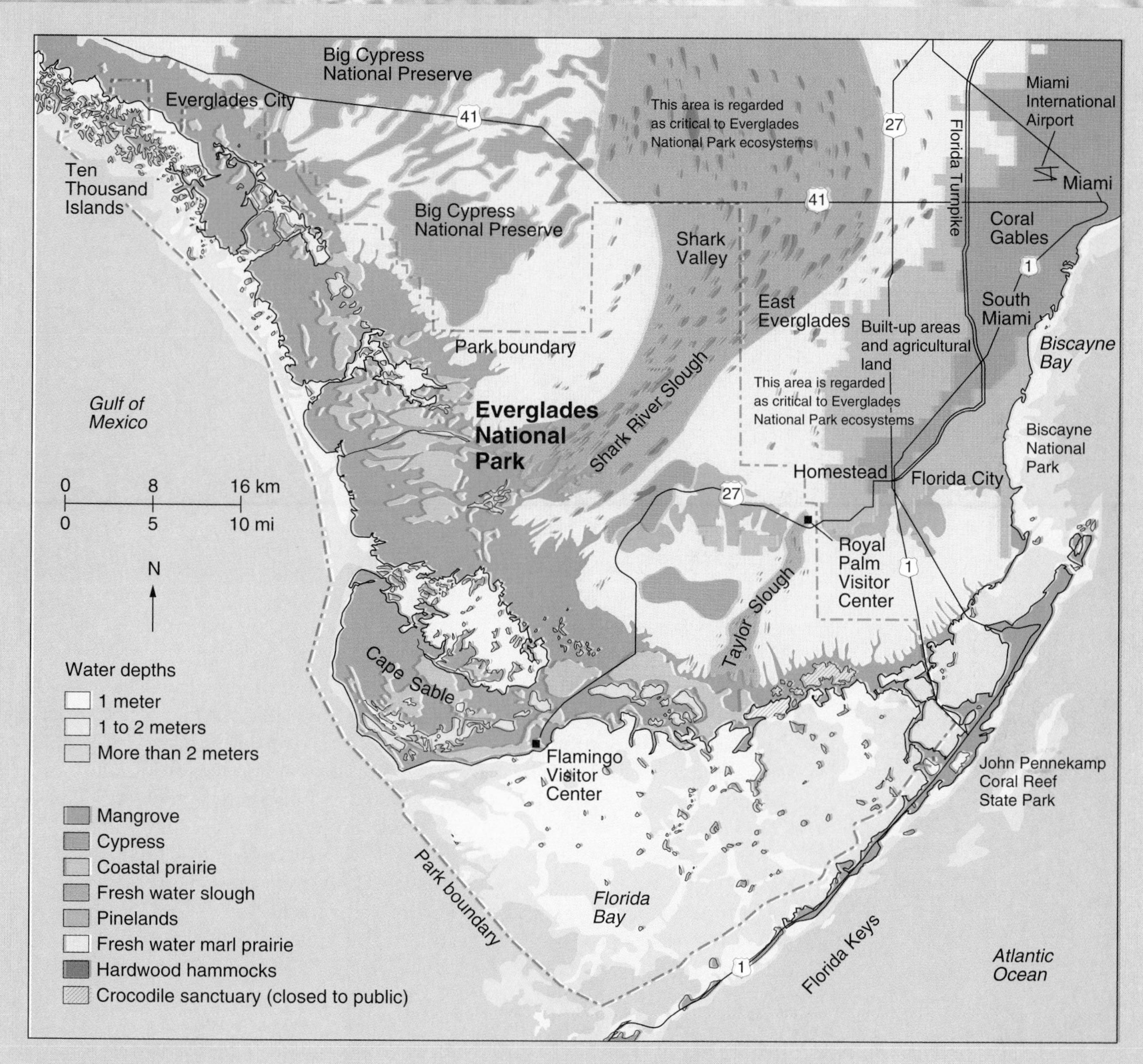


**FIGURE 1**
Freshwater and saltwater wetland environments in southern Florida. The diversity of ecosystems is affected by reduced water flow from the north, caused by freshwater diversion to coastal cities and agricultural areas. Also, agricultural development is adding nutrients to the southward-flowing water, decreasing its quality. (Modified from map of U.S. National Park Service.)

**FIGURE 2**
A shallow "river" of water flows through the Everglades. (Courtesy of Everglades National Park, National Park Service.)

The farm land would be converted into freshwater marshes that would help remove phosphorus from the freshwater flow before it reaches the Everglades to the south. It was proposed to remove 16,000 hectares (40,000 acres) from production. The conversion would cost growers $465 million. Pollution would be reduced by adopting new plant varieties, new fertilization techniques, and new water-pumping practices.

Some agreement will no doubt be reached eventually and at least part of the problem will be ameliorated before the Everglades, a very precious environmental resource, is lost for future generations.

## Even Reef Custodians Are Threatened

Thousands of aquatic species live in the varied habitat of a reef/bay complex. Some of these, such as the sea urchin *Diadema,* are important in preserving reef quality. These spiny creatures work the reef, scraping off encrusting algae for food. These algae are not the zooxanthellae so essential to the life of the coral, but instead are algae that proliferate over the reef. A Key Largo National Marine Sanctuary scientist notes that urchins are absolutely critical to maintaining the coral-algae balance, literally keeping the algae crop back so they don't overgrow the reef.

Healthy *Diadema* have spines that radiate upward in all directions (Figure 14–19a). But in 1991, divers just south of Key West noted urchins with limp, flat spines (Figure 14–19b) on the seafloor. Within four days, hundreds of urchins died. Earlier,

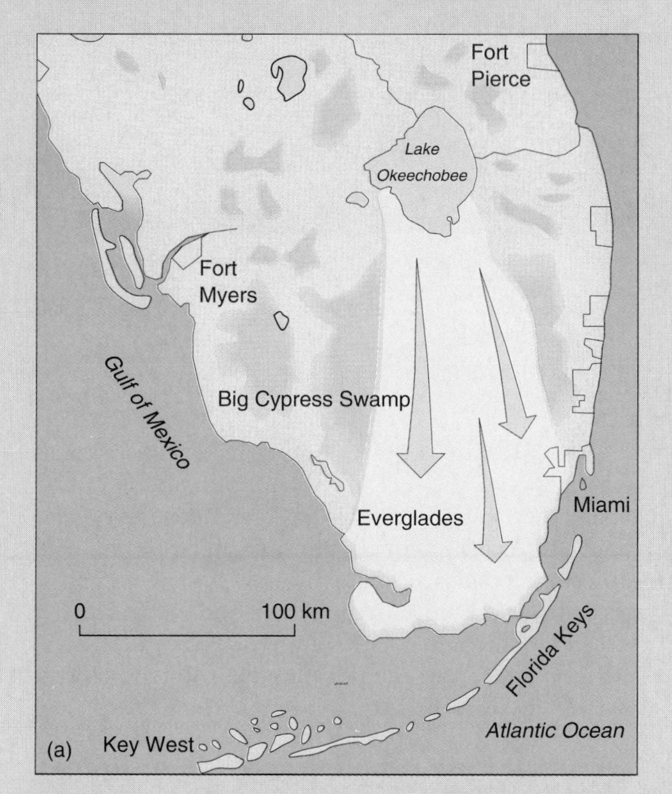


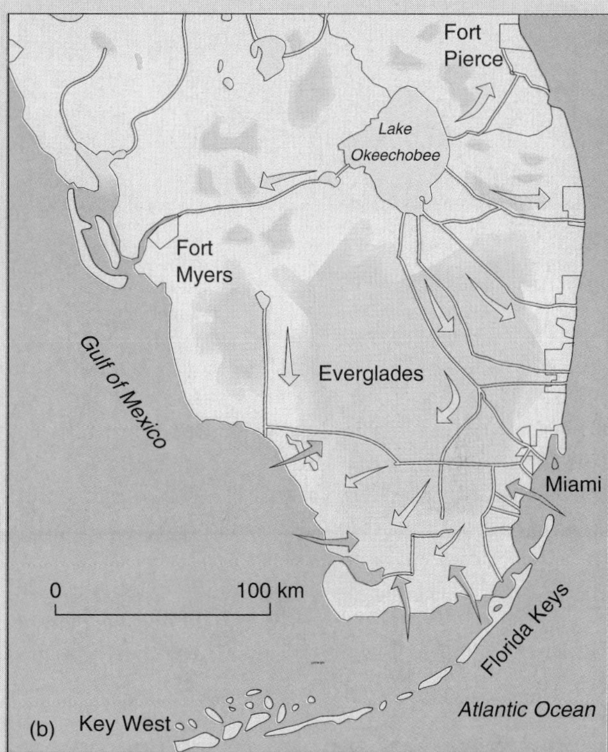


**FIGURE 3**
Effect of drainage changes in south Florida since 1871. (a) In 1871, drainage from Lake Okeechobee spread southward in a broad sheet only a few centimeters deep. This kept the water table very close to the surface, maintaining swampy conditions. (b) Today, drainage is largely controlled by engineering structures. Canals are diverting water flow, lowering the water table. The swamp has been destroyed in some areas, and saltwater is encroaching in coastal wells. (Adapted with the permission of Macmillan College Publishing Company from *Earth's Dynamic Systems,* 6th ed., by W. Kenneth Hamblin. Copyright © 1992 by Macmillan College Publishing Company.)

during 1983 and 1984, a similar event occurred throughout the Caribbean, including Florida's reefs. What happened?

Environmentalists suspect that the urchins were killed by a bacterium or virus related to human activity. However, scientists have not yet identified the exact cause of the sea urchin mortality. Also unknown is the degree to which increasing water temperature, salinity, and sewage pollution may have made these important organisms more susceptible to disease.

## REEFS AS HURRICANE BARRIERS

A healthy reef provides the best hurricane or storm protection possible for a coast. The living reef is built upon a thick, well-cemented mass of

(a)

(b)

**FIGURE 14–19**
(a) A healthy black sea urchin *(Diadema)* on a reef in the Florida Keys. Note the symmetrical distribution and lengths of the spines compared with (b), a sick sea urchin. (Photo (a) by Steven C. Cook; (b) courtesy of John Halas, Florida Keys National Marine Sanctuary.)

limestone composed largely of cemented coral debris. Storm surges may destroy large areas of living coral on a reef flat (Figure 14–20a), but coral quickly become reestablished on the debris. Within a few years, a new layer of living coral covers the storm debris (Figure 14–20b). The periodic sweeping of a reef by storms can help maintain a healthy reef by removing algae that cover the corals.

This scenario requires one thing—healthy corals to regenerate the reef. If urbanization and pollution make the area inhospitable to coral regeneration, the reef will die. Subsequent storms will erode the reef, lowering its surface. This will allow storms and hurricane waves to move over the reef and across the bay to cause greater damage to population centers, such as Key West and other communities on the Florida Keys. A healthy reef is their best hurricane protection.

(a)

(b)

**FIGURE 14–20**
Effects of hurricanes on coral communities in the Florida Keys. (a) A few months after Hurricane Donna (1961), a brain coral is surrounded by broken *Acropora cervicornis* (left) and *Acropora palmata* (top left). Note new growth of *Acropora cervicornis* directly behind the brain coral. (b) Four years later, *Acropora cervicornis* has grown prolifically around the brain coral. After five more years, the brain coral had been completely overgrown by the *Acropora cervicornis.* (Photos courtesy of Gene Shinn, U.S. Geological Survey.)

## The Threat to Southern U.S. Coastal Environments

It is no mystery why the population continues to swell in the southern United States, especially along southern coastal areas: warm climate, scenic vistas, beaches, wetlands, bays, and reefs. Unfortunately, we may be loving these areas to death. Unless we respect the fragility of these ecosystems and their great contribution to the health and wealth of the oceans, we are headed for a serious loss.

Fortunately, most threats to mangrove wetlands and coral reefs can be corrected. It will require more money for pollution control, more restrictive zoning, and some sacrifice. It is essential to preserve these important ecosystems and scenic marine areas.

## LOOKING AHEAD

From the reefs and mangrove wetlands that fringe the southern U.S. coast, we move in the following chapter to hazards and problems of more temperate coasts. We will study the effects of waves, storms, rising sea level, and people on the coastal environment worldwide.

## SUMMARY

Mangrove wetlands differ from temperate wetlands because the warmer environment supports robust mangrove trees that can live in saline waters. Mangroves are important nutrient-producers that aid in land formation in tropical and subtropical areas. Mangroves also provide a sheltered habitat for widely varied aquatic animals. Mangrove wetlands are diminishing in area as urbanization encroaches and wetlands are dredged or filled and built upon.

Finger-fill development creates appealing home sites, but the canals between homes become stagnant and allow stormwater ready access to flood the low-lying homes. Causeways built across bays for access to new building sites restrict tidal exchange and bay waters become more stagnant. The reduced oxygen level in the bay then becomes insufficient to oxidize the organic wastes that seep into the bay and eutrophication may result.

Corals are colonial marine animals that live symbiotically with algae. For both to survive, the water must be warm, of normal salinity, clear so that light can penetrate for photosynthesis in the algae, and driven by waves and currents to transport organic material to the corals. Reefs are well developed in southern Florida, Hawaii, Puerto Rico, and the U.S. Virgin Islands. The reefs of southern Florida are mainly barrier reefs, separated from land by a shallow shelf that contains some patch reefs. Fringing reefs, developed on steep submarine slopes around islands, are characteristic of the U.S. Virgin Islands, Puerto Rico, and Hawaii.

Coral reefs of the Florida-Caribbean type show distinct zonation. Within the lagoon or bay are patch reefs, moundlike structures composed of corals, sponges, and other sea plants and animals. Farther toward the ocean is the back reef, populated by massive corals, especially staghorn coral. The fore reef is the most oceanward part of the reef. Its shallow water is populated largely by elkhorn coral.

The reef provides storm protection for the coast behind it. Coral destruction may be substantial during a hurricane, but new coral animals soon re-establish themselves on the rubble of the reef. However, as pollution kills more and more coral, regeneration may not be possible. Subsequent storms will erode and lower the reef, removing this natural hurricane protection for coastal communities.

Coral reefs are being destroyed in many parts of the world. Boat anchors smash into coral heads and divers' feet trample them, destroying living coral. Organic waste from sewage treatment plants may result in algal blooms that cover the reef and deplete oxygen dissolved in the water. At the same time, animals that eat the algae, such as the sea urchin *Diadema*, are being killed off. This mortality may be traced in part to increased pollution.

Water temperatures can be increased by urban runoff and thermal pollution from industry and power plants. Salinity is increasing where surface water is being diverted for use on land, rather than being discharged into the ocean. Construction is supplying more sediment to coastal waters. Sediment clouds the water and reduces light penetration, which can smother coral polyps.

## KEY TERMS

| | |
|---|---|
| atoll | fringing reef |
| barrier reef | hermatypic |
| coral reef | mangroves |
| finger-fill development | polyps |

## REVIEW QUESTIONS

1. Describe mangrove zonation with respect to salinity.
2. **a.** How do mangroves aid in the creation of new land along the coast?
   **b.** How do mangroves provide nutrients to estuaries and the ocean?

3. What problems develop as wetlands are eliminated?
4. a. What are finger-fill developments?
   b. What problems do they create?
   c. What problems are created when causeways are built across bays to provide access to new housing and business developments?
5. Describe the symbiotic relationship between coral polyps and the plants (algae) that live within the outer parts of the coral structure.
6. List five requirements for coral reef development and explain why each is necessary.
7. Describe the zonation of corals in a typical Florida-Caribbean reef.
8. a. Why does a coral reef provide good protection against hurricanes?
   b. How does the reef regenerate after a storm? How long does this take?
   c. How does urbanization interfere with such recovery after storms?
9. What determines whether a barrier reef or a fringing reef develops along a coast?
10. Describe how coral reefs are being destroyed by each of the following agents: (a) swimmers/divers, (b) boaters, (c) organic-rich effluent, (d) changes in water temperature and salinity, and (e) construction and agriculture on land.
11. Describe three reasons the loss of U.S. coastal wetlands is most severe in Louisiana.

## FURTHER READINGS


Flanagan, R., 1993, Corals under siege: Earth, v. 2, no. 3 (May), p. 26–35.

Mairson, A., 1994, The Everglades: Dying for help, National Geographic, v. 185, no. 4, (April) p. 2–35.

Neigel, J. E., and Aviso, J. C., 1984, On a coral reef, it's a hard knock life: Natural History, v. 93, no. 12, p. 58–64.

Snedaker, J., 1987, Mangrove mythology: Florida Naturalist (Fall), p. 7–8.

Snedaker, S. C., 1989, Overview of ecology and information needs for Florida Bay: Bull. Marine Science, v. 44, no. 1, p. 341–347.

# 15

# Coastal Problems

Coastal areas, where the land meets the ocean or a large lake, are among the most beautiful areas on Earth. Popular for vacation and year-round living, these areas grow more developed each year. In the United States, over half the population now lives within a day's drive of a shoreline, either oceanic or Great Lakes. However, increasing coastal development worries many geologists because they understand clearly the hazards of living along a shoreline and within the coastal zone.

For much of Earth's existence, coastal zones have been its most dynamic and changing areas. Over millions of years, the land has been submerged repeatedly beneath the ocean and exposed once again as the sea retreats. On a much briefer scale (years to decades), coastal areas undergo great erosion from waves, storms, and coastal flooding. These natural geologic processes posed no problem until people began to live along shorelines to take advantage of mild climates, abundant food, easy transportation, and enjoyable living.

Every summer, vacationers swell coastal populations as they crowd the beaches. But more ominous trends are the conversion of summer homes to year-round use and the expansion of coastal cities into formerly pristine coastal environments (examples: Galveston, Texas; Atlantic City, New Jersey; and Long Beach, New York). Year-round use places more people in the coastal zone during fall, winter, and spring—times when coastal storms are far more frequent.

Several lines of evidence, such as records from tidal gauges and submergence of archeological sites that once were above sea level, indicate that sea level is rising. This causes a landward shift of the shoreline—in other words, a loss of real estate. Consequently, building on the shoreline is hazardous to property, because it places structures in the path of a migrating shoreline, ensuring their eventual destruction. Engineering attempts to stop such erosion have been partially successful locally, but have often caused displaced problems (Chapter 1) further along the shoreline.

Coastal storms and hurricanes are an inherent geologic hazard along most U.S. shorelines. As our coasts are increasingly developed, storm damage has

**Severe beach erosion at Westhampton Beach, New York, has placed this home in the surf zone. (Photo by author.)**

risen dramatically each decade. Many coasts are affected by multiple hazards. Large waves continually undermine cliffs, eventually causing landslides. Homes built on these cliffs collapse into the surf when landslides undermine the slope. Storm winds raise the water level and create massive waves that flood low-lying barrier islands and coastal plains, washing away many of the homes. The winds from storms and hurricanes may break open structures, thus increasing water damage from precipitation.

Predicting future coastal change is always controversial because there are so many variables of weather, world climate, nature of the shoreline, and human activity. However, many scientists see evidence that sea level is rising, that storms will continue to occur as frequently as they do now, and that hurricanes may become even *more* common in the coming decades (Chapter 16). If present patterns and rates of coastal development continue, our urbanized coasts are on a collision course with disaster. It is important to consider the integrated nature of a shoreline and the inherent natural hazards in the coastal zone as future development is planned in the coastal zone.

## COASTAL PROCESSES

Any shoreline, whether along the ocean or along a lake, experiences the dynamic processes of erosion and deposition. Wave activity is the most important process in all coastal zones. Currents generated by rising and falling tides are also important erosional/depositional agents along oceanic shorelines.

### Waves

A **wave** is energy in motion. The energy is traveling through water, and the wave is simply the ***mechanical expression*** of the moving energy. On a clear, windless day at the beach, large waves may crash ashore and boats offshore rock back and forth. Such wave activity in the absence of local winds seems puzzling, until you realize that these waves were generated by distant storms. The strong winds in such storm centers exert a frictional drag on the water surface, initiating waves.

As the waves travel through the water, away from the direct influence of the winds that spawned them, they become waves of uniform length called **swells.** Swells moving away from a storm center can travel thousands of kilometers across deep water before the energy of the wave is released by crashing onto the coast.

Symmetrical deep-water swells are described by the terms shown in Figure 15–1. The highest part of the wave is the **crest,** and the lowest part is the **trough.** The vertical distance between crest and trough is the **wave height.** The horizontal distance between any two similar points on successive waves (for example, two crests or two troughs) is the **wavelength.**

A line drawn along a wave crest is called the **wave front.** The direction in which the wave is moving is the **wave normal** (a line perpendicular to the wave fronts). The time for one wavelength to pass a given point is called the **wave period** (in seconds).

The height of waves and their period depend on three things: wind speed, wind duration, and the **fetch** (the distance over which the wind blows). For

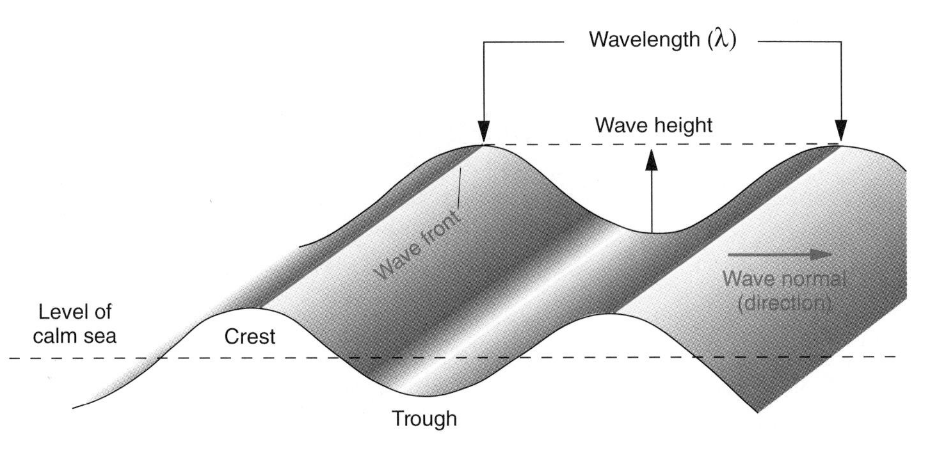


**FIGURE 15–1**
Terms used to describe waves. Wavelength is symbolized by the Greek letter lambda (λ). (Reprinted with the permission of Macmillan College Publishing company from *Physical Geology* by Nicholas K. Coch and Allan Ludman. Copyright © 1992 by Macmillan College Publishing Company, Inc.)

example, compare waves on a lake and on the ocean when wind velocity and duration are identical. Waves will be higher on the ocean because the fetch is far greater than on the lake.

Waves are a mechanism by which energy is transferred along the water surface. Although *waves* visibly travel many kilometers over the surface, the *water* itself does not. Except near shore, the water remains in place, and is merely a *medium* for the wave energy to travel through.

In the open water, water particles in a wave have only a local orbital motion. You can experience this by floating at the surface seaward of the breaking waves along the coast. As long as there is no local current or wind, you will only bob up and down endlessly, in a circular motion. Experimental studies and field observations show that the orbital motion of the water decreases with depth. The orbital motion ceases entirely at a depth equal to about one-half of the wavelength of the waves at the surface (Figure 15–2).

This bottom limit of orbital water movement is the **wave base** (Figure 15–2). Above the wave base is a zone of agitated water, while the water below it is not influenced by wave motion. On the left side of Figure 15–2, the water depth is greater than the wave base, and thus the bottom is undisturbed by the passage of waves overhead. Toward the center of the figure, where waves reach shallow water, the wave base intersects the seafloor. We say that the wave begins to "feel the bottom" in this zone of shoaling (shallowing) waves. The upper part of the wave continues to move landward unimpeded, but the lower part is slowed by friction with the seafloor.

From this point landward, the character of the waves changes markedly; they grow taller and less symmetrical (Figure 15–2). The zone in which these changes occur is called the **breaker zone.** Eventually, these oversteepened waves topple within the **surf zone.** Here, actual forward movement of the water itself occurs within the wave. It is in the surf zone that the energy stored in the wave causes erosion, transport, and deposition of sediment along the shoreline.

## Wave Refraction

If the bottom were uniformly deep along a coastline, approaching waves would slow uniformly as they approach. But the bottom normally is uneven. The part of a wave front that passes over a shallow area begins to slow. This happens around a headland, where land juts out into the ocean (Figure

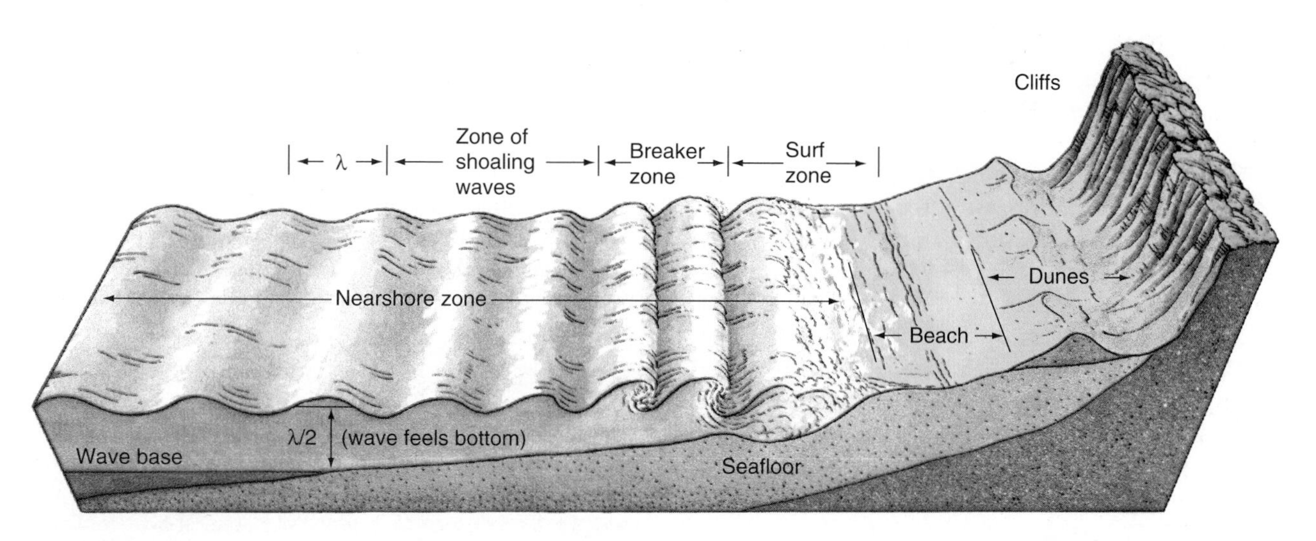


**FIGURE 15–2**
Cross section showing the changes in waves as they approach a coast. When the water depth equals half the wavelength, the waves begin to break. The wave height increases, the wavelength decreases, and the wave becomes asymmetrical. The waves eventually tumble over in the surf zone. (Reprinted with the permission of Macmillan College Publishing company from *Physical Geology* by Nicholas K. Coch and Allan Ludman. Copyright © 1992 by Macmillan College Publishing Company, Inc.)

**FIGURE 15–3**
Diagram showing aerial view of wave refraction along an irregular shoreline. (Reprinted with the permission of Macmillan College Publishing company from *Physical Geology* by Nicholas K. Coch and Allan Ludman. Copyright © 1992 by Macmillan College Publishing Company, Inc.)

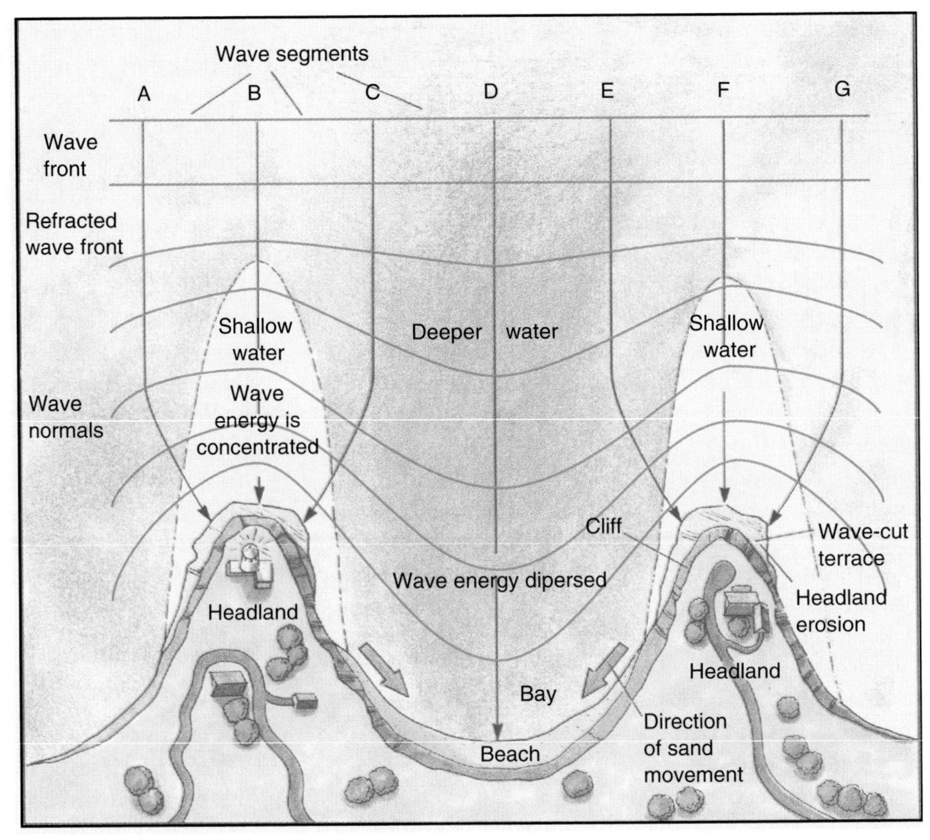


15–3). Parts of the wave that are in deeper water continue to move landward without slowing. This bends the wave fronts, a phenomenon called **refraction,** as the waves advance toward shore (Figure 15–3).

Wave refraction concentrates wave energy on protruding headlands, as shown in Figure 15–3. Equal segments of the deep-water wave shown in this diagram (A–B, B–C, C–D, etc.) have equal amounts of energy. However, as the waves refract, more of this energy is dispersed across the intervening bays, as shown by the pattern. The result of continued wave refraction is that waves erode headlands and smooth out the shoreline. Eroded sediment is deposited offshore, on beaches or in bays.

## Longshore Drift

When waves break at an angle to the shoreline, they move large volumes of sediment along the coast. You can see the process by observing the movement of a single sand grain as successive waves strike the shoreline and recede (Figure 15–4).

Onrushing waves move the sand grain up the beach in the wave-normal direction (A–A′). The retreating water carries the grain down the steepest slope of the beach face (A′–B). This pair of movements transported the sand grain a small *net* distance along the shoreline (A–B). At B, it is picked up by the next incoming wave, and the process repeats, moving the grain even farther along the shoreline (B–C). This movement of sediment along the shoreline by successive waves is called **longshore drift.**

## Tidal Currents

As noted, wind-driven waves are the major erosion agents along oceanic shorelines. However, currents induced by the tides can be much more significant in certain locations, such as sheltered bays between the mainland and offshore islands, and at openings along the shoreline called **tidal inlets** (Figure 15–5). The role of tidal currents in estuarine ecosystems was presented in Chapter 13. In this chapter, we will concentrate on the role of tidal currents in erosion and deposition.

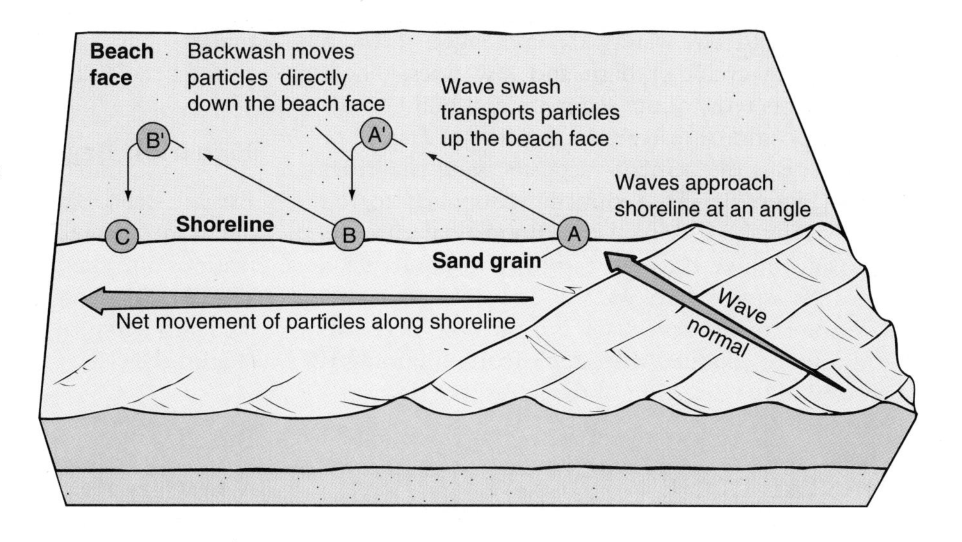


**FIGURE 15–4**
Longshore drift mechanism. Wave fronts advancing at an angle to the shoreline move particles of sediment along the shoreline. (Reprinted with the permission of Macmillan College Publishing company from *Physical Geology* by Nicholas K. Coch and Allan Ludman. Copyright © 1992 by Macmillan College Publishing Company, Inc.)

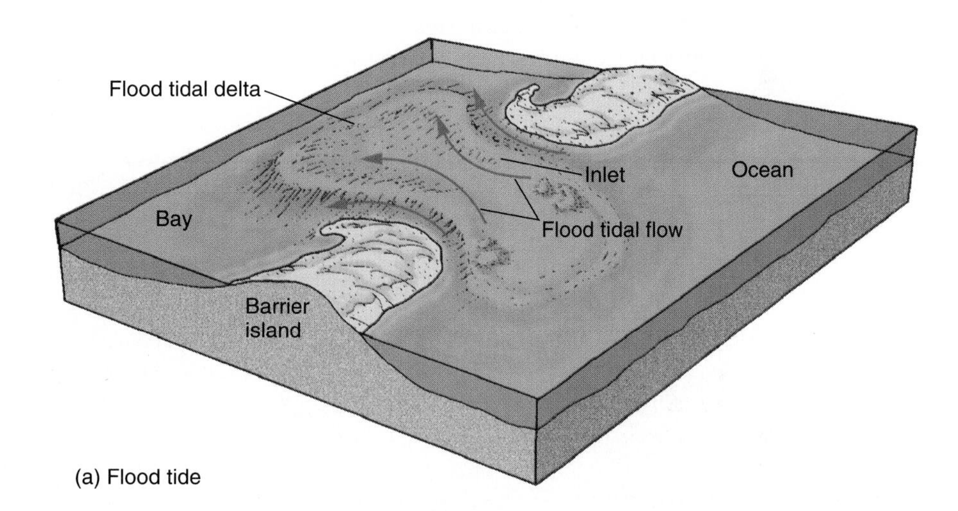


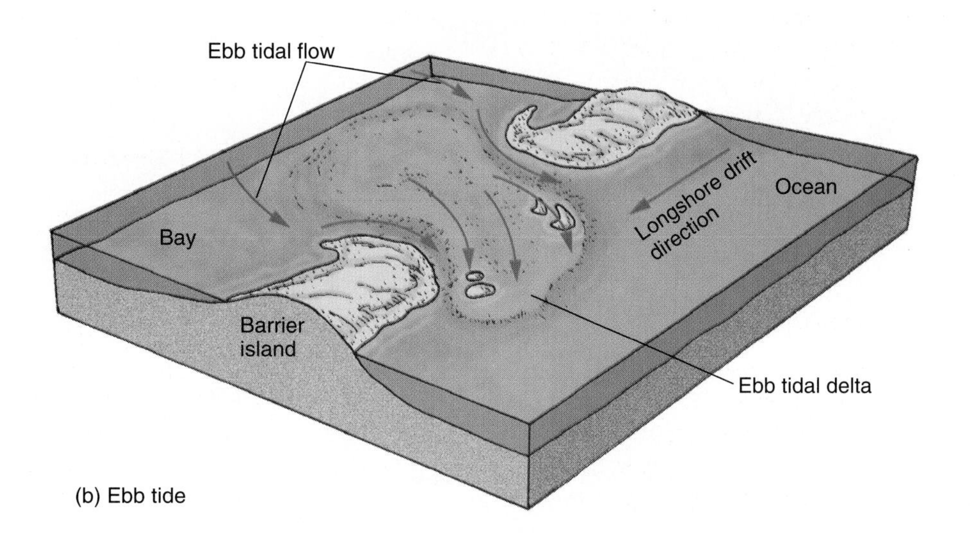


**FIGURE 15–5**
Water and sediment movement at tidal inlets. (a) Flood tide. (b) Ebb tide.

Oceanic and bay waters are exchanged through inlets during cycles of high and low tides. These tidal currents redistribute sediment and build a variety of depositional features in the coastal zone. For example, this movement deposits sand in **tidal deltas** (Figure 15–5). Sediment is supplied to the inlet by longshore drift. Then, flood tidal currents (high tide) move through the inlet to construct a tidal delta in the bay. As the tide ebbs, currents move some of the sand back out into the ocean to build a more subdued tidal delta that commonly is displaced in the direction of the longshore drift.

## COASTAL ZONE

The **coastal zone** includes the area from just beyond the point where waves first break offshore to the limit of high tide inland. Within this coastal zone are several important environments such as estuaries, bays, tidal wetlands (marshes), beaches, cliffed headlands, and barrier islands (Figure 15–6).

### Estuaries, Bays, and Tidal Wetlands

Estuaries, bays, and tidal wetlands are discussed in detail in Chapters 13 and 14, which describe these features and their role in coastal ecosystems (Figure 13–20). It is important to reiterate here that tidal wetlands serve to reduce coastal flooding in storms (Figure 13–17).

### Beaches

A **beach** is a sediment accumulation along part of the coastal zone that is exposed to wave action. In width, a beach extends from the low-water line (low tide) inland to where a change occurs in landforms, sediment type, or vegetation. In general, beaches extend inland to cliffs of rock or sediment, or to

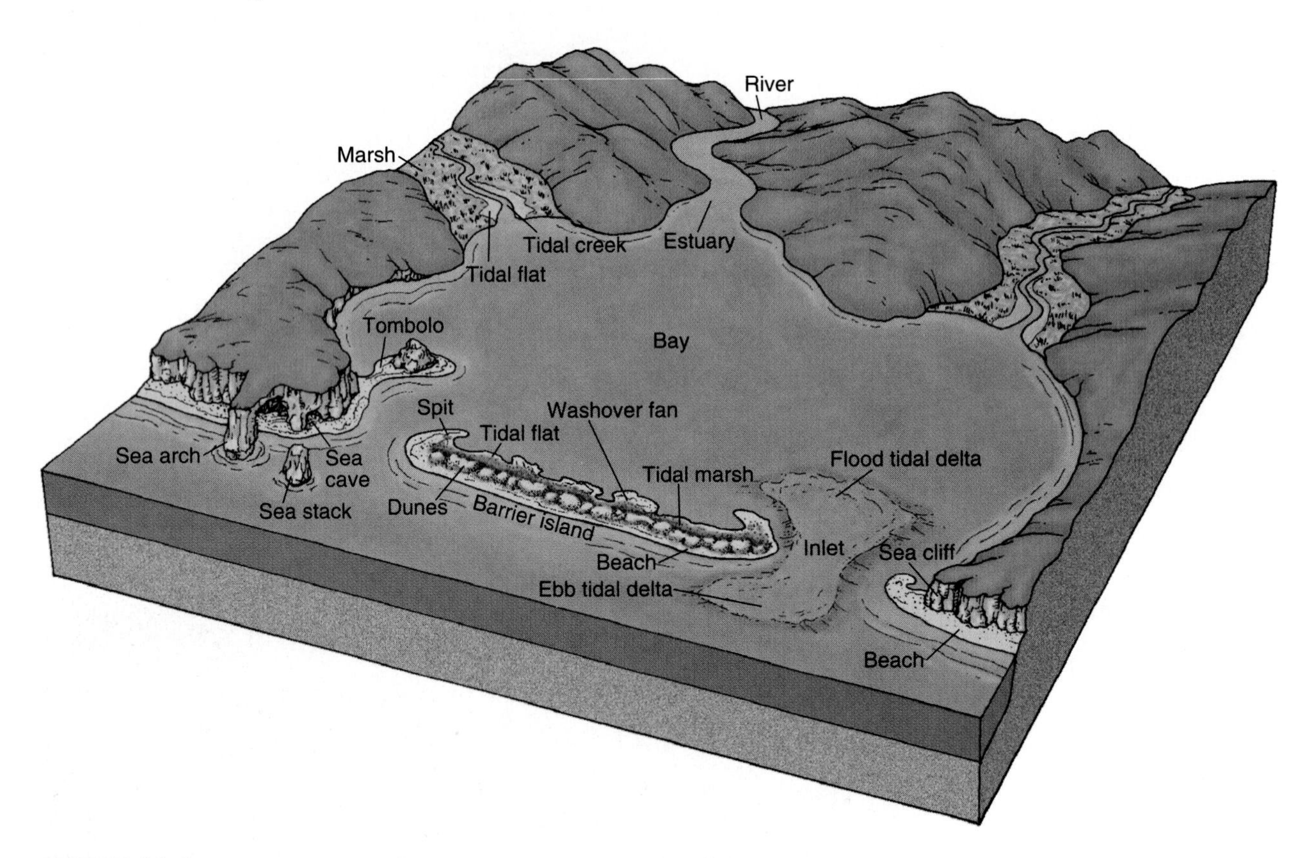


**FIGURE 15–6**
Features of coastal zones described in the text. No location has all of these features together in one place. (Reprinted with the permission of Macmillan College Publishing company from *Physical Geology* by Nicholas K. Coch and Allan Ludman. Copyright © 1992 by Macmillan College Publishing Company, Inc.)

dunes, or to the seaward limit of permanent vegetation.

Beaches occur in various forms (Figure 15–6). Small beaches form along concave portions of cliffed shorelines. More extensive beaches form along **spits,** which are elongate sand bodies attached to land or islands and built into the bay by sediment eroded from headlands or sandy islands. Beaches also form on **tombolos,** which are bars of sand and gravel that connect islands to the mainland or to each other.

Beaches continually build and erode, depending on the balance among sediment deposition, longshore drift, and removal of sand by storm waves. Erosion and deposition occur simultaneously along all shorelines. However, the one that dominates in forming shoreline features varies from coast to coast. Wherever wave erosion removes more material than is being deposited, *net erosion* occurs, and the beach is reduced. Wherever more sediment is being deposited than eroded, *net deposition* occurs, and the beach builds seaward.

## Cliffed Coasts

Some shorelines have high cliffs built of sediment (Figure 15–7). Wave erosion undermines the cliff base, causing the cliff to fail through a variety of mass movements. Material eroded from the cliff by earthfalls, earthflows, and slumping (Chapter 9) then is subjected to wave erosion. In the slumped mass, very fine material (silt and clay) is transported seaward. Fine material (generally sand) is moved along the shoreline to form a beach (Figure 15–7). Coarser material (pebble gravel, cobbles, and boulders) remains behind, forming a submarine **wave-built terrace.** This terrace slopes gently seaward from the base of the cliff into deeper water.

Landslides along the coastal cliffs cause them to recede landward. Some cliffs show no apparent change for a few years and then are cut back tens of meters by extensive mass movements during a single storm. Because of this episodic erosion, cliff retreat rates are usually given as *average* recession rates, such as 0.3 meters (almost 1 foot) per year. This inevitable recession jeopardizes the many homes that are built along cliffed shorelines for their ocean view (Figure 15–8).

The rate of wave erosion is rapid along sandy coasts. Wave erosion is of course much slower along rocky cliffs, but it is just as effective. Both chemical and physical weathering processes (Chapter 6) weaken the rocks on the cliff face, facilitating wave erosion. For example, water can freeze or salt crystals can grow in rock crevices and fractures in the cliffs. Both exert considerable pressure, widening these crevices and fractures. Then, breaking waves force water under great pressure into the widened crevices. Repeated thousands of times, this action wedges out blocks of rock, which fall into the surf. The rock debris is picked up by the waves and thrown repeatedly against the cliff, forming a wave-

**FIGURE 15–7**
Mass movements along this cliffed shoreline result in landslide material deposited on the beach. After waves wash the material, the coarser parts (sand and boulders) remain on the beach while the finer parts (silt and clay) are washed offshore. (Photo by author.)

**FIGURE 15–8**
Cliff recession at Montauk Point, New York. The lighthouse was built far in from the cliff edge during the administration of President George Washington. Cliff erosion in the last twenty years undermined a block-house at the top of the cliff. The remains of that structure are visible in the surf zone at right. (Photo by author.)

cut notch at its base. This may undermine the cliff sufficiently to cause a rockfall and an accumulation of talus (Chapter 9) at the base.

## Barrier Islands

**Barrier islands** are long, narrow accumulations of sand that are parallel to the mainland and separated from it by a bay, sound, or lagoon. Barrier islands differ from spits because they are much longer and wider (Figure 15–6). However, spits can form at the ends of barrier islands as sand is transported into inlets by longshore and tidal currents.

Barrier islands make up most of the Atlantic and Gulf Coast shorelines of the United States. Several major U.S. cities are built on barrier islands, including Miami Beach (Florida), Atlantic City (New Jersey), and Galveston (Texas).

Barrier islands are composed of various sediments. Large changes in grain size occur between adjacent areas, reflecting different agents of transport and different energy levels. For example, coarse sand, gravel, and broken shells may be deposited in the surf zone at the same time that wind is depositing fine sand in dunes only tens of meters away.

Figure 15–9 shows a cross section of a typical barrier island. When you walk from the ocean (right) across the barrier island toward the bay (left), you traverse several sedimentary environments: foreshore, beachface, berm, foredunes, and the vegetated (backdune) area. Figure 15–9 shows the sedimentary nature of these barrier island com-

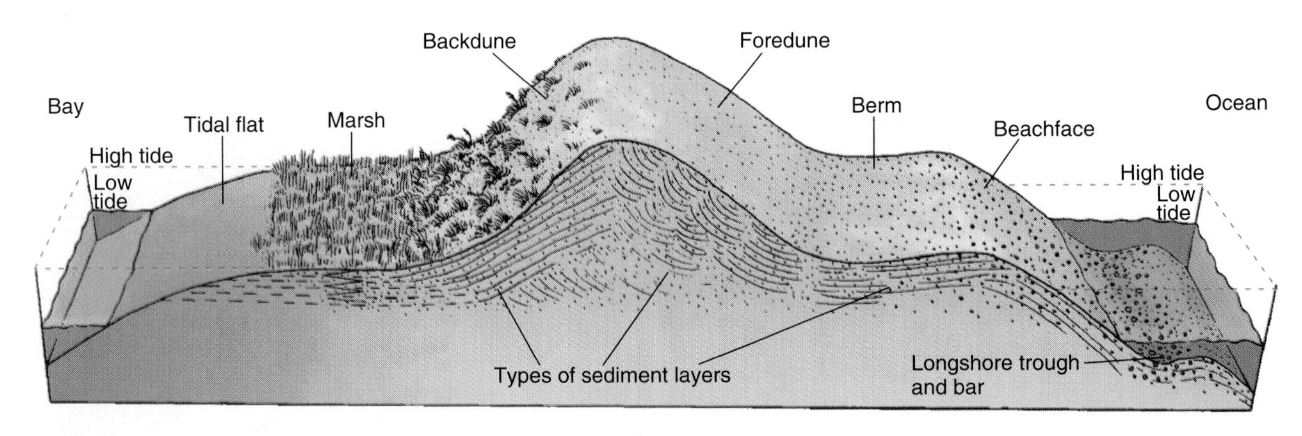


**FIGURE 15–9**
Cross section of a typical barrier island.

ponents. Note how the backdune area grades into the tidal wetland complex (tidal marsh, tidal flat, and bay) described in Chapter 13.

## COASTAL STORMS

In the long, quiet periods between storms, coastal processes redistribute sediment and restore the beach. However, coastal geologists have come to appreciate how much change can occur during a short-lived but powerful storm. In this chapter we will look at the general nature of coastal storms and the changes they cause in coastal areas. These storms are discussed in greater detail in Chapter 16.

### Characteristics of Coastal Storms

Coastal storms are of two types: *tropical* hurricanes and *extratropical* low-pressure cells. Hurricanes spawn in the tropics and migrate northward and westward into temperate regions, particularly the Atlantic and Gulf coasts of the United States. Hurricanes usually occur only during summer to early fall (Chapter 16). The extratropical lows, however, are part of the year-round eastward migration of air masses across North America. Extratropical storms affect the Pacific Coast, the continental interior, and the Atlantic and Gulf coasts. They can occur any time, but are more likely during late fall to early spring. In this chapter, we will use the term **coastal storm** in reference to extratropical storms.

Extratropical coastal storms have wind and waves less powerful than a hurricane's. However, they can cause damage approaching that of a hurricane because of their duration—several days in some cases. A coastal storm has low pressure at its center, surrounded by winds that move counterclockwise—a cyclonic wind pattern. As such a storm approaches a coast, the central low pressure causes the sea surface below it to rise. In addition, its winds act similar to a bulldozer, pushing water ahead of the storm and raising the ocean surface further. While the winds are raising huge waves on the surface, tides may raise the level of the water even more.

The elevated water surface that results from all these factors is called a **storm surge.** As a storm approaches shore, its surge commonly adds 2 to 3 meters (6.5 to 10 feet) to the normal tide level. This is well above mean sea level. A hurricane's storm surge may raise the ocean surface more than 6 meters (20 feet) above normal tide level (Chapter 16).

The effect of an individual storm on a coast is related not only to its strength but to several independent factors, such as nature of the coast and human activity (Table 15–1).

### Beach and Dune Recession

High waves from storms erode large quantities of sediment from beach and dune areas. One of the

**TABLE 15–1**
Factors influencing the severity of coastal storms.

| Factor | Effect |
|---|---|
| Wind velocity | The higher the wind velocity the greater the damage. |
| Storm surge height | The higher the storm surge the greater the damage. |
| Coastal shape | Concave shoreline sections sustain more damage because the water is driven into a confined area by the advancing storm, thus increasing storm surge height and storm surge flooding. |
| Storm center velocity | The slower the storm moves, the greater the damage. The worst possible situation is a storm that stalls along a coast, through several high tides. |
| Nature of coast | Rocky coasts are the least disturbed. Cliffed sedimentary coasts can retreat by slumping or rockfalls, but damage is most severe on low-lying barrier island shorelines because they are easily overwashed by storm waves and storm surges. |
| Previous storm damage | A coast weakened by even a minor previous storm will be subject to proportionally greater damage in a subsequent storm. |
| Human activity | With increased development, property damage increases and more floating debris becomes available to knock down other structures. |

most serious coastal storms to affect the Atlantic shore was the "Ash Wednesday" storm of March 5 through 8, 1962. It moved northward and became stalled against the middle Atlantic coast through five high tides. The U.S. Army Corps of Engineers documented the erosion that occurred at Virginia Beach, Virginia (Figure 15–10). Approximately 30% of the beach and dune sand was removed. The crest of the dunes at Virginia Beach was reduced from an elevation of 4.9 to 3.4 meters (16 to 11 feet), enabling future storm surges to rise over the lowered dunes and flood inland areas.

### Dune Breaching and Washovers

Much material can be eroded from a beach and dunes during a major storm. Some is shifted along the shore by accelerated longshore currents and some moved offshore. In addition, storm waves may wash sediment through low areas between the dunes and onto the back side of the island. This **overwash** is important because it maintains the barrier island's width as its front is eroded. **Washover fans** are lobe-shaped deposits of sand eroded from the ocean side of a shoreline and deposited in the bays behind the barrier island (Figure 15–11).

### Barrier Island Breaching and Inlet Formation

Severe storms can cut through a barrier island to produce a tidal inlet and tidal delta (Figure 15–5) in the bay behind the barrier island. A storm-produced tidal inlet often is short-lived. It closes naturally in a few weeks as longshore drift deposits sand across the inlet and fills it. The filling of a tidal inlet produces a flat, sandy area that invites housing development. However, this area will be the first to flood in a future storm! Time and time again, barrier islands have been breached in storms at the exact place where they were cut through many years before. For this reason, the sites of former tidal inlets should never be considered as building sites.

## COASTAL RECOVERY AFTER STORMS

The process of coastal recovery begins immediately after a storm. Natural processes slowly resume building the beach and dunes. However, the pace often is too slow to satisfy humans, so artificial methods may be used to "restore" beaches and dunes faster.

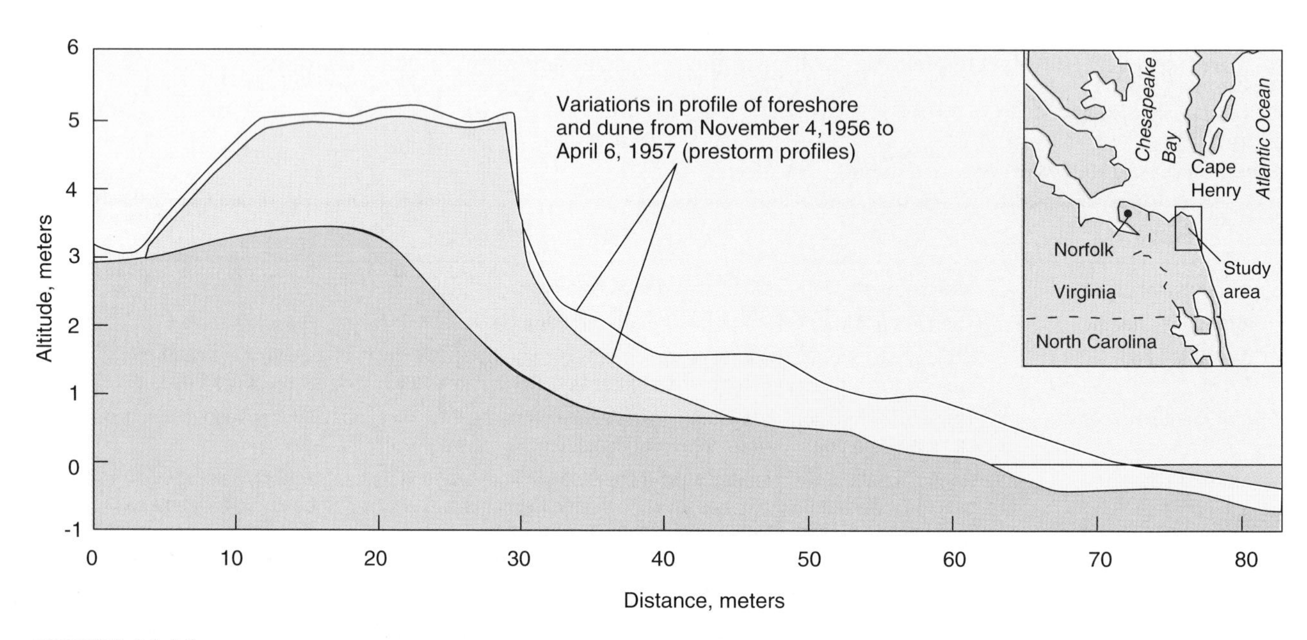


FIGURE 15–10
Beach and dune erosion profile from 1962 storm at Virginia Beach, Virginia. (Modified from Miscellaneous Paper 6 64, U.S. Army Corps of Engineers.)

## Natural Beach Recovery

You may have noticed that beaches look very different in winter and summer. In winter, they generally are narrower, steeper sloping, and covered by coarser sediment. This is because storms are more frequent in the winter and sand is removed faster than it can be supplied by streams or offshore sand bars. During summer, storms are fewer. Generally, wave frequency is less and wave height is lower. These conditions promote onshore deposition of sand and beach growth.

It has long been known that sand stripped from a beach is returned eventually, to some degree. However, only recently have we understood *how* this occurs. Detailed studies of Lake Michigan beaches after storms have shown that sand returns through formation of an offshore ridge of sand and its migration landward (Figure 15–12). During its landward migration, the ridge is separated from the beach by a water-filled trough, called a runnel (together, they are called a **ridge and runnel**).

The migrating ridge shown in Figure 15–12 has "welded" itself to the damaged beach, restoring its width in little more than a week. Ridge and runnel formation is an important beach-restoration process that is visible on many beaches after a storm (Figure 15–13). It is important to remember that, although the width of the beach may have been restored, the front of the beach, and sometimes the dunes, now

**FIGURE 15–11**
Aerial view of washover fans at Cape Romain, South Carolina, in 1987. In 1989 Hurricane Hugo completely removed this sand, exposing older peat beds below. (Photo by author.)

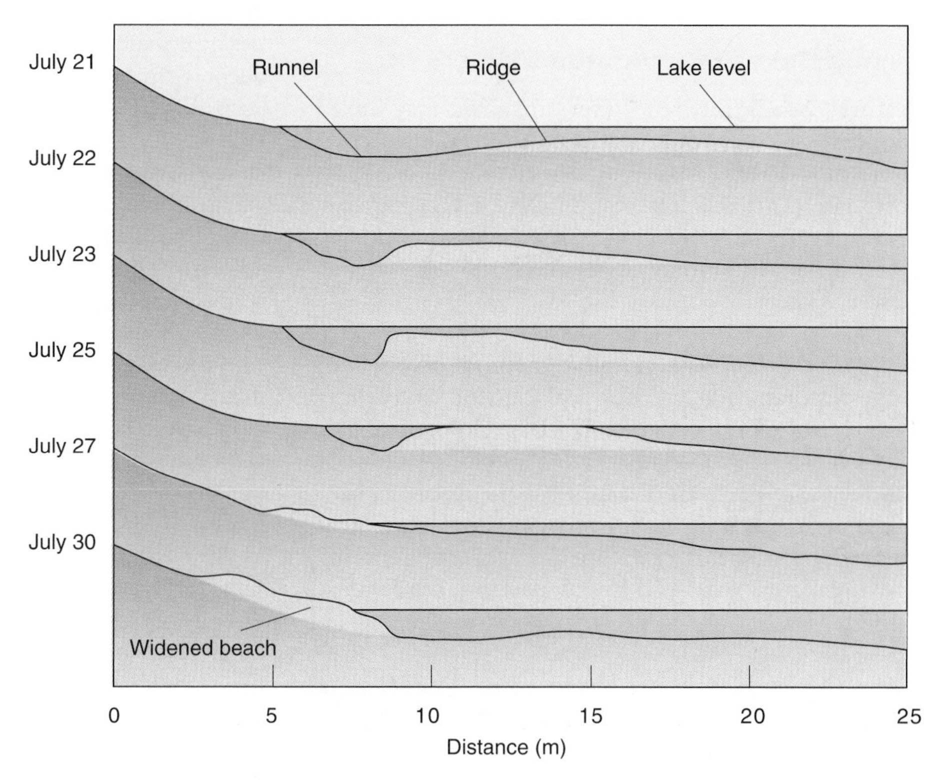


**FIGURE 15–12**
Diagram showing ridge and runnel development. An offshore ridge migrates gradually landward and eventually reaches the beach, increasing its width. During its landward migration, the ridge is separated from the beach face by a water-filled runnel. (After Davis and Fox, Natural beach restoration on Lake Michigan: Journal of Sedimentary Petrology, v. 42, p. 416, Fig. 4. Modified with permission of Society of Economic Paleontologists and Mineralogists.)

**FIGURE 15–13**
Ridge and runnel formation on the south shore of Long Island, New York. (Photo by author.)

may be located further inland than before. Thus, *net* erosion occurs along many coasts. Its environmental significance is discussed later in this chapter.

## Artificial Beach Restoration Methods

Natural beach recovery after a storm may take months or years, but beaches can be "restored" more quickly by a variety of artificial methods. Unfortunately, artificial restoration can be very expensive and only temporary, and may cause other problems. Four commonly used methods are briefly described here.

***Sand Dumping.*** The simplest method of beach/dune restoration is to truck replacement sand from inland to the beach (Figure 15–14). Unfortunately, this sand is very loosely packed. Therefore, dumped sand is more easily eroded in subsequent storms because it has less erosion resistance than sand deposited by a shear force, such as that present in moving water or air. (To review packing, see Figure 5–22.)

***Sand Pumping.*** Another method is to dredge sand from nearshore areas and tidal inlets. The sand is pumped from the dredging site directly onto the eroding beach (Figure 15–15). This method serves two purposes: keeping inlets open for navigation and restoring eroding beaches. However, nearshore dredging must not be done too near the shoreline, because that may steepen the submarine profile and deepen the water, resulting in waves not breaking until they reach closer to shore (Figure 15–2).

***Beach Scraping.*** Sometimes the most expedient way to rebuild an eroded beach is to scrape sand up onto the beach at low tide with bulldozers (Figure 15–16). This method simply reshapes what is already there and raises the beach level. It should be used only when a beach is in imminent danger, because it may steepen the lower beach profile in the nearshore area, which may cause waves to break closer to the shore in future storms. However, sometimes it is the only viable choice if it is necessary to raise the beach surface quickly after a major storm.

An example is in South Carolina, where beaches were stripped by Hurricane Hugo in 1989. Exceptionally high tides were forecast 2 to 3 weeks after the hurricane, and officials were very concerned about the stability of structures remaining on the beach. Extensive sand scraping restored the beach and dunes and helped to prevent erosion from the high tides.

**FIGURE 15–14**
Sand dumping for beach restoration on Tampa Bay, Florida. (Photo by author.)

**FIGURE 15–15**
Beach replenishment by pumping sand from offshore onto the eroded beach. Note the narrow unrestored beach beyond the outflow pipe (background). (Photo by author.)

**FIGURE 15–16**
Beach scraping operations on the South Carolina coast following Hurricane Hugo. (Photo by author.)

***Sand Fencing and Grass Planting.*** This method builds dunes by using sand fencing to trap sand that blows across the beach and then anchoring the sand with vegetation. The sand is retained by choosing plants with extensive root systems, such as beach grass (*Ammophila,* Figure 15–17). It takes time to build dunes in this fashion, but the resulting landforms are more storm resistant than the dune-like ridges of sand built artificially by dumping or bulldozing. This is one of the most effective artificial methods of building beaches and dunes.

## CHANGING SEA LEVEL

Up to now, we have referred to sea level as if it were constant. In fact, it is not, varying daily with the tides. But the average sea level is important to anyone living near the ocean, so maps of coastal areas show "mean sea level," the average of the daily tidal rise and fall. Storm surges raise sea level, but only temporarily. Thus, in practical terms, sea level over a period of years essentially is constant.

However, major *long-term* changes in sea level occur and have done so time and again throughout Earth's history, as recorded in the sedimentary rock record. The present slow increase in sea level is alarming because many large cities may be threatened.

**FIGURE 15–17**
Beach restoration utilizing sand fencing and grass planting in St. Petersburg, Florida. (Photo by author.)

### Causes of Sea Level Change

Global warming is slowly melting Earth's large ice deposits—glaciers and the ice caps over Greenland and Antarctica. This contributes meltwater to the ocean, and sea level is rising steadily. For the past few thousand years, it has risen at about 0.3 meter (1 foot) per century. The rate has been predicted to increase to 1 to 2 meters per century (3 to 6.5 feet). Increasing or decreasing the amount of water in the ocean causes variations in the ocean surface elevation called **eustatic** sea level changes. (Apparent sea level rise also can occur where the land is sinking, a situation discussed later.)

The greenhouse effect (Chapter 11) may be accelerating sea level rise. As the atmosphere gradually warms, it increases the rate of glacier melting and sea level rise. Whether this is happening, and how fast, is controversial. If this effect is real (and some scientists feel it is not), several major coastal cities may become gradually inundated in the next century.

Rising sea level is causing landward migration of the shoreline wherever Earth's crust is stable or where it is subsiding, as along the Gulf Coast (see Box 14–1). Detailed studies of Atlantic and Gulf coastal barrier islands have shown that net long-term erosion is occurring along those shorelines.

Geologist Chris Kraft of the University of Delaware has projected the shoreline that would result from the present rate of sea level rise (Figure 15–18). Many great coastal cities eventually could be submerged. Kraft's projection does not include the accelerated sea level rise attributed to the greenhouse effect. Adding this effect to the rise caused by postglacial melting, and we see that coastal flooding could be higher and come earlier than forecast in Figure 15–18.

In addition to the eustatic change of sea level due to addition of more water from melting ice, Earth forces can elevate or lower (submerge) coastal areas. In the northwestern United States, the shoreline is not migrating inland as sea level rises, because the *land is being uplifted* at a faster rate. In the 1964 Alaska earthquake, parts of the coast either were warped upward or subsided beneath the sea (Figure 5–27). Sea level rise or fall resulting from uplift or subsidence of coastal areas is called *tectonic* sea level change, as opposed to *eustatic*.

The combined effect of tectonic and eustatic changes in sea level makes it difficult to determine the *net* effect of change in many areas. For example, subsidence in the Mississippi Delta region makes the sea level appear to rise faster, whereas tectonic uplift along the Oregon coast makes it seem that the sea level is not rising at all. In both cases, there is an *actual* rise in the ocean surface. Regulatory agencies and developers are increasingly including the factor of rising sea level in planning coastal developments.

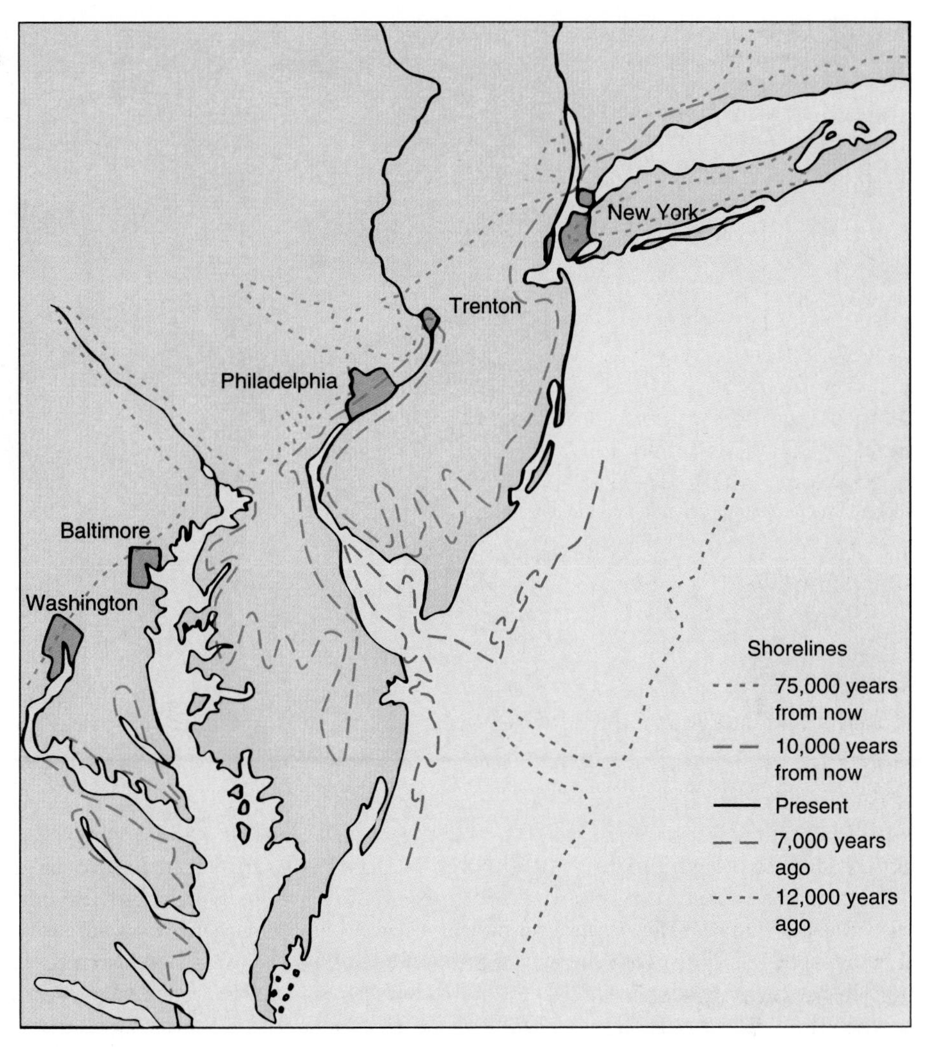


**FIGURE 15–18**
Location of past and future shoreline positions in the northeastern United States. (After J. C. Kraft, 1973, Coastal geomorphology, Fig. 22, p. 352; redrawn with permission of Donald R. Coates.)

## Results of Sea Level Change

Sea level rise along stable and subsiding coasts is causing shoreline erosion further inland. In general, the rate of shoreline retreat is greater along low, gently sloping coasts composed of sediment and slower along high, rocky, cliffed coasts. As sea level rises, it allows storm waves to reach areas not inundated previously. The erosion that accompanies sea level rise can cause cliff retreat and the migration of barrier islands.

***Cliff Retreat.*** Rising sea level elevates waves so they can cut into cliffs at a higher level, initiating mass movements in poorly consolidated sediment (Chapter 9). A cliff may remain relatively stable but then recede many meters in a single event. This tendency for episodic slumping makes building homes near the edges of cliffs an especially risky venture in a time of rising sea level (Figure 15–19).

***Coastal Submergence.*** As sea level rises, it will inundate present coastal areas (Figure 15–18). Unprotected lowlying areas of coastal cities—Boston, New York, Norfolk, Charleston, Miami, San Diego, Los Angeles—will suffer serious damage to their waterfront areas. City planners should prepare for this inevitable flooding now. Vital structures could be moved to higher ground or protected by dikes.

***Barrier Rollover.*** Sea level rise is a special problem along the low-lying, sandy, barrier islands of much of the Atlantic and Gulf coasts. For example, on a

**FIGURE 15–19**
The foundation of this lakefront home was undermined by bluff erosion. (Courtesy of Pennsylvania Department of Environmental Resources.)

gently sloping coastal plain, a 0.3-meter (1 foot) rise in sea level can result in a 60- to 90-meter (200 to 300 foot) landward movement of the shoreline (Figure 15–20). Thus, even houses that are elevated on pilings away from the shore today will be destroyed as water eventually reaches the pilings and scours away the sand that supports them.

Putting a beach home on pilings to minimize damage in present storms is a good short-term strategy (see Chapter 16), but not a long-term solution. The home eventually will be destroyed by wave erosion, as the rising sea level elevates wave action that undermines the house. Pilings are designed so that the sand around them supports part of the weight of the home. As that sand is removed by wave scour, the support becomes increasingly inadequate and reinforcement is necessary. However, by that time, the shoreline may be at, or under the home (Figure 15–21).

Barrier islands respond to rising sea level by erosion on their oceanic sides and deposition of the material by overwash on their back sides. Thus, with time, the ocean front recedes and the back side of the island grows landward in a process called **barrier rollover.** The rollover of a barrier island is shown in Figure 15–20. Note that the island retains its width, but the entire island is displaced landward with time. Note also that the bay width remains the same, because the landward shoreline of the bay also recedes as sea level rises. This will damage structures there too, much to the surprise of the inhabitants who thought living there would protect them from wave damage!

This barrier-migration scenario applies only to coasts where engineering structures have not been built to control erosion. Home and business owners who have ocean or bay frontage may pressure government to build walls and other structures to protect their valuable property. However, this action prevents the overwash that allows the back side of the island to build landward as sea level rises (Figure 15–22).

In short, a barrier island that is protected by engineered structures will only get narrower with time. Ironically, the only way to preserve barrier islands as sea level rises is to let their fronts erode, and allow the sediment to overwash across the island to build out the bayward side. Of course, this natural solution is not popular with property owners.

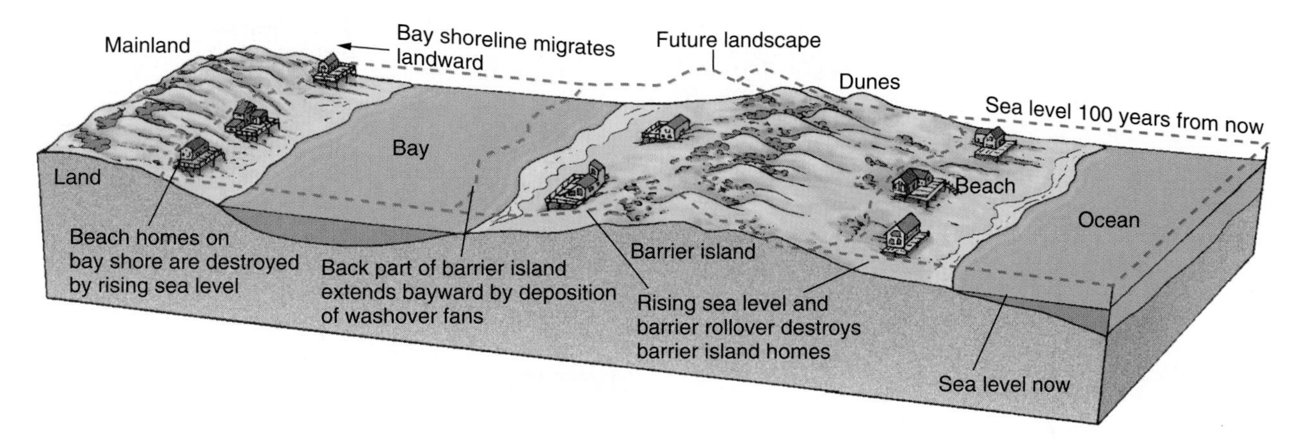


**FIGURE 15–20**
Effect of barrier rollover on a gently sloping coastal plain. In the foreground is the scene today, with homes elevated on pilings. The dashed line shows the same coast as it might look 100 years from now. Note that the width of the barrier island and bay have remained the same, but both have been *displaced* landward. (Vertical scale is somewhat exaggerated.)

## HUMAN INTERFERENCE WITH COASTAL PROCESSES

Accelerating beach erosion has led to engineering structures to prevent erosion and save homes. Although these structures do control erosion in one place, they frequently increase it in another. This is because structures built within the coastal zone disturb the longshore drift. This building alters the dynamic equilibrium along the coast by cutting off the sand supply needed to nourish beaches.

In discussing these structures, we use the terms *updrift* and *downdrift* along a shoreline to describe the direction of sediment transport. *Updrift* is the direction from which the sediment is coming (analogous to "upstream"). *Downdrift* is the direction toward which it is moving (analogous to "downstream").

**FIGURE 15–21**
Severe erosion downdrift from a groin at Westhampton Beach resulted in exposure of the pilings supporting several homes. The homes were subsequently destroyed in storms. (Photo by author.)

FIGURE 15–22
Coastal changes resulting from rising sea level.

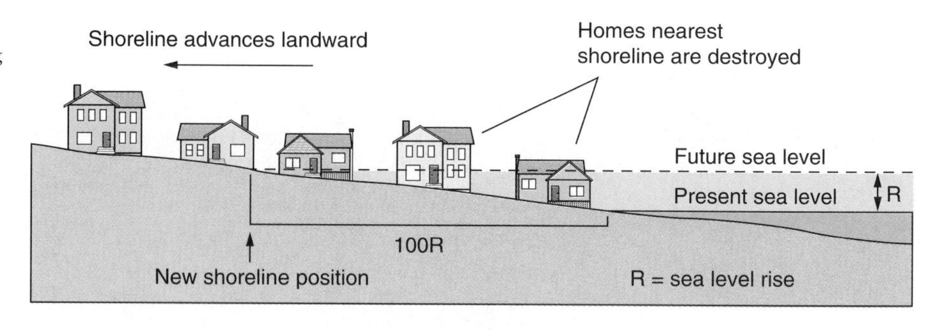


## Dune Elimination

The best barrier to storm-surge intrusion is high, wide, vegetated, natural dunes. However, these dunes are being destroyed by human activities. As people walk across vegetated dunes toward the beach, they kill the grass. The grass roots hold the sand in place and protect it from wind erosion, or **deflation.** Without the grass, wind eventually erodes shallow depressions called **blowouts** in the dunes (Figure 15–23). These depressions lower dune height and make it easier for a storm surge to breach the dunes and overwash the area behind them.

Once a storm-surge channel is established, the dunes erode much faster (Figure 15–24) because they now erode *laterally* from storm surges as well as *vertically* from the wind. Eventually, these dunes become reduced to a low sandy area, offering little surge protection. In many areas, wooden walkways have been built across the dunes for beach access, allowing stabilizing vegetation to be maintained.

## Seawalls

Damage to coastal structures can be prevented by constructing **seawalls** parallel to the shore (Figure 15–25). These massive, expensive structures can protect a coastal segment for several years. However, a seawall eventually *accelerates* beach erosion. This occurs because incoming wave energy is *reflected* off the seawall rather than being dissipated across a wide area of beach. This reflected energy is directed downward in part, causing beach erosion in front of the seawall. The beach then becomes usable only at low tide (Figure 15–25). Eventually, the seawall itself is undermined and fails, and a new one must be built in a more landward position.

Seawalls also prevent the normal erosion of cliffs and beaches, which is essential to maintaining the longshore drift system along the coast. Seawalls also can accentuate the flooding problems on barrier islands and spits during coastal storms. Al-

FIGURE 15–23
Deflation basins in dune systems on the south shore of Long Island, New York. (Photo by author.)

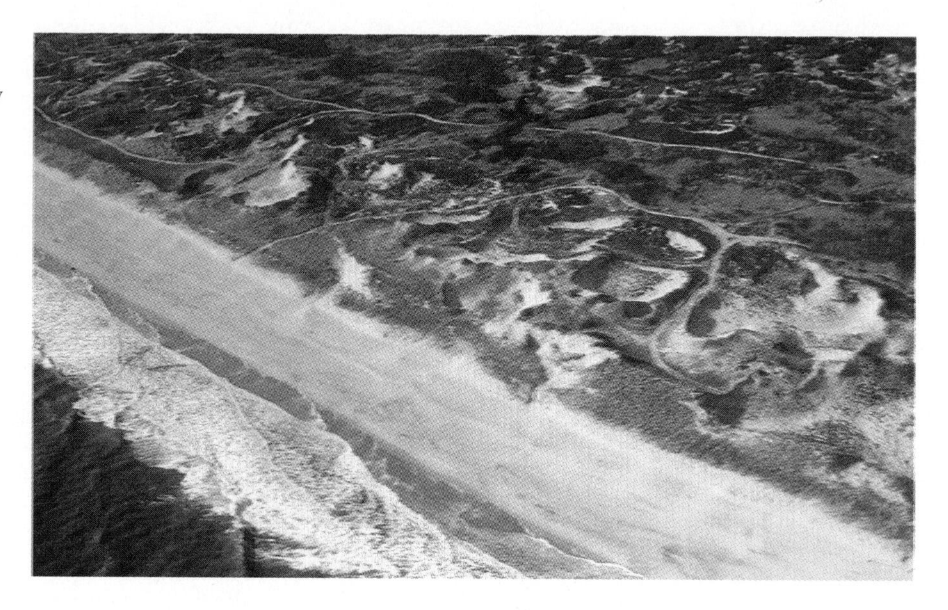

FIGURE 15–24
Surge channel cut through a dune at Southampton, New York. Storm waters broke through a low point in the dune and deposited sand as washover fans on the marsh behind the dune. (Photo by author.)

though seawalls prevent direct destruction by waves, high waves can pump water continually over the seawall. As the storm subsides, this water cannot return to the ocean because of the seawall. Nor can it flow bayward, because the bay level is frequently elevated by rain-swollen estuaries for a time after a storm.

## Groins

Walls of rock, concrete, or wood, called **groins,** commonly are built at right angles to the shoreline to trap sand and replenish a beach (Figure 15–26). The updrift beaches do indeed grow by net deposition. However, beaches downdrift of the groin are eroded, because the groin traps the longshore-drift sand that would have replenished them.

## Jetties

Sediment transport along a coast is also disturbed by the construction of long walls called **jetties** on either side of tidal inlets. These structures prevent sand from filling the inlet, thus maintaining boat ac-

FIGURE 15–25
Seawall at Sea Bright, New Jersey. (Photo by author.)

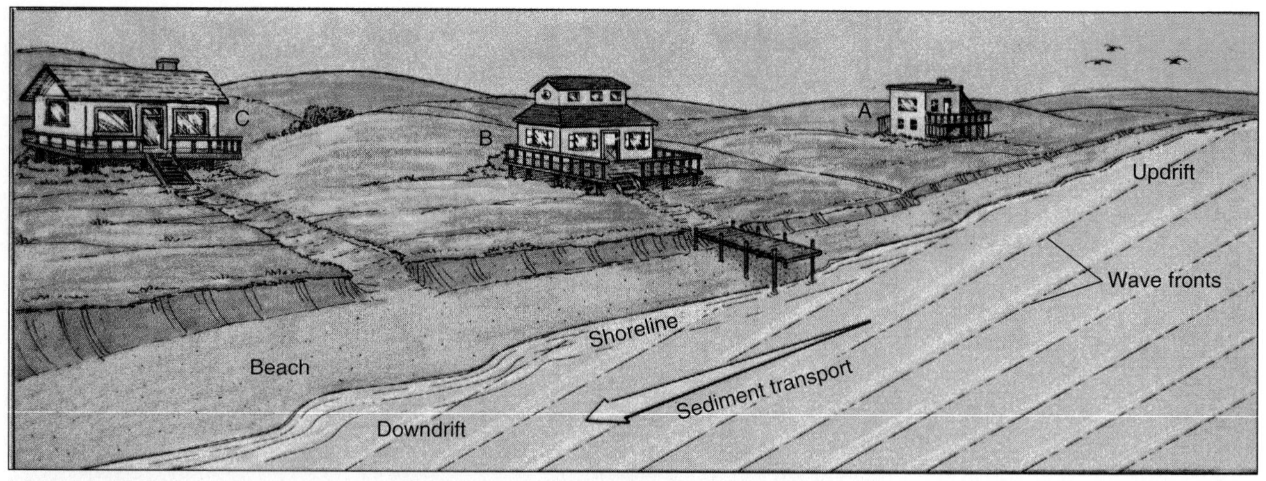


(a)

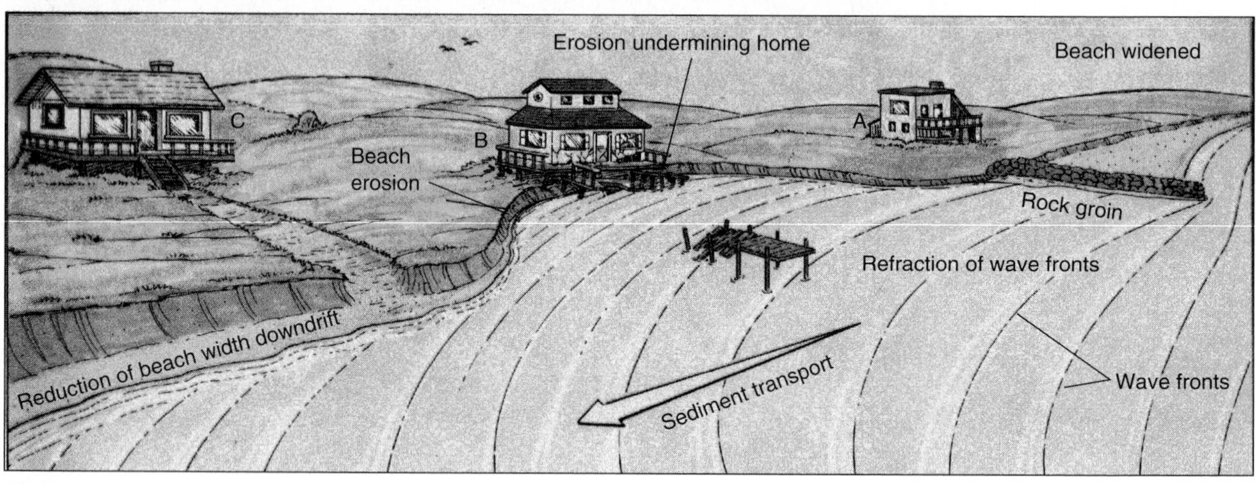


(b)

**FIGURE 15–26**
Effects of groin construction on coastal erosion and deposition. (a) Homeowner A decides to widen the beach by building a rock groin into the surf at the edge of his property. The beach widens at A by sand deposition updrift of the groin. However, the groin has prevented the beach downdrift from being replenished, and net erosion occurs there, undermining the home at B. (Reprinted with the permission of Macmillan College Publishing company from *Physical Geology* by Nicholas K. Coch and Allan Ludman. Copyright © 1992 by Macmillan College Publishing Company, Inc.)

cess between the ocean and the bay. Such inlets are **stabilized inlets** (Figure 15–27).

However, inlet stabilization by jetty construction causes beach erosion downdrift. The updrift jetty traps most of the sediment being carried along the beach. Sediment that passes around the updrift jetty does not make it across the inlet to nourish the beaches on the downdrift side. Instead, most of the sediment is swept into the inlet by tidal currents and is deposited on tidal deltas in the bay (Figure 15–5).

## ARE AMERICA'S BEACHES ALL WASHED UP?

Beach erosion is increasing because sea level is rising (postglacial melting and the greenhouse effect), because of storms and hurricanes, and because human modifications along some coasts have accelerated erosion. Unfortunately, people are building more fixed structures along migrating shorelines and are unaware of the dynamic balance between coastal erosion and deposition.

**FIGURE 15–27**
Shinnecock Inlet at Southampton, New York. The direction of littoral drift is from right to left on this photograph. Note the deposition of the updrift (right) side of the inlet and the erosion on the downdrift side. (Photo by author.)

No coast is spared. Miami Beach must be restored periodically by sand pumped from offshore. In Louisiana the land is subsiding as sea level is rising (Box 14–1). Many of California's beaches are eroding (Figure 15–28) because the river-supplied sand that normally replenishes them is being trapped behind dams in flood-impoundment reservoirs.

Eliminating wetlands for development removes the natural flood protection (Chapter 13), and storm-swollen estuaries now flood barrier islands from the bay side as storms move inland. Where dunes were removed, the most effective barrier to storm waters has been lost. Groins built to trap sand and widen updrift beaches have caused serious problems downdrift.

Coastal development increases loss of life and property in storms. Warning systems and evacuation have dramatically reduced fatalities in storms and hurricanes, but property loss has increased yearly. Although many states now require new homes to be built on pilings 3 to 5 meters (10 to 16 feet) above sea level, this may be of little use if debris transported by the storm surge batters down the supports!

So, are America's beaches all washed up? *No, but the structures on many of them are.* Serious problems can be minimized only when we understand the dynamics of shorelines and the inevitable consequence

**FIGURE 15–28**
Wave erosion has undermined the bluff under this California home. Note the exposed pipe extending out of the bluff at the top right of the photo. (Courtesy of Raymond Pestrong, San Francisco State University.)

**FIGURE 15–29**
High density public use of the coastal zone. Cape Cod National Seashore, Massachusettes. (National Park Service.)

of the gradual sea level rise. Zoning can limit development in coastal areas that are subject to erosion.

Many geologists believe that, instead of rebuilding eroding beaches, we should return more of them to public use when they are devastated by a storm. Examples of such public coastal resources are the National Seashores, which include Cape Cod in Massachusetts (Figure 15–29), Assateague in Maryland, Padre Island in Texas, Hatteras in North Carolina, and Point Reyes in California.

National Seashores are kept in as natural a condition as possible for all to enjoy. Storms will continue to damage their roads, parking lots, bathhouses, and concessions, but these can be rebuilt inexpensively, compared to the cost of restoring high-rise hotels, condominium complexes, and beach homes.

## LOOKING AHEAD

The closing chapter looks at severe weather hazards and their effect on coasts as well as inland areas.

## SUMMARY

Waves are formed as wind shears the water surface. Effective motion in a wave ceases at a depth equal to one-half its wavelength. As a wave enters shallow water, its wave base intersects the bottom and the wave begins to break. Most of the energy carried within a wave is then released within the surf zone.

Waves refracting around a headland cause cliff recession on the headland and sand accumulation on beaches. Sediment is moved along the shoreline by longshore drift and into and out of tidal inlets by tidal currents. Beaches are accumulations of sand along the shoreline. If net sediment supply exceeds erosion, then beaches grow wider. When net erosion exceeds sediment supply, the front of the beach is reduced and sand is removed offshore.

Cliffed coasts recede by rockfall or slumping as waves attack their bases. Barrier islands are linear sand accumulations parallel to the shoreline that are separated from the mainland by bays, and from each other by inlets.

Coastal storms cause major shoreline changes. These extratropical low-pressure systems can remain over a coast for days and result in damage like that from a hurricane. The low pressure and high winds raise the ocean surface into a dangerous storm surge. The severity of a coastal storm depends on wind velocity, storm surge height, coastal shape, storm center velocity, nature of the coast, previous storm damage, and population density.

Storms result in severe dune erosion and beach recession. They can lead to dune breaching, washover fan development, inland flooding, and the formation of new tidal inlets. Beaches restore themselves naturally after a storm by the onshore movement of sand ridges (ridge and runnel). Beaches can be artificially restored by sand dumping, sand pumping, beach scraping, and sand fencing and planting of stabilizing vegetation.

Rising sea level is resulting in a landward retreat of the shoreline in stable and subsiding areas. On barrier islands, the process of barrier rollover results in a gradual landward movement of the barrier island with time. Consequently, any fixed structures eventually will be destroyed and submerged as the shoreline migrates landward.

Human interference with coastal processes and landform development is causing serious problems. Walking across dunes kills the vegetation, which allows blowouts to form. Deflation of the exposed sand lowers dunes and allows storm surges to overtop the dunes more easily. Seawall construction protects structures against wave attack, but causes beach erosion and eventual collapse of the structure as reflected wave energy undermines the wall. Groins cause the updrift beach to grow, but the downdrift beaches erode. Jetties at stabilized inlets cause erosion of the beaches downdrift of the inlet.

## KEY TERMS

| | |
|---|---|
| barrier islands | spits |
| barrier rollover | stabilized inlets |
| beach | storm surge |
| blowout | surf zone |
| breaker zone | swells |
| coastal storm | tidal deltas |
| coastal zone | tidal inlets |
| crest | tombolos |
| deflation | trough |
| eustatic | washover fans |
| fetch | wave |
| groins | wave base |
| jetties | wave-built terrace |
| longshore drift | wave front |
| overwash | wave height |
| refraction | wavelength |
| ridge and runnel | wave normal |
| seawalls | wave period |

## REVIEW QUESTIONS

1. How does wave refraction shape an irregular shoreline?
2. Describe the processes that move sand along a shoreline and through tidal inlets.
3. What conditions determine whether a beach will be eroded or built up?
4. Describe the variations in sediment type at various places across a barrier island.
5. Describe the processes that cause recession on (a) rocky cliffs and (b) bluffs of sediment.
6. Describe five factors that determine the severity of damage in coastal storms.
7. How do beaches recover naturally after a storm?
8. Describe artificial methods by which beaches and dunes have been restored after storms. Then analyze the long-term effectiveness of each method.
9. Eustatic sea level is rising worldwide, yet along some coasts the effect of sea level rise is not visible. Explain.
10. Describe the process of barrier rollover and predict the consequences for present coastal structures.
11. Describe the environmental problems associated with each of these engineering structures: (a) seawalls, (b) groins, and (c) stabilized inlets.

## FURTHER READINGS


Davis, R. A., Jr. (ed.), 1978, Coastal sedimentary environments: New York, Springer-Verlag, 420 p.

Dolan, R., Lins, H., and Stewart, J., 1980, Geographical analysis of Fenwick Island, Maryland, a Middle Atlantic coast barrier island: U.S.G.S. Professional Paper 1177-A, 24 p.

Federal Emergency Management Agency, 1986, Coastal construction manual: 104 p.

Fox, W. T., and Davis, R. A., 1976, Weather patterns and coastal processes, *in* Davis, R. A., Jr., and Ethington, R. L. (eds.): Beach and nearshore sedimentation: Tulsa, S.E.P.M., Spec. Pub. No. 24, p. 1–23.

Kaufman, W., and Pilkey, O. H., 1979, The beaches are moving—The drowning of America's shoreline: Garden City, NY, Anchor/Doubleday, 326 p.

Leatherman, S. P., 1988, Barrier island handbook: College Park, University of Maryland Laboratory for Coastal Research, 931 p.

Leatherman, S. P. (ed.), 1979, Barrier islands from the Gulf of St. Lawrence to the Gulf of Mexico: New York, Academic, 325 p.

MacLeish, W. H., 1980, Our barrier islands are the key issue in 1980, the "Year of the coast": Natural History v. 11, no. 6 (Sept.), p. 46–58.

Morton, R. A., Pilkey, O. H., Jr., Pilkey, O. H., Sr., and Neal, W., 1983, Living with the Texas shore: Durham, NC, Duke Univ. Press, 190 p.

National Committee on Property Insurance, 1988, America's vanishing coastlines—A new concern for the voluntary and residual property insurance markets: 44 p.

South Carolina Coastal Council, 1990, Understanding our coastal environment: Charleston, SC, 40 p.

Time Magazine, 1987, Where's the beach?, August 10, 1987, p. 38–47.

U.S. Army Corps of Engineers, 1981, Low cost shore protection: 36 p.

U.S. Army Corps of Engineers, North Central Div., 1978, Help yourself—A discussion of erosion problems on the Great Lakes and alternative methods of shore protection: 25 p.

Williams, S. J., Dodd, K., and Gohn, K. K., 1990, Coasts in crisis: U.S.G.S. Circular 1075, 32 p.

# 16

# *Severe Weather Hazards*

In Texas, a tornado funnel snakes down from a dark, towering cloud to cut a narrow path of devastation across the landscape (Figure 16–1). A 1000-kilometer (620-mile) stretch of the Atlantic coast is pummeled by a northeaster, a storm whose gale-force winds and massive waves continue for several days without let-up (Figure 16–2). A hurricane devastates an area 170 kilometers (105 miles) wide on the South Carolina coast and continues destruction inland for some 200 kilometers (over 120 miles), flattening forests and releasing torrential rains. The weakening storm triggers floods and landslides across several states (Figure 16–3). These three atmospheric phenomena—tornadoes, northeasters, and hurricanes—are all *severe weather hazards* that cause major damage in the United States each year.

Of all the catastrophes for which one can buy insurance, severe weather hazards have been the most costly, according to data summarized by the New York Times in 1993 (Table 16–1). Of the 11 most costly catastrophes occurring during the 14 years between September 1979 and March 1993, all were due to severe weather except two (the Los Angeles riots and the Loma Prieta earthquake). Even the 1991 fire in Oakland, California, which was initiated by arson, was made possible by another form of severe weather—prolonged, hot, dry winds that blew from arid inland areas to desiccate plants, making them much more flammable.

Other severe weather hazards, such as ice storms, wind, and lightning, are often related to the major weathermakers presented in this chapter. However, they generally are much smaller in scale and result in less damage and little geologic change, so we will not consider them here.

Each severe weather hazard has a specific origin, scale, damage pattern, area of common occurrence, monthly frequency, and a prevailing time of year when conditions encourage its development. For example, most tornadoes occur from March to July, most hurricanes occur from June to November, and most northeasters occur from October through April. Thus, severe weather hazards occur one place or another throughout the year. However, they have one thing in common—*they cause more damage overall than any other natural hazard in the United States* (Table 16–1).

**Wooden plank driven through a palm tree by the high winds of Hurricane Andrew at Homestead, Florida. (National Audio Visual Center.)**

**FIGURE 16–1**
Tornado extending down from a cloud and sweeping across the surface. (National Severe Storm Laboratory.)

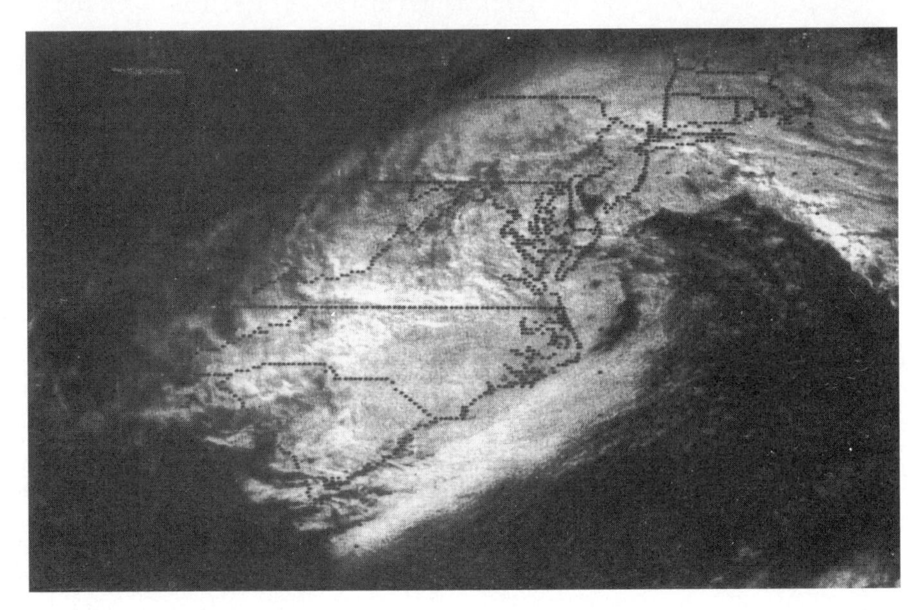

**FIGURE 16–2**
Satellite view of northeaster storm centered at mouth of Chesapeake Bay. The effects of the storm were felt from New England to Florida. (National Weather Service, courtesy of R. Dolan.)

**FIGURE 16-3**
Satellite view of hurricane centered on the coast of southern North Carolina. The strong effects of the storm were felt from Virginia south to northern Florida. (National Weather Service, courtesy of R. Dolan.)

**TABLE 16-1**
Largest insurance payouts for catastrophes in the United States, from September 1979 to July 1993.

| Date | Event | Insured loss (Less than total loss, which is much higher) |
|---|---|---|
| August 1992 | Hurricane Andrew | $16,500,000,000 |
| June-July 1993 | Flooding—Missouri-Mississippi River basin | 12,000,000,000 (estimated; mostly covered by federal flood and crop insurance) |
| September 1989 | Hurricane Hugo | 4,195,000,000 |
| March 1993 | Winter storms in 24 states | |
| October 1991 | Fire—Oakland, California | 1,175,000,000 |
| September 1992 | Hurricane Iniki | 1,600,000,000 |
| October 1989 | Earthquake—Loma Prieta, California | 960,000,000 |
| December 1983 | Winter storms in 41 states | 880,000,000 |
| April-May 1992 | Riots—Los Angeles | 775,000,000 |
| April 1992 | Wind, hail, tornadoes, floods—Texas and Oklahoma | 760,000,000 |
| September 1979 | Hurricane Frederic | 753,000,000 |

*Source:* Compiled by *The New York Times* from data supplied by Florida Department of Insurance (Hurricane Andrew) and Property Claim Services, a division of American Insurance Services Group (all others).

We will begin our tour of severe weather with tornadoes and northeasters and then examine the greatest storms on Earth—hurricanes.

## TORNADOES

Thunderstorms generate a number of weather hazards, including lightning strikes, damaging winds, hailstorms, heavy rain, flash floods, and tornadoes. **Tornadoes** are bodies of air that have a funnel shape, usually rotate counterclockwise (viewed from above), have very low interior pressure, and have very high velocities in their "walls." How a tornado is generated within a thunderstorm is shown in Figure 16–4.

The tornado generated within a severe thunderstorm extends down from the parent cloud, sweeping along the surface as the cloud travels above (Figure 16–4c). Most tornadoes move at an average speed of about 50 kilometers (30 miles) per hour, although speeds varying from stationary to over 110 kilometers (68 miles) per hour have been reported. They generally travel in a southwest-to-northeast direction, but direction can be erratic and may change abruptly.

Tornadoes are serious local hazards because of their extraordinarily low air pressure and very high winds. Pressure can drop up to 100 millibars (3 inches of mercury) inside a tornado's funnel. Their winds usually are less than 230 kilometers (140 miles) per hour but can attain up to 60% greater speed—370 kilometers (230 miles) per hour. Each year tornadoes kill about a hundred people and cost hundreds of million dollars of damage.

Compared to other weather hazards like hurricanes and northeasters, tornadoes damage the smallest area. Within those small areas, however, damage is usually total. A tornado's low pressures and fast rotary winds typically destroy whatever they touch. Tornado diameters range from a few tens of meters up to 1,500 meters (0.9 mile). Their average destructive track is about 7 kilometers (4 mile) long. Tornados are much less predictable than hurricanes and northeasters, striking with little warning, which is why they are so terrifying.

Tornado frequency, state by state, is quite variable (Figure 16–5). According to the National Weather Service, only Alaska does not experience tornadoes. The south-central United States/Gulf Coast region has more tornadoes than anywhere else. Northern Texas, Oklahoma, Kansas, and southern Nebraska have the greatest frequency.

The most dangerous tornado event is when a *tornado swarm* breaks out along a weather front. A "super swarm" occurred during April 3 through 4, 1974, as tornado after tornado was generated along a weather front that moved through Alabama,

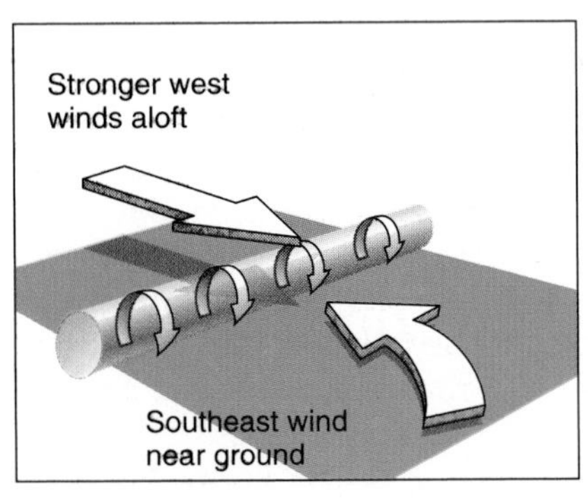


(a)

(b)

(c)

**FIGURE 16–4**
Formation of tornadoes from thunderstorms. (a) Prior to a thunderstorm developing, wind-direction shifts and wind speed increases with increasing height, which initiates horizontal spinning low in the atmosphere. This is not normally visible. (b) As a thunderstorm forms, its rising air "lifts" the rotating air, tilting it from horizontal to vertical. (c) The rotating air, 3 to 10 kilometers (2 to 6 miles) across, extends through the storm. Most strong tornadoes form within this zone. (After National Weather Service, 1992, Tornadoes . . . nature's most violent storms. A preparedness guide. p. 4.)

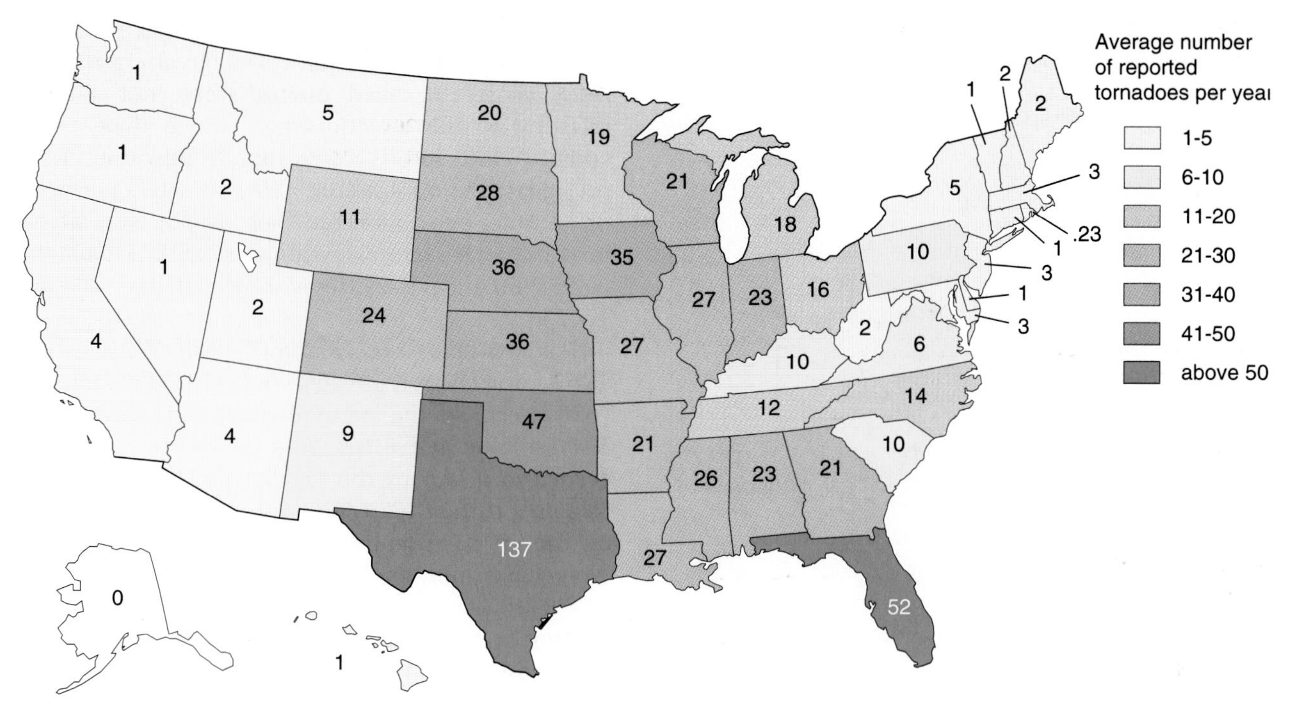


**FIGURE 16–5**
Map of tornado incidence by state. (After National Weather Service, 1992, Tornadoes . . . nature's most violent storms. A preparedness guide. p. 7.)

Georgia, Tennessee, Kentucky, Indiana, and Ohio (Figure 16–6). Most of the tornado tracks were 30 to 50 kilometers (20 to 30 miles) long, but varied from less than 6 up to 240 kilometers (4 to 150 miles). When the storms had passed, 148 separate tornado tracks were identified. During the 16 hours of this outbreak, tornadoes cut through 13 states, killing 307, injuring 6,000, and causing property damage estimated at $600 million.

Tornadoes cause damage in multiple ways. The rapid, rotating winds knock down weaker structures. Extremely low pressure inside a tornado creates such great differences in pressure between the inside and outside of structures that roofs are lifted and removed by the resulting high winds. The wind force and high pressure pick up smaller objects (cars, small structures, animals, people) and can transport them up to hundreds of meters. A Mississippi tornado in 1975 carried a home freezer over 2 kilometers (about a mile).

Tremendous amounts of debris are mobilized in a tornado, and this debris becomes shrapnel that destroys other structures. When debris breaks windows in otherwise strong structures, it permits extreme air pressure to develop on the opposite sides of walls, increasing the chances for structural failure. It is important to remember that in the high winds of a tornado, normally innocuous debris such as cardboard, asphalt roofing tiles, sticks, and fine gravel can become lethal objects.

The U.S. Weather Service issues a *tornado watch* when tornado-breeding thunderstorms are recognized and their arrival is expected within a few hours. A *tornado warning* is issued when tornadoes are spotted or when Doppler radar identifies distinctive "hook-shaped" areas within a local partition of a thunderstorm line that is likely to form a tornado (Figure 16–7).

Tornadoes are classified by the **Fujita Scale** according to their rotational wind velocity and damage (Table 16–2). Survival in a tornado involves sheltering oneself from the high winds and flying debris. This is best achieved by going to a basement or reinforced area in the interior of the lowest floor. If you are caught in the open or in a car, the best advice is to lie below the ground surface in a drainage ditch, or at least flat on the ground.

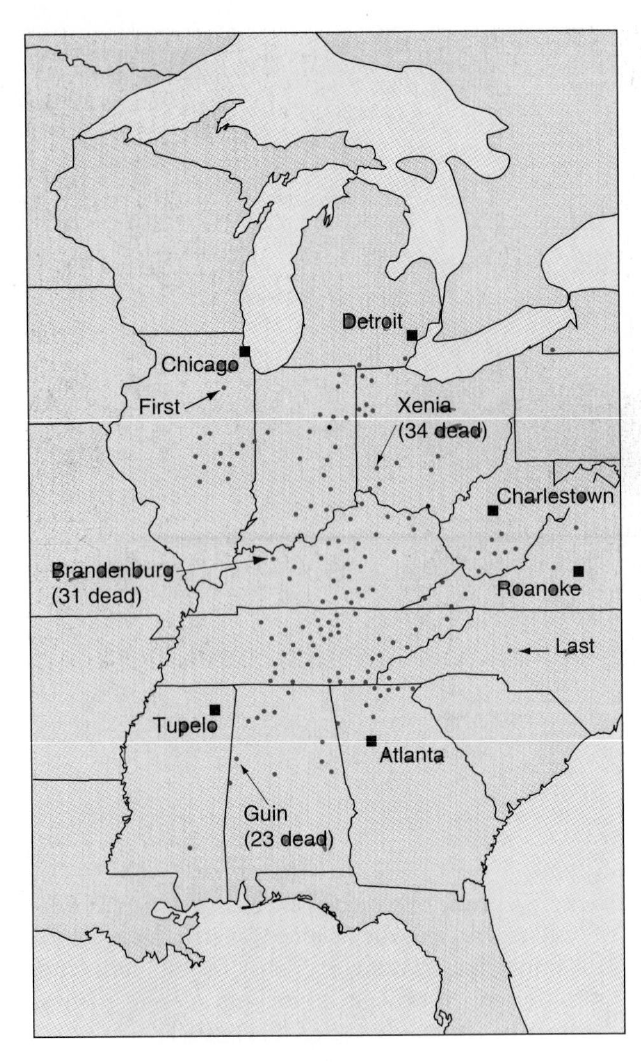


**FIGURE 16–6**
Map showing distribution of tornadoes (dots) during the tornado swarm of April 3 through 4, 1974. (National Weather Service.)

## NORTHEASTERS

When you think of truly large storms, you probably think of hurricanes. Hurricanes are *tropical* storms, spawned in tropical latitudes. But a hurricane is just one type of massive storm. There also are *extratropical* storms that originate in the mid-latitudes, move eastward, and eventually track up the Atlantic coast. Extratropical storms are massive low-pressure systems with cyclonic flow (counterclockwise winds).

As these storms track up the Atlantic coast, their counterclockwise winds flow from the northeast, off the ocean, and onto the land. Extratropical storms that cause significant damage along the Atlantic coast are called **northeasters** (nor'easters), referring to the direction from which their winds come. Wind speeds rarely attain hurricane level (designated as a minimum 119 kilometers per hour or 74 miles per hour). But their massive size affects broad coastal sections as long as 1,500 kilometers (over 930 miles), usually for several days (Figure 16–2).

For example, the March 7 (Ash Wednesday), 1962, northeaster affected over 1,000 kilometers (620 miles) of the Atlantic coast and caused over $300 million in damage. And this storm was far exceeded in power by the "Halloween" storm of October 29 through November 2, 1992, which affected the entire Atlantic coast, but hit hardest from New Jersey into New England. Experts R. Dolan and R. E. Davis, who have developed a rating scale for northeasters, rate the "Halloween" storm as the most powerful on record. Deep-water wave heights during that storm reached 11 meters (36 feet), and the storm's 114-hour duration caused great damage along the Atlantic coast.

The Davis-Dolan scale separates northeasters into five classes, based on their power and effect (Table 16–3). Over 50% of northeasters are weak systems (class 1), with wave heights of 2.5 meters (8.2 feet) and an average duration of 18 hours. Only 3% of all northeasters are class 4 and 5 events, capable of great damage:

- Severe northeasters (class 4) make up 2.4%. They have average wave heights of 5 meters (16.4 feet) and a duration of 63 hours, ensuring that these high waves pound the coast over several tidal cycles.
- Extreme northeasters (class 5) fortunately are very rare (0.6%) and powerful. Their deep-water wave height averages 7 meters (23 feet) and their duration averages 96 hours.

Northeasters originate in various locations, wherever unstable air produces significant temperature and pressure differences. Northeasters of continental origin move west to east across the United States, whereas those forming on the Atlantic coast move northward (Figure 16–8). What they all have in common is a final track that moves them along the Atlantic coast, as northeasterly winds cover the coastal area with rain or snow and massive waves

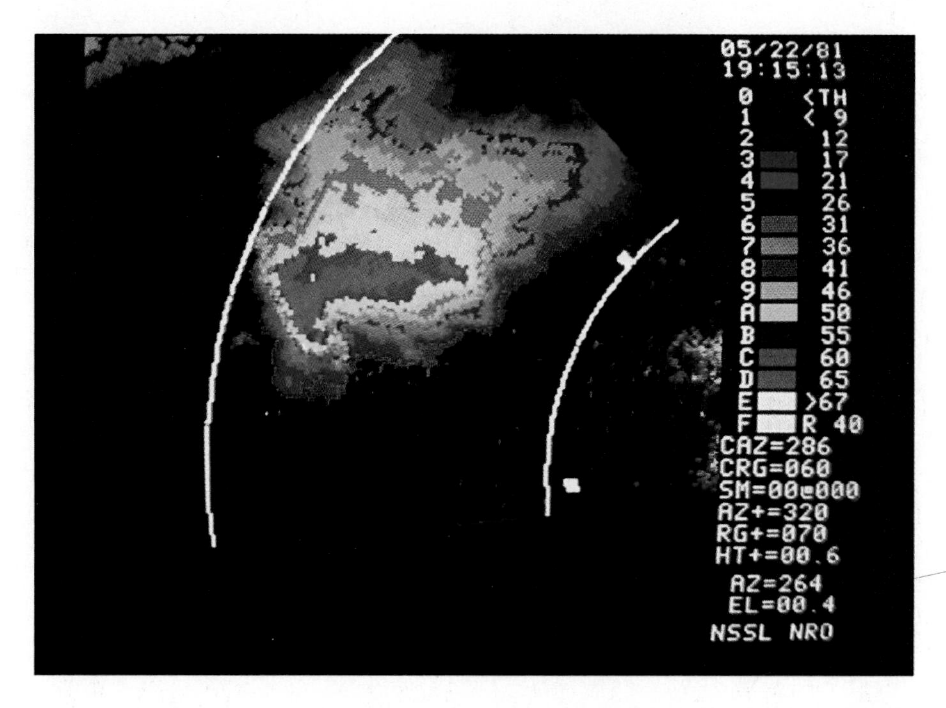


**FIGURE 16–7**
Doppler radar view of an intense thunderstorm showing "hook" shaped protuberance (bottom extension of red area in center of photo). Such patterns indicate the beginning of development of a tornado. (National Severe Storm Laboratory/NOAA.)

**TABLE 16–2**
Fujita Scale of wind intensity, based on wind velocity and resulting damage.

| Scale | Category | Speed km/hr | Expected Damage |
|---|---|---|---|
| F0 | Weak | 64-116 km/hr (40-72 miles/hr) | Branches broken off trees; shallow-rooted trees pushed over; some windows broken. |
| F1 | Moderate | 117-180 km/hr (73-112 miles/hr) | Trees snapped; surfaces peel off roofs; mobile homes pushed off foundations. |
| F2 | Strong | 181-253 km/hr (113-157 miles/hr) | Large trees snapped or uprooted; mobile homes destroyed; roofs torn off frame houses. |
| F3 | Severe | 254-332 km/hr (158-206 miles/hr) | Most trees uprooted; cars overturned; roofs and walls removed from well-constructed buildings. |
| F4 | Devastating | 333-418 km/hr (207-260 miles/hr) | Well constructed houses destroyed; structures blown off foundations; cars thrown; trees uprooted and carried some distance away. |
| F5 | Incredible | 419-512 km/hr (261-318 miles/hr) | Structures the size of autos moved over 90 meters (200 feet); strong frame houses lifted off foundations and disintegrate; auto-sized missiles carried short distances; trees debarked. |

*Source:* T. T. Fujita, Department of Geophysical Sciences, University of Chicago.

**TABLE 16–3**
Dolan/Davis northeaster scale.

| Storm Class | Beach Erosion | Dune Erosion | Overwash | Property Damage |
|---|---|---|---|---|
| Class 1 (Weak) | Minor Changes | None | No | No |
| Class 2 (Moderate) | Modest: mostly to lower beach | Minor | No | Modest |
| Class 3 (Significant) | Erosion: extends across beach | Can be significant | No | Loss of many structures at local scale |
| Class 4 (Severe) | Severe beach erosion and recession | Severe dune erosions or destruction | On low beaches | Loss of structures at community scale |
| Class 5 (Extreme) | Extreme beach erosion | Dunes destroyed over extensive areas | Massive in sheets and channels | Extensive at regional scale: millions of dollars |

*Source:* Davis & Dolan, 1993, American Scientist, v. 81, p. 428-39, Sept-Oct.

pound the coast, destroying structures and causing massive beach and dune erosion.

Analysis of northeaster frequency by Dolan and Davis reveals fewer of these storms during the 1980s. However, the frequency of major northeasters (classes 4 and 5) has increased in recent years. In the period from 1987 to 1993, at least one class 4 or 5 storm has occurred each year, a situation duplicated only once in the past 50 years. Of the eight storms in class 5 during the period studied, seven (88%) have occurred since 1960.

Will serious northeasters become more frequent in the near future? The evidence is unclear at present. However, if they do become more common, they will have serious consequences for our increasingly developed Atlantic shorelines (Figure 16–9).

## HURRICANES

Hurricanes are among nature's most powerful phenomena. They are the most areally destructive weather hazard along the Gulf and Atlantic coasts. In fact, the greatest loss of life from *any* geological hazard in U.S. history is still the 6,000-plus lives lost in the Galveston, Texas, hurricane of 1900. The U.S. Weather Service was then in its infancy, and warning systems developed since that time have greatly reduced the loss of life in hurricanes. However, structural damage from hurricanes has increased as more of the Gulf and Atlantic coasts are developed.

Coastal urbanization has increased markedly since 1970, a time during which the frequency of powerful hurricanes was far less than in the previous twenty years. Recent research suggests that the 10 to 20 year period from 1994 onward will have far more frequent major hurricanes that make landfall along the heavily built-up Gulf and Atlantic coasts. Although present high-technology monitoring systems provide us with adequate warnings, ***evacuation*** is the key problem in heavily developed areas that have limited outside access, such as the Florida Keys, the Outer Banks of North Carolina, or Long Island, New York.

The hurricane hazard is not only the storm itself, with its powerful winds and waves, but the ***response of our coasts*** to this hazard, with their high density of development and numerous coastal engineering structures. This phenomenon-response mechanism is what is making hurricanes that have the same intensities far more hazardous each year. In addition, the greenhouse effect (Chapter 11) may be warming the oceans sufficiently to provide future hurricanes with even more energy and thus increasing the ability to penetrate farther north and inland than ever before.

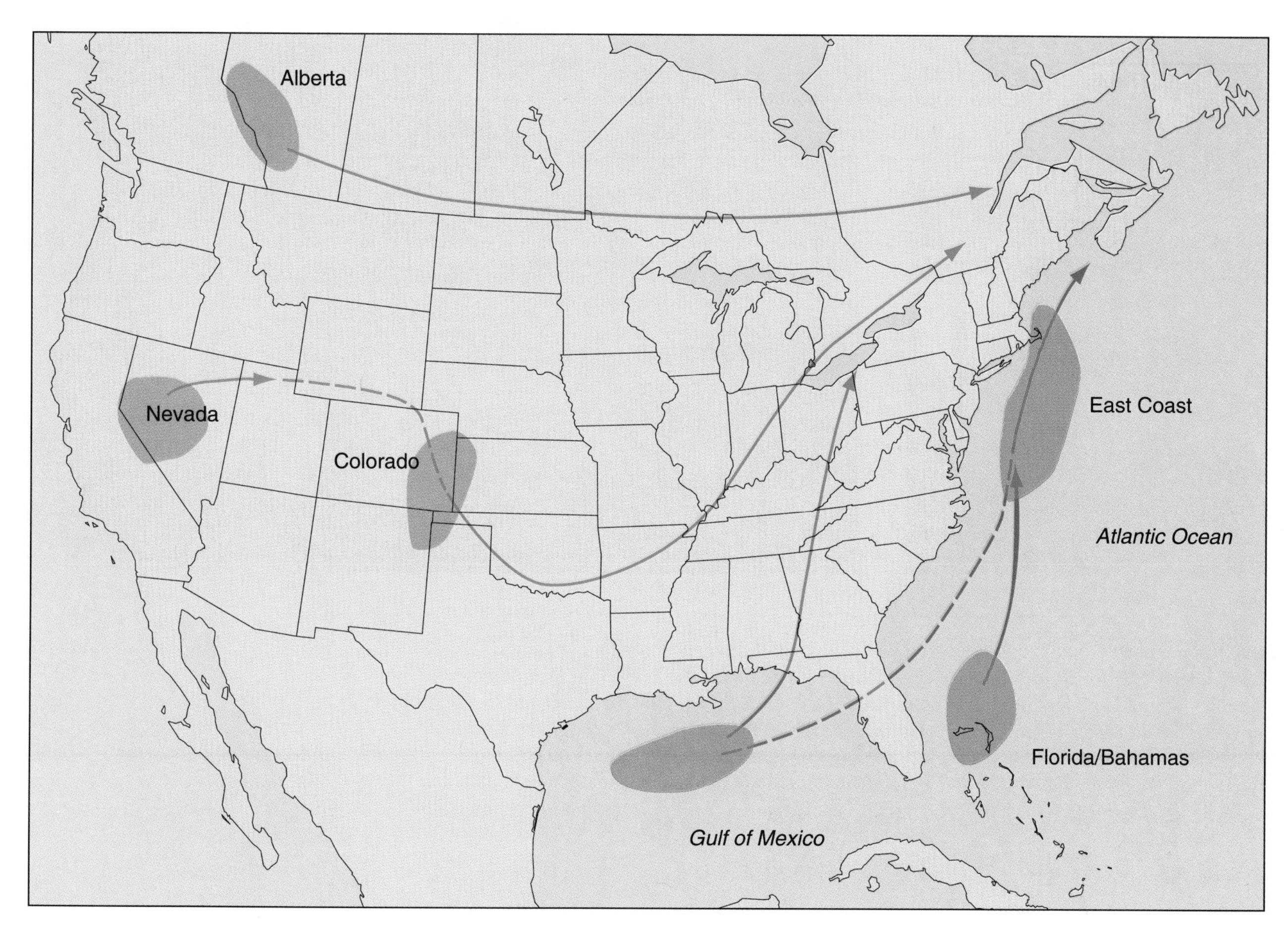


**FIGURE 16–8**
Map of areas where northeaster storms develop. (After Davis, R. E., and Dolan, R., 1993, Northeasters, American Scientist, v. 81, (September-October), p. 428-439, Fig. 16.5, p. 432. Used with permission.)

A **hurricane** is a massive tropical storm system with rotary winds that exceed 119 kilometers per hour (74 miles per hour, blowing counterclockwise around a relatively calm central area of very low pressure (Figure 16–3). Similar storms in the northwestern Pacific are called **typhoons,** or *cyclones* in the Indian Ocean. Hurricanes generally occur during "hurricane season" from June to November in the continental United States, with most concentrating between August and early October. The tracks of the major hurricanes described in this chapter are shown in Figure 16–10.

## Origin of Hurricanes

Many of the most dangerous hurricanes form in the eastern Atlantic off the coast of Africa and are called *Cape Verde storms,* named for the islands in that area. The specific origin and development of hurricanes is beyond the scope of our discussion here; the general discussion that follows is based on information from the National Oceanic and Atmospheric Administration (NOAA) and the National Hurricane Center (NHC).

Hurricanes begin with a disturbance in the westward-flowing air not far north of the Equator. This sets up a vertical air movement that draws heat-laden water vapor upward from the warm ocean below. As the vapor rises higher, it condenses, a process that releases great amounts of heat energy. In a single day, a hurricane can release heat energy equivalent to the fusion of several hundred hydrogen bombs! Put another way, one day's released energy converted to electricity could supply the elec-

(a)

(b)

**FIGURE 16–9**
December 1992 northeaster damage at Sea Bright, New Jersey. (a) Before the storm: a cabana complex with deck extending out in front of protective bulkhead (left). (b) After the storm: waves removed the deck, bulkhead, cabanas, and part of the parking lot. (Photos by author.)

**FIGURE 16–10**
Map showing tracks of hurricanes mentioned in the text.

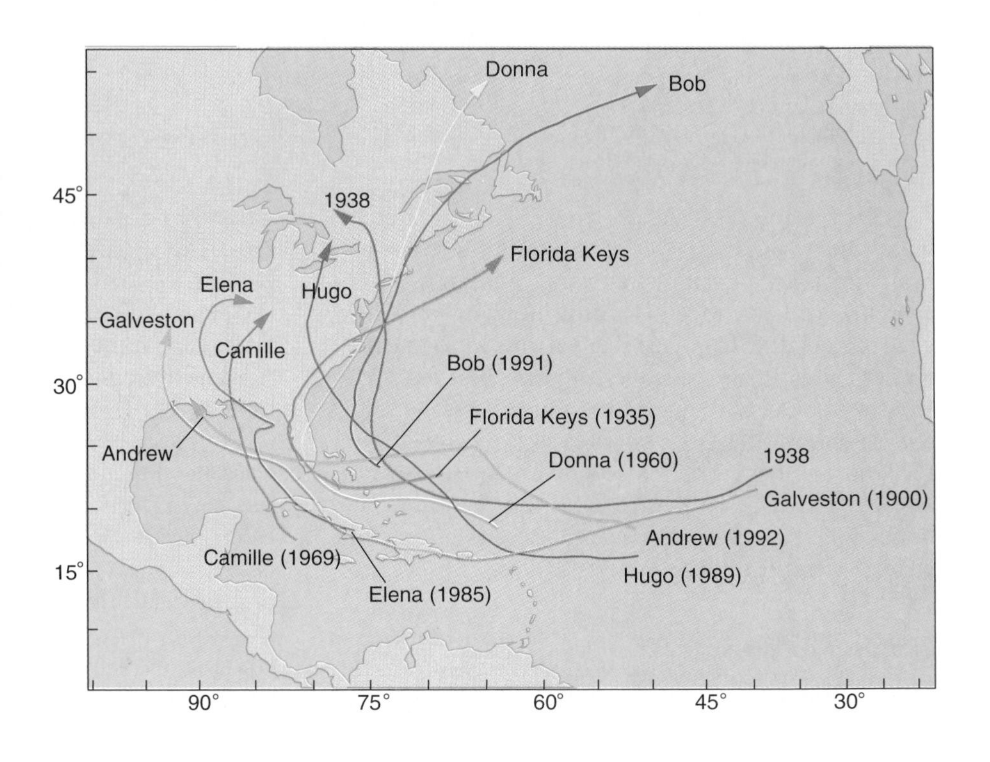

trical needs of the United States for six months. The availability of such enormous energy in a hurricane gives a clue to the vast destructive potential of a fully developed storm.

The initial disturbance in the westward-moving air is called a *tropical wave.* Many waves occur each year, and some develop into a *tropical disturbance,* defined as a moving mass of tropical thunderstorms that maintains its identity for 24 hours or more. Meteorologists at the National Hurricane Center in Coral Gables, Florida, constantly monitor the ocean surface, via ship reports and satellite images, to detect areas of developing low pressure combined with sustained thunderstorm activity. Some tropical disturbances develop further into *tropical storms,* with distinct rotary circulation and faster winds between 62 and 119 kilometers per hour (39–74 miles per hour). If a tropical storm develops a strong rotary circulation with winds exceeding 119 kilometers per hour (74 miles per hour), it is classed as a *hurricane.*

This heat energy drives the wind system into a cyclonic (counterclockwise) spiral. As the storm intensifies and becomes more organized, it develops the familiar form of a hurricane, with its spiralling arms of thunderstorms (Figure 16.3).

The storm is monitored intently for further development as it migrates westward. At this point, anything can happen. Because a hurricane derives its energy from warm ocean water, it loses power as it crosses over land but it may regenerate if it passes over the ocean once again. It may make significant changes in track as it encounters other air masses that block its passage, causing it to stall, or divert it to either side of its predicted track.

Since the early 1950s, names have been assigned to tropical storms for easy reference. Each year, an international committee develops an alphabetical list of alternating male and female names. Names can be reused in a subsequent hurricane season, except those of especially damaging hurricanes, which are "retired" to avoid confusion. Examples include Camille (1969), Hugo (1989), and Andrew (1992).

## Hurricane Structure

A mature hurricane has a series of *rainbands,* which are spiral bands of high wind and torrential rain that surround a relatively calm area of low pressure called the hurricane's **eye** (Figure 16–11). In the **eyewall** are the greatest winds in a hurricane. The eyewall is a zone of extremely turbulent air that can extend 16 to 40 kilometers (10 to 25 miles) outward from the edge of the eye.

A hurricane's winds travel very fast in a counterclockwise direction around the eye at a speed called the *hurricane-wind velocity.* At the same time, the entire storm is being pushed by regional winds at a speed called the *storm-center velocity.* The storm-center velocity is slowest in tropical regions (8 to 24 kilometers per hour or 5 to 15 miles per hour). It speeds up as the storm penetrates temperate regions, where it becomes influenced by other weather systems. In temperate latitudes, hurricanes easily can move 50 kilometers per hour (31 miles per hour), and have traveled 100 kilometers per hour (62 miles per hour) in the case of the 1938 New England Hurricane (Figure 16–10).

Thus, a hurricane has two velocities: the forward speed of the storm as it moves across the ocean or land, and the rotational speed of the winds circling the eye. The combined effect of these two velocities creates a significant difference in the effective wind speed on opposite sides of the storm.

For example, consider a storm moving north at 32 kilometers (20 miles) per hour with rotary winds around the eye of 160 kilometers (100 miles) per hour. On the right side of the storm (Figure 16–12), the velocities add approximately to the sum of the hurricane-wind velocity *plus* the storm-center velocity. However, on the left side of the storm, the rotary winds are moving counter to the direction of the storm center. Thus, the net velocity on this side is the hurricane-wind velocity *minus* the storm-center velocity, or 128 kilometers per hour (80 miles per hour).

This leads to an important generalization—*the winds are always stronger, and the destruction greater, on the right side of a moving hurricane.* If a hurricane stalls, this is not true, because the net wind velocity is the same on all sides of the storm.

## Hurricane Effects on the Ocean

A hurricane not only affects the atmosphere but the ocean as well. Hurricane winds can extend more than 120 kilometers (75 miles) out from the eye, and gale-force winds extend out even farther. These winds generate massive *sea swells,* smooth long-peri-

**FIGURE 16–11**
Cross section of a hurricane showing features described in the text. (Reprinted with the permission of Macmillan College Publishing company from Introductory Oceanography, 6th ed., by Harold V. Thurman. Copyright © 1991 by Macmillan College Publishing Company, Inc.)

od waves that move out in all directions from the storm center. Thus, depending on the storm center velocity, hurricane-generated waves can begin damaging coastal areas 6 to 12 hours before the storm makes landfall.

This highlights the need to evacuate low-lying coastal areas promptly, well before the actual storm reaches shore. The gale-driven advance waves may destroy causeways, embankments, and bridge supports, thus preventing evacuation. Strong advance winds may break or uproot trees and block evacuation routes as well.

The low pressure in the eye actually raises the sea surface, which contributes to the storm surge, which is the elevation of the ocean surface as a hurricane makes landfall.

## Hurricane Intensity and Frequency

The intensity of a hurricane is measured by the numbered categories in the **Saffir-Simpson Hurricane Scale**. Factors considered in this rating scale are wind velocity, height to which the sea rises, and the type of destruction expected (Table 16–4).

The frequency of hurricanes decreases with intensity. Category 1 storms are common each year, but category 5 storms occur far less often. In predicting the damage from a storm of a given category, it is important to realize that the degree of destruction will vary from place to place. Damage occurring at a specific site depends on which side of the storm the locality is, how far it is from the eye, and the degree of hurricane development. The

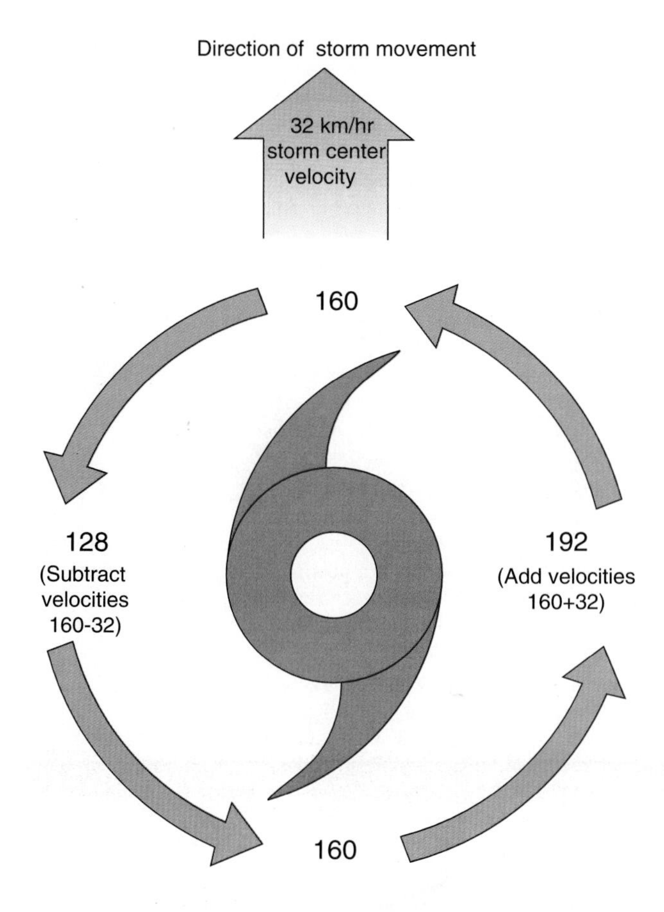


**FIGURE 16–12**
Top view of hurricane showing velocities in kilometers per hour (miles in parentheses) and directions of winds.

damage from a hurricane increases dramatically with each hurricane category, as shown in Table 16–5.

## Monitoring and Tracking Hurricanes

The great reduction in deaths from hurricanes is largely due to the ability of federal agencies to detect embryonic hurricanes and monitor their paths from inception to their dissipation. Ships at sea transmit weather reports that help meteorologists pinpoint centers of low pressure and heavy thunderstorm activity. Scientists at the National Hurricane Center then scan satellite images of Earth produced every 30 minutes to examine a disturbance and watch for its growth (Figure 16–13). If the storm matures and threatens land, observation planes are sent to examine it close up and land-based radars examine the structure of the storm as it approaches land.

Each of these methods provides valuable information. But none can be as detailed as that provided by "hurricane hunters" who fly directly through the storm. An Air Force unit flies from Biloxi, Mississippi, and NOAA flies reconnaissance missions from Miami (Figure 16–14). These planes are capable of staying in the air for 12 hours or more and making a wide variety of measurements, which are sent directly to the National Hurricane Center from the aircraft.

The critical measurements are of wind velocity, pressure, temperature, and the location of the hurricane's eye. Flight altitude is about 3,000 meters (10,000 feet) or lower. The plane drops devices called *dropsondes,* which transmit meteorological data as they fall to the ocean below. This provides a vertical picture of the storm.

The plane flies straight through the wall cloud and into the hurricane's eye. The ride through the wall cloud is bumpy, with heavy rainfall obscuring visibility. When the aircraft penetrates the eyewall into the relatively calm eye, the view is breathtaking (Figure 16–15). The ocean surface below is clearly visible; in the daytime, the sun shines brightly in a clear, blue sky; and at night, the moon and stars are clear.

The aircraft then leaves the eye, penetrating the eyewall at a different place, continuing to obtain data until it reaches the edge of the hurricane. This flight path provides a horizontal picture of the storm (Figure 16–16a). Multiple passes through the storm may be made during a flight. Flight data then can be synthesized by hurricane researchers into a three-dimensional picture of the storm. The radar units in the plane's belly (Figure 16–16a) and tail (Figure 16–16b) provide exceptional pictures of the internal structure of the storm.

## Changes in Hurricane Tracks and Intensities

Hurricanes may move in straight lines, curves, loops, or any combination (Figure 16–10). This is because a hurricane, although massive, is influenced by the large-scale air masses it encounters. A hurricane is like a spinning top on a windy day. The wind propels the spinning top in a certain direction and speed (the storm-center velocity). Any change in wind direction will steer the top, and any change in wind speed will speed or slow the top's forward movement.

**TABLE 16–4**
The Saffir-Simpson Scale of Hurricane Intensity.[1]

| Category | Wind Velocity, Storm Surge Height,[2] and Damage |
|---|---|
| 1 | **Winds 119-153 kilometers per hour (74-95 miles/hr), or storm surge 1.2-1.5 meters (4-5 Feet) above normal.** No real damage to building structures. Damage primarily to unanchored mobile homes, shrubbery, and trees. Also, some coastal road flooding and minor pier damage. |
| 2 | **Winds 154-177 kilometers per hour (96-110 miles/hr), or storm surge 1.8-2.4 meters (6-8 feet) above normal.** Some damage to roofing material, and door and window damage to buildings. Considerable damage to vegetation, mobile homes, and piers. Coastal and low-lying escape routes flood 2 to 4 hours before arrival of hurricane eye. Small craft in unprotected anchorages break moorings. |
| 3 | **Winds 178-209 kilometers per hour (111-130 miles/hr), or storm surge 2.7-3.6 meters (9-12 feet) above normal.** Some structural damage to small residences and utility buildings with a minor amount of curtainwall failures. Mobile homes are destroyed. Flooding near the coast destroys smaller structures, with larger structures damaged by floating debris. Terrain continuously lower than 1.5 meters (5 feet) above sea level may be flooded inland as far as 9.6 kilometers (6 miles). |
| 4 | **Winds 210-249 kilometers per hour (131-155 miles/hr), or storm surge 3.9-5.5 meters (13-18 feet) above normal.** More extensive curtainwall failures with erosion of beach areas. Major damage to lower floors of structures near the shore. Terrain continuously below 3 meters (10 feet) above sea level may be flooded, requiring massive evacuation of residential areas inland as far as 9.6 kilometers (6 miles). |
| 5 | **Winds greater than 249 kilometers (155 miles/hr), or storm surge greater than 5.5 meters (18 feet) above normal.** Complete roof failure on many residences and industrial buildings. Some complete building failures, with small utility buildings blown over or away. Major damage to lower floors of all structures located less than 4.5 meters (15 feet) above sea level and within 457 meters (500 yards) of the shoreline. Massive evacuation of low areas on low ground within 8-16 kilometers (510 miles) of the shoreline may be required. |

[1]NOAA table, 1990 version.
[2]Actual storm surge values vary considerably, depending on coastal configuration and other factors discussed in this chapter.

**TABLE 16–5**
Comparative destructive level of hurricanes, by Saffir-Simpson category.

| Category | Relative Hurricane Destruction Potential |
|---|---|
| 1 | 1 (reference level) |
| 2 | 4 times the damage of a Category 1 hurricane |
| 3 | 40 times the damage |
| 4 | 120 times the damage |
| 5 | 240 times the damage |

*Note:* Based on empirical analysis over the last 42 years by Dr. William M. Gray, Colorado State University meteorologist.

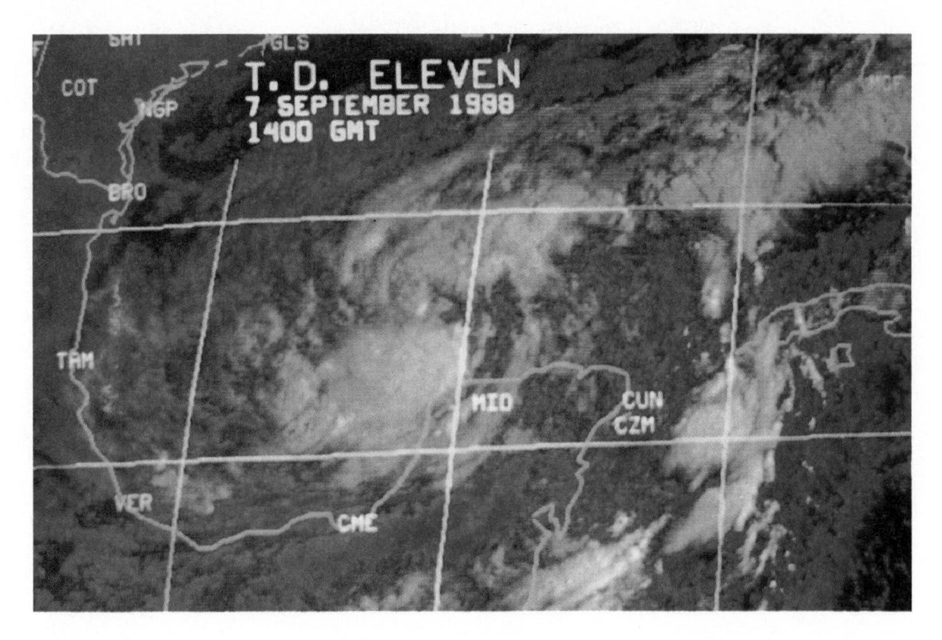


**FIGURE 16–13**
Satellite photograph of Tropical Depression 11 forming just north of the Yucatán peninsula on September 7, 1988. The depression subsequently evolved into Tropical Storm Florence and then into Hurricane Florence, which made landfall on the central Gulf Coast on September 9, 1988. (Courtesy of Robert Sheets, National Hurricane Center.)

The erratic nature of some hurricane tracks is well illustrated by the behavior of Hurricane Elena, a category 3 storm that caused chaos in the Gulf region in 1985. The westbound storm rounded Florida and then turned northward, toward an expected landfall in Mississippi (Figure 16–10). Coastal areas of several Gulf states were evacuated. But Elena slowed and was pushed eastward by local air masses toward the central Florida Gulf coast. It stalled for a while offshore, resulting in evacuation along coastal Florida. Heavy waves and flooding pounded the Florida shore and offshore islands. Elena then turned westward and northward, triggering evacuation of the coastal Gulf States once again. Finally, the hurricane made landfall near Biloxi, Mississippi, and caused $1.4 billion in damage (1989 dollars).

The erratic path of Hurricane Elena points up the necessity for coastal dwellers to be constantly aware of changes in a hurricane's track. The exact landfall site is hard to predict until just a few hours before the storm reaches the coast.

Because hurricanes gain their energy from the warm ocean below, they lose power rapidly as they pass onto land. However, hurricanes sometimes pass over islands or continental masses, lose energy, and then regain it as they pass back over the ocean again. Hurricane Donna (1960), a category 4 storm, had such a track (Figure 16–10). Donna caused great devastation as it moved westward across the Florida Keys. The storm turned abruptly northward, crossed Florida, and moved back out over the Atlantic near Jacksonville, Florida. It re-

**FIGURE 16–14**
Hurricane tracker plane. Radar in the bottom and tail of the planes provide a three-dimensional picture of the storm. Instruments dropped from the plane and measurements made form the plane provide the meteorological characteristics of the storm. (Courtesy Robert Sheets, National Hurricane Center.)

**FIGURE 16–15**
Inside the eye of Hurricane Allen (August, 1980) the air is clear and the winds are low. In the distance are the turbulent clouds marking the eyewall. (Courtesy of Robert Sheets, National Hurricane Center.)

newed its energy from the warm ocean waters as it travelled north to make a second landfall on Long Island, New York. Donna crossed Long Island and Long Island Sound to make yet another landfall in eastern Connecticut and then moved northward into New Hampshire (Figure 16–10). The damage from this hurricane was about $1.8 billion (1990 dollars).

In a more recent example (1992), Hurricane Andrew devastated southern Florida (about $20 billion damage), regained energy from the moist air over the Gulf, and then plowed into Louisiana ($1.5 billion damage) (see Box 16–1).

In the lower latitudes toward the equator, Atlantic hurricanes are pushed and steered by easterly (trade) winds, and usually travel slowly (around 8 to 32 kilometers per hour, or 5 to 20 miles per hour). As they move northward of the Carolinas, they typically accelerate to 50 kilometers per hour (30 miles per hour) or faster. This may seem puzzling, because northern waters are colder than southern waters, which should reduce the energy available to a hurricane. How can a northward-moving hurricane accelerate to such speeds?

The answer is that the hurricane begins to encounter other weather masses as it moves north of Cape Hatteras, North Carolina. A semipermanent area of high pressure exists near Bermuda, called the *Bermuda High*. Other high-pressure masses continually move eastward across the northern United States. Thus, a northward-moving hurricane may encounter a high-pressure area on either side, with a low-pressure trough in between. The hurricane is steered into the trough, and its speed accel-

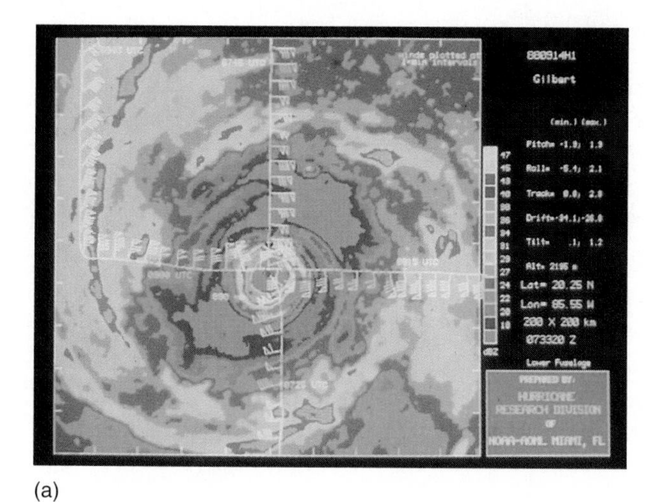


(a)

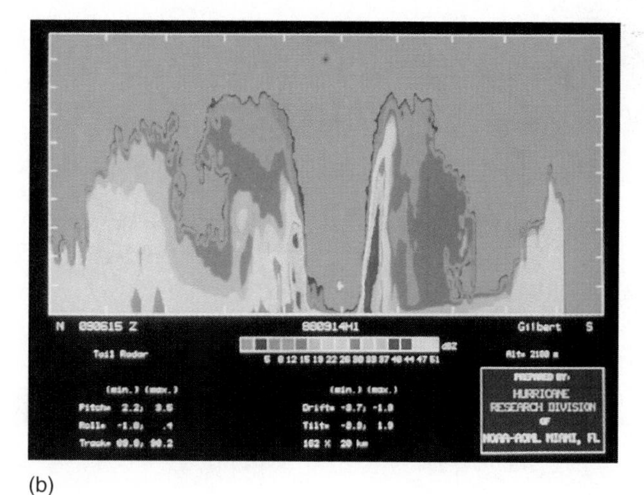


(b)

**FIGURE 16–16**
Radar cross sections of Hurricane Gilbert, September 1988. This hurricane was the largest (1,600 km or 1000 miles across) and lowest pressure (888 mp or 26.2 in of mercury) to hit the Western Hemisphere. It was a category 5 hurricane with peak winds exceeding 320 km/hr or 200 mi/hr. This powerful storm crossed the Yucatán Peninsula of Mexico, weakened, and made landfall on the Gulf Coast of Mexico. (a) Horizontal cross section (bottom radar). The spiral swirls outline the rainbands within the storm. (b) Vertical cross section (tail radar). The eye of the storm is in the low area in the center of the image. The eyewall is the high area bordering the eye. The red areas within the eyewall reflect areas of extreme convection. (Photos courtesy of Robert Sheets, National Hurricane Center.)

erates in the process. Furthermore, according to the National Hurricane Center, if a hurricane has a forward speed greater than about 55 kilometers per hour (34 miles per hour), it has sufficient *momentum* to reach the northeastern United States with little drop in strength. This happened in September 1938, leading to the most devastating hurricane to hit Long Island and New England in over 100 years (Figure 16–10).

## Angle of Hurricane Approach to Coast

As a hurricane tracks toward a coastline, the angle at which it approaches contributes significantly to the resulting damage. To understand this more fully, we refer back to Figure 16–12, which describes the difference in damage intensity on the right and left sides of a hurricane. Hurricane tracks fall into two basic categories: coast-parallel and coast-normal.

A **coast-parallel hurricane track** is offshore and parallel to the land. Hurricane Donna had a coast-parallel track in its migration from Florida to New York (Figure 16–10). As the hurricane moves northward, it keeps its weaker left side over land (Figure 16–17a). This produces less damage, and the effects of the storm at any point are brief. The stronger, faster right side is over the ocean and threatens only ships at sea.

As the storm approaches and passes a given area, it has quite different effects on the coastal waters. Note that, as the storm approaches an area, the strong onshore winds drive water onto the land (the *flood surge*), causing coastal flooding from the ocean (Figure 16–18a). As the hurricane leaves the area, the offshore winds drive impounded water from estuaries and bays over the coast from the land side, a phenomenon called *ebb surge* (Figure 16–18b). Most people know about the approaching storm surge, but many are unaware of the ebb surge.

A **coast-normal hurricane track** brings a hurricane onto the coast at nearly a right angle (Figure 16-17b), such as Hurricane Donna's path across the Florida Keys and that of Hurricane Hugo in South Carolina in 1989 (Figure 16–10). The passage of a powerful coast-normal hurricane over a coast guarantees severe damage on the right side of the storm, and lesser damage on the left.

For example, the 1938 New England Hurricane (Figure 16–10) made landfall in the middle of Long

**BOX 16–1**

## ANDREW, OMAR, AND INIKI—WORLD-CLASS WEATHER HAZARDS OF 1992

The year 1992 was memorable for severe storms in both the Atlantic and Pacific. Two destructive category 4 hurricanes and several typhoons struck the United States and one of its territories, Guam. These storms were of similar power but caused distinctly different patterns of destruction.

### HURRICANE ANDREW

Hurricane Andrew came first, reaching south Florida on August 24. The storm moved slowly northward until a high-pressure air mass to the north blocked its movement and forced it westward, over southern Florida (Figure 1). It made landfall at Homestead, south of Miami, with a storm surge 5 meters (16.4 feet) high and sustained winds of 233 kilometers per hour (145 miles per hour). Fortunately, much of the coast affected by the surge was lightly developed and protected by mangrove forest (Chapter 14), so surge damage was minimized.

However, Andrew's winds roared inland to cause record damage. Although Dade County (Miami) has one of the toughest building codes

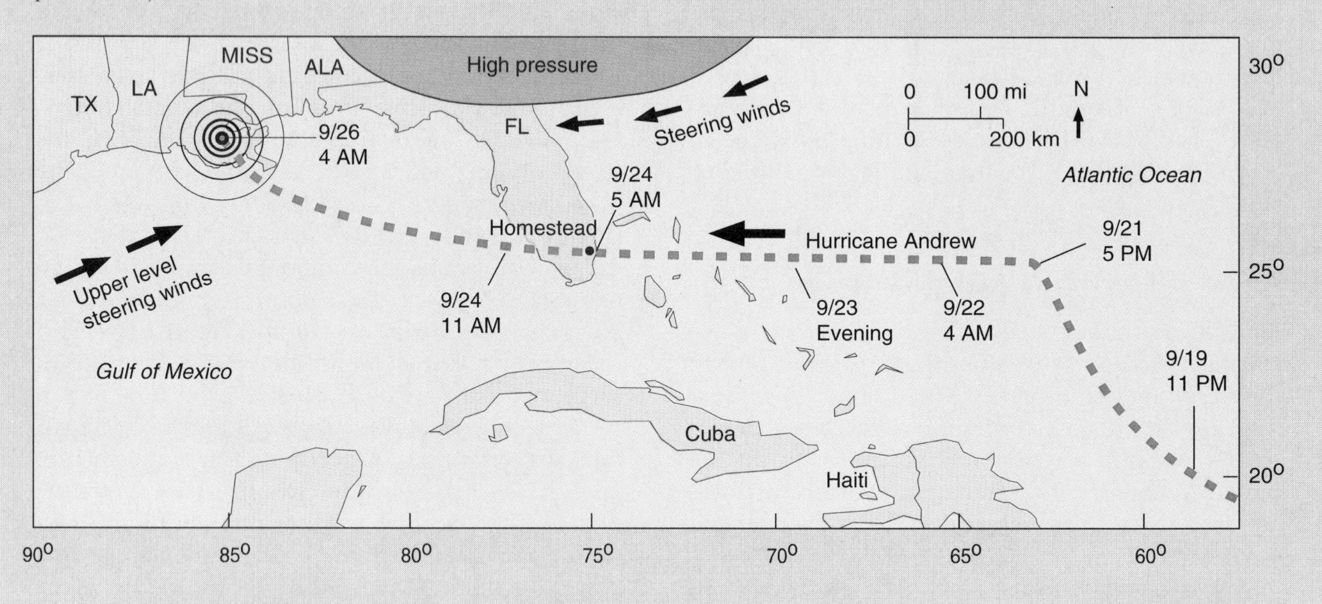


**FIGURE 1**
Track of Hurricane Andrew on August 19 through 26, 1992, across Florida and the Gulf of Mexico, into Louisiana. (Reprinted with permission from Coch, N. K., 1994, Hurricane hazards along the Atlantic coast of the northeastern U.S. *in* C. Finkl (ed.): Coastal hazards—Perception, susceptibility and mitigation: Journal of Coastal Research Special issue 12, p. 115–147.)

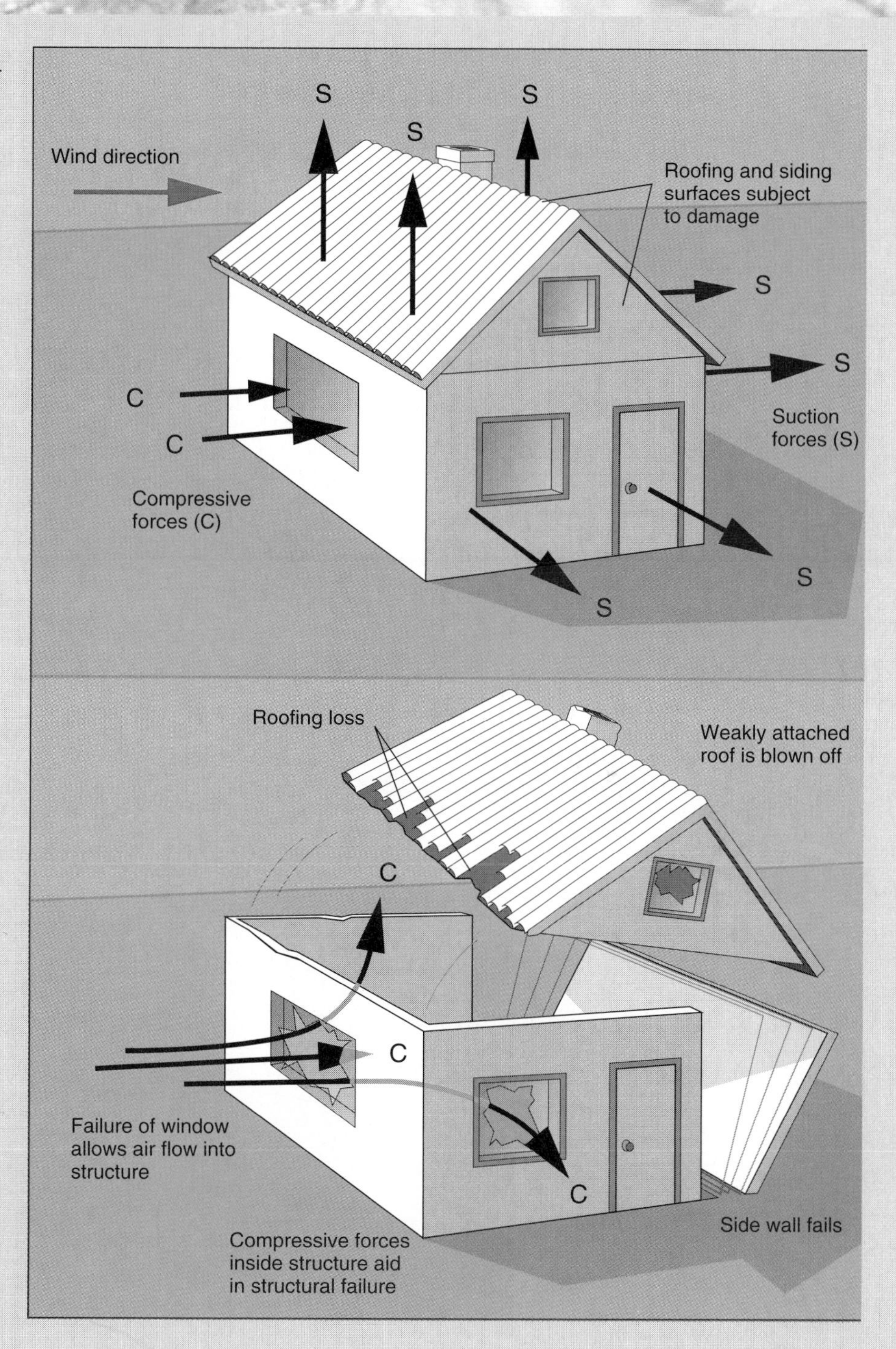


**FIGURE 2**
Mechanism of wind disintegration of structures. (a) Initial conditions, with wind causing low-pressure buildup at points indicated. As long as the roof is held on with hangers, the house may survive. (b) Debris carried by winds breaks windows, allowing the wind to enter the structure. If not properly attached, the roof may fly off and the walls may collapse. (From Coch, N. K., Geologic effects of hurricanes: Geomorphology, v. 10, no. 7-4, p. 37–63. Reprinted with permission of Elsevier Co.)

anywhere, many structures were destroyed because codes were not enforced or because windborne debris opened them, allowing pressure differences and wind force to disintegrate them (Figure 2).

The storm savaged two National Parks (Everglades and Biscayne), caused major agricultural and structural damage, and left thousands of people without homes, power, water, and phones for weeks (Figure 3). Andrew crossed the Florida Peninsula and traveled westward, gaining new energy over the warm waters of the Gulf of Mexico. Forecasters feared a possible landfall along the urbanized Texas Coast. As Andrew moved westward, however, upper-level steering winds diverted it northward toward Louisiana and a landfall along the sparsely populated western portion of the Mississippi Delta (Figure 1).

Andrew made headlines around the world for the spectacular damage it caused. However, the picture could have been far worse, had its track been slightly different. Had Andrew made landfall only 32 kilometers (20 miles) to the north, over the highly urbanized and heavily populated Miami–Fort Lauderdale area, the storm would have been the long-anticipated "Big One" that some day will cause a catastrophe in Florida.

**FIGURE 3**
Surge and wind damage from Hurricane Andrew along Biscayne Bay at Perinne, Florida. Note the flattened mangroves and trees, crushed houses, and the vessel on the lawn in the middle ground. Wind caused extensive damage to the inland homes in the background. (Photo by author.)

## TYPHOON OMAR

In the Atlantic and eastern Pacific, tropical storms are called hurricanes. In the western Pacific, they are called *typhoons*. Those that are steered northward eventually lose energy and dissipate over colder water. But prolonged movement over warm Pacific water, unimpeded by land masses, can generate super-typhoons. They cause great damage when they finally make landfall on Japan, the Philippines, small western Pacific Islands, or on mainland China.

During 1992, the U.S. island of Guam experienced a record five typhoons between August 28 and November 24. The worst was Typhoon Omar (August 28), whose eye moved directly across Guam (Figure 4). Winds ranged from 178 to 217 kilometers per hour (111 to 135 miles per hour), gusting to 241 kilometers per hour (150 miles per hour). The storm surge attained 3 meters (10 feet). Torrential rains of 30 to 48 centimeters (12 to 19 inches) lasted three days. Most typhoons reach Guam during the night, but Omar savaged the island during daylight, providing inhabitants with a spectacular display of flying debris.

Omar caused about $457 million damage on Guam, destroying 2,158 homes and leaving 3,000 people homeless (about 3% of the population). Subsequently, Omar passed over Taiwan, killing two, and causing massive flooding and power disruption. Omar finally dissipated over southeastern China, its torrential rains causing local flooding as far west as Hong Kong.

**FIGURE 4**
Satellite false-color image of Typhoon Omar on August 28, 1992. (Photo courtesy of Joint Typhoon Warning Center, Guam)

## HURRICANE INIKI

Hurricane Iniki, a category 4 hurricane, started from a disturbance in the Caribbean that crossed Central America and intensified into a hurricane in the Pacific. The storm's westward track was safely south of the Hawaiian Islands. Iniki gained energy from the warm waters of the Central Pacific and evolved into a strong hurricane.

Then, unexpectedly, Iniki began to veer northward toward Hawaii. Recalculating its track, forecasters thought the storm would pass between the islands of Kauai and Oahu (Figure 5). This would take the right eyewall of the storm over densely populated Oahu, the island with the state's capital of Honolulu and major air and naval bases ("projected path" in Figure 6).

**FIGURE 5**
Track of Hurricane Iniki through the Hawaiian Islands on September 11, 1992. Times shown are Hawaii Daylight Time.

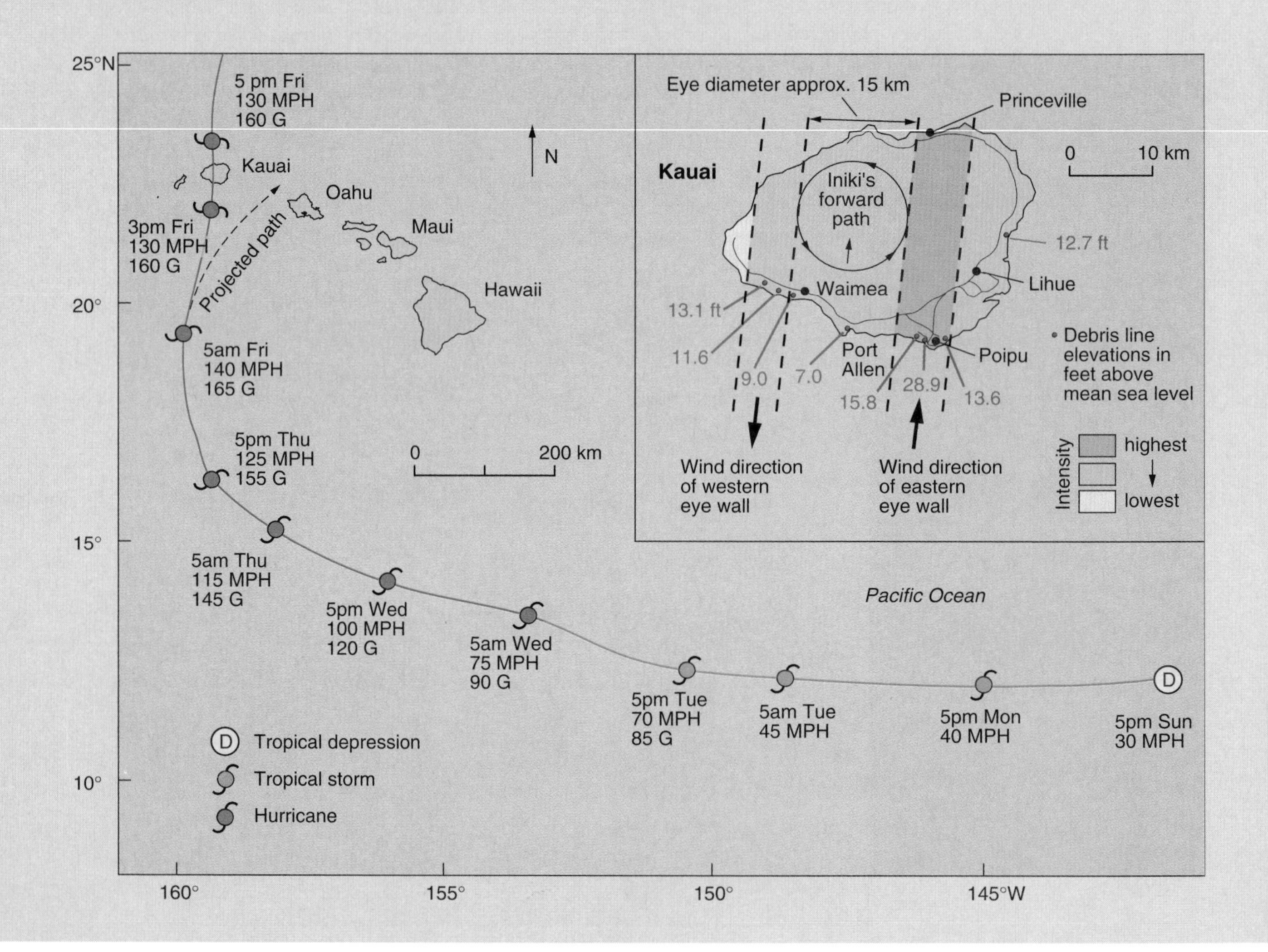

Instead, Iniki passed directly over the less populous island of Kauai, damaging the island's heavily developed southern, eastern, and northern coasts, with its large resorts.

Although Iniki was a category 4 storm, its surge rose only about 2 meters (6.5 feet) because the offshore slope is so steep. In general, the steeper the offshore slope, the lower the surge. However, the steeper the offshore slope, the higher is the wave height. Consequently, Iniki's waves on Kauai reached 10 meters (33 feet) in some places. These gigantic waves pounded seaside structures, reducing them to rubble (Figure 6). Debris was carried inland to form massive debris dams.

Iniki's winds peaked at 177 kilometers per hour (110 miles per hour), not an extreme value. But the mountainous topography of the island increased the wind hazard to the north side of the island. As Iniki's winds roared ashore on the south side, they were funneled through passes in the central mountains, accelerating the wind speed because the flow was restricted. The winds roared out of the mountains to cause unexpectedly severe damage on the north side of the island (Figure 7).

Iniki offers a few lessons. First, on an island with a steep offshore slope, wave damage can exceed surge damage. Second, communities on the leeward side of a mountainous island may experience wind damage as heavy as the windward side. Damage from Iniki crippled Kauai's tourist industry for over a year. The good side is that residents of Kauai learned a valuable lesson, for rebuilding observed by the author is much more resistant to waves, surge, and wind (Figure 8).

**FIGURE 6**
Structural damage from waves in Hurricane Iniki (1992) on the Poipu coast of the island of Kauai, Hawaii. (Photo by author.)

**FIGURE 7**
Wind damage to structures at Princeville, Kauai, in Hurricane Iniki (1992). (Photo by author.)

**FIGURE 8**
Hurricane resistant construction on rebuilt home in the Poipu area of Kauai, Hawaii after Hurricane Iniki (1992). The home was raised up on cross-braced piers anchored into concrete to minimize future surge and wave damage. Metal hangers and straps were used to tie all the levels of the structure together for wind resistance. (Photo by author.)

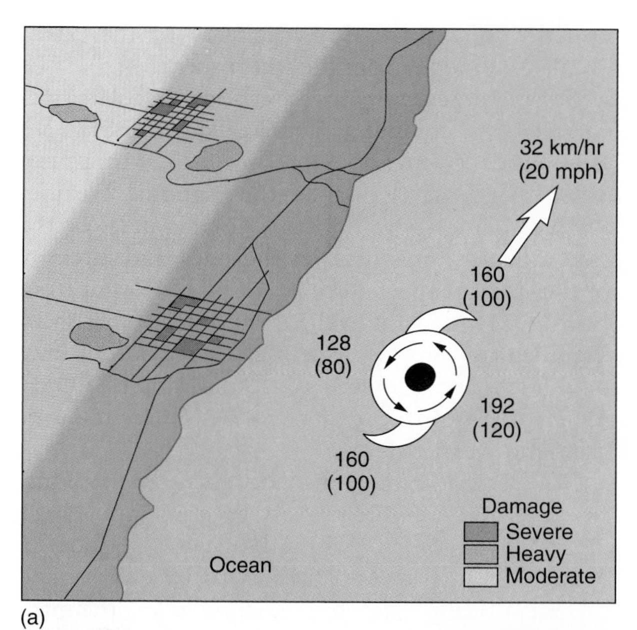


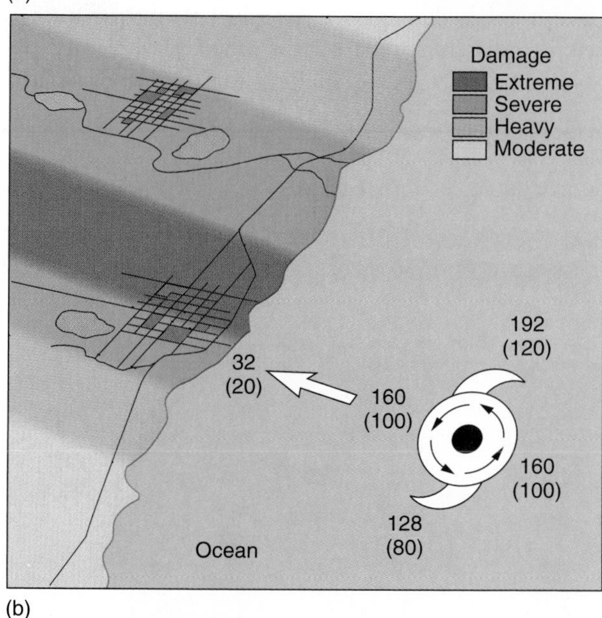


**FIGURE 16–17**
Diagrams showing tracks and wind velocities relative to the coast in (a) coast-parallel and (b) coast-normal hurricanes. Note that the damage diminishes rapidly inland from the coast in a coast-parallel track, whereas belts of extreme, heavy, and severe damage extend far inland in a coast-normal track. Velocities are those perceived by a stationary observer on the ground and are shown in kilometers per hour, with miles per hour in parentheses.

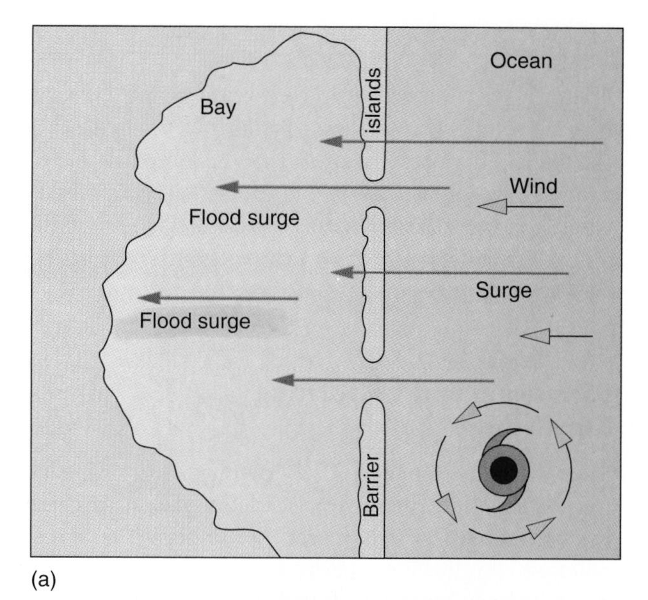


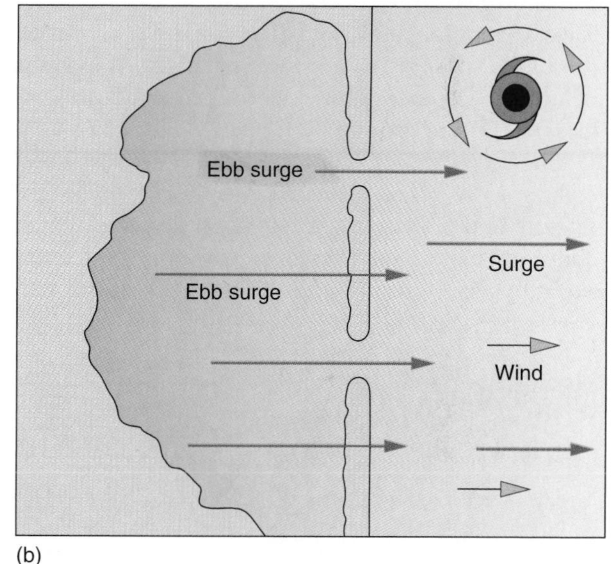


**FIGURE 16–18**
Diagrams showing the differences in coastal water movements and winds as a coast-parallel hurricane (a) enters an area and (b) leaves it.

Island. It devastated eastern Long Island, while the damage in western Long Island and New York City was restricted to some coastal flooding and tree damage. Hurricane Donna (category 4) is interesting because it caused coast-normal damage in Florida and New England, but coast-parallel track damage along much of the Atlantic seaboard in between those points.

## HURRICANE STORM SURGE

Historically, nine out of ten people who are killed in hurricanes are the victims of injury or drowning in the elevated coastal waters that accompany hurricanes. The elevated ocean surface accompanying the passage or landfall of a hurricane is called the storm surge. Storm surge causes most of the structural damage at the shoreline.

### Measuring and Calculating Storm Surge Levels

Storm surge is measured following a hurricane by noting the elevation above sea level of features that indicate "high water mark," or where the water reached. High water marks are easy to see, for the rising waters carry vegetation and sediment, which leave vegetation and stain lines on building structures. As the water recedes from its maximum height, it often strands debris on structures, in trees, or on the ground.

The surge waters carry pieces of debris. These collide with other structures to break them and scratch and abrade their surfaces, leaving telltale abrasion marks. Such features are found and measured by survey teams, and the synthesis of this data provides an accurate areal map showing the extent of inland flooding from the storm surge.

Storm surge levels now are being calculated accurately by a computer model called **SLOSH** (*S*ea *L*ake and *O*verland *S*urges from *H*urricanes). SLOSH was developed by the National Weather Service. It incorporates oceanographic data for the area modeled, plus data on the track and intensity of the storms expected to hit the area. Output from the SLOSH model is in the form of maps that show the calculated storm surge levels with contour lines or color patterns (Figure 16–19). Multiple SLOSH model maps can be generated for different hurricanes hitting that area.

SLOSH calculations made by the National Hurricane Center of surge levels in Hurricane Hugo (1989) were accurate within 0.3 meter (1 foot) of measured surge levels. These maps have proven to be exceptionally useful in predicting where the maximum surge damage will occur. This information is vital to disaster-preparedness officials, who must decide where to locate shelters for people evacuated from the shoreline.

### Causes of Storm Surge

How high the water surface becomes elevated in a hurricane storm surge is controlled by several factors:

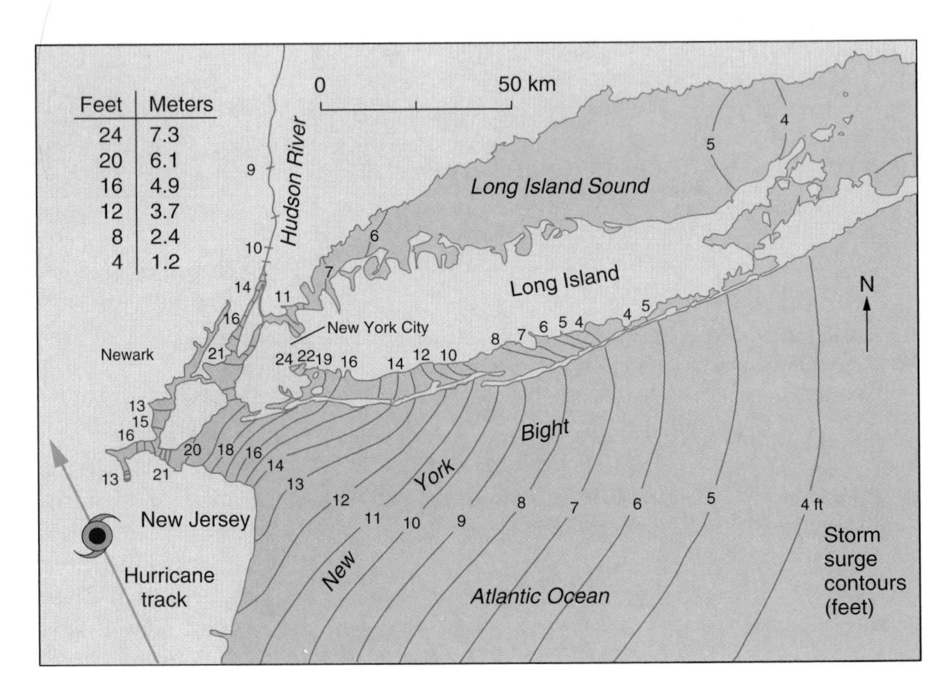


**FIGURE 16–19**
SLOSH map for the New York-New Jersey juncture (New York Bight). The contours (in feet) show expected surge levels for a category 3 hurricane moving northwest at 64 kilometers per hour (40 miles per hour) over northern New Jersey. (SLOSH contour data courtesy of Brian Jarvinen, National Hurricane Center.)

- The greater the storm's intensity (category), the higher the surge, because the greater wind in a stronger hurricane drives more water across shallower depths on the continental shelf, literally "piling up" the water. The gentler the slope, the shallower the water, and the greater the surge height.
- Higher-category storms have lower central pressure, which allows the ocean surface to rise, adding to surge height.
- Tidal stage is important because high tide raises the ocean surface so the storm surge can penetrate farther inland.
- Surge levels increase along concave coasts and in narrowing bays as hurricane winds force water into them. This is shown quite strikingly in Figure 16–19.

Surge levels rise steadily as a hurricane approaches and begins to pass over the coast. In general, the surge rises faster as the eye passes over the coast. Waves atop the surge can greatly increase the damage.

## Two Surge Types—Flood and Ebb

Most people think of storm surge as coming from the ocean onto the land, but, as explained, an *ebb surge* happens as well. Heavy rain may fall on the land for days before a hurricane hits a coast (Figure 16–20b). Consequently streams, lakes, and estuaries may be swollen before the hurricane arrives. These rising waters are prevented from flowing seaward by advancing hurricane winds and the rising storm surge (Figure 16–20c). The rising surge can flood causeways and bridges hours before a hurricane hits. This is one reason why early evacuation is a good idea.

As the hurricane eye passes over the coast, the **flood surge** moves across the shore, destroying structures and carrying sediment and debris landward (Figure 16–20d). After the hurricane passes inland, there is little resistance to the oceanward flow of the impounded flood surge waters and earlier rainfall. This water now flows rapidly seaward as an **ebb surge** (Figure 16–20e).

Ebb surge may overwash barrier islands from the bay side and cause additional erosion and structural failure. Roads and driveways perpendicular to the coast help to channel the flow, and this results in the erosion of **ebb-surge channels** across the shore (Figure 16–21).

## Storm Surge Damage

It is hard to imagine the power of moving water unless you have been hit by a wave in the surf zone. Consequently, many structures that are seriously weakened as the flood surge moves inland are "finished off" later as the ebb surge flows seaward (Figure 16–22).

The great structural damage from the flood surge is caused by three separate components (Figure 16–23):

1. The hammering force of the waves themselves.
2. The upward pressure (hydraulic lift) exerted by water and waves moving under elevated structures.
3. The wave energy reflected from any engineering structures such as groins and jetties (Chapter 15). This reflected wave energy can be concentrated onto the nearest structures and cause their destruction.

Also, and far from least, is the battering that structures receive from debris carried by the water. Fences, wooden walkways, and gazebos can become battering rams that knock down structures in a hurricane. Homes that are poorly anchored to their foundations or built on concrete slabs can be washed inland to collide with other structures (Figure 16–24).

Flood surge results in serious erosion to barrier islands. The surge erodes the beach and dunes, cuts surge channels across the coast, and washes the sand landward in overwash fans (Figure 16–25). The beach lowering and formation of surge channels through the dunes makes erosion by subsequent storms much easier, unless the beaches and dunes are replenished (Chapter 15).

In some cases, storm surge can even cut through a barrier island to form a tidal inlet (Figure 16–26). Some tidal inlets are cut during the flood surge, whereas others are formed when ebb surge waters are channeled back across a barrier island by natural or structural features. In general, more inlet breaching occurs when the hurricane is a high-category event, when wide bays exist behind the barrier islands, or when heavy rain has occurred before hurricane landfall.

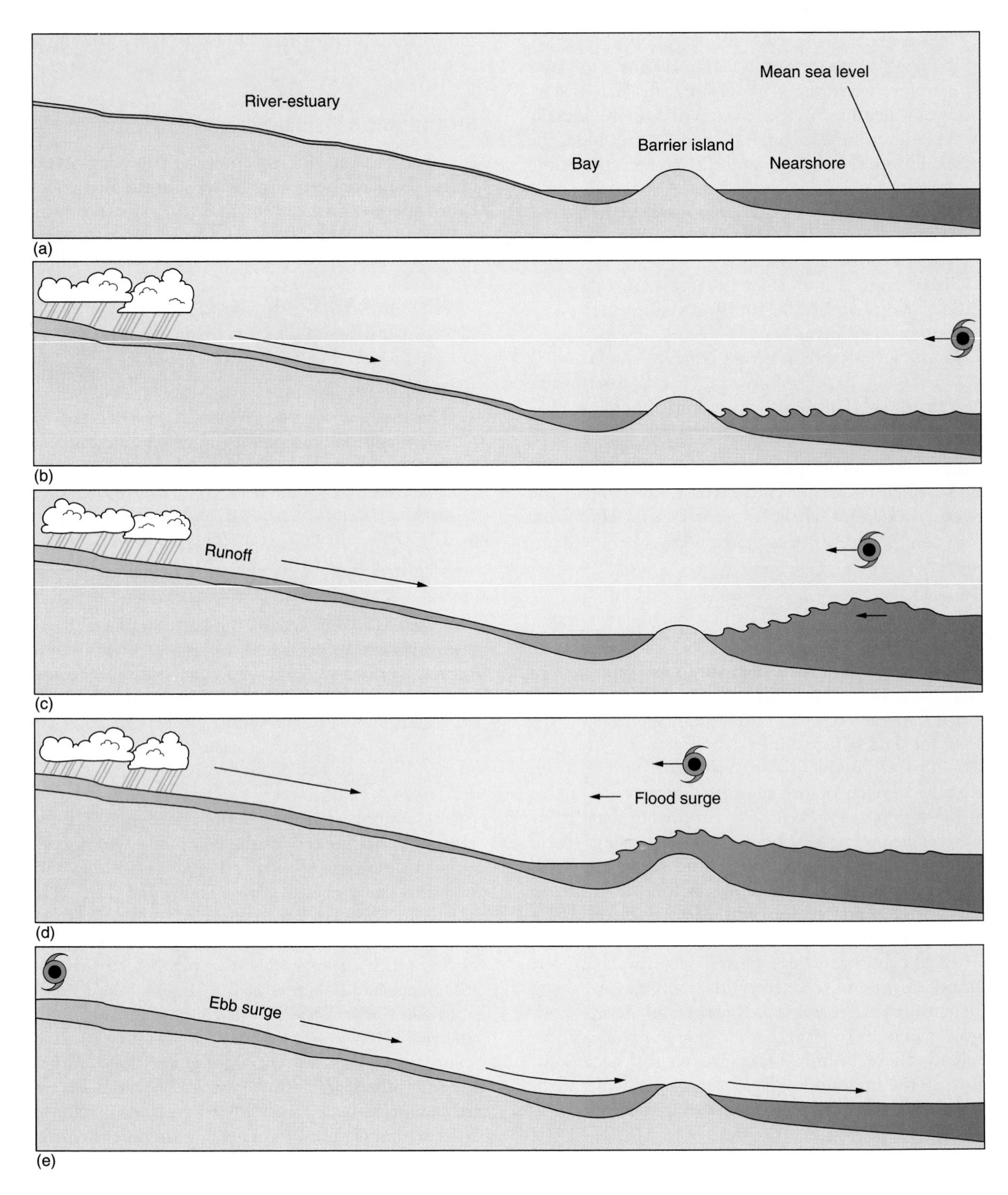


**FIGURE 16–20**
Cross sections showing development of flood surge and ebb surge as a hurricane crosses a coast and moves inland.

**FIGURE 16–21**
Ebb surge channel from Hurricane Hugo (1989) on the South Carolina Coast. (Photo by author.)

Most of these storm-cut inlets close within a few weeks as longshore drift and ridge and runnel formation (Chapter 15) builds bars across their mouths. In some cases, these inlets are stabilized by jetties to allow navigation between the ocean and the bays. However, stabilization of inlets causes another set of environmental problems, as described in Chapter 15.

## HURRICANE WIND DAMAGE

Hurricane winds can cause extensive damage far inland. They may paralyze a region as lines of communication are severed and transportation routes are blocked by flying debris and fallen trees. Recall that winds are stronger and more turbulence occurs in the right eyewall of the hurricane. This can gen-

**FIGURE 16–22**
Bulkhead collapse of Hurricane Hugo (1989) ebb surge at Garden City, South Carolina. Note that the bulkhead collapsed out toward the ocean. Several ebb surge channels are visible in front of the collapsed bulkhead. (Photo by author.)

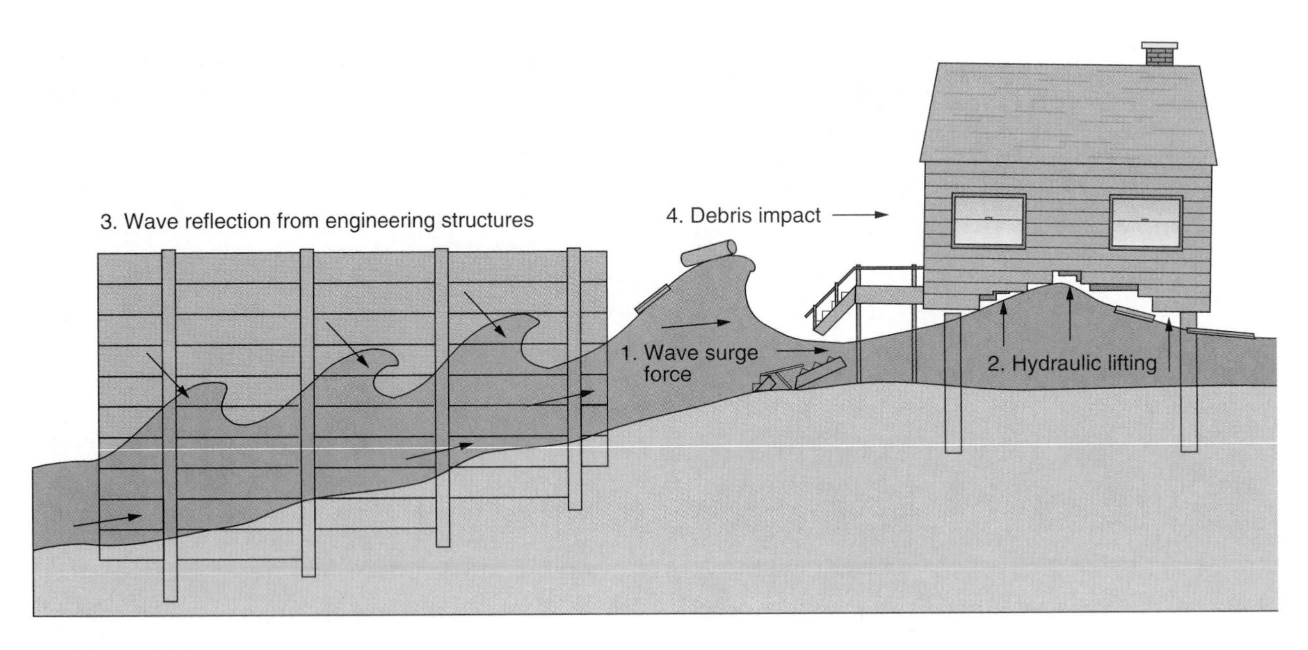


**FIGURE 16–23**
Diagram showing modes of surge damage.

**FIGURE 16–24**
Homes in Garden City, South Carolina washed inland by flood surge of Hurricane Hugo (1989). The empty spaces on the right were sites of trailers that were destroyed by the hurricane's winds. (Photo by author.)

**FIGURE 16–25**
Flood surge damage from Hurricane Hugo (1989) at Cape Romain, South Carolina. Storm surge and waves broke numerous palm trees, eroded sand from the beach and deposited the sand in massive washover fans across the marsh area. (Photo by author.)

erate small, short-lived, but intense tornadoes, affecting the ground beneath the outer rain bands.

Wind velocity increases with height, subjecting high-rise buildings to greater damage. A special problem in urbanized coastal areas where the wind hits a line of tall buildings is that the wind speed increases as it is forced between the buildings. This causes considerable damage to any structures behind the openings in what is called the *wind-tunnel effect.* Many buildings have strong internal structures but weak surface coverings that are easily removed by high winds. As openings develop in the outer structure, it becomes easier for rain to enter during and after the storm. Experience from hurricanes Hugo (1989) and Andrew (1992) shows that subsequent water damage may equal or exceed the direct damage of a hurricane in many cases.

Flying debris in hurricanes can destroy structures and kill and injure people. The Florida Keys Labor Day 1935 Hurricane was one of two category 5 hurricanes to strike the United States (the other was Camille in 1969). Survivors of that storm reported that flying sand stripped clothing and flesh from people exposed to the winds, which exceeded 240 kilometers per hour (150 miles per hour).

In the 1900 Galveston Hurricane (Figure 16–10), a category 4 storm, many people were seriously cut and killed by flying roof tiles. This was an ironic twist of good intentions. A disastrous fire had swept the city earlier, feeding on the many wood-shingled roofs, so laws were passed requiring that new roofs be covered by slate tiles. The new roof tiles became deadly missiles in the 1900 hurricane.

Hurricanes also cause large-scale environmental problems because they blow salt spray far inland and force saltwater into normally brackish or freshwater portions of estuaries. This increase in salinity affects aquatic organisms, and the salt spray can kill, or seriously damage, fresh water vegetation, fish eggs, and crops.

One of the worst effects of high wind is tree destruction. The damage depends on a tree's location relative to the eye, the type of tree, and the saturation of the ground. A tree may topple if its roots are shallow or if the ground has been softened by days of rain. A tree's strength decreases upward as the trunk thins, and wind velocity increases upward, so at some point the wind is sufficient to snap the trunk.

Hurricane Hugo (1989), a category 4 coast-normal storm (Figure 16–10), devastated South Carolina's forests (Figure 16–27). Broken trees, numbering in the hundreds of thousands, were more than a loss to the forestry industry. Broken trees cut power and phone lines and blocked highways, hampering relief operations and service restoration. Uprooted trees broke underground utility lines. High winds and tree damage also impacted wildlife, which are killed by flying debris and falling trees. The loss of trees also means a loss of habitat for many bird species.

## HURRICANE DISRUPTION OF VITAL SERVICES

Hurricanes, especially ones with coast-normal tracks (Figure 16–17), cause r*egional* disruption of vital

**FIGURE 16–26**
Inlet cut by Hurricane Hugo (1989) ebb surge at Pawleys Island, South Carolina. (Photo by author.)

**FIGURE 16–27**
Trees snapped by winds of Hurricane Hugo (1989) in the Bulls Bay area of coastal South Carolina. (Photo by author.)

services due to their high winds and large areal extent. Hurricane winds may extend 120 kilometers (75 miles) outward from the eye, but gale-force winds extend far beyond that.

Coastal and estuarine facilities such as sewage treatment plants and power plants are easily knocked out by storm surge. Loss of electricity prevents gas stations from pumping gasoline, which is essential for vehicles, portable generators, and chain saws. In addition, electric loss prevents water pumps from working, causing shortages of potable water. Survivors of Hurricane Hugo told the author that the most prized items after the storm were gasoline, chain saws, bottled water, plastic sheeting, food, and hand tools. Thus, these are items that should be stocked in hurricane shelters.

## PREDICTING HURRICANE FREQUENCY AND INTENSITY

We have come a long way in our ability to provide reliable hurricane warnings. Once, the only sign of a storm might have been a radio message from a ship at sea or from boats returning to the safety of port in gale-driven seas. Today, satellites spot the birth of a hurricane (Figure 16–13) and follow it until it blows out. We still cannot predict, however, how many hurricanes a particular season will yield, nor where they are likely to strike. Recent research now provides important new information: the African connection.

### Hurricane Frequency, Intensity, and the African Connection

Research by meteorologist William Gray of Colorado State University has provided new insight into hurricane frequency. Dr. Gray has reported a striking correlation between the *frequency of strong Atlantic and Gulf hurricanes* and *West African rainfall* (Figure 16–28). Dr. Gray believes that

**FIGURE 16–28**
These illustrations show the accumulated tracks of all intense hurricanes (category 3, 4, 5) that occurred under two opposite conditions: (a) following the ten *wettest* years in the western Sahel region of Africa, and (b) following the ten *driest* years in that area of Africa. The period sampled was the 42-year span from 1949 to 1990. (Modified with permission from Gray, W. M., and Landsea, C. W., 1991, Predicting U.S. hurricane spawned destruction from West African rainfall, background information for talk by William M. Gray to the 13th Annual National Hurricane Conference, Miami, 40 p.)

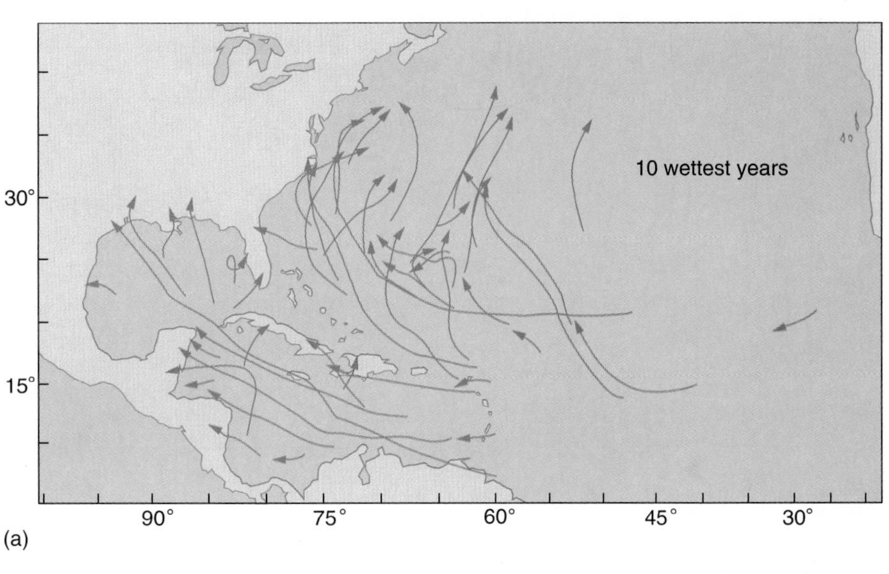


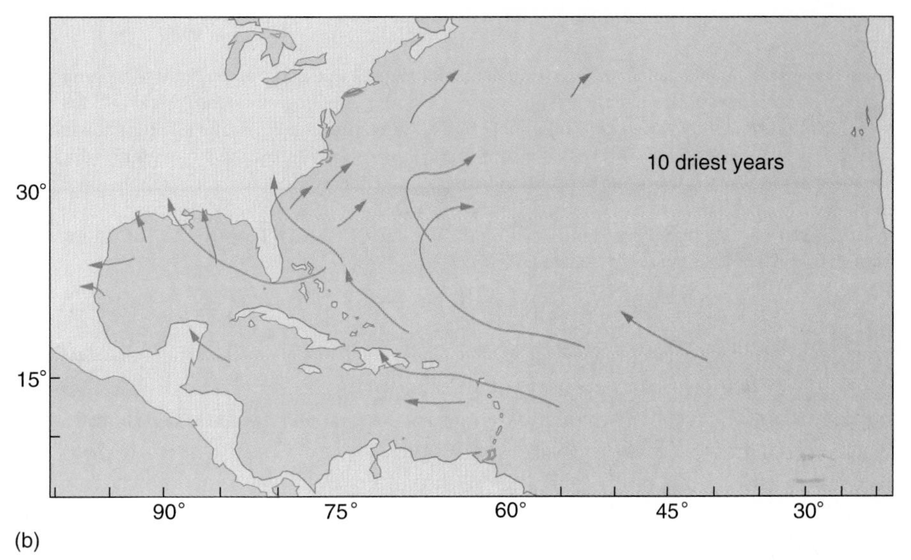


greater African rainfall results in more vegetation and enhanced evapotranspiration from plants. This creates more convection (upward air movement) that interferes with the easterly waves moving across Western Africa into the Atlantic, thus favoring more hurricane development.

Extrapolation of the Florida and Atlantic coast hurricane records back a century shows that *past* wet and dry cycles in West Africa also correlate well with increases and decreases in U.S. hurricane destructiveness (Figure 16–29). If you look at the pattern, you can see that we may be near the end of a "dry" cycle in which powerful hurricanes along the Atlantic coast have been far less frequent than in the previous 26 years.

Importantly, this lull in major hurricane activity coincides with the greatest period of development of our shorelines in U.S. history. The low frequency of hurricanes in recent years has decreased the "hurricane awareness" in people that is crucial to reducing loss of life and property. What will happen to all these structures (and their occupants) as we again enter a "wet" period, in which the frequency of major hurricanes is expected to increase markedly?

And one more thought: A corollary of the greenhouse effect (Chapter 11) is that, as the atmosphere

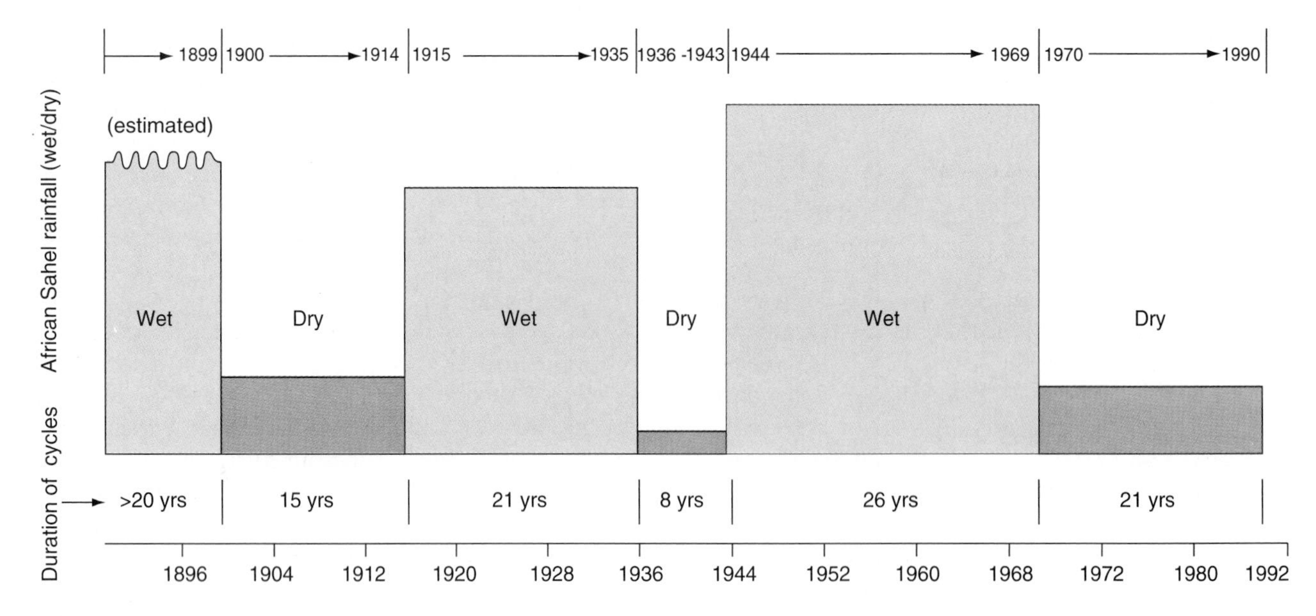


**FIGURE 16–29**
Destructiveness of hurricanes making landfall in Florida and the East Coast. Note the significant differences in hurricane destructiveness between dry and wet periods in West Africa. (Values shown are Hurricane Destruction Potential (HDP), a measure of a hurricane's potential wind and storm surge destruction. HDP is the sum of the squares of a hurricane's maximum wind speed, in knots$^2$ for each 6-hour period of a storm's existence.) (Modified with permission from Gray, W. M., and Landsea, C. W., 1991, Predicting U.S. hurricane spawned destruction from West African rainfall, background information for talk by William M. Gray to the 13th Annual National Hurricane Conference, Miami, 40 p.)

warms, so will the ocean's surface. As ocean waters warm, they provide more energy to develop more frequent and powerful hurricanes. In addition, the warming of more northerly waters may allow more frequent, more powerful hurricanes to travel farther northward to make landfall on these highly developed coasts.

## REDUCING HURRICANE DAMAGE

The destructiveness of past great hurricanes presents many lessons for us to learn, for we seem to repeat the same mistakes, over and over. These lessons are adequate warning and evacuation, maintaining beach width and dune height, engineering solutions, construction codes, and zoning/land use practices. We will look briefly at each.

### Adequate Warning and Evacuation

Obviously, the best place to be in a hurricane is somewhere else. This simple fact highlights the number-one problem in managing the hurricane hazard: *people's failure to evacuate in time.* The eye of a hurricane may be 500 kilometers (300 miles) away, so it does not seem a threat. However, hurricane winds may extend more than 120 kilometers (75 miles) from the eye, and gale-force winds extend even farther. And hurricanes with coast-normal tracks can cause high winds, torrential rains, flooding, and landslides as they penetrate far inland.

Modern forecasting provides a good safety margin for evacuation. If a hurricane is 500 kilometers (300 miles) distant, with a forward speed of 50 kilometers per hour (30 miles per hour), landfall will happen in only 10 hours. It is important to evacuate coastal areas, especially barrier islands, at least 12 hours before landfall. Evacuation becomes very difficult in gale-force winds flooding of low-lying "choke points" on evacuation routes.

### Maintaining Beach Width and Dune Height

A wide beach and high, wide dunes provide the best natural protection against storm surge. Although both will be severely damaged in the storm, they will reduce the damage to structures landward of them. Beaches can be maintained by artificial replenishment, and dunes can be built up both naturally and artificially (Chapter 15).

## Engineering Solutions

The tendency in the past has been to rely on engineering structures to protect against waves and storm surge. However, you saw in Chapter 15 how seawalls sometimes cause flooding behind themselves and must be rebuilt periodically at great expense.

Galveston, Texas, was nearly destroyed by hurricane waves and storm surge on the night of September 8, 1900. Loss of life was at least 6,000. The magnitude of this disaster helped Galveston obtain money to "hurricane-proof" the city. This involved two operations—literally elevating the city by as much as 3.4 meters (11 feet), and protecting the elevated city with a massive seawall.

Every building, roadway, rail line, and utility line was raised and supported, and then the space below was filled with sand dredged from Galveston Bay (Figure 16–30). This took almost seven years. The massive seawall was completed in only two years, rising 5 meters (16 feet) above low tide, higher by a third of a meter than the high-water mark of the 1900 hurricane. The seawall was extended and strengthened later, and these engineering features have protected the city during all subsequent hurricanes.

The effectiveness of the Galveston seawall in preventing erosion is nowhere more apparent than beyond where the wall ends, where waves and storms have caused significant beach recession (see Figure 1–9). This level of engineering effort had never been done before, and the cost makes it unlikely to be repeated elsewhere.

Another engineering structure is a **flood barrier,** a moveable wall or series of gates that can close off a waterway to protect it from hurricane surge. The Army Corps of Engineers has constructed such a barrier across the mouth of Narraganset Bay in Rhode Island to protect the city of Providence. Providence has experienced disastrous surge damage during many previous hurricanes, especially the great New England hurricane of 1938. The flood barrier last protected the city from a storm surge from Hurricane Bob in 1991.

## Construction Codes

One way to reduce storm damage is to rigorously enforce building codes. Some codes require *structural elevation* (Figure 16–31). The structure is

**FIGURE 16–30**
The raising of Galveston, Texas, after the hurricane of 1900. Every home, road, and utility line was raised up at the same time that a protective seawall was built along the ocean. The area was then filled by sand dredged up and pumped through the open spaces under the elevated structures. This massive construction project has protected the city from hurricanes since that time. (Photo from Rosenberg Archives, Galveston.)

**FIGURE 16–31**
Homes in McClellanville, South Carolina, that were rebuilt and raised up on pilings after Hurricane Hugo (1989) caused severe damage in this area. Flood surge reached levels of about 6 meters (20 feet) in this area. (Photo by author.)

supported on pilings, which are buried deeply in the sand and anchored solidly to the building to prevent it from being washed off its foundation (Figure 16–24). The space beneath is kept as clear as possible to allow surge waters to pass through. However, moving debris may knock down the structural supports, or water may scour sand from around the pilings (see Figure 15–21).

Wind damage can be reduced with home shapes that promote smoother windflow and thus offer less wind resistance (gabled rather than flat roof, rounded rather than angular architecture). Each component (foundation, floor, walls, roof) must be anchored *to the others* with metal straps.

## Zoning/Land Use Practices

The only truly effective ways to reduce hurricane damage are to have strict zoning to limit coastal development, or to eliminate coastal development altogether. Limiting development helps because less building density means fewer people to evacuate and fewer structures to be knocked down and to provide debris for destroying others. Many communities are using *setback lines* to prevent people from building in appealing but especially hazardous areas, such as in front of the dune line.

Most coastal managers and geologists feel that *hurricane devastation of coastal zones is inevitable.* The problem is not only the hurricanes, but the danger to people and their structures when hurricanes strike. Many coastal planners feel that the most cost-effective, recreationally useful, and environmentally sound use of coastal zones is to compensate property owners and convert their land to parks, beaches, and seashores, although the compensation could prove very expensive (Chapter 15, Figure 15–29). Repairs to these areas are relatively cheap after a storm, and they preserve natural coastal environments for future generations to enjoy.

## SUMMARY

Severe weather hazards such as tornadoes, northeasters, and hurricanes are the most costly natural hazards in the United States. In general, most tornadoes occur from March to July, hurricanes from June to November, and northeasters from October to April.

Tornadoes are the smallest, but most intense, severe weather hazards. They are funnel-shaped bodies of air that rotate counterclockwise and extend down from thunderstorm clouds. The intense low pressure (a drop of up to 100 millibars) and strong winds up to 370 kilometers per hour (230 miles per hour) cause a swath of destruction from a few tens of meters to 1.5 kilometers wide (1 mile) as a tornado sweeps along a path that averages several kilometers in length. Tornadoes can cause massive regional damage when they form a swarm at the edge of a moving weather front.

Northeasters are extratropical storms that originate in mid-latitudes and move eastward toward the Atlantic coastal region. They are low-pressure systems with winds that rarely attain hurricane levels (minimum 119 kilometers per hour or 74 miles per hour), but their massive size (up to 1,500 kilometer diameter or 932 miles) and long duration (several days) allow them to batter wide areas. Northeasters originate in a number of places, wherever unstable air produces significant temperature and pressure differences. In the end, they all move up the Atlantic coast, and their northeasterly winds cause wind and wave damage along with extreme rainfall or snow, depending on the time of year and the storm track.

A hurricane is a massive tropical storm system with counterclockwise winds, exceeding 119 kilometers per hour (74 miles per hour), rotating around a calm central eye. Hurricanes are created from tropical disturbances in the air overlying the warm ocean just north of the equator. The storm center moves westward across the Atlantic and may turn northward along the Atlantic coast or proceed into the Gulf of Mexico. Winds on the right side of the storm are stronger because they are the approximate sum of the velocity of the rotary winds plus the forward velocity of the storm.

Hurricanes are rated as categories 1 through 5 on the Saffir-Simpson Scale, based on their wind speed and expected surge height. Most storms that turn northward along the Atlantic increase speed with latitude as they become affected by other air masses. Hurricane tracks may change considerably. Hurricanes weaken as they pass over land, which cut them off from their source of energy—oceanic moisture. Hurricanes with coast-parallel tracks keep their weaker left sides along the coast, whereas those with coast-normal tracks devastate wide areas on their right sides as they make landfall.

The low pressure and high winds of a hurricane raise the ocean surface to add to developing a storm surge. Storm surge height now can be predicted by a Weather Service computer model, SLOSH. Flood surge moves onto the land as the hurricane makes landfall, causing great structural destruction and erosion of beaches and dunes. Ebb surge results when the hurricane passes over the area and water impounded inland by the advancing storm surges back into the ocean. Ebb surge completes the destruction of weakened structures and carves ebb surge channels across coastal areas. Hurricane winds damage buildings by removing unanchored roofs and eroding the materials covering their sides. Debris carried by the wind also contributes to the destruction.

Hurricane frequencies and West African rainfall show a close correlation. Wet periods in West Africa correlate with a greater frequency of strong hurricanes that make landfall along the Atlantic coast. The reverse is true for dry periods. The greenhouse effect may increase the frequency of strong hurricanes by increasing ocean temperatures. Warmer ocean waters may allow hurricanes to penetrate farther northward, with undiminished strength, to landfall on the most urbanized coastline in America.

Hurricane damage can be minimized. Coastal areas should be evacuated twelve hours before an expected hurricane landfall, before gale-force winds flood escape routes. The most effective protection against storm surge is to maintain wide beaches and high, wide dunes. Natural methods should be used wherever possible, but engineering solutions can be used in critical areas, although care must be taken not to create other serious problems. Construction codes should ensure that homes are elevated above expected surge heights and that all parts of a structure are securely anchored together. Zoning laws can reduce damage and loss of life when they reduce building density in coastal areas and prevent building in dangerous areas, such as seaward of the dune line.

## KEY TERMS

coast-normal hurricane track
coast-parallel hurricane track
ebb surge
ebb-surge channels
eye
eyewall
flood barrier
flood surge
Fujita Scale
hurricane
northeaster
Saffir-Simpson Hurricane Scale
SLOSH
storm surge
tornadoes
typhoons

## REVIEW QUESTIONS

1. Distinguish among tornadoes, northeasters, and hurricanes on the basis of (a) size, (b) wind speed, and (c) geographic occurrence.
2. Why do tornadoes strike with much less warning than northeasters or hurricanes?
3. What precursors (visual and instrumental) suggest that a tornado may be imminent?
4. Northeasters have winds that rarely reach hurricane force, yet these storms can cause great destruction. Why?
5. Compare northeasters and hurricanes on the basis of (a) the area where they form, (b) their duration, and (c) their path.
6. How and where do hurricanes form? How is their development and movement tracked?
7. Why are the destruction patterns so different on the left and right sides of a hurricane when it makes landfall?

8. What factors are used to assign a category to a given hurricane on the Saffir-Simpson Hurricane Scale? How does the potential destruction increase with hurricane category?
9. What factors affect the path and the intensity of a moving hurricane?
10. What factors determine storm surge height for a given hurricane landfall? How can storm surge height be measured? How can it be predicted?
11. Contrast the origin and the effects of flood surge and ebb surge.
12. What factors determine the wind damage to structures in a given place at hurricane landfall? What biological affects does hurricane wind damage have?
13. Describe how a hurricane landfall impacts vital public services.
14. How can we predict the frequency of hurricane impacts along the Atlantic coast?
15. How might the greenhouse effect influence hurricane frequency and severity?
16. Why is it important to evacuate coastal areas at least twelve hours before a hurricane hits?
17. How can we reduce hurricane loss of life and property damage by the following: (a) maintaining beach width and dune height; (b) engineering structures; (c) construction codes, and (d) zoning codes and land use practices?

## FURTHER READINGS


Christopherson, R. W., 1994, Geosystems: An introduction to physical geography (second edition): Macmillan, Chapter 8 (Weather), p. 228–39.

Davis, R. E., and Dolan, R., 1993, Nor'easters: American Scientist, v. 81, p. 428–39.

Dolan, R., and Davis, R. E., 1992, Rating northeasters: Mariners Weather Log, National Oceanic Data Center, v. 36, no. 1, p. 4–11.

Funk, B., 1980, Hurricane: National Geographic (September), p. 346–79.

Gore, R., 1993, Andrew aftermath: National Geographic (April), p. 2–37.

Gray, W. M., 1990, Strong association between West African climate and U.S. landfall of intense hurricanes: Science, v. 249, p. 1251–56.

Ludlum, D. M., 1963, Early American hurricanes 1492–1870: Boston, American Meteorological Society.

Miller, P., 1987, Tracking tornadoes: National Geographic, v. 171, no. 6, p. 690–715.

Simpson, R. H., and Riehl, H., 1981, The hurricane and its impact: Baton Rouge, Louisiana State University Press.

Snow, J. T., 1984, The tornado: Scientific American, v. 250, no. 4, p. 86–96.

# *Appendixes*

APPENDIX A
Conversion of Metric and English Units

APPENDIX B
Soil Classification

**TABLE A–1**
Conversion of metric and English units.

| Metric–English | | | English–Metric | | |
|---|---|---|---|---|---|
| To convert | to | Multiply by | To convert | to | Multiply by |
| | | Units of Length | | | |
| centimeters (cm) | inches (in) | 0.3937 | in | cm | 2.54 |
| meters (m) | feet (ft) | 3.2808 | ft | m | 0.3048 |
| m | yards (yd) | 1.0936 | yd | m | 0.9144 |
| kilometers (km) | miles (mi) | 0.6214 | mi | km | 1.6093 |
| | | Units of Area | | | |
| $cm^2$ | $in^2$ | 0.1550 | $in^2$ | $cm^2$ | 6.452 |
| $m^2$ | $ft^2$ | 10.764 | $ft^2$ | $m^2$ | 0.0929 |
| $m^2$ | $yd^2$ | 1.196 | $yd^2$ | $m^2$ | 0.8361 |
| $km^2$ | $mi^2$ | 0.3861 | $mi^2$ | $km^2$ | 2.590 |
| $m^2$ | acres | $2.471 \times 10^{-4}$ | acres | $yd^2$ | 4.840 |
| | | | acres | $m^2$ | 4047. |
| | | Units of Volume | | | |
| $cm^3$ | $in^3$ | 0.0610 | $in^3$ | $cm^3$ | 16.3872 |
| $m^3$ | $ft^3$ | 35.314 | $ft^3$ | $m^3$ | 0.02832 |
| $m^3$ | $yd^3$ | 1.3079 | $yd^3$ | $m^3$ | 0.7646 |
| liters (l) | U.S. quart | 1.0567 | U.S. quart | l | 0.9463 |
| l | U.S. gallon | 0.2642 | U.S. gallon | l | 3.7853 |
| | | Units of Mass | | | |
| grams (g) | Avoirdupois ounces | 0.03527 | Avoirdupois ounces | g | 28.3495 |
| g | Troy ounces | 0.03215 | Troy ounces | g | 31.1042 |
| kilograms (kg) | Avoirdupois pounds | 2.2046 | Avoirdupois pounds | kg | 0.4536 |
| | | Units of Density | | | |
| $g/cm^3$ | $lb/ft^3$ | 62.4280 | $lb/ft^3$ | $g/cm^3$ | 0.01060 |
| | | Units of Pressure | | | |
| $kg/cm^2$ | $lb/in^2$ | 14.2233 | $lb/in^2$ | $kg/cm^2$ | 0.0703 |
| bars | atmospheres | 0.98692 | atmospheres | bars | 1.01325 |
| $kg/cm^2$ | atmospheres | 0.95784 | | | |
| $kg/cm^2$ | bars | 0.98067 | | | |
| | | Units of Velocity | | | |
| km/h | mi/h | 0.6214 | mi/h | km/h | 1.6093 |
| km/h | cm/s | 27.78 | mi/h | in/s | 17.60 |
| | | Units of Temperature | | | |
| °C | °F | (9/5)(°C) + 32 | °F | °C | (5/9)(°F – 32) |

**FIGURE A–1**
Temperature scale of degrees Fahrenheit and Celsius.

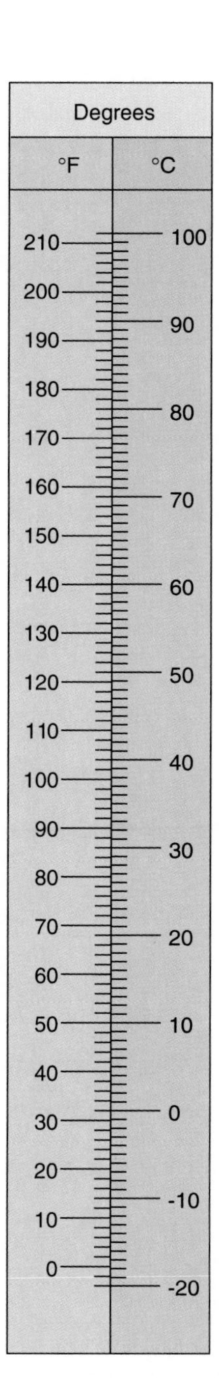

**TABLE B–1**
Soil classification by orders and groups.

| ARID, SEMIARID, AND SUBHUMID CLIMATES (PEDOCALS) | | | | HUMID CLIMATES (PEDALFERS) | |
|---|---|---|---|---|---|
| Azonal Order[1] | Intrazonal Order[2] | Zonal Order[3] (Increasing aridity →) | (Transitional) (Increasing temperatures →) | Zonal Order[3] | Intrazonal Order[2] |
| Lithosols<br>Stony, thin mountain soils that either have had little time to develop or are on slopes steep enough so that thicker sections of soil cannot accumulate.<br>Regosols<br>Very poorly developed soils with no horizon development that have formed on recently deposited sediments. | Saline Soils<br>Soils without horizons that contain an excess of soluble minerals. Formed during the desiccation of fine-grained sediments and saline waters in basins with internal drainage. | Chernozem Soils<br>Dark surficial layer (A) of highly organic soil derived from decay of grass parts. Underlain by a B layer which is brown to yellowish brown. Reduced rainfall results in little leaching and formation of calcium carbonate nodules in the B horizon. Common in the eastern parts of the Dakotas, Nebraska, central Kansas, and western Oklahoma and Texas.<br>Chestnut Soils<br>Similar to chernozem soils but form in a drier climate and are less organic, lighter in color, and have more abundant calcium carbonate nodules in the B horizon. Common in eastern Montana, the western parts of the Dakotas and Nebraska, northeastern Colorado, and western Kansas.<br>Brown Soils<br>Drier versions of the chestnut soils. Have a light-brown color and a B zone with a distinctive columnar structure. Common on the plains abutting the Rocky Mountains.<br>Gray Desert Soils<br>Sandy, pale grayish to reddish-gray soils with little organic material. Form in cooler mid-latitude desert areas of Nevada, Arizona, Utah, and New Mexico. Grayish-red soils characterize subtropical deserts such as those in southern Arizona and New Mexico.<br>Red Desert Soils<br>Red-colored soils in tropical deserts. | Prairie Soils<br>Soils transitional between the pedocals (chernozems) and pedalfers (podzols). Among the most fertile soils because they contain sufficient organic material, such as in the chernozems, and form in sufficient rainfall, such as in the podzols. Common in northern Illinois, Iowa, northwestern Missouri, eastern Kansas, Oklahoma, and north-central Texas. | Podzols<br>Soils with organic-rich accumulations in the top of the A zone underlain by a strongly leached, ash-white horizon in the lower part of the A zone. The B zone is clay-rich. The type of soil that forms in humid subarctic climates. Common in northern Wisconsin, Minnesota, Michigan, and northern New England.<br>Gray-brown Podzols<br>Similar to the podzol but less intensely leached. The base of the A zone is grayish brown rather than ash-white as in podzols. Thick, dark-brown B zone rich in clay minerals. Widespread in the humid temperate northeastern United States.<br>Red-yellow Podzols<br>Similar to gray-brown podzols but has a B zone enriched in hydroxides of iron and aluminum. Common in the humid subtropical climates of the southeastern and Gulf Coastal states.<br>Latosols<br>Deep brownish-red surface deposits with only a thin cover of organic debris. Soil lacks horizons and becomes lighter in color with depth. Forms by intense chemical weathering in hot humid climates. Silica is leached out, and the soil becomes enriched in hydroxides of iron, aluminum, or manganese, depending on the composition of the parent material. | Bog Soils<br>Dark-brown water-saturated and partially decomposed peaty material. Can form in arctic, temperate, or tropic climates wherever plant material accumulates under standing water.<br>Meadow Soils<br>Dark, organic-rich upper layers beneath which is a bluish-gray clay horizon.<br>Planosols<br>Thick and very dark organic-rich soil formed on flat surfaces between stream valleys where soil erosion is limited. |
| | | | | Tundra Soils<br>Soils which form in areas underlain by permafrost. Mixture of organic material and physically weathered rock debris. Chaotic structure resulting from perennial freezing and thawing. Common in arctic and subarctic climates. | |

[1]Soil horizons are poorly developed or absent.
[2]Soil characteristics are determined by local conditions such as poor drainage.
[3]Well-developed soil horizons corresponding to the climatic and vegetative zones in which they are found.
*SOURCE:* Modified with permission from A. N. Strahler, *Physical Geography,* New York, Wiley, 1960.

**TABLE B–2**
Soil classification (Seventh Approximation).

| Soil Order | Characteristics | Some Areas Where It Is Common | Approx. Equiv. in Great Soil Groups |
|---|---|---|---|
| Entisols | Soils without horizons; soil-forming processes have not had sufficient time to produce horizons. | Wide geographic range, from desert sand dunes to frozen ground of subarctic zones. | Azonal soils |
| Inceptisols | Weakly developed soil horizons; soil horizon A is developed. | Wide geographic range wherever soils have just begun to develop on newly deposited or exposed parent materials such as volcanic or glacial deposits. | Lithosols (mountain soils) Regosols (recently deposited sediments) |
| Spodosols | Humid forest soils with a gray leached A horizon and a B horizon enriched in iron or organic material leached from above. Commonly under coniferous forests. | New England, northern Minnesota, and Wisconsin. | Podzolic and brown podzolic soils (pedalfer) |
| Alfisols | Soils with clay enrichment in the B horizon. Lower organic content than mollisols. Medium to high base supply. Commonly under deciduous forests. | Western Ohio, Indiana, lower Wisconsin, northwestern New York, Central Colorado, western Montana. | Gray-brown podzolic soils (pedalfer) |
| Mollisols | Grassland soils with a thick, dark organic-rich surface layer. High base supply (calcium, sodium, and potassium). | Widespread in central and northern Texas, Oklahoma, Kansas, Nebraska, North and South Dakota, and Iowa. | Chestnut, chernozem, and prairie soils (pedocal) |
| Aridisols | Desert and semiarid soils; low organic content along with concentration of soluble salts within soil profile. | Widespread in desert and semiarid areas of Nevada, California, Arizona, New Mexico. | Desert soils (pedocal) |
| Ultisols | Deeply weathered red and orange clay-enriched soils on surfaces that have been exposed for a long time. | Humid temperate to tropical soils. Widespread in southeastern United States east of Mississippi Valley. | Red and yellow podzolic soils, certain lateritic soils (pedalfer) |
| Oxisols | Intensely weathered soils consisting largely of kaolin, hydrated iron and aluminum oxide. Bauxite forms in these soils. | Warm tropical areas with high rainfall. | Most lateritic soils (pedalfer) |
| Histosols | Organic soils and peat. | Mississippi delta, Louisiana; Everglades, Florida; local bogs in many areas. | Bog soils (pedalfer) |
| Vertisols | Swelling soils with high clay content which swell when wet and crack deeply when dry. | Southeast Texas; local areas | Swelling clays |

*SOURCE:* U. S. Department of Agriculture Soil Conservation Service, 1960.

# *Glossary*

*Note:* Number in parentheses indicates the chapter in which the term appears.

**Aa.** The Hawaiian term for more viscous and slow moving lava flows that have a jagged and blocky surface. (4)

**Acid drainage.** Surface and underground drainage that has been acidified through chemical reactions between water and minerals in the rocks. (12)

**Acid precipitation.** Precipitation laced with secondary pollutants, such as sulfuric and nitric acids. (11)

**Acid rain.** Acid precipitation that falls as rain. (11)

**Acid snow.** Acid precipitation that falls as snow. (11)

**Active volcano.** A volcano that has erupted at least once in historical time. (4)

**Adsorption.** Process by which fine-grained sediment particles hold contaminants on their surfaces. (13)

**Aeration.** A method of treating organic-rich waste by infusing it with large volumes of air through submersible pumps. (12)

**Aerobic reaction.** A reaction in which organic material breaks down in the presence of oxygen to produce carbon dioxide. (12)

**Aftershocks.** Smaller seismic events that persist for days or weeks following an earthquake (5)

**A horizon (leached horizon).** Composed of both mineral particles and organic material. Water with dissolved carbon dioxide can remove soluble material such as calcium carbonate as it goes through the layer. (6)

**Amplitude.** The vertical distance between the crest and the trough of a wave form. (5)

**Anaerobic.** A process in which material continues to decay in the absence of oxygen. (12)

**Angle of draw.** The vertical angle between the edge of the mined-out area and the edge of surface subsidence. (10)

**Angle of repose.** The maximum angle at which granular materials can be piled. (9)

**Anthropogenic.** Human activities that accelerate or alter a normally benign process into a problem. (1)

**Aquiclude.** Impermeable rock or sediment or one that transmits groundwater too slowly to be of use to people. (8)

**Aquifer.** A permeable rock or sediment capable of transmitting groundwater rapidly enough to recharge a well or spring. (8)

**Artesian well.** A well in which water rises above the aquifer without pumping. (8)

**Asthenosphere.** A region in the upper mantle, 100 to 250 kilometers below Earth's surface, that is composed of materials of lower rigidity than those above and below. (2)

**Atoll.** A circular coral reef that develops on the flanks of a subsiding island volcano. (4)

**Avalanche.** An extremely rapid form of mass movement on steep slopes. (9)

**Avalanche chutes.** Narrow, steep channels through which avalanches flow. (9)

**Back barrier bays.** Linear estuaries behind barrier islands. (13)

**Bankfull discharge**. A level of discharge that fills the stream channel. (7)

**Barrier islands.** Elongate bodies of sand separated from the mainland by a body of water. (15)

**Barrier reef.** A linear reef parallel to a coast and separated from it by a wide lagoon or bay. (14)

**Barrier rollover.** Response by barrier islands to rising sea level through erosion on their oceanic sides and deposition of the material by overwash on their back sides. (15)

**Basalt.** A fine-grained extrusive igneous rock with the same composition as gabbro. (2)

**Base flow.** That portion of the stream discharge that is supplied by groundwater. (3)

**Base level.** The lowest level to which a stream may erode its channel. (3)

**Bays.** Water bodies in embayments along the shoreline. (13)

**Beach.** A linear zone of coarse sediment accumulation along part of the coastal zone that is exposed to wave action. (15)

**Bedding.** The layering that is characteristic of sedimentary rocks and which is due to changes in sedimentary processes, sources, or conditions as sediments accumulate in the depositional basin. (2)

**Bed load.** That portion of the sediment load that is transported in constant or intermittent contact with the bottom. (3)

**Bench.** A step-like excavation into a slope. (6)

**Benioff zone.** A steeply dipping fault zone along which a plate is subducted. (5)

**Bentonite.** Clay minerals that have the ability to absorb great quantities of water to swell up to eight times their original volume. (10)

**B horizon (accumulation horizon).** The soil zone into which carbonates, clay minerals, and iron oxides are transported by percolating water and in which they accumulate. (6)

**Bioagglomeration.** The process in which aquatic organisms ingest water and particles and excrete the latter as fecal pellets. (13)

**Biodegradable.** Organic wastes which will decompose and decrease in volume over time as they are degraded by biological agents. (12)

**Biological oxygen demand (B.O.D).** The amount of oxygen required for respiration and organic breakdown in a volume of water or bottom sediment. (12)

**Blowout.** Depressions formed by wind in sandy areas where the vegetation has been removed. (15)

**Body waves.** Waves that penetrate Earth and travel through it. (5)

**Bogs.** Local depressions that are generally peat-filled and from where standing water and vegetation accumulate. (13)

**Brackish water.** Water that is intermediate in salinity between fresh and salty. (13)

**Braided channel.** A stream channel that resembles braided hair or a rope. (7)

**Breaker zone.** The zone in which the character of waves changes markedly; they grow taller and less symmetrical. (15)

**Brittle.** A material that eventually breaks when the stress applied exceeds its strength. (2)

**Brittle zone.** The upper zone of a glacier in which motion of ice produces breakage. (3)

**Caldera.** Immense crater formed when a large and particularly violent eruption literally blows the top off of a volcano leaving only the lower part. (4)

**Catchment system.** Paved-over areas or roofs that catch as much rainwater as possible and divert it downward toward an underground cistern. (8)

**Cell.** In a sanitary landfill, the mass of solid waste, covered by sediment, which represents one day's accumulation. (12)

**Cementation.** The process in which sediments are cemented together by minerals deposited in the open spaces between the grains. (2)

**Central vent.** An eruption site at the central, highest, part of a volcano. (4)

**Channelization.** Modification of a stream channel by straightening, clearing, deepening, widening, and lining with concrete or boulders. (7)

**Chemical reaction.** Interactions between atoms involving their outer electrons. (2)

**Chemical weathering.** The compositional change that occurs in minerals that are under chemical attack. (3)

**Chlorofluorocarbons (CFCs).** Chlorine-containing compounds used most commonly in aerosol cans and as foaming agents for styrofoam and insulation. These compounds, when released into Earth's atmosphere, destroy its ozone layer. (11)

**C horizon (partially weathered horizon).** Composed of highly weathered rock or sediment which may preserve some characteristics of the parent material. (6)

**Cinder cones.** Volcanoes composed entirely of tephra. No lava flows are involved. (4)

**Cirque.** A bowl-shaped rock basin which was eroded by a mountain glacier above the snowline. (3)

**Cisterns.** Underground water storage chambers fed by a roof catchment system. (8)

**Clay.** Sedimentary particles with diameters less than 1/256 millimeters. (3)

**Closed system.** A system which neither gains nor loses significant material—it is merely transferred and later recycled. (1)

**Cluster development.** Many structures are grouped together in "villages." The land between is kept in its natural state for walking, picnicking, and bicycling. (6)

**Coastal aquifers.** Aquifers that are recharged inland and extend under coastal plains into the ocean. The pore spaces in their lower ends are commonly filled with saline waters. (8)

**Coastal storm.** Extratropical storms which affect the Pacific Coast, the continental interior, and the Atlantic and Gulf coasts. (15)

**Coastal zone.** Area from just beyond the point where waves first break offshore to the limit of tidal action inland. (15)

**Coast-normal hurricane track.** A hurricane track that is at nearly a right angle to the coast. (16)

**Coast-parallel hurricane track.** A hurricane track that is offshore and parallel to the land. (16)

**Cohesion.** Ability of particles to attract and hold each other. (6, 9)

**Collapse.** A rapid movement of the surface downward into a cavity below. (10)

**Collapse sinkholes.** Steep, rock-walled sinkholes that form rapidly due to sudden collapse of the roof of a solution cavity. (10)

**Combined sewers.** Sewer systems in which sanitary and storm sewers are combined. (12)

**Compaction.** The process in which the pressure of overlying sediments forces the grains together, expelling the water and air that was trapped between the particles. (2)

**Composting.** An aerobic process in which bacterial degradation and oxygenation break down organic materials. (12)

**Compressibility.** The degree to which the volume of a material decreases when normal stress is applied. (6)

**Compressional stress.** Stress resulting from materials being squeezed by forces acting towards one another, like a piece of wood clamped in a vise. (2)

**Condensation.** The process in which cooling air loses its capacity to hold moisture. The vapor begins to change back into micro-droplets of water. (3)

**Conduction.** Heat energy transferred directly from atom to atom in solids. (2)

**Cone of depression.** A depression in the water table caused by excess discharge. (8)

**Confined aquifer.** An aquifer in which the water is under a greater than atmospheric pressure. (8)

**Consumption.** Water use in which the water changes state so that it cannot directly recharge the source. For example, in withdrawal of groundwater for irrigation, the irrigation water is subsequently evaporated and there is a net loss to the water table below. (8)

**Containment structures.** Devices erected around surface storage facilities of pollutants. They prevent a spill from infiltrating into the water table, and facilitate the clean up process. (8)

**Contaminant plume.** Pollutant masses flowing by gravity within the groundwater supply. (8)

**Continental crust.** The portion of Earth's outermost layer that makes up the continents. It has a gross overall composition of granite and ranges from 35 kilometers to 60 kilometers thick. (2)

**Continental drift.** Alfred Wegener's theory that the continents were once joined as a super continent he called *Pangaea*. The super continent subsequently broke apart, resulting in continents that migrated across Earth's surface. (2)

**Continental ice sheets.** A mass of glacier ice that covers most of the topographic features of a region of continental or subcontinental proportions. (3)

**Convection.** Heat energy transferred through the movement of liquids and gasses. (2)

**Convection cell.** The circular path followed by materials transferring heat. (2)

**Convergent margins.** Plate margins where compressional stress moves plates toward each other. (2)

**Coral reefs.** A ridgelike or moundlike structure composed of corals and other aquatic organisms, occurring in shallow water along some subtropical and tropical shorelines. (14)

**Core.** The innermost region of Earth. It is believed to be about 3,570 kilometers thick from its center to its outer edge. It has a solid inner portion and a liquid outer portion, and is composed mainly of metallic nickel and iron. (2)

**Cover-collapse sinkholes.** Sinkholes that form where a thick section of sand overlies a clay layer on top of limestone. (10)

**Cover-subsidence sinkholes.** Sinkholes formed when overlying loose sand moves continually and slowly downward to fill a developing solution cavity. (10)

**Creep.** Slow downslope displacement of sediment, rock or structures under the influence of gravity. (9)

**Crest.** The highest part of a wave. (15)

**Crevasses.** Fractures that form in the brittle zone of a glacier. (3)

**Cross sectional area (stream).** The width multiplied by the mean depth. (3)

**Crust.** The outermost and thinnest layer of Earth. Two different types of crust exist, continental and oceanic. (2)

**Crystal.** Regular geometric shapes formed by minerals under favorable conditions due to the orderly internal arrangement of atoms. (2)

**Cubic packing.** A loose form of particle arrangement where the grains are positioned so their centers are directly above those of grains below. This type of packing has the highest porosity. (5, 9)

**Darcy's law.** (8) Discharge =

$$\frac{(\text{cross-sectional area}) \times (\text{permeability}) \times (\text{vertical distance})}{(\text{distance of transport})}.$$

**Debris avalanches.** The most rapidly flowing, sliding, and falling mass movements. (9)

**Debris basins.** Sediment traps located upstream from reservoirs. (6)

**Debris flows.** Very rapid downslope movement of rock and regolith. (9)

**Deflation.** Wind erosion of sediments. (3, 15)

**Deflation basin.** A depression formed as wind blows sand from a surface unprotected by vegetation. (3)

**Delta.** A depositional plain formed by a river as it enters a standing body of water, such as a lake or the ocean. (3)

**Depth (stream).** The vertical distance between the water surface and the stream bed, averaged from several places across the steam. (3, 7)

**Dilation.** Increase in rock volume due to an increase in stress which causes cracks to develop. (5)

**Discharge.** The volume of water passing a point over a period of time—usually expressed as cubic meters per second or cubic feet per second. (3, 7)

**Discharge (groundwater).** The volume of water that is removed from an aquifer in a given time. (8)

**Displaced problem.** A solution to a geologic environmental problem in one area that results in the same problem or a different problem developing in another area. (1)

**Dissolved ions.** Ions that are dissolved in fluids. (3)

**Dissolved load.** Dissolved ions that are carried by a stream. (3)

**Dissolved oxygen (D.O.).** A measure of the concentration of oxygen in an aquatic environment. (12)

**Divergent margins.** Plate margins where tensional stress moves one plate away from another. (2)

**Diversion culverts.** Channels that are excavated along the top of a slope to intercept water flow and protect the lower slope from erosion. (6)

**Dormant volcano.** A volcano that has not erupted within historic time but is capable of erupting in the future. (4)

**Driving forces.** Forces that promote movement. (9)

**Ductile.** Describes a material under stress that can be deformed or twisted into any shape. (2)

**Dunes.** Streamlined hills composed of sand-size particles deposited by wind. (3)

**Dynamic equilibrium.** When change occurs in one part of a system, it is balanced by change in another part in order to preserve the overall system equilibrium. (1, 7)

**Earthquake.** A shaking of the ground usually caused by rocks rupturing under stress. (5)

**Ebb surge.** The return of flood surge waters, stream flow and precipitation back into the ocean as a hurricane moves inland. (16)

**Ebb-surge channels.** Roads, driveways, and channels perpendicular to the coast that channel hurricane flood flow across the shore back into the ocean. (16)

**Ebb tidal delta.** A mass of coarse material deposited by ebb tidal currents into the ocean. (15)

**Ebb tide.** Seaward flow of estuarine water with the fall of the tide. (13)

**Effluent.** Any liquid discharge from an industrial plant. (12)

**Effluent streams.** Streams that receive most of their discharge from the ground water table. (8)

**Elastic rebound theory.** An explanation for earthquakes in which built-up strain is released as a rock breaks. (5)

**Elastic strain.** The condition in which when a stretching (strain) begins, it is proportional to the stress. (2)

**Energy.** The ability to do work. The work can be constructive, like heating the air, or destructive, like an earthquake. (2)

**Ephemeral streams.** Streams that are dry most of the year because the water table is far beneath their channels. They carry water only for a short period after a rainfall. (8)

**Epicenter.** The point on Earth's surface directly above an earthquake's focus. (5)

**Equilibrium.** A state in which a change in one factor results in a change in the other, so that the system remains the same. (3)

**Erodibility.** The ease with which a soil can be eroded. (6)

**Erosion.** The removal of regolith and rock by the action of streams, glaciers, and coastal waves. (3)

**Eruptive center.** An opening through which lava or tephra is ejected. (4)

**Estuary.** A branch of the ocean characterized by brackish to saline water and tidal action. (13)

**Eustatic.** A worldwide sea level change caused by increasing or decreasing the amount of water in the ocean. (15)

**Eutrophication.** A condition which develops when the D.O. level in a body of water is not enough to satisfy the B.O.D. The organisms within the water die and accumulate on the bottom with the dark colored sediment typical of that environment. (12)

**Evaporation.** The process in which radiant energy from the sun heats the upper portion of bodies of water. This makes water molecules so active that some of them rise into the atmosphere as vapor, against the pull of Earth's gravity. (3)

**Extinct volcano.** A volcano that is not expected to erupt again. (4)

**Eye.** The center of a hurricane, an area of relative calm and very low pressure. (16)

**Eyewall.** The area just outside of the eye of a hurricane, which is the location of the greatest turbulence and highest winds. (16)

**Failure.** Breaking (rocks) or loss of cohesion (sediments) under stress. (2)

**Falls.** A type of mass movement in which rock or regolith falls directly from a cliff onto the ground. (9)

**Faults.** Rock fractures along planar surfaces along which there is movement. (2)

**Fertility.** The ability of a soil to supply the nutrients required for plant growth. (6)

**Fertilizers.** Materials rich in nitrogen, phosphorus, and potassium. (6)

**Fetch.** The distance over which wind blows. (15)

**Finger-fill development.** A pattern of land development that is created by dredging to construct elongated, elevated building sites that are separated by water-filled canals. (14)

**Fissure eruptions.** The ejection of lava from a crack rather than a vent. (4)

**Flocculation.** A process in which ions in saltwater attract electrically charged clay particles, making them aggregate or clump together. (13)

**Flood.** When the volume of water in a stream exceeds the channel's capacity, the excess water spills out of the channel across the adjacent area. (3)

**Flood barrier.** A moveable wall or series of gates that can close off a waterway to protect it from hurricane storm surge. (16)

**Flood frequency recurrence curve.** A graph that expresses in years the likelihood of development of a flood of a given height. (7)

**Flood hazard map.** This map shows areas that would be flooded by stream discharges of a given magnitude (for example, a "50-year flood"). (7)

**Flood impoundment dam.** A containment structure that holds rainwater so it can be released slowly after a flood. (7)

**Floodplain.** The flat plain on either side of a stream that is flooded periodically. (3, 7)

**Flood surge.** As a hurricane eye passes over the coast, the flood surge moves across the shore, destroying structures and carrying sediment and debris landward. (16)

**Flood tidal delta.** A mass of coarse material deposited by flood tidal currents into a coastal bay. (15)

**Flood tide.** Landward flow of estuarine water with the rise of the incoming tide. (13)

**Floodwalls.** Reinforced concrete structures parallel to the river banks which prevent floodwaters from inundating the settled areas behind them. (7)

**Floodways.** Areas of the floodplain of a stream where no new structures or homes are permitted. (7)

**Flows.** Mass movement in which the material behaves like a viscous fluid. (9)

**Focus.** The place within Earth where the rock breaks in an earthquake. It may be anywhere from the surface to about 700 kilometers deep. (5)

**Foreshocks.** Small seismic events caused by minor breaks in strained rocks, especially along tributary faults. (5)

**Frequency.** How often events of a given magnitude occur or are expected to occur. (1)

**Freshwater wetlands.** Areas characterized by specific soil types and vegetation which develop when fresh water covers the ground for long periods. (13)

**Fringing reef.** A reef built directly against the coast. (14)

**Frost wedging.** A weathering process in which water enters rock openings and freezes. The frozen water expands, exerting powerful tensional forces sufficient to wedge the rocks apart. (3, 9)

**Fujita scale.** A scale used to categorize the strength of a tornado. (16)

**Gauging stations.** Facilities at which stream discharge can be measured. (7)

**Geohazard (geologic hazard).** An Earth process that is harmful to humans and/or their property. (1)

**Glacier.** A long-lived mass of ice that forms from accumulated snow on land. It flows very slowly, like thick plastic material, under the force of gravity. (3)

**Gneiss.** A metamorphic rock with separate layers of platy and granular minerals. (2)

**Gradient.** The vertical drop of a stream channel over a given horizontal distance. (3)

**Granite.** A coarse-grained intrusive igneous rock typically containing three minerals—quartz, mica, and feldspar—along with minor amounts of darker minerals. (2)

**Gravity.** A force between two objects that depends on their masses and the distance between them. The greater the mass, or the closer they are, the greater is the gravitational force between them. (3)

**Gravity fault (normal fault).** A fault in which tension pulls the rock apart. One block drops down the fault plane relative to the other block. (2)

**Greenhouse effect.** The process in which increases in natural and anthropogenic carbon dioxide and other atmospheric gases have trapped more and more of Earth's heat, thus increasing atmospheric temperatures (11).

**Groins.** Walls of rock, concrete, or wood built perpendicular to the land into the surf zone to trap sand and build up beaches. (15)

**Groundwater.** Water that infiltrates the surface and moves underground by gravity. (3)

**Gullies.** Rills that deepen and interconnect. (6)

**Hard water.** Water in which chemical reactions between groundwater and limestone have increased the concentration of calcium ions. (8)

**Hazard map.** A map showing areas that are affected by a particular hazard such as lava flows or stream flooding. (4)

**Headward erosion.** Heads of gullies which over time extend themselves by eroding upslope. (6)

**Hermatypic.** Reef-building corals in which the coral animals live in a symbiotic association with algae. (14)

**Hot spot.** A stationary heat source beneath the lithosphere, from which magma rises. (4)

**Hurricane.** A massive low pressure system of tropical origin with rotary winds that exceed 119 kilometers per hour (74 miles per hour) blowing counterclockwise around a relatively calm central area called the eye. (16)

**Hydrologic cycle.** The circulation of water in its three states through the atmosphere, lithosphere, and hydrosphere. (3)

**Hydrolysis.** The process in which polar water molecules react with oppositely charged ions in a mineral structure, dissolving the mineral. (3)

**Hydrosphere.** The liquid outer covering of Earth, encompassing oceans, rivers and lakes. (3)

**Ice caps.** Glaciers so thick they cover most of the topographic features of an entire region. (3)

**Igneous rocks.** Rocks formed from the cooling of molten rock both underground and on Earth's surface. (2)

**Infiltration.** Movement of water into the ground. (3)

**Influent streams.** Streams that supply water into the ground. (8)

**Insoluble residue.** Fine-grained residue left when impure limestone is dissolved. (10)

**Intensity.** The power of an earthquake as perceived by humans. (5)

**Intermittent streams.** Streams that flow only part of the year, when the groundwater table rises to fill their channels. (8)

**Intraplate earthquakes.** Earthquakes that are not associated with plate boundaries but occur within plates. (5)

**Inversion.** Local or regional weather conditions that trap colder air near the surface and warmer air above. (11)

**Ions.** The charged particles that result when an atom gains or loses an electron in a chemical reaction. (2)

**Island arc.** An arc-shaped chain of volcanic islands formed on the concave (landward) side of an oceanic trench. (2)

**Isotopes.** Atoms of an element having different atomic masses are called isotopes of that element. (2)

**Jetties.** Long rock or concrete structures built into the surf zone on either side of a tidal inlet to prevent the inlet from closing through deposition of sand carried along the coast by longshore drift. (15)

**Joints.** Rock fractures along which no movement has occurred. (2)

**Kinetic energy.** The ability of a moving object to induce activity in other objects. (2)

**Lag time (earthquakes).** The difference in arrival time of P and S waves. (5)

**Lag time (streams).** The time difference between peak rainfall and peak stream discharge. (7)

**Lahar.** Volcanic mudflows. (4)

**Lateral erosion.** The sideward erosion by a stream as it migrates over its floodplain with time. (7)

**Lava.** Molten rock that has extruded on to Earth's surface. (2)

**Lava dome.** A volcano built up of viscous lava that does not flow far from the vent before it solidifies. (4)

**Lava flow.** Magma that reaches the surface non-explosively and spreads out. (4)

**Leachate.** A foul smelling liquid produced by the rainwater and groundwater that enter a landfill and react with the solid wastes. (12)

**Leaching.** The solution of subsurface materials by groundwater. (6)

**Levee.** A ridge of sediment deposited alongside a stream as floodwaters rising out of the channel lose energy and deposit their coarser load. (3, 7)

**Limestone.** A biogenic rock formed by the deposition and subsequent compaction and cementation of crystals of calcium carbonate. (2)

**Liquefaction.** Sediment converted to a flowing sediment/water mass through the movement of water upward through a sediment deposit. (5)

**Liquid limit.** The water content at which a soil or sediment can flow. (6)

**Lithosphere.** The uppermost 100 kilometers of Earth, composed of crustal rock and the portion of the mantle above the asthenosphere. (2, 3)

**Lithospheric plates.** The masses of lithosphere that move over the asthenosphere. (2)

**Loess.** Deposits of wind blown silt. (3)

**Longshore drift.** Movement of sediment along a shoreline by successive waves. (15)

**Longwall mining.** A method of mining in which the resource is completely recovered as mining advances. A rotating cutter wheel travels along a "long wall" of coal, shearing off the material being mined. (10)

**Love waves.** A special kind of transverse wave. (5)

**Magma.** Molten rock beneath Earth's surface. (2)

**Magnitude.** The power of a destructive event such as an earthquake or hurricane; magnitude is a measure of how much energy is released. (1, 5)

**Mangrove.** A type of tree that flourishes along coasts that lie between the equator and approximately latitude 30° in both the Northern and Southern Hemispheres. Some species of mangrove are capable of living in salt water, while others have less salinity tolerance. (14)

**Mantle.** The middle layer of Earth. It is about 2,740 kilometers thick and is composed of high density silicate minerals. (2)

**Marble.** A metamorphic rock produced by the recrystallization of limestone or dolomite. (2)

**Marshes.** Shallow water areas characterized by soft-stemmed plants and a lack of trees. They develop in both salt water and fresh water areas. (13)

**Mass movement.** Downslope displacement of masses of regolith and rock. (9)

**Mature soils.** Soils that have developed over enough time (hundreds or thousands of years) to form distinct horizons. (6)

**Meandering channel.** A river pattern consisting of a series of curves. (7)

**Mercalli scale.** A scale to measure the intensity of an earthquake as perceived by humans. (5)

**Metallic resources.** Earth materials from which metals can be extracted. (1)

**Metamorphic rocks.** Rocks that have undergone a change in characteristics through recrystallization resulting from changes in temperature, pressure, the action of hot solutions, or any combination of these factors. (2)

**Metamorphism.** The process by which crystal size, shape, and composition, as well as the alignment of the minerals in a rock, may be significantly altered. (2)

**Mineral.** A substance that fits the following criteria: (1) it must be a solid; (2) it must occur naturally; (3) it must be inorganic; (4) it must have an orderly arrangement of its atoms; and (5) it must have a specific chemical composition, or vary within a well-established range of composition. (2)

**Moho (Mohorovičić discontinuity).** The boundary between the crust and the mantle. (2)

**Monitoring well.** Well drilled into a landfill and in various places around it so that water samples can be extracted and the water quality determined on a regular basis. (8, 12)

**Moraine.** A land form, usually composed largely of till, which is deposited at the margin of a glacier. (3)

**Mudflows.** Mud containing significant water (up to 30 percent) and a large proportion of fine-grained material. (6, 9)

**Nonbiodegradable.** Wastes that undergo minimal breakdown over time and lose little of their original volume. (12)

**Nonmetallic resources.** Useful and abundant Earth materials from which nonmetals are extracted. (1)

**Nonrenewable resources.** Resources that can be replenished only on a geologic scale of millions of years. (1)

**Northeasters (nor'easters).** Immense extratropical storms that can persist over several tidal cycles and thus cause significant damage along long stretches of the Atlantic Coast. The name refers to the direction from which their winds come. (16)

**Nuclear fusion.** A nuclear reaction in which two nuclei join. (3)

**Nuée ardente (pyroclastic flow).** Blasts of gassy incandescent pyroclasts. (4)

**Oceanic crust.** The portion of Earth's outermost layer which makes up the sea floor. It averages about 5 kilometers thick and is made up of basalt. (2)

**O horizon (organic horizon).** Uppermost soil layer, characterized by the accumulation of organic material and the absence of mineral material. (6)

**Outwash.** Sorted and stratified coarse sediments deposited by glacier meltwater streams. (3)

**Outwash plain.** Flat and gently sloping plain underlain by meltwater stream deposits built beyond the margins of a glacier. (3)

**Overwash.** Sediment washed by storm waves through low areas between dunes and onto the back side of a barrier island or inland across the mainland. (15)

**Oxbow lake.** A crescent-shaped lake formed by the cutting through and abandonment of a former meander curve. (7)

**Oxidation.** A process of chemical weathering in which water and oxygen react with minerals in a rock. (3)

**Ozone hole.** An opening in Earth's ozone layer. (11)

**Packing.** Arrangement of the particles in a deposit. (5, 9)

**Pahoehoe.** The Hawaiian term for low-viscosity, "runny" lava that solidifies into a distinctive ropy texture. (4)

**Pancaking.** The process in which poured concrete floors separate from their corner fastenings and fall, floor by floor, onto each other as a result of seismic shaking. (5)

**Paved drop chutes.** Special channels excavated into slopes to remove water immediately. (6)

**Pedalfer.** Clay minerals and iron oxides accumulate in the B horizon in humid climates: *ped* for soil, *al* for aluminum clays and *fer* for ferrum or iron. (6)

**Pedocal.** Soil developed in dry areas where there is insufficient water to remove soluble compounds, so they accumulate within the soil profile; *ped* for soil, *cal* from the Latin for lime. (6)

**Perched water table.** A local water table existing above the level of the regional one because infiltrating water encounters an impermeable layer above the regional water table. (8)

**Perennial streams.** Streams that flow year round because of a high water table and humid climate. (8)

**Period.** The number of waves passing a point per second. (5)

**Permafrost.** Permanently frozen ground. (9)

**Permeability.** The capacity of a rock or sediment to transmit a fluid. (8)

**pH.** The measurement of a solution's concentration of hydrogen ions. (11)

**Physical weathering.** The disintegration of rock into smaller pieces, with no change in the chemical makeup of the minerals. (3)

**Piedmont glacier.** A large glacier formed when two or more trunk glaciers merge at the foot of a mountain range and flow across the adjacent plains. (3)

**Plastic deformation.** A permanent change in the shape or volume of a rock. (2)

**Plasticity index.** The numerical difference between the liquid and plastic limits. (6)

**Plastic limit.** The water content that causes solid soil to become moldable. (6)

**Plastic zone.** The lower part of a glacier which is under extreme pressure due to its own weight. This portion of the glacier flows slowly like an extremely viscous plastic material. (3)

**Plate margins.** Edges where lithospheric plates make contact with other plates. (2)

**Plate tectonics.** A theory that explains the movement and deformation of parts of the outer Earth. It involves the movement of rigid lithospheric slabs, called plates, over a less rigid layer (the asthenosphere). (2)

**Plutonic (intrusive) rock.** Rock formed by the cooling and solidification of magma underground. (2)

**Pollution.** The unfavorable alteration of our surroundings, wholly or largely as a by-product of human action. (1)

**Polyps.** Coral animals of pinhead to dime size. (14)

**Pores.** Open spaces between particles in a rock or sediment. (2)

**Porosity.** The volume of pore space within a rock. (2)

**Potential energy.** Energy in storage prior to use. (2)

**Precipitation.** Liquid particles of rain or solid ones of snow. (3)

**Precursors.** Minor events that indicate an impending major event; measurable movement or other events that may indicate an impending earthquake. (1, 5)

**P wave (primary wave).** The faster of the two body waves. (5)

**Pyroclastic material.** Mixtures of gas and fragments that are expelled from a volcano. It includes both fine airborne material, which can travel some distance, and material that travels near the ground as a mixture of hot gas and fragments. (4)

**Quartzite.** A rock formed from the metamorphism of a quartzose sandstone. (2)

**Quickclays.** Solid clays which become unstable under certain conditions. (5, 9)

**Quicksand.** A sand/water mixture that is fluid because water flows upward through the deposit and exerts pressure on sand grains, keeping them from touching each other. (5)

**Radon.** A radioactive gas. (11)

**Rayleigh waves.** Waves that cause an elliptical or rotating motion in Earth materials. (5)

**Reactivity.** Soils that can react chemically with metal objects placed in the ground. The degree to which the solid portion of our wastes reacts with oxygen, water, and soil bacteria. (6, 12)

**Recharge.** Replenishment of groundwater by infiltration of water from the surface. (8)

**Recharge basins.** Large artificial surface depressions that catch rainwater and channel it to the aquifers below. (8)

**Recurrence interval.** The expected time between geologic events—statistical probabilities based on averages over a long period of time; the average interval in years at which a discharge of a given level is expected. (1, 7)

**Red mangrove.** The most abundant species of mangrove along the seaward fringe of sheltered subtropical and tropical shorelines, along tidal channels, on tidal flats, and on small islands within bays. (14)

**Refraction.** A change in the direction of a wave as it encounters a bottom of unequal depth along the wave front. (15)

**Regolith.** Loose material on Earth's surface. (3)

**Renewable resources.** Resources that can be replenished within a lifetime. (1)

**Residual regolith.** Loose material that remains in place above the underlying material from which it was formed by weathering. (3)

**Resisting forces.** Forces that deter movement. (9)

**Resource.** Any natural material that is both useful to people and available in sufficient amounts. (1)

**Rhombohedral packing.** Particle arrangement in which the centers of grains are located over the spaces between grains below. This type of packing has the lowest porosity. (5, 9)

**Richter scale.** The scale on which the magnitude of an earthquake is measured by instruments. (5)

**Ridge and runnel.** A low, elongate, barely emergent sand bar (ridge) separated by a shallow water-filled trough (runnel) from the beach. (15)

**Rift valleys.** Depressions that form when tensional forces pull rocks apart to create a rift. (5)

**Rills.** Shallow, temporary channels formed in an area during the early stages of stream system development. (6)

**River estuary.** An estuary that develops in an existing river valley. (13)

**Rock cycle.** The process by which one type of rock (igneous, sedimentary, metamorphic) may be converted into other types. (2)

**Rockfall.** The fall of rock particles from the face of a cliff directly on to the land surface. (9)

**Rockslide.** Downslope movement of rock along planar surfaces. (9)

**Room-and-pillar method.** A mining procedure in which most of the coal or ore is removed, but pillars of the resource are left to support the upper strata. (10)

**Roundness.** The degree to which clastic sedimentary particles develop rounded surfaces. (3)

**Runoff.** Water that flows across the land surface. (3)

**Saffir-Simpson scale.** Used to describe the relative power of a hurricane on a rating scale (category) of 1 to 5. The scale utilizes wind velocity and storm surge heights to assign a category to a given storm. (16)

**Salinity.** Concentration of dissolved salts in water. (13)

**Saltation load.** That portion of the sediment load that moves by bouncing along the bed. (3)

**Saltwater encroachment.** Saltwater penetration into an aquifer. (8)

**Salt-wedge estuary.** Develops where the tidal range is moderate and stream discharge dominates over tidal action. The fresh water overlies the salt water in a wedge in the downstream portion of the estuary. (13)

**Sand.** Sedimentary particles with diameters between 1/16 and 2 millimeters. (3)

**Sand boil.** Fluidized sand that erupts from an opening in the land surface. (5, 7)

**Sand dikes.** Fluidized sand that flows upward, cutting through and disrupting the strata above. (5)

**Sand ridges.** Fluidized sand that forms linear features on the surface. (5)

**Sanitary landfill.** A solid waste landfill in which each day's accumulation is covered by a blanket of sediment. (12)

**Sanitary sewers.** An underground system of sewers that collect domestic wastes and deliver them directly to sewage treatment facilities. (12)

**Satellite volcanoes.** Smaller volcanoes that develop on a volcano's flanks. (4)

**Saturated zone.** The groundwater zone in which all the pore spaces are filled with water. (8)

**Schist.** Strongly foliated metamorphic rock composed mainly of platy minerals. (2)

**Sea floor spreading.** The action of convection currents spreading apart the sea floor at the mid-ocean ridge. Lava extruded along the ridge spreads the plates in opposite directions on either side of the rift line. (2)

**Seamount.** An extinct volcano that was worn down and submerged below sea level. (4)

**Seawalls.** A wall built parallel to a coast to protect it from wave erosion. (15)

**Sediment.** Loose material made up of particles from the weathering of rocks, crystals formed by evaporation of surface water, and crystals produced by plants and animals. (2)

**Sedimentary rocks.** Rocks produced by the deposition, compaction, and cementation of sediments at or near Earth's surface. (2)

**Sediment pollution.** The environmental damage to humans, plants, and animals resulting from sediment deposition. (6)

**Segmental pollution.** Alternating clean and polluted areas in a water body depending on the relative balance between B.O.D. and D.O. levels. (13)

**Seismic gaps.** Areas along a fault where strain-releasing seismic events have occurred in the past, but not recently. (5)

**Seismogram.** The chart paper with a tracing of ground motion from a seismometer. (5)

**Seismograph.** A detecting instrument that times and records incoming waves during an earthquake. (5)

**Seismology.** A branch of geology that studies earthquakes and the passage of earthquake waves through Earth. (5)

**Septic tank.** An individual treatment system for the domestic wastes from a home. Sewage is digested in the septic tank. The treated effluent is dispersed through the soil from a leaching field behind the home. (12)

**Sewage sludge.** The solid material left after bacteria break down sewage particles. (12)

**Shear stress.** Stress that acts along planes parallel to the deforming forces, tending to twist the material in opposite directions. (2)

**Sheet flow.** The excess water that flows over the land surface in a layer when the rate of rainfall exceeds the infiltration rate. (3)

**Shield volcanoes.** Volcanoes that take their name from their resemblance to the curved shield of a warrior. Their gentle slope results from the greater distances that the more fluid basaltic lava can flow downslope before it solidifies. (4)

**Silt.** Sedimentary particles with a diameter between 1/256 and 1/16 millimeters. (3)

**Sinkholes.** Circular depressions in the land surface that form as the result of solution of the underlying rocks. (10)

**Slide.** A mass movement process in which rock or sediment moves downslope along a planar surface. (9)

**Slope failure.** A loss of cohesion in slope materials, followed by gravity transport of the material down the slope. (9)

**SLOSH.** *S*ea, *L*ake and *O*verland *S*urges from *H*urricanes. A computer model of the National Weather Service that is used to predict the height of hurricane storm surge. (16)

**Slump.** The sliding of a mass of sediment or poorly consolidated rock downward along a curved surface. (9)

**Smog (smoke + fog).** Local urban air pollution trapped above a city by local weather conditions and topographic features. (11)

**Snow avalanches.** Rapidly moving fluidized masses of snow. (9)

**Snowline.** The lowest altitude of year-round snow cover. (3)

**Soil.** Regolith that has weathered enough to become enriched in the elements needed to support plant life. (3)

**Soil horizons.** Layers in the soil profile that are distinguished by color, organic content, mineralogy, or grain size as a result of weathering of the parent material. (6)

**Soil water zone.** The thin belt of partially saturated pores where water is retained within the root systems of plants. (8)

**Solifluction.** A slow downslope movement of thawed permafrost and water saturated regolith. (9)

**Solution.** The dissolving of a solid in a liquid. (3)

**Solution sinkholes.** Saucer-shaped depressions formed by solution of surface limestone that has only a thin cover of soil or sediment. (10)

**Sorting.** The degree of uniformity of particle sizes in a deposit. (3)

**Spillway.** A surface constructed at a given level so that water rising above that level spills over and into a floodway. (7)

**Spit.** An elongate and commonly curved bar of coarse material built into bays by longshore drift from headlands or sandy islands. (15)

**Spontaneous combustion.** A process in which combustible waste materials ignite due to heat generated by decomposing organic matter. (12)

**Spring.** Any natural discharge of groundwater on Earth's surface. (8)

**Stabilized inlets.** Inlets that are prevented from closing through jetties on either side of the inlet. (15)

**Stage height.** The level to which a stream has risen. (7)

**Storm drains and sewers.** Structures that remove water immediately to prevent flooding. (6)

**Storm sewers.** A system of sewers that collect storm runoff from the land and streets and deliver it directly to nearby lakes, streams, or the ocean. (12)

**Storm surge.** An elevation of the ocean surface resulting from the compound effects of water being pushed shoreward by wind across decreasing depths on the continental shelf, low pressure at the sea surface, tides raising the water level, and winds raising the ocean surface in addition. (15, 16)

**Storm water waste storage facilities.** Facilities that store the excess flow of storm water until it can be sent to a sewage treatment plant. (12)

**Straight channel.** A stream that flows nearly straight as a result of having a channel of resistant rock, well developed linear fractures, or a steep gradient. (7)

**Strain.** The reaction of a material, such as a change in volume or shape, when put under stress. (2)

**Stratovolcanoes.** Volcanoes composed of layers of pyroclastic debris and lava flows. (4)

**Stream.** Flowing surface water that is confined to a channel. (3)

**Stream hydrograph.** A graph that shows the change in stream discharge with time. (7)

**Strength.** The ability of a rock to resist failure. (2)

**Stress.** The internal forces experienced by a material when it is not free to move. Stress is measured as the force applied to a unit area, such as kilograms per square meter ($kg/m^2$). (2)

**Subduction.** The process in which, when two plates converge, one moves beneath the other. (2)

**Subsidence.** Sinking of the ground surface due to the removal of large quantities of water or petroleum from the pores of underlying sediments or rocks. (10)

**Subsidence faults.** Faults in the land surface caused by the withdrawal of fluids. As a consequence, the underlying materials compact and parts of the surface slip down along faults. (10)

**Subsidence fissures.** Fissures on the land surface that are caused when fluid is withdrawn. As a consequence, the surface subsides and tensional forces pull it apart. (10)

**Surface mining.** Excavation of surface materials in order to mine a resource that occurs close to the land surface. (6)

**Surface tension.** A force acting parallel to the water surface which enables fluids to remain as thin films within the pores of rocks and sediments. (9)

**Surface water.** Oceans, lakes, and streams. (3)

**Surface waves.** Waves that travel along Earth's surface. (5)

**Surf zone.** The zone in which over-steepened waves topple. (15)

**Suspended load.** The portion of the sediment load carried within the body of a transporting agent. (3)

**Swamps.** Areas that contain woody plants and trees and are sometimes covered with standing water. (13)

**S wave (secondary wave).** The slower of the two body waves. (5)

**Swells.** Waves of uniform length which travel through the water away from the direct influence of the winds that spawned them. (15)

**Talus slope.** The fan-shaped pile of rock fragments deposited by rockfall at the base of a cliff. (9)

**Tensional stress.** The stress that develops from opposite forces that tend to pull a material apart. (2)

**Tephra.** Airborne lava fragments. (4)

**Thermal pollution.** When temperature is raised in a body of water, the water is unable to hold as much oxygen. This oxygen deprivation often results in fish kills. (12)

**Thrust fault (reverse fault).** A fault in which compressional stress pushes the rock on one side of the fault plane upward relative to the other side. (2)

**Tidal circulation.** Water movement patterns in an estuary as a result of tidal action. (13)

**Tidal deltas.** A deposit of coarse sediment built into a bay or into the ocean as a result of sediment transport by flood or ebb currents, respectively. (15)

**Tidal inlets.** Openings along the shoreline through which ebb and flood currents flow between the ocean and bay. (15)

**Tidal wetlands.** Areas that develop where estuarine waters cover the land. (13)

**Tides.** The rise and fall of the ocean surface as a result of the gravitational force between the Moon and Earth. (3)

**Till.** Unstratified and poorly sorted sediment deposited by actively moving glacial ice. (3)

**Tombolos.** Bars of coarse sediment that connect an island to the mainland or islands to each other. (15)

**Tornadoes.** Funnel-shaped bodies of air that rotate rapidly in a clockwise direction (viewed from above), have very low internal pressures, and have high wind velocities within their walls.

**Tract development.** Typical American home style in which the dwelling is at the center of a lot or tract. (6)

**Transform fault margin.** A plate margin in which the plates neither diverge nor converge but slide horizontally past each other under shearing stress. (2).

**Transpiration.** A breathing-like process in which plants return a portion of the water they absorb to the atmosphere. (3)

**Transported regolith.** Loose material that has been moved and deposited by streams, wind, or glaciers. (3)

**Trough.** The lowest part of a wave. (15)

**Trunk glacier.** A type of mountain glacier formed by the merging of two smaller valley glaciers. (3)

**Tsunami.** Seismic sea waves generated by a major disturbance of the sea floor and overlying water. (5)

**Typhoons.** Hurricanes that persist west of Hawaii in the Pacific Ocean. (16)

**Unconfined aquifer.** An aquifer with no aquiclude above it. Its upper surface is exposed to atmospheric pressure through the interconnected pores of the aquifer. (8)

**Unsaturated zone.** The groundwater zone in which pores are filled partially with air and partially with water that is traveling between the surface and the layer below. (8)

**Urban concentration factor.** The amplification of small amounts of pollutants through use by large populations to create a serious problem. (1)

**Valley glacier.** A glacier that forms in mountainous areas and extends down pre-existing stream valleys. (3)

**Velocity.** The speed with which water flows. (7)

**Vent.** The point at which magma breaks through the oceanic or continental crust. (4)

**Viscosity.** The thickness of a liquid or magma which indicates its resistance to flow. (2, 4)

**Vog.** Volcanic fog formed when sulfur dioxide gas ($SO_2$) is converted to sulfuric acid ($H_2SO_4$) a short distance from an eruption site. (4)

**Volcanic ash.** Sand-sized particles of airborne lava fragments. (4)

**Volcanic bombs.** Airborne boulder-sized pieces of rapidly cooled and solidified lava. (4)

**Volcanic (extrusive) rock.** Rock formed by the cooling of lava on Earth's surface. (2)

**Volcanic glass.** Molten rock that cooled very quickly so that no crystals developed. (2)

**Volcano.** A mountain formed from the eruption of tephra or lava. (4)

**Washover fans.** Lobe-shaped deposits of sand eroded from the ocean side of a shoreline and deposited in the bays behind the barrier island. (15)

**Water content.** The weight percent of water in a soil. (6)

**Water table.** The contact between the unsaturated zone and the saturated zone. (8)

**Wave.** The mechanical expression of energy travelling through water. (15)

**Wave base.** The lower limit of orbital water movement in a wave. (15)

**Wave-built terrace.** Coarse material (pebble gravel, cobbles, and boulders) eroded from a cliff that is subjected to wave action and which remains behind forming a submarine terrace. (15)

**Wave front.** A line drawn along the crest of a wave. (15)

**Wave height.** The vertical distance between the crest and trough of a wave. (15)

**Wavelength.** The horizontal distance between a similar point on two waves, such as between two wave crests. (15)

**Wave normal.** A line perpendicular to the wave fronts. (15)

**Wave period.** The time for one wavelength to pass a given point. (15)

**Weathering.** Physical and chemical changes that occur in sediments and rocks as they are exposed to the atmosphere and biosphere. (2)

**Weathering atmosphere.** Atmosphere conducive to physical and chemical changes in rocks. (3)

**Wetland.** Area inundated or saturated by water for periods long enough to support vegetation that is adapted to saturated soil conditions. (13)

**Width (stream).** The distance measured along the stream surface from one bank to the other. (3, 7)

# Index